Lecture Notes in Computer Scie

T0238150

Commenced Publication in 1973
Founding and Former Series Editors:
Gerhard Goos, Juris Hartmanis, and Jan van Leeuwen

José Luiz Fiadeiro Neil Harman
Markus Roggenbach Jan Rutten (Eds.)

Algebra and Coalgebra in Computer Science

First International Conference, CALCO 2005
Swansea, UK, September 3-6, 2005
Proceedings

Springer

Volume Editors

José Luiz Fiadeiro
University of Leicester, Department of Computer Science
University Road, Leicester LE1 7RH, UK
E-mail: jose@fiadeiro.org

Neil Harman
Markus Roggenbach
University of Wales Swansea, Department of Computer Science
Singleton Park, Swansea SA2 8PP, UK
E-mail: {n.a.harman, m.roggenbach}@swansea.ac.uk

Jan Rutten
Vrije Universiteit Amsterdam, Centre for Mathematics and Computer Science (CWI)
Department of Software Technology
Kruislaan 413, P.O. Box 94079, 1090 GB Amsterdam, The Netherlands
E-mail: janr@cwi.nl

Library of Congress Control Number: 2005931121

CR Subject Classification (1998): F.3.1, F.4, D.2.1, I.1

ISSN 0302-9743
ISBN-10 3-540-28620-9 Springer Berlin Heidelberg New York
ISBN-13 978-3-540-28620-2 Springer Berlin Heidelberg New York

Springer is a part of Springer Science+Business Media

springeronline.com

Printed in Germany

Typesetting: Camera-ready by author, data conversion by Scientific Publishing Services, Chennai, India
Printed on acid-free paper SPIN: 11548133 06/3142 5 4 3 2 1 0

Preface

In April 2004, after one year of intense debate, CMCS, the International Workshop on Coalgebraic Methods in Computer Science, and WADT, the Workshop on Algebraic Development Techniques, decided to join their forces and reputations into a new high-level biennial conference. CALCO, the Conference on Algebra and Coalgebra in Computer Science, was created to bring together researchers and practitioners to exchange new results related to foundational aspects, and both traditional and emerging uses of algebras and coalgebras in computer science. A steering committee was put together by merging those of CMCS and WADT: Jiri Adamek, Ataru Nakagawa, Michel Bidoit, José Fiadeiro (co-chair), Hans-Peter Gumm, Bart Jacobs, Hans-Jörg Kreowski, Ugo Montanari, Larry Moss, Peter Mosses, Fernando Orejas, Francesco Parisi-Presicce, John Power, Horst Reichel, Markus Roggenbach, Jan Rutten (co-chair), and Andrzej Tarlecki.

CALCO 2005 was the first instance of this new conference. The interest that it generated in the scientific community suggests that it will not be the last. Indeed, it attracted as many as 62 submissions covering a wide range of topics roughly divided into two areas:

Algebras and Coalgebras as Mathematical Objects: Automata and languages; categorical semantics; hybrid, probabilistic, and timed systems; inductive and coinductive methods; modal logics; relational systems and term rewriting.

Algebras and Coalgebras in Computer Science: Abstract data types; algebraic and coalgebraic specification; calculi and models of concurrent, distributed, mobile, and context-aware computing; formal testing and quality assurance; general systems theory and computational models (chemical, biological, etc); generative programming and model-driven development; models, correctness and (re)configuration of hardware/middleware/architectures; re-engineering techniques (program transformation); semantics of conceptual modelling methods and techniques; semantics of programming languages; validation and verification.

Every submission received three or four reviews, which were generally of excellent quality. We want to thank all reviewers, the list of which is at the end of this preface, for carrying out their task with competence and precision, but also with the enthusiasm that comes from contributing to the birth of a new conference. Decisions were made during two weeks of animated e-mail discussion. In the end, a total of 25 papers were selected, the revised versions of which can be found in this volume. We were also lucky to have invited talks by three expert researchers: Samson Abramsky, Gordon Plotkin and Vladimiro Sassone. We are very grateful to the three of them.

The technical programme of the conference was preceded by a Young Researchers Workshop, CALCO-jnr, dedicated to presentations by PhD students and by those who had completed their doctoral studies in recent years, thus following on one of the traditional features of WADT. A technical report collects contributions selected from the presentations. CALCO-jnr was organized by Peter Mosses, John Power and Monika Seisenberger. A meeting of the IFIP WG1.3 – Foundations of System Specification – took place immediately after the conference and was hosted by Peter Mosses.

The project of hosting this first edition of CALCO was seized with both hands by a young and enthusiastic team led by Neal Harman and Markus Roggenbach from the University of Wales Swansea. The organizers would like to thank John V. Tucker and Monika Seisenberger for their invaluable advice and support, and IT Wales and the support staff, especially Sue Phillips, for making the event possible. CALCO 2005 received generous contributions from the Welsh Development Agency (WDA), IFIP, BCS-FACS, Digita and IT Wales.

Alfred Hofmann and his team at Springer lent us their support from Day –1 by agreeing to publish this volume. The work of the PC was supported by the Conference Online Service; Tiziana Margaria, Bernhard Steffen and their team deserve all our applause; it was very reassuring to feel that Martin Karusseit was available 24 hours a day, 7 days a week, but the truth is that we only needed him to sort out our own silly mistakes ...

We would like to reserve our final words of thanks to all the authors who have contributed such good quality papers to CALCO 2005.

June 2005

José Luiz Fiadeiro and Jan Rutten
Program Co-chairs

Neal Harman and Markus Roggenbach
Organizing Co-chairs

Organization

Program Committee

Luca Aceto, Aalborg University, Denmark and Reykjavík University, Iceland
Jiri Adamek, University of Braunschweig, Germany
Christel Baier, University of Bonn, Germany
Michel Bidoit, CNRS, Cachan, France
Jules Desharnais, Laval University, Canada
José Luiz Fiadeiro, University of Leicester, UK (co-chair)
Marie-Claude Gaudel, LRI-CNRS, Paris, France
Reiko Heckel, University of Leicester, UK
Hans-Peter Gumm, Philipps University, Marburg, Germany
Ugo Montanari, University of Pisa, Italy
Larry Moss, Indiana University, USA
Peter Mosses, University of Wales Swansea, UK
Fernando Orejas, Politechnical University of Catalunia, Barcelona, Spain
Francesco Parisi-Presicce, George Mason University, Fairfax, USA
John Power, University of Edinburgh, UK
Horst Reichel, Technical University of Dresden, Germany
Jan Rutten, CWI & Free University Amsterdam, The Netherlands (co-chair)
Eugene Stark, Stony Brook University, New York, USA
Andrzej Tarlecki, Warsaw University, Poland
John Tucker, University of Wales Swansea, UK
Martin Wirsing, Ludwig Maximilian University, Muenchen, Germany

Additional Reviewers

Thorsten Altenkirch
Serge Autexier
Denis Bechet
Gerd Behrmann
Marco Bernardo
Rocco De Nicola
Stéphane Demri
Josée Desharnais
Wan Fokkink
Magne Haveraaen
Rolf Hennicker
Wolfram Kahl
Alexander Kurz
Slawek Lasota
Mohamed Mejri
Stefan Milius
Marco Pistore
Steffen Priebe
Markus Roggenbach
Grigore Rosu
Marie-Christine Rousset
Emil Sekerinski
Pawel Sobocinski
Pascal Tesson
Hendrik Tews
Robert Walters
James Worrell

Table of Contents

Invited Talks

Contributed Papers

Abstract Scalars, Loops, and Free Traced and Strongly Compact Closed Categories

Samson Abramsky

Oxford University Computing Laboratory,
Wolfson Building, Parks Road, Oxford OX1 3QD, U.K
http://web.comlab.ox.ac.uk/oucl/work/samson.abramsky/

Abstract. We study structures which have arisen in recent work by the present author and Bob Coecke on a categorical axiomatics for Quantum Mechanics; in particular, the notion of *strongly compact closed category*. We explain how these structures support a notion of *scalar* which allows quantitative aspects of physical theory to be expressed, and how the notion of strong compact closure emerges as a significant refinement of the more classical notion of compact closed category.

We then proceed to an extended discussion of free constructions for a sequence of progressively more complex kinds of structured category, culminating in the strongly compact closed case. The simple geometric and combinatorial ideas underlying these constructions are emphasized. We also discuss variations where a prescribed monoid of scalars can be 'glued in' to the free construction.

1 Introduction

In this preliminary section, we will discuss the background and motivation for the technical results in the main body of the paper, in a fairly wide-ranging fashion. The technical material itself should be essentially self-contained, from the level of a basic familiarity with monoidal categories (for which see e.g. [20]).

1.1 Background

In recent work [4,5], the present author and Bob Coecke have developed a categorical axiomatics for Quantum Mechanics, as a foundation for high-level approaches to quantum informatics: type systems, logics, and languages for quantum programming and quantum protocol specification. The central notion in our axiomatic framework is that of *strongly compact closed category*. It turns out that this rather simple and elegant structure suffices to capture most of the key notions for quantum informatics: compound systems, unitary operations, projectors, preparations of entangled states, Dirac bra-ket notation, traces, scalars, the Born rule. This axiomatic framework admits a range of models, including of course the Hilbert space formulation of quantum mechanics.

Additional evidence for the scope of the framework is provided by recent work of Selinger [25]. He shows that the framework of completely positive maps acting

J.L. Fiadeiro et al. (Eds.): CALCO 2005, LNCS 3629, pp. 1–29, 2005.

on generalized states represented by density operators, used in his previous work on the semantics of quantum programming languages [24], fits perfectly into the framework of strongly compact closed categories.[1] He also showed that a simple construction (independently found and studied in some depth by Coecke [9]), which can be carried out completely generally at the level of strongly compact closed categories, corresponds to passing to the category of completely positive maps (and specializes exactly to this in the case of Hilbert spaces).

1.2 Multiplicatives and Additives

We briefly mention a wider context for these ideas. To capture the branching structure of measurements, and the flow of (classical) information from the result of a measurement to the future evolution of the quantum system, an additional *additive* level of structure is required, based on a functor $\oplus$, as well as the *multiplicative* level of the compact closed structure based around the tensor product (monoidal structure) $\otimes$. This delineation of additive and multiplicative levels of Quantum Mechanics is one of the conceptually interesting outcomes of our categorical axiomatics. (The terminology is based on that of Linear Logic [12] — of which our structures can be seen as 'collapsed models'). In terms of ordinary algebra, the multiplicative level corresponds to the multilinear-algebraic aspect of Quantum Mechanics, and the additive level to the linear-algebraic. But this distinction is usually lost in the sea of matrices; in particular, it is a real surprise how much can be done purely with the multiplicative structure.

It should be mentioned that we fully expect an *exponential* level to become important, in the passage to the multi-particle, infinite dimensional, relativistic, and eventually field-theoretic levels of quantum theory.

We shall not discuss the additive level further in this paper. For most purposes, the additive structure can be regarded as freely generated, subject to arithmetic requirements on the scalars (see [4]).

1.3 Explicit Constructions of Free Structured Categories

Our main aim in the present paper is to give explicit characterizations of free constructions for various kinds of categories-with-structure, most notably, for traced symmetric monoidal and strongly compact closed categories. We aim to give a synthetic account, including some basic cases which are well known from the existing literature [20,19]. We will progressively build up structure through the following levels:

(1) Monoidal Categories
(2) Symmetric Monoidal Categories
(3) Traced Symmetric Monoidal Categories

[1] Selinger prefers to use the term 'dagger compact closed category', since the notion of adjoint which is formalized by the dagger operation $()^\dagger$ is a separate structure which is meaningful in a more general setting.

(4) Compact Closed Categories
(5) Strongly Compact Closed Categories
(6) Strongly Compact Closed Categories with prescribed scalars

Of these, those cases which have not, to the best of our knowledge,, appeared previously are (3), (5) and (6). But in any event, we hope that our account will serve as a clear, accessible and useful reference.

Our constructions also serve to *factor* the Kelly-Laplaza construction [19] of the free compact closed category through the $\mathcal{G}$ or Int construction [14,2] of the compact closed category freely generated by a traced symmetric monoidal category, which is a central part of (the mathematically civilised version of) the so-called 'Geometry of Interaction' [12,3].

It should be emphasized that constructions (1)–(4) are free over *categories*, (5) over categories with involutions, and (6) over a comma category of categories with involution with a specified evaluation of scalars. We note that Dusko Pavlovic has give a free construction of traced categories over *monoidal categories* [21]. His construction is elegant, but abstract and less combinatorial/geometric than ours: perhaps necessarily so, since in our situation the monoidal structure, which itself has some spatial content, is added freely. Another reference is by Katis, Sabadini and Walters [16]. They construct a free 'feedback category', which is a trace minus the Yanking axiom — which is very important for the dynamics of the trace — over a monoidal category, and then formally quotient it to get a traced category. A treatment in the same style as the present paper of free traced, compact closed and strongly compact closed categories over a monoidal category remains a topic for future investigation.

Furthermore, we will work entirely with the *strict* versions of the categories-with-structure we will study. Since in each case, every such category is monoidally equivalent to a strict one, this does not really lose any generality; while by greatly simplifying the description of the free constructions, it makes their essential content, especially the geometry that begins to emerge as we add traces and compact closure (paths and loops), much more apparent.

1.4 Diagrammatics

Our free constructions have immediate diagrammatic interpretations, which make their geometric content quite clear and vivid. Diagrammatic notation for tensor categories has been extensively developed with a view to applications in categorical formulations of topological invariants, quantum groups, and topological quantum field theories [15]. Within the purely categorical literature, a forerunner of these developments is the early work of Kelly on coherence [17,18]; while the are also several precursors in the non-categorical literature, notably Penrose's diagrammatic notation for abstract tensors [22].

Diagrammaticnotationhasplayedanimportantroleinourownworkwith Coecke on applying our categorical axiomatics to quantum informatics, e.g. to quantum protocols [4]. For example, the essence of the verification of the teleportation protocol is the diagrammatic equality shown in Figure 1. For details, see [4,5].

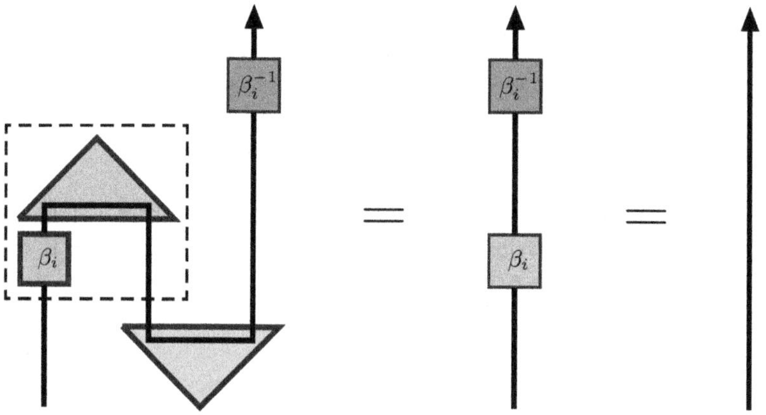

Fig. 1. Diagrammatic proof of teleportation

1.5 Categorical Quantum Logic

The diagrammatics of our constructions leads in turn to the idea of a *logical* formulation, in which the diagrammatic representation of a morphism in the free category is thought of as a *proof-net*, in the same general sense as in Linear Logic [11].

More precisely, morphisms in the free category will correspond to proof nets in normal form, and the definition of composition in the category gives a direct construction for normalizing a cut between two such proof nets. One advantage of the logical formulation is that we get an explicit syntactic description of these objects, and we can decompose the normalization process into cut-reduction steps, so that the computation of the normal form can be captured by a rewriting system. This provides an explicit computational basis for deciding equality of proofs, which corresponds in the categorical context to verifying the *commutativity of a diagram.*

In the categorical approach to quantum informatics [4], verifying the correctness of various quantum protocols is formulated as showing the commutativity of certain diagrams; so a computational theory of the above kind is directly applicable to such verifications.

In a joint paper with Ross Duncan [6], we have developed a system of Categorical Quantum Logic along these lines, incorporating additive as well as multiplicative features. This kind of logic, and its connection with Quantum Mechanics, is very different to the traditional notion of 'Quantum Logic' [8]. Duncan is continuing to develop this approach in his forthcoming thesis.

1.6 Overview

The further structure of the paper is as follows. In Section 2 we explore the abstract notion of scalar which exists in any monoidal category. As we will see, scalars play an important role in determining the structure of free traced and strongly compact closed categories, as they correspond to the values of *loops*. In Section 3, we review the notions of compact closed and strongly compact closed categories. The need for the notion of strong compact closure, to capture the structure of the complex spaces arising in Quantum Mechanics, is explained. In Section 4, we turn to the free constructions themselves.

Notation. We set up some notation which will be useful. We define $[n] := \{1, \ldots, n\}$ for $n \in \mathbb{N}$. We write $S(n)$ for the symmetric group on $[n]$. If $\pi \in S(n)$ and $\sigma \in S(m)$, we define $\pi \otimes \sigma \in S(n+m)$ by

$$\pi \otimes \sigma(i) = \begin{cases} \pi(i), & 1 \leq i \leq n \\ \sigma(i-n)+n, & n+1 \leq i \leq n+m. \end{cases}$$

Given $\lambda : [n] \to X$, $\mu : [m] \to X$, we define $[\lambda, \mu] : [n+m] \to X$ by

$$[\lambda, \mu](i) = \begin{cases} \lambda(i), & 1 \leq i \leq n \\ \mu(i-n), & n+1 \leq i \leq n+m. \end{cases}$$

We write $\mathcal{M}(X)$ for the free commutative monoid generated by a set X. Concretely, these are the finite multisets over X, with the addition given by multiset union, which we write as $S \uplus T$.

2 Scalars in Monoidal Categories

The concept of a *scalar* as a basis for quantitative measurements is fundamental in Physics. In particular, in Quantum Mechanics complex numbers α play the role of *probability amplitudes*, with corresponding probabilities $\alpha\bar{\alpha} = |\alpha|^2$.

A key step in the development of the categorical axiomatics for Quantum Mechanics in [4] was the recognition that the notion of scalar is meaningful in great generality — in fact, in any monoidal (not necessarily symmetric) category.[2]

Let $(\mathcal{C}, \otimes, \mathrm{I})$ be a strict monoidal category . We define a *scalar* in $\mathcal{C}$ to be a morphism $s : \mathrm{I} \to \mathrm{I}$, *i.e.* an endomorphism of the tensor unit.

Example 1. In $\mathbf{FdVec}_{\mathbb{K}}$, the category of finite-dimensional vector spaces over a field $\mathbb{K}$, linear maps $\mathbb{K} \to \mathbb{K}$ are uniquely determined by the image of 1, and hence correspond biuniquely to elements of $\mathbb{K}$; composition corresponds to multiplication of scalars. In **Rel**, there are just two scalars, corresponding to the Boolean values 0, 1.

---

[2] Susbsequently, I became aware through Martin Hyland of the mathematical literature on Tannakian categories [23,10], stemming ultimately from Grothendiek. Tannakian categories embody much stronger assumptions than ours, in particular that the categories are abelian as well as compact closed, although the idea of strong compact closure is absent. But they certainly exhibit a consonant development of a large part of multilinear algebra in an abstract setting.

The (multiplicative) monoid of scalars is then just the endomorphism monoid $\mathcal{C}(\mathrm{I},\mathrm{I})$. The first key point is the elementary but beautiful observation by Kelly and Laplaza [19] that this monoid is always commutative.

Lemma 1. *$\mathcal{C}(\mathrm{I},\mathrm{I})$ is a commutative monoid*

Proof.

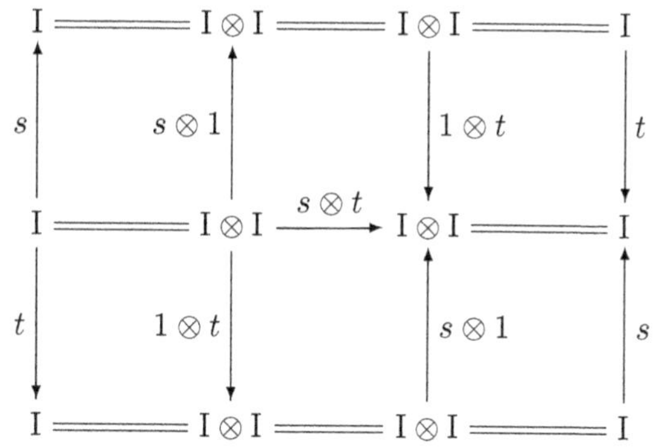

□

We remark that in the non-strict case, where we have unit isomorphisms

$$\lambda_A : \mathrm{I} \otimes A \to A \qquad \rho_A : A \otimes \mathrm{I} \to A$$

the proof makes essential use of the coherence axiom $\lambda_{\mathrm{I}} = \rho_{\mathrm{I}}$.

The second point is that a good notion of *scalar multiplication* exists at this level of generality. That is, each scalar $s : \mathrm{I} \to \mathrm{I}$ induces a natural transformation

$$s_A : A \xrightarrow{\simeq} \mathrm{I} \otimes A \xrightarrow{s \otimes 1_A} \mathrm{I} \otimes A \xrightarrow{\simeq} A \,.$$

with the naturality square

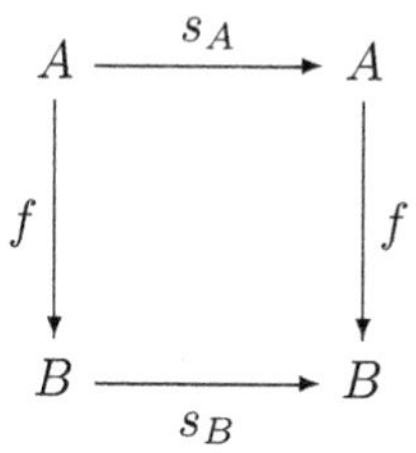

We write $s \bullet f$ for $f \circ s_A = s_B \circ f$. Note that

$$1 \bullet f = f \tag{1}$$
$$s \bullet (t \bullet f) = (s \circ t) \bullet f \tag{2}$$
$$(s \bullet g) \circ (t \bullet f) = (s \circ t) \bullet (g \circ f) \tag{3}$$
$$(s \bullet f) \otimes (t \bullet g) = (s \circ t) \bullet (f \otimes g) \tag{4}$$

which exactly generalizes the multiplicative part of the usual properties of scalar multiplication. Thus scalars act globally on the whole category.

3 Strongly Compact Closed Categories

A *compact closed category* is a symmetric monoidal category in which to each object A a *dual* A^*, a *unit* $\eta_A : \mathrm{I} \to A^* \otimes A$ and a *counit* $\epsilon_A : A \otimes A^* \to \mathrm{I}$ are assigned in such a way that the following 'triangular identities' hold:

$$A \xrightarrow{1_A \otimes \eta_A} A \otimes A^* \otimes A \xrightarrow{\epsilon_A \otimes 1_A} A \quad = \quad 1_A$$

$$A^* \xrightarrow{\eta_A \otimes 1_A} A^* \otimes A \otimes A^* \xrightarrow{1_A \otimes \epsilon_A} A^* \quad = \quad 1_{A^*}$$

Viewing monoidal categories as bicategories with a single 0-cell, this amounts to the axiom:

Every object (1-cell) has an adjoint

We can also view compact closed categories as *-autonomous categories [7] for which $\otimes = \invamp$, and hence as 'collapsed' models of Linear Logic [11].

3.1 Examples

– $(\mathbf{Rel}, \times)$: Sets, relations, and cartesian product. Here $\eta_X \subseteq \{*\} \times (X \times X)$ and we have
$$\eta_X = \epsilon_X^c = \{(*, (x, x)) \mid x \in X\}.$$
– $(\mathbf{FdVec}_{\mathbb{K}}, \otimes)$: Vector spaces over a field $\mathbb{K}$, linear maps, and tensor product. The unit and counit in $(\mathbf{FdVec}_{\mathbb{C}}, \otimes)$ are
$$\eta_V : \mathbb{C} \to V^* \otimes V :: 1 \mapsto \sum_{i=1}^{i=n} \bar{e}_i \otimes e_i$$
$$\epsilon_V : V \otimes V^* \to \mathbb{C} :: e_j \otimes \bar{e}_i \mapsto \langle \bar{e}_i \mid e_j \rangle$$
where n is the dimension of V, $\{e_i\}_{i=1}^{i=n}$ is a basis for V and $\bar{e}_i$ is the linear functional in V^* determined by $\bar{e}_j(e_i) = \delta_{ij}$.

3.2 Duality, Names and Conames

For each morphism $f : A \to B$ in a compact closed category we can construct a *dual* $f^* : B^* \to A^*$:

$$f^* \quad = \quad B^* \xrightarrow{\eta_A \otimes 1} A^* \otimes A \otimes B^* \xrightarrow{1 \otimes f \otimes 1} A^* \otimes B \otimes B^* \xrightarrow{1 \otimes \epsilon_B} A^*$$

a *name*

$$\ulcorner f \urcorner : \mathrm{I} \to A^* \otimes B \quad = \quad \mathrm{I} \xrightarrow{\eta} A^* \otimes A \xrightarrow{1 \otimes f} A^* \otimes B$$

and a *coname*

$$\llcorner f \lrcorner : A \otimes B^* \to \mathrm{I} \quad = \quad A \otimes B^* \xrightarrow{f \otimes 1} B \otimes B^* \xrightarrow{\epsilon} \mathrm{I}$$

The assignment $f \mapsto f^*$ extends $A \mapsto A^*$ into a contravariant endofunctor with $A \simeq A^{**}$. In any compact closed category, we have

$$\mathcal{C}(A \otimes B^*, \mathrm{I}) \simeq \mathcal{C}(A, B) \simeq \mathcal{C}(\mathrm{I}, A^* \otimes B).$$

For $R \in \mathbf{Rel}(X, Y)$ we have

$$\ulcorner R \urcorner = \{(*, (x, y)) \mid xRy, x \in X, y \in Y\}$$
$$\llcorner R \lrcorner = \{((x, y), *) \mid xRy, x \in X, y \in Y\}$$

and for $f \in \mathbf{FdVec}_{\mathbb{K}}(V, W)$ with matrix (m_{ij}) in bases $\{e_i^V\}_{i=1}^{i=n}$ and $\{e_j^W\}_{j=1}^{j=m}$ of V and W respectively:

$$\ulcorner f \urcorner : \mathbb{K} \to V^* \otimes W :: 1 \mapsto \sum_{i,j=1}^{i,j=n,m} m_{ij} \cdot \bar{e}_i^V \otimes e_j^W$$
$$\llcorner f \lrcorner : V \otimes W^* \to \mathbb{K} :: e_i^V \otimes \bar{e}_j^W \mapsto m_{ij}.$$

3.3 Why Compact Closure Does Not Suffice

In inner-product spaces we have the *adjoint*:

$$\frac{A \xrightarrow{f} B}{A \xleftarrow{f^\dagger} B} \qquad \langle f\phi \mid \psi \rangle_B = \langle \phi \mid f^\dagger \psi \rangle_A$$

This is *not* the same as the dual — the types are different. In "degenerate" CCC's in which $A^* = A$, e.g. **Rel** or real inner-product spaces, we have $f^* = f^\dagger$. In complex inner-product spaces such as Hilbert spaces, the inner product is *sesquilinear*

$$\langle \psi \mid \phi \rangle = \overline{\langle \phi \mid \psi \rangle}$$

and the isomorphism $A \simeq A^*$ is not linear, but *conjugate linear*:

$$\langle \lambda \bullet \phi \mid - \rangle \quad = \quad \bar{\lambda} \bullet \langle \phi \mid - \rangle$$

and hence does not live in the category **Hilb** at all!

3.4 Solution: Strong Compact Closure

We define the *conjugate space* of a Hilbert space $\mathcal{H}$: this has the same additive group of vectors as $\mathcal{H}$, while the scalar multiplication and inner product are "twisted" by complex conjugation:

$$\alpha \bullet_{\bar{\mathcal{H}}} \phi := \bar{\alpha} \bullet_{\mathcal{H}} \phi \qquad \langle \phi \mid \psi \rangle_{\bar{\mathcal{H}}} := \langle \psi \mid \phi \rangle_{\mathcal{H}}$$

We can define $\mathcal{H}^* = \bar{\mathcal{H}}$, since $\mathcal{H}$, $\bar{\mathcal{H}}$ have the same orthornormal bases, and we can define the counit by

$$\epsilon_{\mathcal{H}} : \mathcal{H} \otimes \bar{\mathcal{H}} \to \mathbb{C} :: \phi \otimes \psi \mapsto \langle \psi \mid \phi \rangle_{\mathcal{H}}$$

which is indeed (bi)linear rather than sesquilinear!

The crucial observation is this: $()^*$ has a *covariant* functorial extension $f \mapsto f_*$, which is essentially identity on morphisms; and then we can *define*

$$f^\dagger = (f^*)_* = (f_*)^*.$$

3.5 Axiomatization of Strong Compact Closure

In fact, there is a more concise and elegant axiomatization of strongly compact closed categories, which takes the adjoint as primitive [5]. It suffices to require the following structure on a (strict) symmetric monoidal category $(\mathcal{C}, \otimes, I, \tau)$:

- A strict monoidal involutive assignment $A \mapsto A^*$ on objects.
- An identity-on-objects, contravariant, strict monoidal, involutive functor $f \mapsto f^\dagger$.
- For each object A a unit $\eta_A : I \to A^* \otimes A$ with $\eta_{A^*} = \tau_{A^*,A} \circ \eta_A$ and such that either the diagram

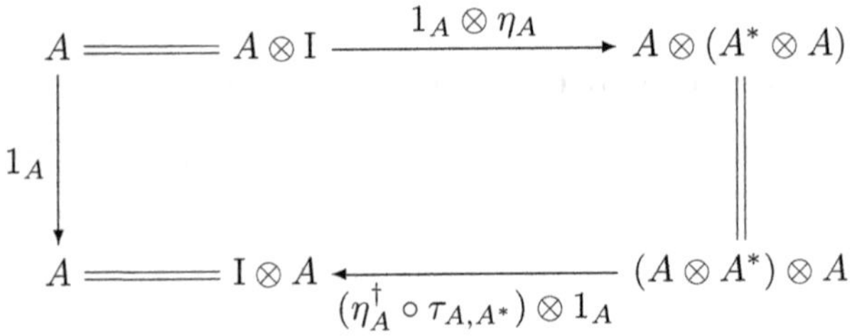

$$
\begin{array}{ccccc}
A & \!\!=\!\!= & A \otimes I & \xrightarrow{\;1_A \otimes \eta_A\;} & A \otimes (A^* \otimes A) \\
\Big\downarrow{\scriptstyle 1_A} & & & & \Big\| \\
A & \!\!=\!\!= & I \otimes A & \xleftarrow[(\eta_A^\dagger \circ \tau_{A,A^*}) \otimes 1_A]{} & (A \otimes A^*) \otimes A
\end{array}
\tag{5}
$$

or the diagram

$$
\begin{array}{ccccc}
A & \!\!=\!\!= I \otimes A & \xrightarrow{\;\eta_A \otimes 1_A\;} (A^* \otimes A) \otimes A & \!\!=\!\! & A^* \otimes (A \otimes A) \\
\Big\downarrow{\scriptstyle 1_A} & & & & \Big\downarrow{\scriptstyle 1_{A^*} \otimes \tau_{A,A}} \\
A & \!\!=\!\!= I \otimes A & \xleftarrow[\eta_A^\dagger \otimes 1_A]{} (A^* \otimes A) \otimes A & \!\!=\!\! & A^* \otimes (A \otimes A)
\end{array}
\tag{6}
$$

commutes, where $\tau_{A,A} : A \otimes A \simeq A \otimes A$ is the twist map.
- Given such a functor $()^\dagger$, we define an isomorphism α to be *unitary* if $\alpha^{-1} = \alpha^\dagger$. We additionally require that the canonical natural isomorphism for symmetry given as part of the symmetric monoidal structure on $\mathcal{C}$ is (componentwise) unitary in this sense.

While diagram (5) is the analogue to (3) with $\eta_A^\dagger \circ \tau_{A,A^*}$ playing the role of the counit, diagram (6) expresses *Yanking* with respect to the canonical trace of the compact closed structure.[3] We only need one commuting diagram as compared

---

[3] In fact, we have used the 'left trace' here rather than the more customary 'right trace' which we shall use in our subsequent discussion of traced monoidal categories. In the symmetric context, the two are equivalent; we chose the left trace here because, given our other notational conventions, it requires less use of symmetries in stating the axiom.

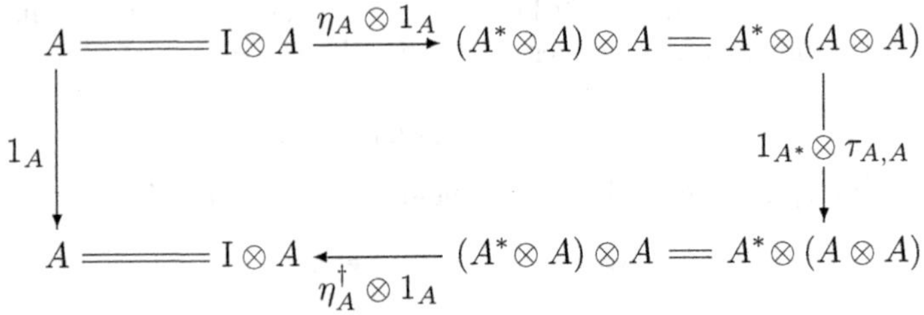

to (3) and (3) in the definition of compact closure, since due to the strictness assumption (i.e. $A \mapsto A^*$ being involutive) we were able to replace the second diagram by $\eta_{A^*} = \tau_{A^*,A} \circ \eta_A$.

Standard Triangular Identities Diagrammatically

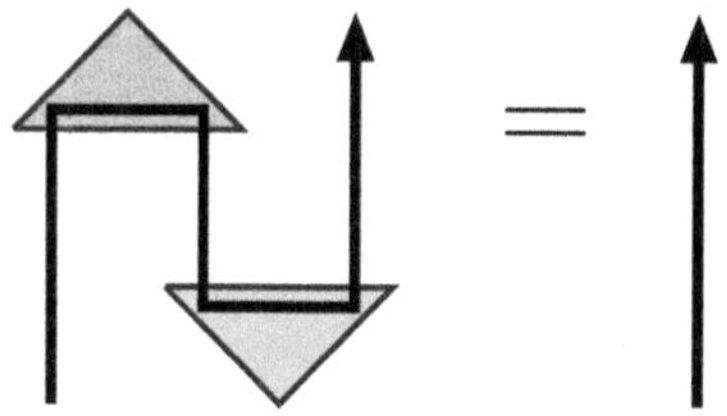

$(\epsilon_A \otimes 1_A) \circ (1_A \otimes \eta_A) = 1_A$

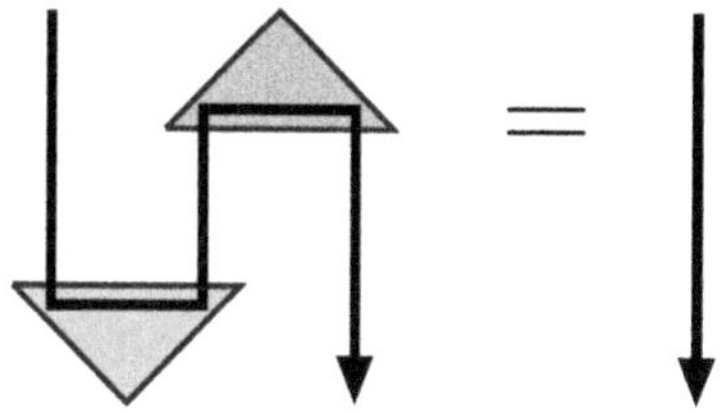

$(1_{A^*} \otimes \epsilon_A) \circ (\eta_A \otimes 1_{A^*}) = 1_{A^*}$

Yanking Diagrammatically

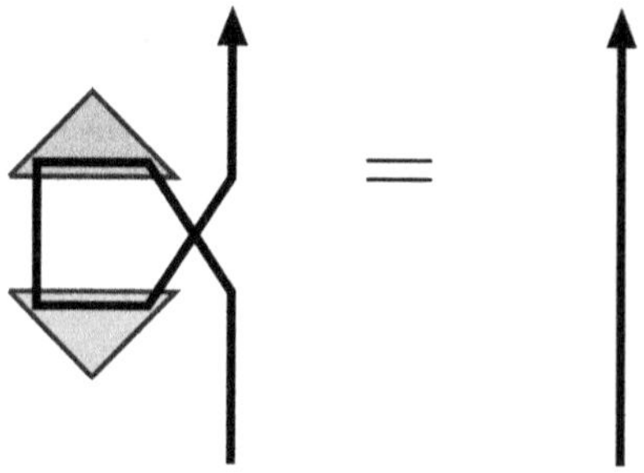

$$(\eta_A^\dagger \otimes 1_A) \circ (1_{A^*} \otimes \tau_{A,A}) \circ (\eta_A \otimes 1_A) = 1_A.$$

4 Free Constructions

We will now give detailed descriptions of free constructions for a number of types of category-with-structure. We shall consider the following cases:

(1) Monoidal Categories
(2) Symmetric Monoidal Categories
(3) Traced Symmetric Monoidal Categories
(4) Compact Closed Categories
(5) Strongly Compact Closed Categories
(6) Strongly Compact Closed Categories with prescribed scalars

For cases (1)–(4), we shall consider adjunctions of the form

$$\mathbf{Cat} \underset{U_S}{\overset{F_S}{\rightleftarrows}} \mathbf{S\text{-}Cat} \quad (F_S \dashv U_S)$$

where S ranges over the various kinds of structure. Specifically, we shall give explicit descriptions in each case of $F_S(\mathcal{C})$ for a category $\mathcal{C}$. This explicit description — not algebraically by generators and relations, but giving direct combinatorial definitions of the normal forms and how they compose, thus solving the **word problem** over these categories— is the strongest form of coherence theorem available for notions such as compact closure and traces. In these cases, cyclic structures arise, violating the compatibility requirements for stronger forms of coherence developed in [17,18]. This point is discussed in the concluding section of [19].

In case (5), we consider an adjunction

$$\mathbf{InvCat} \underset{U_{\mathsf{SCC}}}{\overset{F_{\mathsf{SCC}}}{\rightleftarrows}} \mathbf{SCC\text{-}Cat} \quad (\bot)$$

where **InvCat** is the category of categories with a specified involution, (what Selinger calls 'dagger categories' in [25]), and functors which preserve the involution. Finally, in (6) we consider an adjunction with respect to a comma category, which allows us to describe the free strongly compact slosed category generated by a category $\mathcal{C}$, together with a prescribed multiplicative monoid of scalars.

Our treatment will be incremental, reflecting the fact that in our sequence (1)–(6), each term arises by adding structure to the previous one. Each form of structure is reflected conceptually by a new feature arising in the corresponding free construction:

M	monoidal	lists
SM	symmetric monoidal	permutations
Tr	traced symmetric monoidal	loops
CC	compact closed	polarities
SCC	strong compact closed	reversals

We will also begin to see a primitive graph-theoretic geometry of *points*, *lines* and *paths* begin to emerge as we progress through the levels of structure. There is in fact more substantial geometry lurking here than might be apparent: the elaboration of these connections must be left to future work.

Finally, we mention a recurring theme. To form a 'pure' picture of each construction, it is useful to consider the case $F_S(\mathbf{1})$ explicitly, where $\mathbf{1}$ is the category with (one object and) one morphism (*i.e.* one generator, no relations).

4.1 Monoidal Categories

We begin with the simple case of monoidal categories. The objects of $F_{\mathsf{M}}(\mathcal{C})$, the free monoidal category generated by the category $\mathcal{C}$, are lists of objects of $\mathcal{C}$. The (strict) monoidal structure is given by concatenation; the tensor unit I is the empty sequence.

Arrows:

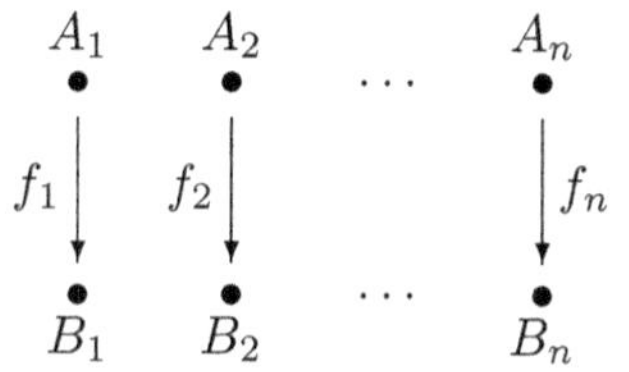

$$f_i : A_i \to B_i$$

An arrow from one list of objects to another is simply a list of arrows of $\mathcal{C}$ of the appropriate types. Note that there can only be an arrow between lists of the same length. Composition is performed pointwise in the obvious fashion.

Formally, we set $[n] := \{1, \ldots, n\}$, and define an object of $F_{\mathsf{M}}(\mathcal{C})$ to be a pair (n, A), where $n \in \mathbb{N}$, and $A : [n] \to \mathsf{Ob}\,\mathcal{C}$. Tensor product of objects is defined by $(n, A) \otimes (m, B) = (n + m, [A, B])$. The tensor unit is $\mathrm{I} = (0, !)$, where ! is the unique function from the empty set.

A morphism $\lambda : (n, A) \to (m, B)$ can only exist if $n = m$, and is specified by a map $\lambda : [n] \to \mathsf{Mor}\,\mathcal{C}$, satisfying

$$\lambda_i : A_i \longrightarrow B_i.$$

Arrows in $F_{\mathsf{M}}(\mathcal{C})$ are thus simply those expressible in the form

$$f_1 \otimes \cdots \otimes f_k : A_1 \otimes \cdots \otimes A_k \longrightarrow B_1 \otimes \cdots \otimes B_k.$$

Unicity of the monoidal functor to a monoidal category $\mathcal{M}$ extending a given functor $F : \mathcal{C} \to U_{\mathsf{M}}\mathcal{M}$ is then immediate.

Note that

$$F_{\mathsf{M}}(\mathbf{1}) = (\mathbb{N}, =, +, 0).$$

4.2 Symmetric Monoidal Categories

The objects of $F_{\mathsf{SM}}(\mathcal{C})$ are the same as in the monoidal case.

An arrow $(n, A) \longrightarrow (n, B)$ is given by (π, λ), where $\pi \in S(n)$ is a permutation, and $\lambda_i : A_i \to B_{\pi(i)}$, $1 \leq i \leq n$.

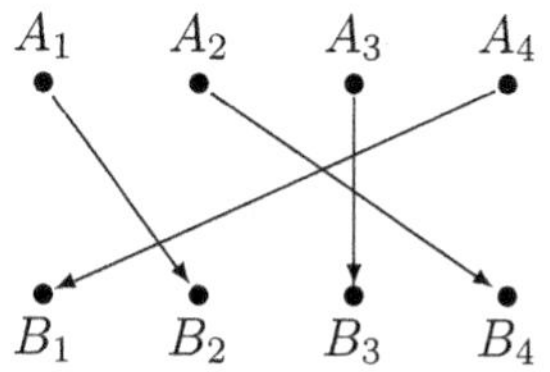

Composition in $F_{\mathsf{SM}}(\mathcal{C})$ is described as follows. Form paths of length 2, and compose the arrows from $\mathcal{C}$ labelling these paths.

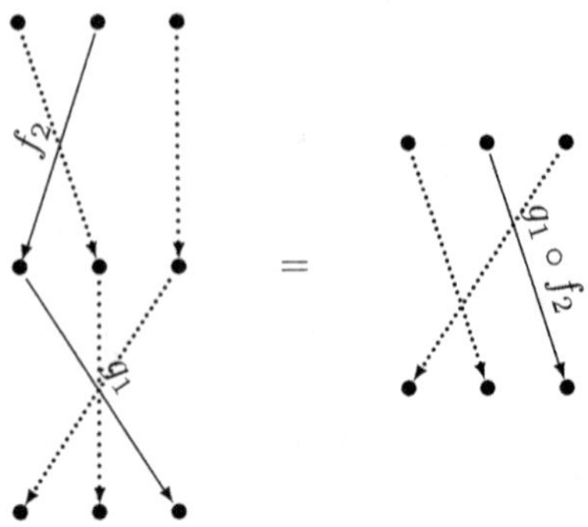

Note that $F_{\mathsf{SM}}(\mathbf{1}) = \coprod_n S(n)$ (coproduct of categories). Thus the free monoidal category on the trivial generating category comprises (the disjoint union of) all the finite symmetric groups.[4]

Let $\mathcal{M}$ be a symmetric monoidal category, and consider a tensor product $A_1 \otimes \cdots \otimes A_n$. Each element $\pi \in S(n)$ of the symmetric group $S(n)$ induces an isomorphism, which by abuse of notation we also write as π:

$$\pi : A_1 \otimes \cdots \otimes A_n \xrightarrow{\cong} A_{\pi(1)} \otimes \cdots \otimes A_{\pi(n)}.$$

Now note that under the above concrete description of $F_{\mathsf{SM}}(\mathcal{C})$, arrows

$$(\pi, \lambda) : (n, A) \longrightarrow (n, B)$$

can be written as

$$\pi^{-1} \circ \bigotimes_{i=1}^{n} f_i \ : \ \bigotimes_{i=1}^{n} A_i \longrightarrow \bigotimes_{i=1}^{n} B_i. \tag{7}$$

Again, the freeness property follows directly. The main observation to be made is that such arrows are closed under composition:

$$(\sigma^{-1} \circ \bigotimes_{i=1}^{n} g_i) \circ (\pi^{-1} \circ \bigotimes_{i=1}^{n} f_i) = (\sigma \circ \pi)^{-1} \circ \bigotimes_{i=1}^{n} (g_{\pi(i)} \circ f_i) \tag{8}$$

and tensor product:

$$(\pi^{-1} \circ \bigotimes_i f_i) \otimes (\sigma^{-1} \circ \bigotimes_i g_i) \ = \ (\pi \otimes \sigma)^{-1} \circ \left(\bigotimes_i f_i \otimes \bigotimes_i g_i \right), \tag{9}$$

where if $\pi \in S(n)$, $\sigma \in S(m)$, $\pi \otimes \sigma \in S(n+m)$ is the evident concatenation of the two permutations, as defined in the Introduction.

The above closed form expression for composition requires the 'naturality square':

[4] At this point, a possible step towards geometry presents itself. If we considered free braided monoidal categories, we would find a similar connection to the braid groups [15]. However, we shall not pursue that here.

$$\bigotimes_i B_{\pi(i)} \xrightarrow{\quad \pi^{-1} \quad} \bigotimes_i B_i$$

$$\bigotimes_i g_{\pi(i)} \downarrow \qquad\qquad\qquad \downarrow \bigotimes_i g_i$$

$$\bigotimes_i C_{\sigma \circ \pi(i)} \xrightarrow{\quad \sigma \circ (\sigma \circ \pi)^{-1} \quad} \bigotimes_i C_{\sigma(i)}$$

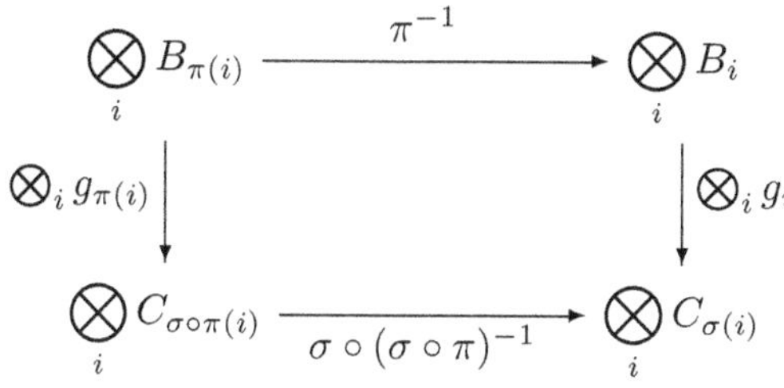

4.3 Traced Symmetric Monoidal Categories

We now come to a key case, that of traced symmetric monoidal categories. Much of the structure of strongly compact closed categories in fact appears already at the traced level. This is revealed rather clearly by our incremental development of the free constructions.

We begin by recalling the basic notions. let $(\mathcal{C}, \otimes, I, \tau)$ be a symmetric monoidal category. Here $\tau_{A,B} : A \otimes B \xrightarrow{\cong} B \otimes A$ is the symmetry or twist natural isomorphism. A *trace* on $\mathcal{C}$ is a family of functions

$$\mathsf{Tr}_{A,B}^U : \mathcal{C}(A \otimes U, B \otimes U) \longrightarrow \mathcal{C}(A, B)$$

for objects A, B, U of $\mathcal{C}$, satisfying the following axioms:

- **Input Naturality:**

$$\mathsf{Tr}_{A,B}^U(f) \circ g = \mathsf{Tr}_{A',B}^U(f \circ (g \otimes 1_U))$$

  where $f : A \otimes U \to B \otimes U$, $g : A' \to A$,
- **Output Naturality:**

$$g \circ \mathsf{Tr}_{A,B}^U(f) = \mathsf{Tr}_{A,B'}^U((g \otimes 1_U) \circ f)$$

  where $f : A \otimes U \to B \otimes U$, $g : B \to B'$,
- **Feedback Dinaturality:**

$$\mathsf{Tr}_{A,B}^U((1_B \otimes g) \circ f) = \mathsf{Tr}_{A,B}^{U'}(f \circ (1_A \otimes g))$$

  where $f : A \otimes U \to B \otimes U'$, $g : U' \to U$,
- **Vanishing (I,II):**

$$\mathsf{Tr}_{A,B}^I(f) = f \quad \text{and} \quad \mathsf{Tr}_{A,B}^{U \otimes V}(g) = \mathsf{Tr}_{A,B}^U(\mathsf{Tr}_{A \otimes U, B \otimes U}^V(g))$$

  where $f : A \otimes I \to B \otimes I$ and $g : A \otimes U \otimes V \to B \otimes U \otimes V$.
- **Superposing:**

$$g \otimes \mathsf{Tr}_{A,B}^U(f) = \mathsf{Tr}_{W \otimes A, Z \otimes B}^U(g \otimes f)$$

  where $f : A \otimes U \to B \otimes U$ and $g : W \to Z$.
- **Yanking:** $\qquad\qquad \mathsf{Tr}_{U,U}^U(\tau_{U,U}) = 1_U.$

Diagrammatically, we depict the trace as feedback:

$$\frac{A \otimes U \xrightarrow{\ f\ } B \otimes U}{A \xrightarrow{\mathsf{Tr}^U_{A,B}(f)} B}$$

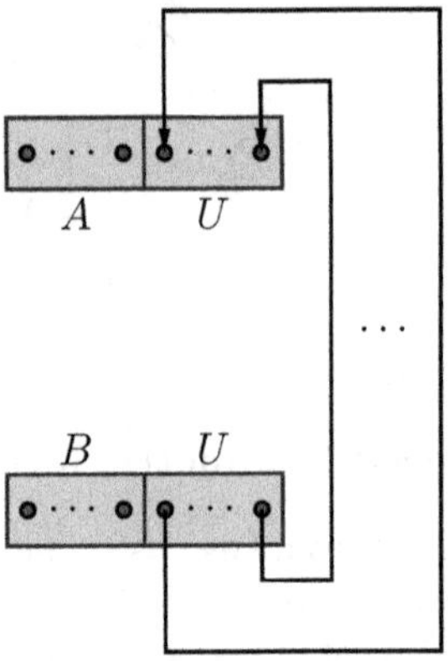

It corresponds to *contracting indices* in traditional tensor calculus.

We now consider the free symmetric monoidal category generated by $\mathcal{C}$, $F_{\mathsf{SM}}(\mathcal{C})$, as described in the previous section. Recall that morphisms in $F_{\mathsf{SM}}(\mathcal{C})$ can be written as

$$\pi^{-1} \circ \bigotimes_{i=1}^{n} f_i \ : \ \bigotimes_{i=1}^{n} A_i \longrightarrow \bigotimes_{i=1}^{n} B_i .$$

Our first observation is that *this category is already canonically traced.* Understanding why this is so, and why $F_{\mathsf{SM}}(\mathcal{C})$ is *not* the free traced category, will lay bare the essential features of the free construction we are seeking.

Note firstly that, if there is an arrow $f : (n, A) \otimes (p, U) \to (m, B) \otimes (p, U)$ in $F_{\mathsf{SM}}(\mathcal{C})$, then we must have $n+p = m+p$, and hence $n = m$. Thus we can indeed hope to form an arrow $A \to B$ in $F_{\mathsf{SM}}(\mathcal{C})$. Now we consider the 'geometry' arising from the permutation π, together with the diagrammatic feedback interpretation of the trace. We illustrate this with the following example.

Example. Consider the arrow $f = \pi^{-1} \circ \bigotimes_{i=1}^{4} f_i$, where $\pi = (2, 4, 3, 1)$, and $f_i : A_i \to B_{\pi(i)}$. Suppose that $A_i = U_i = B_i$, $2 \leq i \leq 4$, and write $U = \bigotimes_{i=2}^{4} U_i$. We wish to compute $\mathsf{Tr}^U_{A_1,B_1}(f)$. The geometry is made clear by the following figure.

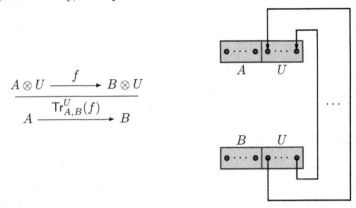

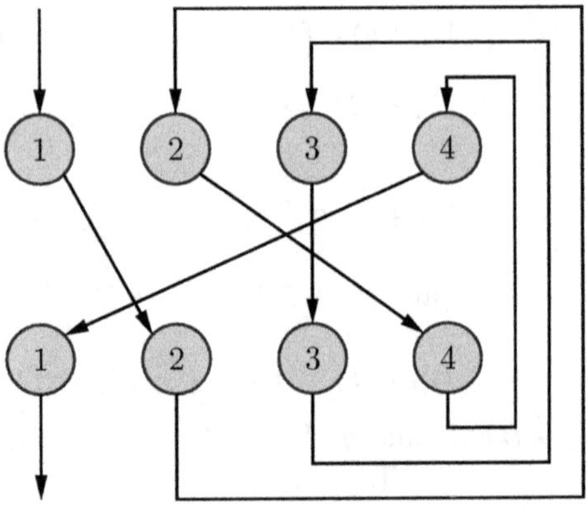

We simply follow the path leading from A_1 to B_1:

$$1 \to 2 \to 4 \to 1$$

composing the arrows which label the arcs in the path: thus

$$\mathsf{Tr}^U_{A_1,B_1}(f) = f_4 \circ f_2 \circ f_1$$

in this case. A similar procedure can always be followed for arrows in the form (7), which as we have seen is general for $F_{\mathsf{SM}}(\mathcal{C})$. (It is perhaps not immediately obvious that a path from an input will always emerge from the feedback zone into an output. See the following Proposition 1). Moreover, this assignment does lead to a well-defined trace on $F_{\mathsf{SM}}(\mathcal{C})$. However, this is *not* the free traced structure generated by $\mathcal{C}$.

To see why this construction does not give rise to the free interpretation of the trace, note that in our example, the node U_3 is involved in a cycle $U_3 \to U_3$, which does not appear in our expression for the trace of f. In fact, note that if we trace an endomorphism $f : A \to A$ out completely, *i.e.* writing $f : \mathrm{I} \otimes A \to \mathrm{I} \otimes A$ we form $\mathsf{Tr}^A_{\mathrm{I},\mathrm{I}}(f) : \mathrm{I} \to \mathrm{I}$, then we get a *scalar*. Indeed, the importance of scalars in our context is exactly that they give the values of loops. Now in $F_{\mathsf{SM}}(\mathcal{C})$, the tensor unit is the empty list, and there is only one scalar — the identity. It is exactly this collapsed interpretation of the scalars which prevents the trace we have just (implicitly) defined on $F_{\mathsf{SM}}(\mathcal{C})$ from giving the free traced category on $\mathcal{C}$.

We now turn to a more formal account, culminating in the construction of $F_{\mathsf{Tr}}(\mathcal{C})$.

Geometry of Permutations. We begin with a more detailed analysis of permutations $\pi \in S(n+m)$, with the decomposition $n+m$ reflecting our distinction between the visible (input-output) part of the type, and the hidden (feedback) part, arising from the application of the trace.

We define an *n-path* (or if n is understood, an *input-output path*) of π to be a sequence

$$i, \pi(i), \pi^2(i), \ldots, \pi^{k+1}(i) = j$$

where $1 \leq i, j \leq n$, and for all $0 < p < k$, $\pi^p(i) > n$. We write $P_\pi(i)$ for the n-path starting from i, which is clearly unique if it exists, and also $p_\pi(i) = j$. We write $P^0_\pi(i)$ for the set of elements of $\{n+1, \ldots, n+m\}$ appearing in the sequence. A *loop* of π is defined to be a cycle

$$j, \pi(j), \ldots, \pi^{k+1}(j) = j$$

where $n < j \leq n + m$. We write $\mathcal{L}(\pi)$ for the set of all loops of π.

Proposition 1. *The following holds for any permutation $\pi \in S(n+m)$:*

1. *For each i, $1 \leq i \leq n$, $P_\pi(i)$ is well-defined.*
2. $p_\pi \in S(n)$.

3. *The family of sets*

$$\{P^0_\pi(i) \mid 1 \leq i \leq n\} \cup \mathcal{L}(\pi)$$

form a partition of $\{n+1, \ldots, n+m\}$.

Proof.

1. Consider the sequence
$$i, \pi(i), \pi^2(i), \ldots$$
Either we reach $\pi^{k+1} = j \leq n$, or there must be a least l such that
$$\pi^{k+1}(i) = \pi^{l+1}(i) > n, \qquad 0 \leq k < l.$$
(Note that the fact that $i \leq n$ allows us to write the left hand term as $\pi^{k+1}(i)$). But then, applying π^{-1}, we conclude that $\pi^k(i) = \pi^l(i)$, a contradiction.
2. If $p_\pi(i) = p_\pi(j)$, then $\pi^{k+1}(i) = \pi^{l+1}(j)$, where say $k \leq l$. Applying $(\pi^{-1})^{k+1}$, we obtain $i = \pi^{l-k}(j) \leq n$, whence $l = k$ and $i = j$.
3. It is standard that distinct cycles are disjoint. We can reason similarly to part (2) to show that if $P^0_\pi(i)$ meets $P^0_\pi(j)$, then $i = j$. Similar reasoning to (1) shows that $P^0_\pi(i) \cap L = \varnothing$, for $L \in \mathcal{L}(\pi)$. Finally, iterating π^{-1} on $j > n$ either forms a cycle, or reaches $i \leq n$; in the latter case, $j \in P^0_\pi(i)$.

□

We now give a more algebraic description of the permutation p_π. Firstly, we extend our notation by defining $[n{:}m] := \{n+1, \ldots, m\}$. Now we can write $[n{+}m] = [n] \sqcup [n{:}n{+}m]$, where $\sqcup$ is disjoint union. We can use this decomposition to express $\pi \in S(n{+}m)$ as the disjoint union of the following four maps:

$$\pi_{1,1} : [n] \longrightarrow [n] \qquad \pi_{1,2} : [n] \longrightarrow [n{:}n{+}m]$$

$$\pi_{2,1} : [n{:}n{+}m] \longrightarrow [n] \qquad \pi_{2,2} : [n{:}n{+}m] \longrightarrow [n{:}n{+}m]$$

We can view these maps as *binary relations* on $[n{+}m]$ (they are in fact injective partial functions), and use relational algebra (union $R \cup S$, relational composition $R; S$ and reflexive transitive closure R^*) to express p_π in terms of the π_{ij}:

$$p_\pi \;=\; \pi_{1,1} \;\cup\; \pi_{1,2}; \pi^*_{2,2}; \pi_{2,1}.$$

We can also characterize the elements of $\mathcal{L}(\pi)$:

$$j \in \bigcup \mathcal{L}(\pi) \quad \Longleftrightarrow \quad \langle j, j \rangle \in \pi^*_{2,2} \cap \mathsf{id}_{[n:n+m]}.$$

Loops. We follow Kelly and Laplaza [19] in making the following basic definitions. The *loops* of a category $\mathcal{C}$, written $\mathcal{L}[\mathcal{C}]$, are the endomorphisms of $\mathcal{C}$ quotiented by the following equivalence relation: a composition

$$A_1 \xrightarrow{f_1} A_2 \xrightarrow{f_2} \cdots \quad A_k \xrightarrow{f_k} A_1$$

is equated with all its cyclic permutations. A *trace function* on $\mathcal{C}$ is a map on the endomorphisms of $\mathcal{C}$ which respects this equivalence. We note in particular the following standard result [14]:

Proposition 2. *If $\mathcal{C}$ is traced, then the trace applied to endomorphisms:*

$$g : A \to A \quad \longmapsto \quad \mathsf{Tr}^{\mathrm{I}}_{A,A}(f) : \mathrm{I} \to \mathrm{I}$$

is a (scalar-valued) trace function.

Traces of Decomposable Morphisms. We now turn to a general proposition about traced categories, from which the structure of the free category will be readily apparent. It shows that whenever a morphism is decomposable into a tensor product followed by a permutation (as all morphisms in $F_{\mathsf{SM}}(\mathcal{C})$ are), then the trace can be calculated explictly by composing over paths.

Proposition 3. *Let $\mathcal{C}$ be a traced symmetric monoidal category, and consider a morphism of the form*

$$f = \pi^{-1} \circ \bigotimes_{i=1}^{n+m} f_i : C \longrightarrow D$$

where $C = \bigotimes_{i=1}^{n} A_i \otimes \bigotimes_{j=n+1}^{n+m} U_j$, $D = \bigotimes_{i=1}^{n} B_i \otimes \bigotimes_{j=n+1}^{n+m} U_j$, $\pi \in S(n+m)$, and $f_i : C_i \to D_{\pi(i)}$. Then

$$\mathsf{Tr}^{U}_{C,D}(f) = \left(\prod_{l \in \mathcal{L}(\pi)} s_l \right) \bullet (p_{\pi}^{-1} \circ \bigotimes_{i=1}^{n} g_n)$$

where for each $1 \leq i \leq n$, with n-path

$$P_{\pi}(i) = i, p_1, \ldots, p_k, j$$

g_i is the composition

$$A_i \xrightarrow{\ f_i\ } U_{p_1} \xrightarrow{\ f_{p_1}\ } \cdots\cdots \quad U_{p_k} \xrightarrow{\ f_{p_k}\ } B_j$$

and for $l = p_1, \cdots, p_k, p_1 \in \mathcal{L}(\pi)$, $s_l = \mathsf{Tr}^{U_{p_1}}_{\mathrm{I,I}}(f_{p_k} \circ \cdots \circ f_{p_1})$. The product $\prod_l s_l$ refers to multiplication in the monoid of scalars, which we know by Proposition 1 to be commutative.

Taken together with the following instance of Superposing:

$$\mathsf{Tr}^{U}_{A,B}(s \bullet f) = s \bullet \mathsf{Tr}^{U}_{A,B}(f) \tag{10}$$

this Proposition yields a closed form description of the trace on expressions of the form:

$$s \bullet (\pi^{-1} \circ \bigotimes_j f_j) \ : \ \bigotimes_j A_j \ \longrightarrow \ \bigotimes_j B_j . \tag{11}$$

We approach the proof of this Proposition via a number of lemmas.

Firstly, a simple consequence of Feedback Dinaturality:

Lemma 2. *Let* $U = \bigotimes_{i=1}^{n} U_i$, *and* $\sigma \in S(n)$. *Let* $\sigma U = \bigotimes_{i=1}^{n} U_{\sigma(i)}$. *Then*

$$\mathsf{Tr}^{U}_{A,B}(f) = \mathsf{Tr}^{\sigma(U)}_{A,B}((1_A \otimes \sigma) \circ f \circ (1_B \otimes \sigma^{-1})).$$

Lemma 3.

$$\mathsf{Tr}^{U}_{A,B}(f) \otimes \mathsf{Tr}^{V}_{C,D}(g) = \mathsf{Tr}^{V\otimes U}_{A\otimes C,B\otimes D}((1_A \otimes \tau_{U,D\otimes V}) \circ (f \otimes g) \circ (1_A \otimes \tau_{C\otimes V,U})).$$

The proof is in the Appendix.

We now show how the trace is evaluated along cyclic paths of any length. We write $\sigma_{k+1} = \begin{pmatrix} 1\,2 \cdots & k & k+1 \\ 2\,3 \cdots & k+1 & 1 \end{pmatrix}$, the cyclic permutation of length $k+1$. Note the useful recursion formula:

$$\sigma_{k+1} = (\tau \otimes 1) \circ (1 \otimes \sigma_k). \tag{12}$$

Suppose we have morphisms $f_i : A_i \to A_{i+1}$, $1 \leq i \leq k+1$. We write $U = \bigotimes_{i=2}^{k+1} A_i$, and $V = \bigotimes_{i=3}^{k+1} A_i$. (By convention, a tensor product over an empty range of indices is taken to be the tensor unit I).

Lemma 4. *For all* $k \geq 0$*:*

$$\mathsf{Tr}^{U}_{A_1,A_{k+2}}(\sigma_{k+1} \circ \bigotimes_{i=1}^{k+1} f_i) = f_{k+1} \circ f_k \circ \cdots \circ f_1.$$

The proof is relegated to the Appendix. This lemma simultaneously generalizes Vanishing I ($k = 0$) and Yanking ($k = 1$, $A_1 = A_2 = A_3$, $f_1 = f_2 = 1_{A_1}$), and also the Generalized Yanking of [3]. The geometry of the situation is made clear by the following diagram.

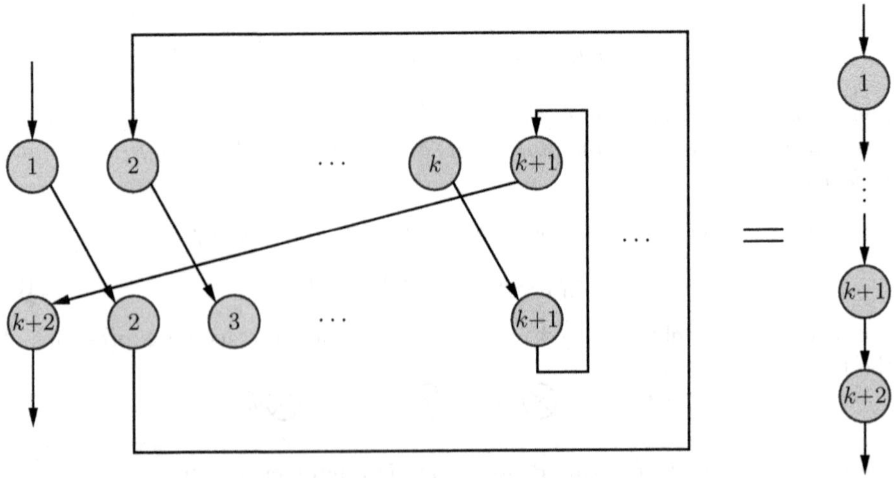

Proof of Proposition 3. Note that for $k = 0$, this is just Vanishing I. Up to conjugation by some permutation σ, we can express π as the tensor product of its n-paths and loops:

$$\pi = \sigma \circ \left(\bigotimes_{i=1}^{n} P_\pi(i) \otimes \bigotimes_{L \in \mathcal{L}[\pi]} L \right) \circ \sigma^{-1}.$$

Using Lemmas 3 and 2, we can express the trace of f in terms of the traces of the morphisms corresponding to the n-paths and loops of π. The trace of each n-path is given by Lemma 4. □

Description of $F_{\mathsf{Tr}}(\mathcal{C})$ The objects are as for $F_{\mathsf{SM}}(\mathcal{C})$. A morphism now has the form (S, π, λ), where (π, λ) are as in $F_{\mathsf{SM}}(\mathcal{C})$, and S is a *multiset of loops* in $\mathcal{L}[\mathcal{C}]$, *i.e.* an element of $\mathcal{M}(\mathcal{L}[\mathcal{C}])$, the free commutative monoid generated by $\mathcal{L}[\mathcal{C}]$.

Note that such a morphism

$$(S, \pi, \lambda) : (n, A) \longrightarrow (n, B)$$

can be written as

$$\left(\prod_{[s_i : A \to A]_\sim \in S} \mathsf{Tr}^{A}_{\mathrm{I},\mathrm{I}}(s_i) \right) \bullet (\pi^{-1} \circ \bigotimes_{i=1}^{n} \lambda_i) \;:\; \bigotimes_i A_i \longrightarrow \bigotimes_i B_i \tag{13}$$

in the language of traced symmetric monoidal categories. This will be our closed-form description of morphisms in the free traced category. It follows from Proposition 3, together with equations (1)–(4), (8), (9), (10), that this is indeed closed under the traced monoidal operations.

We define the main operations on morphisms.

Composition

$$(T, \sigma, \mu) \circ (S, \pi, \lambda) = (S \uplus T, \sigma \circ \pi, i \mapsto \mu_{\pi(i)} \circ \lambda_i)$$

Tensor Product

$$(S, \pi, \lambda) \otimes (T, \sigma, \mu) = (S \uplus T, \pi \otimes \sigma, [\lambda, \mu])$$

Trace

$$\mathsf{Tr}^{m}_{n,n}(S, \pi, \lambda) = (S \uplus T, p_\pi, \mu)$$

where

$$T = \{[\lambda_{\pi^l(j)} \circ \cdots \circ \lambda_j]_\sim \mid \pi^{l+1}(j) = j \in \mathcal{L}(\pi)\},$$

$$\mu : i \mapsto \lambda_{p_\pi(i)} \circ \lambda_{\pi^k(i)} \circ \cdots \circ \lambda_i.$$

Note that $F_{\mathsf{Tr}}(\mathcal{C})(\mathrm{I},\mathrm{I}) = \mathcal{M}(\mathcal{L}[\mathcal{C}])$. Also,

$$F_{\mathsf{Tr}}(\mathbf{1}) = (\coprod_{n \in \mathbb{N}} S(n)) \times (\mathbb{N}, +, 0).$$

That is, the objects in this free category are the natural numbers; a morphism is a pair (π, n), where π is a permutation, and n is a natural number counting the number of loops.

4.4 Compact Closed Categories

The free construction for compact closed categories was characterized in the pioneering paper by Kelly and Laplaza [19]. Their construction is rather complex. Even when simplified to the strict monoidal case, several aspects of the construction are bundled in together, and it can be hard to spot what is going one. (For example, the path construction we gave for the trace in the previous section is implicit in their paper — but not easy to spot!). We are now in a good position to disentangle and clarify their construction. Indeed, we have already explictly constructed $F_{\mathsf{Tr}}(\mathcal{C})$, and there is the $\mathcal{G}$ or Int construction of Joyal, Street and Verity [14][5], which is developed in the symmetric monoidal context with connections to Computer Science issues and the Geometry of Interaction in [2]. This construction gives the free compact closed category generated by a traced monoidal category. Thus we can recover the Kelly-Laplaza construction as the composition of these two adjunctions:

$$\mathbf{Cat} \underset{U_{\mathsf{Tr}}}{\overset{F_{\mathsf{Tr}}}{\underset{\longleftarrow}{\overset{\longrightarrow}{\bot}}}} \mathsf{Tr}\text{−}\mathbf{Cat} \underset{U}{\overset{\mathcal{G}}{\underset{\longleftarrow}{\overset{\longrightarrow}{\bot}}}} \mathsf{CC}\text{−}\mathbf{Cat}$$

Adjoints compose, so $F_{\mathsf{CC}}(\mathcal{C}) = \mathcal{G} \circ F_{\mathsf{Tr}}(\mathcal{C})$. This factorization allows us to 'rationally reconstruct' the Kelly-Laplaza construction.

The main notion which has to be added to those already present in $F_{\mathsf{Tr}}(\mathcal{C})$ is that of *polarity*. The ability to distincguish between positive and negative occurrences of a variable will allow us to transpose variables from inputs to outputs, or vice versa. This possibility of transposing variables means that we no longer have the simple situation that morphisms must be between lists of generating objects of the same length. However, note that in a compact closed category, $(A \otimes B)^* \simeq A^* \otimes B^*$, so any object constructed from generating objects by tensor product and duality will be isomorphic to one of the form

$$\bigotimes_i A_i \otimes \bigotimes_j B_j^*.$$

Moreover, any morphism

$$f : \bigotimes_i A_i \otimes \bigotimes_j B_j^* \longrightarrow \bigotimes_k C_k \otimes \bigotimes_l D_l^* \tag{14}$$

[5] Prefigured in [1], and also in some unpublished lectures of Martin Hyland [13].

will, after transposing the negative objects, be in biunique correspondence with one of the form

$$\bigotimes_i A_i \otimes \bigotimes_l D_l \longrightarrow \bigotimes_k C_k \otimes \bigotimes_j B_j. \tag{15}$$

A key observation is that *in the free category, this transposed map (15) will again be of the closed form (13)* which characterizes morphisms in $\mathcal{F}_{\mathsf{Tr}}(\mathcal{C})$, as we saw in the previous section. From this, the construction of $\mathcal{F}_{\mathsf{CC}}(\mathcal{C})$ will follow directly.

Objects. The objects in $\mathcal{F}_{\mathsf{CC}}(\mathcal{C})$ are, following the $\mathcal{G}$ construction applied to $\mathcal{F}_{\mathsf{Tr}}(\mathcal{C})$, pairs of objects of $\mathcal{F}_{\mathsf{Tr}}(\mathcal{C})$, hence of the form (n, m, A^+, A^-), where

$$A^+ : [n] \longrightarrow \mathsf{Ob}\,\mathcal{C} \qquad A^- : [m] \longrightarrow \mathsf{Ob}\,\mathcal{C}.$$

Such an object can be read as the tensor product

$$\bigotimes_{i=1}^{n} A_i^+ \otimes \bigotimes_{j=1}^{m} (A_j^-)^*.$$

This is equivalent to the Kelly-Laplaza notion of signed set, under which objects have the form (n, A, sgn), where $\mathsf{sgn} : [n] \to \{+, -\}$.

Operations on Objects. The tensor product is defined componentwise on the positive and negative components. Formally:

$$(n, m, A^+, A^-) \otimes (p, q, B^+, B^-) = (n + p, m + q, [A^+, B^+], [A^-, B^-]).$$

The duality simply interchanges positive and negative components:

$$(n, m, A^+, A^-)^* = (m, n, A^-, A^+).$$

Note that the duality is involutive, and distributes through tensor:

$$A^{**} = A, \qquad (A \otimes B)^* = A^* \otimes B^*.$$

Morphisms. A morphism has the form

$$(S, \pi, \lambda) : (n, m, A^+, A^-) \longrightarrow (p, q, B^+, B^-)$$

where we require $n + q = k = m + p$, $\pi \in S(k)$, and $\lambda : [k] \to \mathsf{Mor}\,\mathcal{C}$, such that

$$\lambda_i : [A^+, B^-]_i \longrightarrow [B^+, A^-]_{\pi(i)}.$$

S is a multiset of loops, just as in $\mathcal{F}_{\mathsf{Tr}}(\mathcal{C})$. Note that (S, λ, π) can indeed be seen as a morphism in $\mathcal{F}_{\mathsf{Tr}}(\mathcal{C})$ in the transposed form (15), as discussed previously.

We now describe the compact closed operations on morphisms.

Composition. Composition of a morphism $f : A \to B$ with a morphism $g : B \to B$ is given by feeding 'outputs' by f from the positive component of B as inputs to g (since for g, B occurs negatively, and hence the positive and negative components are interchanged); and symmetrically, feeding the g outputs from the negative components of B as inputs to f. This symmetry allows the strong form of duality present in compact closed categories to be interpreted in a very direct and natural fashion.

This general prescription is elegantly captured algebraically in terms of the trace, which co-operates with the duality to allow symmetric interaction between the two morphisms which are being composed. This is illustrated by the following diagram, which first appeared in [1]:

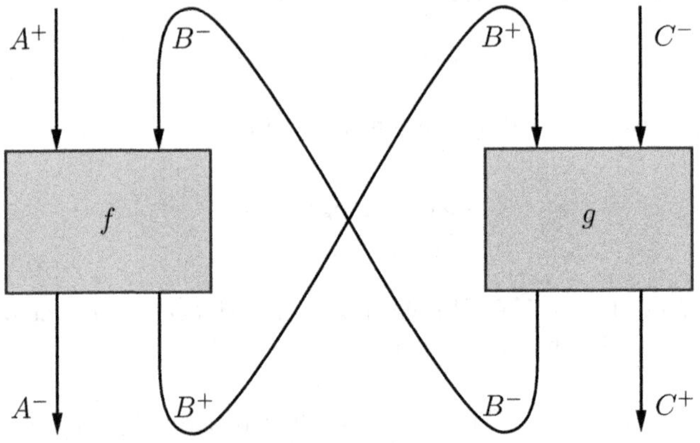

A concrete account for $F_{\mathsf{CC}}(\mathcal{C})$ follows directly from our description of the trace in $F_{\mathsf{Tr}}(\mathcal{C})$: chase paths, and compose (in $\mathcal{C}$) the morphisms labelling the paths to get the labels. In general, loops will be formed, and must be added to the multiset. Formally, given arrows

$$f : A \longrightarrow B, \qquad g : B \longrightarrow C$$

where

$$f = (S, \pi, \lambda) : (n, m, A^+, A^-) \longrightarrow (p, q, B^+, B^-)$$
$$g = (T, \sigma, \mu) : (p, q, B^+, B^-) \longrightarrow (r, s, C^+, C^-)$$

we form the composition

$$(S \uplus T \uplus U,\ \mathrm{EX}(\pi, \sigma),\ \rho) : (n, m, A^+, A^-) \longrightarrow (r, s, C^+, C^-).$$

There is an algebraic description of the permutation component $\mathrm{EX}(\pi, \sigma)$, which can be derived from our algebraic description of p_π in the previous section. In the same manner as we did there, we can decompose each of π and σ into four components, which we write as matrices:

$$\pi = \begin{pmatrix} \pi_{A^+A^-} & \pi_{A^+B^+} \\ \pi_{B^-A^-} & \pi_{B^-B^+} \end{pmatrix} \qquad \sigma = \begin{pmatrix} \sigma_{A^+A^-} & \sigma_{A^+B^+} \\ \sigma_{B^-A^-} & \sigma_{B^-B^+} \end{pmatrix}$$

Now if we write

$$\mathrm{EX}(\pi, \sigma) = \theta = \begin{pmatrix} \theta_{A^+A^-} & \theta_{A^+C^+} \\ \theta_{C^-A^-} & \theta_{C^-C^+} \end{pmatrix}$$

then we can define

$$\begin{array}{rcl} \theta_{A^+A^-} & = & \pi_{A^+A^-} \cup \pi_{A^+B^+};\sigma_{B^+B^-};(\pi_{B^-B^+};\sigma_{B^+B^-})^*;\pi_{B^-A^-} \\ \theta_{A^+C^+} & = & \pi_{A^+B^+};(\sigma_{B^+B^-};\pi_{B^-B^+})^*;\sigma_{B^+C^+} \\ \theta_{C^-A^-} & = & \sigma_{C^-B^-};(\pi_{B^-B^+};\sigma_{B^+B^-})^*;\pi_{B^-A^-} \\ \theta_{C^-C^+} & = & \sigma_{C^-C^+} \cup \sigma_{C^-B^-};\pi_{B^-B^+};(\sigma_{B^+B^-};\pi_{B^-B^+})^*;\sigma_{B^+C^+}. \end{array}$$

This is essentially the 'Execution formula' [12] — see also [14] and [2]; it appears implicitly in [19] as a coequaliser.

Similarly, we can characterize the loops formed by composing π and σ, $\mathcal{L}(\pi, \sigma)$, by

$$\begin{array}{rcl} j & \in & \bigcup \mathcal{L}(\pi, \sigma) \iff \\ \langle j, j \rangle & \in & ((\pi_{B^-B^+};\sigma_{B^+B^-})^* \cap \mathsf{id}_{B^-}) \cup ((\sigma_{B^+B^-};\pi_{B^-B^+})^* \cap \mathsf{id}_{B^+}). \end{array}$$

The labelling function ρ simply labels $\mathrm{EX}(\pi, \sigma) : i \mapsto j$ with the arrow in $\mathcal{C}$ formed by composing the arrows labelling the arcs in the path from i to j described by the above 'flow matrix'. Similarly, U is the multiset of loops labelling the cycles in $\mathcal{L}(\pi, \sigma)$.

One can give algebraic descriptions of ρ and U by reformulating λ and μ as *graph homomorphisms* into (the underlying graph of) $\mathcal{C}$. One can then form a homomorphism $\nu : G \to U_{\mathbf{Graph}}\mathcal{C}$ from a combined graph G, which gives an 'intensional description' of the composition. This combined graph will comprise the disjoint union of the graphs corresponding to the two arrows being composed, together with explicit *feedback arcs*, labelled by ν with identity arrows in $\mathcal{C}$. One then considers the *path category* G^* freely generated from this graph [20, 2.VII]; the above flow expressions for $\mathrm{EX}(\pi, \sigma)$ yield a description of the paths in this category when relational composition is reinterpreted as concatenation of paths. We can then read off ρ and U from the unique functorial extension of ν to this path category.

Tensor Product. This is defined componentwise as in $F_{\mathsf{Tr}}(\mathcal{C})$, with appropriate permutation of indices in order to align positive and negative components correctly.

Units and Counits. Firstly, we describe the identity morphisms explicitly:

$$\mathsf{id} = (\varnothing, \mathsf{id}_{[n+m]}, i \mapsto 1_{[A,B]_i}) : (n, m, A^+, A^-) \longrightarrow (n, m, A^+, A^-).$$

We join each dot in the input to the corresponding one in the output, and label it with the appropriate identity arrow.

Now consider the unit $\eta_A : \mathrm{I} \to A^* \otimes A$. Once we have unpacked the definition of what comprises an arrow of this type, we see that we can make exactly the same definition as for the identity! The unit is just *the right transpose of the identity*. Similarly, the counit is *the left transpose of the identity.*

Thus identities, units and counits are essentially all the same, except that the polarities allow variables to be transposed freely between the domain and codomain.

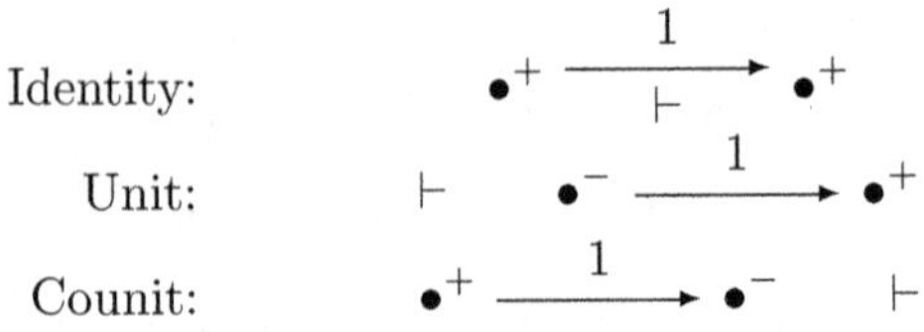

4.5 Strongly Compact Closed Categories

We now wish to analyze the new notion of strongly compact closed category in the same style as the previous constructions. Fortunately, there is a simple observation which makes this quite transparent. *Provided that the category we begin with is already equipped with an involution* (but no other structure), then this involution 'lifts' through all our constructions, yielding the free 'dagger version' (in the sense of [25]) of each of our constructions. In particular, our construction of $\mathcal{F}_{\mathsf{CC}}(\mathcal{C})$ in the previous section in fact gives rise to the free strongly compact closed category.

More precisely, we shall describe an adjunction

$$\mathbf{InvCat} \; \underset{U_S CC}{\overset{F_S CC}{\rightleftarrows}} \; \mathbf{SCC\text{-}Cat} \qquad (\bot)$$

where **InvCat** is the category of categories with a specified *involution, i.e.* an identity on objects, contravariant, involutive functor; and functors preserving the involution.

Our previous construction of $\mathcal{F}_{\mathsf{CC}}(\mathcal{C})$ lifts directly to this setting. The main point is that we can define an involution $()^\dagger$ on $\mathcal{F}_{\mathsf{CC}}(\mathcal{C})$, under the assumption that we are given a primitive $()^\dagger$ on the generating category $\mathcal{C}$. The dagger on $\mathcal{F}_{\mathsf{CC}}(\mathcal{C})$ will endow it with the structure of a strongly compact closed category (for which the compact closed part will coincide with that already described for $\mathcal{F}_{\mathsf{CC}}(\mathcal{C})$).

Given

$$(S, \pi, \lambda) : (n, m, A^+, A^-) \longrightarrow (p, q, B^+, B^-),$$

we can define

$$(S, \pi, \lambda)^\dagger = (\{[s^\dagger]_\sim \mid [s]_\sim \in S\}, \pi^{-1}, j \mapsto \lambda^\dagger_{\pi^{-1}(j)}).$$

In short, we reverse direction on the arrows connecting the dots (including reversing the direction of loops), and label the reversed arrows with the reversals of the original labels. This contrasts with the dual f^*, which by the way types are interpreted in this free situation, is essentially *the same combinatorial object* as f, but with a different 'marking' by polarities — there are no reversals involved. Thus, if we had a labelling morphism

$$\lambda_i = [A^+, B^-]_i \xrightarrow{f_i} [B^+, A^-]_{\pi(i)=j}$$

then we will get

$$(\lambda^\dagger)_j = [B^+, A^-]_j \xrightarrow{f_i^\dagger} [A^+, B^-]_{\pi^{-1}(j)=i}.$$

It is easy to see that $\eta_A = \epsilon_A^\dagger$, so this is compatible with our previous construction of $F_{\mathsf{CC}}(\mathcal{C})$.

4.6 Parameterizing on the Monoid

So far, the scalars have arisen intrinsically from the loops in the generating category $\mathcal{C}$. However, we may wish for various reasons to be able to 'glue in' a preferred multiplicative monoid of scalars into our traced, compact closed,or strongly compact closed category, For example, we may wish to consider only a few generating morphisms, but to take the complex numbers $\mathbb{C}$ as scalars. We will present a construction which accomodates this, as a simple refinement of the previous ones. There are versions of this construction for each of the traced, compact closed, and strongly compact closed cases: we shall only discuss the last of these.

Firstly, we note that there is a functor

$$\mathcal{L} : \mathbf{InvCat} \longrightarrow \mathbf{InvSet}$$

which sends a category to its sets of loops. The dagger defines an involution on the set of loops. Involution-preserving functors induce involution-preserving functions on the loops.

Now let **InvCMon** be the category of commutative monoids with involution, and involution-preserving homomorphisms. There is an evident forgetful functor $U_{\mathbf{InvCMon}} \longrightarrow \mathbf{InvSet}$. We can form the comma category $(\mathcal{L} \downarrow U_{\mathbf{InvCMon}})$, whose objects are of the form $(\mathcal{C}, \varphi, M)$, where φ is an involution-preserving map from $\mathcal{L}[\mathcal{C}]$ to the underlying set of M. Here we can think of M as the prescribed monoid of scalars, and φ as specifying how to evaluate loops from $\mathcal{C}$ in this monoid.

There is a forgetful functor $U_{\mathcal{V}} : \mathsf{SCC}\text{−}\mathbf{Cat} \longrightarrow \mathcal{V}$

$$U_{\mathcal{V}} : \mathcal{C} \longmapsto (U_{\mathsf{SCC}}(\mathcal{C}),\ f : A \to A \longmapsto \mathsf{Tr}^A_{\mathrm{I,I}}(f),\ \mathcal{C}(\mathrm{I},\mathrm{I})).$$

Our task is to construct an adjunction

$$\mathcal{V} \underset{U_{\mathcal{V}}}{\overset{F_{\mathcal{V}}}{\underset{\longleftarrow}{\overset{\longrightarrow}{\perp}}}} \mathbf{SCC\text{−}Cat}$$

which builds the free SCC on a category *with prescribed scalars.* This is a simple variation on our previous construction of $F_{\mathsf{SCC}}(\mathcal{C})$, which essentially acts by composition with the loop evaluation function φ on $F_{\mathsf{SCC}}(\mathcal{C})$. We use the prescribed monoid M in place of $\mathcal{M}(\mathcal{L}[\mathcal{C}])$. Thus a morphism in $F_{\mathcal{V}}(\mathcal{C})$ will have the form (m, π, λ), where $m \in M$. Multiset union is replaced by the monoid operation of M. The action of the dagger functor on elements of M is by the given involution on M. When loops in $\mathcal{C}$ arise in forming compositions in the free category, they are evaluated in M using the function φ.

The monoid of scalars in this free category will of course be M.

References

1. S. Abramsky, R. Jagadeesan, New Foundations for the Geometry of Interaction, *Information and Computation,* 111(1):53-119, 1994. Conference version appeared in LiCS '92.
2. S. Abramsky. Retracing some paths in process algebra. *CONCUR 96: Proceedings of the Seventh International Conference on Concurrency Theory*, LNCS **1119**, 1–17, 1996.
3. S. Abramsky and E. Haghverdi and P. J. Scott. Geometry of Interaction and Linear Combinatory Algebras. *Mathematical Structures in Computer Science* 12:625–665, 2002.
4. S. Abramsky and B. Coecke. *A categorical semantics of quantum protocols.* Proceedings of the 19th Annual IEEE Symposium on Logic in Computer Science (LiCS'04), IEEE Computer Science Press, 415–425, 2004. (extended version at `arXiv: quant-ph/0402130`)
5. S. Abramsky and B. Coecke. Abstract Physical Traces. *Theory and Applications of Categories*, 14:111–124, 2005.
6. S. Abramsky and R. W. Duncan. *Categorical Quantum Logic.* In the Proceedings of the Second International Workshop on Quantum Programming Languages, 2004.
7. M. Barr. **-autonomous Categories.* Springer-Verlag 1979.
8. G. Birkhoff and J. von Neumann. The logic of quantum mechanics. *Annals of Mathematics* 37, 823–843, 1937.
9. B. Coecke. Delinearizing Linearity. Draft paper, 2005.
10. P. Deligne. Catégories Tannakiennes. In *The Grothendiek Festschrift, Vol. II*, 111–195. Birkhauser 1990.
11. J.-Y. Girard, Linear Logic. *Theoretical Computer Science* 50(1):1-102, 1987.
12. J.-Y. Girard, Geometry of Interaction I: Interpretation of System F, in: *Logic Colloquium '88*, ed. R. Ferro, et al. North-Holland, pp. 221-260, 1989.
13. Martin Hyland. Personal communication, July 2004.
14. A. Joyal, R. Street and D. Verity, Traced monoidal categories. *Math. Proc. Camb. Phil. Soc.* 119, 447–468, 1996.
15. C. Kassel. *Quantum Groups.* Springer-Verlag 1995.
16. P. Katis, N. Sabadini and R. F. C. Walters. Feedback, trace and fixed point semantics. Proceedings of FICS01: Workshop on Fixed Points in Computer Science, 2001. Available at `http://www.unico.it/ walters/papers/index.html`
17. G. M. Kelly. Many-variable functorial calculus I. *Springer Lecture Notes in Mathematics 281*, 66–105, 1972.
18. G. M. Kelly. An abstract approach to coherence. *Springer Lecture Notes in Mathematics 281*, 106–147, 1972.

19. G. M. Kelly and M. L. Laplaza. Coherence for compact closed categories. *Journal of Pure and Applied Algebra* **19**, 193–213, 1980.
20. S. Mac Lane. *Categories for the Working Mathematician.* Springer-Verlag 1971.
21. Dusko Pavlovic. *A semantical approach to equilibria, adaptation and evolution.* Unpublished manuscript, November 2004.
22. R. Penrose. Applications of negative-dimensional tensors. In *Combinatorial Mathematics and Its Applications*, ed. D. J. Welsh, 221-244, Academic Press 1971.
23. N. S. Rivano. *Catégories Tannakiennes.* Springer-Verlag 1972.
24. P. Selinger. Towards a quantum programming language. *Mathematical Structures in Computer Science* 14(4), 527–586, 2004.
25. P. Selinger. Dagger compact closed categories and completely positive maps. To appear in *Proceedings of the 3rd International Workshop on Quantum Programming Languages*, 2005.

Appendix

The following equational proofs involve some long typed formulas. To aid in readability, we have annotated each equational step (reading down the page) by *underlining* each redex, and *overlining* the corresponding contractum.

Proof of Lemma 3

Proof.

$$\underline{\mathsf{Tr}^U_{A,B}(f) \otimes \mathsf{Tr}^V_{C,D}(g)}$$

$$= \{\text{Superposing}\}$$

$$\overline{\mathsf{Tr}^V_{A\otimes C,B\otimes D}(\underline{\mathsf{Tr}^U_{A,B}(f) \otimes g})}$$

$$= \{\text{Naturality of } \tau\}$$

$$\mathsf{Tr}^V_{A\otimes C,B\otimes D}(\overline{\tau_{D\otimes V,B} \circ (\underline{g \otimes \mathsf{Tr}^U_{A,B}(f)}) \circ \tau_{A,C\otimes V}})$$

$$= \{\text{Superposing}\}$$

$$\mathsf{Tr}^V_{A\otimes C,B\otimes D}(\underline{\tau_{D\otimes V,B}} \circ \overline{\mathsf{Tr}^U_{C\otimes V\otimes A,D\otimes V\otimes B}(g \otimes f)} \circ \underline{\tau_{A,C\otimes V}})$$

$$= \{\text{Input/Output Naturality}\}$$

$$\mathsf{Tr}^V_{A\otimes C,B\otimes D}(\mathsf{Tr}^U_{A\otimes C\otimes V,B\otimes D\otimes V}(\underline{\overline{(\tau_{D\otimes V,B} \otimes 1_U)} \circ (g \otimes f) \circ \overline{(\tau_{A,C\otimes V} \otimes 1_U)}}))$$

$$= \{\mathsf{SM}\text{ Coherence}\}$$

$$\underline{\mathsf{Tr}^V_{A\otimes C,B\otimes D}(\mathsf{Tr}^U_{A\otimes C\otimes V,B\otimes D\otimes V}}(\overline{(1_A \otimes \tau_{U,D\otimes V}) \circ (f \otimes g) \circ (1_A \otimes \tau_{C\otimes V,U})}))$$

$$= \{\text{Vanishing II}\}$$

$$\overline{\mathsf{Tr}^{V\otimes U}_{A\otimes C,B\otimes D}}((1_A \otimes \tau_{U,D\otimes V}) \circ (f \otimes g) \circ (1_A \otimes \tau_{C\otimes V,U})).$$

□

Proof of Lemma 4

Proof. Note that for $k = 0$, this is just Vanishing I. We now reason inductively when $k > 0$.

$$
\begin{aligned}
& \underline{\mathsf{Tr}^{U}_{A_1,A_{k+2}}}(\sigma_{k+1} \circ \textstyle\bigotimes_{i=1}^{k+1} f_i) \\
= {} & \{\text{Vanishing II}\} \\
& \overline{\mathsf{Tr}^{A_2}_{A_1,A_{k+2}}(\mathsf{Tr}^{V}_{A_1\otimes A_2,A_{k+2}\otimes A_2}}(\underline{\sigma_{k+1} \circ \textstyle\bigotimes_{i=1}^{k+1} f_i})) \\
= {} & \{(12)\} \\
& \mathsf{Tr}^{A_2}_{A_1,A_{k+2}}(\mathsf{Tr}^{V}_{A_1\otimes A_2,A_{k+2}\otimes A_2}(\overline{\underline{(\tau \otimes 1)} \circ (1 \otimes \sigma_k) \circ \textstyle\bigotimes_{i=1}^{k+1} f_i})) \\
= {} & \{\text{Naturality in } A_{k+2} \otimes A_2\} \\
& \mathsf{Tr}^{A_2}_{A_1,A_{k+2}}(\overline{\tau} \circ \mathsf{Tr}^{V}_{A_1\otimes A_2,A_2\otimes A_{k+2}}((1 \otimes \sigma_k) \circ \underline{\textstyle\bigotimes_{i=1}^{k+1} f_i})) \\
= {} & \{\text{Bifunctoriality of } \otimes\} \\
& \mathsf{Tr}^{A_2}_{A_1,A_{k+2}}(\tau \circ \mathsf{Tr}^{V}_{A_1\otimes A_2,A_2\otimes A_{k+2}}((1 \otimes \sigma_k) \circ \overline{(1 \otimes \textstyle\bigotimes_{i=2}^{k+1} f_i) \circ (\underline{f_1 \otimes 1_U})})) \\
= {} & \{\text{Naturality in } A_1 \otimes A_2\} \\
& \mathsf{Tr}^{A_2}_{A_1,A_{k+2}}(\tau \circ \mathsf{Tr}^{V}_{A_2\otimes A_2,A_2\otimes A_{k+2}}((1 \otimes \sigma_k) \circ (1 \otimes \textstyle\bigotimes_{i=2}^{k+1} f_i)) \circ (\underline{\overline{f_1 \otimes 1_{A_2}}})) \\
= {} & \{\text{Naturality in } A_1\} \\
& \mathsf{Tr}^{A_2}_{A_2,A_{k+2}}(\tau \circ \mathsf{Tr}^{V}_{A_2\otimes A_2,A_2\otimes A_{k+2}}(\underline{(1 \otimes \sigma_k) \circ (1 \otimes \textstyle\bigotimes_{i=2}^{k+1} f_i)})) \circ \overline{f_1} \\
= {} & \{\text{Bifunctoriality of } \otimes\} \\
& \mathsf{Tr}^{A_2}_{A_2,A_{k+2}}(\tau \circ \underline{\mathsf{Tr}^{V}_{A_2\otimes A_2,A_2\otimes A_{k+2}}(\overline{1 \otimes (\sigma_k \circ \textstyle\bigotimes_{i=2}^{k+1} f_i)})}) \circ f_1 \\
= {} & \{\text{Superposing}\} \\
& \mathsf{Tr}^{A_2}_{A_2,A_{k+2}}(\tau \circ \overline{(1 \otimes \underline{\mathsf{Tr}^{V}_{A_2,A_{k+2}}(\sigma_k \circ \textstyle\bigotimes_{i=2}^{k+1} f_i)})}) \circ f_1 \\
= {} & \{\text{Induction hypothesis}\} \\
& \mathsf{Tr}^{A_2}_{A_2,A_{k+2}}(\underline{\tau \circ (1 \otimes (\overline{f_{k+1} \circ \cdots \circ f_2}))}) \circ f_1 \\
= {} & \{\text{Naturality of } \tau\} \\
& \mathsf{Tr}^{A_2}_{A_2,A_{k+2}}(\overline{\underline{((f_{k+1} \circ \cdots \circ f_2) \otimes 1)} \circ \tau}) \circ f_1 \\
= {} & \{\text{Naturality in } A_{k+2}\} \\
& \overline{(f_{k+1} \circ \cdots \circ f_2)} \circ \underline{\mathsf{Tr}^{A_2}_{A_2,A_2}(\tau)} \circ f_1 \\
= {} & \{\text{Yanking}\} \\
& (f_{k+1} \circ \cdots \circ f_2) \circ \overline{1_{A_2}} \circ f_1 \\
= {} & f_{k+1} \circ \cdots \circ f_2 \circ f_1 .
\end{aligned}
$$

□

Labels from Reductions: Towards a General Theory

Bartek Klin, Vladimiro Sassone, and Paweł Sobociński

Warsaw University, University of Sussex, and PPS – Université Paris VII

Abstract. We consider open terms and parametric rules in the context of the systematic derivation of labelled transitions from reduction systems.

1 Introduction

Since the seminal ideas of logicians of the early 20th century, it has become customary to encapsulate the dynamics of computation in terse and elegant *rewrite* calculi. For instance, the essence of conventional computation is condensed in Church's beta-reduction rule $(\lambda x.M)N \rightarrow M\{x := N\}$, while the mechanics of π-calculus interaction is captured by the rule

$$\overline{a}\langle n\rangle.P \mid a(x).Q \longrightarrow P \mid Q\{x := n\} .$$

The beauty and power of such formalisms can hardly be overestimated: they centre our models on the essential, and help us focus our reasoning on fundamental principles. However, models are normally used not only to describe, but also to design, specify, analyse, and – most importantly – as the foundations for advanced, ground-breaking techniques. A well consolidated, relevant example is 'model checking,' where simple tools used at a suitable abstraction level and driven by powerful ideas have afforded spectacular results. Similarly, notions revolving around semantic equivalences and coinduction have had a strong, lasting impact.

Several such ideas rely on relatively lower-level models based on *transition systems*. Intuitively, these describe individual steps of computing entities, rather than providing an overall picture of the computational primitives of the model as such. For instance, in the case of π-calculus terms a transition $\overline{a}\langle n\rangle.P \xrightarrow{\overline{a}n} P$ would express that the system is ready to evolve to P by engaging in action $\overline{a}$ and offering it to (potential partners in) the environment. There would then be a dual rule $a(x).Q \xrightarrow{a(n)} Q\{x := n\}$ for message receivers, and finally an inference rule would dictate how dual actions can meet in the environment and complete each other to yield finished interactions.

$$\frac{A \xrightarrow{\overline{a}n} A' \qquad B \xrightarrow{a(n)} B'}{A \mid B \longrightarrow A' \mid B'}$$

Although the resulting term-transformation systems are equivalent, the differences between these approaches are significant, and are better not dismissed hastily by a simple 'matter-of-taste' argument. The fundamental point of a 'labelled-transition' semantics is that it is compositional: it explains the behaviour of complex systems by extrapolating it from the behaviour of their components. This is in sharp contrast with a 'reduction' semantics, where $\overline{a}\langle n\rangle.P$ and $a(x).Q$ are completely inert, have no meaning of their own. This distinction is of paramount importance for applications like model

J.L. Fiadeiro et al. (Eds.): CALCO 2005, LNCS 3629, pp. 30–50, 2005.

checking and bisimulation, that rely on the information afforded by labels and transitions to analyse system components in isolation.

Influenced by Plotkin's successful 'structural operational semantics' [10], many formalisms in the seventies and eighties had been originally equipped exclusively with a labelled-transition semantics, including CCS and the π-calculus. In recent years however it has become increasingly important for complex computational models to have both a reduction semantics, to explain their mechanics in intuitive, self-justifying terms, and a labelled-transition semantics, to serve as basis for semantic analysis. In particular, several papers have been devoted to identify characterisations of reduction-based equivalences in terms of labels and bisimulations. This is the context of the present work: is it possible, and how, to *derive labelled transition systems from reductions* so as to equip calculi with rich and treatable semantics theories? And to what extent can this be done parametrically, i.e., independently on the specific calculus at hand? Questions like these gained momentum as work on 'universal' models emerged from the field of concurrency, as e.g. action calculi [7], tile systems [2], and, more recently, bigraphs [8,3]. Such models are meant to provide general frameworks independent of specific models, such that several calculi can be recast and understood as fragments therein. In ambitious terms, one could think of these frameworks as semantic universes which individual models can be instantiated from. The question therefore arose as to how to associate meaning and reasoning techniques to such 'universal' meta-models.

Much progress has been made since, mainly by Robin Milner and his collaborators. The rest of this introduction will revisit the main ideas underlying the approach, whilst the main body of paper will present the technical details in a slightly novel fashion, and try to accommodate in the theory the idea of parametric rules and open terms.

The central technical challenge is thus how to associate labelled transitions to terms from reduction systems. Peter Sewell[16] exploited the intuition that labels in labelled transition systems express the compositional properties of terms, i.e., the extend to which a term is amenable to engage in interactions with the environment, and how. Thus, if term a when inserted in a *context* $c[-]$ can perform a reduction, say $c[a] \to a'$, then $c[-]$ is a strong candidate as a label for a transition $a \xrightarrow{c} a'$. (This spells out as: 'a is ready to interact with context $c[-]$, and a' would be the result of such potential interaction.') This intuition is very suggestive indeed; the devil however is as usual the details: in order for this idea to give a sensible bisimulation, it is fundamental to select carefully which contexts to consider: certainly not all, but only those c which are the '*smallest*' to trigger a given reduction. Failing to do so would give rise to a 'garbled' semantics, as the excess transitions would convey misleading information as to what term a is ready to engage with and what the environment is expected to contribute.

The need to formalise the notion of 'smallest' leads to *category theory*, where it is possible to express such universal properties in term of uniqueness of certain 'arrow' factorisation. For instance, in categorical terms the fact that C is the disjoint union (the so-called coproduct) of sets A and B is expressed by saying that all pairs of maps (arrows) $f : A \to X$ and $g : B \to X$ factor uniquely via injections into C and a map $[f, g] : C \to X$. In complete analogy, a context $c[-]$ is the 'smallest' to create redex l in a, if all contexts $c'[-]$ that create l factor as $c'[-] = e[c[-]]$ for a unique context $e[-]$. For instance, in the λ-calculus

$$\lambda x.x \xrightarrow{(-)y} y, \qquad \text{but not} \qquad \lambda x.x \xrightarrow{(-)yz} yz,$$

as $(-)yz$ arises uniquely as the composition of $(-)z$ and $(-)y$.

The first step to rephrase our notion of 'smallness' as a problem of unique arrow factorisation is to recast terms as arrows in categories. This can be done following Lawvere's seminal approach to algebraic theories, that here we instantiate using MacLane's notion of 'product and permutation' category (PROP) – roughly speaking, 'linear' Lawvere theories – that we recall in §2. An arrow $f : n \to m$ in a PROP represents a m-tuple of contexts containing altogether n 'holes;' i.e., when f is fed with n terms to plug its holes, it yields a tuple of m terms. The question as to whether or not term a in context c manifests a redex l becomes now whether there exists a suitable context d such that $ca = dl$. This allows us to express the minimality of c by ranging over all equations of the kind $c'a = d'l$, seeking for unique ways to factor c' through c. In §3 we recall how such universal property is elegantly expressed by the notion of *idem-relative-pushout*, a breakthrough due Leifer and Milner [4]. Remarkably, such formalisation supports the central '*congruence theorem*' that bisimulation on the labelled transition systems derived following the theory is a congruence, i.e., it is closed under all contexts. Due to the generality of the framework, such a result has already been applied to a variety of different models [3,13,9,12,1,14].

This paper's original contribution concerns our initial ideas on the treatment of *open terms* and *parametric rules* in the above framework.

For the sake of illustration, let us consider on the CCS rule for interaction. To express such a rule as a collection of ground rules $\overline{a}P \mid aQ \longrightarrow P \mid Q$ is not entirely satisfactory in this setting: even in the simplest cases, we have infinitely many rewrite rules to deal with, and these give rise to infinitely many higher-order labels, e.g., of the kind $\overline{a}P \xrightarrow{-|aQ} P \mid Q$. This appears to make a poor use of the generality, elegance and succinctness of theory of relative pushouts. Ideally, the rule should be expressed parametrically, as in[1]

$$a.1 \mid \overline{a}.2 \longrightarrow 1 \mid 2$$

and the labels should be derivable for open terms with universal property imposed both on the contexts and the parameters. A label should thus consist both of a smallest context and the most general parameter which makes a reduction possible; for example

$$\langle a.P, 1\rangle \xrightarrow{1|\overline{a}.2} P \mid 1 \qquad \text{and} \qquad a.P \mid 1 \xrightarrow[\overline{a}.1]{} P \mid 1,$$

where the label above the transition denotes a context with two holes (to insert the left-hand pair in), and the label below a transition denotes a parameter (to fill the left-hand side open term with).

As it turns out, the framework is robust enough to adapt easily to the new question. Rather than investigating ('square') equations such as $ca = dl$, we now face '*hexagonal*' equations $cap = dlq$ in order to establish the universal property that, at the same time, identifies the smallest context c as well as the largest parameter p that unearth redex l in term a. The main technical device we introduce to that purpose is to pair the notion

[1] In the paper we use natural numbers to denote context parameters ('holes').

of slice pushout (a rephrasing of relative pushouts) with a dual notion of coslice pullback: the role of the pushout is to determine c as before, while the pullback of course ascertains p. Such coupling of universal properties gives rise to the new notion of '*lux*' (locally universal hexagon), introduced in §4. These have been considered previously by Peter Sewell, who referred to them as hex-RPOs. In fact, much of our technical development has been foreshadowed in his unpublished notes [15].

Our main technical results are a characterisation of categories with luxes in terms of slice pushouts and coslice pullbacks (Theorem 1) and, of course, the fundamental *congruence theorem* for the labelled transition systems derived using our theory of luxes (Theorem 3). Most of the ideas presented here are work in progress, and in the concluding section we discuss merits and shortcomings of our proposal, as well as identifying some of the main avenues of future work on luxes.

Structure of the Paper. In §2 we recall the notion of PROP and the construction of categories of terms. §3 illustrates the existing theory based on slice pushouts, and its extension to a bicategorical setting. §4-6 contain the main body of the paper, with our definition of luxes, their properties, and the congruence theorem. Finally, §7 discusses the shortcomings of the current theory and points forward to open issues and future research.

We assume the reader to have a basic knowledge of category theory, as can be acquired from any graduate textbook. Throughout the paper we use standard categorical notations, where $\circ$ denotes (right-to-left) composition and is most often omitted.

2 PROPs as Categories of Terms

A 'product and permutation' category [5], PROP, can be described, roughly, as a linear Lawvere theory; more accurately, PROPs are one-sorted symmetric monoidal theories whereas Lawvere theories are one-sorted finite product theories. We recall a straightforward definition below.

Definition 1 (PROP). A PROP is a category $\mathbf{C}$ where:

- objects are the natural numbers (here denoted $\underline{0}, \underline{1}, \underline{2}, \ldots$);
- for each n, the group of permutations of n elements, $S(n)$, is a subgroup of all the invertible elements of the homset $[\underline{n}, \underline{n}]$. The identity permutation corresponds to the identity $1_{\underline{n}} : \underline{n} \to \underline{n}$;
- there is a functor $\otimes : \mathbf{C} \times \mathbf{C} \to \mathbf{C}$ which acts as addition on the objects, i.e., $\underline{m} \otimes \underline{n} = \underline{m+n}$, and additionally:
  - is associative: $(f \otimes f') \otimes f'' = f \otimes (f' \otimes f'')$;
  - given $\sigma \in S(n)$ and $\sigma' \in S(n')$, we have $\sigma \otimes \sigma' = \sigma \times \sigma' : \underline{n+n'} \to \underline{n+n'}$, where $\times$ denotes the product of permutations;
  - for any two natural numbers n, n', let $\gamma_{\underline{n},\underline{n'}} : \underline{n+n'} \to \underline{n+n'}$ be the permutation which swaps the two blocks of n and n'. Then for any maps $f : \underline{m} \to \underline{n}$ and $f' : \underline{m'} \to \underline{n'}$ we have $\gamma_{\underline{n},\underline{n'}}(f \otimes f') = (f' \otimes f)\gamma_{\underline{m},\underline{m'}}$.

Example 1. For any algebraic signature (i.e., set of operator names with finite arities) Σ, the *free PROP* $\mathbf{P}_\Sigma$ over Σ has n-tuples of terms over Σ that altogether contain m distinct holes, as arrows $t : \underline{m} \to \underline{n}$. Permutations in $[\underline{n}, \underline{n}]$ are tuples built solely of holes, $\otimes$ acts on arrows as tuple juxtaposition, and arrow composition is the standard composition of terms.

Example 2. PROPs can also be induced from signatures modulo term equations. Consider the signature $\Sigma = \{\mathtt{nil} : 0,\ a. : 1,\ \overline{a}. : 1,\ | : 2\}$ corresponding to the grammar:

$$P ::= \mathtt{nil} \mid aP \mid \overline{a}P \mid P \,|\, P\,,$$

where a ranges over some fixed set A of actions, and the associativity equation:

$$P \mid (Q \mid R) = (P \mid Q) \mid R\,.$$

The PROP **PAP** (Prefix and Associative Parallel composition) is built of terms over Σ quotiented by the associativity equation, with permutations, $\otimes$ and composition defined as in Example 1. Additionally, one can quotient terms by the commutativity equation

$$P \mid Q = Q \mid P\,;$$

the resulting PROP will be called **PACP** (Prefix and Associative, Commutative Parallel Composition).

3 Labelled Transitions for Ground Reductions

This section introduces the background material we need in later sections. First, we briefly recall Leifer and Milner's notion of idem-relative-pushout (IPO) as well as its dual, the idem-relative-pullback (IPB). Following a brief informal and discussion on how IPOs have been used in order to generate labelled transition systems (LTS) for calculi with ground reduction rules, we shall demonstrate that IPOs and IPBs can be conveniently studied in a category of factorisations, where they are easily seen to be coproducts and products, respectively. We conclude with a short note on how to generalise the theory to G-categories [11,13].

3.1 Pushouts in Slices

Let $\mathbf{C}$ be a category and V, W objects of $\mathbf{C}$. The *slice category* $\mathbf{C}/W$ has as objects pairs $\langle X, a\rangle$, where $a : X \to W$ is an arrow of $\mathbf{C}$, while its arrows $f\colon \langle X, a\rangle \to \langle X', a'\rangle$ are arrows $f\colon X \to X'$ in $\mathbf{C}$ such that $a'f = a$. The dual notion of a *coslice category* $V/\mathbf{C}$ consists of the pairs $\langle b, X\rangle$, where $b : V \to X$ and of maps $f\colon \langle b, X\rangle \to \langle b', X'\rangle$ for $f\colon X \to X'$ such that $fb = b'$.

Let $r\colon V \to W$ be an arrow of $\mathbf{C}$. A *pushout* of $f\colon \langle V, r\rangle \to \langle C, c\rangle$ and $g\colon \langle V, r\rangle \to \langle D, d\rangle$ in the slice category $\mathbf{C}/W$ is, equivalently, a *coproduct* in $\langle V, r\rangle\,/(\mathbf{C}/W)$. Spelling this definition out, the span of f and g identifies a commutative square $cf = r = dg$ in $\mathbf{C}$, while the pushout diagram $h\colon \langle C, c\rangle \to \langle E, e\rangle$ and $k\colon \langle D, d\rangle \to \langle E, e\rangle$ determines a universal set of arrows such that $hf = kg$, $eh = c$ and $ek = d$, as in the diagram below. We shall say that a category has *slice pushouts* when it has pushouts in all slices.[2]

[2] Leifer and Milner [4] use the term relative pushouts, or RPOs, to refer to pushouts in slices.

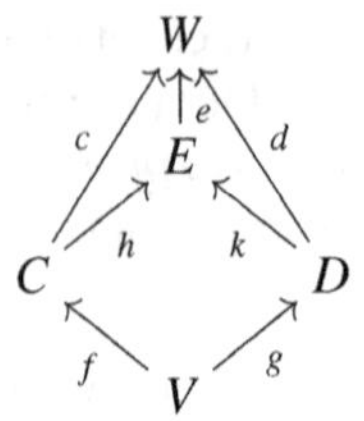

Lemma 1. *Free PROPs have slice pushouts.*

Proof (sketch). A diagram

$$\langle \underline{k}, c\rangle \xleftarrow{\;a\;} \langle \underline{m}, t\rangle \xrightarrow{\;b\;} \langle \underline{l}, d\rangle$$

in $\mathbf{P}_\Sigma/\underline{n}$ is an arrow $t : \underline{m} \to \underline{n}$ in $\mathbf{P}_\Sigma$ with its two decompositions:

$$ca = t = db$$

As usual, a *position* ρ in a given term t is a finite sequence of numbers which encodes a path downward from a root node of t. The set of positions in t, with the standard prefix ordering, is denoted S_t.

It is straightforward to check that decompositions of a given arrow t into p arrows are in 1-1 correspondence to monotonic functions from S_t to the set $\{1, \ldots, p\}$ with the natural ordering. Consider such functions Λ_{ca} and Λ_{db} corresponding to the two decompositions above, and define

$$\Lambda_1(\rho) = \begin{cases} 1 \text{ if } \Lambda_{ca}(\rho) = 1 \text{ and } \Lambda_{db}(\rho) = 1 \\ 2 \text{ if } \Lambda_{ca}(\rho) = 1 \text{ and } \Lambda_{db}(\rho) = 2 \\ 3 \text{ if } \Lambda_{ca}(\rho) = 2 \end{cases}$$

$$\Lambda_2(\rho) = \begin{cases} 1 \text{ if } \Lambda_{ca}(\rho) = 1 \text{ and } \Lambda_{db}(\rho) = 1 \\ 2 \text{ if } \Lambda_{ca}(\rho) = 2 \text{ and } \Lambda_{db}(\rho) = 1 \\ 3 \text{ if } \Lambda_{db}(\rho) = 2 \end{cases}$$

Λ_1 and Λ_2 are monotonic, hence they correspond to two decompositions of t:

$$x_1 y_1 z_1 = t = x_2 y_2 z_2$$

Moreover, $x_1 = x_2$, $z_1 = a$ and $z_2 = b$. Let the domain of $x = x_1 = x_2$ be $\underline{q}$. The square

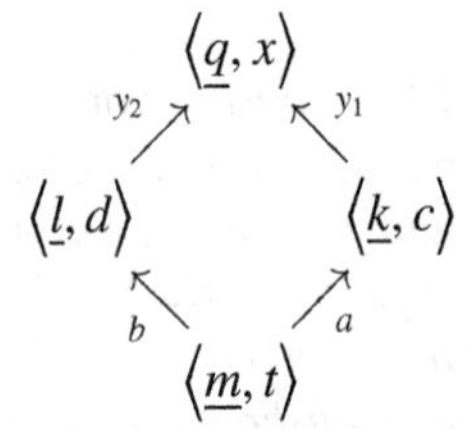

is a pushout in $\mathbf{P}_\Sigma/\underline{n}$.

Lemma 2. *PAP has slice pushouts.*

Proof (sketch). Arrows in **PAP** can be represented as tuples of finite, ordered trees with nodes of any degree, where an immediate child of a node of degree higher than 1 must have degree at most 1. Additionally, nodes of degree 1 are labelled with elements of A. Leaves of such trees correspond to occurrences of the constant `nil`, nodes of degree 1 to applications of prefix composition operators, and nodes of higher degree to term fragments built solely of the associative parallel composition operator. On this representation of arrows, a pushout construction very similar to that of Lemma 1 can be made.

Interestingly, **PACP** does not have slice pushouts. Indeed, there is no unique mediation between the squares in the slice of $\underline{1}$:

$$\begin{array}{ccc} & \langle \underline{1}, \mathtt{nil} \mid 1\rangle & \\ {}^{id}\nearrow & & \nwarrow^{id} \\ \langle \underline{1},\mathtt{nil}\mid 1\rangle & & \langle \underline{1},\mathtt{nil}\mid 1\rangle \\ {}_{\mathtt{nil}}\nwarrow & & \nearrow_{\mathtt{nil}} \\ & \langle \underline{0},\mathtt{nil}\mid \mathtt{nil}\rangle & \end{array} \qquad \begin{array}{ccc} & \langle \underline{2}, 1 \mid 2\rangle & \\ {}^{\langle 1,\mathtt{nil}\rangle}\nearrow & & \nwarrow^{\langle \mathtt{nil},1\rangle} \\ \langle \underline{1},\mathtt{nil}\mid 1\rangle & & \langle \underline{1},\mathtt{nil}\mid 1\rangle \\ {}_{\mathtt{nil}}\nwarrow & & \nearrow_{\mathtt{nil}} \\ & \langle \underline{0},\mathtt{nil}\mid \mathtt{nil}\rangle & \end{array}$$

By an idem-relative-pushout [4] we mean the (square) diagram in **C** obtained by applying the forgetful functor $U_W : \mathbf{C}/W \to \mathbf{C}$ (which projects $\langle V, r\rangle$ to V) to a pushout diagram in $\mathbf{C}/W$. Let $\mathcal{I}$ denote the class of IPOs in **C**.

In categories with slice pushouts, it makes sense to talk about IPOs without worrying about particular slices, as the conclusion of the following lemma implies:

Lemma 3. *If* **C** *has slice pushouts and a diagram D in $\mathbf{C}/X$ maps via the forgetful functor U_X to an IPO (i.e., $U_X D \in \mathcal{I}$) then D is a pushout diagram in $\mathbf{C}/X$.*

Moreover, in categories with slice pushouts, IPOs behave somewhat like ordinary pushouts, as demonstrated by the following lemma.

Lemma 4. *Suppose that* **C** *has slice pushouts and the left square is an IPO. Then the entire diagram is an IPO iff the right square is an IPO.*

$$\begin{array}{ccccc} A & \longrightarrow & B & \longrightarrow & C \\ \downarrow & & \downarrow & & \downarrow \\ D & \longrightarrow & E & \longrightarrow & F \end{array}$$

3.2 Pullbacks in Coslices

Dually, a pullback of $f\colon \langle a, A\rangle \to \langle r, W\rangle$ and $g\colon \langle b, B\rangle \to \langle r, W\rangle$ in the coslice category $V/\mathbf{C}$ is, equivalently, a product in $(V/\mathbf{C})/\langle r, W\rangle$. We say that a category has *coslice pullbacks* when it has pullbacks in all of its coslices.

Lemma 5. *Free PROPs have coslice pullbacks.* ***PAP*** *has coslice pullbacks.*

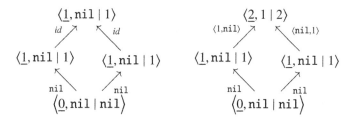

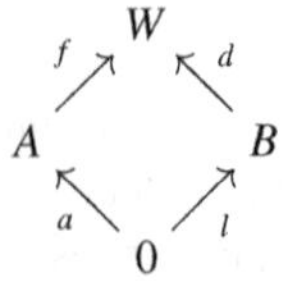

Fig. 1. An IPO corresponding to a label

Proof. Proceed exactly as in Lemmas 1 and 2.

By an idem-relative-pullback (IPB), we mean the (square) diagram obtained from a pullback diagram in a coslice category under the image of the forgetful functor to **C**. We immediately obtain dual versions of Lemmas 3 and 4, the latter of which we state below.

Lemma 6. *Suppose that* **C** *has coslice pullbacks and the right square is an IPB. Then the entire diagram is an IPB iff the left square is an IPB.*

$$\begin{array}{ccccc} A & \longrightarrow & B & \longrightarrow & C \\ \downarrow & & \downarrow & & \downarrow \\ D & \longrightarrow & E & \longrightarrow & F \end{array}$$

3.3 Labels

As we mentioned in the Introduction, IPOs have been used by Leifer and Milner to derive labelled transition systems for calculi equipped with a reduction semantics derived from a set of ground rules. Here we give a brief overview of the technique.

Leifer and Milner's framework of choice is their notion of '*reactive system*,' which consists of a category of contexts with a chosen object 0, a subcategory of evaluation contexts which satisfies certain additional axioms and a set of reduction rules $\mathcal{R}$. The arrows with domain 0 are thought of as closed terms. We shall not give a formal definition here; instead we refer the reader ahead to Definition 3, which deals with a more general situation where reduction rules may be open – to obtain a (closed) reactive system from that definition one needs to assume additionally that the domains of l and r in every rule $\langle l, r\rangle \in \mathcal{R}$ are 0.

Sewell's central idea [16] which guides the definition of the derived LTS is that labels should be certain contexts – more accurately, $a \xrightarrow{f} a'$ when fa (a in the context of f) can perform a single reduction and result in a'. Moreover, as explained in the Introduction, f must be the 'smallest' such context. The notion of IPO gives us a precise way to measure when a context is the smallest. Indeed, consider Fig. 1, where a is an arbitrary term, l is the left hand side of a reduction rule $\langle l, r\rangle \in \mathcal{R}$ and d is an evaluation context. The fact that the diagram is commutative implies that fa can perform a reduction resulting in dr, where the redex l has been replaced by r, the right-hand side of the rule. Requiring the diagram to be an IPO results in an elegant formalisation of the fact that f does not contain redundant material, not necessary for the reduction. The LTS determined in this way can be shown to be well-behaved. In particular, if the underlying category has slice pushouts then bisimilarity is a *congruence*, in the sense that $a \sim b$ implies that $ca \sim cb$ for al c in $\mathbf{C}$.

3.4 Category of Factorisations

The category of factorisations of an arrow provides a convenient setting for studying slice pushouts and coslice pullbacks which we shall use in the rest of the paper.

Definition 2 (Factorisations). The category $\mathrm{Fact}(\mathbf{C}, r)$ of factorisations of an arrow $r\colon V \to W$ in $\mathbf{C}$ is consists of objects and arrows as defined below.

- objects: commutative diagrams in $\mathbf{C}$ of the form

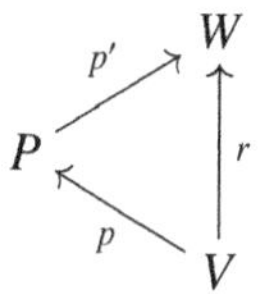

- arrows: an arrow from $\langle P, p, p' \rangle$ to $\langle Q, q, q' \rangle$ is a commutative diagram in $\mathbf{C}$ of the form

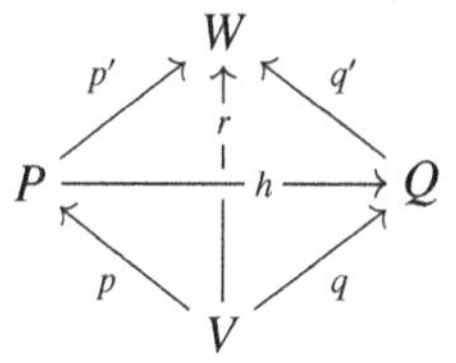

- Composition and identities are obvious.

The following fact is immediate.

Proposition 1. $(V/\mathbf{C})/\langle r, W \rangle \cong \mathrm{Fact}(\mathbf{C}, r) \cong \langle V, r \rangle /(\mathbf{C}/W)$.

Such categories of factorisations form a convenient universe to speak about slice pushouts from $\langle V, r \rangle$ and coslice pullbacks from $\langle r, W \rangle$ in $\mathbf{C}$, since the former are precisely the coproducts and the latter are the products in $\mathrm{Fact}(\mathbf{C}, r)$.

3.5 Generalisation to G-Categories

In [11,13] the second and third author generalised Leifer and Milner's theory to a 2-categorical setting, where structural congruence axioms (usually involving the commutativity of parallel composition) are replaced by invertible 2-cells. The extra structure is necessary, because simply quotienting terms results in structures where IPOs do not exist, as in the case of **PACP** defined in Example 2 (cf. also [11]).

The problem is alleviated by working with G-categories – 2-categories with invertible 2-cells – and considering GIPOs. The latter are the natural bicategorical generalisation of IPOs: namely, rather than pushouts in slice categories, one considers bi-pushouts in pseudo-slice categories. One can, equivalently, define a category of pseudo-factorisations and consider bicoproducts (obtaining GIPOs) and biproducts (obtaining GIPBs).

Example 3. A *G-PROP* is a PROP with the underlying category carrying the structure of a G-category, i.e., a 2-category with all 2-cells invertible. As an example, consider the PROP **PAP** from Example 2 (see the proof of Lemma 2 for an explicit representation of the arrows of **PAP**). Additionally, a 2-cell from a term t to a term t' is a family, indexed by the nodes of t, of permutations on the sets of their immediate children, such that the application of all these permutations to t yields t'. Note how such 2-cells induce bijections between the sets S_t and S'_t of positions respectively in t and t'. Clearly, all such 2-cells are invertible, hence the theory of GIPOs described in [11,13] applies. In particular, the lack of slice pushouts in the PROP **PACP** is avoided here: pseudo-slice bipushouts exist in the above G-PROP, which in the following will be denoted **PA2CP**.

4 Hexagons and Universality

In this section we set out on a path to extend the technique of LTS derivation to systems where reduction rules are open in the sense that they can be instantiated with arbitrary parameters. In such a setting, we would also like generate labels for possibly open terms. The basic idea is that instead of considering simply the smallest context which allows a reduction, we would like to calculate both a smallest context and the most general parameter at the same time. We discuss a reasonable universal property, the *locally universal hexagon*, or *lux*, referred to by Sewell [15] as hex-RPO. These can be used to generate a labelled transition system with information about both contexts and parameters, reminiscent of work on tile systems [2].

In order to understand this universal property, we consider its relationship with slice pushouts and coslice pullbacks in the underlying category. It is convenient to work in slices of the so-called twisted arrow category. We show in Theorem 1 that a category has luxes if and only if it has slice pushouts, coslice pullbacks and these 'commute.' This result allows us isolate sufficient conditions for luxes to exist. Assuming that the underlying category has mono arrows, we show that bisimilarity is a congruence. Finally, we examine how the theory generalises to G-categories.

Definition 3 (Open reactive system). An open reactive system $\mathbb{C}$ is a triple $\langle \mathbf{C}, \mathbf{D}, \mathcal{R} \rangle$ consisting of:

- a category $\mathbf{C}$ with a distinguished object 0 – we shall usually refer to its arrows as contexts and, specifically, to the arrows with domain 0 as terms;
- a composition reflecting subcategory $\mathbf{D}$ of $\mathbf{C}$ – the arrows of $\mathbf{D}$ are termed *evaluation* contexts;
- a set $\mathcal{R}$ of pairs of arrows of $\mathbf{C}$, so that if $\langle l, r \rangle \in \mathcal{R}$, then the domains and codomains of l and r are equal – we shall refer to $\mathcal{R}$ as the set of reduction rules.

Given an open reactive system $\mathbb{C}$, one can define a reduction relation on the terms of $\mathbb{C}$ as follows: $a \longrightarrow\!\!\!\!\triangleright\, a'$ if $a = dlx$ and $a' = drx$ for some $d \in \mathbf{D}$, $x \in \mathbf{C}$, and $\langle l, r \rangle \in \mathcal{R}$.

We shall refer to commutative diagrams such as the one illustrated if Figure 2 as commutative hexagons, or simply *hexagons*. The following universal property defines locally universal hexagons, or luxes.

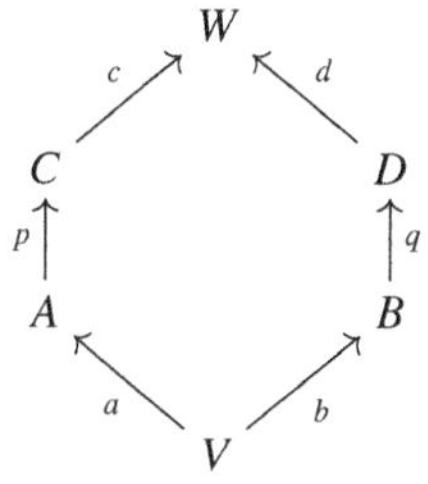

Fig. 2. A hexagon

Definition 4 (Luxes). A *locally universal hexagon* (*lux*) for the hexagon of Fig. 2 is a hexagon that factors through it (cf. diagram *i*), and that additionally satisfies a universal property:

> for any other such hexagon (cf. diagram *ii*), there exist unique $h'' : Y \to Y'$ and $z'' : X' \to X$ such that diagram (*iii*) is commutative, $h = h'h''$ and $z = z''z'$.

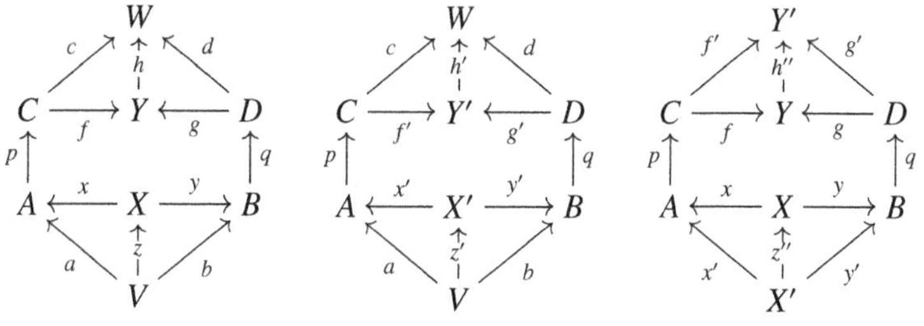

We denote the lux of diagram (*i*) above as $(p)^{fg}_{xy}(q)$.

We shall say that a category **C** has luxes if every hexagon has a lux. As was the case for IPOs, in categories with luxes one does not need to know which hexagon is a lux for: if a hexagon is a lux, then it is such for all hexagons through which it factors. This property, analogous to Lemma 3, will be proved formally as Lemma 7 below.

In order obtain first intuitions about luxes, let us consider a very simple example. Below we denote string concatenation by; noted in diagrammatic order (i.e., left-to-right).

Example 4. Consider a free monoid over an alphabet Σ viewed as a category **S** with one object. Then luxes exist in **S**. Consider a hexagon as in Fig. 2, where $a; p; c = b; q; d$. Let h be the largest suffix common to c and d, then there exist words f and g so that $c = f; h$ and $d = g; h$. Similarly, let z be the largest prefix common to a and b, then there exist words x and y so that $z; x = a$ and $z; y = b$. Clearly $x; p; f = t; q; g$ and it is straightforward to check that the universal property holds. We now fix $\Sigma = \{a, b\}$ and illustrate several examples of luxes in **S** in Fig. 3.

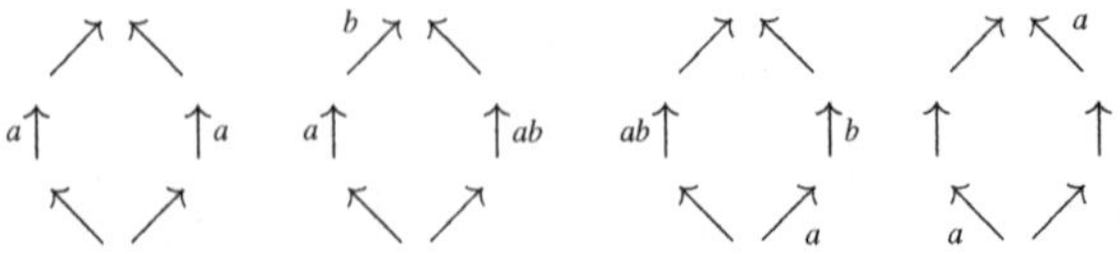

Fig. 3. Luxes in a free monoid

Armed with the notion of lux, we are ready to define a labelled transition system on possibly open terms.

Definition 5 (LTS). Given an open reactive system $\mathbb{C}$, an LTS can be derived as follows:

- nodes are arbitrary arrows $a : A \to C$ – i.e., domain does not need to be 0: terms are possibly open;
- there is a transition $a \xrightarrow{x} b$ whenever there exist $\langle l, r \rangle \in \mathcal{R}$ and a lux

$$
\begin{array}{ccc}
 & f \nearrow \nwarrow g & \\
a \uparrow & & \uparrow l \\
 & x \nwarrow \nearrow y &
\end{array}
$$

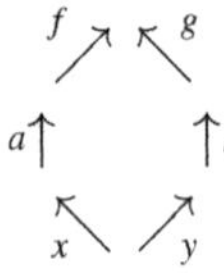

such that $b = gry$ – thanks to Lemma 7 below, this is well given.

5 Properties of Locally Universal Hexagons

In order to study the properties of luxes, it is convenient to work in *twisted arrow categories*, that we introduce below. Here we give their definition from [6], and examine some of its basic properties. We mention that the category can be concisely described as the category of elements for the homfunctor $\mathbf{C}(-, -) : \mathbf{C}^{\mathrm{op}} \times \mathbf{C} \to \mathbf{Set}$.

Definition 6 (Twisted Arrow Categories). Given a category $\mathbf{C}$, the twisted arrow category $\mathrm{Tw}(\mathbf{C})$ has

- objects: the arrows of $\mathbf{C}$;
- arrows: an arrow from $f : A \to B$ to $f' : A' \to B'$ consists of arrows $p : B \to B'$ and $q : A' \to A$ such that $pfq = f'$; in other words, an arrow from f to f' is a factorisation of f' through f, as in the diagram below.

$$
\begin{array}{ccc}
B & \xrightarrow{c} & B' \\
f \uparrow & & \uparrow f' \\
A & \xleftarrow{a} & A'
\end{array}
$$

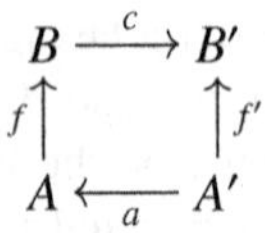

In symbols, we shall use $f \underset{a}{\xrightarrow{c}} f'$ to denote such an arrow of $\mathrm{Tw}(\mathbf{C})$.

As promised, the twisted arrow category gives us a simplified setting in which we may consider the universal property of luxes. Indeed, our first observation is that hexes are in 1-1 correspondence with cospans

$$p \underset{a}{\overset{c}{\longleftrightarrow}} r \underset{b}{\overset{d}{\longleftrightarrow}} q$$

in Tw(**C**), where $cpa = r = dqb$. Secondly, it is easily verified that luxes are precisely the coproduct diagrams in slices of Tw(**C**).

Proposition 2. *A lux is a hexagon in* **C** *that results from a coproduct diagram in the slice category* $\mathrm{Tw}(\mathbf{C})/r$.

Notice that we explicitly talk about the coproduct *diagram* (as opposed to *object*), which includes the cospan formed by the coproduct coprojections.

The following lemma justify us referring to locally universal hexagons (without mentioning which hexagon it is universal with respect to). Thus, when talking about categories with luxes, we shall often abuse notation – in contrast with §3 where we distinguished between slice pushouts and IPOs and coslice pullbacks and IPBs.

Lemma 7. *In a category with luxes, if a hexagon factors through a lux (possibly for another hexagon) then it is a lux for that hexagon.*

Proof. We know that a category has luxes iff every slice $\mathrm{Tw}(\mathbf{C})/r$ has coproducts. It is straightforward to verify that, given an arbitrary category **C**, when every slice has coproducts, if $\langle A, a\rangle \xrightarrow{f} \langle C, c\rangle \xleftarrow{g} \langle D, d\rangle$ is a coproduct diagram in $\mathbf{C}/X$, then for any $X' \in \mathbf{C}$ with $a' : A \to X'$, $b' : B \to X'$ and $c' : C \to X'$ such that $a' = c'f$ and $b' = c'g$ $\langle A, a'\rangle \xrightarrow{f} \langle C, c'\rangle \xleftarrow{g} \langle D, d'\rangle$ is a coproduct diagram in $\mathbf{C}/X'$.

In order to obtain further insights into luxes, we shall explore the relationship between slices of Tw(**C**) and the category of factorisations of Definition 2.

First we notice that there is a faithful functor $\mathcal{I}\colon V/\mathbf{C} \to \mathrm{Tw}(\mathbf{C})$ which is the first projection on objects and takes an arrow $h : \langle p, P\rangle \to \langle q, Q\rangle$ to $p \underset{\mathrm{id}}{\overset{h}{\longleftrightarrow}} q$. Similarly, there is a functor $\mathcal{J}\colon (\mathbf{C}/W)^{\mathrm{op}} \to \mathrm{Tw}(\mathbf{C})$ which takes $h : \langle P, p'\rangle \to \langle Q, q'\rangle$ to $q' \underset{h}{\overset{\mathrm{id}}{\longleftrightarrow}} p'$.

Both $\mathrm{Fact}(\mathbf{C}, r)$ and $\mathrm{Fact}(\mathbf{C}, r)^{\mathrm{op}}$ can be seen as full subcategories of Tw(**C**) via the functors $\mathcal{I}/r\colon \mathrm{Fact}(\mathbf{C}, r) \to \mathrm{Tw}(\mathbf{C})/r$ and $\mathcal{J}/r\colon \mathrm{Fact}(\mathbf{C}, r)^{\mathrm{op}} \to \mathrm{Tw}(\mathbf{C})/r$. Observe that the second functor is well defined, since $(\mathbf{C}/W)^{\mathrm{op}}/\langle V, r\rangle \cong (\langle V, r\rangle/(\mathbf{C}/W))^{\mathrm{op}}$, for all categories **C** and arrows $r\colon V \to W$ in it. We illustrate the actions of $\mathcal{I}/r$ and $\mathcal{J}/r$ in Fig. 4.

Lemma 8. $\mathcal{I}/r$ *and* $\mathcal{J}/r$ *have left adjoints, respectively* $\Phi\colon \mathrm{Tw}(\mathbf{C})/r \to \mathrm{Fact}(\mathbf{C}, r)$ *and* $\Psi\colon \mathrm{Tw}(\mathbf{C})/r \to \mathrm{Fact}(\mathbf{C}, r)^{\mathrm{op}}$.

It is useful for us to examine the functors Φ and Ψ in more detail. The action of Φ on objects and arrows of $\mathrm{Tw}(\mathbf{C})/r$ is shown in Fig. 5. Note that $\Phi \circ I/r = \mathrm{id}_{\mathrm{Fact}(\mathbf{C},r)}$ and $\Psi \circ J/r = \mathrm{id}_{\mathrm{Fact}(\mathbf{C},r)^{\mathrm{op}}}$. In fact, Lemma 8 states that both $\mathrm{Fact}(\mathbf{C}, r)$ and its opposite are full reflective subcategories of $\mathrm{Tw}(\mathbf{C})/r$.

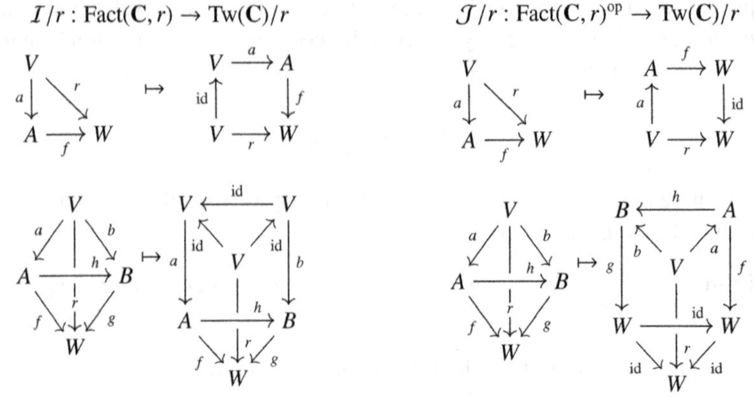
$\mathcal{I}/r : \mathrm{Fact}(\mathbf{C}, r) \to \mathrm{Tw}(\mathbf{C})/r$
$\mathcal{J}/r : \mathrm{Fact}(\mathbf{C}, r)^{\mathrm{op}} \to \mathrm{Tw}(\mathbf{C})/r$

Fig. 4. $\mathcal{I}/r$ and $\mathcal{J}/r$ on objects and arrows

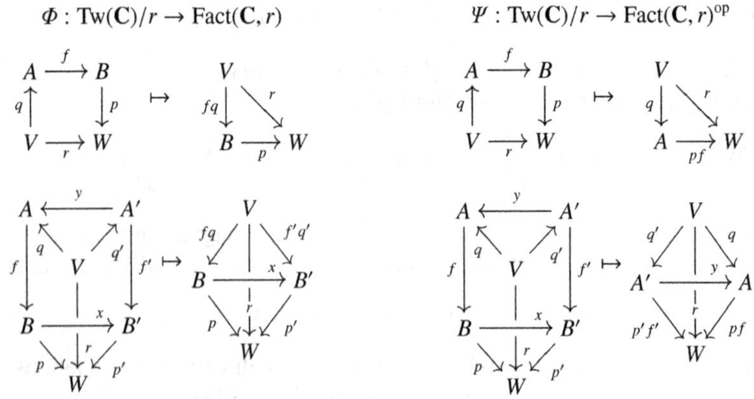
$\Phi : \mathrm{Tw}(\mathbf{C})/r \to \mathrm{Fact}(\mathbf{C}, r)$
$\Psi : \mathrm{Tw}(\mathbf{C})/r \to \mathrm{Fact}(\mathbf{C}, r)^{\mathrm{op}}$

Fig. 5. Φ and Ψ on objects and arrows

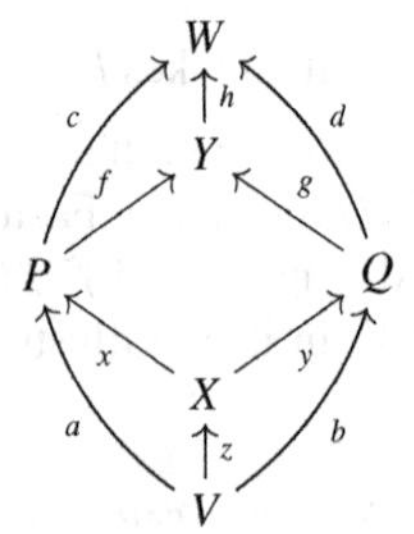

Fig. 6. Commutativity of slice pushouts and coslice pullbacks

Corollary 1. *Coproducts in* $\mathrm{Tw}(\mathbf{C})/r$ *map via* Φ *to coproducts in* $\mathrm{Fact}(\mathbf{C}, r)$*, and thus to coproducts in* $\langle V, r\rangle / (\mathbf{C}/W)$*, which are pushouts in* $\mathbf{C}/W$.

Corollary 2. *Coproducts in* Tw(*catC*)/*r map via* Ψ *to products in* Fact($\mathbf{C}$, *r*)*, and thus to products in* $(V/\mathbf{C})/\langle r, W\rangle$ *which are pullbacks in* $V/\mathbf{C}$.

Lemma 9. *A diagram (i) is a coproduct diagram of* $p \underset{a}{\overset{c}{\longleftrightarrow}} r$ *and* $q \underset{b}{\overset{d}{\longleftrightarrow}} r$ *in* Tw($\mathbf{C}$)/*r iff (1) diagram (ii) is a pushout in* $\mathbf{C}/W$*, and (2) diagram (iii) is a pullback in* $V/\mathbf{C}$.

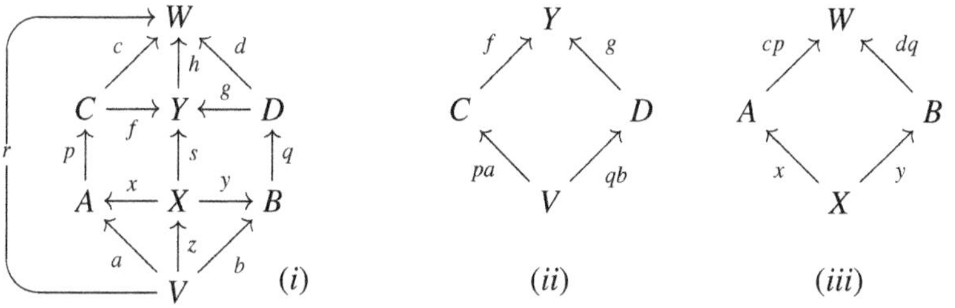

Proof. The only if direction is given by Corollaries 1 and 2. The if direction is easily verified.

Note that Lemma 9 explicitly assumes that the resulting hex is commutative. Consider for instance the following diagram in **Set**:

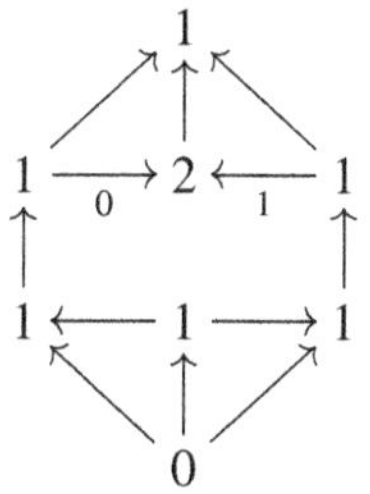

The lower hexagon results from calculating a local pushout of $0 \to 1$ with itself in $\mathbf{Set}/1 \cong \mathbf{Set}$, while the upper one from a local pullback of $1 \to 1$ with itself in $0/\mathbf{Set} \cong \mathbf{Set}$. Notice that the resulting inner hexagon is *not* commutative.

We shall say that slice pushouts and coslice pullbacks *commute* when, given a commutative square (the outside of Fig. 6), constructing a pushout of *a* and *b* in $\mathbf{C}/W$ and a pullback of *c* and *d* in $V/\mathbf{C}$ results in an inner commutative diagram ($fx = gy$).

Theorem 1. *A category* $\mathbf{C}$ *has luxes iff it has slice pushouts, coslice pullbacks and these commute.*

Proof. If $\mathbf{C}$ has slice pushouts, coslice pullbacks and these commute, then one can explicitly construct a lux, using the conclusions of Lemma 9, since it is easy to show that the commutativity property ensures the commutativity of the resulting hexagon.

Conversely, if $\mathbf{C}$ has luxes then it is easy to show that it has slice pushouts, i.e., coproducts in Fact($\mathbf{C}$, *r*) and coslice pullbacks, i.e., products in Fact($\mathbf{C}$, *r*). Indeed, it is enough to calculate the lux of the hexagon below:

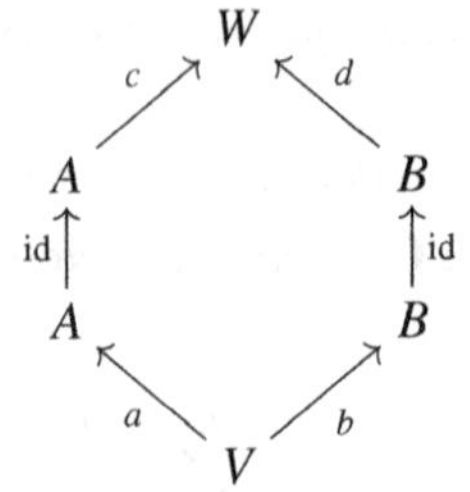

Using the fact that Φ and Ψ preserve coproducts, the resulting lux maps via Φ to the slice pushout of a and b in $\mathbf{C}/W$ and via Ψ to the coslice pullback of c and d in $V/\mathbf{C}$. The commutativity property follows directly.

As an immediate consequence, it follows that **Set** does not have luxes, since the commutativity property is not satisfied.

When working in categories with luxes, we can use the conclusions of Lemma 9 to obtain a characterisation of luxes in terms of IPOs and IPBs.

Lemma 10. *In a category with luxes, a commutative diagram (i) is a lux iff diagram (ii) is an IPO and diagram (iii) is an IPB.*

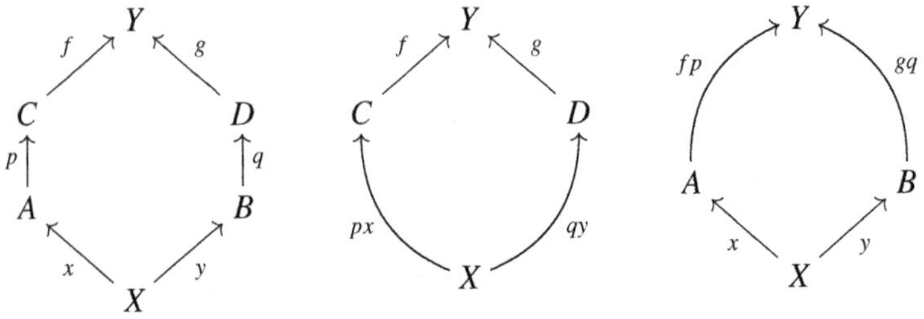

(i) (ii) (iii)

It is useful to consider properties of **C** that ensure that slice pushouts and coslice pullbacks commute. One obvious such property is that either all arrows of **C** are mono, another is that all arrows of **C** are epi.

Corollary 3. *The following conditions are each sufficient for the existence of luxes in category* **C**.

1. **C** *has slice pushouts, slice pullbacks and all arrows are mono;*
2. **C** *has slice pushouts, slice pullbacks and all arrows are epi.*

Theorem 2. *Free PROPs have luxes.* ***PAP*** *has luxes.*

Proof. It is easy shown by induction that all arrows in free PROPs, and all arrows in **PAP**, are mono. This means that no two different terms can be made equal by putting them in the same context. Then use Lemmas 1, 2, and 5.

Theorem 3 (Congruence). *Suppose that $\mathbb{C}$ is an open reactive system. Let $\sim$ denote bisimilarity on the LTS introduced in Definition 5.*

If **C** *has luxes and all arrows of* **C** *are mono, then $\sim$ is a congruence, in the sense that if $p \sim q$, then $cp \sim cq$ for all contexts c in* **C**.

Proof. It is enough to show that $\{\langle cp, cq\rangle \mid p \sim q,\ c \in \mathbf{C}\}$ is a bisimulation.

Indeed, suppose that $p \sim q$ and $cp \xrightarrow[x]{f} p'$. Then we can find a lux, illustrated as the outside of diagram (i) below, where $\langle l, r\rangle \in \mathcal{R}$ and $p' = gry$.

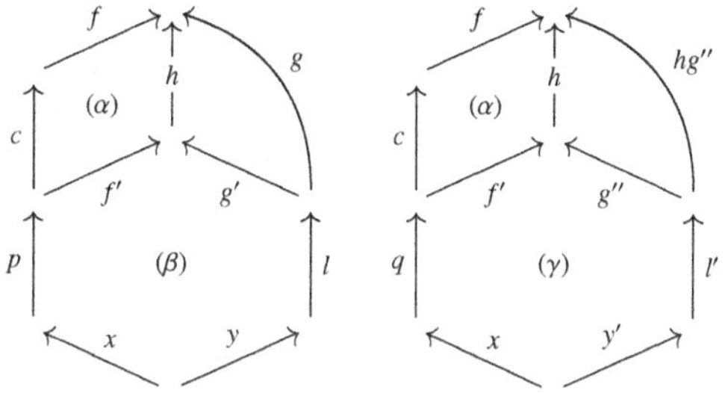

(*i*) (*ii*)

We now calculate a slice pushout of px and ly in diagram (*i*), resulting in f', g' and h such that $hf' = fc$ and $hg' = g$. Then (β) is an IPB and an IPO in the sense of Lemma 10, using the Lemma yields that (β) is a lux. We obtain $p \xrightarrow[x]{f'} g'ry$

Since $p \sim q$, also $q \xrightarrow[x]{f'} q'$ where $q' \sim g'ry$. Let (γ) be a lux responsible for the transition, so that $\langle l', r'\rangle \in \mathcal{R}$ and $q' = g''r'y'$. Pasting the IPO (α) results in a hexagon which is an IPO. Using the fact that h is mono, it is also an IPB. Thus $cq \xrightarrow[x]{f} hq'$. But $p' = gry = hg'ry$, and since $q' \sim g'ry'$, the proof is complete.

Dually, the following holds.

Proposition 3. *If* **C** *has luxes and all arrows of* **C** *are epi, then $px \sim qx$ for all x in* **C**.

As a consequence of Theorem 2 and the fact that the arrows of free PROPs are mono, the LTS obtained from Definition 5 for any reactive system over a free PROP yields a congruent bisimilarity.

6 Structural Congruence as Invertible 2-Cells

In §3.5 we gave a rough description of how to generalise the concepts of IPOs and IPBs to G-categories. Here we give a brief description of how to generalise the theory of luxes. The definition of G-lux is simple to state.

Definition 7 (G-lux). Given a G-category **C**, the definition of Tw(**C**) can easily be extended to a G-category. The arrows are now twisted squares with a 2-cell, and the 2-cells are 2-cells between the top and bottom components such that everything commutes. A G-lux is a bicoproduct in pseudo-slice category Tw(**C**)/r.

An open G-reactive system is simply an open reactive system on a G-category, the only extra requirement is for the subcategory of evaluation contexts to be full on the 2-dimensional structure.

Given a G-reactive system, it is easy to extend Definition 5 to generate an LTS using G-luxes. One obtains a transition system with possibly open terms as states. It is also possible to consider an LTS where the states are terms quotiented by isomorphism (or, in process calculus terminology, structural congruence) – the congruence theorem holds in both instances; see [17, Ch. 2] for details.

It is fairly straightforward to rework the theory presented in the previous section in this more general setting, but we omit the details here. Using the concepts discussed in §3.5, one obtains generalised versions of Theorem 1, Lemma 10 and Theorem 3. In the latter, the mono requirement is replaced by a 2-categorical version which states that for any arrow f and 2-cells α and β, if $f\alpha = f\beta$ then $\alpha = \beta$.

Proposition 4. *PA2CP (cf. Example 3) has G-luxes.*

Proof (sketch). The proof follows the general structure of those of Theorem 2 and Lemma 2. To show that **PA2CP** has pseudo-slice bipushouts, consider a 2-cell $\alpha : t \Rightarrow t'$ with decompositions $t = ca$, $t' = db$. These decompositions correspond to monotonic functions Λ_{ca}, Λ_{db} as sketched in the proof of Lemma 1. The following function on S_t:

$$\Lambda_1(\rho) = \begin{cases} 1 \text{ if } \Lambda_{ca}(\rho) = 1 \text{ and } \Lambda_{db}(\overline{\alpha}(\rho)) = 1 \\ 2 \text{ if } \Lambda_{ca}(\rho) = 1 \text{ and } \Lambda_{db}(\overline{\alpha}(\rho)) = 2 \\ 3 \text{ if } \Lambda_{ca}(\rho) = 2 \end{cases}$$

(where $\overline{\alpha} : S_t \to S_{t'}$ is the bijection, induced by α, between positions in terms) defines a decomposition $t = xyz$, and moreover $z = a$. Analogously one obtains a decomposition $t' = x'y'z'$, with $z' = b$ and $x' = x$. These decompositions form the 1-cell part of the required pseudo-slice bipushout square; to find the required 2-cells, proceed as in the case of free monoids and permutations in [17].

Example 5. We can construct an open (G-)reactive system on **PA2CP** by letting the set of reduction rules $\mathcal{R}$ be the singleton consisting of the single rule $\langle a.1 \mid \overline{a}.2, 1 \mid 2\rangle$. In the following, we shall use P, Q and X as meta variables which stand for any closed term (arrow $\underline{0} \to \underline{1}$) of **PA2CP**.

The subcategory of evaluation contexts is taken to be the smallest composition reflecting 2-full 2-subcategory which includes arrows of the form $\langle 1 \mid P\rangle : \underline{1} \to \underline{1}$. In more intuitive terms, the non-evaluation contexts are precisely the contexts which have a hole under a prefix.

In Fig. 7, we illustrate several examples of G-luxes, which in turn lead to labels of the induced LTS. Thus, the top left diagram gives a transition $\langle a.P, 1\rangle \xrightarrow{1 \| \overline{a}.2} P \mid 1$, the next diagram leads to a transition $a.P \mid 1 \xrightarrow[\overline{a}.1]{} P \mid 1$. The next transition induced is $a.P \mid \overline{a}.Q \longrightarrow P \mid Q$, which can be seen as internal reduction since no external context or parameter is required. In the second row, the first lux from the left demonstrates the function of the 2-cells in **PA2CP**: here γ is the unique permutation $a.P \mid \overline{a}.Q \mid \overline{a}.1 \to a.P \mid \overline{a}.1 \mid \overline{a}.Q$. The label generated is $\langle a.P \mid \overline{a}.Q, 1\rangle \xrightarrow{1 \| \overline{a}.2} a.P \mid \overline{a}.1 \mid \overline{a}.Q$.

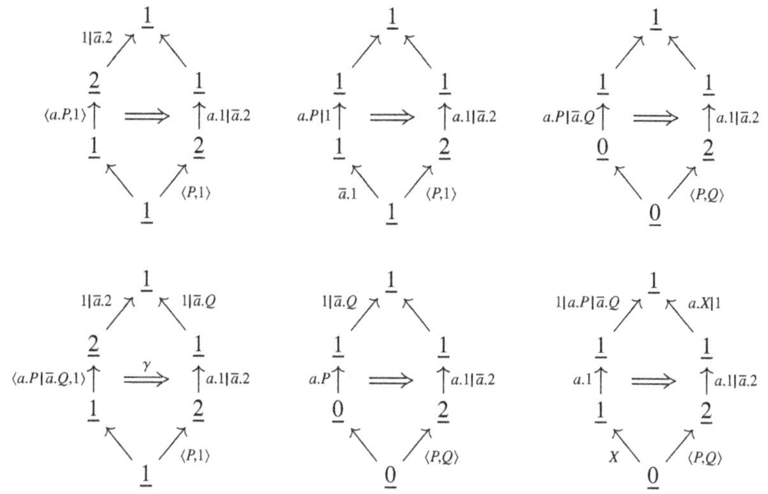

Fig. 7. G-luxes in **PA2CP**

The final two diagrams illustrate what we believe is the main problem with luxes – indeed, the problem can be observed already in the much simpler Example 4. Roughly, while the universal property of luxes ensures that there is no redundant information in the contexts and in the parameters, there may still be some overlap between contexts and parameters. Indeed, consider the middle diagram of the second row of Fig. 7. The lux leads to the transition $a.P \xrightarrow{1|\bar{a}.Q} P \mid Q$. However, the information in Q is not necessary for this reduction, since it appears both in the context on the left and in the parameter on the right. Unfortunately, since Q is arbitrary, this means that the resulting LTS is infinitely branching. The final diagram is even more redundant, since no part of the term is actually necessary for the reduction; now we have X appearing both as a parameter an the left and as part of the context on the right, and P and Q appearing both as the parameters on the right and as part of the context on the left. This diagram induces the transition $a.1 \xrightarrow[X]{1|a.P|\bar{a}.Q} a.X \mid P \mid Q$.

7 Towards a General Theory

Let us consider the following simple property, in order to rule out some of the 'redundant' luxes identified in Example 5.

Definition 8 (Irredundant hexagon). A hexagon is said to be irredundant when there exist $k\colon A \to D$ and $l\colon B \to D$ so that all regions of Fig. 8 are commutative; that is $lb = pa$, $cl = dq$, $ka = qb$ and $dk = cp$.

A lux is said *irredundant* when it is irredundant as a hexagon. The property can be extended to cover G-luxes in the obvious way, that is, instead of commutativity one requires the presence of compatible 2-cells.

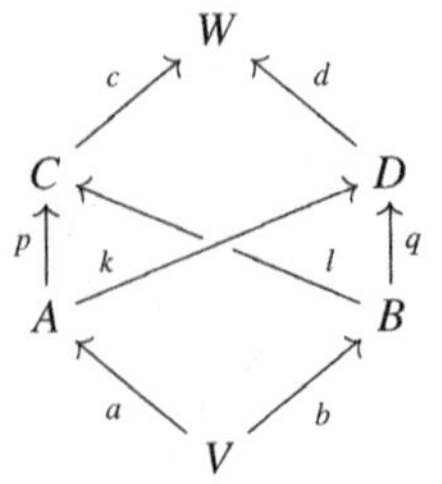

Fig. 8. An irredundant hexagon

Example 6. Consider the luxes illustrated in Fig. 3 and discussed in Example 4. It is easy to show that the first three diagrams are irredundant as hexes, but the final one is not. Now consider the G-luxes illustrated in Fig. 7 and discussed in Example 5. Again, all of the luxes apart from the middle and the right lux of the second row are irredundant.

Thus, by considering irredundant luxes, we eliminate the problematic luxes identified in our case studies. The obvious next steps are to alter the LTS definition so that only irredundant luxes are taken into account, and to study bisimilarity on the resulting structures. Alas, the simple-minded modification to Definition 5 will not do: the technique we use to prove our congruence results (viz., Theorem 3 and Proposition 3) does not stand for irredundant luxes alone, as a lux that is a factor of an irredundant lux need not be irredundant itself. Indeed, bisimilarity in general is *not* a congruence. Recalling e.g. the reactive system of Example 5, and consider the possible labels of transitions with domains of the form $a.P$. If a reduction involves $a.P$, then it must occur in context with an output $\overline{a}.Q$, for some Q; this introduces redundancy, as Q must arise as a parameter that instantiates the reduction rule. Thus $a.P$ has *no* irredundant labels, which implies $a.P \sim b.P$, for all $a \neq b$. But $a.P$ can reduce in the presence of an output on a while $b.P$ cannot, and so congruence is broken. As future work, we plan to study ways of deriving transition systems for open terms that at the same time carry no redundancy in the labels, and are sufficient for coinductive reasoning.

In any case, we feel that the framework for deriving labelled transition systems from reductions is still in its infancy, and requires further development. Our ultimate goal is an abstract method that, when applied to the standard reduction system of a calculus like, say, the π-calculus, yields a labelled transition system on which bisimilarity is a congruence, and moreover: (1) gives rise to feasible coinductive techniques, and (2) is fully abstract with respect to standard equivalences defined in terms of contextual closures (such as barbed congruence). The final theory will necessarily involve a satisfactory treatment of variables, parameters and parametric rules. We believe that luxes and irredundancy, introduced in this paper, shall serve as important tools in our future research on 'labels from reductions.'

References

1. H. Ehrig and B. König. Deriving bisimulation congruences in the DPO approach to graph re writing. In *Proceedings of FoSSaCS'04, 7th International Conference on Foundations of Software Science and Computation Structures*, volume 2987 of *LNCS*, pages 151–166, 2004.

2. F. Gadducci and U. Montanari. The tile model. In G. Plotkin, C. Stirling, and M. Tofte, editors, *Proof, Language and Interaction: Essays in Honour of Robin Milner*, pages 133–166. MIT Press, 2000.
3. O. H. Jensen and R. Milner. Bigraphs and mobile processes. Technical Report 570, University of Cambridge, 2003.
4. J. Leifer and R. Milner. Deriving bisimulation congruences for reactive systems. In *International Conference on Concurrency Theory, CONCUR'00*, volume 1877 of *LNCS*, pages 243–258, 2000.
5. S. Mac Lane. Categorical algebra. *Bulletin of the American Mathematical Society*, 71:40–106, 1965.
6. S. Mac Lane. *Categories for the Working Mathematician*. Graduate Texts in Mathematics. Springer Verlag, 2nd edition, 1992.
7. R. Milner. Calculi for interaction. *Acta Informatica*, 33(8):707–737, 1996.
8. R. Milner. Bigraphical reactive systems. In *International Conference on Concurrency Theory, CONCUR'01*, volume 2154 of *LNCS*, pages 16–35, 2001.
9. R. Milner. Bigraphs for Petri nets. In *Lectures on Concurrency and Petri Nets 2003*, volume 3098 of *Lecture Notes in Computer Science*, pages 686–701, 2004.
10. G. Plotkin. A structural approach to operational semantics. Technical Report DAIMI FN-19, Aarhus University, Computer Science Department, 1981.
11. V. Sassone and P. Sobociński. Deriving bisimulation congruences using 2-categories. *Nordic Journal of Computing*, 10(2):163–183, 2003.
12. V. Sassone and P. Sobociński. A congruence for Petri nets. In *Workshop on Petri Nets and Graph Transformation*, volume 127 of *ENTCS*, pages 107–120, 2005.
13. V. Sassone and P. Sobociński. Locating reaction with 2-categories. *Theoretical Computer Science*, 333(1-2):297–327, 2005.
14. V. Sassone and P. Sobociński. Reactive systems over cospans. *Proceedings of Logics in Computer Science, LICS 2005*, IEEE Press 2005.
15. P. Sewell. Working notes PS15–PS19, 2000. Unpublished notes.
16. P. Sewell. From rewrite rules to bisimulation congruences. *Theoretical Computer Science*, 274(1–2):183–230, 2002.
17. P. Sobociński. *Deriving process congruences from reaction rules*. PhD thesis, BRICS, University of Aarhus, 2004.

Adequacy for Algebraic Effects with State

Gordon Plotkin*

Laboratory for the Foundations of Computer Science,
School of Informatics, University of Edinburgh,
King's Buildings, Edinbursgh EH9 3JZ, Scotland
gdp@inf.ed.ac.uk

Abstract. In previous work we gave an operational semantics and adequacy theorem for algebraic effects in programming languages. This covered finitary algebraic operations, thereby accommodating ordinary and probabilistic nondeterminism, output, and exceptions (but without handling), together with their combinations. With some extra effort, infinitary operations can also be covered, thereby also accommodating input and state. However one does not thereby obtain the natural operational semantics for state, which employs configurations of programs and states. We propose instead to consider the natural coalgebra of states given by the *update* and *lookup* operations; this coalgebra is the final comodel of the Lawvere theory for state. We therefore give an account integrating coalgebras given by comodels of Lawvere theories into the algebraic theory of effects. The coalgebras are used for the dynamics of the state component of the configurations.

* This work was done with the support of EPSRC grants GR/M56333 and GR/S86372/01.

J.L. Fiadeiro et al. (Eds.): CALCO 2005, LNCS 3629, p. 51, 2005.

Bisimilarity Is Not Finitely Based over BPA with Interrupt

Luca Aceto[1,3], Wan Fokkink[4], Anna Ingolfsdottir[1,2], and Sumit Nain[1]

[1] Brics (Basic Research in Computer Science),
Centre of the Danish National Research Foundation,
Department of Computer Science, Aalborg University,
Fr. Bajersvej 7B, 9220 Aalborg Ø, Denmark
{luca, annai, nain}@cs.aau.dk
[2] Department of Computer Science, University of Iceland, 107 Reykjavík, Iceland
annaing@hi.is
[3] School of Computer Science, Reykjavík University, Ofanleiti 2, 103 Reykjavík, Iceland
luca@ru.is
[4] Vrije Universiteit Amsterdam, Department of Computer Science,
Section Theoretical Computer Science, De Boelelaan 1081a,
1081 HV Amsterdam, The Netherlands
wanf@cs.vu.nl

Abstract. This paper shows that bisimulation equivalence does not afford a finite equational axiomatization over the language obtained by enriching Bergstra and Klop's Basic Process Algebra with the interrupt operator. Moreover, it is shown that the collection of closed equations over this language is also not finitely based. In sharp contrast to these results, the collection of closed equations over the language BPA enriched with the disrupt operator is proven to be finitely based.

1 Introduction

Programming and specification languages often include constructs to specify mode switches (see, e.g., [8,11,23,24,26]). Indeed, some form of mode transfer in computation appears in the time-honoured theory of operating systems in the guise of, e.g., interrupts, in programming languages as exceptions, and in the behaviour of control programs and embedded systems as discrete "mode switches" triggered by changes in the state of their environment.

In light of the ubiquitous nature of mode changes in computation, it is not surprising that classic process description languages either include primitive operators to describe mode changes—for example, LOTOS [15,23] offers the so-called *disruption operator*—or have been extended with variations on mode transfer operators. For instance, examples of such operators that may be added to CCS are discussed by Milner in [25, pp. 192–193], and the reference [17] offers some discussion of the benefits of adding one of those, viz. the *checkpointing operator*, to that language.

In the setting of Basic Process Algebra (BPA), as introduced by Bergstra and Klop in [12], some of these extensions, and their relative expressiveness, have been discussed in the early paper [11]. That preprint of Bergstra's has later been revised and extended

J.L. Fiadeiro et al. (Eds.): CALCO 2005, LNCS 3629, pp. 52–66, 2005.

in [7]. There, Baeten and Bergstra study the equational theory and expressiveness of BPA_δ (the extension of BPA with a constant δ to describe "deadlock") enriched with two mode transfer operators, viz. the *disrupt* and *interrupt* operators. In particular, they offer an equational axiomatization of bisimulation equivalence [25,29] over the resulting extension of the language BPA_δ. This axiomatization is finite, if so is the underlying set of actions—a state of affairs that is most pleasing for process algebraists.

However, the axiomatization of bisimulation equivalence offered by Baeten and Bergstra in [7] relies on the use of four auxiliary operators—two per mode transfer operator. Although the use of auxiliary operators in the axiomatization of behavioral equivalences over process description languages has been well established since Bergstra and Klop's axiomatization of parallel composition using the left and communication merge operators [13], to our mind, a result like the aforementioned one always begs the question whether the use of auxiliary operators is necessary to obtain a finite axiomatization of bisimulation equivalence.

For the case of parallel composition, Moller showed in [27,28] that strong bisimulation equivalence is not finitely based over CCS [25] and PA [13] without the left merge operator. (The process algebra PA [13] contains a parallel composition operator based on pure interleaving without communication, and the left merge operator.) Thus auxiliary operators are necessary to obtain a finite axiomatization of parallel composition. But, is the use of auxiliary operators necessary to give a finite axiomatization of bisimulation equivalence over the language BPA enriched with the mode transfer operators studied by Baeten and Bergstra in [7]?

We address the above natural question in this paper. In particular, we focus on BPA enriched with the interrupt operator. Intuitively, "p interrupted by q" describes a process that normally behaves like p. However, at each point of the computation before p terminates, q can interrupt it, and begin its execution. If this happens, p resumes its computation upon termination of q.

We show that, in the presence of two distinct actions, bisimulation equivalence is *not* finitely based over BPA with the interrupt operator. Moreover, we prove that the collection of closed equations over this language is also not finitely based. This result provides evidence that the use of auxiliary operators in the technical developments presented in [7] is indeed necessary in order to obtain a finite axiomatization of bisimulation equivalence.

Our main result adds the interrupt operator to the list of operators whose addition to a process algebra spoils finite axiomatizability modulo bisimulation equivalence; see, e.g., [4,3,14,16,20,30,31] for other examples of non-finite axiomatizability results over process algebras, and some of their precursors in the setting of formal language theory. Of special relevance for concurrency theory are the aforementioned results of Moller's to the effect that the process algebras CCS and PA without the auxiliary left merge operator from [12] do not have a finite equational axiomatization modulo bisimulation equivalence [27,28]. Recently, in collaboration with Luttik, the first three authors have shown in [5] that the process algebra obtained by adding Hennessy's merge operator from [22] to CCS does not have a finite equational axiomatization modulo bisimulation equivalence. This result is in sharp contrast with a theorem established by Fokkink and Luttik in [18] to the effect that the process algebra PA [13] affords an ω-complete axiomatization that is finite if so is the underlying set of actions. Aceto, Ésik and Ingolfsdottir proved in [2] that there is no finite equational axiomatization that is ω-complete for the max-plus algebra of the natural numbers, a result whose process algebraic implications are discussed in [1]. Fokkink and Nain have shown in [19] that no congruence over the language BCCSP, a basic formalism to express finite process behaviour, that is included in possible worlds equivalence, and includes ready trace equivalence, affords a finite ω-complete equational axiomatization.

The paper is organized as follows. We begin by presenting the language BPA with the interrupt operator, its operational semantics and preliminaries on equational logic in Section 2. There we also show that the interrupt operator is not definable in BPA modulo bisimilarity. The general structure of the proof of our main result, to the effect that bisimilarity is not finitely based over the language we consider in this paper, is presented in Section 3. In that section, we also show how to reduce the proof of our main result to that of a technical statement describing a key property of closed instantiations of sound equations that is preserved under equational derivations (Proposition 3). We conclude the paper by showing in Section 4 that, in sharp contrast to the main result of the paper, the use of auxiliary operators is *not* necessary in order to obtain a finite axiomatization of bisimulation equivalence over closed terms in the language obtained by enriching BPA with the disrupt operator from [7].

2 Preliminaries

We begin by introducing the basic definitions and results on which the technical developments to follow are based. The interested reader is referred to [7,12] for more information.

2.1 The Language $\mathrm{BPA_{int}}$

We assume a non-empty alphabet A of atomic actions, with typical elements a, b. The language for processes we shall consider in this paper, henceforth referred to as $\mathrm{BPA_{int}}$, is obtained by adding the interrupt operator from [7] to Bergstra and Klop's BPA [12]. This language is given by the following grammar:

$$t ::= x \mid a \mid t \cdot t \mid t + t \mid t \rhd t \ ,$$

where x is a variable drawn from a countably infinite set V and a is an action. In the above grammar, we use the symbol $\rhd$ for the *interrupt operator*. We shall use the meta-variables t, u, v, w to range over process terms, and write $var(t)$ for the collection of variables occurring in the term t. The *size* of a term is the number of operator symbols in it. A process term is *closed* if it does not contain any variables. Closed terms will be typically denoted by p, q, r, s. As usual, we shall often write tu in lieu of $t \cdot u$, and we assume that $\cdot$ binds stronger than $+$.

A (closed) substitution is a mapping from process variables to (closed) $\mathrm{BPA_{int}}$ terms. For every term t and substitution σ, the term obtained by replacing every occurrence of a variable x in t with the term $\sigma(x)$ will be written $\sigma(t)$. Note that $\sigma(t)$ is closed, if so is σ. In what follows, we shall use the notation $\sigma[x \mapsto p]$, where σ is a closed substitution and p is a closed $\mathrm{BPA_{int}}$ term, to stand for the substitution mapping x to p, and acting like σ on all of the other variables in V.

In the remainder of this paper, we let a^1 denote a, and a^{m+1} denote $a(a^m)$. Moreover, we consider terms modulo associativity and commutativity of $+$. In other words, we do not distinguish $t + u$ and $u + t$, nor $(t+u)+v$ and $t+(u+v)$. This is justified because $+$ is associative and commutative with respect to the notion of equivalence we shall consider over $\mathrm{BPA_{int}}$. (See axioms A1, A2 in Table 3.) In what follows, the symbol $=$ will denote equality modulo associativity and commutativity of $+$.

We say that a term t has $+$ as head operator if $t = t_1 + t_2$ for some terms t_1 and t_2. For example, $a + b$ has $+$ as head operator, but $(a + b)a$ does not.

For $k \geq 1$, we use a *summation* $\sum_{i \in \{1,\dots,k\}} t_i$ to denote $t_1 + \cdots + t_k$. It is easy to see that every $\mathrm{BPA_{int}}$ term t has the form $\sum_{i \in I} t_i$, for some finite, non-empty index set I, and terms t_i $(i \in I)$ that do not have $+$ as head operator. The terms t_i $(i \in I)$ will be referred to as the *(syntactic) summands* of t. For example, the term $(a + b)a$ has only itself as (syntactic) summand.

The operational semantics for the language $\mathrm{BPA_{int}}$ is given by the labelled transition system

$$\left(\mathrm{BPA_{int}}, \left\{ \xrightarrow{a} \mid a \in A \right\}, \left\{ \xrightarrow{a} \checkmark \mid a \in A \right\}\right) ,$$

where the transition relations $\xrightarrow{a}$ and the unary predicates $\xrightarrow{a} \checkmark$ are, respectively, the least subsets of $\mathrm{BPA_{int}} \times \mathrm{BPA_{int}}$ and $\mathrm{BPA_{int}}$ satisfying the rules in Table 1. Intuitively, a transition $t \xrightarrow{a} u$ means that the system represented by the term t can perform the action a, thereby evolving into u. The special symbol $\checkmark$ stands for (successful) termination; therefore the interpretation of the statement $t \xrightarrow{a} \checkmark$ is that the process term t can terminate by performing a. Note that, for every closed term p, there is some action a for which either $p \xrightarrow{a} p'$ holds for some p', or $p \xrightarrow{a} \checkmark$ does.

For terms t, u, and action a, we say that u is an *a-derivative* of t if $t \xrightarrow{a} u$.

Table 1. Transition Rules for $\mathrm{BPA_{int}}$

$$a \xrightarrow{a} \checkmark$$

$$\frac{t \xrightarrow{a} \checkmark}{t + u \xrightarrow{a} \checkmark} \qquad \frac{u \xrightarrow{a} \checkmark}{t + u \xrightarrow{a} \checkmark} \qquad \frac{t \xrightarrow{a} t'}{t + u \xrightarrow{a} t'} \qquad \frac{u \xrightarrow{a} u'}{t + u \xrightarrow{a} u'}$$

$$\frac{t \xrightarrow{a} \checkmark}{t \cdot u \xrightarrow{a} u} \qquad \frac{t \xrightarrow{a} t'}{t \cdot u \xrightarrow{a} t' \cdot u}$$

$$\frac{t \xrightarrow{a} \checkmark}{t \rhd u \xrightarrow{a} \checkmark} \qquad \frac{t \xrightarrow{a} t'}{t \rhd u \xrightarrow{a} t' \rhd u} \qquad \frac{u \xrightarrow{a} \checkmark}{t \rhd u \xrightarrow{a} t} \qquad \frac{u \xrightarrow{a} u'}{t \rhd u \xrightarrow{a} u' \cdot t}$$

The transition relations $\xrightarrow{a}$ naturally compose to determine the possible effects that performing a sequence of actions may have on a $\mathrm{BPA_{int}}$ term.

Definition 1. *For a sequence of actions $a_1 \cdots a_k$ ($k \geq 0$), and* BPA$_{int}$ *terms t, t', we write $t \stackrel{a_1 \cdots a_k}{\longrightarrow} t'$ iff there exists a sequence of transitions*

$$t = t_0 \stackrel{a_1}{\to} t_1 \stackrel{a_2}{\to} \cdots \stackrel{a_k}{\to} t_k = t' \ .$$

Similarly, we say that $a_1 \cdots a_k$ ($k \geq 1$) is a termination trace of a BPA$_{int}$ *term t iff there exists a term t' such that*

$$t \stackrel{a_1 \cdots a_{k-1}}{\longrightarrow} t' \stackrel{a_k}{\to} \surd \ .$$

If $t \stackrel{a_1 \cdots a_k}{\longrightarrow} t'$ holds for some BPA$_{int}$ *term t', or $a_1 \cdots a_k$ is a termination trace of t, then $a_1 \cdots a_k$ is a* trace *of t.*

The depth *of a term t, written $depth(t)$, is the length of the longest trace it affords.*

The norm *of a term t, denoted by $norm(t)$, is the length of its shortest termination trace; this notion stems from [9].*

The depth and the norm of closed terms can also be characterized inductively thus:

$$\begin{aligned} depth(a) &= 1 \\ depth(p+q) &= \max\{depth(p), depth(q)\} \\ depth(pq) &= depth(p) + depth(q) \\ depth(p \rhd q) &= depth(p) + depth(q) \end{aligned}$$

$$\begin{aligned} norm(a) &= 1 \\ norm(p+q) &= \min\{norm(p), norm(q)\} \\ norm(pq) &= norm(p) + norm(q) \\ norm(p \rhd q) &= norm(p) \ . \end{aligned}$$

Note that the depth and the norm of each closed BPA$_{\text{int}}$ term are positive.

In what follows, we shall sometimes need to consider the possible origins of a transition of the form $\sigma(t) \stackrel{a}{\to} p$, for some action a, closed substitution σ, BPA$_{\text{int}}$ term t and closed term p. Naturally enough, we expect that $\sigma(t)$ affords that transition if $t \stackrel{a}{\to} t'$, for some t' such that $p = \sigma(t')$. However, the above transition may also derive from the initial behaviour of some closed term $\sigma(x)$, provided that the collection of initial moves of $\sigma(t)$ depends, in some formal sense, on that of the closed term substituted for the variable x. Similarly, we shall sometimes need to consider the possible origins of a transition of the form $\sigma(t) \stackrel{a}{\to} \surd$, for some action a, closed substitution σ and BPA$_{\text{int}}$ term t.

To fully describe these situations, we introduce the auxiliary notion of configuration of a BPA$_{\text{int}}$ term. To this end, we assume a set of symbols

$$V_d = \{x_d \mid x \in V\}$$

disjoint from V. Intuitively, the symbol x_d (read "during x") will be used to denote that the closed term substituted for variable x has begun executing, but has not yet terminated.

Definition 2. *The collection of* BPA$_{int}$ *configurations is given by the following grammar:*

$$c ::= t \mid x_d \mid c \cdot t \mid c \rhd t \ ,$$

where t is a BPA$_{int}$ *term, and $x_d \in V_d$.*

For example, the configuration $x_d \cdot (a \rhd x)$ is meant to describe a state of the computation of some term in which the (closed term substituted for the) occurrence of variable x on the left-hand side of the $\cdot$ operator has begun its execution (and has not terminated), but the one on the right-hand side has not. Note that each configuration contains at most one occurrence of an $x_d \in V_d$.

We shall consider the symbols x_d as variables, and use the notation $\sigma[x_d \mapsto p]$, where σ is a closed substitution and p is a closed $\mathrm{BPA_{int}}$ term, to stand for the substitution mapping x_d to p, and acting like σ on all of the other variables.

Table 2. SOS Rules for the Auxiliary Transitions $\stackrel{x}{\rightarrow}$, $\stackrel{x_s}{\rightarrow}$ and $\stackrel{x}{\rightarrow}\surd$ $(x \in V)$

$$\frac{}{x \stackrel{x_s}{\rightarrow} x_d} \qquad \frac{}{x \stackrel{x}{\rightarrow}\surd}$$

$$\frac{t \stackrel{x}{\rightarrow} t'}{t + u \stackrel{x}{\rightarrow} t'} \qquad \frac{t \stackrel{x_s}{\rightarrow} c}{t + u \stackrel{x_s}{\rightarrow} c} \qquad \frac{t \stackrel{x}{\rightarrow}\surd}{t + u \stackrel{x}{\rightarrow}\surd}$$

$$\frac{u \stackrel{x}{\rightarrow} u'}{t + u \stackrel{x}{\rightarrow} u'} \qquad \frac{u \stackrel{x_s}{\rightarrow} c}{t + u \stackrel{x_s}{\rightarrow} c} \qquad \frac{u \stackrel{x}{\rightarrow}\surd}{t + u \stackrel{x}{\rightarrow}\surd}$$

$$\frac{t \stackrel{x}{\rightarrow} t'}{tu \stackrel{x}{\rightarrow} t'u} \qquad \frac{t \stackrel{x_s}{\rightarrow} c}{tu \stackrel{x_s}{\rightarrow} cu} \qquad \frac{t \stackrel{x}{\rightarrow}\surd}{tu \stackrel{x}{\rightarrow} u}$$

$$\frac{t \stackrel{x}{\rightarrow} t'}{t \rhd u \stackrel{x}{\rightarrow} t' \rhd u} \qquad \frac{t \stackrel{x_s}{\rightarrow} c}{t \rhd u \stackrel{x_s}{\rightarrow} c \rhd u} \qquad \frac{t \stackrel{x}{\rightarrow}\surd}{t \rhd u \stackrel{x}{\rightarrow}\surd}$$

$$\frac{u \stackrel{x}{\rightarrow} u'}{t \rhd u \stackrel{x}{\rightarrow} u't} \qquad \frac{u \stackrel{x_s}{\rightarrow} c}{t \rhd u \stackrel{x_s}{\rightarrow} ct} \qquad \frac{u \stackrel{x}{\rightarrow}\surd}{t \rhd u \stackrel{x}{\rightarrow} t}$$

The way in which the initial behaviour of a term may depend on that of the variables that occur in it is formally described by three auxiliary transition relations whose elements have the following forms:

- $t \stackrel{x_s}{\rightarrow} c$ (read "t can start executing x and become c in doing so"), where t is a term, x is a variable, and c is a configuration,
- $t \stackrel{x}{\rightarrow} t'$, where t and t' are terms and x is a variable, or
- $t \stackrel{x}{\rightarrow}\surd$, where t is a term.

The first of these types of transitions will be used to account for those transitions of the form $\sigma(t) \stackrel{a}{\rightarrow} p$ that are due to a-labelled transitions of the closed term $\sigma(x)$ that do not lead to its termination. The second will describe the origin of transitions of the form $\sigma(t) \stackrel{a}{\rightarrow} \sigma(t')$ that are due to a-labelled transitions of the closed term $\sigma(x)$ that lead to its termination. Finally, transitions of the third kind will allow us to describe the origin of termination transitions of the form $\sigma(t) \stackrel{a}{\rightarrow}\surd$ that are due to a-labelled termination transitions of the closed term $\sigma(x)$.

The SOS rules defining these transitions are given in Table 2. In those rules, the meta-variables t, u, t' and u' denote $\mathrm{BPA_{int}}$ terms, and c ranges over the collection of configurations that contain one occurrence of a symbol of the form x_d. The attentive reader might have already noticed that the left-hand sides of the rules in Table 2 are always $\mathrm{BPA_{int}}$ terms, and therefore that no transitions are possible from configurations that contain one occurrence of a symbol of the form x_d. This is in line with our aim in defining the auxiliary transition relations $\xrightarrow{x}$, $\xrightarrow{x_s}$ and $\xrightarrow{x}\surd$ ($x \in V$), viz. to describe the possible origins of the *initial* transitions of a term of the form $\sigma(t)$, with t a $\mathrm{BPA_{int}}$ term and σ a closed substitution.

Lemma 1. *For each* BPA_{int} *term t, configuration c and variable x, if $t \xrightarrow{x_s} c$, then x_d occurs in c. Moreover, if $c = x_d$ then x is a summand of t.*

The precise connection between the transitions of a term $\sigma(t)$ and those of t is expressed by the following lemma.

Lemma 2 (Operational Correspondence). *Assume that t is a* BPA_{int} *term, σ is a closed substitution and a is an action. Then the following statements hold:*

1. *If $t \xrightarrow{a}\surd$, then $\sigma(t) \xrightarrow{a}\surd$.*
2. *If $t \xrightarrow{x}\surd$ and $\sigma(x) \xrightarrow{a}\surd$, then $\sigma(t) \xrightarrow{a}\surd$.*
3. *If $t \xrightarrow{x} t'$ and $\sigma(x) \xrightarrow{a}\surd$, then $\sigma(t) \xrightarrow{a} \sigma(t')$.*
4. *Assume that $t \xrightarrow{x_s} c$ and $\sigma(x) \xrightarrow{a} p$, for some closed term p. Then*
$$\sigma(t) \xrightarrow{a} \sigma[x_d \mapsto p](c) \ .$$
5. *If $t \xrightarrow{a} t'$, then $\sigma(t) \xrightarrow{a} \sigma(t')$.*
6. *Assume that $\sigma(t) \xrightarrow{a}\surd$. Then either $t \xrightarrow{a}\surd$ or there is a variable x such that $t \xrightarrow{x}\surd$ and $\sigma(x) \xrightarrow{a}\surd$.*
7. *Assume that $\sigma(t) \xrightarrow{a} p$, for some closed term p. Then one of the following possibilities applies:*
   - *$t \xrightarrow{x} t'$, $\sigma(x) \xrightarrow{a}\surd$ and $p = \sigma(t')$, for some term t' and variable x,*
   - *$t \xrightarrow{a} t'$ for some term t' such that $p = \sigma(t')$, or*
   - *$t \xrightarrow{x_s} c$ and $\sigma(x) \xrightarrow{a} q$, for some variable x, configuration c and closed term q such that $\sigma[x_d \mapsto q](c) = p$.*

In this paper, we consider the language $\mathrm{BPA_{int}}$ modulo bisimulation equivalence [29].

Definition 3. *Two closed* BPA_{int} *terms p and q are bisimilar, denoted by $p \leftrightarrow q$, if there exists a symmetric binary relation $\mathcal{B}$ over closed* BPA_{int} *terms which relates p and q, such that:*

- *if $r\ \mathcal{B}\ s$ and $r \xrightarrow{a} r'$, then there is a transition $s \xrightarrow{a} s'$ such that $r'\ \mathcal{B}\ s'$;*
- *if $r\ \mathcal{B}\ s$ and $r \xrightarrow{a}\surd$, then $s \xrightarrow{a}\surd$.*

Such a relation $\mathcal{B}$ will be called a bisimulation. The relation $\leftrightarrow$ will be referred to as bisimulation equivalence or bisimilarity.

It is well known that $\leftrightarrow$ is an equivalence relation [29]. Moreover, the transition rules in Table 1 are in the 'path' format of Baeten and Verhoef [10]. Hence, bisimulation equivalence is a congruence with respect to all the operators in the signature of $\mathrm{BPA_{int}}$.

Note that bisimilar closed $\mathrm{BPA_{int}}$ terms afford the same finite non-empty collection of (termination) traces, and therefore have the same norm and depth.

Bisimulation equivalence is extended to arbitrary $\mathrm{BPA_{int}}$ terms thus:

Definition 4. *Let t, u be* BPA$_{int}$ *terms. Then $t \leftrightarrow u$ iff $\sigma(t) \leftrightarrow \sigma(u)$ for every closed substitution σ.*

For instance, we have that

$$x \triangleright y \leftrightarrow (x \triangleright y) + yx$$

because, as our readers can easily check, the terms $p \triangleright q$ and $(p \triangleright q) + qp$ have the same set of initial "capabilities", i.e.,

$$p \triangleright q \xrightarrow{a} r \text{ iff } (p \triangleright q) + qp \xrightarrow{a} r \ , \text{ for each } a \text{ and } r, \text{ and}$$
$$p \triangleright q \xrightarrow{a} \checkmark \text{ iff } (p \triangleright q) + qp \xrightarrow{a} \checkmark, \text{ for each } a \ .$$

It is natural to expect that the interrupt operator cannot be defined in the language BPA modulo bisimulation equivalence. This expectation is confirmed by the following simple, but instructive, result:

Proposition 1. *There is no* BPA$_{int}$ *term t such that t does not contain occurrences of the interrupt operator, and $t \leftrightarrow x \triangleright y$.*

Proof. Assume, towards a contradiction, that t is a BPA$_{\mathrm{int}}$ term such that t does not contain occurrences of the interrupt operator, and $t \leftrightarrow x \triangleright y$.

Consider the closed substitution σ_a mapping each variable to a. Since

$$\sigma_a(t) \leftrightarrow a \triangleright a \text{ and } a \triangleright a \xrightarrow{a} \checkmark \ ,$$

we have that $\sigma_a(t) \xrightarrow{a} \checkmark$. Lemma 2(6) yields that either $t \xrightarrow{a} \checkmark$ or there is a variable z such that $t \xrightarrow{z} \checkmark$ and $\sigma_a(z) \xrightarrow{a} \checkmark$. We shall now argue that both of these possibilities imply that $t \not\leftrightarrow x \triangleright y$, contradicting our assumption.

Indeed, using the former possibility and Lemma 2(1), we may infer that

$$\sigma_a[x \mapsto a^2](t) \xrightarrow{a} \checkmark \ .$$

This implies that $t \not\leftrightarrow x \triangleright y$, because $a^2 \triangleright a$ does not have termination traces of length 1.

Assume now that there is a variable z such that $t \xrightarrow{z} \checkmark$ and $\sigma_a(z) \xrightarrow{a} \checkmark$. It is not hard to see that $t \leftrightarrow z + u$ for some term u, since t does not contain occurrences of the interrupt operator and $t \xrightarrow{z} \checkmark$. We claim that

$$\sigma_a[x \mapsto a^2](t) \not\leftrightarrow a^2 \triangleright a \ .$$

If $z \neq x$, our claim follows, because, reasoning as above,

$$\sigma_a[x \mapsto a^2](t) \leftrightarrow a + \sigma_a[x \mapsto a^2](u) \xrightarrow{a} \checkmark$$

whereas $a^2 \triangleright a$ does not have termination traces of length 1.

If $t \leftrightarrow x + u$, then $\sigma_a[x \mapsto a^2](t) \xrightarrow{a} p$ for some $p \leftrightarrow a$. On the other hand, the two a-derivatives of $a^2 \triangleright a$, namely $a \triangleright a$ and a^2, have depth 2, and thus neither of them is bisimilar to a. □

2.2 Equational Logic

An *axiom system* is a collection of equations $t \approx u$ over the language $\mathrm{BPA_{int}}$. An equation $t \approx u$ is derivable from an axiom system E, notation $E \vdash t \approx u$, if it can be proven from the axioms in E using the rules of equational logic (viz. reflexivity, symmetry, transitivity, substitution and closure under $\mathrm{BPA_{int}}$ contexts):

$$t \approx t \qquad \frac{t \approx u}{u \approx t} \qquad \frac{t \approx u \quad u \approx v}{t \approx v} \qquad \frac{t \approx u}{\sigma(t) \approx \sigma(u)}$$

$$\frac{t \approx u \quad t' \approx u'}{t + t' \approx u + u'} \qquad \frac{t \approx u \quad t' \approx u'}{tt' \approx uu'} \qquad \frac{t \approx u \quad t' \approx u'}{t \rhd t' \approx u \rhd u'} \; .$$

Without loss of generality one may assume that substitutions happen first in equational proofs, i.e., that the rule

$$\frac{t \approx u}{\sigma(t) \approx \sigma(u)}$$

may only be used when $(t \approx u) \in E$. In this case, the equation $\sigma(t) \approx \sigma(u)$ is called a *substitution instance* of an axiom in E.

Moreover, by postulating that for each axiom in E also its symmetric counterpart is present in E, one may assume that applications of symmetry happen first in equational proofs. In the remainder of this paper, we shall tacitly assume that our equational axiom systems are closed with respect to symmetry.

It is well-known (see, e.g., Sect. 2 in [21]) that if an equation relating two closed terms can be proven from an axiom system E, then there is a closed proof for it.

Definition 5. *An equation $t \approx u$ over the language BPA_{int} is sound with respect to $\leftrightarrow$ iff $t \leftrightarrow u$. An axiom system is sound with respect to $\leftrightarrow$ iff so is each of its equations.*

An example of a collection of equations over the language $\mathrm{BPA_{int}}$ that are sound with respect to $\leftrightarrow$ is given in Table 3. Those equations stem from [12]. Equations dealing with the interrupt operator using two auxiliary operators are offered in [7].

Table 3. Some Axioms for $\mathrm{BPA_{int}}$

A1	$x + y \approx y + x$
A2	$(x + y) + z \approx x + (y + z)$
A3	$x + x \approx x$
A4	$(x + y)z \approx (xz) + (yz)$
A5	$(xy)z \approx x(yz)$

3 Bisimilarity Is Not Finitely Based over $\mathrm{BPA_{int}}$

Our order of business in the remainder of this paper will be to show the following theorem:

Theorem 1. *Bisimilarity is not finitely based over the language* BPA_{int}*—that is, there is no finite axiom system that is sound with respect to* $\leftrightarrow$*, and proves all of the equations* $t \approx u$ *such that* $t \leftrightarrow u$*. Moreover, the same holds true if we restrict ourselves to the collection of closed equations over* BPA_{int} *that hold modulo* $\leftrightarrow$*.*

The above theorem is an immediate corollary of the following result:

Theorem 2. *Let* E *be a finite collection of equations over the language* BPA_{int} *that is sound modulo* $\leftrightarrow$*. Let* $n > 2$ *be larger than the size of each term in the equations in* E*. Then* E *does not prove the equation* e_n*, where the family of equations* e_n $(n \geq 1)$ *is defined thus:*

$$e_n : \left(\sum_{i=1}^{n} p_i\right) \rhd a \approx b + \sum_{i=2}^{n} b((b^{i-1} + b) \rhd a) + a\sum_{i=1}^{n} p_i \ . \tag{1}$$

In the above family, $p_1 = b$ *and* $p_i = b(b^{i-1} + b)$ *for* $i > 1$*.*

Observe that, for each $n \geq 1$, the closed equation e_n is sound modulo bisimilarity. Indeed, the left-hand and right-hand sides of the equation have isomorphic labelled transitions systems. Therefore, as claimed above, Theorem 1 is an immediate consequence of Theorem 2. In the remainder of this study, we shall offer a proof of Theorem 2. In order to prove this theorem, it will be sufficient to establish the following technical result:

Proposition 2. *Let* E *be a finite axiom system over the language* BPA_{int} *that is sound modulo bisimilarity. Let* $n > 2$ *be larger than the size of each term in the equations in* E*. Assume, furthermore, that*

- $E \vdash p \approx q$,
- $p \leftrightarrow \left(\sum_{i=1}^{n} p_i\right) \rhd a$ *and*
- p *has a summand bisimilar to* $\left(\sum_{i=1}^{n} p_i\right) \rhd a$.

Then q *has a summand bisimilar to* $\left(\sum_{i=1}^{n} p_i\right) \rhd a$*.*

Indeed, assuming Proposition 2, we can prove Theorem 2, and therefore Theorem 1, as follows.

Proof of Theorem 2: Assume that E is a finite axiom system over the language $\mathrm{BPA}_{\mathrm{int}}$ that is sound modulo bisimilarity. Pick $n > 2$ and larger than the size of the terms in the equations in E. Assume that, for some closed term q,

$$E \vdash \left(\sum_{i=1}^{n} p_i\right) \rhd a \approx q \ .$$

Using Proposition 2, we have that q has a summand bisimilar to $\left(\sum_{i=1}^{n} p_i\right) \rhd a$. Note now that the summands of the right-hand side of equation e_n, viz.

$$b + \sum_{i=2}^{n} b((b^{i-1} + b) \rhd a) + a\sum_{i=1}^{n} p_i \ ,$$

are the terms

- b,
- $b((b^{i-1}+b) \rhd a)$, for some $2 \leq i \leq n$, and
- $a \sum_{i=1}^{n} p_i$.

Unlike $\left(\sum_{i=1}^{n} p_i\right) \rhd a$, none of these terms can initially perform both an a and a b action. It follows that no summand of the right-hand side of equation e_n is bisimilar to $\left(\sum_{i=1}^{n} p_i\right) \rhd a$, and thus that

$$q \neq b + \sum_{i=2}^{n} b((b^{i-1}+b) \rhd a) + a \sum_{i=1}^{n} p_i \ .$$

We may therefore conclude that the axiom system E does not prove equation e_n, which was to be shown. □

Our order of business will now be to provide a proof of Proposition 2. Our proof of that result will be proof-theoretic in nature, and will proceed by induction on the depth of equational derivations from a finite axiom system E. The crux in such an induction proof is given by the following proposition, to the effect that the statement of Proposition 2 holds for closed instantiations of axioms in E.

Proposition 3. *Let $t \approx u$ be an equation over the language* BPA$_{int}$ *that holds modulo bisimilarity. Let σ be a closed substitution, $p = \sigma(t)$ and $q = \sigma(u)$. Assume that*

- *$n > 2$ and the size of t is smaller than n,*
- *$p \leftrightarrow \left(\sum_{i=1}^{n} p_i\right) \rhd a$ and*
- *p has a summand bisimilar to $\left(\sum_{i=1}^{n} p_i\right) \rhd a$.*

Then q has a summand bisimilar to $\left(\sum_{i=1}^{n} p_i\right) \rhd a$.

Indeed, let us assume for the moment that the above result holds. Using it, we can prove Proposition 2 thus:

Proof of Proposition 2: Assume that E is a finite axiom system over the language BPA$_{\text{int}}$ that is sound with respect to bisimulation equivalence, and that the following hold, for some closed terms p and q and positive integer $n > 2$ that is larger than the size of each term in the equations in E:

1. $E \vdash p \approx q$,
2. $p \leftrightarrow \left(\sum_{i=1}^{n} p_i\right) \rhd a$, and
3. p has a summand bisimilar to $\left(\sum_{i=1}^{n} p_i\right) \rhd a$.

We prove that q also has a summand bisimilar to $\left(\sum_{i=1}^{n} p_i\right) \rhd a$ by induction on the depth of the closed proof of the equation $p \approx q$ from E. Recall that, without loss of generality, we may assume that applications of symmetry happen first in equational proofs (that is, E is closed with respect to symmetry).

We proceed by a case analysis on the last rule used in the proof of $p \approx q$ from E. The case of reflexivity is trivial, and that of transitivity follows immediately by using the inductive hypothesis twice. Below we only consider the other possibilities.

- CASE $E \vdash p \approx q$, BECAUSE $\sigma(t) = p$ AND $\sigma(u) = q$ FOR SOME EQUATION $(t \approx u) \in E$ AND CLOSED SUBSTITUTION σ. Since $n > 2$ is larger than the size of each term mentioned in equations in E, the claim follows by Proposition 3.
- CASE $E \vdash p \approx q$, BECAUSE $p = p' + p''$ AND $q = q' + q''$ FOR SOME p', q', p'', q'' SUCH THAT $E \vdash p' \approx q'$ AND $E \vdash p'' \approx q''$. Since p has a summand bisimilar to $\left(\sum_{i=1}^{n} p_i\right) \rhd a$, we have that so does either p' or p''. Assume, without loss of generality, that p' has a summand bisimilar to $\left(\sum_{i=1}^{n} p_i\right) \rhd a$. Since p is bisimilar to $\left(\sum_{i=1}^{n} p_i\right) \rhd a$, so is p'. The inductive hypothesis now yields that q' has a summand bisimilar to $\left(\sum_{i=1}^{n} p_i\right) \rhd a$. Hence, q has a summand bisimilar to $\left(\sum_{i=1}^{n} p_i\right) \rhd a$, which was to be shown.
- CASE $E \vdash p \approx q$, BECAUSE $p = p'p''$ AND $q = q'q''$ FOR SOME p', q', p'', q'' SUCH THAT $E \vdash p' \approx q'$ AND $E \vdash p'' \approx q''$. This case is vacuous. In fact, $norm(p) = 1$ by our assumption that $p \leftrightarrow \left(\sum_{i=1}^{n} p_i\right) \rhd a$, whereas the norm of a closed term of the form $p'p''$ is at least 2.
- CASE $E \vdash p \approx q$, BECAUSE $p = p' \rhd p''$ AND $q = q' \rhd q''$ FOR SOME p', q', p'', q'' SUCH THAT $E \vdash p' \approx q'$ AND $E \vdash p'' \approx q''$. The claim is immediate because p and q are their only summands, and E is sound modulo bisimilarity.

This completes the proof. □

In light of our previous discussion, all that we are left to do to complete our proof of Theorem 1 is to show Proposition 3. The rather lengthy proof of that result may be found in [6].

4 BPA with the Disrupt Operator

As mentioned in Sect. 1, in their paper [7], Baeten and Bergstra have given a finite axiomatization of bisimilarity over BPA_δ (the extension of BPA with a constant δ to describe "deadlock") enriched with two mode transfer operators, viz. the *disrupt* and *interrupt* operators, using auxiliary operators. The main result in this paper (Theorem 1) shows that the use of auxiliary operators is indeed necessary in order to obtain a finite axiomatization of bisimulation equivalence over the language $\mathrm{BPA}_{\mathrm{int}}$, and that this holds true even if we restrict ourselves to axiomatizing the collection of closed equations over this language.

A natural question to ask at this point is whether this negative result applies also to the language $\mathrm{BPA}_{\mathrm{dis}}$ obtained by enriching BPA with the disrupt operator. Intuitively, "p disrupted by q"—which we shall write $p \blacktriangleright q$ in what follows—describes a process that normally behaves like p. However, at each point of the computation before p terminates, q can begin its execution. If this happens, q takes over, and p never resumes its computation. This intuition is captured formally by the following transition rules:

$$\frac{t \xrightarrow{a} \surd}{t \blacktriangleright u \xrightarrow{a} \surd} \qquad \frac{t \xrightarrow{a} t'}{t \blacktriangleright u \xrightarrow{a} t' \blacktriangleright u} \qquad \frac{u \xrightarrow{a} \surd}{t \blacktriangleright u \xrightarrow{a} \surd} \qquad \frac{u \xrightarrow{a} u'}{t \blacktriangleright u \xrightarrow{a} u'}$$

It is not hard to see that the following equations are sound modulo bisimilarity over the language $\mathrm{BPA}_{\mathrm{dis}}$:

$$a \blacktriangleright x \approx a + x$$
$$ax \blacktriangleright y \approx a(x \blacktriangleright y) + y \quad \text{and}$$
$$(x + y) \blacktriangleright z \approx (x \blacktriangleright z) + (y \blacktriangleright z) \ .$$

The last of these equations is particularly important, at least as far as obtaining a finite equational axiomatization of bisimilarity over the collection of closed terms in the language $\mathrm{BPA}_{\mathrm{dis}}$ is concerned. (The interested reader may have already noticed that its soundness modulo bisimulation equivalence depends crucially on the fact that transitions due to moves of the second argument of a disrupt discard the first argument.) Indeed, its repeated use in conjunction with the first two laws allows us to eliminate occurrences of the disrupt operator from closed terms. This effectively reduces the problem of finitely axiomatizing bisimilarity over the collection of closed terms in the language $\mathrm{BPA}_{\mathrm{dis}}$ to that of offering a finite axiomatization of bisimilarity over closed BPA terms. As shown by Bergstra and Klop in [12], the five equations in Table 3 suffice to axiomatize bisimilarity over the language BPA.

In sharp contrast to Theorem 1, we therefore have that:

Theorem 3. *The collection of closed equations over* $\mathrm{BPA}_{\mathit{dis}}$ *that hold modulo* $\leftrightarrow$ *is axiomatized by the five equations in Table 3 together with the aforementioned three equations for the disrupt operator, and is therefore finitely based.*

It follows that the use of auxiliary operators is *not* necessary in order to obtain a finite axiomatization of bisimulation equivalence over closed terms in the language $\mathrm{BPA}_{\mathrm{dis}}$.

The axiomatization of bisimilarity over closed terms in the language $\mathrm{BPA}_{\mathrm{dis}}$ offered in the theorem above is not ω-complete. For example, the reader can easily check that the disrupt operator is associative modulo bisimilarity, i.e., that the equation

$$(x \blacktriangleright y) \blacktriangleright z \approx x \blacktriangleright (y \blacktriangleright z)$$

holds modulo $\leftrightarrow$. This equation is not provable using the equations mentioned in Theorem 3. However, we conjecture that bisimilarity also affords a finite ω-complete axiomatization over $\mathrm{BPA}_{\mathrm{dis}}$.

References

1. L. ACETO, Z. ÉSIK, AND A. INGOLFSDOTTIR, *On the two-variable fragment of the equational theory of the max-sum algebra of the natural numbers*, in Proceedings of the 17th International Symposium on Theoretical Aspects of Computer Science, STACS 2000 (Lille), H. Reichel and S. Tison, eds., vol. 1770 of Lecture Notes in Computer Science, Springer-Verlag, Feb. 2000, pp. 267–278.
2. ———, *The max-plus algebra of the natural numbers has no finite equational basis*, Theoretical Comput. Sci., 293 (2003), pp. 169–188.
3. L. ACETO, W. FOKKINK, R. VAN GLABBEEK, AND A. INGOLFSDOTTIR, *Nested semantics over finite trees are equationally hard*, Information and Computation, 191 (2004), pp. 203–232.

4. L. Aceto, W. Fokkink, and A. Ingolfsdottir, *A menagerie of non-finitely based process semantics over BPA*—from ready simulation to completed traces*, Mathematical Structures in Computer Science, 8 (1998), pp. 193–230.
5. L. Aceto, W. Fokkink, A. Ingolfsdottir, and B. Luttik, *CCS with Hennessy's merge has no finite equational axiomatization*, Theoretical Comput. Sci., 330 (2005), pp. 377–405.
6. L. Aceto, W. Fokkink, A. Ingolfsdottir, and S. Nain, *Bisimilarity is not Finitely Based over BPA with Interrupt*, Research report RS-04-24, BRICS, Oct. 2004. Available from `http://www.brics.dk/RS/04/24/index.html`.
7. J. C. Baeten and J. Bergstra, *Mode transfer in process algebra*, Report CSR 00–01, Technische Universiteit Eindhoven, 2000. This paper is an expanded and revised version of [11].
8. J. C. Baeten, J. Bergstra, and J. W. Klop, *Syntax and defining equations for an interrupt mechanism in process algebra*, Fundamenta Informaticae, IX (1986), pp. 127–168.
9. ———, *Decidability of bisimulation equivalence for processes generating context-free languages*, J. Assoc. Comput. Mach., 40 (1993), pp. 653–682.
10. J. C. Baeten and C. Verhoef, *A congruence theorem for structured operational semantics*, in Proceedings CONCUR 93, Hildesheim, Germany, E. Best, ed., vol. 715 of Lecture Notes in Computer Science, Springer-Verlag, 1993, pp. 477–492.
11. J. Bergstra, *A mode transfer operator in process algebra*, Report P8808, Programming Research Group, University of Amsterdam, 1988.
12. J. Bergstra and J. W. Klop, *Fixed point semantics in process algebras*, Report IW 206, Mathematisch Centrum, Amsterdam, 1982.
13. ———, *Process algebra for synchronous communication*, Information and Control, 60 (1984), pp. 109–137.
14. S. Blom, W. Fokkink, and S. Nain, *On the axiomatizability of ready traces, ready simulation and failure traces*, in Proceedings 30th Colloquium on Automata, Languages and Programming—ICALP'03, Eindhoven, J. C. Baeten, J. K. Lenstra, J. Parrow, and G. J. Woeginger, eds., vol. 2719 of Lecture Notes in Computer Science, Springer-Verlag, 2003, pp. 109–118.
15. E. Brinksma, *A tutorial on LOTOS*, in Proceedings of the IFIP Workshop on Protocol Specification, Testing and Verification, M. Diaz, ed., North-Holland, 1986, pp. 73–84.
16. J. H. Conway, *Regular Algebra and Finite Machines*, Mathematics Series (R. Brown and J. De Wet eds.), Chapman and Hall, London, United Kingdom, 1971.
17. A. Dsouza and B. Bloom, *On the expressive power of CCS*, in Foundations of Software Technology and Theoretical Computer Science (Bangalore, 1995), P. S. Thiagarajan, ed., vol. 1026 of Lecture Notes in Computer Science, Springer-Verlag, 1995, pp. 309–323.
18. W. Fokkink and B. Luttik, *An omega-complete equational specification of interleaving*, in Proceedings 27th Colloquium on Automata, Languages and Programming—ICALP'00, Geneva, U. Montanari, J. Rolinn, and E. Welzl, eds., vol. 1853 of Lecture Notes in Computer Science, Springer-Verlag, July 2000, pp. 729–743.
19. W. Fokkink and S. Nain, *On finite alphabets and infinite bases: From ready pairs to possible worlds*, in Proceedings of Foundations of Software Science and Computation Structures, 7th International Conference, FOSSACS 2004, I. Walukiewicz, ed., vol. 2897, Springer-Verlag, 2004, pp. 182–194.
20. J. L. Gischer, *The equational theory of pomsets*, Theoretical Comput. Sci., 61 (1988), pp. 199–224.
21. J. F. Groote, *A new strategy for proving ω–completeness with applications in process algebra*, in Proceedings CONCUR 90, Amsterdam, J. C. Baeten and J. W. Klop, eds., vol. 458 of Lecture Notes in Computer Science, Springer-Verlag, 1990, pp. 314–331.

22. M. HENNESSY, *Axiomatising finite concurrent processes*, SIAM J. Comput., 17 (1988), pp. 997–1017.
23. ISO, *Information processing systems – open systems interconnection – LOTOS – a formal description technique based on the temporal ordering of observational behaviour* ISO/TC97/SC21/N DIS8807, 1987.
24. S. MAUW, *PSF – A Process Specification Formalism*, PhD thesis, University of Amsterdam, Dec. 1991.
25. R. MILNER, *Communication and Concurrency*, Prentice-Hall International, Englewood Cliffs, 1989.
26. R. MILNER, M. TOFTE, R. HARPER, AND D. MACQUEEN, *The Definition of Standard ML (Revised)*, MIT Press, 1997.
27. F. MOLLER, *The importance of the left merge operator in process algebras*, in Proceedings 17th ICALP, Warwick, M. Paterson, ed., vol. 443 of Lecture Notes in Computer Science, Springer-Verlag, July 1990, pp. 752–764.
28. ———, *The nonexistence of finite axiomatisations for CCS congruences*, in Proceedings 5th Annual Symposium on Logic in Computer Science, Philadelphia, USA, IEEE Computer Society Press, 1990, pp. 142–153.
29. D. PARK, *Concurrency and automata on infinite sequences*, in 5th GI Conference, Karlsruhe, Germany, P. Deussen, ed., vol. 104 of Lecture Notes in Computer Science, Springer-Verlag, 1981, pp. 167–183.
30. V. REDKO, *On defining relations for the algebra of regular events*, Ukrainskii Matematicheskii Zhurnal, 16 (1964), pp. 120–126. In Russian.
31. P. SEWELL, *Nonaxiomatisability of equivalences over finite state processes*, Annals of Pure and Applied Logic, 90 (1997), pp. 163–191.

Algebra ∩ Coalgebra = Presheaves

J. Adámek*

Technical University of Braunschweig
J.Adamek@tu-bs.de

Abstract. The intersection of algebra and coalgebra, i.e., the collection of all categories that are varieties as well as covarieties, is proved to consist of precisely the presheaf categories.

1 Introduction

A sequential automaton can be viewed as an algebra, or as a coalgebra. In fact, the main ingredient, the next-state function

$$\delta\colon Q \times I \longrightarrow Q \qquad (I = \text{the input set})$$

defines an algebra of the endofunctor $(-) \times I$ of **Set**, but by currying it

$$\hat{\delta}\colon Q \longrightarrow Q^I$$

one gets a coalgebra of the endofunctor $(-)^I$. Is this a unique such situation, or are there other interesting examples of coalgebras that are algebras? A surprisingly general example was discovered by James Worrell: he proved that for every (not necessarily finitary) signature Σ we can view Σ-coalgebras, i.e., coalgebras of the polynomial endofunctor

$$H_\Sigma Q = \coprod_{\sigma \in \Sigma} Q^n \qquad (n = \text{arity of } \sigma)$$

as a variety of algebras. In fact, the category $\mathbf{Coalg}\, H_\Sigma$ of all Σ-coalgebras is equivalent to a presheaf category $\mathbf{Set}^{\mathscr{A}^{\mathrm{op}}}$ for some small category $\mathscr{A}$, see [W]. Now $\mathbf{Set}^{\mathscr{A}^{\mathrm{op}}}$ is always a variety, in fact, a variety of unary algebras—but not always one-sorted! Thus, the slogan

Σ-coalgebras form a variety of algebras

is, in general, only true if we move from one-sorted sets to many-sorted ones. Therefore, in the present paper we consider algebra and coalgebra of many-sorted sets (given by endofunctors of $\mathbf{Set}^S$ for nonempty sets S).

We are going to describe the intersection of algebra and coalgebra, i.e., those categories which are at the same time varieties of F-algebras and covarieties of

* Support by the grant MSM 6840770014 of the Ministry of Education of the Czech Republic is acknowledged.

J.L. Fiadeiro et al. (Eds.): CALCO 2005, LNCS 3629, pp. 67–82, 2005.

G-coalgebras for endofunctors F and G of many-sorted sets. We consider all these categories as *concrete categories*, i.e., pairs consisting of a category $\mathscr{V}$ and a forgetful functor $V\colon \mathscr{V} \longrightarrow \mathbf{Set}^S$. Given two concrete categories $V_i\colon \mathscr{V}_i \longrightarrow \mathbf{Set}^S$ for $i = 1, 2$ we call them *concretely equivalent* if there exists an equivalence functor $E\colon \mathscr{V}_1 \longrightarrow \mathscr{V}_2$ such that V_1 is naturally isomorphic to $V_2 \cdot E$; notation $\mathscr{V}_1 \simeq \mathscr{V}_2$ (see [P]). We will strengthen the result of James Worrell in several directions:

(1) Considering presheaf categories $\mathbf{Set}^{\mathscr{A}^{\mathrm{op}}}$ as concrete categories over $\mathbf{Set}$, we prove that the category of Σ-coalgebras is *concretely* equivalent to a presheaf category. And conversely: given an endofunctor H of $\mathbf{Set}$ such that $\mathbf{Coalg}\,H$ is concretely equivalent to a presheaf category, then H is a reduction of a polynomial functor—thus $\mathbf{Coalg}\,H$ is the category of Σ-coalgebras for some Σ. ("Reduction" means that the value at the empty set can be changed.)

(2) Considering $\mathbf{Set}^{\mathscr{A}^{\mathrm{op}}}$ as a concrete category over $\mathbf{Set}^S$ (S-sorted sets), where S is the set of objects of $\mathscr{A}$, this category is *always* concretely equivalent to a covariety of coalgebras. And conversely: every many-sorted variety concretely equivalent to a many-sorted covariety is a category of presheaves.

(3) In contrast to (2), only very special small categories $\mathscr{A}$ have the property that $\mathbf{Set}^{\mathscr{A}^{\mathrm{op}}}$ is concretely equivalent to $\mathbf{Coalg}\,H$ over $\mathbf{Set}$: $\mathscr{A}$ has to be equivalent to the Σ-tree category for some signature. This category has all Σ-trees[1] as objects, and morphisms from t' to t are all nodes of t whose subtree (in t) is t'.

For all these results we work with concrete categories over $\mathbf{Set}^S$. For example, every variety of S-sorted coalgebras, or every covariety of S-sorted coalgebras (i.e., algebras and coalgebras over endofunctors of $\mathbf{Set}^S$). The fact that Worrell's result can be strengthened as above makes heavy use of concrete equivalence: we do not know the answer to the

Open Problem. For which endofunctors H of $\mathbf{Set}^S$ is the category $\mathbf{Coalg}\,H$ equivalent to a presheaf category?

Acknowledgement. The author is grateful to Jiří Rosický and Hans Porst for interesting discussions on the topic of this paper.

2 Varieties and Covarieties

Remark 1. What is a many-sorted variety? Each of the following is a reasonable answer, depending on the generality one has in mind:

(a) An equationally presentable category of Σ-algebras, where Σ is a finitary, S-sorted signature, see e.g. [LEW].

(b) As above, but dropping "finitary". Thus, an S-sorted *signature* is a set Σ together with an *arity* of every operation symbol $\sigma \in \Sigma$ of the form

$$\sigma\colon \left(s_i\right)_{i<n} \longrightarrow s$$

[1] By a tree we mean a labelled, ordered tree throughout the paper. And we consider trees always up to isomorphism (of labelled, ordered trees) only.

where s_i and s are sorts, and n is a cardinal number. The concept of an equationally presentable category of Σ-algebras is analogous to the finitary case (a).

(c) A full subcategory of $\mathbf{Alg}\, H$, the category of H-algebras for a *varietor* on $\mathbf{Set}^S$ (i.e., an endofunctor H having free algebras), closed under
    products,
    subalgebras, and
    quotient algebras.
  See e.g. [Re].

(d) A monadic category over $\mathbf{Set}^S$, see e.g. [AP].

Fortunately, our results about algebra meeting coalgebra are independent of the answer one chooses: even if varieties are understood in the most general sense (d), the meet of algebra and coalgebra is given by presheaf categories, a special case of (a).

Notation 2.2. For an endofunctor H of $\mathbf{Set}^S$ we denote by $\mathbf{Alg}\, H$ the category of all *H-algebras*, i.e., pairs (A, α) where A is an S-sorted set and $\alpha\colon HA \longrightarrow A$ an S-sorted function. Morphisms $f\colon (A,\alpha) \longrightarrow (B,\beta)$, called *homomorphisms*, are the S-sorted functions with $f \cdot \alpha = \beta \cdot Hf$.

For example,

$$\mathbf{Alg}\, H_\Sigma$$

is the usual category of Σ-algebras. Here $H_\Sigma\colon \mathbf{Set}^S \longrightarrow \mathbf{Set}^S$ is the *polynomial functor* defined on objects $X = (X_s)_{s\in S}$ by

$$(H_\Sigma X)_s = \coprod_{\sigma\in\Sigma_s} \prod_{i<n} X_{s_i}$$

where the arity of σ is $\sigma\colon (s_i)_{i<n} \longrightarrow \sigma$, and analogously on morphisms.

Remark 2. What is a many-sorted covariety? Each of the following is a reasonable answer, depending on the generality one has in mind:

(a) A coequationaly presentable category of Σ-algebras, where Σ is a finitary, S-sorted signature. That is, a full subcategory of $\mathbf{Alg}\, H_\Sigma$ presentable by subsets D of the cofree coalgebras $C_\Sigma(X)$ (formed by colored trees, see 2.7 below). A coalgebra satisfies D if, under any coloring by colors from X, the tree expansions of all states are trees lying in D.

(b) As above but dropping "finitary".

(c) A full subcategory of $\mathbf{Coalg}\, H$, the category of H-coalgebras for a *covarietor* on $\mathbf{Set}^S$ (i.e., an endofunctor H having cofree coalgebras $C(X)$ on all $X \in \mathbf{Set}^S$), closed under
    coproducts
    subcoalgebras, and
    quotient coalgebras.
  See [Ru].

(d) A comonadic category over $\mathbf{Set}^S$, see e.g. [AP].

Fortunately, our results about algebra meeting coalgebra are independent of the answer one chooses: even if covarieties are understood in the most general sense (d), the meet of algebra and coalgebra is given by presheaf categories, a special case of (a).

Example 1. As observed by Jan Rutten [Ru], deterministic systems having binary input and halting states (not reacting to inputs) are coalgebras of the one-sorted polynomial endofunctor

$$H_\Sigma = X \times X + 1.$$

A cofree coalgebra $C_\Sigma(k)$ is the coalgebra of all binary trees, possibly infinite, colored by k colors. In fact, given a coalgebra

$$\alpha \colon Q \longrightarrow Q \times Q + 1$$

and a coloring $f \colon Q \longrightarrow k$ of its states, every state q produces a k-colored tree $\bar{f}(q)$ which is full tree expansion in the system, and $\bar{f} \colon Q \longrightarrow C_\Sigma(k)$ is the unique homomorphism extending f.

Consider the subset

$$D = C(1) - \{t_0\}$$

of all trees distinct from the trivial, root-only, tree t_0. A coalgebra satisfies D iff it has no halting states. This covariety is also a variety: each such coalgebra is given by a function

$$\alpha \colon Q \longrightarrow Q \times Q$$

corresponding to a pair of endofunctions of Q, or, to the algebra structure

$$\hat{\alpha} \colon Q + Q \longrightarrow Q.$$

Thus the covariety presented by D is nothing else than the category

$$\mathbf{Alg}(\mathrm{Id} + \mathrm{Id})$$

of algebras of the functor Id + Id.

Next, consider the subset

$$D' \subseteq C(1)$$

of all finite trees. A coalgebras satisfies D' iff it always halts in finitely many steps. Is the corresponding covariety a variety? No! It is easy to verify that the covariety presented by D' does not have products on the level of **Set**. (In fact, consider coalgebras A_n and states q_n such that A_n halts in n steps when started in q_n, but not in less steps. Then the cartesian product of these coalgebras contains the state $(q_0, q_1, q_2, \dots)$ in which halting is impossible.) Since varieties of one-sorted algebras have products on the level of **Set**, this finishes the argument.

Notation 2.5. (1) For an endofunctor H of $\mathbf{Set}^S$ we denote by $\mathbf{Coalg}\, H$ the category of all H-*coalgebras*, i.e., pairs (A, α) consisting of an S-sorted set A and an S-sorted function $\alpha\colon A \longrightarrow HA$. Morphisms $f\colon (A,\alpha) \longrightarrow (B,\beta)$, called *homomorphisms*, are the S-sorted functions such that $\beta \cdot f = Hf \cdot \alpha$.

For example, a Σ-coalgebra, i.e., an object of $\mathbf{Coalg}\, H_\Sigma$, consists of an S-sorted set A and an S-sorted function α assigning to every element $a \in A_s$ an n-tuple $(a_i)_{i<n}$ in $\prod_{i<n} A_{s_i}$ for some operation $\sigma\colon (s_i)_{i<n} \longrightarrow s$ of Σ_s. Notation:

$$\alpha(a) = \sigma(a_i)_{i<n}.$$

(2) Given an S-sorted set X (of colors), a *cofree coalgebra* on X is a coalgebra $C(X)$ with a structure map

$$\psi_X\colon C(X) \longrightarrow HC(X)$$

together with a universal coloring

$$C(X) \xrightarrow{\varepsilon_X} X.$$

The universal property means that for every coalgebra $\alpha\colon A \longrightarrow HA$ and every coloring $f\colon A \longrightarrow X$ there exists a unique coalgebra homomorphism $\bar{f}\colon A \longrightarrow C(X)$ such that $f = \varepsilon_X \cdot \bar{f}$.

Remark 2.6. (a) To say that H has cofree coalgebras means that the forgetful functor $U\colon \mathbf{Coalg}\, H \longrightarrow \mathbf{Set}^S$ has a right adjoint. Such functors H are called *covarietors* in [AP].

(b) We can interpret $C(X)$ as the collection of all possible behaviors of systems (represented as coalgebras) whose states are not observable, but their colors are. Thus, if we color the states, using an arbitrary f, we can assign to every state $a \in A$ the corresponding behavior $\bar{f}(a)$ we observe.

Example 2.7. See [AP] For a polynomial functor H_Σ we have the cofree coalgebra $C_\Sigma(X)$ of all X-colored Σ-trees. More detailed:

(1) By a *Σ-tree* τ is meant a tree labelled in Σ in such a way that every node labelled by a symbol σ of arity $\sigma\colon (s_i)_{i<n} \longrightarrow s$ (such nodes are said to *have sort* s) has n children, and the i-th child has sort s_i. The sort of the root is called the *sort of the tree* τ. We denote by T_Σ the S-sorted set of all Σ-trees. (Recall that trees are considered up to isomorphism throughout the paper.) And T_Σ is a terminal Σ-coalgebra whose coalgebra structure is the inverse of tree-tupling.

(2) Given an S-sorted set X of "colors", by an *X-colored Σ-tree* is meant a Σ-tree together with a coloring of its nodes: every node of sort s is colored by an element of X_s. The color of the root is called the *color of the tree* τ. We denote by $C_\Sigma(X)$ the S-sorted set of all X-colored trees.

(3) The coalgebra structure ψ_X of the cofree coalgebra $C_\Sigma(X)$ has sorts

$$(\psi_X)_s\colon C_\Sigma(X)_s \longrightarrow \coprod_{\sigma\in\Sigma_s} \prod_{i<n} C_\Sigma(X)_{s_i}$$

which assign to every tree τ whose root is labelled by σ of arity $\sigma\colon (s_i)_{i<n} \longrightarrow s$ the n-tuple of its children in $\prod_{i<n} C_\Sigma(X)_{s_i}$. The color of tree-roots yields the universal map

$$\varepsilon_X\colon C_\Sigma(X) \longrightarrow X.$$

(4) For every Σ-coalgebra (A, α) and every coloring $f\colon A \longrightarrow X$ the unique homomorphism

$$\bar{f}\colon A \longrightarrow C_\Sigma(X) \qquad \text{with } f = \varepsilon_X \cdot \bar{f}$$

can be described as follows. Given an element $a \in A_s$ then $\alpha(a)$ has the form

$$\alpha(a) = \sigma(a_i)_{i<n} \quad \text{for some } \sigma \in \Sigma_s$$

where the sorts s_i of a_i are such that σ has arity $\sigma\colon (s_i)_{i<n} \longrightarrow s$. This gives us an unfolding tree τ of a: the root of τ is labelled by σ and colored by $f_s(a)$, the children a_i of the root are labelled (recursively) by an operation in Σ and colored by $f_{s_i}(a_i)$, etc. We put

$$\bar{f}_s(a) = \tau.$$

Definition 2.8. (See [Ru].) Let H be a covarietor. Given a subobject

$$d\colon D \hookrightarrow C(X)$$

of a cofree coalgebra, we say that a coalgebra A *satisfies* D provided that for every coloring $f\colon A \longrightarrow X$ the corresponding homomorphism $\bar{f}$ factorizes through d. By a *covariety* of S-sorted coalgebras is meant a full subcategory of **Coalg** H, where H is a covarietor in **Set**, which can be presented by subobjects of cofree coalgebras; or, equivalently, which is closed under coproducts, subcoalgebras, and quotient coalgebras, see [Ru].

Example 2.9. M-sets as a variety. Unary algebras with operations indexed by a set M are just algebras of the functor $X \longmapsto M \times X$. If M is a monoid with unit e, recall that an *M-set* is a unary algebra $\alpha\colon M \times A \longrightarrow A$ such that

$$\alpha(e, x) = x \quad \text{for } x \in A$$

and, whenever $u \cdot v = w$ in M, then

$$\alpha(w, x) = \alpha(u, \alpha(v, x)) \quad \text{for } x \in A.$$

This is, obviously, precisely the variety presented by the equations $e(x) = x$ and $w(x) = u(v(x))$.

Example 2.10. M-sets as a covariety. Here we use the obvious currification, for every unary algebra A:

$$\frac{M \times A \longrightarrow A}{A \longrightarrow A^M}$$

to obtain a coalgebra of the set functor $X \longmapsto X^M$. Observe that this functor is polynomial: consider one M-ary operation. The elements of a cofree coalgebra $C(X)$ are the trees colored in X such that the children of every node are indexed by M. Therefore, the nodes of such a tree are in a bijective correspondence with words in M^*: the empty word is the root, and the children of a word w are the words wm for all $m \in M$. Consequently, to present a member of $C(X)$

means to present a coloring of words in M^* by colors in X. Shortly: we can consider $C(X)$ as the set of all colorings:

$$C(X) = X^{M^*}.$$

Consider the coequation

$$D \subseteq C(2)$$

of all colorings of M^* by two colors such that
(a) e has the same color as the empty word and
(b) if $u \cdot v = w$ in M then w has the same color as the two-letter word uv.
It is easy to verify that D presents precisely the M-sets in $\mathbf{Coalg}(-)^M$.

3 Coalgebras as Presheaves

Assumption 3.1. Unlike the rest of the paper, in the present section we work with one-sorted algebras and coalgebras only. Thus "concrete category" means here a category together with a faithful functor into **Set**. Examples include

(a) $\mathbf{Alg}\, H$ and $\mathbf{Coalg}\, H$ for endofunctors H of **Set**, and
(b) presheaf categories $\mathbf{Set}^{\mathscr{A}^{\mathrm{op}}}$ (where $\mathscr{A}$ is any small category) endowed with the forgetful functor

$$V\colon \mathbf{Set}^{\mathscr{A}^{\mathrm{op}}} \longrightarrow \mathbf{Set}, \quad V(A) = \coprod_{s \in \mathbf{obj}\, \mathscr{A}} A(s).$$

Remark 3.2. For every one-sorted signature Σ each Σ-coalgebra (A, α) defines a graph (with multiple, directed edges) on the set A: the edges from a node b into a node a with $\alpha(a) = \sigma(a_i)_{i<n}$ are precisely all the indices i with $b = a_i$.

For example, the graph of the terminal coalgebra T_Σ, see Example 2.7 (1) has (uncolored) Σ-trees as nodes, and edges from t' to t are all children of t which are (isomorphic to) t'.

Notation 3.3. Let $\mathscr{T}_\Sigma$ be the free category on the graph T_Σ; we can describe it as follows:

- objects are all Σ-trees;
- morphisms from t' to t are all nodes of t such that the corresponding subtree of t is (isomorphic to) t';
- identity morphisms are given by the roots of trees;
- composition is the obvious one, given by the transitivity of the relation "subtree".

In particular, every morphism is a composite of the *basic morphisms* $t' \longrightarrow t$ given by all children of t equal to t'.

Theorem 3.4. For every one-sorted signature Σ the category of Σ-coalgebras is concretely equivalent, over Set, to the presheaf category of $\mathscr{T}_\Sigma$:

$$\mathbf{Coalg}\, H_\Sigma \simeq \mathbf{Set}^{\mathscr{T}_\Sigma^{\mathrm{op}}}.$$

Remark. (1) This theorem is a special case of the following result (the proof of which is completely analogous): given a Σ-coalgebra A, let $\mathscr{A}$ denote the free category on the graph of A. Then the comma-category $\mathbf{Coalg}\, H_\Sigma/A$ is concretely equivalent to the presheaf category of $\mathscr{A}$.

(2) Although we formulated the theorem for one-sorted signatures only, it holds for S-sorted ones: $\mathbf{Coalg}\, H_\Sigma$ is concretely equivalent over $\mathbf{Set}^S$ to $(\mathbf{Set}^S)^{\mathscr{T}_\Sigma^{\mathrm{op}}}$.

(3) The equivalence of $\mathbf{Coalg}\, H_\Sigma$ to a presheaf category was proved by James Worrell [W]; he uses wide-pullbacks preserving functors instead of polynomial functors, but for set functors these are the same.

(4) The proof below is due to Hans Porst and Christian Dzieron [PD], except that the concreteness of the equivalence was not mentioned there.

Proof. For every H_Σ-coalgebra A denote by

$$h_A\colon A \longrightarrow T_\Sigma$$

the unique homomorphism (which takes every element to the tree unfolding in A). Define a functor $E\colon \mathbf{Coalg}\, H_\Sigma \longrightarrow \mathbf{Set}^{\mathscr{T}_\Sigma^{\mathrm{op}}}$ as follows. With every coalgebra A associate a presheaf $EA\colon \mathscr{T}_\Sigma^{\mathrm{op}} \longrightarrow \mathbf{Set}$ defined on objects $t \in T_\Sigma$ by

$$EA(t) = h_A^{-1}(t).$$

Thus, to every tree t we assign the set of all elements of A which unfold to t. For every basic morphism $j\colon t' \longrightarrow t$ (where $t = \sigma(t_i)_{i<n}$ and $t' = t_j$) the function $EA(j)\colon h_A^{-1}(t) \longrightarrow h_A^{-1}(t')$ assigns to every element a with $h_A(t) = a$ the element a_j, where $\alpha(a) = (a_i)_{i<n}$.

With every coalgebra homomorphism $f\colon (A, \alpha) \longrightarrow (B, \beta)$ we associate a natural transformation $Ef\colon EA \longrightarrow EB$ whose t-component is the restriction of f to $h_A^{-1}(t) \longrightarrow h_B^{-1}(t)$. Then E is an equivalence functor, see [PD].

$VE\colon \mathbf{Coalg}\, H_\Sigma \longrightarrow \mathbf{Set}$ takes a coalgebra A to the isomorphic copy of UA one gets from the bijection between UA and $\coprod h_A^{-1}(t)$. This isomorphism $UA \simeq VEA$ is clearly natural in A. □

Remark 3.5. As mentioned in the Introduction, we call a set functor H a *reduction* of H' provided that H and H' coincide on all nonempty sets and nonempty functions. (Example: the constant functor C_1 of value 1 has the reduction C_{10} with $C_{10}\emptyset = \emptyset$.) In that case $\mathbf{Coalg}\, H$ is, obviously, concretely equivalent (in fact, isomorphic) to $\mathbf{Coalg}\, H'$. Thus we have an immediate

Corollary 3.6. For every reduction H of a polynomial functor the category $\mathbf{Coalg}\, H$ is concretely equivalent to a presheaf category.

Example 3.7. (i) Let Σ be the signature with a nullary symbol, then H_Σ has C_1 (the constant functor of value 1) as a summand. Consequently, reductions H of H_Σ can have *any* value at $\emptyset$: define $H\emptyset = R$ (an arbitrary set) and for the empty map $f\colon \emptyset \longrightarrow X$ let $Hf\colon M \longrightarrow H_\Sigma X$ factor through the summand $1 = C_1 X \hookrightarrow H_\Sigma X$ above.

(ii) If Σ has no nullary symbols, then H_Σ has no reductions $H \neq H_\Sigma$: any element of $H\emptyset$ leads to a natural transformation from C_{10} to H_Σ. But $\hom(C_{10}, H_\Sigma) = \emptyset$, thus, $H\emptyset = \emptyset$. This means $H = H_\Sigma$.

Theorem 3.8. *For a set functor H the category* **Coalg** H *is concretely equivalent to a presheaf category iff H is a reduction of a polynomial functor.*

Proof. Sufficiency is Corollary 3.6. For the necessity, let $\mathscr{A}$ be a small category such that **Coalg** H is concretely equivalent to $V\colon \mathbf{Set}^{\mathscr{A}^{\mathrm{op}}} \longrightarrow \mathbf{Set}$ of 3.1 (b). Observe that V is a coproduct of hom-functors: in fact, the functor $A \longmapsto A(s)$ of evaluation at s is, by the Yoneda lemma, representable by the object $\mathscr{A}(-,s)$ of $\mathbf{Set}^{\mathscr{A}^{\mathrm{op}}}$. Since hom-functors preserve limits, and coproducts commute in Set with connected limits, we conclude that

$$V \text{ preserves connected limits.}$$

Consequently, so does the forgetful functor $U\colon \mathbf{Coalg}\, H \longrightarrow \mathbf{Set}$ (since $U \approx V \cdot E$ for some equivalence functor $E\colon \mathbf{Coalg}\, H \longrightarrow \mathbf{Set}^{\mathscr{A}^{\mathrm{op}}}$).

We will prove next that H preserves nonempty wide pullbacks, i.e., limits of cocones $p_i\colon P_i \longrightarrow P$ $(i \in I)$ in Set, provided that the limit cone $q_i\colon Q \longrightarrow Q_i$ $(i \in I)$ has nonempty domain, $Q \neq \emptyset$. In fact, given elements

$$x_i \in HP_i \ (i \in I) \quad \text{and} \quad x \in HP$$

with $Hp_i(x_i) = x$, it is our task to prove that there is a unique $y \in HQ$ with $x_i = Hq_i(y)$ for all $i \in I$. We can endow each P_i with the constant function of value x_i and obtain an H-coalgebra; analogously with P. This turns each p_i into a coalgebra homomorphism. Now **Coalg** H is complete (because $\mathbf{Set}^{\mathscr{A}^{\mathrm{op}}}$ is), thus, it has a wide pullback of that cocone of homomorphisms. Since U preserves this wide pullback, we have a coalgebra structure $\rho\colon Q \longrightarrow HQ$ turning each q_i into a homomorphism (and forming the domain of the wide pullback in **Coalg** H). For every element $z \in Q$ the element $y = \rho(z)$ of HQ has the desired property: for each $i \in I$ we have

$$Hq_i(y) = (Hq_i \cdot \rho)(z) = (\mathrm{const}\, x_i \cdot q_i)(z) = x_i.$$

And y is unique: suppose y' also fulfils $Hq_i(y) = x_i$, then the constant function with value y' turns each q_i into a coalgebra homomorphism, from which it easily follows that ρ is equal to this constant function; thus, $y = y'$.

We are ready to prove that H is a reduction of a polynomial functor, using results of Věra Trnková on the structure of set functors. Express H as a coproduct of functors H_i preserving terminal objects, $H_i 1 \approx 1$, which is possible by [T1], see I.11. It is clear that since H preserves nonempty wide pullbacks, so does each H_i. It is sufficient to prove that, then, H_i is a reduction of a representable functor. Then H is a reduction of a coproduct of representable functors. This concludes the proof: polynomial functors are precisely the coproducts of representable functors.

Since H_i preserves terminal objects and nonempty wide pullbacks, it preserves nonempty products. Then the reduction H_i' of H_i with $H_i'\emptyset = \emptyset$ preserves products. Věra Trnková calls a set functor separating if it preserves disjointness of pairs of subobjects. Every separating functor preserves finite intersections, see Corollary 2.1 in [T1]. If H_i' is separating, then it consequently preserves limits, thus, it is representable. If H_i' is not separating, then it is naturally isomorphic to C_{01} by [T2], IV.4. This is a reduction of C_1, representable by $\emptyset$. Thus, H_i' is a reduction of a representable functor, which implies the same for H_i. □

4 Presheaves as Coalgebras

Notation 4.1. Throughout this section $\mathscr{A}$ denotes a small category, and we consider S-sorted sets for

$$S = \mathbf{obj}\,\mathscr{A}.$$

Thus, unlike 3.1(b), the presheaf category is now endowed with the forgetful functor

$$U\colon \mathbf{Set}^{\mathscr{A}^{\mathrm{op}}} \longrightarrow \mathbf{Set}^S, \quad A \longmapsto (A(s))_{s\in S}.$$

Remark 4.2. The presheaf category is a variety of S-sorted unary algebras: take $\Sigma = \mathbf{mor}\,\mathscr{A}$ as the signature, and given a morphism $u \in \mathscr{A}(t,s)$ let $u\colon s \longrightarrow t$ be the arity of u. Then the following equations present $\mathbf{Set}^{\mathscr{A}^{\mathrm{op}}}$ in $\mathbf{Alg}\,\Sigma$:

$$\mathrm{id}_s(x) = x \quad \text{and} \quad u(v(x)) = w(x)$$

where s is any sort, x is a variable of sort s, and $w \in \mathscr{A}(t,s)$ is a morphism which factorizes as $w = v \cdot u$ in $\mathscr{A}$.

We are now going to show that $\mathbf{Set}^{\mathscr{A}^{\mathrm{op}}}$ is also a covariety of coalgebras.

Observation 4.3. Generalizing Example 2.10 to presheaves

$$A\colon \mathscr{A}^{\mathrm{op}} \longrightarrow \mathbf{Set}$$

observe that A is completely specified by (a) the underlying S-sorted set $UA = (A_s)_{s\in S}$ and (b) the function

$$\mathscr{A}(t,s) \times A_s \longrightarrow A_t \qquad \text{for all} \quad (t,s) \in S \times S$$

which assigns to every pair $u\colon t \longrightarrow s$ in $\mathscr{A}$ and $x \in A_s$ the value $Au(x)$ in A_t. We use, as in 2.10, currification:

$$\frac{\mathscr{A}(t,s) \times A_s \longrightarrow A_t}{A_s \longrightarrow A_t^{\mathscr{A}(t,s)}}$$

and, for a given sort $s \in S$, we collect all these functions together by using the product (over all $t \in S$):

$$A_s \longrightarrow \prod_{t\in S} A_t^{\mathscr{A}(t,s)}.$$

In this sense, every presheaf is a coalgebra of the endofunctor $H^{\mathscr{A}}$ of $\mathbf{Set}^S$ given on objects A by

$$(H^{\mathscr{A}} A)_s = \prod_{t \in S} A_t^{\mathscr{A}(t,s)}, \tag{1}$$

for all $s \in S$. This is a polynomial functor (see Definition 2.2) of the signature Σ which has for every sort s precisely one operation $\sigma_s \in \Sigma_s$ of arity

$$\sigma_s \colon (t\,t\, \ldots\, t\,t'\,t' \ldots\, t' \ldots) \longrightarrow s$$

in which every sort $t \in S$ is repeated precisely $\mathscr{A}(t,s)$-times.

Remark 4.4. Let us describe the cofree coalgebras

$$C(X)$$

for the above polynomial functor $H^{\mathscr{A}}$, see Example 2.7. An element τ of $C(X)_s$ is a tree whose root is colored in X_s and labelled in Σ_s – but since Σ_s has just one element, we can forget the latter label. Thus we only need to remember that the children of the root of sort t correspond to $\mathscr{A}(t,s)$, for any $t \in S$. And we need to know the color, from X_t, of each child of sort t. Analogously, the children of a child of sort t correspond to $\mathscr{A}(t',t)$ for any $t' \in S$ and are colored in $X_{t'}$. Etc. Thus, such a tree τ is completely described by its sort s and a coloring of all finite sequences of composable morphisms

$$t_0 \xrightarrow{u_0} t_1 \xrightarrow{u_1} t_2 \longrightarrow \cdots \xrightarrow{u_{n-1}} t_n = s \tag{2}$$

of $\mathscr{A}$ by colors from X_{t_0}.

Denote by

$$\mathscr{A}\langle t, s\rangle$$

the set of all sequences (2) in $\mathscr{A}$ with $t = t_0$. This includes, for $t = s$, the empty sequence, consisting of s alone. Then we can consider $C(X)$ as the coalgebra of all colorings of sequences:

$$C(X)_s = \prod_{t \in S} X_t^{\mathscr{A}\langle t,s\rangle}$$

for all $s \in S$ (compare with $C(X) = X^{M^*}$ in 2.10). Here $(\varepsilon_X)_s \colon C(X)_s \longrightarrow X_s$ takes each coloring to the color of the empty sequence:

$$(\varepsilon_X)_s \colon \varphi \longmapsto \varphi_s(s) \qquad \text{for all } \varphi = (\varphi_t)_{t \in S}.$$

And the coalgebra structure

$$(\psi_X)_s \colon C(X)_s \longrightarrow (H^{\mathscr{A}} C(X))_s = \prod_{p \in S} \prod_{t \in S} X_t^{\mathscr{A}\langle t,p\rangle \times \mathscr{A}(p,s)}$$

is determined by the obvious concatenation maps $\mathscr{A}\langle t,p\rangle \times \mathscr{A}(t,s) \longrightarrow \mathscr{A}(t,s)$. Thus $(\psi_X)_s$ assigns to a coloring $\varphi = (\varphi_t)_{t \in S}$ of sequences the tuple (indexed by all morphisms $u\colon p \longrightarrow s$ of $\mathscr{A}$) of colorings $\varphi_s(\dots p \xrightarrow{u} s)$ obtained by concatenating u at the end of the sequence, and then using φ_s.

Notation 4.5. The S-sorted set all sorts of which are $\{0,1\}$ is denoted by **bool**.

Proposition 4.6. The presheaf category $\mathbf{Set}^{\mathscr{A}^{\mathrm{op}}}$ is a covariety of $H^{\mathscr{A}}$-coalgebras presented by the subset

$$D \subseteq C(\mathbf{bool})$$

which consists of precisely those colorings $(\varphi_t)_{t\in S}$ such that

(i) $\varphi_s(s \xrightarrow{\mathrm{id}} s) = \varphi_s(s)$

and

(ii) $\varphi_t(t \xrightarrow{u_0} t_1 \xrightarrow{u_1} s) = \varphi_t(t \xrightarrow{u_1\cdot u_0} s)$ for every composable pair u_0, u_1 of morphisms of $\mathscr{A}$.

Proof. (1) Let A be a presheaf. For every map $f\colon UA \longrightarrow \mathbf{bool}$ we prove that $\bar{f}$ factorizes through D. In fact, the homomorphism $\bar{f}\colon UA \longrightarrow C(\mathbf{bool})$ assigns to every element $a \in A_s$ the following coloring $\varphi = (\varphi_t)_{t\in S}$ of sequences (2) above form the composed morphism $v = u_0 \cdot u_1 \cdot \dots \cdot u_{n-1}$ in $\mathscr{A}^{\mathrm{op}}$, then φ_{t_0} colors the sequence by the color which f_{t_0} assigns to $A(v)(a)$. This coloring φ fulfils (i) because A preserves identity morphisms, and (ii) because A preserves composition.

Conversely, let A be a coalgebra of $H^{\mathscr{A}}$ satisfying D. The underlying S-sorted set $UA = (A_s)_{s\in S}$ gives the object-function of a presheaf. The coalgebra structure

$$\alpha_s\colon A_s \longrightarrow \prod_{t\in T} A_t^{\mathscr{A}(t,s)} \qquad (s \in S)$$

defines, for every morphism $u\colon t \longrightarrow s$ of $\mathscr{A}$, a function $A(u)\colon A_s \longrightarrow A_t$ assigning to every element $a \in A_s$ the u-component of $\alpha_s(a)$. This would give the morphism-function of a presheaf, in case we verify the preservation of identity morphisms and composition. In fact:

(i) $A(\mathrm{id}_s)(a) = a$ follows from the property that every coloring $f\colon UA \longrightarrow$ **bool** assigns the same color to a and to $A(\mathrm{id}_s)(a)$. This property is a consequence of (i) in the definition of D and the fact that $\bar{f}_s(a) \in D_s$ because $\bar{f}_s(a)$ colors $s \xrightarrow{\mathrm{id}} s$ with the color of $A(\mathrm{id}_s)(a)$.

(ii) $A(u_0\cdot u_1)(a) = Au_0(Au_1(a))$ follows, for every composable pair $t_0 \xrightarrow{u_0} t_1 \xrightarrow{u_1} s$ in $\mathscr{A}$, from the property that every coloring $f\colon UA \longrightarrow$ **bool** colors both sequences in the same color. This property is a consequence of (ii) in the definition of D and the fact that $\bar{f}_s(a) \in D_s$ because $\bar{f}_s(a)$ colors $t_0 \xrightarrow{u_0} u_1 \xrightarrow{u_1} s$ and $t_0 \xrightarrow{u_1\cdot u_0} s$ with the above two colors. □

5 Algebra ∩ Coalgebra

Remark 5.1. We saw in Section 2 that the category

$$M\text{-}\mathbf{Set} \qquad (M \text{ a monoid})$$

is equal both to a variety of one-sorted algebras and a to covariety of one-sorted coalgebras. Well, "equal" is a bit strong: it only is a variety and a covariety up to a concrete equivalence over Set. In the present section we prove that there are no other such categories, that is, for categories concrete over Set we prove that

$$\text{Algebra} \cap \text{Coalgebra} = M\text{-sets.} \tag{3}$$

Is this not in a contradiction to the result of Section 3 that $\mathbf{Coalg}\, H_\Sigma$ is both a covariety and a presheaf category? No contradiction! If $\mathbf{Set}^{\mathscr{A}^{\mathrm{op}}}$ is viewed as a concrete category over Set, see 3.1 (b), then, unless $\mathscr{A}$ is equivalent to a monoid, the presheaf category is never a variety. (Proof: assuming that V of 3.1 (b) is naturally isomorphic to a forgetful functor of a variety, then V preserves products. Consider the product $1 = 1 \times 1$ of the terminal presheaf 1 with itself: if V preserves this product, then $\mathbf{obj}\,\mathscr{A}$ is a product of two copies of itself. Thus, $\mathbf{obj}\,\mathscr{A}$ has at most one element.)

Theorem 5.2. *(Algebra ∩ Coalgebra=Presheaves.) For every concrete category $\mathscr{V}$ on S-sorted sets the following conditions are equivalent:*

(i) $\mathscr{V}$ is concretely equivalent both to a variety of S-sorted algebras, and to a covariety of S-sorted coalgebras;
(ii) $\mathscr{V}$ is concretely equivalent to $\mathbf{Set}^{\mathscr{A}^{\mathrm{op}}}$ for a small category $\mathscr{A}$ with $S = \mathbf{obj}\,\mathscr{A}$.

Remark. (a) The statement (and the proof) of the above theorem does not depend on the generality of the concept of variety we use: this can mean an equationalty defined class of Σ-algebras, or, more generally, any monadic category over $\mathbf{Set}^S$. The same goes for covariety: this can mean a coequationally defined class of Σ-coalgebras, or, more generally, a comonadic category over $\mathbf{Set}^S$, see Section 2.

(b) In the one-sorted case we get the above equation (3).

Proof. Let $V\colon \mathscr{V} \longrightarrow \mathbf{Set}^S$ be a concrete category concretely equivalent to a variety and concretely equivalent to a covariety. It is sufficient to prove that the monad $\mathbb{T} = (T, \eta, \mu)$ induced by V is isomorphic to the monad induced, for some category $\mathscr{A}$ with $S = \mathbf{obj}\,\mathscr{A}$, by the forgetful functor $U\colon \mathbf{Set}^{\mathscr{A}^{\mathrm{op}}} \longrightarrow \mathbf{Set}^S$ of Notation 4.1. In fact, the functor V is monadic, i.e., the category $\mathscr{V}$ is concretely equivalent to the category of $\mathbb{T}$-algebras. But also U is monadic, so that $\mathbf{Set}^{\mathscr{A}^{\mathrm{op}}}$ is also concretely equivalent to the category of $\mathbb{T}$-algebras.

Since V, due to the coalgebra part, preserves colimits, its composite T with a left adjoint of V also preserves colimits. By Special Adjoint Functor Theorem, see [M], T has a right adjoint, $T \dashv R$. Each component

$$R_s\colon \mathbf{Set}^S \longrightarrow \mathbf{Set} \qquad (s \in S)$$

is, then, also a right adjoint – thus, it is representable by an S-sorted set. Let $M(s,t)$ denote, for $t \in S$, the components of that representing set. Then R is naturally isomorphic to the functor

$$X \longmapsto \left(\prod_{t\in S} X_t^{M(s,t)}\right)_{s\in S}$$

whose left adjoint, T, is given (by currification) as

$$X \longmapsto \Big(\coprod_{t\in S} M(s,t)\times X_t\Big)_{s\in S}.$$

Thus, without loss of generality we can assume that T is defined by

$$(TX)_s = \coprod_{t\in S} M(s,t)\times X_t.$$

More succinctly: let $P_s\colon \mathbf{Set}^S \longrightarrow \mathbf{Set}$ be the s-th projection, then

$$P_s\cdot T = \coprod_{t\in S} M(s,t)\bullet P_t$$

(where $M\bullet$ denotes copowers indexed by M).

The unit of the monad $\mathbb{T}$

$$\eta\colon \mathrm{Id}\longrightarrow T$$

consists of natural transformations

$$P_s\eta\colon P_s \longrightarrow \coprod_{t\in S} M(s,t)\bullet P_t.$$

Since P_s is representable by the S-sorted set whose components are empty except the s-th one which is 1, we see that, by the Yoneda lemma, $P_s\eta$ is just a choice of an element

$$e_s\in M(s,s).$$

Then η_s has components

$$(\eta_s)_X\colon X_s \longrightarrow \coprod_{t\in S} M(s,t)\times X_t$$

which are the coproduct injections corresponding to e_s.

Analogously, the monad multiplication

$$\mu\colon T\cdot T\longrightarrow T$$

consists of natural transformations μ_s from

$$P_s\cdot T\cdot T = \coprod_{t\in S} M(s,t)\bullet P_tT = \coprod_{t,t'\in S} (M(s,t)\times M(t,t'))\bullet P_{t'}$$

into $P_s\cdot T$. By using the Yoneda lemma on the (u,v)-component of μ_s, for any $(u,v)\in M(s,t)\times M(t,t')$:

$$P_{t'}\longrightarrow P_s\cdot T\cdot T\xrightarrow{\ \mu_s\ } P_s\cdot T = \coprod_{t'\in T} M(s,t')\bullet P_{t'},$$

we see that μ chooses for u, v an element in $M(t', s)$, which we denote by $v \cdot u$. Then μ_s has components

$$(\mu_s)_X \colon \coprod_{t,t' \in S} [M(s,t) \times M(t,t')] \bullet P_{t'} \longrightarrow \coprod_{t' \in T} M(s,t') \bullet P_{t'}$$

given, for every (u,v)-copy of $P_{t'}$, by the coproduct injection corresponding to $v \cdot u \in M(s,t')$.

Let us prove that we obtain a category $\mathscr{A}$ with objects S and hom-sets $M(s,t)$ where e_s is the identity morphism and $(u,v) \longmapsto v \cdot u$ is the composition. In fact, the unit law of monads

$$\mu \cdot T\eta = \mu \cdot \eta T = \mathrm{id}$$

tells us precisely that e_s is a unit of the composition. And the associative law of monads

$$\mu \cdot T\mu = \mu \cdot \mu T$$

tells us precisely that the above composition is associative. Thus, we obtained a category $\mathscr{A}$.

It remains to verify that the forgetful functor $U \colon \mathbf{Set}^{\mathscr{A}^{\mathrm{op}}} \longrightarrow \mathbf{Set}^S$ induces the above monad $\mathbb{T}$. In fact, for every S-sorted set X we have an obvious presheaf $X^* \colon \mathscr{A}^{\mathrm{op}} \longrightarrow \mathbf{Set}$ with

$$UX^* = TX:$$

put $X^*(s) = \coprod_{t \in S} \mathscr{A}(s,t) \times X_t$ on objects, and for a morphism $w \colon s' \longrightarrow s$ of $\mathscr{A}$ define $X^*(w) \colon X^*(s) \longrightarrow X^*(s')$ at a pair $(u,a) \in \mathscr{A}(s,t) \times X_t$ to be the pair $(u \cdot w, a) \in \mathscr{A}(s',t) \times X_t$, shortly: $X^*(w)(u,a) = (u \cdot w, a)$. The above natural transformation $\eta_X : X \longrightarrow TX = UX^*$ is clearly a universal arrow for U. Since U is monadic on $\mathbf{Set}^s$, this proves that U induces the monad $\mathbb{T}$. □

6 Conclusions

We have started the present paper by asking whether sequential automata are the only type of systems which have both an algebraic treatement and a coalgebraic one. The answer is esentially affirmative in the realm of one-sorted algebra and coalgebra: the only generalizations of sequential automata living in "algebra ∩ coalgebra" are M-sets for monoids M (where the case of automata corresponds to free monoids). We say "essentially" because our answer depends on viewing categories of algebras and coalgebras as concrete categories over **Set**. For the corresponding question concerning abstract categories we do not know the answer at present.

A surprising fact proved by F. E. J. Linton and R. C. Paré [LP] is that every covariety is equivalent to the dual of a variety. We do not know the answer to the following

Open problem: Is every variety equivalent to the dual of a covariety?

For varieties and covarieties of many-sorted algebras, i.e., monadicity and comonadicity over $\mathbf{Set}^S$ up to concrete equivalence, the categories of presheaves form the intersection of algebra and coalgebra. In other words, the obvious currification (which enables us to view many-sorted unary algebras as coalgebras) describes the whole intersection of algebra and coalgebra. A generalization of Theorem 3.4, the opposite implication to J. Worrell's result about categories of Σ-coalgebras, to many-sorted sets is also left as an open problem.

References

[AP] J. Adámek and H.-E. Porst, On varieties and covarieties in a category, Math. Structures Comput. Sci. 13 (2003), 201–232

[LEW] J. Loeckx, H.-D. Ehrich and M. Wolf, Specification of abstract data types, Wiley and Teubner, Chichester 1996

[LP] F. E. J. Linton and R. C. Paré, Injectives in topoi I: Representing coalgebras as algebras, Lect. Notes Mathem. 719, Springer Verlag 1970, 196–206

[M] S. MacLane, Categories for the working mathematician, Springer Verlag 1971

[P] H.-E. Porst, What is concrete equivalence? Appl. Categorical Structures 2 (1994), 57–70

[PD] H.-E. Porst and C. Dzieron, On coalgebras which are algebras, in Categorical Structures and Applications (Gähler and Preuss, editors) World Scientific Publ. 2004, 227–234

[Re] J. Reiterman, One more cagorical model of universal algebra, Math. Z. 161 (1978), 137–146

[Ru] J. Rutten, Universal coalgebra: a theory of systems, Theor. Comput. Sci. 249 (2000), 3–80

[T1] V. Trnková, Some properties of set functors, Comment Math. Univ. Carolinae 10 (1969), 323–352

[T2] V. Trnková, On descriptive classification of set-functors I, Comment. Math. Univ. Carolinae 12 (1971), 143–174

[W] J. Worrell, A note on coalgebras and presheaves, Electr. Notes Theor. Comput. Sci. 65.1 (2002)

Strong Splitting Bisimulation Equivalence

J.A. Bergstra and C.A. Middelburg

[1] Programming Research Group, University of Amsterdam,
P.O. Box 41882, 1009 DB Amsterdam, The Netherlands
janb@science.uva.nl
[2] Department of Philosophy, Utrecht University,
P.O. Box 80126, 3508 TC Utrecht, The Netherlands
janb@phil.uu.nl
[3] Computing Science Department, Eindhoven University of Technology,
P.O. Box 513, 5600 MB Eindhoven, The Netherlands
keesm@win.tue.nl

Abstract. We present ACP^c, a process algebra with conditional expressions in which the conditions are taken from a Boolean algebra, and extensions of this process algebra with mechanisms for condition evaluation. We confine ourselves to finitely branching processes. This restriction makes it possible to present^c in a concise and intuitively clear way, and to bring the notion of splitting bisimulation equivalence and the issue of condition evaluation in process algebras with conditional expressions to the forefront.

1 Introduction

It is not unusual that process algebras include conditional expressions of some form. Several extensions of ACP [1,2] include conditional expressions of the form $\zeta :\rightarrow p$ or $p \lhd \zeta \rhd q$ (see e.g. [3,4,5,6]). What are considered to be conditions differs from one extension to another. The set of conditions is usually one of the following: (i) a two-valued set, usually called $\mathbb{B}$; (ii) the set of all propositions with a given set of propositional variables and with finite conjunctions and disjunctions; (iii) the domain of a free Boolean algebra over a given set of generators. The third alternative generalizes the first two alternatives. In this paper, we present ACP^c, an extension of ACP with conditional expressions of the form $\zeta :\rightarrow p$ in which the domain of a free Boolean algebra over a given set of generators is taken as the set of conditions. We give the axioms of ACP^c, describe the structural operational semantics of ACP^c, and introduce a variant of bisimulation equivalence, called splitting bisimulation equivalence, for which the axiomatization is sound. In the title, the qualification "strong" is used to indicate that splitting bisimulation equivalence does not provide for abstraction from internal actions. Outside the title, we leave out this qualification.

How conditions are evaluated is usually not considered. The state operators as introduced in [4] allow for a kind of condition evaluation. However, state operators were not especially devised for that purpose. In this paper, we extend

J.L. Fiadeiro et al. (Eds.): CALCO 2005, LNCS 3629, pp. 83–97, 2005.

ACP^c with operators especially devised for condition evaluation and with the state operators from [4]; and show how those extensions are related. Two kinds of operators are devised for condition evaluation, one for the case where condition evaluation is not dependent on process behaviour and the other for the case where condition evaluation is dependent on process behaviour. We show how a theory about the set of atomic conditions can be used for condition evaluation with an operator of the former kind, that the operators of the former kind are superseded by the operators of the latter kind and that those operators are in their turn superseded by the state operators.

The work presented in this paper can easily be adapted to other process algebras based on (strong) bisimulation models, such as the strong bisimulation version of CCS [7]. Adaptation to CSP [8], which is not based on bisimulation models, will be more difficult and in part perhaps even impossible. In some extensions of ACP with conditional expressions, the conditions are propositions of a three-, four- or five-valued propositional logic, see e.g. [9,10]. Such conditions will bring us outside the domain of Boolean algebras.

In [11], we investigated conditional expressions in the setting of ACP more extensively. In that paper, we presented ACP^c for the first time. We also presented its main models, called full splitting bisimulation models. We extended ACP^c with the above-mentioned operators especially devised for condition evaluation, the state operators from [4] and the signal emission operator from [6], which like the state operators allows for a kind of condition evaluation. We also showed how those extensions are related. On purpose to incorporate the past in conditions, we also added a retrospection operator on conditions to ACP^c.

All this fitted in with our intention at the time: to arrive at a well-considered extension of ACP with conditional expressions in which retrospective conditions can be used. Retrospective conditions allow for looking back on conditions under which preceding actions have been performed. Their addition is considered to be a basic way to increase expressiveness. In the full splitting bisimulation models of ACP^c, infinitely branching processes are taken into account. Because the set of atomic conditions is not required to be finite, those models are rather complicated. Moreover, the adaptation of the full splitting bisimulation models to the retrospection operator on conditions is quite substantial. As a result, other interesting matters are pushed to the background in [11].

The current paper can be viewed as an extended abstract of some parts of [11]. Most importantly, the full splitting bisimulation models of ACP^c and the addition of the retrospection operator on conditions to ACP^c are not covered. Moreover, because we confine ourselves to finitely branching processes, the presentation of what is left over has been fairly simplified.

We do not give proofs of the theorems concerning congruence properties of splitting bisimulation equivalence, soundness of axiomatizations for splitting bisimulation equivalence, and uniqueness of solutions of guarded recursive specifications. Those theorems follow from the corresponding theorems in [11] because the structural operational semantics induces a model isomorphic to the full bisimulation model that covers only finitely branching processes.

2 BPA with Conditions

BPA_δ is a subtheory of ACP that does not support parallelism and communication (see e.g. [2]). In this section, we present an extension of BPA_δ with guarded commands, i.e. conditional expressions of the form $\zeta :\rightarrow p$. The extension is called $\mathrm{BPA}_\delta^\mathrm{c}$. In the extension, just as in BPA_δ, it is assumed that a fixed but arbitrary finite set of *actions* A, with $\delta \notin \mathsf{A}$, has been given. Moreover it is assumed that a fixed but arbitrary set of *atomic conditions* C_at has been given.

In $\mathrm{BPA}_\delta^\mathrm{c}$, conditions are taken from the domain of the free Boolean algebra over C_at. We denote this algebra by $\mathcal{C}$. As usual, we identify Boolean algebras with their domain. Thus, we also write $\mathcal{C}$ for the domain of $\mathcal{C}$. It is well known that $\mathcal{C}$ is isomorphic to the Boolean algebra of equivalence classes with respect to logical equivalence of the set of all propositions with elements of C_at as propositional variables and with finite conjunctions and disjunctions (see e.g. [12]).

The algebraic theory $\mathrm{BPA}_\delta^\mathrm{c}$ has two sorts:

- the sort $\mathbf{P}$ of *processes*;
- the sort $\mathbf{C}$ of (*finite*) *conditions*.

The algebraic theory $\mathrm{BPA}_\delta^\mathrm{c}$ has the following constants and operators to build terms of sort $\mathbf{C}$:

- the *bottom* constant $\bot : \mathbf{C}$;
- the *top* constant $\top : \mathbf{C}$;
- for each $\eta \in \mathsf{C}_\mathsf{at}$, the *atomic condition* constant $\eta : \mathbf{C}$;
- the unary *complement* operator $- : \mathbf{C} \rightarrow \mathbf{C}$;
- the binary *join* operator $\sqcup : \mathbf{C} \times \mathbf{C} \rightarrow \mathbf{C}$;
- the binary *meet* operator $\sqcap : \mathbf{C} \times \mathbf{C} \rightarrow \mathbf{C}$.

The algebraic theory $\mathrm{BPA}_\delta^\mathrm{c}$ has the following constants and operators to build terms of sort $\mathbf{P}$:

- the *deadlock* constant $\delta : \mathbf{P}$;
- for each $a \in \mathsf{A}$, the *action* constant $a : \mathbf{P}$;
- the binary *alternative composition* operator $+ : \mathbf{P} \times \mathbf{P} \rightarrow \mathbf{P}$;
- the binary *sequential composition* operator $\cdot : \mathbf{P} \times \mathbf{P} \rightarrow \mathbf{P}$;
- the binary *guarded command* operator $:\rightarrow : \mathbf{C} \times \mathbf{P} \rightarrow \mathbf{P}$.

We use infix notation for the binary operators. The following precedence conventions are used to reduce the need for parentheses. The operators to build terms of sort $\mathbf{C}$ bind stronger than the operators to build terms of sort $\mathbf{P}$. The operator $\cdot$ binds stronger than all other binary operators to build terms of sort $\mathbf{P}$ and the operator $+$ binds weaker than all other binary operators to build terms of sort $\mathbf{P}$.

The constants and operators of $\mathrm{BPA}_\delta^\mathrm{c}$ to build terms of sort $\mathbf{P}$ are the constants and operators of BPA_δ and additionally the guarded command operator. Let p and q be closed terms of sort $\mathbf{P}$ and ζ be a closed term of sort $\mathbf{C}$. Intuitively, the constants and operators to build terms of sort $\mathbf{P}$ can be explained as follows:

Table 1. Axioms of Boolean algebras

$\phi \sqcup \bot = \phi$	BA1	$\phi \sqcap \top = \phi$	BA5
$\phi \sqcup -\phi = \top$	BA2	$\phi \sqcap -\phi = \bot$	BA6
$\phi \sqcup \psi = \psi \sqcup \phi$	BA3	$\phi \sqcap \psi = \psi \sqcap \phi$	BA7
$\phi \sqcup (\psi \sqcap \chi) = (\phi \sqcup \psi) \sqcap (\phi \sqcup \chi)$	BA4	$\phi \sqcap (\psi \sqcup \chi) = (\phi \sqcap \psi) \sqcup (\phi \sqcap \chi)$	BA8

Table 2. Axioms of $\mathrm{BPA}_\delta^\mathrm{c}$

$x + y = y + x$	A1	$\top :\rightarrow x = x$	GC1
$(x + y) + z = x + (y + z)$	A2	$\bot :\rightarrow x = \delta$	GC2
$x + x = x$	A3	$\phi :\rightarrow \delta = \delta$	GC3
$(x + y) \cdot z = x \cdot z + y \cdot z$	A4	$\phi :\rightarrow (x + y) = \phi :\rightarrow x + \phi :\rightarrow y$	GC4
$(x \cdot y) \cdot z = x \cdot (y \cdot z)$	A5	$\phi :\rightarrow x \cdot y = (\phi :\rightarrow x) \cdot y$	GC5
$x + \delta = x$	A6	$\phi :\rightarrow (\psi :\rightarrow x) = (\phi \sqcap \psi) :\rightarrow x$	GC6
$\delta \cdot x = \delta$	A7	$(\phi \sqcup \psi) :\rightarrow x = \phi :\rightarrow x + \psi :\rightarrow x$	GC7

- δ cannot perform any action;
- a first performs action a unconditionally and then terminates successfully;
- $p + q$ behaves either as p or as q, but not both;
- $p \cdot q$ first behaves as p, but when p terminates successfully it continues by behaving as q;
- $\zeta :\rightarrow p$ behaves as p under condition ζ.

Some earlier extensions of ACP include conditional expressions of the form $p \triangleleft \zeta \triangleright q$; see e.g. [4]. This notation with triangles originates from [13]. We treat conditional expressions of the form $p \triangleleft \zeta \triangleright q$, where p and q are terms of sort **P** and ζ is a term of sort **C**, as abbreviations. That is, we write $p \triangleleft \zeta \triangleright q$ for $\zeta :\rightarrow p + -\zeta :\rightarrow q$.

The axioms of $\mathrm{BPA}_\delta^\mathrm{c}$ are the axioms of Boolean Algebras (BA) given in Table 1 and the additional axioms given in Table 2. Axioms A1–A7 are the axioms of BPA_δ. So $\mathrm{BPA}_\delta^\mathrm{c}$ imports the (equational) axioms of both BA and BPA_δ. The axioms of BA given in Table 1 have been taken from [14]. Several alternatives for this axiomatization can be found in the literature. If we use basic laws of BA other than axioms BA1–BA8, such as $\phi \sqcap \phi = \phi$ and $-(\phi \sqcap \psi) = -\phi \sqcup -\psi$, in a step of a derivation, we will refer to them as applications of BA and not give their derivation from axioms BA1–BA8. Axioms GC1–GC7 have been taken from [4], but with the axiom $x \cdot z \triangleleft \phi \triangleright y \cdot z = (x \triangleleft \phi \triangleright y) \cdot z$ (CO5) replaced by $\phi :\rightarrow x \cdot y = (\phi :\rightarrow x) \cdot y$ (GC5).

Example 1. Consider a careful pedestrian who uses a crossing with traffic lights to cross a road with busy traffic safely. When the pedestrian arrives at the crossing and the light for pedestrians is green, he or she simply crosses the street. However, when the pedestrian arrives at the crossing and the light for

Table 3. Transition rules for $\mathrm{BPA}_\delta^{\mathrm{c}}$

$$\frac{}{a \xrightarrow{[\top]\,a} \surd}$$

$$\frac{x \xrightarrow{[\phi]\,a} \surd}{x + y \xrightarrow{[\phi]\,a} \surd} \qquad \frac{y \xrightarrow{[\phi]\,a} \surd}{x + y \xrightarrow{[\phi]\,a} \surd} \qquad \frac{x \xrightarrow{[\phi]\,a} x'}{x + y \xrightarrow{[\phi]\,a} x'} \qquad \frac{y \xrightarrow{[\phi]\,a} y'}{x + y \xrightarrow{[\phi]\,a} y'}$$

$$\frac{x \xrightarrow{[\phi]\,a} \surd}{x \cdot y \xrightarrow{[\phi]\,a} y} \qquad \frac{x \xrightarrow{[\phi]\,a} x'}{x \cdot y \xrightarrow{[\phi]\,a} x' \cdot y}$$

$$\frac{x \xrightarrow{[\phi]\,a} \surd}{\psi :\to x \xrightarrow{[\phi \sqcap \psi]\,a} \surd}\ \phi \sqcap \psi \neq \bot \qquad \frac{x \xrightarrow{[\phi]\,a} x'}{\psi :\to x \xrightarrow{[\phi \sqcap \psi]\,a} x'}\ \phi \sqcap \psi \neq \bot$$

pedestrians is red, he or she first makes a request for green light (e.g. by pushing a button) and then crosses the street when the light has changed. This behaviour can be described in $\mathrm{BPA}_\delta^{\mathrm{c}}$ as follows:

$$\mathit{PED} = \mathit{arrive} \cdot (\mathit{green} :\to \mathit{cross} + \mathit{red} :\to (\mathit{make\text{-}req} \cdot (\mathit{green} :\to \mathit{cross}))) .$$

The careful pedestrian described above does not cross the street if the light for pedestrians does not change from red to green after a request for green light. Whether the change from red to green will ever happen is not described here.

Henceforth, we write $\mathcal{T}_{\mathbf{P}}$ for the set of all closed terms of sort $\mathbf{P}$ and $\mathcal{T}_{\mathbf{C}}$ for the set of all closed terms of sort $\mathbf{C}$. The terms of sort $\mathbf{C}$ are interpreted in $\mathcal{C}$ as usual. Henceforth, we write $\mathcal{C}^-$ for $\mathcal{C} \setminus \{\bot\}$.

We proceed to the presentation of the structural operational semantics of $\mathrm{BPA}_\delta^{\mathrm{c}}$. The following transition relations on $\mathcal{T}_{\mathbf{P}}$ are used:

- for each $\ell \in \mathcal{C}^- \times \mathsf{A}$, a binary relation $\xrightarrow{\ell}$;
- for each $\ell \in \mathcal{C}^- \times \mathsf{A}$, a unary relation $\xrightarrow{\ell} \surd$.

We write $p \xrightarrow{[\alpha]\,a} q$ instead of $(p, q) \in \xrightarrow{(\alpha,a)}$ and $p \xrightarrow{[\alpha]\,a} \surd$ instead of $p \in \xrightarrow{(\alpha,a)} \surd$. The relations $\xrightarrow{\ell} \surd$ and $\xrightarrow{\ell}$ can be explained as follows:

- $p \xrightarrow{[\alpha]\,a} \surd$: p is capable of performing action a under condition α and then terminating successfully;
- $p \xrightarrow{[\alpha]\,a} q$: p is capable of performing action a under condition α and then proceeding as q.

The structural operational semantics of $\mathrm{BPA}_\delta^{\mathrm{c}}$ is described by the transition rules given in Table 3.

Bisimilarity has to be adapted to the setting with guarded actions. In the definition given below, we use a well-known notion from the field of Boolean algebras: the partial order relation $\sqsubseteq$ on $\mathcal{C}$ defined by $\alpha \sqsubseteq \beta$ iff $\alpha \sqcup \beta = \beta$. Moreover, we use the notation $\bigsqcup A$, where $A = \{\alpha_1, \ldots, \alpha_n\} \subseteq \mathcal{C}$, for $\alpha_1 \sqcup \ldots \sqcup \alpha_n$.

A *splitting bisimulation* B between closed terms $p, q \in \mathcal{T}_\mathbf{P}$ is a binary relation on $\mathcal{T}_\mathbf{P}$ such that $B(p, q)$ and for all p_1, q_1 such that $B(p_1, q_1)$:

- if $p_1 \xrightarrow{[\alpha]\,a} p_2$, then there exists a finite set $CT' \subseteq \mathcal{C}^- \times \mathcal{T}_\mathbf{P}$ such that $\alpha \sqsubseteq \bigsqcup \mathrm{dom}(CT')$ and for all $(\alpha', q_2) \in CT'$, $q_1 \xrightarrow{[\alpha']\,a} q_2$ and $B(p_2, q_2)$;
- if $q_1 \xrightarrow{[\alpha]\,a} q_2$, then there exists a finite set $CT' \subseteq \mathcal{C}^- \times \mathcal{T}_\mathbf{P}$ such that $\alpha \sqsubseteq \bigsqcup \mathrm{dom}(CT')$ and for all $(\alpha', p_2) \in CT'$, $p_1 \xrightarrow{[\alpha']\,a} p_2$ and $B(p_2, q_2)$;
- if $p_1 \xrightarrow{[\alpha]\,a} \surd$, then there exists a finite set $C' \subseteq \mathcal{C}^-$ such that $\alpha \sqsubseteq \bigsqcup C'$ and for all $\alpha' \in C'$, $q_1 \xrightarrow{[\alpha']\,a} \surd$;
- if $q_1 \xrightarrow{[\alpha]\,a} \surd$, then there exists a finite set $C' \subseteq \mathcal{C}^-$ such that $\alpha \sqsubseteq \bigsqcup C'$ and for all $\alpha' \in C'$, $p_1 \xrightarrow{[\alpha']\,a} \surd$.

Two closed term $p, q \in \mathcal{T}_\mathbf{P}$ are *splitting bisimulation equivalent* or *splitting bisimilar* for short, written $p \leftrightarrows q$, if there exists a splitting bisimulation B between p and q. Let B be a splitting bisimulation between p and q. Then we say that B is a splitting bisimulation *witnessing* $p \leftrightarrows q$.

The name splitting bisimulation is used because a transition of one of the related processes may be simulated by a set of transitions of the other process. Splitting bisimulation should not be confused with split bisimulation [15].

Splitting bisimilarity is a congruence with respect to alternative composition, sequential composition and guarded command.

Proposition 1 (Congruence). *For all $p, p', q, q' \in \mathcal{T}_\mathbf{P}$ and $\alpha \in \mathcal{C}$, $p \leftrightarrows q$ and $p' \leftrightarrows q'$ implies $p + p' \leftrightarrows q + q'$, $p \cdot p' \leftrightarrows q \cdot q'$ and $\alpha :\to p \leftrightarrows \alpha :\to q$.*

The axioms of $\mathrm{BPA}_\delta^\mathrm{c}$ constitute a sound and complete axiomatization of splitting bisimilarity.

Theorem 1 (Soundness). *For all $p, q \in \mathcal{T}_\mathbf{P}$, $\mathrm{BPA}_\delta^\mathrm{c} \vdash p = q$ implies $p \leftrightarrows q$.*

Theorem 2 (Completeness). *For all $p, q \in \mathcal{T}_\mathbf{P}$, $p \leftrightarrows q$ implies $\mathrm{BPA}_\delta^\mathrm{c} \vdash p = q$.*

Proof. The proof follows the same line as the completeness proof for BPA_δ given in [16]. □

3 ACP with Conditions

In order to support parallelism and communication, we add parallel composition and encapsulation operators to $\mathrm{BPA}_\delta^\mathrm{c}$, resulting in ACP^c.

Like in $\mathrm{BPA}_\delta^\mathrm{c}$, it is assumed that a fixed but arbitrary finite set of *actions* A, with $\delta \notin \mathsf{A}$, and a fixed but arbitrary set of *atomic conditions* C_at has been given. We write A_δ for $\mathsf{A} \cup \{\delta\}$. In ACP^c, it is further assumed that a fixed but arbitrary commutative and associative *communication* function $| : \mathsf{A}_\delta \times \mathsf{A}_\delta \to \mathsf{A}_\delta$, such that $\delta \mid a = \delta$ for all $a \in \mathsf{A}_\delta$, has been given. The function $|$ is regarded to give the result of synchronously performing any two actions for which this is possible, and to be δ otherwise.

The theory ACP^c is an extension of $\mathrm{BPA}_\delta^\mathrm{c}$. It has the constants and operators of $\mathrm{BPA}_\delta^\mathrm{c}$ and in addition:

Table 4. Additional axioms for $\mathrm{ACP^c}$ $(a, b, c \in \mathsf{A}_\delta)$

$x \parallel y = x \mathbin{⌊⌊} y + y \mathbin{⌊⌊} x + x \mid y$	CM1	$\partial_H(a) = a$	if $a \notin H$	D1
$a \mathbin{⌊⌊} x = a \cdot x$	CM2	$\partial_H(a) = \delta$	if $a \in H$	D2
$a \cdot x \mathbin{⌊⌊} y = a \cdot (x \parallel y)$	CM3	$\partial_H(x + y) = \partial_H(x) + \partial_H(y)$		D3
$(x + y) \mathbin{⌊⌊} z = x \mathbin{⌊⌊} z + y \mathbin{⌊⌊} z$	CM4	$\partial_H(x \cdot y) = \partial_H(x) \cdot \partial_H(y)$		D4
$a \cdot x \mid b = (a \mid b) \cdot x$	CM5			
$a \mid b \cdot x = (a \mid b) \cdot x$	CM6	$(\phi :\to x) \mathbin{⌊⌊} y = \phi :\to (x \mathbin{⌊⌊} y)$		GC8
$a \cdot x \mid b \cdot y = (a \mid b) \cdot (x \parallel y)$	CM7	$(\phi :\to x) \mid y = \phi :\to (x \mid y)$		GC9
$(x + y) \mid z = x \mid z + y \mid z$	CM8	$x \mid (\phi :\to y) = \phi :\to (x \mid y)$		GC10
$x \mid (y + z) = x \mid y + x \mid z$	CM9	$\partial_H(\phi :\to x) = \phi :\to \partial_H(x)$		GC11
$a \mid b = b \mid a$	C1			
$(a \mid b) \mid c = a \mid (b \mid c)$	C2			
$\delta \mid a = \delta$	C3			

- the binary *parallel composition* operator $\parallel : \mathbf{P} \times \mathbf{P} \to \mathbf{P}$;
- the binary *left merge* operator $⌊⌊ : \mathbf{P} \times \mathbf{P} \to \mathbf{P}$;
- the binary *communication merge* operator $\mid : \mathbf{P} \times \mathbf{P} \to \mathbf{P}$;
- for each $H \subseteq \mathsf{A}$, the unary *encapsulation* operator $\partial_H : \mathbf{P} \to \mathbf{P}$.

We use infix notation for the additional binary operators as well.

The constants and operators of $\mathrm{ACP^c}$ to build terms of sort $\mathbf{P}$ are the constants and operators of ACP and additionally the guarded command operator.

Let p and q be closed terms of $\mathrm{ACP^c}$. Intuitively, the additional operators can be explained as follows:

- $p \parallel q$ behaves as the process that proceeds with p and q in parallel;
- $p \mathbin{⌊⌊} q$ behaves the same as $p \parallel q$, except that it starts with performing an action of p;
- $p|q$ behaves the same as $p \| q$, except that it starts with performing an action of p and an action of q synchronously;
- $\partial_H(p)$ behaves the same as p, except that it does not perform actions in H.

The axioms of $\mathrm{ACP^c}$ are the axioms of $\mathrm{BPA}_\delta^\mathrm{c}$ and the additional axioms given in Table 4. CM2–CM3, CM5–CM7, C1–C3 and D1–D2 are actually axiom schemas in which a, b and c stand for arbitrary constants of $\mathrm{ACP^c}$ (i.e. $a, b, c \in \mathsf{A}_\delta$). In D1–D4, H stands for an arbitrary subset of A. So, D3 and D4 are axiom schemas as well.

Axioms A1–A7, CM1–CM9, C1–C3 and D1–D4 are the axioms of ACP. So $\mathrm{ACP^c}$ imports the axioms of ACP.

A well-known subtheory of ACP is PA, ACP without communication. Likewise, we have a subtheory of $\mathrm{ACP^c}$, to wit $\mathrm{PA^c}$. The theory $\mathrm{PA^c}$ is $\mathrm{ACP^c}$ without the communication merge operator, without axioms CM5–CM9 and C1–C3, and

Table 5. Additional transition rules for ACP^{c}

$$\frac{x \xrightarrow{[\phi]\,a} \surd}{x \parallel y \xrightarrow{[\phi]\,a} y} \qquad \frac{y \xrightarrow{[\phi]\,a} \surd}{x \parallel y \xrightarrow{[\phi]\,a} x} \qquad \frac{x \xrightarrow{[\phi]\,a} x'}{x \parallel y \xrightarrow{[\phi]\,a} x' \parallel y} \qquad \frac{y \xrightarrow{[\phi]\,a} y'}{x \parallel y \xrightarrow{[\phi]\,a} x \parallel y'}$$

$$\frac{x \xrightarrow{[\phi]\,a} \surd,\; y \xrightarrow{[\psi]\,b} \surd}{x \parallel y \xrightarrow{[\phi \sqcap \psi]\,c} \surd}\; a \mid b = c,\; \phi \sqcap \psi \neq \bot \qquad \frac{x \xrightarrow{[\phi]\,a} \surd,\; y \xrightarrow{[\psi]\,b} y'}{x \parallel y \xrightarrow{[\phi \sqcap \psi]\,c} y'}\; a \mid b = c,\; \phi \sqcap \psi \neq \bot$$

$$\frac{x \xrightarrow{[\phi]\,a} x',\; y \xrightarrow{[\psi]\,b} \surd}{x \parallel y \xrightarrow{[\phi \sqcap \psi]\,c} x'}\; a \mid b = c,\; \phi \sqcap \psi \neq \bot \qquad \frac{x \xrightarrow{[\phi]\,a} x',\; y \xrightarrow{[\psi]\,b} y'}{x \parallel y \xrightarrow{[\phi \sqcap \psi]\,c} x' \parallel y'}\; a \mid b = c,\; \phi \sqcap \psi \neq \bot$$

$$\frac{x \xrightarrow{[\phi]\,a} \surd}{x \mathbin{\|\!\!\lfloor} y \xrightarrow{[\phi]\,a} y} \qquad \frac{x \xrightarrow{[\phi]\,a} x'}{x \mathbin{\|\!\!\lfloor} y \xrightarrow{[\phi]\,a} x' \parallel y}$$

$$\frac{x \xrightarrow{[\phi]\,a} \surd,\; y \xrightarrow{[\psi]\,b} \surd}{x \mid y \xrightarrow{[\phi \sqcap \psi]\,c} \surd}\; a \mid b = c,\; \phi \sqcap \psi \neq \bot \qquad \frac{x \xrightarrow{[\phi]\,a} \surd,\; y \xrightarrow{[\psi]\,b} y'}{x \mid y \xrightarrow{[\phi \sqcap \psi]\,c} y'}\; a \mid b = c,\; \phi \sqcap \psi \neq \bot$$

$$\frac{x \xrightarrow{[\phi]\,a} x',\; y \xrightarrow{[\psi]\,b} \surd}{x \mid y \xrightarrow{[\phi \sqcap \psi]\,c} x'}\; a \mid b = c,\; \phi \sqcap \psi \neq \bot \qquad \frac{x \xrightarrow{[\phi]\,a} x',\; y \xrightarrow{[\psi]\,b} y'}{x \mid y \xrightarrow{[\phi \sqcap \psi]\,c} x' \parallel y'}\; a \mid b = c,\; \phi \sqcap \psi \neq \bot$$

$$\frac{x \xrightarrow{[\phi]\,a} \surd}{\partial_H(x) \xrightarrow{[\phi]\,a} \surd}\; a \notin H \qquad \frac{x \xrightarrow{[\phi]\,a} x'}{\partial_H(x) \xrightarrow{[\phi]\,a} \partial_H(x')}\; a \notin H$$

with axiom CM1 replaced by $x \parallel y = x \mathbin{\|\!\!\lfloor} y + y \mathbin{\|\!\!\lfloor} x$ (M1). In other words, the possibility that actions are performed synchronously is not covered by PA^{c}.

The structural operational semantics of ACP^{c} is described by the transition rules for BPA$_\delta^{\mathrm{c}}$ and the additional transition rules given in Table 5.

Splitting bisimilarity is a congruence with respect to parallel composition, left merge, communication merge and encapsulation.

Proposition 2 (Congruence). *For all $p, p', q, q' \in \mathcal{T}_{\mathbf{P}}$, $p \,\underline{\leftrightarrow}\, q$ and $p' \,\underline{\leftrightarrow}\, q'$ implies $p \parallel p' \,\underline{\leftrightarrow}\, q \parallel q'$, $p \mathbin{\|\!\!\lfloor} p' \,\underline{\leftrightarrow}\, q \mathbin{\|\!\!\lfloor} q'$, $p \mid p' \,\underline{\leftrightarrow}\, q \mid q'$ and $\partial_H(p) \,\underline{\leftrightarrow}\, \partial_H(q)$.*

The axioms of ACP^{c} constitute a sound and complete axiomatization of splitting bisimilarity.

Theorem 3 (Soundness). *For all $p, q \in \mathcal{T}_{\mathbf{P}}$, ACP$^{\mathrm{c}} \vdash p = q$ implies $p \,\underline{\leftrightarrow}\, q$.*

Theorem 4 (Completeness). *For all $p, q \in \mathcal{T}_{\mathbf{P}}$, $p \,\underline{\leftrightarrow}\, q$ implies ACP$^{\mathrm{c}} \vdash p = q$.*

Proof. The proof follows the same line as the completeness proof for ACP given in [16]. □

4 Guarded Recursion

In order to allow for the description of (potentially) non-terminating processes, we add guarded recursion to ACP^{c}.

Table 6. Axioms for recursion

$\langle X \vert E\rangle = \langle t_X \vert E\rangle$	if $X = t_X \in E$	RDP
$E \Rightarrow X = \langle X \vert E\rangle$	if $X \in \mathrm{V}(E)$	RSP

Table 7. Transition rules for recursion

$$\frac{\langle t_X|E\rangle \xrightarrow{[\phi]\,a} \surd}{\langle X|E\rangle \xrightarrow{[\phi]\,a} \surd}\ X = t_X \in E \qquad \frac{\langle t_X|E\rangle \xrightarrow{[\phi]\,a} x'}{\langle X|E\rangle \xrightarrow{[\phi]\,a} x'}\ X = t_X \in E$$

A *recursive specification* over $\mathrm{ACP^c}$ is a set of equations $E = \{X = t_X \mid X \in V\}$ where V is a set of variables and each t_X is a term of $\mathrm{ACP^c}$ that only contains variables from V. We write $\mathrm{V}(E)$ for the set of all variables that occur on the left-hand side of an equation in E. A *solution* of a recursive specification E is a set of processes (in some model of $\mathrm{ACP^c}$) $\{P_X \mid X \in \mathrm{V}(E)\}$ such that the equations of E hold if, for all $X \in \mathrm{V}(E)$, X stands for P_X. Let t be a term of $\mathrm{ACP^c}$ containing a variable X. We call an occurrence of X in t *guarded* if t has a subterm of the form $a \cdot t'$ containing this occurrence of X. A recursive specification over $\mathrm{ACP^c}$ is called a *guarded* recursive specification if all occurrences of variables in the right-hand sides of its equations are guarded or it can be rewritten to such a recursive specification using the axioms of $\mathrm{ACP^c}$ and the equations of the recursive specification.

For each guarded recursive specification E and each variable $X \in \mathrm{V}(E)$, we introduce a constant of sort **P** standing for the unique solution of E for X. This constant is denoted by $\langle X|E\rangle$. We will also use the following notation. Let t be a term of $\mathrm{ACP^c}$ and E be a guarded recursive specification over $\mathrm{ACP^c}$. Then we write $\langle t|E\rangle$ for t with, for all $X \in \mathrm{V}(E)$, all occurrences of X in t replaced by $\langle X|E\rangle$.

The additional axioms for recursion are the equations given in Table 6. Both RDP and RSP are axiom schemas. A side condition is added to restrict the variables, terms and guarded recursive specifications for which X, t_X and E stand. The additional axioms for recursion are known as the recursive definition principle (RDP) and the recursive specification principle (RSP). The equations $\langle X|E\rangle = \langle t_X|E\rangle$ for a fixed E express that the constants $\langle X|E\rangle$ make up a solution of E. The conditional equations $E \Rightarrow X = \langle X|E\rangle$ express that this solution is the only one.

The structural operational semantics for the constants $\langle X|E\rangle$ is described by the transition rules given in Table 7.

Guarded recursive specifications over $\mathrm{ACP^c}$ have unique solutions in the model induced by the structural operational semantics of $\mathrm{ACP^c}$ extended with guarded recursion.

Theorem 5 (Unique solutions). *Let E be a guarded recursive specifications over $\mathrm{ACP^c}$. If $\{p_X \mid X \in \mathrm{V}(E)\}$ and $\{q_X \mid X \in \mathrm{V}(E)\}$ are solutions of E, then $p_X \mathrel{\underline{\leftrightarrow}} q_X$ for all $X \in \mathrm{V}(E)$.*

Table 8. Axioms for condition evaluation ($a \in \mathsf{A}_\delta$, $\eta \in \mathsf{C}_{\mathsf{at}}$, $\eta' \in \mathsf{C}_{\mathsf{at}} \cup \{\bot, \top\}$)

$\mathsf{CE}_h(a) = a$	CE1	$\mathsf{CE}_h(\bot) = \bot$	CE6
$\mathsf{CE}_h(a \cdot x) = a \cdot \mathsf{CE}_h(x)$	CE2	$\mathsf{CE}_h(\top) = \top$	CE7
$\mathsf{CE}_h(x + y) = \mathsf{CE}_h(x) + \mathsf{CE}_h(y)$	CE3	$\mathsf{CE}_h(\eta) = \eta' \quad \text{if } h(\eta) = \eta'$	CE8
$\mathsf{CE}_h(\phi :\rightarrow x) = \mathsf{CE}_h(\phi) :\rightarrow \mathsf{CE}_h(x)$	CE4	$\mathsf{CE}_h(-\phi) = -\mathsf{CE}_h(\phi)$	CE9
$\mathsf{CE}_h(\mathsf{CE}_{h'}(x)) = \mathsf{CE}_{h \circ h'}(x)$	CE5	$\mathsf{CE}_h(\phi \sqcup \psi) = \mathsf{CE}_h(\phi) \sqcup \mathsf{CE}_h(\psi)$	CE10
		$\mathsf{CE}_h(\phi \sqcap \psi) = \mathsf{CE}_h(\phi) \sqcap \mathsf{CE}_h(\psi)$	CE11

5 Evaluation of Conditions

Guarded commands cannot always be eliminated from closed terms of ACP^{c} because conditions different from both $\bot$ and $\top$ may be involved. The condition evaluation operators introduced below, can be brought into action in such cases.

There are unary *condition evaluation* operators $\mathsf{CE}_h{:}\mathbf{P} \rightarrow \mathbf{P}$ and $\mathsf{CE}_h{:}\mathbf{C} \rightarrow \mathbf{C}$ for each endomorphisms h of $\mathcal{C}$.

These operators can be explained as follows: $\mathsf{CE}_h(p)$ behaves as p with each condition ζ occurring in p replaced according to h. If the image of $\mathcal{C}$ under h is $\mathbb{B}$, i.e. the Boolean algebra with domain $\{\bot, \top\}$, then guarded commands can be eliminated from $\mathsf{CE}_h(p)$. In the case where the image of $\mathcal{C}$ under h is not $\mathbb{B}$, CE_h can be regarded to evaluate the conditions only partially.

Henceforth, we write $\mathcal{H}$ for the set of all endomorphisms of $\mathcal{C}$.

The additional axioms for CE_h, where $h \in \mathcal{H}$, are the axioms given in Table 8.

Example 2. We return to Example 1, which is concerned with a pedestrian who uses a crossing with traffic lights to cross a road with busy traffic safely. Let h_g be such that $h_g(\mathit{green}) = \top$ and $h_g(\mathit{red}) = \bot$; and let h_r be such that $h_r(\mathit{green}) = \bot$ and $h_r(\mathit{red}) = \top$. Then we can derive the following:

$$\mathsf{CE}_{h_g}(\mathit{PED}) = \mathit{arrive} \cdot \mathit{cross} \quad \text{and} \quad \mathsf{CE}_{h_r}(\mathit{PED}) = \mathit{arrive} \cdot \mathit{make\text{-}req} \cdot \delta\,.$$

So in a world where the traffic light for pedestrians is green he or she will cross the street without making a request for green light; and in a world where the traffic light for pedestrians is red he or she will become completely inactive after making a request for green light. In reality, the request would cause a change from red to green, but the condition evaluation operators CE_h cannot deal with that. We will return to this issue in Example 3.

The structural operational semantics of ACP^{c} extended with condition evaluation is described by the transition rules for ACP^{c} and the transition rules given in Table 9.

The elements of $\mathcal{C}$ can be used to represent equivalence classes with respect to logical equivalence of the set of all propositions with elements of C_{at} as propositional variables and with finite conjunctions and disjunctions. We write $\mathcal{P}$ for this set of propositions. It is likely that there is a theory $\varPhi$ about the atomic

Table 9. Transition rules for condition evaluation

$\dfrac{x \xrightarrow{[\phi]\,a} \surd}{\mathsf{CE}_h(x) \xrightarrow{[h(\phi)]\,a} \surd}\; h(\phi) \neq \bot$	$\dfrac{x \xrightarrow{[\phi]\,a} x'}{\mathsf{CE}_h(x) \xrightarrow{[h(\phi)]\,a} \mathsf{CE}_h(x')}\; h(\phi) \neq \bot$

conditions in the shape of a set of propositions. Let $\Phi \subset \mathcal{P}$, and let $h_\Phi \in \mathcal{H}$ be such that for all $\alpha, \beta \in \mathcal{C}$:

$$\Phi \vdash \langle\!\langle h_\Phi(\alpha) \rangle\!\rangle \Leftrightarrow \langle\!\langle \alpha \rangle\!\rangle \quad \text{and} \quad h_\Phi(\alpha) = h_\Phi(\beta) \;\text{ iff }\; \Phi \vdash \langle\!\langle \alpha \rangle\!\rangle \Leftrightarrow \langle\!\langle \beta \rangle\!\rangle \qquad (1)$$

where $\langle\!\langle \alpha \rangle\!\rangle$ is a representative of the equivalence class of propositions isomorphic to α. Then we have $h_\Phi(\alpha) = \top$ iff $\langle\!\langle \alpha \rangle\!\rangle$ is derivable from Φ and $h_\Phi(\alpha) = \bot$ iff $\neg \langle\!\langle \alpha \rangle\!\rangle$ is derivable from Φ. The image of $\mathcal{C}$ under h_Φ is $\mathbb{B}$ iff Φ is a complete theory. If Φ is not a complete theory, then h_Φ is not uniquely determined by (1). However, the images of $\mathcal{C}$ under the different endomorphisms satisfying (1) are isomorphic subalgebras of $\mathcal{C}$. Moreover, if both h and h' satisfy (1), then $\Phi \vdash \langle\!\langle h(\alpha) \rangle\!\rangle \Leftrightarrow \langle\!\langle h'(\alpha) \rangle\!\rangle$ for all $\alpha \in \mathcal{C}$.

Below, we show that condition evaluation on the basis of a complete theory can be viewed as substitution on the basis of the theory. That leads us to the use of the following convention: for $\alpha \in \mathcal{C}$, $\underline{\alpha}$ stands for an arbitrary closed term of sort **C** of which the value in $\mathcal{C}$ is α.

Proposition 3 (Condition evaluation on the basis of a theory). *Let $\Phi \subset \mathcal{P}$ be a complete theory and let p be a closed term of* ACP^{c}. *Then $\mathsf{CE}_{h_\Phi}(p) = p'$ where p' is p with, for all $\alpha \in \mathcal{C}$, in all subterms of the form $\underline{\alpha} :\rightarrow q$, $\underline{\alpha}$ replaced by $\top$ if $\Phi \vdash \langle\!\langle \alpha \rangle\!\rangle$ and $\underline{\alpha}$ replaced by $\bot$ if $\Phi \vdash \neg \langle\!\langle \alpha \rangle\!\rangle$.*

Proof. This result follows immediately from the definition of h_Φ and the distributivity of CE_{h_Φ} over all operators of ACP^{c}. □

In μCRL [17], an extension of ACP which includes conditional expressions, we find a formalization of the substitution-based alternative for CE_{h_Φ}.

The substitution-based alternative works properly because condition evaluation by means of a condition evaluation operator is not dependent on process behaviour. Hence, the result of condition evaluation is globally valid. Below, we will generalize the condition evaluation operators introduced above in such a way that condition evaluation may be dependent on process behaviour. In that case, the result of condition evaluation is in general not globally valid.

Remark 1. Let $h \in \mathcal{H}$. Then h induces a theory $\Phi \subset \mathcal{P}$ such that $h = h_\Phi$, viz. the theory Φ defined by

$$\Phi = \{\langle\!\langle h(\alpha) \rangle\!\rangle \Leftrightarrow \langle\!\langle \alpha \rangle\!\rangle \mid \alpha \in \mathcal{C}\} \cup \{\langle\!\langle \alpha \rangle\!\rangle \Leftrightarrow \langle\!\langle \beta \rangle\!\rangle \mid h(\alpha) = h(\beta)\}\,.$$

Consequently, condition evaluation by means of the condition evaluation operators introduced above is always condition evaluation of which the result can be determined from a set of propositions.

Table 10. Axioms for generalized condition evaluation ($a \in \mathsf{A}_\delta$)

$\mathsf{GCE}_h(a) = a$	GCE1
$\mathsf{GCE}_h(a \cdot x) = a \cdot \mathsf{GCE}_{\mathsf{eff}(a,h)}(x)$	GCE2
$\mathsf{GCE}_h(x + y) = \mathsf{GCE}_h(x) + \mathsf{GCE}_h(y)$	GCE3
$\mathsf{GCE}_h(\phi :\rightarrow x) = \mathsf{CE}_h(\phi) :\rightarrow \mathsf{GCE}_h(x)$	GCE4

Table 11. Transition rules for generalized condition evaluation

$$\frac{x \xrightarrow{[\phi]\, a} \surd}{\mathsf{GCE}_h(x) \xrightarrow{[h(\phi)]\, a} \surd}\; h(\phi) \neq \bot \qquad \frac{x \xrightarrow{[\phi]\, a} x'}{\mathsf{GCE}_h(x) \xrightarrow{[h(\phi)]\, a} \mathsf{GCE}_{\mathsf{eff}(a,h)}(x')}\; h(\phi) \neq \bot$$

We proceed with generalizing the condition evaluation operators introduced above. It is assumed that a fixed but arbitrary function $\mathsf{eff} : \mathsf{A} \times \mathcal{H} \rightarrow \mathcal{H}$ has been given.

There is a unary *generalized condition evaluation* operator $\mathsf{GCE}_h : \mathbf{P} \rightarrow \mathbf{P}$ for each $h \in \mathcal{H}$; and there is again the unary operator $\mathsf{CE}_h : \mathbf{C} \rightarrow \mathbf{C}$ for each $h \in \mathcal{H}$.

The generalized condition evaluation operator GCE_h allows, given the function eff, to evaluate conditions dependent of process behaviour. The function eff gives, for each action a and endomorphism h, the endomorphism h' that represents the changed results of condition evaluation due to performing a. The function eff is extended to A_δ such that $\mathsf{eff}(\delta, h) = h$ for all $h \in \mathcal{H}$.

The additional axioms for GCE_h, where $h \in \mathcal{H}$, are the axioms given in Table 10 and axioms CE6–CE11 from Table 8.

Example 3. We return to Example 1, which is concerned with a pedestrian who uses a crossing with traffic lights to cross a road with busy traffic safely. In Example 2, we illustrated that the condition evaluation operators CE_h cannot deal with the change from red light to green light caused by a request for green light. Here, we illustrate that the generalized condition evaluation operators GCE_h can deal with such a change. Let h_g and h_r be as in Example 2; and let eff be such that $\mathsf{eff}(\mathit{make\text{-}req}, h_r) = h_g$ and $\mathsf{eff}(a, h) = h$ otherwise. Then we can derive the following:

$$\mathsf{GCE}_{h_g}(\mathit{PED}) = \mathit{arrive} \cdot \mathit{cross}\ ,$$

$$\mathsf{GCE}_{h_r}(\mathit{PED}) = \mathit{arrive} \cdot \mathit{make\text{-}req} \cdot \mathit{cross}\ .$$

The change from red light to green light is due to interaction between the pedestrian and the traffic lights.

The structural operational semantics of $\mathrm{ACP^c}$ extended with generalized condition evaluation is described by the transition rules for $\mathrm{ACP^c}$ and the transition rules given in Table 11.

We can add both the condition evaluation operators and the generalized condition evaluation operators to $\mathrm{ACP^c}$. However, Proposition 4 stated below makes it clear that the latter operators supersede the former operators.

The equation $\mathsf{CE}_h(\mathsf{CE}_{h'}(x)) = \mathsf{CE}_{h \circ h'}(x)$ is an axiom, but the equation $\mathsf{GCE}_h(\mathsf{GCE}_{h'}(x)) = \mathsf{GCE}_{h \circ h'}(x)$ is not an axiom. The reason is that the latter equation is only valid if eff satisfies $\mathsf{eff}(a, h \circ h') = \mathsf{eff}(a, h) \circ \mathsf{eff}(a, h')$ for all $a \in \mathsf{A}$ and $h, h' \in \mathcal{H}$.

As their name suggests, the generalized condition evaluation operators are generalizations of the condition evaluation operators.

Proposition 4 (Generalization). *We can fix the function* eff *such that* $\mathsf{GCE}_h(x) = \mathsf{CE}_h(x)$ *for all* $h \in \mathcal{H}$.

Proof. Clearly, if $\mathsf{eff}(a, h') = h'$ for all $a \in \mathsf{A}$ and $h' \in \mathcal{H}$, then $\mathsf{GCE}_h(x) = \mathsf{CE}_h(x)$ for all $h \in \mathcal{H}$. □

The state operators that are added to ACP^{c} in Sect. 6 are in their turn generalizations of the generalized condition evaluation operators.

6 State Operators

The state operators make it easy to represent the execution of a process in a state. The basic idea is that the execution of an action in a state has effect on the state, i.e. it causes a change of state. Besides, there is an action left when an action is executed in a state. The operators introduced here generalize the state operators added to ACP in [18]. The main difference with those operators is that guarded commands are taken into account.

It is assumed that a fixed but arbitrary set S of *states* has been given, together with functions $\mathsf{act} : \mathsf{A} \times S \to \mathsf{A}_\delta$, $\mathsf{eff} : \mathsf{A} \times S \to S$ and $\mathsf{eval} : \mathcal{C} \times S \to \mathcal{C}$, where, for each $s \in S$, the function $h_s : \mathcal{C} \to \mathcal{C}$ defined by $h_s(\alpha) = \mathsf{eval}(\alpha, s)$ is an endomorphism of $\mathcal{C}$.

There are unary *state operators* $\lambda_s : \mathbf{P} \to \mathbf{P}$ and $\lambda_s : \mathbf{C} \to \mathbf{C}$ for each $s \in S$.

The state operator λ_s allows, given the above-mentioned functions, processes to interact with a state. Let p be a process. Then $\lambda_s(p)$ is the process p executed in state s. The function act gives, for each action a and state s, the action that results from executing a in state s. The function eff gives, for each action a and state s, the state that results from executing a in state s. The function eval gives, for each condition α and state s, the condition that results from evaluating α in state s. The functions act and eff are extended to A_δ such that $\mathsf{act}(\delta, s) = \delta$ and $\mathsf{eff}(\delta, s) = s$ for all $s \in S$.

The additional axioms for λ_s, where $s \in S$, are the axioms given in Table 12. Axioms SO1–SO3 are the axioms for the state operators added to ACP in [18].

The structural operational semantics of ACP^{c} extended with state operators is described by the transition rules for ACP^{c} and the transition rules given in Table 13.

We can add, in addition to the state operators, the condition evaluation operators and/or the generalized condition evaluation operators from Sect. 5 to ACP^{c}.

The state operators are generalizations of the generalized condition evaluation operators from Sect. 5.

Table 12. Axioms for state operators ($a \in \mathsf{A}_\delta$, $\eta \in \mathsf{C}_{\mathsf{at}}$, $\eta' \in \mathsf{C}_{\mathsf{at}} \cup \{\bot, \top\}$)

$\lambda_s(a) = \mathsf{act}(a, s)$	SO1	$\lambda_s(\bot) = \bot$	SO5
$\lambda_s(a \cdot x) = \mathsf{act}(a, s) \cdot \lambda_{\mathsf{eff}(a,s)}(x)$	SO2	$\lambda_s(\top) = \top$	SO6
$\lambda_s(x + y) = \lambda_s(x) + \lambda_s(y)$	SO3	$\lambda_s(\eta) = \eta' \quad \text{if } \mathsf{eval}(\eta, s) = \eta'$	SO7
$\lambda_s(\phi :\rightarrow x) = \lambda_s(\phi) :\rightarrow \lambda_s(x)$	SO4	$\lambda_s(-\phi) = -\lambda_s(\phi)$	SO8
		$\lambda_s(\phi \sqcup \psi) = \lambda_s(\phi) \sqcup \lambda_s(\psi)$	SO9
		$\lambda_s(\phi \sqcap \psi) = \lambda_s(\phi) \sqcap \lambda_s(\psi)$	SO10

Table 13. Transition rules for state operators

$$\frac{x \xrightarrow{[\phi]\, a} \surd}{\lambda_s(x) \xrightarrow{[\mathsf{eval}(\phi,s)]\, \mathsf{act}(a,s)} \surd}\ \mathsf{act}(a, s) \neq \delta,\ \mathsf{eval}(\phi, s) \neq \bot$$

$$\frac{x \xrightarrow{[\phi]\, a} x'}{\lambda_s(x) \xrightarrow{[\mathsf{eval}(\phi,s)]\, \mathsf{act}(a,s)} \lambda_{\mathsf{eff}(a,s)}(x')}\ \mathsf{act}(a, s) \neq \delta,\ \mathsf{eval}(\phi, s) \neq \bot$$

Proposition 5 (Generalization). *We can fix S,* act*,* eff *and* eval *such that, for some $f : \mathcal{H} \to S$, $\lambda_{f(h)}(x) = \mathsf{GCE}_h(x)$ holds for all $h \in \mathcal{H}$.*

Proof. Clearly, if $S = \mathcal{H}$, f is the identity function on $\mathcal{H}$, and $\mathsf{act}(a, s) = a$, $\mathsf{eff}(a, s) = \mathsf{eff}(a, f^{-1}(s))$ and $\mathsf{eval}(\alpha, s) = f^{-1}(s)(\alpha)$ for all $a \in \mathsf{A}$, $s \in S$ and $\alpha \in \mathcal{C}$, then $\lambda_{f(h)}(x) = \mathsf{GCE}_h(x)$ holds for all $h \in \mathcal{H}$. □

Notice that, in so far as condition evaluation is concerned, the state operators do not add anything to the generalized condition evaluation operators.

7 Concluding Remarks

Conditional expressions of the form $\zeta :\rightarrow p$ are not new. They were added to ACP for the first time in [3]. In [4], it was proposed to take the domain of a free Boolean algebra over a given set of generators as the set of conditions. Splitting bisimilarity is based on a variant of bisimilarity that was defined for the first time in [4]. The formulation given here is closer to the one given in [5]. State operators were added to ACP for the first time in [18]. The condition evaluation operators and the generalized condition evaluation operators were introduced for the first time in [11]. We are not aware of other work studying condition evaluation in a process algebra with conditional expressions.

In ACP^{c}, like in ACPps [6], conditional expressions give rise to the inclusion of conditional transitions in the behaviour being described, whereas in most other process algebraic formalisms that include conditional expressions, they concern the conditional inclusion of unconditional transitions (see e.g. μCRL [19]). ACP^{c}, like ACPps, is a development following ideas from [4]. ACP^{c} is based on a more abstract view on conditions than ACPps, but it lacks signal emission – a mechanism from ACPps that allows for a kind of condition evaluation.

References

1. Bergstra, J.A., Klop, J.W.: Process algebra for synchronous communication. Information and Control **60** (1984) 109–137
2. Baeten, J.C.M., Weijland, W.P.: Process Algebra. Volume 18 of Cambridge Tracts in Theoretical Computer Science. Cambridge University Press, Cambridge (1990)
3. Baeten, J.C.M., Bergstra, J.A., Mauw, S., Veltink, G.J.: A process specification formalism based on static COLD. In Bergstra, J.A., Feijs, L.M.G., eds.: Algebraic Methods II: Theory, Tools and Applications. Volume 490 of Lecture Notes in Computer Science., Springer-Verlag (1991) 303–335
4. Baeten, J.C.M., Bergstra, J.A.: Process algebra with signals and conditions. In Broy, M., ed.: Programming and Mathematical Methods. Volume F88 of NATO ASI Series., Springer-Verlag (1992) 273–323
5. Bergstra, J.A., Ponse, A., van Wamel, J.J.: Process algebra with backtracking. In de Bakker, J.W., de Roever, W.P., Rozenberg, G., eds.: A Decade of Concurrency (Reflections and Perspectives). Volume 803 of Lecture Notes in Computer Science., Springer-Verlag (1994) 46–91
6. Baeten, J.C.M., Bergstra, J.A.: Process algebra with propositional signals. Theoretical Computer Science **177** (1997) 381–405
7. Hennessy, M., Milner, R.: Algebraic laws for non-determinism and concurrency. Journal of the ACM **32** (1985) 137–161
8. Brookes, S.D., Hoare, C.A.R., Roscoe, A.W.: A theory of communicating sequential processes. Journal of the ACM **31** (1984) 560–599
9. Bergstra, J.A., Ponse, A.: Process algebra and conditional composition. Information Processing Letters **80** (2001) 41–49
10. van der Zwaag, M.B.: Models and Logics for Process Algebra. PhD thesis, Programming Research Group, University of Amsterdam, Amsterdam (2002)
11. Bergstra, J.A., Middelburg, C.A.: Splitting bisimulations and retrospective conditions. Computer Science Report 05-03, Department of Mathematics and Computer Science, Eindhoven University of Technology (2005)
12. Monk, J.D., Bonnet, R., eds.: Handbook of Boolean Algebras. Volume 1. Elsevier, Amsterdam (1989)
13. Hoare, C.A.R., Hayes, I.J., He Jifeng, Morgan, C.C., Roscoe, A.W., Sanders, J.W., Sorensen, I.H., Spivey, J.M., Sufrin, B.A.: Laws of programming. Communications of the ACM **30** (1987) 672–686
14. Halmos, P.R.: Lectures on Boolean Algebras. Mathematical Studies. Van Nostrand, Princeton, NJ (1963)
15. Busi, N., van Glabbeek, R.J., Gorrieri, R.: Axiomatising ST-bisimulation semantics. In Olderog, R.R., ed.: PROCOMET'94. Volume 56 of IFIP Transactions A., North-Holland (1994) 169–188
16. Baeten, J.C.M., Verhoef, C.: Concrete process algebra. In Abramsky, S., Gabbay, D.M., Maibaum, T.S.E., eds.: Handbook of Logic in Computer Science. Volume IV. Oxford University Press, Oxford (1995) 149–268
17. Groote, J.F., Ponse, A.: Proof theory for μCRL: A language for processes with data. In Andrews, D.J., Groote, J.F., Middelburg, C.A., eds.: Semantics of Specification Languages. Workshops in Computing Series, Springer-Verlag (1994) 232–251
18. Baeten, J.C.M., Bergstra, J.A.: Global renaming operators in concrete process algebra. Information and Control **78** (1988) 205–245
19. Groote, J.F., Ponse, A.: The syntax and semantics of μCRL. In Ponse, A., Verhoef, C., van Vlijmen, S.F.M., eds.: Algebra of Communicating Processes 1994. Workshops in Computing Series, Springer-Verlag (1995) 26–62

Complete Axioms for Stateless Connectors*

Roberto Bruni, Ivan Lanese, and Ugo Montanari

Computer Science Department, University of Pisa, Italy
{bruni, lanese, ugo}@di.unipi.it

Abstract. The conceptual separation between computation and coordination in distributed computing systems motivates the use of peculiar entities commonly called *connectors*, whose task is managing the interaction among distributed components. Different kinds of connectors exist in the literature, at different levels of abstraction. We focus on a basic algebra of connectors which is expressive enough to model, e.g., all the architectural connectors of CommUnity. We first define the operational, observational and denotational semantics of connectors, then we show that the observational and denotational semantics coincide and finally we give a complete normal-form axiomatization.

1 Introduction

The advent of modern communication technologies shifted the focus of computer science researchers from isolated computing systems to distributed communicating systems, in which *interaction* plays the prominent role. In Milner's words [21], "computing has grown into informatics and Turing's logical computing machines are matched by a logic of interaction". In this perspective, the analysis of global computing systems is facilitated by approaches, techniques and paradigms that exploit a clean conceptual separation between *computation* and *coordination*. This is much evident at several levels of abstraction (architecture, software, processes), where issues like reusability, maintenance, heterogeneity call for modular specifications, theories and models.

When separating coordination from computation, the notion of a *connector* emerges in different contexts, with slightly different meaning, expressiveness and functionalities. The common trait is the role of a connector: a component that mediates the interaction of other computational components and connectors. In particular, connectors have been studied within both *algebraic* and *categorical approaches* to system modeling.

The algebraic approach [14,20] models systems a elements of a suitable term algebra, with constants modeling basic components that can be composed via the other operators, e.g., parallel composition and name restriction. Operational and abstract semantics are usually based on a labelled transition system defined by structural induction.

The categorical approach [13] models systems as objects in a category, with morphisms defining relations such as subsystem or refinement. Complex software architectures can be modeled as diagrams in the category, with universal constructions, such as colimit, building an object in the same category that behaves as the whole system

* Research supported by the FET-GC Project IST-2001-32747 AGILE and by the project HPRN-CT-2002-00275 SEGRAVIS.

J.L. Fiadeiro et al. (Eds.): CALCO 2005, LNCS 3629, pp. 98–113, 2005.

and that is uniquely determined up to isomorphism. The use of architectural connectors within the categorical approach is well exemplified by CommUnity [11,10].

Having rigorous mathematical foundations is crucial for the analysis of coordinated distributed systems. Several different kinds of connectors have been studied in the literature, studying e.g. the observational semantics of process contexts [17,24,12,16], or analysing suitable equational theories and reduction to normal forms [25,6,3,15].

In this paper, we concentrate on the algebraic approach by promoting a small algebra of connectors, for which we define suitable operational, observational and denotational semantics. The first one is expressed using the *Tile Model* [12]. The observational semantics we select is *tile bisimilarity*, that also coincides with *tile trace equivalence* for the algebra under inspection. The denotational semantics is original to this contribution and it is based on (an algebra of) suitable boolean matrices, called tick-tables. We first show that the observational and denotational semantics coincide and then give a complete normal-form axiomatization for them, which is the main result of the paper.

Our connectors are rather simple: they essentially model basic synchronization, mutual exclusion and hiding and they are all stateless. Nevertheless, we think that the analysis of these connectors is quite interesting, since they allow to build a wide range of coordination connectors. For instance, they are expressive enough to model the multiple-action synchronization mechanism of CommUnity which uses morphisms and complex architectural connectors. This is shown in the previous work [2], where an encoding from CommUnity into the Tile Model is defined. One of the main results of [2] is that the translation of a diagram is tile bisimilar to the translation of its colimit.

The above mentioned main result of this paper, namely the complete axiomatization of abstract semantics, improves the work in [2] by showing that, for the part of action coordination, tile bisimilarity can be axiomatized as a suitable equational theory, where equivalence classes have standard representatives. While in the algebraic approach equivalence classes are usually abstract entities, having a normal form gives a concrete representation that matches a nice feature of the categorical approach, namely that the colimit of a diagram is its best concrete representative.

The research initiated in [2] and extended in this paper is a first step towards a more general reconciliation between the categorical and the algebraic approach, of which CommUnity and the Tile Model are just intended to be two selected representatives.

With respect to other approaches to synchronization connectors existing in the literature [3,5,25,6,15], our main contribution is the introduction of the mutual exclusion connector, which allows to specify a wider range of possible synchronization policies. Furthermore, the semantics based on matrices is new and it provides a clean mathematical definition of connectors. Finally, we characterize the classes of matrices that can be specified, both with and without mutual exclusion.

Structure of the paper. § 2 contains some background on symmetric monoidal categories and on the Tile Model. § 3 presents syntax and semantics of connectors, showing the correspondence between the observational and denotational semantics. § 4 contains the main results of the paper, namely the axiomatization of connectors and the theorems for semantic equivalence and normal form: we consider the case without mutual exclusion first (§ 4.1) and the general case later (§ 4.2). Conclusion and future work are in § 5. Due to space constraints, proofs are just sketched: full proofs can be found in [4].

2 Background

Symmetric monoidal categories for connectors. It has been shown in the literature that distributed systems can be conveniently modeled as graphs [8,22,9] that straightforwardly account for the network distribution of processes, mobile agents, etc. The advantage representing configuration graphs as (freely generated) symmetric monoidal categories is three-fold. First, they introduce a suitable notion of (observable) interfaces for configurations. Second, they introduce two key operations for composing graphs, namely sequential and parallel compositions. Third, the natural isomorphism defined by symmetries allows to take graphs up to interface-preserving graph isomorphism.

We recall that a *(strict) monoidal category* [18] $(C, \otimes, e)$ is a category C together with a functor $\otimes \colon C \times C \to C$ called the *tensor product* and an object e called the *unit*, such that for any arrows $\alpha_1, \alpha_2, \alpha_3 \in C$ we have $(\alpha_1 \otimes \alpha_2) \otimes \alpha_3 = \alpha_1 \otimes (\alpha_2 \otimes \alpha_3)$ and $\alpha_1 \otimes id_e = \alpha_1 = id_e \otimes \alpha_1$. Note that, by functoriality of $\otimes$ we have, e.g., $\alpha_1 \otimes \alpha_2 = \alpha_1 \otimes id_{a_2}; id_{b_1} \otimes \alpha_2 = id_{a_1} \otimes \alpha_2; \alpha_1 \otimes id_{b_2}$ for any $\alpha_i \colon a_i \to b_i, i \in \{1,2\}$.

Definition 1. *A* symmetric (strict) monoidal category $(C, \otimes, e, \gamma)$ *is a (strict) monoidal category* $(C, \otimes, e)$ *together with a family of arrows, called* symmetries, $\{\gamma_{a,b} \colon a \otimes b \to b \otimes a\}_{a,b}$ *indexed by pairs of objects in* C *such that for any two arrows* $\alpha_1, \alpha_2 \in C$ *with* $\alpha_i \colon a_i \to b_i$, *we have* $\alpha_1 \otimes \alpha_2; \gamma_{b_1,b_2} = \gamma_{a_1,a_2}; \alpha_2 \otimes \alpha_1$ *(that is* γ *is a natural isomorphism) that satisfies the coherence equalities (for any objects* a, b, c*):*

$$\gamma_{a,b}; \gamma_{b,a} = id_{a \otimes b} \qquad\qquad \gamma_{a \otimes b,c} = id_a \otimes \gamma_{b,c}; \gamma_{a,c} \otimes id_b.$$

The categories we are interested in are those freely generated from an unsorted (hyper)signature Σ, i.e., from a ranked family of operators $\sigma \colon n \to m$. The objects are just natural numbers expressing the arities of the interfaces, i.e., the number of "attach points", with $n \otimes m = n + m$ and $e = 0$. The operators $\sigma \in \Sigma$ are seen as basic arrows with source and target defined accordingly to the arity of σ. Symmetries can be always expressed in terms of the basic symmetry $\gamma_{1,1} \colon 2 \to 2$. Intuitively, symmetries can be used to rearrange the input-output interfaces of graph-like configurations. We call *permutation* any composition of identities and symmetries. A generic arrow can always be expressed as a suitable composition of id_1, $\gamma_{1,1}$ and $\sigma \in \Sigma$.

Lemma 1. *Any arrow* α *can be decomposed as* $id_{n_1} \otimes \sigma_1 \otimes id_{m_1}; \ldots; id_{n_k} \otimes \sigma_k \otimes id_{m_k}$ *for some natural numbers* $k, n_1, \ldots, n_k, m_1, \ldots, m_k$ *and* $\sigma_1, \ldots, \sigma_k \in \{\gamma_{1,1}\} \cup \Sigma$.

An arrow expressed using only identities and (possibly multiple instances of) one particular $\sigma \in \{\gamma_{1,1}\} \cup \Sigma$ is called a *layer of* σ. For example, a permutation is a layer of $\gamma_{1,1}$.

Tile Model. In this paper, we choose the Tile Model for defining the operational and observational semantics of connectors. In fact, tile configurations are suitable to represent connectors, which include input and output interfaces where actions can be observed and that can be used to compose configurations and to coordinate their behaviours.

The Tile Model [12] is a rule-based framework whose main ingredients are rewrite rules with side effects, called *basic tiles* that combine inspirations from SOS rules [23],

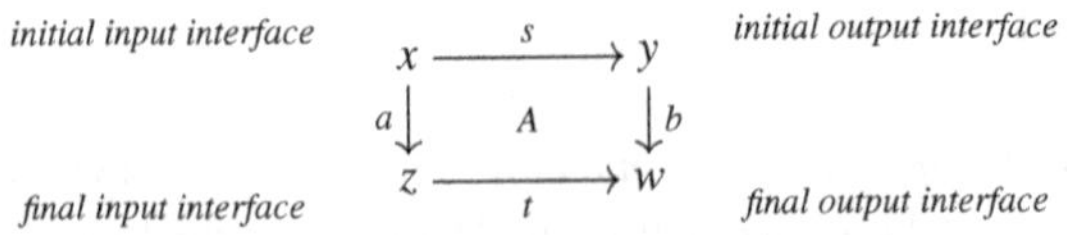

Fig. 1. Graphical representation of a tile A

context systems [17], *structured transition systems* [7] and *rewriting logic* [19]. A tile $A : s \xrightarrow[b]{a} t$ is a rewrite rule stating that the *initial configuration* s can evolve to the *final configuration* t via A, producing the *effect* b; but the step is allowed only if the 'arguments' of s can produce a, which acts as the *trigger* of A (see Figure 1). Triggers and effects are called *observations* and tile vertices are called *interfaces*. The operational semantics of concurrent systems can be expressed via tiles if system configurations form a monoidal category $\mathcal{H}$, and observations form a monoidal category $\mathcal{V}$ with the same set of objects as $\mathcal{H}$. Abusing the notation, we denote by $\_ \otimes \_$ both monoidal functors of $\mathcal{H}$ and $\mathcal{V}$ and by $\_;\_$ both sequential compositions in $\mathcal{H}$ and $\mathcal{V}$.

Definition 2. *A* tile system *is a tuple $\mathcal{R} = (\mathcal{H}, \mathcal{V}, N, R)$ where $\mathcal{H}$ and $\mathcal{V}$ are monoidal categories with the same set of objects $O_{\mathcal{H}} = O_{\mathcal{V}}$, N is the set of rule names and $R \colon N \to \mathcal{H} \times \mathcal{V} \times \mathcal{V} \times \mathcal{H}$ is a function such that for all $A \in N$, if $R(A) = \langle s, a, b, t \rangle$, then the arrows s, a, b, t can form a tile like in Figure 1.*

Tiles can be composed horizontally, in parallel, and vertically to generate larger steps. Horizontal composition $A;B$ coordinates the evolution of the initial configuration of A with that of B, yielding the 'synchronization' of the two rewrites. Horizontal composition is possible only if the initial configurations of A and B interact cooperatively: the effect of A must provide the trigger for B. The parallel composition $A \otimes B$ builds concurrent steps. Vertical composition $A * B$ is sequential composition of computations.

Tiles can be seen as sequents of *tile logic*: the sequent $s \xrightarrow[b]{a} t$ is *entailed* by the tile logic associated with $\mathcal{R}$, written $\mathcal{R} \vdash s \xrightarrow[b]{a} t$, if it can be obtained by composing some basic tiles in R (possibly using also auxiliary tiles, like identities $id \xrightarrow[a]{a} id$ propagating observations). The "borders" of composed sequents are defined in Figure 2.

The main feature of tiles is their double labeling with triggers and effects, that allows to observe the input-output behaviour of configurations. By taking $\langle$trigger, effect$\rangle$ pairs as labels one can see tiles as a labeled transition system. In this context, the usual notion of bisimilarity is called *tile bisimilarity*.

Definition 3. *Let $\mathcal{R} = (\mathcal{H}, \mathcal{V}, N, R)$ be a tile system. A symmetric relation $\sim_{\mathrm{t}}$ on configurations is called* tile bisimulation *if whenever $s \sim_{\mathrm{t}} t$ and $\mathcal{R} \vdash s \xrightarrow[b]{a} s'$, then t' exists such that $\mathcal{R} \vdash t \xrightarrow[b]{a} t'$ and $s' \sim_{\mathrm{t}} t'$.*

The maximal tile bisimulation is called *tile bisimilarity* and it is denoted by $\simeq_{\mathrm{t}}$. Note that $s \simeq_{\mathrm{t}} t$ only if s and t have the same input-output interfaces.

$$\frac{s \xrightarrow[b]{a} t \quad h \xrightarrow[c]{b} f}{s;h \xrightarrow[c]{a} t;f} \text{ (hor)} \qquad \frac{s \xrightarrow[b]{a} t \quad h \xrightarrow[d]{c} f}{s \otimes h \xrightarrow[b \otimes d]{a \otimes c} t \otimes f} \text{ (par)} \qquad \frac{s \xrightarrow[b]{a} t \quad t \xrightarrow[d]{c} h}{s \xrightarrow[b;d]{a;c} h} \text{ (ver)}$$

Fig. 2. Inference rules for tile logic

A syntactic property on tiles guaranteeing that $\simeq_t$ is a congruence, i.e. that the semantics is compositional, is the so-called *basic source format* [12], which amounts to require that $\mathcal{H}$ is generated from a (hyper)signature Σ and that the initial configuration of each basic tile consists of a basic operator in Σ.

We shall focus on tile systems of *stateless* connectors, meaning that in all basic tiles the final configuration is equal to the initial one. Operatively, this means that the behaviour of a connector is history independent. An easy consequence is that $\simeq_t$ coincides with tile trace equivalence.

3 Algebra of Connectors

We present here a rich algebra of connectors for action coordination. We have developed such an algebra to model systems where multiple actions can be executed at each time, either independently or synchronized. Connectors are used to guarantee the global consistency of local evolutions. For instance, in the translation of CommUnity [2], connectors are used in conjunction with other operators representing the computational entities. Roughly these have n attach points associated with actions and according to the computed action they emit 1 tick (action performed) and $n-1$ unticks (forced inactivity).

We remark that all structures that we are going to present are based on the symmetric strict monoidal structure given by symmetries γ, tensor product $\otimes$ and unit 0.

The complete list of connectors is in Figure 3. The ordinary basic connectors are in the leftmost part of the table, while their duals are on the right (symmetry is self-dual). The term mex stands for "mutual exclusion". We also speak about synch connectors (∇ and Δ), choice connectors ($\blacktriangledown$ and $\blacktriangle$), hiding connectors (! and ¡) and inaction connectors ($\mathbf{0}$ and $\overline{\mathbf{0}}$). This set of connectors has been used in [2] to model action coordination in CommUnity, an architectural design language which has the extreme separation between computation and coordination as distinctive feature.

We now define the tile semantics for our connectors. As usual for tiles, we first fix the categories of configurations and of observations and then we give the basic tiles.

As explained in Section 2, the objects of our categories are natural numbers.

The horizontal category of configurations is the free symmetric (strict) monoidal category generated by the basic connectors. The basic connector γ is the symmetry $\gamma_{1,1}$. We call *connector* any arrow in the horizontal category. Given a connector $\alpha: n \to m$ we denote by $\alpha^c: m \to n$ its dual, defined in the obvious way for basic connectors (see Figure 3) and then inductively by $(\alpha;\beta)^c = \beta^c;\alpha^c$ and $(\alpha \otimes \beta)^c = \alpha^c \otimes \beta^c$.

The vertical category is the free monoidal category generated by the arrows $\mathsf{tick}: 1 \to 1$ and $\mathsf{untick}: 1 \to 1$.

Ordinary structure			Dual structure		
name	symbolic	graphical	name	symbolic	graphical
symmetry	$\gamma\colon 2 \to 2$		**symmetry**	$\gamma\colon 2 \to 2$	
duplicator	$\nabla\colon 1 \to 2$		**coduplicator**	$\Delta\colon 2 \to 1$	
bang	$!\colon 1 \to 0$		**cobang**	$¡\colon 0 \to 1$	
mex	$\dot{\nabla}\colon 1 \to 2$		**comex**	$\dot{\Delta}\colon 2 \to 1$	
zero	$\mathbf{0}\colon 1 \to 0$		**cozero**	$\bar{\mathbf{0}}\colon 0 \to 1$	

Fig. 3. Syntax of basic connectors

$$\gamma \xrightarrow[y\otimes x]{x\otimes y} \gamma \text{ where } x,y \in \{\mathsf{tick},\mathsf{untick}\} \qquad \nabla \xrightarrow[\mathsf{tick}\otimes\mathsf{tick}]{\mathsf{tick}} \nabla \qquad \nabla \xrightarrow[\mathsf{untick}\otimes\mathsf{untick}]{\mathsf{untick}} \nabla \qquad \mathbf{0} \xrightarrow[id_0]{\mathsf{untick}} \mathbf{0}$$

$$! \xrightarrow[id_0]{\mathsf{tick}} ! \qquad ! \xrightarrow[id_0]{\mathsf{untick}} ! \qquad \dot{\nabla} \xrightarrow[\mathsf{untick}\otimes\mathsf{untick}]{\mathsf{untick}} \dot{\nabla} \qquad \dot{\nabla} \xrightarrow[\mathsf{tick}\otimes\mathsf{untick}]{\mathsf{tick}} \dot{\nabla} \qquad \dot{\nabla} \xrightarrow[\mathsf{untick}\otimes\mathsf{tick}]{\mathsf{tick}} \dot{\nabla}$$

Fig. 4. Basic tiles for ordinary connectors

The tiles defining the semantics of ordinary connectors are in Figure 4. The first rule specifies that a symmetry can accept any input pair which is swapped in the output. Then there are the two rules for duplicator, where the constraint is that all the actions must coincide. Last rule in the first row defines the only allowed behaviour for zero, which admits just untick on its interface. Rules in the second row specify the behaviour of bang, which hides any action on its interface, and mex: if the trigger is untick, then the effects are two unticks, otherwise the trigger tick is propagated to exactly one effect.

Dual connectors have symmetric tiles. For instance, the tiles for Δ are:

$$\Delta \xrightarrow[\mathsf{tick}]{\mathsf{tick}\otimes\mathsf{tick}} \Delta \qquad \Delta \xrightarrow[\mathsf{untick}]{\mathsf{untick}\otimes\mathsf{untick}} \Delta$$

From the tile system we can derive an observational semantics using tile bisimilarity. This semantics is compositional, as proved by the following theorem.

Theorem 1. *In all the tile systems built using only the above tiles for connectors, $\simeq_t$ is a congruence (w.r.t. parallel and sequential composition).*

Proof. Trivial, since all the basic tiles satisfy the basic source property. □

It is worth noting that the axioms for symmetry "bisimulate", in the sense that the left hand side and the right hand side of each axiom are tile bisimilar.

id	0	1
0	✓	
1		✓

γ	00	01	10	11
00	✓			
01			✓	
10		✓		
11				✓

∇	00	01	10	11
0	✓			
1				✓

!	∅
0	✓
1	✓

$\blacktriangledown$	00	01	10	11
0	✓			
1		✓	✓	

0	∅
0	✓
1	

Fig. 5. Denotational semantics

The coordination policy of a connector $\alpha: n \to m$ can be represented as a $2^n \times 2^m$ tick-table whose cells contain boolean values. Each row (resp. column) represents a combination of tick/untick values (denoted as 1 or 0 in the tick-tables) for the n inputs (resp. m outputs). If a cell is true (i.e., marked), then the corresponding combination of inputs and outputs is admissible, otherwise (the cell is false, i.e., empty, unmarked) the corresponding combination of inputs and outputs is forbidden. The tick-tables for basic connectors are in Figure 5 (dual basic connectors have transposed tables).

We denote with $T(\alpha)$ the tick-table associated to connector α. Furthermore, given a position $[i, j]$ in a tick-table T we denote with $d_T([i, j])$ its *domain*, that is the set of elements in its input and output interfaces on which tick actions are performed.

A connector $\alpha: n \to m$ can be seen as an hypergraph where basic connectors are edges and elements of interfaces are nodes. The solution of the network of constraints S associated with α is the set of consistent assignments of tick/untick values to all the nodes appearing in the graph denoted by α in such way that a corresponding "tiling" can be found. However this semantics is too *concrete* when one is not interested in knowing the way in which all constraints of the network are satisfied. A more abstract semantics of α is the solution of the network, where all the internal nodes have been existentially quantified. Thus tick-tables can be seen as the *denotational semantics* of connectors.

Next lemma shows the effect on tick-tables of the operations on connectors.

Lemma 2. *For any two connectors $\alpha: n \to h, \beta: h \to m$, $T(\alpha;\beta)$ is the product matrix $T(\alpha) \times T(\beta)$, i.e., $T(\alpha;\beta)[i, j] = \bigvee_k (T(\alpha)[i,k] \wedge T(\beta)[k, j])$. For any two connectors $\alpha: n \to h, \beta: l \to m$, $T(\alpha \otimes \beta)$ is obtained by refining each marked entry of $T(\beta)$ by a copy of $T(\alpha)$, and each unmarked entry of $T(\beta)$ by the empty table with the same dimension as $T(\alpha)$. Moreover, for any connector α, $T(\alpha^c)$ is the transposition of $T(\alpha)$.*

The denotational semantics of connectors given by tick-tables agrees with the observational semantics defined by tile bisimilarity, that is two connectors are tile bisimilar iff they have the same associated tick-table.

Theorem 2. *For each pair of connectors α and β, $\alpha \simeq_t \beta$ iff $T(\alpha) = T(\beta)$.*

Proof. Since all connectors are stateless, two connectors are tile bisimilar iff their allowed combinations of ticks and unticks on the interfaces are equal, i.e., iff their tables are equal. □

4 Normal Form

We first show an axiomatization of connectors which is correct and complete w.r.t. their denotational semantics, and then we show an algorithm to derive a standard representative for each equivalence class, i.e., a normal form. From the categorical point of view, this corresponds to compute the colimit of a diagram (such as a CommUnity one).

Several axioms for connectors have been proposed, studied and applied in the literature, see e.g. [1,5,25,6,3]. Axioms over connectors are usually aimed at characterizing a category of *links between objects* as the equational term algebra freely generated from a restricted set of basic connectors. Usually the axioms have just to consider the few possible ways in which two or three basic connectors can be composed together. However, our algebra is very rich and thus a few more complex patterns need to be considered.

The consistency of all the axioms we are going to present w.r.t. the denotational semantics can be checked just by looking at the tables associated to each term. More precisely, for each axiom $\alpha = \beta$ that we propose, it is easy to check that $T(\alpha) = T(\beta)$.

Notation. Given a set of connectors S we denote with $\mathbf{CC}(S)$ the class of connectors generated by connectors in S. Note that symmetries are always included in $\mathbf{CC}(S)$, even when $S = \varnothing$.

Two edges are *adjacent* if they share a node in the graph representation of the connector. An edge is adjacent to any node in its interfaces. A *path* in the graph is a sequence of nodes $\{n_i | i \in \{1,\ldots,n\}\}$ such that for each $i \in \{1,\ldots,n-1\}$ n_i is an element of the input interface of a basic connector and n_{i+1} is an element of the output interface of the same basic connector if the connector is not a symmetry. As suggested by their graphical representation, for symmetries the path can only enter in the first element of the input interface and exit from the second one in the output interface, or enter from the second one and exit from the first one. The *components* of a path are all its nodes and all the edges traversed. We say that two components of a graph are *linearly connected* iff there exists a path of which they are both components. The relation of *connectedness* is the transitive closure of the relation of linear connectedness.

We let ∇^n denote the "tree" of ∇ connectors with n leaves, inductively defined as $\nabla^0 = !$ and $\nabla^{n+1} = \nabla; id \otimes \nabla^n$. Note that $\nabla^1 = id$. We also define connectors for structured objects in terms of connectors defined for smaller objects:

$$\nabla_0 = id_0 \qquad \nabla_{n+1} = \nabla \otimes \nabla_n; id \otimes \gamma_{1,n} \otimes id_n \qquad !_0 = id_0 \qquad !_{n+1} = ! \otimes !_n$$

Note that $\nabla_1 = \nabla$ and $!_1 = !$. Similar notations are used for the other connectors.

4.1 Connectors for Synchronization

First, we focus on the class of connectors $\mathbf{CC}(\nabla, \Delta, !, ¡)$. The tick-tables associated to these connectors can be characterized as below.

Proposition 1. *Let* $\alpha \in \mathbf{CC}(\nabla, \Delta, !, ¡)$. *Then* $T(\alpha)$ *satisfies the following properties.*

- $T(\alpha)[0,0] = \checkmark$;
- *suppose* $T(\alpha)[i_1, j_1] = \checkmark$ *and* $T(\alpha)[i_2, j_2] = \checkmark$;
  - *if* $d_{T(\alpha)}([i,j]) = d_{T(\alpha)}([i_1,j_1]) \cup d_{T(\alpha)}([i_2,j_2])$ *then* $T(\alpha)[i,j] = \checkmark$
  - *if* $d_{T(\alpha)}([i,j]) = d_{T(\alpha)}([i_1,j_1]) \cap d_{T(\alpha)}([i_2,j_2])$ *then* $T(\alpha)[i,j] = \checkmark$
  - *if* $d_{T(\alpha)}([i,j]) = d_{T(\alpha)}([i_1,j_1]) \setminus d_{T(\alpha)}([i_2,j_2])$ *then* $T(\alpha)[i,j] = \checkmark$
  - *if* $d_{T(\alpha)}([i,j]) = \overline{d_{T(\alpha)}([i_1,j_1])}$ *then* $T(\alpha)[i,j] = \checkmark$

Proposition 1 says that the cell with empty domain is always enabled and that table entries are closed under domain union, intersection, difference and complement.

Intuitively, the last four properties are true because connectors built of synch and hiding connectors individuate equivalence classes on the elements of the interfaces (connected elements are in the same class), different equivalence classes act independently and domains are unions of such classes. It is an easy consequence of the proposition that, for instance, for any $\alpha \in \mathbf{CC}(\nabla, \Delta, !, ¡)$, $T(\alpha)[1,1] = \checkmark$.

We call *synch-tables* the tables that satisfy these properties.

Definition 4 (Base). *Given a synch-table T its base $b(T)$ is the set of the domains of its marked cells that are minimal w.r.t. set inclusion.*

The synch-tables are uniquely identified by their bases.

Lemma 3. *Let T_1 and T_2 be any two synch-tables with the same dimension. Then $T_1 = T_2$ iff $b(T_1) = b(T_2)$.*

Proof. The if part is trivial. The only if part follows from Proposition 1. □

Analogous structures have been already studied in the literature [5,25,6,3]. If we inspect which equalities are satisfied among those in [3], then according to the terminology therein, we have a gs-monoidal structure $(\nabla, !)$, a cogs-monoidal structure $(\Delta, ¡)$, a match-share structure (∇, Δ) and a new-bang structure $(!, ¡)$. The whole structure is called a p-monoidal structure. Interestingly, the p-monoidal axioms characterize exactly tile bisimilarity and allow for normal-form reduction. This is explained below in detail.

As far as the gs-structure is concerned there are three axioms expressing the "associativity", "commutativity" and "unit" for the ∇ (with ! as "unit").

$$\nabla;(id \otimes !) = id \qquad \nabla;\gamma = \nabla \qquad \nabla;(\nabla \otimes id) = \nabla;(id \otimes \nabla)$$

A cogs-monoidal structure is just a gs-monoidal structure in the dual category. Therefore the axioms are obtained by reversing the order of composition.

The axioms of match-share categories have been proposed in [3], where the free algebra of match-share connectors has been shown to model partition relations between non-empty source and target objects. There are three match-share axioms:

$$\nabla;\Delta = id \qquad \Delta;\nabla = (id \otimes \nabla);(\Delta \otimes id) \qquad \Delta;\nabla = (\nabla \otimes id);(id \otimes \Delta)$$

The leftmost axiom essentially says that the multiplicity of connections between two objects is not important. The other two axioms (which are in fact equivalent, thus one of them can be dropped) say that the path connecting two objects is not important.

The new-bang categories just contain the axiom $¡;! = id_0$ which represents garbage-collection of isolated nodes.

We want to use the axioms to reduce any connector in a suitable normal form. We start by defining a *sorted form* that forces a standard order on connector layers.

Definition 5 (Sorted form). *A connector $\alpha \in \mathbf{CC}(\nabla, \Delta, !, ¡)$ is in sorted form iff*

$$\alpha \equiv \alpha_{¡};\alpha_{\gamma};\alpha_{\Delta};\beta_{\nabla};\beta_{\gamma};\beta_{!}$$

where α_σ and β_σ are layers of σ and $\equiv$ is syntactic identity.

Proposition 2. *Any connector* $\alpha \in \mathbf{CC}(\nabla, \Delta, !, ¡)$ *can be transformed in sorted form using the axioms.*

Proof. The proof is by induction on the construction of the connector. Essentially, we have to prove that given a connector α in sorted form, we can transform in sorted form any connector $id_{n_1} \otimes \sigma \otimes id_{n_2}; \alpha$. For each σ one can find axioms that make it to commute with all the other connectors that it must traverse to reach its final position. □

We want now to define for connectors a *normal form* which is strictly related to tick-tables. We first need an auxiliary definition.

Definition 6 (Central point). *A* central point *is any element of interface shared by layers* α_Δ *and* β_∇.

Definition 7 (Normal form). *A connector* $\alpha \in \mathbf{CC}(\nabla, \Delta, !, ¡)$ *is in normal form iff:*

1. *it is in sorted form;*
2. *hiding connectors have central points as interface;*
3. *each central point is linearly connected to at least an external interface.*

Theorem 3. *Any connector* $\alpha \in \mathbf{CC}(\nabla, \Delta, !, ¡)$ *can be transformed in normal form using the axioms.*

Proof. Trivial, using Proposition 2. □

The theorems below show the connection between normal forms and synch-tables.

Theorem 4. *For each synch-table* T*, we can build a connector* $\alpha \in \mathbf{CC}(\nabla, \Delta, !, ¡)$ *in normal form such that* $T = T(\alpha)$*. Moreover the construction is unique up to the axioms of symmetric monoidal categories and of associativity and commutativity of synch connectors.*

Proof. Let $b(T)$ be the base of T. We build α in the following way:

- we create a central point P_b for each element $b \in b(T)$;
- we build a tree Δ^n (resp. ∇^m) on the left (resp. right) of each central point P_b, where n (resp. m) is the number of elements in b that are in the left (resp. right) interface;
- we add permutations to connect the trees Δ^n and ∇^m to the corresponding elements of the interfaces. □

Theorem 5. *We have a bijective correspondence between synch-tables and connectors in* $\mathbf{CC}(\nabla, \Delta, !, ¡)$ *up to the axioms.*

Proof. The proof is done by showing that the function from synch-tables to connectors up to the axioms defined in Theorem 4 is bijective. □

$$\blacktriangledown;\blacktriangle = id \tag{1}$$
$$\nabla;\blacktriangle = \mathbf{0};\overline{\mathbf{0}} \tag{2}$$
$$\blacktriangle;\nabla = \nabla_2;\blacktriangle \otimes \blacktriangle \tag{3}$$
$$\blacktriangle;\mathbf{0} = \mathbf{0}\otimes\mathbf{0} \tag{4}$$
$$\nabla;id\otimes\mathbf{0} = \mathbf{0};\overline{\mathbf{0}} \tag{5}$$
$$\Delta;\mathbf{0} = \mathbf{0}\otimes\mathbf{0} \tag{6}$$
$$\blacktriangledown;!_2 = ! \tag{7}$$
$$\nabla;\blacktriangledown\otimes id = \blacktriangledown;\nabla\otimes\nabla;id\otimes\blacktriangle\otimes id;id\otimes\gamma \tag{8}$$
$$\blacktriangledown;\nabla\otimes id = \nabla;\blacktriangledown\otimes\blacktriangledown;id\otimes\Delta\otimes id;id\otimes\gamma \tag{9}$$
$$\blacktriangle;\blacktriangledown = \blacktriangledown_2;\nabla\otimes\nabla\otimes\nabla\otimes\nabla;id\otimes\blacktriangle\otimes(\blacktriangle;!)\otimes\blacktriangle\otimes id;\gamma\otimes\gamma;id\otimes(\blacktriangle;!)\otimes id \tag{10}$$
$$\text{¡};\blacktriangledown;\blacktriangledown\otimes id = \text{¡}_3;\blacktriangledown\otimes\blacktriangledown\otimes\blacktriangledown;id\otimes\gamma\otimes\gamma\otimes id;\Delta\otimes\Delta\otimes\Delta \tag{11}$$
$$id_2 = \blacktriangledown\otimes\blacktriangledown;id\otimes\Delta\otimes id;id\otimes\nabla\otimes id;\blacktriangle\otimes\blacktriangle \tag{12}$$
$$id_2 = \blacktriangledown\otimes(\text{¡};\blacktriangledown)\otimes\blacktriangledown;id\otimes\gamma\otimes\gamma\otimes id;\Delta\otimes\Delta\otimes\Delta;id\otimes\nabla\otimes id;\blacktriangle\otimes\blacktriangle \tag{13}$$
$$!_n = id_n\otimes\text{¡};id_n\otimes\nabla^n;\blacktriangle_n;!_n \tag{14}$$

Fig. 6. Axioms for mutual exclusion, textually

4.2 Adding the Mutual Exclusion Connector

As we have already seen, connectors in $\mathbf{CC}(\nabla,\Delta,!,\text{¡})$ allow to specify only a small class of tick-tables. In particular, we can express synchronization constraints but not mutual exclusion constraints. This is proved by the fact that the class $\mathbf{CC}(\nabla,\Delta,!,\text{¡})$ has limited expressiveness. For instance, it is not expressive enough to model all CommUnity connectors. To solve that problem we will add the mutual exclusion connector $\blacktriangledown : 2 \to 1$.

Following the analogy with Section 4.1, one may think that also the dual connector $\blacktriangle$ must be explicitly introduced, but this is not strictly required since the complex term $id\otimes(\text{¡};\nabla)\otimes id;id\otimes\blacktriangledown\otimes id_2;id_2\otimes\gamma\otimes id;\Delta\otimes id\otimes\Delta;!\otimes id\otimes!$ exhibits the same behaviour as $\blacktriangle$. Similarly both inaction connectors $\mathbf{0}$ and $\overline{\mathbf{0}}$ can be derived as auxiliary connectors. In fact we have for instance $T(\mathbf{0}) = T(\blacktriangledown;\Delta;!)$.

One may start considering just the axiomatization of choice and inaction connectors, separately w.r.t. synch and hiding connectors. Thus one individuates a gs-monoidal structure $(\blacktriangledown,\mathbf{0})$, a cogs-monoidal structure $(\blacktriangle,\overline{\mathbf{0}})$ and a new-bang structure $(\mathbf{0},\overline{\mathbf{0}})$. Unluckily no simple axiomatization can be found for $(\blacktriangledown,\blacktriangle)$, since they form neither a match-share category since $T(\blacktriangle;\blacktriangledown) \neq T(\blacktriangledown\otimes id;id\otimes\blacktriangle)$ nor an r-monoidal [3] category since $T(\blacktriangle;\blacktriangledown) \neq T(\blacktriangledown\otimes\blacktriangledown;id\otimes\gamma\otimes id;\blacktriangle\otimes\blacktriangle)$.

Thus we resort to a complex axiomatization that deals with all the four classes of connectors at the same time. The axioms are textually written in Figure 6. For simplicity, dual axioms are omitted. Axioms 1–9 are quite simple. The other ones, which are more complex, are depicted in Figure 7 and commented below. The last one, which is actually an axiom scheme, is drawn only for $n = 3$. Axiom 10 deals with commutation of $\blacktriangle$ and $\blacktriangledown$, but w.r.t. the conceptually similar axiom 3, we have to force mutual exclusion on all the paths. Axiom 11 shows that mutual exclusion on three actions can be enforced by imposing mutual exclusion separately on each pair of actions. Axiom 12 shows that given two independent actions we can freely add an action for their synchronized execution and in that case axiom 13 says that we can also force mutual exclusion on the two paths corresponding to the asynchronous execution of the two starting actions. Finally, axiom 14 means that if all the elements of the interfaces of a connector are adjacent to a node in the interface of a connector $\boldsymbol{\Delta};!$ (or of the dual form), then for each denotation we can obtain a concrete correct behaviour by performing an untick on each internal node, thus there is no real constraint on the behaviour of the elements of the interfaces, which can be considered disconnected and closed by an hiding connector.

We present here some useful equivalence lemmas.

Lemma 4. $\boldsymbol{\nabla}_n;\Delta^n \otimes \Delta^n = \Delta^n;\boldsymbol{\nabla}$.

Proof. The proof is by induction on n. □

Lemma 5. *For each connector $\alpha : m \to n$ let $\alpha^c : n \to m$ be its dual connector. We have $\alpha \otimes id_n;\Delta_n;!_n = id_m \otimes \alpha^c;\Delta_m;!_m$.*

Proof. The proof is by induction on the number of basic connectors in α. □

Also for $\mathbf{CC}(\nabla,\Delta,!,\text{¡},\boldsymbol{\nabla})$ we can define a sorted form and a normal form.

Definition 8 (Sorted form). *A connector $\alpha \in \mathbf{CC}(\nabla,\Delta,!,\text{¡},\boldsymbol{\nabla})$ is in sorted form iff:*

$$\alpha \equiv \alpha_{\text{¡}};\alpha_{\bar{\mathbf{0}}};\alpha_{\boldsymbol{\nabla}};\alpha_{\gamma};\alpha_{\boldsymbol{\Delta}};\beta_{\nabla};\beta_{\gamma};\beta_{\boldsymbol{\Delta}};\beta_{\mathbf{0}};\beta_{!}$$

where α_σ and β_σ are layers of σ and $\equiv$ is syntactic identity.

Note that the definition of central point (Definition 6) can be applied also to this new sorted form. Central points can be linearly connected to both *free variables* (i.e., external interfaces) and *hidden variables* (i.e., interfaces of hiding connectors).

Proposition 3. *Any connector $\alpha \in \mathbf{CC}(\nabla,\Delta,!,\text{¡},\boldsymbol{\nabla})$ can be transformed in sorted form using the axioms.*

Proof. The proof strategy is the same of Proposition 2, but here when moving connectors to their final layer, other basic connectors may be created, and in this case we have to check that the procedure indeed terminates.

In most cases the proof can be done by induction on the width of layer that must be traversed, that is on the maximum number of connectors in all the paths of the layer. The complex case is the one dealing with connectors ∇ and $\boldsymbol{\Delta}$.

The only risk to cycle is when a ∇ connector while traversing layer $\alpha_{\boldsymbol{\nabla}}$ creates a $\boldsymbol{\Delta}$ connector that while traversing layer α_{Δ} creates again a $\boldsymbol{\nabla}$ connector that goes back to the original ∇ connector. One can see that this may happen only if there are two different paths starting from the same ∇ connector, one for each element of its right interface, which arrive to the same Δ connector. These paths can be deleted by isolating either the connector $\nabla;\boldsymbol{\nabla} \otimes \boldsymbol{\nabla};id \otimes \Delta \otimes id$ or the connector $\boldsymbol{\nabla};\Delta$, which can be transformed using the axioms respectively into $\boldsymbol{\nabla};\nabla \otimes id;id \otimes \gamma$ and into $\mathbf{0};\bar{\mathbf{0}}$. □

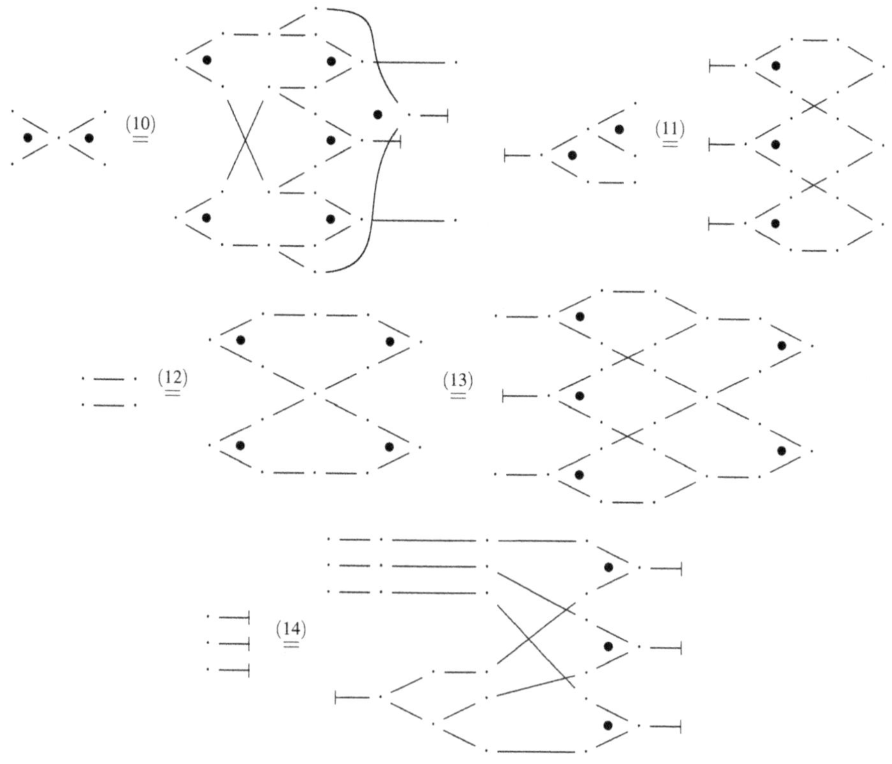

Fig. 7. Complex axioms for mutual exclusion, graphically

Definition 9 (Normal form). *A connector* $\alpha \in \mathbf{CC}(\nabla, \Delta, !, ¡, \blacktriangledown)$ *is in normal form if and only if:*

1. *α has the form $\alpha_{¡};\alpha_{\mathbf{0}};\alpha_{\blacktriangledown};\alpha_{\gamma};\alpha_{\Delta};\beta_{\nabla};\beta_{\gamma};\beta_{\blacktriangle};\beta_{\bar{\mathbf{0}}};\beta_{!}$;*
2. *hiding connectors are adjacent to either roots of mex trees or central points;*
3. *there exists at most one path between a fixed central point and a fixed variable;*
4. *no two central points are linearly connected to exactly the same set of variables;*
5. *each central point is linearly connected to at least one free variable;*
6. *each hidden variable is linearly connected to at most two central points;*
7. *no two hidden variables are linearly connected to the same set of central points;*
8. *each pair of central points associated with disjoint sets of free variables is linearly connected to an hidden variable;*
9. *hidden variables are on the left of central points, unless they are adjacent to them.*

Theorem 6. *Any connector* $\alpha \in \mathbf{CC}(\nabla, \Delta, !, ¡, \blacktriangledown)$ *can be transformed in normal form using the axioms.*

Proof. Thanks to Proposition 3, α can be transformed in sorted form. Then the conditions can be satisfied one at the time using the axioms. □

Again, there is a precise correspondence between normal forms and tick-tables.

Theorem 7. *For any* tick-*table T with $T[0,0] = \checkmark$, we can build a connector α in normal form such that $T(\alpha) = T$. Moreover the construction is unique up to the axioms of symmetric monoidal categories and of associativity and commutativity of synch and choice connectors.*

Proof. Given a tick-table T, the connector α is realized in the following way:

- variables that always have value 0 are connected to $\mathbf{0}$ or $\overline{\mathbf{0}}$ connectors;
- for other input (resp. output) variables, we build a tree of ▼ (resp. ▲) with as many leaves as the number of checked cells that have that variable in the domain;
- for each pair of central points with disjoint sets of free variables, we create an hidden variable for them;
- for each checked cell in the table, we create a central point with two outgoing trees of synch connectors. The number of leaves is the number of free variables (input on the left, output on the right) in the domain, plus (on the left) the number of hidden variables associated to the central point;
- we connect leaves of synchronization trees with leaves of mutual exclusion trees using permutations, connecting each central point to the associated variables. □

The theorem below establishes a bijective correspondence between denotations and standard implementations of connectors.

Theorem 8. *We have a bijective correspondence between* tick-*tables T with $T[0,0] = \checkmark$ and connectors in* $\mathbf{CC}(\nabla, \Delta, !, \text{¡}, ▼)$ *up to the axioms.*

Proof. Analogously to Theorem 5, the proof is done by showing that the function from tick-tables to connectors up to the axioms defined in Theorem 7 is bijective. □

These results can be used to extend the research in [2], where a mapping from CommUnity to the Tile Model was presented, and the main result was that the translation of a CommUnity diagram is tile bisimilar to the translation of its colimit. Using the correspondence between observational semantics and connectors up to the axioms we can state that the (synchronization part of the) translation of a CommUnity diagram is equal up to the axioms to the (synchronization part of the) translation of its colimit. More in general, colimit computation in the categorical approach is now strongly related to normalization using suitable axioms in the algebraic approach.

5 Conclusion and Future Work

We have presented different classes of connectors and we have shown how they can be analyzed from different points of view: their concrete structures can be described by graphs, their operational and observational semantics are given using tiles and tile bisimilarity, while the representation based on tick-tables provides a denotational semantics. We have proved that there is a bijective correspondence among connectors up to axioms, classes of bisimilar connectors and denotations. This allows to extend the result of [2] proving that there is a correspondence between colimit computation in a categorical framework and normalization up to the axioms in an algebraic framework.

Our work leaves an open problem: we argue that the axiom schema 14 (see Figure 7) is needed for the completeness of the axiomatization, but we do not know whether a different finite axiomatization of $\mathbf{CC}(\nabla, \Delta, !, ¡, \mathbb{\nabla})$ exists or not.

As future work we plan to study the complexity of our reduction to normal form. Furthermore we want to generalize our connectors to a setting where we have a richer set *Act* of actions ruled by a synchronization algebra, instead of just two possible observations (tick/untick). Another interesting extension is given by probabilistic connectors.

References

1. J.A. Bergstra, C.A. Middelburg, and G. Stefanescu. Network algebra for asynchronous dataflow. *International Journal of Computer Mathematics*, 65:57–88, 1997.
2. R. Bruni, J.L. Fiadeiro, I. Lanese, A. Lopes, and U. Montanari. New insights on architectural connectors. In *Proc. IFIP TCS 2004*, pp. 367–379. Kluwer Academics, 2004.
3. R. Bruni, F. Gadducci, and U. Montanari. Normal forms for algebras of connections. *Theoret. Comput. Sci.*, 286(2):247–292, 2002.
4. R. Bruni, I. Lanese, and U. Montanari. Normal forms for stateless connectors. Tech. Rep. TR-05-11, Computer Science Department, University of Pisa, Italy.
5. V.E. Cazanescu and G. Stefanescu. Towards a new algebraic foundation of flowchart scheme theory. *Fundamenta Informaticae*, 13:171–210, 1990.
6. A. Corradini and F. Gadducci. An algebraic presentation of term graphs, via gs-monoidal categories. *Applied Categorical Structures*, 7:299–331, 1999.
7. A. Corradini and U. Montanari. An algebraic semantics for structured transition systems and its application to logic programs. *Theoret. Comput. Sci.*, 103:51–106, 1992.
8. P. Degano and U. Montanari. A model for distributed systems based on graph rewriting. *Journal of the ACM*, 34(2):411–449, 1987.
9. H. Ehrig, M. Pfender, and H. J. Schneider. Graph grammars: an algebraic approach. In *Proc. IEEE Conference on Automata and Switching Theory*, pp. 167–180, 1973.
10. J. L. Fiadeiro. *Categories for Software Engineering*. Springer, 2004.
11. J.L. Fiadeiro, A. Lopes, and M. Wermelinger. A mathematical semantics for architectural connectors. In *Generic Programming*, *LNCS* 2793, pp. 190–234. Springer, 2003.
12. F. Gadducci and U. Montanari. The tile model. In *Proof, Language and Interaction: Essays in Honour of Robin Milner*, pp. 133–166. MIT Press, 2000.
13. J. A. Goguen. Categorical foundations for general systems theory. In *Advances in Cybernetics and Systems Research*, pp. 121–130. Transcripta Books, 1973.
14. C.A.R. Hoare. *CSP – Communicating Sequential Processes*. International Series in Computer Science. Prentice-Hall, 1985.
15. P. Katis, N. Sabadini, and R.F.C. Walters. Bicategories of Processes. *Journal of Pure and Applied Algebra*, 115:141–178, 1997.
16. Y. Lafont. Interaction combinators. *Inform. and Comput.*, 137(1):69–101, 1997.
17. K.G. Larsen and L. Xinxin. Compositionality through an operational semantics of contexts. In *Proc. ICALP'90*, *LNCS* 443, pp. 526–539. Springer, 1990.
18. S. MacLane. *Categories for the Working Mathematician*. Springer, 1971.
19. J. Meseguer. Conditional rewriting logic as a unified model of concurrency. *Theoret. Comput. Sci.*, 96:73–155, 1992.
20. R. Milner. *A Calculus of Communicating Systems*, *LNCS* 92. Springer, 1989.

21. R. Milner. Turing, computation and communication. Turing anniversary lecture, 1997.
22. R. Milner. Bigraphical reactive systems. In *Proc. CONCUR 2001*, *LNCS* 2154, pp. 16–35. Springer, 2001.
23. G.D. Plotkin. A structural approach to operational semantics. Tech. Rep. DAIMI FN-19, Aarhus University, 1981.
24. A. Rensink. Bisimilarity of open terms. *Inform. and Comput.*, 156(1/2):345–385, 2000.
25. G. Stefanescu. *Network Algebra.* Discrete Math. and Theoret. Comp. Sci., Springer, 2000.

On the Semantics of Coinductive Types in Martin-Löf Type Theory

Federico De Marchi*

Department of Mathematics, University of Utrecht,
Utrecht, P.O. Box 80010, 3508 TA Utrecht, The Netherlands

Abstract. There are several approaches to the problem of giving a categorical semantics to Martin-Löf type theory with dependent sums and products and extensional equality types. The most established one relies on the notion of a type-category (or category with attributes) with Σ and Π types. We extend such a semantics by introducing coinductive types both on the syntactic level and in a type-category. Soundness of the semantics is preserved.

As an example of such a category, we prove that the type-category built over a locally cartesian closed category $\mathcal{C}$ admits coinductive types whenever $\mathcal{C}$ has final coalgebras for all polynomial functors.

1 Introduction

The problem of finding a categorical semantics to Martin-Löf type theory has given rise to a substantial amount of very interesting research, over the years. Most of it was inspired by Seely's first attempt to use locally cartesian closed categories [20], which however was proved slightly inaccurate. In fact, he had glossed over the need to have a choice of pullbacks that compose on the nose in the category, which is essential in order to interpret substitution. In order to fix the problem, Cartmell devised the notion of a contextual category [7], on which Streicher based his semantics for a dependent type theory with dependent products and sums, and extensional equality [21]. However, the axioms for a contextual category are very "uncategorical" in spirit, since they assume a well-founded order on objects; for this reason, this notion was later replaced by the more abstract one of category with attributes [11], or type-category, as Pitts calls them [19].

From a fibrational point of view, a type-category is just a split fibration. Adapting an argument of Bénabou [6], Hofmann could show that every locally cartesian closed category $\mathcal{C}$ gives rise to a split fibration which is equivalent (in a suitable higher-order sense) to the canonical indexing of $\mathcal{C}$. In this way, he could fix the bug in Seely's paper by interpreting type theory in the type-category built out of any locally cartesian closed one.

* This research is supported by NWO Grant n. 613.000.222.
Email: marchi@math.uu.nl

J.L. Fiadeiro et al. (Eds.): CALCO 2005, LNCS 3629, pp. 114–126, 2005.

In [23], Benno van den Berg and the present author suggested that such a semantics could be extended to one for a theory with coinductive types, provided the original locally cartesian closed category had M-types (i.e. final coalgebras of polynomial functors). It is the purpose of this paper to make that statement precise.

The problem of adding infinite objects to Martin-Löf type theory has been considered by several authors over the years [8,16,17]. The main source of trouble, in this case, is that infinite (non-well-founded) objects in a type might have infinitely long reductions, therefore making a full description of their normal form impossible to achieve. When in presence of well-founded types, one can give a description of an infinite object by its finite approximations [17,10,14,23,1]; however, in our setting we want to avoid using well-founded types. In Section 2 we shall introduce a system of rules for coinductive types. These are very close to the categorical formulation of the properties of the final coalgebra for a polynomial functor. In particular, our way of introducing terms of a coinductive type is by unfolding a coalgebra at a particular state. This is analogue to the concept of a productive definition of a term as discussed by Coquand in [8]. The guardedness he requires there is given in our context by the polynomial functor itself.

In Section 3 we recall the concept of a type-category and introduce that of a coinductive type therein, and we show how these categories provide a sound categorical semantics for the aforementioned type theory. Finally, in Section 4 we show that, following Hofmann's construction [11], any locally cartesian closed category with final coalgebras of polynomial functors gives rise to a type-category with coinductive types, thus providing a wide class of examples.

Acknowledgements. The author would like to thank Benno van den Berg for the several useful discussions, and to Dr. Thorsten Altenkirch for his valuable suggestions.

2 Coinductive Types in Martin-Löf Type Theory

We consider a version of Martin-Löf type theory with Σ-types, Π-types, and extensional equality, as presented in [15,18]. When one views the theory as a programming language, according to the "types-as-specification" paradigm, it may be desirable to allow for some programs to have an infinite computation. When using type theory to study constructive mathematics, coinductive types can help modelling some non-well-founded sets [2]. In presence of inductive types, infinite programs can be fully described by the collection of their finite (but arbitrarily long) approximations [17,16,9,14]. Alternatively, one can describe the elements of a non-well-founded set by means of a recursive definition, provided this is guarded, or productive [8,10]. In our axioms, we shall resort to the second method, since we do *not* assume to have W-types.

Categorically, it is clearly understood that non-well-founded terms over a signature can be collected into the final coalgebra for a polynomial functor associated to it [13,22,5,3]. In order to be closer to the categorical semantics we intend to present, we give axioms here, which closely resemble those of a final coalgebra.

It is to be remarked that our definitions make equality undecidable in the system. However, the fact that we take extensional equality into account already breaks decidability, and we choose to accept this drawback. We are aware of recent work by Altenkirch on "tracking the proofs" of equality of various terms within a version of intensional type theory which he calls "observational type theory" [4]. This seems to be a promising area for further developments in the direction of actual implementations of the theory.

Now, we come to the axioms. The introduction rule takes the obvious form

$$[M\text{-FORM}] \qquad \frac{A \ \mathit{type} \qquad B(a) \ \mathit{type} \ [a : A]}{M(A,B) \ \mathit{type}}$$

We should think of the elements of type A as term constructors, the arity of $a : A$ being given by the type $B(a)$. Note that, following the conventions of [18], we are omitting those contexts which are not discharged by the rule, and we are omitting the obvious substitution rules that should come together with the introduction of a new type constructor. Given types A and $B(a)$ $[a : A]$ as above, we shall often write $P_b(X)$ for the type $(\Sigma a : A)(B(a) \to X)$.

Elements of the coinductive type $M(A,B)$ are defined corecursively. Whenever we have a type X and an element of the function type $f : X \to P_b(X)$, we can think of it as a way of describing the *evolution* of the elements of X according to the signature described by A and B. In particular, every element $x : X$ is *productive*, since $f(x).1$ is an element of type A (i.e. a term constructor of the specified signature) and $f(x).2$ takes any $b : B(f(x).1)$ to another element in X, from which we can reiterate the procedure indefinitely. In this way, we can associate to each element $x : X$ a (possibly non-well-founded) tree whose nodes are labelled by elements of type A and branches departing from a node labelled by a are labelled by elements of type $B(a)$. This is the *unfolding* of f at x, or the *behaviour* of x under the evolution f.

$$[M\text{-INTRO}] \qquad \frac{x : X \qquad f : X \to P_b(X)}{\mathsf{unfold}(f,x) : M(A,B)}$$

When we are given an element t of the coinductive type $M(A,B)$, which we think of as a tree, we can extract its *root*, which is an element of A, and its *branching function*, which has type $B(\mathsf{root}(t)) \to M(A,B)$.

$$[M\text{-ELIM}] \qquad \frac{t : M(A,B)}{\mathsf{root}(t) : A} \qquad \frac{t : M(A,B)}{\mathsf{br}(t) : B(\mathsf{root}(t)) \to M(A,B)}$$

In order to understand how root and br act on an element $t : M(A,B)$, suppose t is the tree representing the behaviour of some $x : X$ under $f : X \to P_b(X)$; that is, $t = \mathsf{unfold}(f,x)$. Then, $f(x)$ is an element of $P_b(X) \equiv (\Sigma a : A)(B(a) \to X)$, which is a pair (a,s). These are precisely the root and the branching function of the original t. This explanation, justifies the following equality rules:

$$[M\text{-EQ}]\quad \frac{x : X \qquad f : X \to P_b(X)}{\mathsf{root}(\mathsf{unfold}(f,x)) = f(x).1 \,:\, A}$$

$$\frac{x : X \qquad\qquad f : X \to P_b(X)}{\mathsf{br}(\mathsf{unfold}(f,x)) = (b)\mathsf{unfold}(f,(f(x).2)b) \,:\, B(f(x).1) \to M(A,B)}$$

$$\frac{t : M(A,B)}{t = \mathsf{unfold}((y)(\mathsf{root}(y),\mathsf{br}(y)),t) \,:\, M(A,B)}$$

By the definitions of root and br, it follows immediately that the abstraction $m\equiv(y)(\mathsf{root}(y),\mathsf{br}(y))$ has type $M(A,B) \to P_b(M(A,B))$. Therefore, for any $t : M(A,B)$ we can unfold m at t. The third equality rule above is stating precisely that $\mathsf{unfold}(m,t) = t$. This rule is essential in proving that, given any function $f : X \to P_b(X)$, the function $(x)\mathsf{unfold}(f,x)$ is the *unique* one to $M(A,B)$ which preserves the root and branching functions.

3 Type-Categories and Coinductive Types

Now that we have introduced the axioms of our type theory, we can approach the question of giving it a categorical semantics. It has already been mentioned that Seely's idea of using locally cartesian closed categories (lccc's) for modelling dependent types is very insightful, but not correct [20]. His idea was to interpret contexts by objects in the category, and a judgement of the form $A(x_1,\ldots,x_n)$ *type* $[x_1 : X_1,\ldots,x_n : X_n]$ by an arrow $\alpha : A \longrightarrow X$ over the object interpreting the context $[x_i : X_i]$. Given another context $[y_1 : Y_1,\ldots,y_m : Y_m]$ interpreted by an object Y, and an n-tuple of terms $f_i(y_1,\ldots,y_m) : X_i$, this is interpreted by an arrow $f : Y \longrightarrow X$, and the interpretation of the substituted term $A[f_i/x_i]$ is the pullback of α along f. Now, supposing we are given two composable substitutions, interpreted by maps

$$Z \xrightarrow{g} Y \xrightarrow{f} X,$$

the interpretation of the type $(A[f_i/x_i])[g_j/y_j]$ obtained by first performing the substitution f and then the substitution g, should be the same as the interpretation of the type $A[f_i[g_j/y_j]/x_i]$, obtained by performing the composite substitution on A. In other words, the pullback of α first along f and then along g ought to be the same as the pullback along the composite gf. Unfortunately, in general we cannot make a coherent choice of pullbacks in an lccc which is closed under pullback pasting. Hence, the need for a more refined model, in which substitution can be traced more accurately.

The first attempt in this direction was that of Cartmell [7], who proposed in his PhD thesis the notion of a contextual category. This has been further studied by Streicher [21]; however, Cartmell himself, and later other authors, found that contextual categories have a rather technical and cumbersome definition, that could be left aside, in favour of what have been called *categories with attributes*, or *type-categories* [19,11].

A type-category, in the notation of Pitts, is specified by a category $\mathcal{C}$ with a terminal object 1, together with the following extra structure:

- for each object X in $\mathcal{C}$, a set $Type_{\mathcal{C}}(X)$ of *X-indexed types*;
- for each X in $\mathcal{C}$, a map $p : Type_{\mathcal{C}}(X) \longrightarrow \mathrm{Ob}(\mathcal{C}/X)$ which takes an X-indexed type A to the *canonical projection*

$$\pi_A : X \ltimes A \longrightarrow X$$

  from the *total object* $X \ltimes A$ of A to X itself;
- for each map $f : Y \longrightarrow X$ in $\mathcal{C}$, an operation assigning to each X-indexed type A a Y-indexed type f^*A, called the *pullback of A along f*, together with a morphism

$$f \ltimes A : Y \ltimes f^*A \longrightarrow X \ltimes A$$

  making the following into a pullback:

$$\begin{array}{ccc} Y \ltimes f^*A & \xrightarrow{f \ltimes A} & X \ltimes A \\ {\scriptstyle \pi_{f^*A}} \downarrow & & \downarrow {\scriptstyle \pi_A} \\ Y & \xrightarrow[f]{} & X. \end{array} \tag{1}$$

These data are subject to the following coherence conditions, for $A \in Type_{\mathcal{C}}(X)$, $f : Y \longrightarrow X$ and $g : Z \longrightarrow Y$:

$$\begin{array}{ccc} \mathsf{id}_X^* A = A & \text{and} & \mathsf{id}_X \ltimes A = \mathsf{id}_{X \ltimes A}; \\ g^*(f^*A) = (fg)^*A & \text{and} & (f \ltimes A)(g \ltimes f^*A) = (fg) \ltimes A. \end{array}$$

Example (The syntactic category of a theory). We shall provide a wide class of examples of type-categories in the next section. For the time, it is useful to notice that any dependent type theory $\mathbb{T}$ gives rise to a type-category $\mathcal{T}$. Objects in $\mathcal{T}$ are equivalence classes of well-formed contexts in the theory, modulo the relation determined by provable equality of two contexts. An arrow f from (the equivalence class of) a context $Y = [y_j : Y_j]$ $(j = 1, \ldots, m)$ to $X = [x_i : X_i]$ $(i = 1, \ldots, n)$ consists of the equivalence class (again, modulo provable equality) of an n-tuple of terms $f_i(y_1, \ldots, y_m) : X_i$. The final object in $\mathcal{T}$ is clearly given by the empty context. The family $Type_{\mathcal{T}}(X)$ consists of all those types A for which the judgement A *type* $[x_i : X_i]$ is derivable in $\mathbb{T}$, the canonical projection of such a type being the projection

$$(x_i)_{i=1,\ldots,n} : [x_1 : X_1, \ldots, x_n : X_n, x : A] \longrightarrow [x_1 : X_1, \ldots, x_n : X_n].$$

The reindexing along a context morphism $f : Y \longrightarrow X$ of a type $A(x_1, \ldots, x_n)$ depending on the context X, is the type obtained by substituting the x_i's by the terms specified by f:

$$f^*A(y_1, \ldots, y_m) = A[f_1/x_1, \ldots, f_n/x_n].$$

It is clear that these data satisfy the conditions for a type-category, which is called the *syntactic category* built over $\mathbb{T}$.

When interpreting a type theory in a type-category $\mathcal{C}$, objects of $\mathcal{C}$ are used to represent well-formed contexts of the theory, whereas arrows are used to interpret substitution; that is, tuples of terms of the appropriate types, which depend on the variables defined in the domain. If we interpret a context X by an object (which we denote again by X, abusing the notation), then a judgement of the form A *type* $[X]$ is interpreted by an element $A \in Type_{\mathcal{C}}(X)$. If $f : Y \longrightarrow X$ is a substitution, then the pullback f^*A will interpret the type A with the variables substituted according to f. The coherence conditions expressed above ensure that composite substitutions are correctly interpreted. The total object of a type A depending on a context X interprets the context $[X, a : A]$, with the obvious projection onto X (see [19,11] for further details). Finally, terms of a given type are interpreted by sections of the canonical projection; that is, a judgement $t : A\ [X]$ is interpreted by a morphism $t : X \longrightarrow X \ltimes A$ in $\mathcal{C}$ such that $\pi_A t = \mathsf{id}_X$. Given a substitution $f : Y \longrightarrow X$ between contexts, and a term t as above, the term obtained by substituting all the variables in t according to f is interpreted by the unique section f^*t of π_{f^*A} such that $(f \ltimes A) f^*t = tf$, which is determined by the pullback (1).

Note that the map $p : Type_{\mathcal{C}}(X) \longrightarrow \mathrm{Ob}(\mathcal{C}/X)$ induces the structure of a category on the collection of types over X, in an obvious way: maps between two elements A and B are maps in the slice category $\mathcal{C}/X$ between the canonical projections π_A and π_B. The map p then becomes a full and faithful functor from $Type_{\mathcal{C}}(X)$ to $\mathcal{C}/X$. Moreover, for a $\mathcal{C}$-morphism $f : Y \longrightarrow X$ the pullback functor $f^* : \mathcal{C}/X \longrightarrow \mathcal{C}/Y$ restricts to a functor $f^* : Type_{\mathcal{C}}(X) \longrightarrow Type_{\mathcal{C}}(Y)$, whose action on objects is precisely the one specified by the type-category structure. The association $X \mapsto Type_{\mathcal{C}}(X)$ and $f \mapsto f^*$ defines a functor $\mathcal{C} \to \mathcal{C}at$. Functoriality is ensured by the coherence conditions for the pullback functors, and it is precisely the condition needed in order for the substitution to be correctly interpreted. We could not use the slice categories $\mathcal{C}/X$ because they give rise to a *pseudo*-functor, and the action of the pullback functors composes only up to isomorphism.

Now, suppose the left adjoint Σ_f to the pullback functor $f^* : \mathcal{C}/X \longrightarrow \mathcal{C}/Y$ restricts to categories of types as well. Then, it is possible to interpret dependent sums in our model. If A *type* $[X]$ and $B(a)$ *type* $[X, a : A]$ are judgements in the theory, interpreted by objects A in $Type_{\mathcal{C}}(X)$ and B in $Type_{\mathcal{C}}(X \ltimes A)$, then the composite $\Sigma_{\pi_A} \pi_B = \pi_A \pi_B$ is the canonical projection of an object in $Type_{\mathcal{C}}(X)$, which we define to be the interpretation of the type $\Sigma(A, B)$ in the context X.

Likewise, we can interpret dependent products in $\mathcal{C}$ provided the pullback functors have a right adjoint. More specifically, we need that for A in $Type_{\mathcal{C}}(X)$ and B in $Type_{\mathcal{C}}(X \ltimes A)$ there is an indexed type $\Pi(A, B)$ in $Type_{\mathcal{C}}(X)$ and a morphism $ap_{A,B} : \pi_A^* \Pi(A, B) \longrightarrow B$ in $Type_{\mathcal{C}}(X \ltimes A)$ with the obvious universal property. Stability of the interpretation under substitution is ensured by the further requirements that, for any $f : Y \longrightarrow X$ in $\mathcal{C}$,

$$f^* \Pi(A, B) = \Pi(f^*A, f^*B) \quad \text{and} \quad (f \ltimes A)^* ap_{A,B} = ap_{f^*A, (f \ltimes A)^* B}.$$

We refer the reader to Streicher's monograph [21] for a treatment of extensional equality types.

Once we have defined an interpretation of the type-valued and term-valued function symbols of a theory into a type-category, we can use the rules for dependent products and coproducts in order to inductively define an interpretation of all well-formed contexts, of types depending on a context, and of their terms. We shall then say that a judgement of the form A *type* $[X]$ is satisfied by the model if A is interpreted by an element of $Type_{\mathcal{C}}(X)$, and similarly for one of the form $a : A\ [X]$. Equality judgements will be satisfied when the two sides of the equality have the same interpretation in $\mathcal{C}$. The properties of dependent products and coproducts ensure that this model is sound, in the sense that any judgement derivable in the type theory is satisfied by any interpretation (see [19] for more details).

Example (Interpretation in the syntactic category). Given a type theory $\mathbb{T}$, this has an obvious interpretation into its syntactic category $\mathcal{T}$. A context is interpreted by its equivalence class, a judgement of the form A *type* $[X]$ is interpreted by the element A in $Type_{\mathcal{T}}(X)$, and a term $t : A\ [X]$ is interpreted by the section

$$(x_1, \ldots, x_n, t(x_1, \ldots, x_n)) : [x_1 : X_1, \ldots, x_n : X_n] \longrightarrow [x_1 : X_1, \ldots, x_n : X_n, x : A]$$

of the canonical projection of A. It is clear that a judgement is provable in $\mathbb{T}$ if an only if it is satisfied by the model.

We now proceed to specify the amount of structure needed in order to interpret coinductive types. Unsurprisingly, the properties closely resemble the type theoretic rules described in Section 2.

Definition 1. A type-category $\mathcal{C}$ with dependent sums, dependent products and extensional equality (in the sense of [11]), *has coinductive types* if for any A in $Type_{\mathcal{C}}(X)$ and B in $Type_{\mathcal{C}}(X \ltimes A)$ there is a type

$$M(A, B) \text{ in } Type_{\mathcal{C}}(X)$$

with the following properties:

- for any type Y in $Type_{\mathcal{C}}(X)$ and sections y of π_Y and g of $\pi_{Y \to P_b(Y)}$ there is a section
$$unfold(g, y) \ \text{ of } \ \pi_{M(A,B)};$$
- for any section t of $\pi_{M(A,B)}$ there are sections
$$root(t) \ \text{ of } \ \pi_A \ \text{ and } \ br(t) \ \text{ of } \ \pi_{root(t)^* B \to M(A,B)}$$
- and the following are equal sections of π_A, $\pi_{(g(y).1)^* B \to M(A,B)}$ and $\pi_{M(A,B)}$, respectively:

$$\begin{aligned} root(unfold(g, y)) &= g(y).1 \\ br(unfold(g, y)) &= (b)unfold(g, (g(y).2)b) \\ unfold((s)(root(s), br(s)), t) &= t; \end{aligned}$$

– for a morphism $f : X' \longrightarrow X$ in $\mathcal{C}$, the following coherence condition holds:

$$f^* M(A, B) = M(f^* A, f^* B),$$

together with the analogous conditions for the aforementioned sections.

It is immediate from the definition how to interpret a Martin-Löf type theory with coinductive types in any type-category which has coinductive types, and soundness readily extends.

Theorem 2 (soundness). *With the notion of satisfaction described above, the collection of judgements satisfied by a model in a type-category with coinductive types is closed under the rules of Σ-types, Π-types, extensional equality as well as those given in Section 2 for coinductive types.*

Moreover, it is an immediate consequence of the definitions that the syntactic category of a type theory $\mathbb{T}$ with coinductive types has the structure of a type-category with coinductive types, which provides a complete semantics for $\mathbb{T}$, in the sense that a judgement is provable in the theory if and only if it is satisfied by the model.

4 From M-Types to Coinductive Types

As we mentioned in the introduction, the reason for introducing type-categories was that of overcoming the problem of not being able to make a coherent choice of pullbacks in a locally cartesian closed category, in such a way that the pullback functors between the slice categories would compose on the nose (instead of up to isomorphism). Another way to phrase the problem is in terms of fibrations. The association $X \mapsto \mathcal{C}/X$ defines a pseudofunctor $\mathcal{C}^{\mathrm{op}} \to \mathit{Cat}$ (it is a pseudofunctor precisely because composition of the pullback functors is possible only up to coherent isomorphisms), which is also called the *canonical indexing* of $\mathcal{C}$. It is a way to view $\mathcal{C}$ as an indexed category, or, equivalently, as a cloven fibration (for definitions, see for example [12]). Moreover, the strict functoriality on the composition of the pullback amounts exactly to saying that this fibration is split. So, locally cartesian closed categories fail to support an interpretation of type theory because their canonical indexing is not a split fibration.

However, Bénabou had described a method of associating to any fibration an equivalent split one (i.e. a split fibration on the same base category, whose fibres are pointwise equivalent to those of the given one) [6]. Hofmann found an application of his work in the present context [11], and described an explicit way of associating a type-category $\widehat{\mathcal{C}}$ to any lccc $\mathcal{C}$, in such a way that $\mathit{Type}_{\widehat{\mathcal{C}}}(X)$ is equivalent to $\mathcal{C}/X$ for every object X in $\mathcal{C}$.

The underlying category of $\widehat{\mathcal{C}}$ is again $\mathcal{C}$. For an object X in $\mathcal{C}$, the collection $\mathit{Type}_{\widehat{\mathcal{C}}}(X)$ consists of those functors $F : \mathcal{C}/X \longrightarrow \mathcal{C}^{\rightarrow}$ which take any morphism in $\mathcal{C}/X$ to a pullback square, and such that the codomain of $F(f)$ is the domain of f for any object f in $\mathcal{C}/X$. Here, the notation $\mathcal{C}^{\rightarrow}$ indicates the category whose objects are arrows in $\mathcal{C}$ and morphisms are commuting squares.

Given an object X in $\mathcal{C}$ and a functor $F \in Type_{\widehat{\mathcal{C}}}(X)$, we define its canonical projection $p(F)$ to be $\pi_F = F(\mathsf{id}_X)$, which is a map over X. Its domain will be the total object $X \ltimes F$. Finally, given a morphism $f : Y \longrightarrow X$ in $\mathcal{C}$, the pullback of F along f is given by the functor f^*F which takes an object g over Y to the arrow $F(fg)$. The map $f \ltimes F : Y \ltimes f^*F \longrightarrow X \ltimes F$ is the top arrow in the following square

$$\begin{array}{ccc} Y \ltimes f^*F & \xrightarrow{f \ltimes F} & X \ltimes F \\ {\scriptstyle F(f)=\pi_{f^*F}} \downarrow & & \downarrow {\scriptstyle \pi_F = F(\mathsf{id}_X)} \\ Y & \xrightarrow[f]{} & X, \end{array}$$

which is a pullback because it is the action of the functor F on the arrow f in $\mathcal{C}/X$. It is not hard to show that these data satisfy the necessary coherence conditions, therefore they define a type-category, which has the universal property that $Type_{\widehat{\mathcal{C}}}(X)$ is equivalent to $\mathcal{C}/X$ for every object X in $\mathcal{C}$, one direction of the equivalence being the canonical projection functor $p : Type_{\widehat{\mathcal{C}}}(X) \longrightarrow \mathcal{C}/X$.

Furthermore, any extra structure on $\mathcal{C}$ induces some structure in $\widehat{\mathcal{C}}$, with the exception of impredicative universes, as Hofmann explains. In particular, the presence of left and right adjoints to the pullback functors in $\mathcal{C}$ ensures that $\widehat{\mathcal{C}}$ has dependent sums and products, respectively, whereas the existence of equalisers determines the existence of extensional equality types. For example, for elements A in $Type_{\widehat{\mathcal{C}}}(X)$ and B in $Type_{\widehat{\mathcal{C}}}(X \ltimes A)$, the total objects for $\Sigma(A,B)$ and $\Pi(A,B)$ are

$$\pi_{\Sigma(A,B)} = \Sigma_{\pi_B}\pi_A = X \ltimes A \ltimes B \xrightarrow{\pi_B} X \ltimes A \xrightarrow{\pi_A} X$$

and

$$\pi_{\Pi(A,B)} = \Pi_{\pi_A}\pi_B : X \ltimes \Pi(A,B) \longrightarrow X.$$

Having dependent products and coproducts, we can form, for any Y in $Type_{\widehat{\mathcal{C}}}(X)$ and A and B as above, the element

$$P_b(Y) = \Sigma(A, \Pi(B, \pi_B^* \pi_A^* Y))$$

which sits again in $Type_{\widehat{\mathcal{C}}}(X)$.

Lemma 3. *Given A and B as above, the association $Y \mapsto P_b(Y)$ defines a polynomial functor on $Type_{\widehat{\mathcal{C}}}(X)$.*

Proof. Let us first remind that, given a map $f : B \longrightarrow A$ in a locally cartesian closed category $\mathcal{C}$, the polynomial endofunctor on $\mathcal{C}$ associated to f is defined by

$$P(X) = \Sigma_A(A \times X \xrightarrow{\pi_A} A)^{(B \xrightarrow{f} A)},$$

where the exponential is taken in the slice category $\mathcal{C}/A$. We have already mentioned that the functor $p : Type_{\widehat{\mathcal{C}}}(X) \longrightarrow \mathcal{C}/X$ defines an equivalence. In particular, this means that $Type_{\widehat{\mathcal{C}}}(X)$ is locally cartesian closed; hence, it makes sense to talk about polynomial functors on it. Moreover, if we show that the canonical projection of $P_b(Y)$ is a polynomial expression over π_Y, the same will hold in $Type_{\widehat{C}}(X)$, because of the equivalence, and the result will be proved.

Using the descriptions above for dependent products and coproducts, we get the following chain of equalities in $\mathcal{C}/X$:

$$\begin{aligned} p(\Sigma(A, \Pi(B, \pi_B^* \pi_A^* Y))) &= \Sigma_{\pi_A}(p(\Pi(B, \pi_B^* \pi_A^* Y))) \\ &= \Sigma_{\pi_A} \Pi_{\pi_B}(\pi_B^* \pi_A^* p(Y)) \\ &= \Sigma_{\pi_A}((\pi_A^* p(Y))^{\pi_B}). \end{aligned}$$

Now, note that π_A is the unique map from π_A to the terminal object id_X in $\mathcal{C}/X$, and $\pi_A^* p(Y)$ is the first projection from $\pi_A \times p(Y)$; hence, we can rewrite $p(\Sigma(A, \Pi(B, \pi_B^* \pi_A^* Y)))$ as

$$\Sigma_{\pi_A}((\pi_A \times p(Y) \to \pi_A)^{\pi_B}) = P_{\pi_B}(p(Y)).$$

Functoriality of this expression in Y is an easy check, which closes the proof. □

Using the previous result, we can now deduce the existence of coinductive types in the type category associated to an lccc with *M-types* (i.e. final coalgebras of polynomial functors, [23]).

Theorem 4. *The type category $\widehat{\mathcal{C}}$ associated to a locally cartesian closed category $\mathcal{C}$ with M-types has coinductive types.*

Proof. Let $A \in Type_{\widehat{C}}(X)$ and $B \in Type_{\widehat{C}}(X \ltimes A)$. Then, by Lemma 3, the mapping

$$Y \mapsto P_b(Y)$$

defines a polynomial functor over $Type_{\widehat{C}}(X) \simeq \mathcal{C}/X$. By assumption, this functor has a final coalgebra

$$m : M(A, B) \longrightarrow \Sigma(A, B \to \pi_A^*(M(A, B))), \tag{2}$$

whose domain is to interpret our coinductive type. In fact, here we are implicitly using the fact that locally cartesian closed pretoposes with M-types are closed under slicing, as proved in [23].

In order to give a concrete description of the functor $M(A, B) : \mathcal{C}/X \longrightarrow \mathcal{C}^{\to}$, we can reason as follows. For the functor to satisfy the coherence condition of Definition 1, it is necessary that the equation $f^* M(A, B) = M(f^* A, f^* B)$ holds for any $f : X' \longrightarrow X$ in $\mathcal{C}$. In particular, this means that, for an arrow $g : Y \longrightarrow X'$, one must have

$$M(A, B)(f \circ g) = M(f^* A, f^* B)(g)$$

and choosing $g = \mathsf{id}_{X'}$ we get that $M(A, B)(f) = p(M(f^* A, f^* B))$ for any f in $\mathcal{C}/X$. Therefore, in order to describe $M(A, B)$, we first make a choice of final coalgebras for the functor $P_{\pi_{f^* B}}$ in each slice category $\mathcal{C}/X'$ (for $f : X' \longrightarrow X$), and then define $M(A, B)(f)$ to be the carrier of the chosen coalgebra for $P_{\pi_f * B}$. The coherence condition is then automatically fulfilled.

Given sections $y : X \longrightarrow X \ltimes Y$ of π_Y and $g : X \longrightarrow X \ltimes (Y \to P_b(Y))$, g determines a map $\tilde{g}$ over X:

$$
\begin{array}{ccc}
X \ltimes Y & \xrightarrow{\quad \tilde{g} \quad} & X \ltimes P_b(Y) \\
 & \pi_Y \searrow \quad \swarrow \pi_{P_b(Y)} & \\
 & X. &
\end{array}
$$

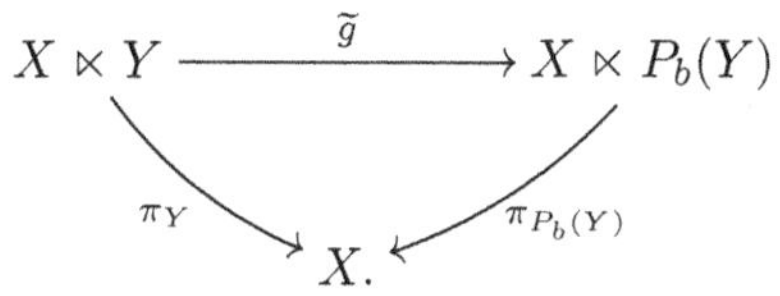

This is a P_b-coalgebra in $Type_{\widehat{C}}(X)$, hence there is a unique coalgebra morphism

$$
\begin{array}{ccc}
X \ltimes Y & \xrightarrow{\quad \overline{g} \quad} & X \ltimes M(A, B) \\
 & \pi_Y \searrow \quad \swarrow \pi_{M(A,B)} & \\
 & X. &
\end{array}
$$

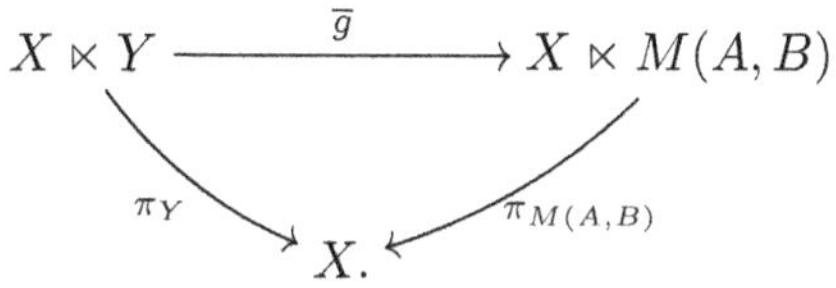

Post-composition with $\overline{g}$ takes y to a section of $\pi_{M(A,B)}$:

$$unfold(g, y) = \overline{g}y.$$

Given a section t of $\pi_{M(A,B)}$, post-composition with the map m of (2) gives a section of $p(P_b(M(A, B))) = \Sigma_{\pi_A} p(B \to \pi_A^* M(A, B))$. Projection on A then determines a section

$$root(t) = (mt).1 \quad \text{of} \quad \pi_A,$$

whereas the second projection is a section

$$br(t) = (mt).2 \quad \text{of} \quad \pi_{root(t)^* B \to M(A,B)}.$$

The various equations of Definition 1 are obviously satisfied by the data we have just defined, because of the finality of $M(A, B)$.   □

Remark. The choice of a collection of final coalgebras made in the proof is needed in order to ensure coherence of coinductive types under pullback. The situation is analogous to that of Hofmann [11], where he needs a choice of pullbacks and equalisers in order to describe the identity types.

5   Conclusions

We have extended the extensional type theory of Martin-Löf with Σ-types and Π-types, by adding coinductive types. These allow for the specification of infinite programs, and for the study of non-well-founded structures.

A sound categorical semantics has been given, in terms of type-categories with coinductive types, which extends the well-known semantics presented in [19,11]. We also prove that any locally cartesian closed category with M-types gives rise to a type-category with coinductive types. This formalised the statement in [23] saying that coinductive pretoposes give a semantics of type theories with coinductive types.

From a computational point of view, the choice of working with extensional equality (as well as the axioms we adopted for M-types in the type theory) are rather unpleasant, since they make equality undecidable. We heard from Altenkirch that the so-called "observational types" he introduced in [4] might help handling infinite objects in an intensional setting, and that some research in this direction has been undertaken.

In [23], the structure of a coinductive pretopos is studied, and that is not only a locally cartesian closed category with M-types, but also exact (and with distributive sums). These further properties have been ignored in the present setting. They might be of use in giving a semantics to the study of setoids built out of a type theory with conductive types, and this topic is clearly related to the study of non-well-founded set theory [2].

References

1. Michael Abbott, Thorsten Altenkirch, and Neil Ghani. Representing strictly positive types. Presented at APPSEM annual meeting, invited for submission to Theoretical Computer Science, 2004.
2. Peter Aczel. *Non-Well-Founded Sets.* Center for the Study of Language and Information, Stanford University, 1988. CSLI Lecture Notes, Volume 14.
3. Peter Aczel, Jiří Adámek, Stefan Milius, and Jiří Velebil. Infinite trees and completely iterative theories: a coalgebraic view. *Theoretical Computer Science*, 300:1–45, 2003.
4. T. Altenkirch. Extensional equality in intensional type theory. In *14th Symposium on Logic in Computer Science (LICS'99)*, pages 412–421. IEEE, 1999.
5. Michael Barr. Terminal coalgebras for endofunctors on sets. Available from `ftp://www.math.mcgill.ca/pub/barr/trmclgps.zip`, 1999.
6. Jean Bénabou. Fibered categories and the foundations of naive category theory. *J. Symbolic Logic*, 50(1):10–37, 1985.
7. John Cartmell. Generalised algebraic theories and contextual categories. *Ann. Pure Appl. Logic*, 32(3):209–243, 1986.
8. Thierry Coquand. Infinite objects in type theory. In *Types for proofs and programs (Nijmegen, 1993)*, volume 806 of *Lecture Notes in Comput. Sci.*, pages 62–78. Springer, 1994.
9. Veronica Gaspes. Infinite objects in type theory, 1997.
10. Lars Hallnäs. On the syntax of infinite objects: an extension of Martin-Löf's theory of expressions. In *COLOG-88 (Tallinn, 1988)*, volume 417 of *Lecture Notes in Comput. Sci.*, pages 94–104. Springer, 1990.
11. Martin Hofmann. On the interpretation of type theory in locally cartesian closed categories. In *Computer science logic (Kazimierz, 1994)*, volume 933 of *Lecture Notes in Comput. Sci.*, pages 427–441. Springer, 1995.
12. Bart Jacobs. *Categorical logic and type theory*, volume 141 of *Studies in Logic and the Foundations of Mathematics.* North-Holland Publishing Co., Amsterdam, 1999.

13. Bart Jacobs and Jan Rutten. A tutorial on (co)algebras and (co)induction. *Bulletin of the EATCS*, 62:222–259, 1996.
14. Ingrid Lindström. A construction of non-well-founded sets within Martin-Löf's type theory. *Journal of Symbolic Logic*, 54(1):57–64, 1989.
15. Per Martin-Löf. *Intuitionistic type theory*, volume 1 of *Studies in Proof Theory. Lecture Notes*. Bibliopolis, Naples, 1984.
16. Per Martin-Löf. Mathematics of infinity. In *COLOG-88 (Tallinn, 1988)*, volume 417 of *Lecture Notes in Comput. Sci.*, pages 146–197. Springer, 1990.
17. N. P. Mendler, P. Panangaden, and R. L. Constable. Infinite objects in type theory. In *Symposium on Logic in Computer Science (LICS '86)*, pages 249–257. IEEE Computer Society Press, 1986.
18. Bengt Nordström, Kent Petersson, and Jan M. Smith. *Programming in Martin-Löf's type theory*, volume 7 of *International Series of Monographs on Computer Science*. The Clarendon Press Oxford University Press, 1990.
19. Andrew M. Pitts. Categorical logic. In *Handbook of logic in computer science, Vol. 5*, Oxford Sci. Publ., pages 39–128. Oxford Univ. Press, 2000.
20. R. A. G. Seely. Locally cartesian closed categories and type theory. *Math. Proc. Cambridge Philos. Soc.*, 95(1):33–48, 1984.
21. Thomas Streicher. *Semantics of type theory*. Progress in Theoretical Computer Science. Birkhäuser, 1991. Correctness, completeness and independence results.
22. Daniele Turi and Jan Rutten. On the foundations of final coalgebra semantics: non-well-founded sets, partial orders, metric spaces. *Mathematical Structures in Computer Science*, 8(5):481–540, October 1998.
23. Benno van den Berg and Federico De Marchi. Non-well-founded trees in categories. (submitted). Available online at `http://arxiv.org/abs/math.CT/0409158`.

Look: Simple Stochastic Relations Are Just, Well, Simple

Ernst-Erich Doberkat*

Chair for Software Technology, University of Dortmund
doberkat@acm.org

Abstract. Simple systems cannot decomposed further. Algebraically, simple systems have only isomorphisms as epis. We characterize simple stochastic relations through different forms of bisimulations for the case that the underlying spaces are Polish, and analytic, respectively. This requires a closer investigation of bisimulations, congruences and their mutual relationship. We provide a complete characterization of simple stochastic relations for analytic spaces.

1 Introduction

An algebraic structure which is isomorphic to each of its non-trivial factor spaces is called simple. Take e.g. a simple and non-trivial group G and an epimorphism $\phi : G \to H$, then ϕ is an isomorphism, because the factor system $G/\mathsf{ker}\,(\phi)$ is isomorphic to G, thus the kernel $\mathsf{ker}\,(\phi)$ is trivial, cp. [8, p. 104]. Thus a system S is simple if each epimorphism $S \to T$ is an isomorphism. On the other hand, the very close connection between simple systems and trivial bisimulations is well known in the theory of coalgebras: a system is simple iff it has only trivial bisimulations.

We show in this paper that the investigation of simple stochastic relations through bisimulations is fruitful as well. While in coalgebras heavy use is being made of weak pullbacks, this is not possible for stochastic relations, since they are not available there — in fact, one is glad to have semi-pullbacks [6]. Hence one has to bypass this difficulty at the cost of some rather technical constructions. A further technical point to be considered concerns the structure of the base space. Stochastic relations can be defined on top of arbitrary measurable spaces, but the probabilistic structure of these spaces is too poor to be of much use to us. Thus we resort to a richer structure, viz., Polish spaces and their Borel images, analytic spaces. A closer look will reveal that a careful distinction between these spaces will be required, since stochastic relations on them are different in subtle ways, as we will see.

Bisimulations are usually defined through spans of morphisms, and it turns out that we need to capture different conditions on equivalence relations through

* Research funded in part by Deutsche Forschungsgemeinschaft, grant DO 263/8-1, *Algebraische Eigenschaften stochastischer Relationen.*

J.L. Fiadeiro et al. (Eds.): CALCO 2005, LNCS 3629, pp. 127–141, 2005.

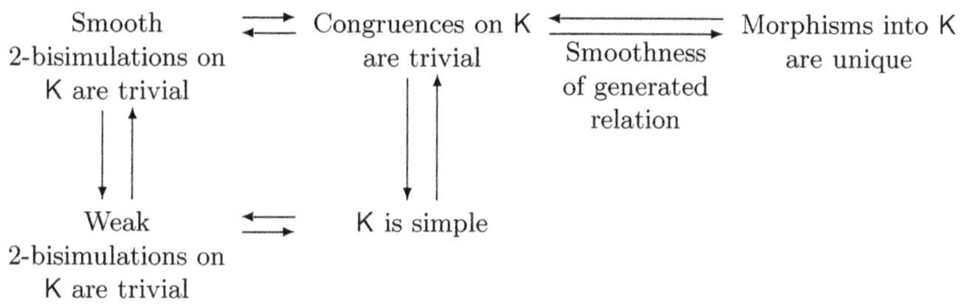

Fig. 1. Simple systems: the Polish case

bisimulations in order to appreciate all properties of simple relations. We introduce 2-bisimulations as those bisimulations that are based on a set-theoretic relation on the space involved, so that the morphisms are just the projections. For our discussion those 2-bisimulations are of interest that are defined through the kernels of morphisms, or, equivalently, through congruences; we call them smooth. There is a very close relationship between congruences and such smooth 2-bisimulations, since we show that each congruence gives rise to such a smooth 2-bisimulation (this is easy in the theory of coalgebras, it turns out to be rather hard work for the stochastic case). This observation yields immediately that a stochastic relation has only trivial congruences iff it has only trivial smooth 2-bisimulations.

This characterization is provided for the case that the spaces on which the stochastic relations are built are Polish. Going a step further to analytic spaces (hence to Borel images of Polish spaces) indicates that we need a further kind of 2-bisimulations that are called *weak* (they focus on Borel sets that are invariant under the congruence and leave other Borel sets alone). In the Polish case we can show that there is no difference, but in the analytic case this is presumably not the case. We prove that a stochastic relation for relations over Polish spaces is simple iff it has only the trivial congruence iff smooth as well as weak 2-bisimulations are trivial. It is shown that if there can be at most one morphism into a simple stochastic relation, then the relation is simple (the converse holds as well, but under a restrictive condition); Figure 1 gives an overview. This is essentially the situation for the analytic case, too, but the equivalence of weak and smooth 2-bisimulations is a bit weaker; Figure 2 provides a pictorial summary for this case as well. All this leads to a complete characterization of simple analytic relations by injective measurable maps into the unit interval of the reals. It implies that final systems do not exist unless the system is truly probabilistic: then there is exactly one.

Final coalgebras are used by Rutten [10, 11] for establishing a calculus of coinduction. Since the structure of simple systems is much poorer for stochastic relations, such an endeavor cannot be expected to be as fruitful as in the general coalgebraic case, but we indicate that the identification of simple relations may occasionally be helpful nevertheless. We derive in the full paper [5] an explicit representation of the number of heaps that is central to the analysis of Williams'

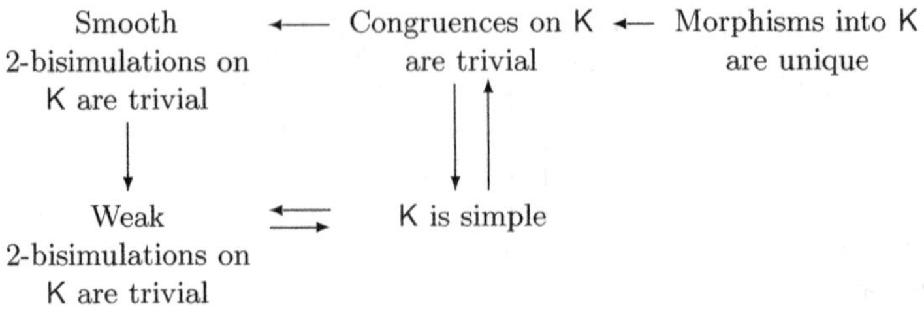

Fig. 2. Simple systems: the analytic case

algorithms for heap construction from a continuous representation, and we show how the inversions of an array can be counted. These examples illustrate possible applications of the results presented here.

The paper is organized as follows: Section 2 defines the category of stochastic relations and the basic version of bisimulations. Since smooth equivalence relations play an important role for the discussion at the paper's core, they are studied separately in Section 3, and some techniques associated with them are introduced and illustrated. Section 4 applies this to congruences, introduces smooth and weak 2-bisimulations, and studies briefly the relationship between them. Simple relations are formally introduced in Section 5, characterizations both for the Polish and the analytic case are given. Section 6 mentions related work and gives indications for further studies.

2 Stochastic Relations

Let $(X, \mathcal{A})$ be a measurable space, then $\mathbf{S}(X, \mathcal{A})$ denotes the set of all subprobability measures on $(X, \mathcal{A})$. This is made into a measurable space through the weak-*-σ-algebra, which is the smallest σ-algebra $\mathcal{A}^*$ on $\mathbf{S}(X, \mathcal{A})$ that renders the evaluation map $\mu \mapsto \mu(A)$ measurable for all $A \in \mathcal{A}$.

Definition 1. *A stochastic relation* $\mathsf{K} = ((X, \mathcal{A}), (Y, \mathcal{B}), K)$ *with* $K : (X, \mathcal{A}) \rightsquigarrow (Y, \mathcal{B})$ *is a map* $K : X \to \mathbf{S}(Y, \mathcal{B})$ *that is* $\mathcal{A}$-$\mathcal{B}^*$*-measurable.*

Thus a stochastic relation $K : (X, \mathcal{A}) \rightsquigarrow (Y, \mathcal{B})$ has the following properties:

1. $B \mapsto K(x)(B)$ is a subprobability measure on $\mathcal{B}$ for each $x \in X$,
2. $x \mapsto K(x)(B)$ is a $\mathcal{A}$-measurable map from X to the unit interval $[0, 1]$ for each $B \in \mathcal{B}$.

Suppose that the base spaces $(X, \mathcal{A})$ and $(Y, \mathcal{B})$ are identical, then a stochastic relation may be interpreted as a coalgebra for the subprobability functor. This is an endofunctor in the category of measurable spaces with measurable maps as morphisms. This point of view appears quite as attractive, because it suggests to fit stochastic relations tightly under the roof of coalgebras, making tried and tested approaches available for investigating problems of stochastic relations.

Unfortunately, this proposal will work only partially. There are two reasons for that: First, it is well known that the subprobability functor has some idiosyncratic properties making work with it sometimes a little strenuous (for example, there do not exist weak pullbacks). Second, a coalgebra $\langle X, c\rangle$ for functor $\mathbf{F}$ is defined as a morphism $c : X \to \mathbf{F}X$, so the codomain of morphism c is just the image of its domain under $\mathbf{F}$. This is rather restrictive, both structurally and regarding applications. "Unfolding" domain and codomain into two independent objects provides much needed maneuverability, see e.g. [4, 6].

Let K_1 and K_2 be stochastic relations with $\mathsf{K}_i = ((X_i, \mathcal{A}_i), (Y_i, \mathcal{B}_i), K_i), i = 1, 2$, then $\mathsf{f} : \mathsf{K}_1 \to \mathsf{K}_2$ is a *morphism* between K_1 and K_2 iff $\mathsf{f} = (\phi, \psi)$ is a pair of surjective maps such that $\phi : X_1 \to X_2$ is $\mathcal{A}_1 - \mathcal{A}_2$-measurable, $\psi : Y_1 \to Y_2$ is $\mathcal{B}_1 - \mathcal{B}_2$-measurable with $K_2 \circ \phi = \mathbf{S}(\psi) \circ K_1$. Here $\mathbf{S}(\psi) : \mathbf{S}(Y_1, \mathcal{B}_1) \to \mathbf{S}(Y_2, \mathcal{B}_2)$ is the $\mathcal{B}_1^* - \mathcal{B}_2^*$ measurable map defined through $\mathbf{S}(\psi)(\mu)(B_2) := \mu(\psi^{-1}[B_2])$.

We want the maps underlying a morphism be onto in order to make sure that each element in the range can be traced back to at least one element from the domain, so that there are no unrelated elements in the range.

A *bisimulation* between the stochastic relations K_1 and K_2 is a stochastic relation M (which is sometimes called the *mediating object*) together with two morphisms

$$\mathsf{K}_1 \xleftarrow{\ \mathsf{f}_1\ } \mathsf{M} \xrightarrow{\ \mathsf{f}_2\ } \mathsf{K}_2.$$

If K_1 and K_2 coincide, this is called a *bisimulation on* K_1. Thus a bisimulation is a span of morphisms. The notion of a bisimulation can be refined: let $\mathsf{M} = ((A, \mathcal{X}), (B, \mathcal{Y}), M)$ be the mediating object with suitable σ-algebras $\mathcal{X}$ and $\mathcal{Y}$ on A resp. B. If A and B are measurable subsets of $X_1 \times X_2$ resp. $Y_1 \times Y_2$, and if $\mathsf{f} = (\pi_{1,X_1}, \pi_{1,Y_1}), \mathsf{f}' = (\pi_{2,X_2}, \pi_{2,Y_2})$ — π indicating projections —, then the bisimulation is called a *2-bisimulation.* Thus a 2-bisimulation renders this diagram commutative:

$$\begin{array}{ccccc}
X_1 & \xleftarrow{\ \pi_{1,X_1}\ } & A & \xrightarrow{\ \pi_{2,X_2}\ } & X_2 \\
\downarrow{\scriptstyle K_1} & & \downarrow{\scriptstyle M} & & \downarrow{\scriptstyle K_2} \\
\mathbf{S}(Y_1, \mathcal{A}_1) & \xleftarrow[\mathbf{S}(\pi_{1,Y_1})]{} & \mathbf{S}(B, \mathcal{Y}) & \xrightarrow[\mathbf{S}(\pi_{2,Y_2})]{} & \mathbf{S}(Y_2, \mathcal{B}_2)
\end{array}$$

We will discuss special versions of bisimulations in Section 4.

General measurable spaces are far too general for obtaining results of interest. We will deal with Polish and analytic spaces instead which are much more fruitful. A topological space X is called a *Polish space* iff there exists a metric for the topology on X which is second countable and which is complete. Thus the topology has a countable base (or, equivalently, a countable dense subset), and each Cauchy sequence converges. The Borel sets $\mathcal{B}(X)$ on X are the smallest σ-algebra on X that contains the open sets. A measurable space $(X, \mathcal{A})$ is called *analytic* iff there exists a Polish space X_0 with $X \subseteq X_0$ and (slightly abusing notation) $\mathcal{A} = \mathcal{B}(X) := \mathcal{B}(X_0) \cap X$ such that $X = f[Y]$ for some Polish space Y and a Borel measurable map $f : Y \to X_0$; Y and f define the analytic structure on X. We will write down Polish resp. analytic spaces without their σ-algebra whenever this does not lead to confusion.

$\mathbf{S}(X)$ will always carry the weak-*-σ-algebra $\mathcal{B}(X)^*$. If X is Polish, $\mathbf{S}(X)$ is Polish as well: the weak topology, i.e., the smallest topology that makes $\mu \mapsto \int_X f\, d\mu$ continuous for all bounded and continuous $f : X \to \mathbb{R}$, is metrizable and has a countable dense subset. Similarly, if X is Polish or analytic, $\mathbf{S}(X)$ shares this respective property. We will denote by $\mathbf{P}(X)$ the subspace of all probability measures.

General stochastic relations with their morphisms form a category, and we are mostly interested in the full subcategories of objects (X, Y, K) in which both X and Y are Polish or both are analytic. We will not name these categories but rather mention that we talk about Polish objects or about analytic objects, depending on which case applies. Note that when a weak 2-bisimulation on an analytic object is investigated, we do so in the category of all stochastic relations, contrary to the usual custom of defining bisimulations through morphisms in the category one investigates the object in.

We will make use of set-valued maps when investigating 2-bisimulations. Suppose that we have Polish spaces V and W and a map $R : V \to \mathcal{P}(W)$ assigning each $v \in V$ a nonempty subset $R(v)$ of W. A measurable map $f : V \to W$ is called a *measurable selector for* R iff $f(v) \in R(v)$ holds for each $v \in V$, so that f picks for each $v \in V$ an element of $R(v)$ in a measurable way. The existence of a measurable selector for a set-valued map can by no means always be guaranteed (in fact, without the condition of measurability the existence of a selector is equivalent to the axiom of choice). The Himmelberg-Van Vleck Theorem [13, Theorem 4.2.e] provides a sufficient condition:

Proposition 1. *Assume that V and W are Polish spaces, and that $R : V \to \mathcal{P}(W)$ is a set-valued map such that*

1. *$R(v)$ is a non-empty closed subset of W for each $v \in V$,*
2. *the weak inverse $\exists R(C) := \{v \in V \mid R(v) \cap C \neq \emptyset\}$ is a Borel set for each compact subset C of W.*

Then there exists a measurable selector for R.

3 Smooth Equivalences

Fix an analytic space X. Smooth equivalence relations are introduced and some properties are demonstrated. This is discussed separately for the reader's convenience.

Definition 2. *An equivalence relation α on X is called* smooth *iff there exists a Borel measurable map $f : X \to Q$ into a Polish space Q such that α is just the kernel of f, thus $\alpha = \mathsf{ker}(f) := \{\langle x, x' \rangle \in X \times X \mid f(x) = f(x')\}$.*

An equivalent definition [12] is to postulate the existence of a sequence $(A_n)_{n\in\mathbb{N}}$ of Borel sets in X such that $x \ \alpha \ x'$ iff $\forall n \in \mathbb{N} : [x \in A_n \Leftrightarrow x' \in A_n]$. This formulation is useful when it comes to define e.g. equivalent states in a Kripke model $\mathcal{M}$ for a modal logic: denote for a formula φ by $A_\varphi := \{s \mid \mathcal{M}, s \models \varphi\}$ the set of all states that satisfy φ, then the equivalence relation defined through $s \sim s'$ iff $\forall \varphi : [s \in A_\varphi \Leftrightarrow s' \in A_\varphi]$ is smooth, because there are only countably many formulas. It is the equivalence underlying the Hennessy-Milner Theorem, see [2, 3].

Let X be a set with $A \subseteq X \times X$ a relation. Denote by $\ell(A)$ the smallest equivalence relation containing A. Denote for an equivalence relation α on X by $[x]_\alpha$ the equivalence class of $x \in X$.

This definition is central for the development:

Definition 3. *Let α be a smooth equivalence relation on the analytic space X.*

1. *A subset $B \subseteq X$ is called* α-invariant *iff B is the union on α-classes.*
2. *The σ-algebra of α-invariant Borel sets is denoted by $\mathcal{I}(\mathcal{B}(X), \alpha)$.*

It can be shown [4] that $\mathcal{I}(\mathcal{B}(X), \alpha) = \sigma(\{A_n \mid n \in \mathbb{N}\})$, provided α is defined through the sequence $(A_n)_{n\in\mathbb{N}}$, as indicated above. Since we can always find such a sequence [12, Lemma 3.1.6, Exercise 5.1.10] for a smooth relation, the σ-algebra of α- invariant Borel sets is countably generated, and vice versa. A smooth relation α on X constitutes a measurable subset $\alpha \subseteq X \times X$.

The factor space X/α is made into a measurable space by endowing it with the largest σ-algebra $\mathcal{B}(X)/\alpha$ that renders the canonic projection $\eta_\alpha : x \mapsto [x]_\alpha$ measurable. It is well known that $(X/\alpha, \mathcal{B}(X)/\alpha)$ is an analytic space, in particular that $\mathcal{B}(X/\alpha) = \mathcal{B}(X)/\alpha$ holds, see [12, Exercise 5.1.14].

We list some important and helpful properties of $\mathcal{I}(\mathcal{B}(X), \alpha)$ and related structures for later reference.

Lemma 1. *Let α be a smooth equivalence relation. Then*

1. $\mathcal{I}(\mathcal{B}(X), \alpha) = \eta_\alpha^{-1}[\mathcal{B}(X/\alpha)]$,
2. *If $P \in \mathcal{I}(\mathcal{B}(X), \alpha)$, then $(P \times X) \cap \alpha = (X \times P) \cap \alpha = (P \times P) \cap \alpha$*
3. $\otimes[X, \alpha] := \{(P \times X) \cap \alpha \mid P \in \mathcal{I}(\mathcal{B}(X), \alpha)\}$ *is a σ-algebra on α,*
4. *If $\mu \in \mathbf{S}(X, \mathcal{I}(\mathcal{B}(X), \alpha))$ is a subprobability measure on the α-invariant Borel sets of X, then $\mu^\bullet((P \times X) \cap \alpha) := \mu(P)$ defines a subprobability measure on $\otimes[X, \alpha]$.*

Property 1 states that the Borel sets $\mathcal{B}(X/\alpha)$ are essentially the α-invariant Borel sets in X. The Borel sets in the factor space look somewhat inaccessible, hence this representation will be a great help, since we now have a specific handle on them (via the factor map η_α). Property 3 looks at the trace the invariant Borel sets leave on α: these sets form a σ-algebra, when the pairs forming the equivalence relation are considered as a space in its own right, and property 4 strengthens this view, because also a measure may be transported in this way. It will be most helpful in the sequel to see that we can view measures that are defined on the invariant Borel sets of X as measures on the somewhat strange looking σ-algebra $\otimes [X, \alpha]$.

The interplay between smooth relations and measurable maps is further illustrated by the technique of transporting a smooth relation backwards along a measurable map. The proof makes use of Souslin's famous characterization of Borel sets as both analytic and complements of analytic sets.

Lemma 2. *Let α be a smooth equivalence relation on the analytic space A so that $\alpha = \mathsf{ker}(h)$ for some measurable map $h : A \to W$, W being a Polish space. Define for the Polish space X and the Borel map $f : X \to A$ on X the smooth relation $\alpha_f := \mathsf{ker}(h \circ f)$. If $E \subseteq X$ is a α_f-invariant Borel set, then*

1. *$f[E]$ is a α-invariant Borel set in A,*
2. *$E = f^{-1}[f[E]]$.*

Consequently, the invariant Borel sets of α_f are just the inverse images of the invariant Borel set of α under f, viz., $\mathcal{I}(\mathcal{B}(X), \alpha_f) = f^{-1}[\mathcal{I}(\mathcal{B}(A), \alpha)]$.

Because usually the image of a Borel set will not be a Borel set again, property 1 is somewhat surprising, (exactly the observation that the images of Borel sets under Borel maps are no longer Borel had led to the development of the theory of analytic sets). Property 2 permits an explicit representation of the α_f-invariant sets in terms of their f-images. This is a rather strong and unusual statement as well, indicating that smoothness is rather strong a property, in particular when combined with invariance.

When investigating different morphisms for a stochastic relation, a relation derived from the images will be helpful. Abbreviate for two maps $g_1, g_2 : V \to W$ with common domain V and common range W (V and W are for the time being arbitrary sets) the common product image by

$$\lfloor g_1 \| g_2 \rfloor := \{\langle g_1(v), g_2(v)\rangle \mid v \in V\}.$$

Intuitively, g_1 and g_2 correspond to two processes; they are run in parallel, and we have a look at the equivalence relation spawned by them: they may be viewed as the events common to both morphisms.

Lemma 3. *Assume B is a Polish space, and let $\psi_1, \psi_2 : B \to Y$ be surjective Borel maps. Assume further that $\ell(\lfloor \psi_1 \| \psi_2 \rfloor)$ is smooth. Let*

$$\mathcal{C} := \{C \in \mathcal{B}(Y) \mid \psi_1^{-1}[C] = \psi_2^{-1}[C]\}$$

be the σ-algebra of common events. Then the common events are exactly the $\ell(\lfloor \psi_1 \| \psi_2 \rfloor)$-invariant Borel sets, thus $\mathcal{C} = \mathcal{I}(\mathcal{B}(Y), \ell(\lfloor \psi_1 \| \psi_2 \rfloor))$.

4 Congruences

Let $\mathsf{K} = (X, Y, K)$, be a stochastic relation over the analytic spaces X and Y. A *congruence* $\mathsf{c} = (\alpha, \beta)$ on K is a pair of smooth equivalence relations α and β on

X resp. Y such that $K(x_1)(B) = K(x_2)(B)$ holds whenever $\langle x_1, x_2\rangle \in \alpha$, and B is a β-invariant Borel set in Y. Thus if α cannot separate x_1 from x_2, and if the elements of B cannot be separated by β, then the probabilities for the respective transitions are equal. Note that the kernel $\mathsf{ker}(\mathsf{f})$ of a morphism $\mathsf{f} : \mathsf{K} \to \mathsf{K}'$ is a congruence on K, where $\mathsf{ker}(\phi, \psi)$ is defined as the pair $(\mathsf{ker}(\phi), \mathsf{ker}(\psi))$, see [4]. This observation will be used occasionally.

We have defined bisimulations and 2-bisimulations in Section 2. Congruences permit specializing the notion of a bisimulation further, and these specializations will be used later on when characterizing simple systems.

Definition 4. *Let α and β be smooth equivalence relations on X resp. Y.*

1. *A 2-bisimulation $\mathsf{M} = (\alpha, \beta, M)$ on K is called a* smooth 2-bisimulation *on K.*
2. *If for the stochastic relation $\mathsf{N} = ((\alpha, \mathcal{B}(\alpha)), (\beta, \otimes[Y, \beta]), N)$*

$$(\mathbf{S}(\pi_{i,Y}) \circ N(a_1, a_2))(E) = K(a_i)(E)$$

holds for $i = 1, 2$ whenever $\langle a_1, a_2\rangle \in \alpha$ and E is a β-invariant Borel set of Y, then N is called a weak 2-bisimulation *on K.*

Smooth 2-bisimulations correspond to the bisimulation equivalences studied in coalgebras [9], as we will see soon. Weak 2-bisimulations restrict their attention to the β-invariant Borel sets of Y (rather than on all Borel sets), $N(a)((B \times Y) \cap \beta)$ is defined for $a \in \alpha$ and for the Borel set $B \in \mathcal{I}(\mathcal{B}(Y), \beta)$, see Lemma 1, part 3. This look of course much more restrictive than for a smooth 2-bisimulation: Clearly a smooth 2-bisimulation is a weak one, but we will show in Proposition 2 that we can produce a smooth 2-bisimulation from a weak one, provided the relation K is a Polish object.

We will begin with a fairly deep connection between congruences, smooth and weak 2-bisimulations, that says among others that in Polish spaces congruences and 2-bisimulations are just two sides of the same medal. Fix for the discussion that follows the stochastic relation $\mathsf{K} = (X, Y, K)$ and a pair $\mathsf{c} = (\alpha, \beta)$ of smooth equivalence relations on the analytic spaces X resp. Y.

Proposition 2. *Consider the following conditions:*

a. $\mathsf{c} = (\alpha, \beta)$ is a congruence on K.
b. There exists a weak 2-bisimulation $((\alpha, \mathcal{B}(\alpha)), (\beta, \otimes[Y, \beta]), N)$ on K.
c. There exists a smooth 2-bisimulation (α, β, M) on K.

Then the following holds:

1. *$c \Rightarrow b \Rightarrow a$ is true for the analytic spaces X and Y,*
2. *If both X and Y are Polish, then all conditions are equivalent.*

Proof. 0. "$c \Rightarrow b$" is quite obvious, since each smooth 2-bisimulation is a weak one, so for the general case the implication $b \Rightarrow a$, and for the Polish case the implication $a \Rightarrow c$ needs to be established.

1. "$b \Rightarrow a$" Let $C \in \mathcal{I}(\mathcal{B}(Y), \beta)$ be a β-invariant Borel subset of Y, then $(C \times Y) \cap \beta$ equals both $Y \times C) \cap \beta$ and $(C \times C) \cap \beta$. (Lemma 1, part 2). Thus we obtain for $\langle x, x' \rangle \in \alpha$ from $((\alpha, \mathcal{B}(\alpha)), (\beta, \otimes [Y, \beta]), N)$ being a 2-bisimulation: $K(x)(C) = N(x, x')((C \times Y) \cap \beta) = N(x, x')((Y \times C) \cap \beta) = K(x')(C)$.

2. "$a \Rightarrow c$" This part is much harder. We need to construct a stochastic relation $M : \alpha \rightsquigarrow \beta$ so that (α, β, M) forms a 2-bisimulation. The plan goes as follows: we show that the problem can be considered a selection problem. For this, we define on α a suitable set-valued map Γ that takes on closed sets of measures on β and that satisfies the conditions of Proposition 1 for the existence of a selector. The main difficulty will lie in showing that Γ takes in fact non-empty values, and here invariant sets come in. An analysis of the topological situation [12, Corollary 3.2.6] shows that we may select suitable Polish topologies on X and Y with the same Borel sets as before such that both α and β are closed, hence Polish subsets of X resp. Y, and that $K : X \to \mathbf{S}(Y)$ may assumed to be continuous.

Given $\langle x_1, x_2 \rangle \in \alpha$, the set

$$\Gamma(x_1, x_2) := \{\mu \in \mathbf{S}(\beta) \mid \mathbf{S}(\pi_{1,Y})(\mu) = K(x_1), \mathbf{S}(\pi_{2,Y})(\mu) = K(x_2)\}$$

will be scrutinized with the goal of finding a measurable selector for Γ. It is immediate that it is a closed subset of $\mathbf{S}(\beta)$, because the projections induce continuous maps on the respective spaces of subprobabilities. Whenever $C \subseteq \mathbf{S}(\beta)$ is compact, the weak inverse $\exists\Gamma(C) := \{\langle x_1, x_2 \rangle \in \alpha \mid \Gamma(x_1, x_2) \cap C \neq \emptyset\}$ of C is a closed subset of α. This is established through a standard sequential compactness argument.

We want to show first that $\Gamma(x_1, x_2) \neq \emptyset$, whenever $\langle x_1, x_2 \rangle \in \alpha$. For this the techniques developed in [6] are used. Put $Z := Y/\beta$ with $\mathcal{C} := \mathcal{B}(Y/\beta)$, then $(Z, \mathcal{C})$ is an analytic space, hence it is known to be separable. The map $\psi : y \mapsto [y]_\beta$ is measurable from Y onto Z, and we have $S := \{\langle y_1, y_2 \rangle \mid \psi(y_1) = \psi(y_2)\} = \beta$. We know moreover from Lemma 1, part 1 that $\eta_\beta^{-1}[\mathcal{C}] = \mathcal{I}(\mathcal{B}(Y), \beta)$ holds. Now fix $\langle x_1, x_2 \rangle \in \alpha$ and put $\nu_1 := K(x_1), \nu_2 := K(x_2)$, then a measure θ_1 on the σ-algebra $\otimes [Y, \beta]$ is defined through $\theta_1((B \times B) \cap \beta) = \nu_1(B) \, (= \nu_2(B))$, see Lemma 1, part 4. The use of [6, Lemma 3] yields an extension of θ_1 to a measure θ which is defined on all of $\mathcal{B}(S)$. Thus we have now $\theta \in \mathbf{S}(S)$ such that

$$\forall E_i \in \psi^{-1}[\mathcal{C}] : \mathbf{S}(\pi_{i,Y})(\theta)(E_i) = \nu_i(E_i), i = 1, 2.$$

From [6, Proposition 5] we obtain a measure $\mu \in \mathbf{S}(S)$ such that

$$\forall E_i \in \mathcal{B}(S) : \mathbf{S}(\pi_{i,Y})(\mu)(E_i) = \nu_i(E_i), i = 1, 2.$$

But this means that $\Gamma(x_1, x_2) \neq \emptyset$, thus we can apply the Himmelberg-Van Vleck Selection Theorem (Proposition 1) and obtain a measurable selector M for Γ, consequently, $M : \alpha \rightsquigarrow \beta$. Thus $\mathsf{M} := (\alpha, \beta, M)$ is a stochastic relation. From M being a selector to Γ one sees that M is a 2-bisimulation for K, since

$$(\mathbf{S}(\pi_{1,Y}) \circ M)(x_1, x_2) = K(x_1) \text{ and } (\mathbf{S}(\pi_{2,Y}) \circ M)(x_1, x_2) = K(x_2)$$

is true for all $\langle x_1, x_2 \rangle \in \alpha$.

Thus we have established a very close relationship between congruences and 2-bisimulations for stochastic relations. The basic idea has been to extend a stochastic relation that is defined on a small and fairly easy to handle σ-algebra to a larger one. But this is complicated, because we do not have direct access to the Borel sets, when we need it: the Borel sets are defined in terms of a closure operation and not through some explicit procedure, so we cannot put a handle on them directly. Hence we have to walk a by-path: we show through a selection argument that such a measure must exist. This argument works essentially as follows: we know that the situation is easily managed on a the small σ-algebra which we start from (this is like the begin of a proof by induction: the picture is nice and clear in the beginning). We know also that our request for an extension is not unreasonable, since our map Γ has some reasonable properties (this is like the induction hypothesis); from this we conclude that we can find an extension through a selector (this is much like the inductive step itself).

Quite apart from the somewhat involved technical development, this close relationship is somewhat akin to general coalgebras. The situation cannot be mirrored, however, since for coalgebras one usually requires a functor which preserves weak pullbacks, see [9]. The structure for the subprobability functor $\mathbf{S}$ is, however, slightly more involved: the existence of semi-pullbacks can be established, examples show that the hope for establishing weak pullbacks is vain, see [6]. Consequently it seems to be difficult to fit general coalgebras and stochastic relations too tightly under one common roof.

Anyway, Proposition 2 provides us with a considerable degree of freedom, because we can select the proper instrument in investigating simplicity without having to be afraid that we loose important properties. This holds at least in the Polish case. In the case of an analytic object we have to be a bit careful, but the proposition tells us as well where to install watch dogs.

A partial converse to Proposition 2 is furnished through

Lemma 4. *Let α and β be smooth equivalence relations on X resp. Y. Assume that $\mathsf{M} := ((\alpha, \mathcal{B}(\alpha)), (\beta, \otimes[Y, \beta]), M)$ is a weak 2-bisimulation on K. Then (α, β) is a congruence of K.*

Consequently, both smooth and weak 2-bisimulations for a stochastic relation K are defined on congruences.

We leave this Section now, well armed with bisimulations of different sorts for an investigation of simple relations.

5 Simple Relations

We are now ready to characterize simple systems both for Polish and analytic spaces. We deal first with the Polish case which is a bit easier to handle, and turn then to the analytic case. We develop a technique for reducing the analytic to the Polish case, so that we may capitalize on previous results. A complete characterization of simple relations will then be given for the general analytic case.

Call a congruence $\mathsf{c} = (\alpha, \beta)$ on X and Y *trivial* iff both equivalence relations are the identity, viz., iff both $\alpha = \Delta_X$ and $\beta = \Delta_Y$ hold. Similarly, call a smooth or weak 2-bisimulation *trivial* iff the underlying congruence is trivial.

Definition 5. *A stochastic relation* K *is called* simple *iff each morphism with domain* K *is an isomorphism.*

This definition looks a bit stronger than usual, since usually epimorphisms emanating from a simple structure are assumed to be isomorphisms. But since we deal only with surjective maps, the common definition applies in this context.

We characterize simple systems if the spaces on which the relation is defined both are Polish:

Theorem 1. *Consider these statements for the Polish object* K

(a). K *is simple.*
(b). Each smooth 2-bisimulation on K *is trivial.*
(c). Each weak 2-bisimulation on K *is trivial.*
(d). Let $\mathsf{f}_1, \mathsf{f}_2 : \mathsf{M} \to \mathsf{K}$ *be morphisms, where* M *is a Polish object, then* $\mathsf{f}_1 = \mathsf{f}_2$.
(e). Each congruence on K *is trivial.*

Then

1. *These implications hold always: (a)* $\Leftrightarrow$ *(b)* $\Leftrightarrow$ *(c)* $\Leftrightarrow$ *(e)* $\Leftarrow$ *(d).*
2. *Let in (d)* $\mathsf{f}_i = (\phi_i, \psi_i)$. *If both* $\ell(\lfloor\phi_1\|\phi_2\rfloor)$ *and* $\ell(\lfloor\psi_1\|\psi_2\rfloor)$ *are smooth, then (e)* $\Rightarrow$ *(d) holds as well.*

A visual overview is provided through Figure 1 in the Introduction. The **proof** for Theorem 1 is broken into several pieces:

"(e) $\Rightarrow$ *(a)"* Let $\mathsf{f} : \mathsf{K} \to \mathsf{L}$ be a morphism, then f can be factored through $\mathsf{K}/\mathsf{ker}(\mathsf{f})$ as $\mathsf{f} = \mathsf{f}' \circ \eta_{\mathsf{ker}(\mathsf{f})}$ with an isomorphism f' [4, Corollary 3]. $\mathsf{ker}(\mathsf{f})$ is a congruence which is trivial by assumption. Thus f is an isomorphism.

"(a) $\Rightarrow$ *(e)"* If c is a congruence on K, then $\eta_{\mathsf{c}} : \mathsf{K} \to \mathsf{K}/\mathsf{c}$ is a morphism.

"(b) $\Leftrightarrow$ *(e)"* This is a special case of Proposition 2.

"(d) $\Rightarrow$ *(b)"* Let $\mathsf{M} := (A, B, M)$ be a smooth bisimulation on K, then

$$(\pi_{1,X}, \pi_{1,Y}), (\pi_{2,X}, \pi_{2,Y}) : \mathsf{M} \to \mathsf{K}$$

are morphisms which are equal by assumption.

This settles the proof of part 1. Turning to the proof of part 2, assume that (e) holds in addition to $(\ell(\lfloor\phi_1\|\phi_2\rfloor), \ell(\lfloor\psi_1\|\psi_2\rfloor))$ being smooth. We note from the proof of Lemma 3 that a $\ell(\lfloor\psi_1\|\psi_2\rfloor)$-invariant Borel set $D \subseteq Y$ has the property that $\psi_1^{-1}[D] = \psi_2^{-1}[D]$ holds, hence that D is an event common to ψ_1 and ψ_2. Now define the equivalence relation $R_D := \{\langle x_1, x_2\rangle \mid K(x_1)(D) = K(x_2)(D)\}$, then $\lfloor\psi_1\|\psi_2\rfloor \subseteq R_D$ follows from $\mathsf{f}_1, \mathsf{f}_2 : \mathsf{M} \to \mathsf{K}$ being morphisms: suppose $\langle x_1, x_2\rangle = \langle\phi_1(a), \phi_2(a)\rangle$, and $E = \psi_1^{-1}[D] = \psi_2^{-1}[D]$, we obtain

$$K(x_1)(D) = (K \circ \phi_1)(a)(D) = M(a)(E) = K(x_2)(D).$$

Since R_D is an equivalence relation for each D, and $\ell(\lfloor\phi_1\|\phi_2\rfloor)$ is the smallest equivalence relation containing $\lfloor\phi_1\|\phi_2\rfloor$, this implies

$$\ell(\lfloor\phi_1\|\phi_2\rfloor) \subseteq \bigcap\{R_D \mid D \in \mathcal{I}(\mathcal{B}(Y), \ell(T))\}$$

which in turn yields that $(\ell(\lfloor\phi_1\|\phi_2\rfloor), \ell(\lfloor\psi_1\|\psi_2\rfloor))$ is a congruence on K. This congruence is trivial by assumption, yielding $\mathsf{f}_1 = \mathsf{f}_2$, as desired.

Turning to the analytic case, we will reduce this case to the one to the Polish one, and we have seen that we can move smooth equivalence relations along arrows (albeit reversing the direction) in Lemma 2. This will be used now to move congruences.

Proposition 3. *Let $\mathsf{K} = (X, Y, K)$ be a Polish object, $\mathsf{L} = (A, B, L)$ be an analytic object, assume that $\mathsf{f} = (\phi, \psi) : \mathsf{K} \to \mathsf{L}$ is a morphism, and that $\mathsf{c} = (\alpha, \beta)$ is a congruence on L. Then $\mathsf{c}_\mathsf{f} := (\alpha_\phi, \beta_\psi)$, is a congruence on K.*

For a characterization of simple stochastic relations analogous to Theorem 1 we fix an analytic object $\mathsf{K} = (X, Y, K)$ together with Polish spaces X_0, Y_0 and surjective Borel maps $f : X_0 \to X$ and $g : Y_0 \to Y$ which define the analytic structure on X resp. Y. We establish for K the following property:

Proposition 4. *These conditions are equivalent for K:*

1. *Each weak 2-bisimulation on K is trivial.*
2. *Each congruence on K is trivial.*

Proof. 0. Since each weak 2-bisimulation is defined on a congruence, the implication *2* $\Rightarrow$ *1* is obvious from Lemma 4. In order to establish the other implication, we will construct from a given congruence $\mathsf{c} = (\alpha, \beta)$ on K together with the derived pair $\mathsf{c}_{f,g} := (\alpha_f, \beta_g)$ a stochastic relation $\mathsf{K}_0 := (X_0, Y_0, K_0)$ on which $\mathsf{c}_{f,g}$ is a congruence, then construct a smooth 2-bisimulation $\mathsf{M}_0 = (\alpha_f, \beta_g, M_0)$ on K_0, and use this for constructing a weak 2-bisimulation $\mathsf{M} = (\alpha, \beta, M)$ on K.

1. The relations α_f and β_g are smooth equivalence relations on X_0 resp. Y_0. Define for $E \in \mathcal{I}(\mathcal{B}(Y), \beta)$ and $x_0 \in X_0$ $K_0'(x_0)(g^{-1}[E]) := K(f(x_0))(E)$, then we see from Lemma 2 that $K_0' : (X_0, \mathcal{B}(X_0)) \rightsquigarrow (Y_0, \mathcal{I}(\mathcal{B}(Y_0), \beta_g))$ is a stochastic relation, so by [6, Proposition 6] we can find a stochastic relation $K_0 : (X_0, \mathcal{B}(X_0)) \rightsquigarrow (Y_0, \mathcal{B}(Y_0))$ extending K_0'. Then $\mathsf{c}_{f,g}$ is a congruence on K_0: let $\langle x_0, x_1\rangle \in \alpha_f$, and $E_0 \in \mathcal{I}(\mathcal{B}(Y_0), \beta_g)$ be an invariant Borel set in Y_0. We know then that $\langle f(x_0), f(x_1)\rangle \in \alpha$, and that $E_0 = g^{-1}[g[E_0]]$ with $g[E_0] \in \mathcal{I}(\mathcal{B}(Y), \beta)$. This yields $K_0(x_0)(E_0) = K_0(x_0)(g^{-1}[g[E_0]]) = K_0(x_1)(E_0)$. From Proposition 2 we get a smooth 2-bisimulation $\mathsf{M}_0 = (\alpha_f, \beta_g, M_0)$ on K_0. We show that this implies $M_0(x_0, x_1)((P \times Y_0) \cap \beta_g) = M_0(x_0', x_1')((P \times Y_0) \cap \beta_g)$, provided $P \in \mathcal{I}(\mathcal{B}(Y_0), \beta_g)$ is a β_g-invariant Borel set in Y_0, and $\langle x_0, x_1\rangle, \langle x_0', x_1'\rangle \in \alpha_f$ with $f(x_0) = f(x_0')$ or $f(x_1) = f(x_1')$ using the bisimulation property for M_0: assume that $f(x_0) = f(x_0')$, then

$$\begin{aligned} M_0(x_0, x_1)((P \times Y_0) \cap \beta_g) &= K_0(x_0)(P) \\ &\overset{(*)}{=} K_0(x_0')(P) \\ &= M_0(x_0', x_1')((P \times Y_0) \cap \beta_g) \end{aligned}$$

Eq. (*) follows from the observation that $f(x_0) = f(x_0')$ implies $\langle x_0, x_0' \rangle \in \alpha_f$ (note that x_1, x_1' are used to make sure that the respective arguments lie in the domain of M_0).

Now introduce the stochastic relation $\mathsf{M} = ((\alpha, \mathcal{B}(\alpha)), (\beta, \otimes [Y, \beta]), M)$ by defining for $\langle a, a' \rangle = \langle f(x_0), f(x_0') \rangle \in \alpha$ and for $B \in \mathcal{I}(\mathcal{B}(Y), \beta)$ the subprobability

$$M(a, a')((B \times Y) \cap \beta) := M_0(x_0, x_0')((g^{-1}[B] \times Y_0) \cap \beta_g).$$

Then the discussion above shows that M is well defined, provided we can establish that $(g^{-1}[B] \times Y_0) \cap \beta_g \in \otimes [Y_0, \beta_g]$ is true. But we know that $g^{-1}[B] \in \mathcal{I}(\mathcal{B}(Y_0), \beta_g)$ holds.

2. It remains to show that M is indeed a weak 2-bisimulation. Let $\langle a, a' \rangle \in \alpha$ with $a = f(x_0), a' = f(x_0')$, and take a β-invariant Borel set $E \subseteq Y$. Then $\pi_{1,Y}^{-1}[E] = (E \times Y) \cap \beta$, thus putting all this together, we obtain

$$\begin{aligned} M(a, a')(\pi_{1,Y}^{-1}[E]) &= M_0(x_0, x_0')((g^{-1}[E] \times Y) \cap \beta_g) \\ &= K_0(x_0)(g^{-1}[E]) \\ &= K(a)(E) \end{aligned}$$

We obtain as a consequence the analogue to Theorem 1 for analytic objects, Figure 2 in the Introduction provides a pictorial summary.

Theorem 2. *Consider these statements for the analytic object* K

(a). K *is simple.*
(b). Each smooth 2-bisimulation on K *is trivial.*
(c). Each weak 2-bisimulation on K *is trivial.*
(d). Let $\mathsf{f}_1, \mathsf{f}_2 : \mathsf{M} \to \mathsf{K}$ *be morphisms, where* M *is an analytic object, then* $\mathsf{f}_1 = \mathsf{f}_2$.
(e). Each congruence on K *is trivial.*

Then these implications hold: (d) $\Rightarrow$ *(a)* $\Leftrightarrow$ *(e)* $\Rightarrow$ *(b)* $\Rightarrow$ *(c)* $\Leftarrow$ *(e).*

We are now in a position to characterize simple systems over analytic spaces completely. Let $\mathbb{1} := \{*\}$ be the one-element space with the discrete topology (which is Polish) and $\mathcal{P}(\mathbb{1})$ as its Borel sets. This space plays a distinguished rôle:

Proposition 5. *The analytic objects* $(X, \mathbb{1}, K)$ *such that* $x \mapsto K(x)(\mathbb{1})$ *is injective are exactly the simple analytic objects.*

Thus the simple objects in the category of stochastic relations over analytic spaces are in one-to-one correspondence with the injective Borel maps from analytic spaces to the unit interval.

Call finally an object F *final* iff given another object M there exists exactly one morphism $\mathsf{f} : \mathsf{M} \to \mathsf{F}$. In view of Theorem 1, a final object is simple. The category of stochastic relations does not have final objects: Being simple, a final object would have the shape $\mathsf{F} = (X, \mathbb{1}, F)$ according to Proposition 5. But X cannot have more than one element, thus $\mathsf{F} = (\mathbb{1}, \mathbb{1}, F)$ with $F(*)(\mathbb{1}) = r$ for some

$r, 0 \leq r \leq 1$. But then there would be a unique morphism $(\mathbb{1}, \mathbb{1}, K) \to (\mathbb{1}, \mathbb{1}, K')$ with $K'(*)(\mathbb{1}) = r' \neq r$. This is evidently impossible.

We have, however, the following positive result:

Corollary 1. *The full subcategory of stochastic relations* (X, Y, K) *such that* $K(x)(Y) = 1$ *holds for all* $x \in X$ *has a final object* $(\mathbb{1}, \mathbb{1}, F)$.

6 Conclusion

We have identified simple stochastic relations over analytic spaces by providing a complete characterization for them: they are essentially the injective Borel maps from the space to the unit interval. It turns out that, save the case of truly probabilistic systems, there does not exist a final system. Essential tools are bisimulations in various forms. We develop the idea of a bisimulation along the lines originally proposed by Milner and later taken up e.g. by Rutten [9]: bisimulations are based there on set-theoretic relations, and these relations can be restricted further, e.g. to be an equivalence relation. In this way we get a hierarchy of bisimulations: those that are general and are based on the notion of a morphism, those that are based on projections, and finally those that are based on equivalence relations. A result that may be of independent interest is the observation that a congruence on a stochastic relation may be extended to a bisimulation, provided the stage for this relation is made up from Polish spaces.

The characterization of simple systems through bisimulations and congruences has been undertaken by Rutten for coalgebras under the assumption that the functor on which the coalgebra is based preserves weak pullbacks [9]. We know, however, that the probability functor does not have this appealing property [3, 6], so that a recourse to the coalgebraic methods employed by Rutten is not possible.

We show how a characterization of simple systems can be used for the average case analysis of algorithms. This is — again — somewhat similar to the coalgebraic case, but the non-existence of final systems limits the applicability of this method severely. In this case we show that simple systems can be used for obtaining a new result for counting heaps.

Some further work should be done here, for example it may be interesting to see whether other structures based on probabilistic relations (like stochastic Petri nets, for example) offer themselves for a similar treatment by exploiting fully the power of bisimulations for a version of these systems that are based on stochastic relations. It could be shown in [7] that stochastic relations and their congruences are a helpful tool in the investigation of problems related to model checking [1].

References

[1] C. Baier, B. Haverkort, H. Hermanns, and J.-P. Koert. Model-checking algorithms for continuous time Markov chains. *IEEE Trans. Softw. Eng.*, 29(6):524 – 541, June 2003.

[2] J. Desharnais, A. Edalat, and P. Panangaden. Bisimulation of labelled Markov-processes. *Information and Computation*, 179(2):163 – 193, 2002.
[3] E.-E. Doberkat. Semi-pullbacks and bisimulations in categories of stochastic relations. In *Proc. ICALP'03*, volume 2719 of *Lecture Notes in Computer Science*, pages 996 – 1007, Berlin, 2003. Springer-Verlag.
[4] E.-E. Doberkat. Factoring stochastic relations. *Information Processing Letters*, 90(4):161 – 166, May 2004.
[5] E.-E. Doberkat. Look: simple stochastic relations are just, well, simple. Technical Report 152, Chair for Software Technology, University of Dortmund, November 2004.
[6] E.-E. Doberkat. Semi-pullbacks for stochastic relations over analytic spaces. *Math. Struct. Comp. Sci.*, 2005. (in print).
[7] E.-E. Doberkat. Zeno paths, congruences and bisimulations for continuous-time stochastic logic. Technical Report 155, Chair for Software Technology, University of Dortmund, March 2005.
[8] S. Lang. *Algebra*. Addison-Wesley, Reading, Mass., 1965.
[9] J. J. M. M. Rutten. Universal coalgebra: a theory of systems. *Theoretical Computer Science*, 249(1):3 – 80, 2000. Special issue on modern algebra and its applications.
[10] J. J. M. M. Rutten. Bisimulation in enumerative combinatorics. *ENTCS*, 65(1):1 – 19, October 2002.
[11] J. J. M. M. Rutten. Behavioral differential equations: a coinductive calculus of streams, automata and power series. *Theor. Comp. Sci.*, (308):1 – 53, 2003.
[12] S. M. Srivastava. *A Course on Borel Sets*. Graduate Texts in Mathematics. Springer-Verlag, Berlin, 1998.
[13] D. H. Wagner. A survey of measurable selection theorems. *SIAM J. Control Optim.*, 15(5):859 – 903, August 1977.

Modelling Fusion Calculus Using HD-Automata*

Gianluigi Ferrari[1], Ugo Montanari[1], Emilio Tuosto[1],
Björn Victor[2], and Kidane Yemane[2]

[1] Dipartimento di Informatica, Università di Pisa, Italy
[2] Dept. of Information Technology, Uppsala University, Sweden

Abstract. We propose a coalgebraic model of the Fusion calculus based on HD-automata. The main advantage of the approach is that the partition refinement algorithm designed for HD-automata is easily adapted to handle Fusion calculus processes. Hence, the transition systems of Fusion calculus processes can be minimised according to the notion of observational semantics of the calculus. As a beneficial side effect, this also provides a bisimulation checker for Fusion calculus.

1 Introduction

Nominal calculi, process calculi with primitive mechanisms for local name generation, name exchange and scoping rules, have been successfully applied to specify and verify properties of *global computing systems*. Names provide a suitable abstraction to describe a variety of different computational phenomena such as mobility, localities, distributed object systems, security keys, session identifiers and so on. For instance, the π-calculus has been exploited for modelling and verifying a finite instance of the Handover protocol of the Public Land Mobile Network [19]. Several properties of cryptographic protocols have been naturally expressed through spi-calculus specifications [1]. Nominal calculi also provide a basic programming model that has been incorporated in novel programming languages (see e.g. [6, 2]) and workflow languages for Web Service coordination [4, 15].

Verification via semantic equivalence provides a well established framework to reason about the behaviour of systems specified using nominal calculi. In this approach, checking behavioural properties is reduced to the problem of contrasting two system abstractions in order to determine whether their behaviours coincide with respect to a suitable notion of semantic equivalence. However, in the case of nominal calculi verification via semantic equivalence is intrinsically difficult. Indeed, when an unbound number of new names can be generated during execution, models of nominal calculi (e.g. labelled transition systems) tend to be infinite even in the simplest cases unless explicit mechanisms are introduced to deal with names.

Symbolic semantics [14, 3, 16] is a well established approach to finite state verification of nominal calculi. Symbolic semantics takes a *syntax-based* approach and generalises standard operational semantics by keeping track of equalities among names: transitions are derived in the context of such constraints. The main advantage of the

* Work supported by the PROFUNDIS FET-GC project.

J.L. Fiadeiro et al. (Eds.): CALCO 2005, LNCS 3629, pp. 142–156, 2005.

symbolic semantics is that it yields a smaller transition system. The idea of symbolic semantics has been exploited to provide a convenient characterisation of *open bisimilarity* [24] and in the design of the corresponding bisimulation checker, the *Mobility WorkBench* (MWB) [26].

An alternative class of models for nominal calculi are the so-called *syntax-free* models where names are explicitly dealt with regardless of the syntactic structure of the calculi. *Indexed LTSs* [5] and *History Dependent automata* (HD-automata in brief) [17,21, 18] are examples of syntax-free models of nominal calculi developed following the approach based on name permutations. HD-automata have an added value with respect to indexed LTS because they are equipped with powerful verification techniques. Here we focus on HD-automata, which encompass the main features of nominal calculi, namely creation and deallocation of names, and accounts for a compact representation of process behaviour by collapsing states differing only for the renaming of local names. Basically, the "history" of the names appearing in the computation is explicitly represented so that it is possible to reconstruct the associations that have led to a given state. Clearly, if a state is reached in two different computations, different histories are assigned to its names [17,21]. In [18], states of HD-automata have been equipped with name symmetries which further reduces the size of the automata and guarantee the existence of the minimal realization. The computation of the minimal automata is derived by exploiting a coalgebraic presentation of the partition refinement algorithm [7]. The minimisation algorithm for the early semantics of π-calculus has been implemented in the Mihda toolkit [10]. Since Mihda does not rely on a symbolic semantics, the number of states of the minimal HD-automaton is unnecessarily large due to the existence of different input transitions for different instantiations of the input variable. Hence, the integration of symbolic techniques and syntax-free models would provide more powerful verification methods. Notice that minimisation algorithms for syntax-based models have been already developed (e.g., for the open semantics of the π-calculus[22]).

In this paper we introduce a novel symbolic semantics for the Fusion calculus [20]. The Fusion calculus is a nominal calculus which extends the π-calculus with the capability of *fusing* names. When two names are fused then they can be used interchangeably. The Fusion calculus has been introduced as a simplification and generalisation of the π-calculus. Apart from the theoretical interest Fusion calculus seem to arise naturally in the implementation of distributed systems, e.g. workflow languages for service coordination [23].

The main technical contribution of this paper is the development of a coalgebraic framework for the Fusion calculus equipped with the symbolic semantics. We remark that this is the first coalgebraic semantics for Fusion calculus. Moreover, our results allows the minimisation algorithm for finite HD-automata to be smoothly extended to the Fusion calculus. The coalgebraic semantics behaves in accordance with the symbolic semantics of Fusion calculus, e.g. bisimilar processes are mapped together by the morphisms yielding the minimal HD-automaton. Finally, this provides an algorithm for checking hyperbisimulation of finitary Fusion calculus agents.

The paper is structured as follows: In section 2 we describe HD-automata as a coalgebra over a category of *named sets* and the minimisation procedure for the HD-automata and, in Section 3, we provide a brief overview of Fusion calculus together with a new symbolic semantics. This will be followed by the HD-automata for Fusion calculus and how minimisation works for Fusion calculus HD-automata in Section 4, and we conclude in Section 5 with some conclusion, related work and some ideas for future work.

2 History Dependent Automata

Verification of concurrent and mobile systems specified using nominal calculi is intrinsically difficult since the state space can easily become extremely large or even infinite. *History Dependent automata* (HD-automata in brief) [21, 17, 18, 7] are an operational model for nominal calculi designed to address finite state verification. HD-automata can be seen as automata enriched by equipping states and transition with names. This permits to model name creation/deallocation or name extrusion which are typical linguistic mechanisms of nominal calculi.

A noteworthy fact is that names in states of HD-automata have *local meaning*, hence a compact representation of agent behaviour can be achieved by collapsing states that differ only for renaming of local names. Following [10], we provide a formal definition of HD-automata which basically differs from definitions of [10] in avoiding usage of dependent types. Indeed, the presentation in [10] aims at showing how the formal definition of the partition refinement algorithm for HD-automata guides and corresponds to its implementation Mihda [8]. Here, we build named sets on top of *permutation algebras*, namely we give a more abstract definition which focuses on the main mathematical ingredients necessary to describe the coalgebraic model and the related minimisation algorithm for Fusion calculus without taking into account implementation details.

Before giving the formal definitions, it is worth to collect some notations.

We consider a set $\mathcal{N}_\star$ made of a countable set of names $\mathcal{N}$ and a distinguished element $\star \notin \mathcal{N}$. We let $\mathrm{Aut}(\mathcal{N})$ be the set of bijective endofunctions on $\mathcal{N}$, i.e., the permutations of $\mathcal{N}$. Given a permutation ρ such that $\mathrm{dom}(\rho) \subset \mathcal{N}$ (where $\mathrm{dom}(\rho)$ is the domain of ρ), $\underline{\rho}$ is the automorphism on $\mathcal{N}_\star$ obtained by extending ρ on $\mathcal{N}_\star$ so that $\underline{\rho}(x) = x$ for any $x \in \mathcal{N}_\star \setminus \mathrm{dom}(\rho)$. The application of a function on names γ to an element e is written as $e\gamma$ and, when e is a set, it stands for the point-wise application of γ to the elements of e.

Definition 1 (Permutation algebras). *A* permutation algebra *is an algebra* $\langle S, O\rangle$, *where* S *is the carrier of the algebra and the set of operations* O *contains unary operators* $\{\widehat{\rho} \mid \rho \in \mathrm{Aut}(\mathcal{N})\}$ *such that the following axioms hold.*

$$\forall x \in S.\, x\widehat{id} = x, \qquad \forall x \in S \forall \rho_1, \rho_2 \in \mathrm{Aut}(\mathcal{N}).\, x\widehat{\rho_1;\rho_2} = (x\widehat{\rho_1})\widehat{\rho_2}.$$

Given a finite set of names N, $\mathrm{sym}(N)$ is the set defined as $\mathrm{sym}(N) = \{\rho \in \mathrm{Aut}(\mathcal{N}) \mid \forall x \notin N.\rho(x) = x\}$. Notice that $\mathrm{sym}(N)$ is a subgroup of $\mathrm{Aut}(\mathcal{N})$.

Following [18], we can see the Fusion calculus as a permutation algebra: the carrier is the set of processes of Fusion calculus (up to structural congruence) and the operations are name permutations interpreted as substitutions.

We introduce the notions of *named sets* and *named functions* which form the category **NS** of named sets, in terms of permutation algebras. Then we study the structure of **NS**.

Definition 2 (Named sets). *A* named set *(ns) is a pair* $\langle Q, \mathrm{g}\rangle$ *where;*

1. Q *is a permutation algebra;*
2. $\mathrm{g} : Q \to \bigcup_{N \in \wp_{fin}(\mathcal{N})} \{\mathrm{sym}(N)\}$ *assigns to any* $q \in Q$ *a group of permutations* $\mathrm{g}(q)$ *over a finite set of names such that* $q = q\widehat{\rho}$, *for any* $\rho \in \mathrm{g}(q)$. *The* names of q, *written as* $|q|$, *are defined as the domain of the permutations* $\rho \in \mathrm{g}(q)$ *while* $\|q\|$ *is the cardinality of* $|q|$.

We let D, E, F range over nss and, given $D = \langle Q, \mathrm{g}\rangle$, we write Q_D (resp. g_D) to denote Q (resp. g).

Definition 3 (Isomorphism of named sets). *Two nss D and E are* isomorphic *if there exists an* isomorphism between D and E, *namely a bijective function* $s : Q_D \to Q_E$ *such that, for any* $d \in Q_D$ *there is a bijective correspondence* $n : |d| \to |s(d)|$ *such that* $\mathrm{g}_D(d) = \mathrm{g}_E(s(d)); n$.

Therefore, we consider nss up-to isomorphism so that two nss are considered equal whenever they have the same "structure" despite of having different underlying sets and different names associated to each element. It only matters the number of names and the symmetry associated to each element, namely, names are local to elements of nss.

Named functions basically are functions that preserve the structure of nss.

Definition 4 (Named functions). *Given two named sets D and E, a* named function *(nf)* $H : D \to E$ *is a pair* $\langle h, \Sigma\rangle$ *where* $h : Q_D \to Q_E$ *and* $\Sigma : Q_D \to \wp_{fin}(Q_E \times \mathcal{N}_\star^{\mathcal{N}})$ *are such that, for all* $q \in Q_D$ *and* $(e, \sigma) \in \Sigma(q)$,

1. σ *is injective,* $\sigma(|e|) \subseteq |q| \cup \{\star\}$ *and* $\sigma(x) = x$, *for any* $x \in \mathcal{N} \setminus |e|$;
2. $\sigma; \mathrm{g}_D(q) \subseteq \Sigma_2(q)$, *where* $\Sigma_2(q) = \bigcup_{(e,\sigma)\in\Sigma(q)} \{\sigma\}$ *and the permutations in* $\mathrm{g}_D(q)$ *are all meant to act as the identity on* $\star$;
3. $\mathrm{g}_E(h(q)); \sigma = \Sigma_2(q)$.

Named functions are ranged over by H, K and J. We write h_H and Σ_H for denoting the first and the second components of H. The intuition behind conditions 1, 2 and 3 naturally emerges when nfs are exploited to describe the transitions out of a certain state. Intuitively, elements in $\mathrm{h}_H(q)$ are the transitions out of q and $\Sigma_H(q)$ contains the mappings of names of target states of those transitions to names of q. Condition 1 ensures that any name in $|q|$ has a unique "meaning" along each transition in $\mathrm{h}_H(q)$ (injectivity of σ) and establishes that names in target states are either mapped on names of the source state or to $\star$, the distinguished name representing the generation of a new name along a transition. Condition 2 states that the group of the starting state q does not *generate* transitions which are not in $\Sigma_H(q)$. Finally, condition 3 states that any permutation is in the symmetry of $\mathrm{h}_H(q)$ iff, when applied to any σ mapping names of the transitions to those of q, it yields a map in $\Sigma_H(q)$.

Definition 5 (Composition of named functions). *The* composition $H; K$ *of two nfs* $H : D \to E$ *and* $K : E \to F$ *is* $\langle \mathrm{h}_H; \mathrm{h}_K, \Sigma_H; \Sigma_K\rangle$ *where* $\Sigma_H; \Sigma_K : Q_D \to \wp_{fin}(Q_F \times \mathcal{N}_\star^{\mathcal{N}})$ *is such that*

$$\Sigma_H; \Sigma_K : q \mapsto \bigcup_{(e,\sigma)\in\Sigma_H(q)} \{(f, \sigma; \sigma') \mid (f, \sigma') \in \Sigma_K(\mathrm{h}_H(q))\}.$$

Proposition 1. *In Definition 5, $H;K$ is a nf. Composition of nfs is associative and has identities.*

Proof. The proof proceeds as the corresponding proof in [9]. □

Definition 6 (Category of named sets). *The* category **NS** *has nss as objects and nfs as morphisms.*

The basic characteristics of **NS** are collected in Proposition 2.

Proposition 2 (Structure of NS). *The category* **NS** *has an initial object, a terminal object, and finite powerset functor defined as follows:*

1. *the initial object is given by $\bot = \langle \emptyset, \emptyset \rangle$;*
2. *the terminal object is given by $I = \langle \{*\}, * \mapsto \emptyset \rangle$;*
3. *the powerset functor on* **NS** *is obtained by lifting the covariant powerset functor $\wp_{fin}(\_)$ on* **Set**, *namely $\wp_{fin}(D) = \langle \wp_{fin}(D_Q), \mathrm{g} \rangle$, where, given $Q \subseteq Q_D$, $\mathrm{g}(Q) = \{\rho \mid \rho \text{ is a permutation over } \bigcup_{q \in Q} |q|\} \wedge Q\rho = Q$.*

Definition 7 (Pairing of named sets). *Given two nss D and E, the* pairing $D \otimes E$ of D and E *is defined as $D \otimes E = \langle Q_D \times Q_E, \mathrm{g} \rangle$ where*

- $\mathrm{g} : Q_D \times Q_E \to \bigcup_{N,M \in \wp_{fin}(\mathcal{N})} \{\mathrm{sym}(N) + \mathrm{sym}(M)\}$ *is such that* $\mathrm{g}(d,e) = \{\rho_1 + \rho_2 \mid \rho_1 \in \mathrm{g}_D(d) \wedge \rho_2 \in \mathrm{g}_E(e)\}$.

Formally, $D \otimes E$ is not a ns because the range of $\mathrm{g}_{D \otimes E}$ is not the union of symmetries over a finite set of names. However, observing that for any $(d,e) \in Q_D \times Q_E$, $\mathrm{g}(d,e)$ is a symmetry on $|d| + |e|$, we can find a ns whose group function maps (d,e) to a symmetry over as many names as in $|d| + |e|$ and whose permutations correspond bijectively to the permutations in $\mathrm{g}(d,e)$. Notice that, since nss are considered up-to isomorphism, it does not matter which set of names is chosen for any pair (d,e).

Definition 8 (HD-automata). *Fixed a ns of labels L, a* HD-automaton *over L is a coalgebra for $T_L(D) = \wp_{fin}(L \otimes D)$.*

We emphasise that nfs provide the formal mean to describe a generic step of the iterative minimisation algorithm. Intuitively, nfs map states of the automaton in equivalence classes containing those states considered equivalent.

Definition 9 (Kernel of named functions). *The* kernel of a nf $H : E \to F$ *(written as* $\ker H$*) is the ns D such that:*

1. *$Q_D = \ker \mathrm{h}_H$ considered as permutation algebra where for all $A \in Q_D$ and $\rho \in \mathrm{Aut}(\mathcal{N})$, $A\rho$ is the element-wise application of ρ to A;*
2. *the group of $A \in \ker \mathrm{h}_H$ is $\mathrm{g}_F(\mathrm{h}_H(a))$, for $a \in A$.*

In [7, 10] the normalisation functor for the early semantics of π-calculus has been introduced. In this context, the concept of *redundancy* relies on the concept of *active names* because of the presence of freshly generated names. We generalise this concept by means of *redundant transitions*. Generally, redundant transitions are transitions describing behaviours that can be matched by other transitions in the bisimulation game.

Redundant transitions occur when HD-automata are built out of a nominal calculus. During this phase, it is not possible to decide which are the redundant transitions[1]. Therefore, all the transitions are taken when HD-automata are built and redundant ones are removed during the minimisation. This is achieved by means of *normalisation* which basically gets rid of redundant transitions.

Definition 10 (Normalisation functor). *A* normalisation functor N *is any functor such that* $N(D)$ *is isomorphic to a subset of* D.

The minimisation algorithm on a T_L coalgebra $(D, K : D \to T_L(D))$ is specified by the equations 1 and 2 below.

$$H_{(0)} \stackrel{\text{def}}{=} \langle q \mapsto \bot, q \mapsto \emptyset \rangle, \quad \text{where } \mathrm{dom}(H_{(0)}) = D \tag{1}$$

$$H_{(i+1)} \stackrel{\text{def}}{=} K; N(T(H_{(i)})), \tag{2}$$

where N is a normalisation functor and $T : \mathbf{NS} \to \mathbf{NS}$ is the functor defined as

$$T(D) = \begin{cases} T_L(D) & D \in \mathrm{obj}(\mathbf{NS}) \\ \langle \mathrm{h}, \Sigma \rangle & D = \langle h_D, \Sigma_D \rangle \in \mathbf{NS}(E, F) \text{ for } E, F \in \mathrm{obj}(\mathbf{NS}) \end{cases}$$

where, given $B \in \wp_{\text{fin}}(L \otimes E)$,

$$\begin{aligned} \mathrm{h}(B) &= \{\langle l, \mathrm{h}_D(q) \rangle \mid \langle l, q \rangle \in B\} \\ \Sigma(B) &= \{\langle l, \mathrm{h}_D(q), \sigma; \sigma' \rangle \mid \langle l, q, \sigma' \rangle \in B \wedge \langle l, q', \sigma' \rangle \in \Sigma_D(q)\}. \end{aligned}$$

All the states of automaton K are initially considered equivalent, indeed, $\ker H_0$ gives rise to a single equivalence class containing the whole $\mathrm{dom}(K)$. At the generic $(i+1)$-th iteration, as specified in (2), the image through $H_{(i)}$ of the i-th iteration is composed with K as stated by the definition of functor T, then the normalisation functor removes the redundant transitions. The algorithm builds the minimal realisation $\overline{H}$ of (finite) HD-automata by constructing (an approximation of) the final coalgebra morphism. The kernel of $\overline{H}$ yields the equivalence classes where equivalent states are grouped in the same class.

The proof of the convergence of the algorithm is based on the observation that T is a monotonic functor over finite chains. In order to establish this, we must give a way of saying when two nfs are the same.

Definition 11 (Equivalence of named functions). *Two nfs* $H : D \to E$ *and* $K : D' \to E'$ *are* equivalent *when* $\ker H$ *is isomorphic to* $\ker K$ *via the bijections* n *and* s *(see Definition 3) and for all* $q \in D$, $\mathrm{g}_E(\mathrm{h}_H(q)); n = \mathrm{g}_{E'}(\mathrm{h}_K(s(q)))$.

Definition 12 (Order of named functions). *Let* $H : D \to E$ *and* $K : D \to F$ *be two nfs,* H is less than or equal to K *(written as* $H \preceq K$*) if, and only if,*

- $Q_{\ker H}$ *is coarser than* $Q_{\ker K}$ *and* $\forall A \in Q_{\ker H}. \forall B \in Q_{\ker K}.\ B \subseteq A \Rightarrow \mathrm{g}_{\ker H}(A) \subseteq \mathrm{g}_{\ker K}(B)$.
- $\forall A \in Q_{\ker H}. \forall B \in Q_{\ker K}.\ \forall q \in A \cap B. \Sigma_H(q) \subseteq \Sigma_K(q)$.

[1] In general, to decide redundancy is as difficult as deciding bisimilarity.

Proposition 3. *Relation $\preceq$ is a preorder and $H \preceq K \wedge K \preceq H$ implies H* and *K* equivalent.

Proof. The first condition of Definition 12 and $H \preceq K \wedge K \preceq H$ imply that $\ker H$ is isomorphic to $\ker K$. It remains to prove that the hypothesis implies the last condition of Definition 11. Assume that there is $q \in \mathrm{dom}(H)$ such that $g_{\mathrm{cod}(H)}(\mathrm{h}_H(q)) \neq g_{\mathrm{cod}(K)}(\mathrm{h}_K(q))$. Then, for all $\sigma \in \Sigma_H(q)$,

$$g_{\mathrm{cod}(H)}(\mathrm{h}_H(q));\sigma \neq g_{\mathrm{cod}(K)}(\mathrm{h}_K(q));\sigma. \qquad (3)$$

By Definition 12, $\Sigma_H(q) = \Sigma_K(q)$ since $H \preceq K \wedge K \preceq H$. Moreover, conditions on nfs (Definition 4) imply that $g_{\mathrm{cod}(H)}(\mathrm{h}_H(q));\sigma = \Sigma_{\mathrm{h}_H}(q)$ and $g_{\mathrm{cod}(K)}(\mathrm{h}_K(q));\sigma = \Sigma_{\mathrm{h}_K}(q)$ that contradicts (3). □

Monotonicity is preserved by composition of named functions:

Lemma 1. *Let H and K be two nfs such that $H \preceq K$. For any ns J, if* $\mathrm{cod}(J) = \mathrm{dom}(H) = \mathrm{dom}(K)$ *then $J;H \preceq J;K$.*

Finally, we can prove the convergence of the iterative algorithm:

Theorem 1 (Convergence). *If the normalisation functor N is monotone on nfs then the iterative algorithm described by (1) and (2) converges on finite state HD-automata.*

Proof. By construction, $\wp_{\mathrm{fin}}(\_)$ is monotone, hence T is monotone because it is the composition of two monotone functors. By monotonicity of T and Lemma 1, map $\mathcal{M} : H \mapsto K;T(H)$ is monotone and finite. Finally, all nfs chains having finite domain are finite, hence, the algorithm in (1) and (2) converges to the maximal fix-point of $\mathcal{M}$. □

The proof of Theorem 1 mimics that in [10]. The only difference is that the theorem in [10] is proved only for the case of the early semantics of π-calculus, while here, the result is extended to the general case of finite HD-automata, with the only additional assumption that the normalisation functor is monotone.

3 Syntax and Semantics of the Fusion Calculus

We briefly recollect the syntax and operational semantics of the Fusion calculus referring the reader to [20] for further details. In Section 3.1 we present a new canonical symbolic semantics.

Definition 13 (Fusion calculus syntax). *The* free actions *ranged over by* α, fusion action *ranged over by* φ, actions *ranged over by* γ, *and the* agents *ranged over by* $P, Q, \ldots$, *are defined by*

$$\begin{array}{llll} \alpha ::= u\tilde{x} \mid \overline{u}\tilde{x} & & P ::= \mathbf{0} \mid \gamma.Q \mid Q+R \mid Q \mid R \\ \gamma ::= \alpha \mid \varphi & & \quad\;\; \mid (x)Q \mid [x=y]Q \mid A\langle\tilde{x}\rangle, \end{array}$$

where $x, y, u, v \ldots$ range over $\mathcal{N}$ and represent communication channels, which are also the values transmitted. An input action $u\tilde{x}$ means "input objects along the port u and replace $\tilde{x}$ with these objects". Note that input does not entail binding. The output action

Table 1. Transition rules for the Fusion calculus

$$\text{PREF}\ \frac{-}{\alpha.P \xrightarrow{\alpha} P} \qquad \text{SUM}\ \frac{P \xrightarrow{\gamma} P'}{P+Q \xrightarrow{\gamma} P'} \qquad \frac{P \xrightarrow{\gamma} P'}{P \mid Q \xrightarrow{\gamma} P' \mid Q}\ \text{PAR}$$

$$\text{COM}\ \frac{P \xrightarrow{u\tilde{x}} P',\ Q \xrightarrow{\overline{u}\tilde{y}} Q',\ |\tilde{x}| = |\tilde{y}|}{P \mid Q \xrightarrow{\{\tilde{x}=\tilde{y}\}} P' \mid Q'} \qquad \frac{P \xrightarrow{\varphi} P',\ z\varphi x,\ z \neq x}{(z)P \xrightarrow{\varphi\backslash z} P'\{x/z\}}\ \text{SCOPE}$$

$$\text{PASS}\ \frac{P \xrightarrow{\gamma} P',\ z \notin \mathrm{n}(\gamma)}{(z)P \xrightarrow{\gamma} (z)P'} \qquad \frac{P \xrightarrow{(\tilde{y})a\tilde{x}} P',\ z \in \tilde{x} - \tilde{y},\ a \notin \{z, \bar{z}\}}{(z)P \xrightarrow{(z\tilde{y})a\tilde{x}} P'}\ \text{OPEN}$$

$$\text{MATCH}\ \frac{P \xrightarrow{\gamma} P'}{[x=x]P \xrightarrow{\gamma} P'} \qquad \frac{P \equiv P' \quad P \xrightarrow{\gamma} Q \quad Q \equiv Q'}{P' \xrightarrow{\gamma} Q'}\ \text{STRUCT}$$

$\overline{u}\tilde{x}$ means "output the objects $\tilde{x}$ along the port u". A fusion action $\{\tilde{x} = \tilde{y}\}$ represents an obligation to make $\tilde{x}$ and $\tilde{y}$ equal everywhere, limited by the scope of the names involved.

A Fusion calculus process is either the void process **0**, or a process prefixed by an action $\gamma.P$ that is ready to perform γ and continue as P, or non-deterministic choice $P+Q$, or the parallel composition of two processes $Q \mid R$, or a process with a restricted name $(x)Q$, or a process preceded by a check for the equality of names $[x=y]Q$, or else the invocation of a process definition $A\langle\tilde{x}\rangle$, where A is a process identifier and we assume that for each process identifier there is a single defining equation $A\langle\tilde{x}\rangle \stackrel{\text{def}}{=} P$ such that all the names in $\mathrm{fn}(P)$ appear in $\tilde{x}$.

Definition 14 (Fusion calculus semantics). *The labelled transition system of Fusion calculus is the least relation satisfying the inference rules in Table 1.*

Use of the SCOPE rule entails a substitution of the scoped name z for a nondeterministically chosen name x related to it by φ. For the purpose of the equivalence defined below it will not matter which such x replaces z. The only rule dealing with bound actions is OPEN. Using structural congruence, pulling the relevant scope to top level, we can still infer e.g. $P \mid (x)ayx.Q \xrightarrow{(x)ayx} P \mid Q$ using PREF and OPEN (provided $x \notin \mathrm{fn}(P)$, otherwise an alpha-conversion is necessary).

Definition 15 (Hyperbisimulation [20]). *A* fusion bisimulation *is a binary symmetric relation $\mathcal{S}$ between agents such that $(P,Q) \in \mathcal{S}$ implies:*
If $P \xrightarrow{\gamma} P'$ with $\mathrm{bn}(\gamma) \cap \mathrm{fn}(Q) = \emptyset$, then $Q \xrightarrow{\gamma} Q'$ and $(P'\sigma_\gamma, Q'\sigma_\gamma) \in \mathcal{S}$. Agents P and Q are fusion bisimilar, *written $P \dot{\sim} Q$, if $(P,Q) \in \mathcal{S}$ for some fusion bisimulation $\mathcal{S}$. A* hyperbisimulation *is a substitution closed fusion bisimulation.*

The interesting point in this definition is the treatment of fusion actions. Indeed, if γ is a fusion action, it only makes sense to relate P' and Q' when a substitution σ_γ, induced by γ, has been performed. For instance, a fusion $\{x=y\}$ induces a substitution $\sigma_{\{x=y\}}$, i.e. $\{y/x\}$ or $\{x/y\}$.

Theorem 2. [20] *Hyperequivalence is the largest congruence in fusion bisimilarity.*

Table 2. Canonical symbolic transition system for the Fusion calculus

$$\text{PREF}\ \frac{-}{\alpha\,.P \xrightarrow{\emptyset,\alpha} P\sigma_\alpha} \qquad \text{SUM}\ \frac{P \xrightarrow{M,\gamma} P'}{P+Q \xrightarrow{M,\gamma} P'} \qquad \frac{P \xrightarrow{M,\gamma} P'}{P \mid Q \xrightarrow{M,\gamma} P' \mid Q\sigma_M\sigma_\gamma}\ \text{PAR}$$

$$\text{SCOPE}\ \frac{P \xrightarrow{M,\varphi} P',\ z\varphi x,\ z \neq x,\ z \notin \mathrm{n}(M)}{(z)P \xrightarrow{M,\varphi\backslash z} P'\{x/z\}} \qquad \frac{P \xrightarrow{M,\gamma} P' \qquad M' = M[x=y]}{[x=y]P \xrightarrow{M',\gamma\sigma_{[x=y]}} P'\sigma_{M'}}\ \text{MATCH}$$

$$\text{PASS}\ \frac{P \xrightarrow{M,\gamma} P',\ z \notin \mathrm{n}(M,\gamma)}{(z)P \xrightarrow{M,\gamma} (z)P'}$$

$$\frac{P \xrightarrow{M,(\tilde{y})a\tilde{x}} P',\ z \in \tilde{x}-\tilde{y},\ a \notin \{z,\bar{z}\},\ z \notin \mathrm{n}(M)}{(z)P \xrightarrow{M,(z\tilde{y})a\tilde{x}} P'}\ \text{OPEN}$$

$$\text{COM}\ \frac{P \xrightarrow{M,u\tilde{x}} P',\ \ Q \xrightarrow{N,\bar{v}\tilde{y}} Q',\ \ |\tilde{x}| = |\tilde{y}|,\ \ L = MN[u=v],\ \varphi = \{\tilde{x}=\tilde{y}\}\sigma_L}{P \mid Q \xrightarrow{L,\varphi} (P' \mid Q')\sigma_L\sigma_\varphi}$$

3.1 Canonical Symbolic Semantics of Fusion Calculus

Having briefly presented Fusion calculus syntax together with its concrete semantics, we provide a new symbolic semantics of Fusion calculus which lend itself to coalgebraic modelling through HD-automata. The *canonical* symbolic semantics for the Fusion calculus is defined along the lines of symbolic semantics for the π-calculus [24, 22]. Symbolic semantics are often used to give efficient characterisations of bisimulation equivalences for value-passing calculi.

In Table 2 we present the symbolic transition system where structurally equivalent agents are considered the same. Like in [24] a symbolic transition is of the form $P \xrightarrow{M,\gamma} Q$, where M is the enabling condition of the action γ in the sense that M represents the equalities a minimal substitution σ_M must make true in order for $P\sigma_M$ to perform the corresponding action in the original labelled transition system. σ_M is the *substitutive effect* of M: an idempotent substitution s.t. $\sigma_M(x) = \sigma_M(y)$ iff $M \Rightarrow x = y$. We generalise substitutive effects to actions, where σ_α is the identity substitution.

In Table 2 we write MN for denoting the concatenation of M and N. Following Pistore and Sangiorgi's work [22], our transition rules apply substitutions to the continuation of a transition: like [22], a substitution σ_M, making the condition for the transition true, and in addition a substitution σ_γ, the substitutive effect of the action, is applied to the right-hand side of the transition. The motivation for this is to make the definition of bisimulation simpler and more in line with the algorithms used in the HD framework (see Section 4). We show later in this section that bisimulation using the symbolic semantics coincides with the original non-symbolic version.

Using canonical substitutions gives us pleasant properties like the following:

Lemma 2. *If* $P \xrightarrow{M,\gamma} P'$*, then* $\gamma = \gamma\sigma_M$ *and* $P' = P'\sigma_M = P'\sigma_\gamma = P'\sigma_M\sigma_\gamma$.

The definition of symbolic hyperbisimulation is similar to that of symbolic open bisimulation [24,22], but does not have the complication of distinctions.

Definition 16 (Symbolic hyperbisimulation). *A binary symmetric process relation* $\mathcal{S}$ *is a* symbolic hyperbisimulation *if* $(P,Q) \in \mathcal{S}$ *implies:*
If $P \overset{M,\gamma}{\longmapsto} P'$ *with* $\mathrm{bn}(\gamma) \cap \mathrm{fn}(Q) = \emptyset$ *then* $Q \overset{N,\gamma'}{\longmapsto} Q'$ *such that*
$M \Rightarrow N$, $\gamma = \gamma'\sigma_M$, *(note* $\gamma = \gamma\sigma_M$*)*
and $(P', Q'\sigma_M) \in \mathcal{S}$ *(note* $P' = P'\sigma_M$*).*
P is symbolically hyperequivalent *to Q, written* $P \simeq Q$, *if* $(P,Q) \in \mathcal{S}$ *for some symbolic hyperbisimulation* $\mathcal{S}$.

Since the symbolic semantics applies the substitutive effects, we can leave most of that out of the bisimulation definition. It is still necessary to apply substitution corresponding to the stronger condition, σ_M, to the label and continuation of the transition of Q. (Note that $Q'\sigma_M = Q'\sigma_M\sigma_\gamma$.)

Theorem 3. $P \sim Q$ *iff* $P \simeq Q$

Proof. See appendix.

4 From Fusion Calculus to HD-Automata

This section describes how agents of Fusion calculus can be mapped onto HD-automata and what normalisation means for the HD-automata for Fusion calculus. We first introduce labels and transitions and then define the normalisation functor for Fusion calculus. Let us remark that the monotonicity of this functor guarantees the convergence of the minimisation algorithm on finite HD-automata that correspond to Fusion calculus agents. We conclude the section with an informal discussion on the correspondence between hyperequivalence and minimisation. In order to keep the coalgebraic presentation as simple as possible, we limit to a monadic version of Fusion calculus where tuples in communication actions carry a single name.

Though not increasing the expressiveness of the calculus, polyadicity would obscure the main picture of the coalgebraic presentation with cumbersome technical details. (A mapping of the polyadic Fusion calculus to HD-automata is given in [25].)

The labels of the canonical symbolic semantics of Fusion calculus consists of enabling conditions and actions; both of them can be represented as nss.

Let M be the ns $\langle\{\bullet\},\mathrm{g}\rangle$ where $\mathrm{g} = \{id_{x,y}, exch_{x,y}\}$, namely, g contains the identity and the exchanging permutation on the two names in M i.e, $|\bullet| = \{x,y\}$.

Definition 17 (Matching named set). *A* matching named set *is a ns of the form* $\mathtt{M} = \underbrace{M \otimes \ldots \otimes M}_{n \geq 0}$, *also written as* M^n *(recall, that pairing treats names in a component as distinguished from those in other components;* $\mathtt{M}_i$ *is the i-th component of* $\mathtt{M}$*).*

Given a name substitution $\sigma \in \mathcal{N}_\star{}^{\mathcal{N}}$, *the* interpretation of $\mathtt{M}$ in σ *is* $[\![\mathtt{M}]\!]_\sigma$ *and is defined as*

$$\sigma(x_1) = \sigma(y_1) \wedge \cdots \wedge \sigma(x_n) = \sigma(y_n),$$

where, for any $i = 1,\ldots,n$, $\{x_i, y_i\}$ *are the names in* $\mathtt{M}_i$.

As notation, M^0 is the ns $\langle\{\bullet\}, \bullet \mapsto \emptyset\rangle$ namely, the singleton ns where $|\bullet| = \emptyset$. Basically, enabling conditions are represented by tupling matching nss each representing a fusion of two names. The interpretation of $\mathtt{M}$ under σ is the statement constraining the names x_i and y_i of any component $\mathtt{M}_i$ to be identified once they are interpreted through σ. Notice that the interpretation of M^0 under any substitution always holds and, indeed, it represents the trivial condition $[x = x]$. (Any substitution σ such that $[\![\mathtt{M}]\!]_\sigma$ holds true is said to be compatible with $\mathtt{M}$.)

Definition 18 (Labels for Fusion calculus). *Let* $\mathtt{M}$ *be a matching ns and Lab be the set* $\{\mathtt{tau}, \mathtt{in}, \mathtt{out}, \mathtt{fuse}\}$. *The* nss of labels for Fusion calculus *are the nss* $\mathtt{M} \otimes \langle Lab, \mathrm{g}\rangle$ *where* $\mathrm{g} : Lab \to \mathcal{N}_\star$ *is such that*

$$\|\mathtt{tau}\| = 0$$
$$\|\mathtt{fuse}\| = 2 \wedge \mathrm{g}(\mathtt{fuse}) \text{ has the identity and the exchanging permutations}$$
$$\|\mathtt{in}\|, \|\mathtt{out}\| \leq 2 \wedge \mathrm{g}(\mathtt{in}),\ \mathrm{g}(\mathtt{out}) \text{ have only the identity permutation.}$$

$\mathtt{M}$ *is called the* enabling part *and* g *is called the* action part *of labels for Fusion calculus.*

Hereafter, L stands for the ns of labels for Fusion calculus and $K : D \to T_L(D)$ is an HD-automaton. (Roughly, L represents labels of transitions of the canonical symbolic semantics of Fusion calculus.)

Let P be a Fusion calculus agent, $K[P]$ denotes the coalgebraic specification of the HD-automaton associated to P, namely a T_L-coalgebra such that $\mathrm{dom}(K[P]) = D[P]$ and $\mathrm{cod}(K[P]) = T_{NS}(D[P])$.

Let $Q_{D[P]}$ denote the set of Fusion calculus processes reachable from P, namely

$$Q_{D[P]} \stackrel{\text{def}}{=} \{P\} \cup \bigcup_{P \stackrel{M,\gamma}{\longmapsto} P'} \{P'\} \cup Q_{D[P']}.$$

It is trivial to equip $Q_{D[P]}$ with a named set structure. Indeed, for any $q \in Q_{D[P]}$, the group component $\mathrm{g}_{D[P]}(q)$ is the identity on $\mathrm{fn}(q)$.

Function $\mathrm{h}_{K[P]}$ associates, to each state, its outgoing transitions and is defined as

$$\mathrm{h}_{K[P]}(q) = \{\langle l, q', \sigma\rangle \mid q \stackrel{M,\gamma}{\longmapsto} q' \wedge l \text{ corresponds to } M, \gamma\}$$

where, for any $\langle l, q', \sigma\rangle \in \mathrm{h}_{K[P]}(q)$ σ maps the names in $\mathrm{fn}(q')$ that correspond to those in $\mathrm{fn}(q)$ and names generated in the transition to $\star$ (and similarly for the names of l). Recall, indeed, that the names in q' and l are *local*, hence, even though they are syntactically equal in the Fusion calculus transition system, they must be considered different in HD-automata states.

The HD-automaton obtained by this definition is a T_L-coalgebra by construction. Observe that infinite HD-automata can be obtained using the construction above, however, there are interesting classes of Fusion calculus agents that generate finite HD-automata: this is the case of *finitary* agents. The *degree of parallelism* $\deg(P)$ of a Fusion calculus agent P is defined as follows:

$$\begin{array}{rlrl}
\deg(\mathbf{0}) &= 0 & \deg(\alpha\,.\,P) &= 1 \\
\deg((x)P) &= \deg(P) & \deg(P \mid Q) &= \deg(P) + \deg(Q) \\
\deg([x = y]P) &= \deg(P) & \deg(P + Q) &= \max(\deg(P), \deg(Q))
\end{array}$$
$$\deg(A\langle y_1, \ldots, y_n\rangle) = \deg(P[^{y_1,\ldots,y_n}/_{x_1,\ldots,x_n}]), \quad \text{if } A(x_1, \ldots, x_n) \stackrel{\text{def}}{=} P$$

Agent P is *finitary* if $\max\{\deg(P') \mid P \overset{M_1,\gamma_1}{\longmapsto} \cdots \overset{M_i,\gamma_i}{\longmapsto} P'\} < \infty$. In [18,21] the following result has been proved:

Theorem 4 (Theorem 47 of [18]). *Let P be a finitary agent. Then the HD-automaton $K[P]$ is finite.*

The class of finitary agents is expressive enough for non-trivial specification, as witnessed by e.g. the Handover protocol [19] in π-calculus, which can trivially be specified also in the finitary Fusion calculus.

Let us now define the normalisation functor for Fusion calculus.

Definition 19 (Redundancy of labels). *Let $\langle l,q,\sigma\rangle$ and $\langle l',q,\sigma'\rangle$ be two hdt of K. Assuming that the matching ns of l (resp. l') is* $\mathtt{M}$ *(resp.* $\mathtt{M}'$*), l* is redundant wrt *l' iff l and l' have the same action part and $[\![M]\!]_\sigma$ logically implies $[\![M']\!]_{\sigma'}$ but not vice versa.*

Definition 20 (Redundant transitions). *Let $q \in Q_D$ be a state of K, an hdt $\langle l_1,q_1,\sigma_1\rangle$ is* redundant (abbreviated as rhdt) for *q if there is $\langle l_2,q_1,\sigma_2\rangle$ in $\Sigma_K(q)$ such that l_1 is redundant wrt l_2 and, for a substitution σ accomplishing with the interpretation of the enabling part of l_1, $\sigma_2;\sigma = \sigma_1$.*

The intuition is that $t = \langle l_1,q_1,\sigma_1\rangle$ is dominated by another transition $t' = \langle l_2,q_1,\sigma_2\rangle$ reaching the same target state as t and with the same label but having

- enabling conditions weaker than those of t and,
- under the conditions of t, the names associated to the label of t' are the same as those of t.

Definition 21 (Normalisation functor for Fusion calculus). *The* normalisation functor for Fusion calculus *denoted by* $\mathtt{N} : \mathbf{NS} \to \mathbf{NS}$*, is defined as follows:*

$$\mathtt{N}(D) = \begin{cases} \langle \mathtt{h},\Sigma\rangle & D = \langle h_D,\Sigma_D\rangle \in \mathbf{NS}(\wp_{fin}(L\otimes E), \wp_{fin}(L\otimes F)) \text{ for } E,F \in obj(\mathbf{NS}) \\ D & \text{otherwise.} \end{cases}$$

where, for $B \in \wp_{fin}(L\otimes E)$,

$$\begin{aligned} \mathtt{h}(B) &= \{\langle l,q\rangle \mid \nexists\langle l,q,\sigma\rangle \text{ rhdt in } \Sigma(B)\} \\ \Sigma(B) &= \{\langle l,q,\sigma\rangle \in B \mid \langle l,q,\sigma\rangle \text{ not rhdt in } \} \end{aligned}$$

Basically, N filters those transitions out of a given state q that are redundant because of the presence of another transition having weaker conditions on names.

Proposition 4. *The functor* $\mathtt{N}$ *is monotonic on nfs.*

Theorem 5. *The minimisation algorithm described in Section 2 converges on finite HD-automata for Fusion calculus.*

Proof. By the monotonicity of T and $\mathtt{N}$ and Theorem 1.

Let us remark that normalisation through $\mathtt{N}_H$ is based on Definition 16 of symbolic hyperbisimulation. Indeed, redundancy conditions for Fusion calculus simply are the conditions in Definition 16 relating the enabling part and the action part of bisimilar transitions. Hence, a tight relationship can be established between Fusion calculus hyperbisimulation and the outcome of the minimisation algorithm.

Theorem 6 (Minimisation and hyperbisimulation). *Two finitary Fusion calculus processes are hyperbisimilar iff they have the same minimal realisation.*

Proof. (Sketch.) On the one hand, given two bisimilar Fusion calculus agents P and Q, if corresponding HD-automata $K[P]$ and $K[Q]$ are finite, their minimal realisations, say $\overline{K}[P]$ and $\overline{K}[Q]$, achieved by the minimisation algorithm are equivalent. Namely, $\ker \overline{K}[P] = \ker \overline{K}[Q]$ which implies that their corresponding classes have the same symmetries. On the other hand, if $K[P]$ and $K[Q]$ are finite and they both have the same minimal realisation, say $\overline{K}$, then P and Q are bisimilar. Basically, this is due to the fact that the transitions out of a state in $K[P]$ (resp. $K[Q]$) have a corresponding transition in the behaviour of P (resp. Q). By construction of K all the possible transitions of P have a matching non rhdt in H, hence can be matched by Q as well and vice versa. □

5 Conclusions

Related Work. This work is related to the work of Ferrari et al [7, 8, 10], and Cattani and Sewell [5] which both follow syntax-independent model approach to the operational semantics of process calculi. The former goes further to introduce a minimisation procedure of transition systems for nominal calculi in a coalgebraic setting but only treated an early semantics of π-calculus. Another related work worth noting is the work of Fiore and Staton [11], and Gadducci et al [12] where they provide a formal comparison of several operational semantics of nominal calculi.

In this paper we take the approach of Ferrari et al [7, 8, 10] further in order to give the same functionality for the Fusion calculus. Hyperbisimulation in Fusion calculus is more sophisticated than early bisimulation of π-calculus which was studied in the above mentioned work because of the closure under all substitution required by hyperbisimulation. We solve these technical challenges by providing a new symbolic semantics of Fusion calculus and conservatively extending HD-automata. The presentation of the minimisation algorithm given in Section 2 differs from those [7, 10] because we neatly distinguish between two phases. The first phase takes the initial HD-automaton to the current iteration of the algorithm, while the second phase applies a suitable normalisation functor for removing redundant transitions.

In the future we wish to take this research further and to deal with open semantics of π-calculus. Open semantics of π-calculus is complicated because extruded names need to be recorded and kept distinct from all other names under renaming. On the theoretical side, it would be interesting to study how this work is related to a presheaf model of open semantics of π-calculus as in [13] and other approaches as in [11, 12].

References

1. M. Abadi and A. Gordon. A Calculus for Cryptographic Protocols: The Spi Calculus. *Information and Computation*, 148(1):1–70, January 1999.
2. N. Benton, L. Cardelli, and C. Fournet. Modern Concurrency Abstractions for C#. *ACM Transactions on Programming Languages and Systems*, 26(5):269–304, Sept. 2004.
3. M. Boreale and R. De Nicola. A Symbolic Semantics for the π-calculus. *Information and Computation*, 126(1):34–52, April 1996.

4. R. Bruni, H. Melgratti, and U. Montanari. Theoretical Foundations for Compensations in Flow Composition Languages. In *Annual Symposium on Principles of Programming Languages POPL*, pages 209–220, New York, NY, USA, 2005. ACM Press.
5. G. L. Cattani and P. Sewell. Models for Name-Passing Processes: Interleaving and Causal (Extended Abstract). In *Proceedings of the Fifteenth Annual IEEE Symposium on Logic in Computer Science, LICS 2000*, pages 322–333. IEEE Computer Society Press, 2000.
6. S. Conchon and F. Le Fessant. Jocaml: Mobile Agents for Objective-Caml. In *International Symposium on Agent Systems and Applications*, pages 22–29, Palm Springs, California, Oct. 1999.
7. G. Ferrari, U. Montanari, and M. Pistore. Minimizing Transition Systems for Name Passing Calculi: A Co-algebraic Formulation. In M. Nielsen and U. Engberg, editors, *Foundations of Software Science and Computation Structures*, volume 2303 of *Lecture Notes in Computer Science*, pages 129–143. Springer-Verlag, 2002.
8. G. Ferrari, U. Montanari, and E. Tuosto. From Co-algebraic Specifications to Implementation: The Mihda toolkit. In F. de Boer, M. Bonsangue, S. Graf, and W. de Roever, editors, *Symposium on Formal Methods for Components and Objects*, volume 2852 of *Lecture Notes in Computer Science*, pages 319 – 338. Springer-Verlag, November 2002.
9. G. Ferrari, U. Montanari, and E. Tuosto. Modular Verification of Systems via Service Coordination. In *Monterey Workshop 2004*, October 2004. To appear on the workshop post-proceedings.
10. G. Ferrari, U. Montanari, and E. Tuosto. Coalgebraic Minimisation of HD-automata for the π-Calculus in a Polymorphic λ-Calculus. *Theoretical Computer Science*, 331:325–365, 2005.
11. M. Fiore and S. Staton. Comparing Operational Models of Name-Passing Process Calculi. In J. Adamek, editor, *Proc. CMCS'04*, ENTCS. Elsevier, 2004.
12. F. Gadducci, M. Miculan, and U. Montanari. About permutation algebras and sheaves (and named sets, too!). Technical Report UDMI/26/2003/RR, Department of Mathematics and Computer Science, University of Udine, 2003.
13. N. Ghani, B. Victor, and K. Yemane. Relationally Staged Computation in the π-calculus. In *Procedings of CMCS 2004*, number 106, 11 in ENTCS, pages 105–120, 2004.
14. M. Hennessy and H. Lin. Symbolic Bisimulations. *Theoretical Computer Science*, 138(2):353–389, February 1995.
15. C. Laneve and G. Zavattaro. Foundations of Web Transactions. In *Foundations of Software Science and Computation Structures*, Lecture Notes in Computer Science, 2005. To appear.
16. H. Lin. Complete Inference Systems for Weak Bisimulation Equivalences in the π-Calculus. *Information and Computation*, 180(1):1–29, January 2003.
17. U. Montanari and M. Pistore. History Dependent Automata. Technical report, Computer Science Department, Università di Pisa, 1998. TR-11-98.
18. U. Montanari and M. Pistore. π-Calculus, Structured Coalgebras, and Minimal HD-Automata. In M. Nielsen and B. Roman, editors, *Mathematical Foundations of Computer Science*, volume 1983 of *Lecture Notes in Computer Science*, pages 569–578. Springer-Verlag, 2000. An extended version will be published on Theoretical Computer Science.
19. F. Orava and J. Parrow. An Algebraic Verification of a Mobile Network. *Formal Aspects of Computing*, 4(5):497–543, 1992.
20. J. Parrow and B. Victor. The Fusion Calculus: Expressiveness and Symmetry in Mobile Processes. In *Proceedings of LICS '98*, pages 176–185. IEEE, Computer Society Press, July 1998.
21. M. Pistore. *History Dependent Automata*. PhD thesis, Computer Science Department, Università di Pisa, 1999.
22. M. Pistore and D. Sangiorgi. A Partition Refinement Algorithm for the π-Calculus. *Information and Computation*, 164(2):467–509, 2001.

23. U. Roxburgh. BizTalk Orchestration: Transactions, Exceptions, and Debugging, 2001. Microsoft Corporation. Available at http://msdn.microsoft.com/library/en-us/dnbiz/html/bizorchestr.asp.
24. D. Sangiorgi. A Theory of Bisimulation for the π-Calculus. *Acta Informatica*, 33(1):69–97, 1996.
25. E. Tuosto, B. Victor, and K. Yemane. Polyadic History-Dependent Automata for the Fusion Calculus. Technical Report 2003-62, Department of Information Technology, Uppsala, Sweden, December 2003. Available at http://www.it.uu.se/research/reports/.
26. B. Victor and F. Moller. The Mobility Workbench — A Tool for the π-Calculus. In D. Dill, editor, *Computer Aided Verification*, volume 818 of *Lecture Notes in Computer Science*, pages 428–440. Springer-Verlag, 1994.

A Proof Sketches for Section 3

Lemma 3. *For any Fusion calculus agent P, if $P \xrightarrow{\gamma} P'$, then $P\sigma \xrightarrow{\gamma\sigma} P'\sigma$, for any substitution σ.*

In the remainder of this section we establish the correspondence between symbolic hyperequivalence (Definition 16) and the standard hyperequivalence (Definition 15) by proving Theorem 3: $P \sim Q$ iff $P \simeq Q$.

Lemma 4.

1. $\sigma\sigma_{R\sigma} = \sigma_R\rho$, *for any substitution σ and some ρ, where R is an equivalence relation.*
2. *If $M \Rightarrow N$ then for any substitution σ, $M\sigma = N\sigma\rho$, for some substitution ρ.*
3. $\sigma_R\sigma_{S\sigma_R} = \sigma_S\sigma_R$, *where R and S are equivalence relations.*

Lemma 5.

1. *If $P \overset{M,\gamma}{\longmapsto} P'$, then $P\sigma \overset{M\sigma,\gamma\sigma}{\longmapsto} P'\sigma\sigma_{M\sigma}\sigma_{\gamma\sigma}$.*
2. *if $P\sigma \overset{N,\gamma'}{\longmapsto} P'$, then $P \overset{M,\gamma}{\longmapsto} P''$ with $M\sigma \Leftrightarrow N$, $\gamma\sigma = \gamma'$, and $P' = P''\sigma\sigma_N\sigma_{\gamma'}$.*

Proof. By transition induction, using Lemma 4. □

Lemma 6. *$P \simeq Q$ implies $P\sigma \simeq Q\sigma$, for any substitution σ.*

Proof. Straightforward diagram chasing, using Lemmas 4 and 5. □

Lemma 7.

1. *If $P \overset{M,\gamma}{\longmapsto} P'$, then $P\sigma_M \xrightarrow{\gamma} P''$ s.t. $P' = P''\sigma_\gamma$;*
2. *if $M \Rightarrow N$ and $P\sigma_M \xrightarrow{\gamma} P'$, then $P \overset{N,\gamma'}{\longmapsto} P''$ such that $\gamma = \gamma'\sigma_M$ and $P'\sigma_\gamma = P''\sigma_M$.*

Proof. Again by transition induction, using Lemmas 4 and 3. □

Proof of Theorem 3:

$\Rightarrow$: by showing $s = \{(P,Q) : P \sim Q\}$ is a symbolic hyperbisimulation, using Lemmas 7 and 4. □

$\Leftarrow$: We already have closure under substitution (Lemma 6), and show that $s = \{(P,Q) : P \simeq Q\}$ is a fusion bisimulation using Lemmas 7 and 4. □

An Algebraic Framework for Verifying the Correctness of Hardware with Input and Output: A Formalization in HOL

Anthony Fox

Computer Laboratory, University of Cambridge

Abstract. The HOL-4 proof system has been used to implement an algebraic framework for verifying the correctness of hardware with input and output. Implementations and specifications are modelled as *iterated maps*, with input and output modelled using *streams*. The correctness model supports three types of abstraction: temporal abstraction (with *immersions*), data abstraction, and stream abstraction.

This work has been used to formally verify the ARM6 microprocessor. This paper discusses this processor's input and output behaviour and shows how this has been modelled and verified in HOL. The verification is believed to be the first complete formal verification of a commercial off-the-shelf (COTS) processor. The definition of correctness given here is new – it is suited to verifying ARM's block data transfer instructions, these load and store sets of registers.

1 Introduction

This paper describes an algebraic framework for the specification and verification of hardware with input and output. The framework is based on an approach developed by Harman and Tucker [10,11], which was later refined and extended by Fox [5]. This work addresses the following questions:

> *How suited is the* HOL-*4 proof system to formalizing the approach of Harman, Tucker and Fox?*
>
> *Can the approach be used to formally verify a commercial processor design?*
>
> *Does the approach need to be modified in order to handle the input and output behaviour of the ARM6?*

The correctness framework is formalized in Section 2, and Section 3 introduces some *one-step* theorems, which provide a mechanism to formally verify correctness. The input and output behaviour of the ARM6 is *briefly* described in Section 4. Although the formalization presented here derives from a major case study (the mechanical verification of commercial pipelined processor), the concepts are generic and applicable in a wide variety of settings, both in terms

J.L. Fiadeiro et al. (Eds.): CALCO 2005, LNCS 3629, pp. 157–174, 2005.

of the tools used and the system under consideration. By not using an overly bespoke model of correctness, one can: (*i*) make objective comparisons between different verifications; (*ii*) benefit from using an established method for structuring specifications and proofs; and (*iii*) avoid introducing bugs when defining correctness.

Consider the device shown in Figure 1: it has an internal state, which evolves over time; input is accepted from the environment; and output is sent to the environment. Input and output are modelled using *streams*; these are maps of the form $\{0, 1, \ldots\} \to D$, where D is a set of data values. The state space and rate of state evolution determines the level of abstraction for the device's model. The correctness of a (concrete) model with respect to a more abstract model is established with the use of abstraction maps; there are maps for: time, state, input, and output.

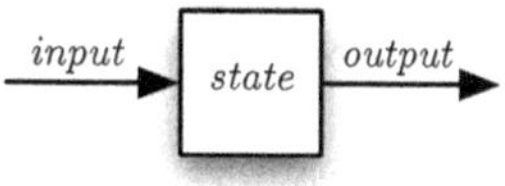

Fig. 1. A device with IO

The HOL-4 proof system has been used to:

1. formalize the correctness framework;
2. model the ARM instruction set architecture (ISA) and ARM6 processor; and
3. formally verify the correctness of the processor model.

Initially, a closed-system version of the processor was verified i.e. there was no input and output. Later, a more realistic processor model was verified: this covers the exception and memory interface behaviour. The processor:

- accepts signals for resets, memory aborts and interrupts (fast and slow);
- receives input and sends output along a data bus; and
- communicates with a memory unit using a set of flags and an address bus.

The memory for the ARM6 was originally treated as part of the processor's internal state i.e. the memory was a map (from 30-bit addresses to 32-bit words), capable of being updated on each clock cycle. This had the advantage that data transfer instructions (memory loads and stores) could be verified in the same way as all of the other instructions. However, this is a naïve model. Processor cores are paired with different memory units; these are complex devices and they may vary in behaviour e.g. they may differ through caching, aborts and update latencies. In order to model and verify the ARM6, modifications were required to the correctness model of [5] – these changes will be discussed in this paper.

1.1 Background and Related Work

The HOL-4 system (`hol.sf.net`) has its origins in the LCF system of Milner [9], which formed the basis for the higher-order logic of Gordon [8].

This work derives from research carried out at Swansea on the algebraic specification of hardware and software. A correctness model for microprocessors was developed by Harman and Tucker [10]: an important part of their approach was in the treatment of temporal abstraction, with the explicit use of *clocks* and *retiming* maps. This work forms part of the more general subject of *synchronous concurrent algorithms* (SCAs) [16] and stream processing [17,15]. The current HOL formalization builds on a version without input and output [6].

Many notions of correctness have been used in processor verifications; for example, Aagaard *et. al* [1] classify a selection of definitions that have been used to verify processors with out-of-order execution. The model of correctness presented here is not suited to verifying out-of-order designs, nor is it based on the *flushing* correctness model of Burch and Dill [2]. Manolios [12] identifies some problems with the Burch and Dill model (it does not exclude trivial implementations), and he provides a better definition of correctness.

One approach that has been used in verifying pipelined designs is to check a set of safety and liveness properties e.g. prove that there are no deadlocks, and that control and data hazards are correctly handled. This approach is not explicitly used in verifying the ARM6 – instead the correctness definition is in the form of a *commuting equation* – this is sufficient (but perhaps not necessary) to ensure that the usual pipeline correctness properties hold. Ideally, a set of liveness and safety properties should be sufficient with respect to the correctness definition presented here, but this is not necessarily the case (one might just rule out particular classes of bugs).

The work presented here is unique with respect to its treatment of input and output. Processor verifications are usually carried out in the absence of external interrupts, and the main memory is usually considered to be part of the state space. A notable exception to this is the work of Sawada and Hunt [13]. They use an instruction set that is much simpler than that of the ARM architecture (e.g. there are no block data transfers), but the FM9801's micro-architecture is significantly more complex than the ARM6's. The processor's complexity has meant that their *top-level* correctness definition does not guarantee the *complete* correctness of their implementation.

The ARM6 verification is believed to be the first complete formal verification of a COTS processor. In earlier work, either the processor was not a widely available commercial product (in many cases designed with verification in mind) or the verification has been in some respect incomplete i.e. not verifying interrupts/exceptions or the entire instruction set. ARM processors are prevalent in embedded (and, in particular, low power) systems. Documentation on the ARM architecture can be found in [7] and [14].

2 Algebraic Models

Section 2.1 introduces iterated map state-output functions; these are used to model hardware with input and output. In Section 2.3, correctness is defined, using abstraction maps, which are discussed in Section 2.2. Some important stream operations are provided in Section 2.4. In Section 2.5 this work is placed in the context of algebraic specification.

To aid readability, the symbols listed in Table 1 will be used when writing HOL code (which will be in typewriter font).

Table 1. Symbols used in HOL

Symbol	ASCII	Use	Symbol	ASCII	Use
$\vdash_{def}$, $\vdash$	|-	proof turnstile	$\neg$	~	logical negation
α,β,...	'a,'b,...	type variables	$\wedge$	/\	conjunction
$\rightarrow$	->	function type constructor	$\vee$	\/	disjunction
$\times$	#	tuple type constructor	$\Rightarrow$	=>	implication
$\forall$	!	for all	$\in$	IN	set membership
$\exists$	?	there exists	$\langle\|$	<|	record opening parenthesis
λ	\	lambda abstraction	$\|\rangle$	|>	record closing parenthesis

2.1 State-Output Functions and Iterated Maps

State-output functions are used to model hardware with input and output; in [5] these are maps of the form:

$$f : \mathbb{N} \times A \times [\mathbb{N} \rightarrow B] \rightarrow A \times C$$

where A is a set of states (the *state space*), $[\mathbb{N} \rightarrow B]$ is a set of input streams (i.e. time indexed data values) and C is a set of output values. Each set is required to be non-empty. Time is modelled using the natural numbers. Given an initial state a and input stream i, if $f(t, a, i) = (b, c)$ then the state at time t is b and the output is c.

In HOL, the following record datatypes are used:

```
state_inp = ⟨| state : α; inp : num → β |⟩
state_out = ⟨| state : α; out : β |⟩
```

A state-output function is any function of the form:

```
f : num → (α,β) state_inp → (α,γ) state_out .
```

The state space is represented by the type variable α, the input space has type `num` $\rightarrow$ β and the output has type γ. (It is more convenient to use curried functions in HOL.) The HOL-4 system ensures that all types have at least one member, therefore, these type variables can be freely instantiated with any concrete type. For clarity and convenience, records are used instead of tuples (pairs) – the HOL-4 system automatically defines projection and update functions for record types e.g. `state_out_inp x` is the input component of record `x` (this can be written as `x.inp`), and one can also write `x with inp := y`. The notation *name* := *exp* is used to assign a value to a named component.

Iterated map state-output functions are characterised by the predicate `IMAP`:

```
⊢def IMAP f init next out =
       (∀x. (f 0 x).state = init x.state) ∧
       (∀t x. (f (t + 1) x).state = next (f t x).state (x.inp t)) ∧
       ∀t x. (f t x).out = out (f t x).state
⊢def IS_IMAP f = ∃init next out. IMAP f init next out
⊢def IS_IMAP_INIT f init = ∃next out. IMAP f init next out
```

In an iterated map specification, the state at time zero is a function (`init`$:\alpha\to\alpha$) of the *pre-initial* state `x.state`. Each successive state, at time $t+1$, is a function (`next`$:\alpha\to\beta\to\alpha$) of the state at time t and the stream input `x.inp` at time t. For example, the state at cycle three is:

```
next (next (next (init x.state) (x.inp 0)) (x.inp 1)) (x.inp 2) .
```

The output is always a function (`out`$:\alpha\to\gamma$) of the current state.

The predicate `IS_IMAP` asserts that `f` is an iterated map i.e. that `f` can be defined with *initialisation*, *next state* and *output* functions. If such functions exist then they will be unique, however, for convenience, HOL-4's quantifier for unique existence is not used here. The predicate `IS_IMAP_INIT` is used to assert that `f` has a particular initialisation function.

Example 1. Assume ⊢ `IMAP f I (λa i. a + i) (λa. ODD a)`, where `I` is the identity map and `ODD n` is true if, and only if, the natural number `n` is odd. Let `x = ⟨|state := 4; inp := (λt. t MOD 3)|⟩`. The iterated map `f` evaluates as follows:

```
⊢ f 0 x = ⟨|state := 4; out := F|⟩      ⊢ f 3 x = ⟨|state := 7; out := T|⟩
⊢ f 1 x = ⟨|state := 4; out := F|⟩      ⊢ f 4 x = ⟨|state := 7; out := T|⟩
⊢ f 2 x = ⟨|state := 5; out := T|⟩      ⊢ f 5 x = ⟨|state := 8; out := F|⟩ .
```

Example 2. Let `x = ⟨|state := 0; inp := (λt. 1)|⟩`, and let the functions `f` and `g` evaluate as follows:

```
⊢ f 0 x = ⟨|state := 0; out := F|⟩      ⊢ g 0 x = ⟨|state := 0; out := F|⟩
⊢ f 1 x = ⟨|state := 0; out := F|⟩      ⊢ g 1 x = ⟨|state := 0; out := T|⟩
⊢ f 2 x = ⟨|state := 1; out := F|⟩ .
```

It is clear that ⊢ `¬IS_IMAP f ∧ ¬IS_IMAP g`. The function `g` is not an iterated map because the output is not a function of the state; `f` is not an iterated map because the next state cannot be a function of the previous state and input i.e. the state at cycle three differs from that at cycle two, even though the previous state and input values are the same.

2.2 Data, Temporal and Stream Abstraction

State-output functions are used to model systems at different levels of abstraction. When comparing an *implementation* with a *specification*, four abstractions are used, these are for: time, state, input and output.

A data abstraction maps the states of an implementation to the states of a specification. The following property is used to assert the validity of a data abstraction:

```
⊢def DATA_ABSTRACTION abs initi inits = SURJ abs (RANGE initi) (RANGE inits)
⊢def SURJ f s t = (∀x. x ∈ s ⇒ f x ∈ t) ∧ ∀y. y ∈ t ⇒ ∃x. x ∈ s ∧ (f x = y)
```

A data abstraction is valid provided that it is a surjective mapping between the initial state spaces i.e. all initial states of the implementation must map to an initial state of the specification, and all initial states of the specification must be mapped to by some initial state of the implementation. When used in Section 2.3, this condition ensures that the implementation captures all of the behaviour of a specification.

Temporal abstraction is achieved with the use of an *immersion*:

```
⊢def IMMERSION imm = ∀(x :(α,β) state_inp). FREE_IMMERSION (imm x)
⊢def FREE_IMMERSION f = (f 0 = 0) ∧ ∀t1 t2. t1 < t2 ⇒ f t1 < f t2
```

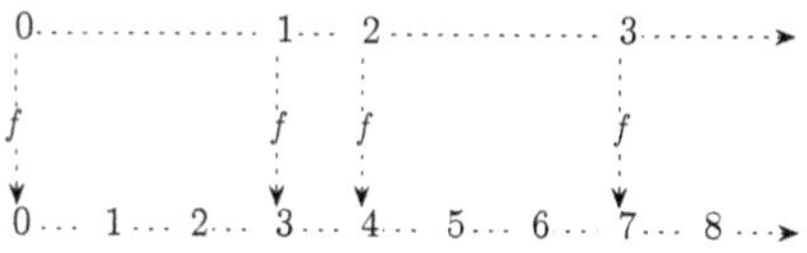

Fig. 2. A free immersion

A *free immersion* is any monotonic increasing map (over the natural numbers) with identity at cycle zero e.g. see Figure 2. Free immersions map abstract cycles at the specification level to concrete cycles at the implementation level. A state- and input-dependent immersion (or just, an immersion) is any map from a state-input record to a free immersion. Thus, the temporal abstraction is a function of the implementation's pre-initial state and input.

The validity of a stream abstraction is asserted with the following predicate:

```
⊢def STREAM_ABSTRACTION smpl sstrm istrm =
       (∃i. i ∈ istrm) ∧ ∀x. x.inp ∈ istrm ⇒ smpl x ∈ sstrm
```

Here `sstrm` and `istrm` are the sets of all valid input streams at the specification and implementation levels respectively.[1] The stream abstraction property is not as strong as the data abstraction property i.e. it is not necessary to cover all of the valid input streams at the specification level. This weaker condition is needed in the context of the ARM6 verification – the set of valid exception sequences depends on the state of the processor's core.

2.3 Correctness

The correctness of an implementation with respect to a specification is asserted with the following predicate:

```
⊢def CORRECT spec impl imm abs osmpl ismpl sstrm istrm =
       IMMERSION imm ∧
       DATA_ABSTRACTION abs (state_out_state ∘ impl 0) (state_out_state ∘ spec 0) ∧
       STREAM_ABSTRACTION ismpl sstrm istrm ∧
       ∀x. x.inp ∈ istrm ⇒
         (let y = ⟨|state := abs x.state; inp := ismpl x|⟩ in
            (∀t. (spec t y).state = abs (impl (imm x t) x).state) ∧
            (OUTPUT spec y = osmpl (x,OUTPUT impl x)))
⊢def OUTPUT g x t = (g t x).out
```

[1] In HOL-4, predicate sub-typing is not built into the logic, which is why predicates are needed to specify the domains of the input stream. Initialisation functions are used to restrict the domain of the state space.

The state-output functions are `spec` and `impl`, and the corresponding input stream spaces are `sstrm` and `istrm`. There are four abstractions: a data abstraction `abs`; an immersion `imm`; an input stream abstraction `ismpl`; and an output stream abstraction `osmpl`. The predicates from Section 2.2 are used to ensure these maps have the required properties.[2] The function `OUTPUT` gives the output stream for a state-output function.

Two equations are used: one asserts that the states of the two systems correspond with respect to the abstractions; and the other asserts that the output streams correspond through abstraction. The equations must hold for all cycles at the specification level, states at the implementation level, and for all input streams that are in `istrm`. Thus, all of the states produced by the specification `spec` can be obtained (in the same temporal order) through abstraction, using the implementation `impl`. Likewise, the output of the specification is a function of the output of the implementation. In the context of pipelined processors, this means that if the processor fails to resolve a hazard (a data or control dependency) or if the processor deadlocks (when, at the specification level, the state continues to evolve) then correctness will not hold because it will be impossible to satisfy the commutativity equation.

The output stream abstraction, `osmpl`, takes two arguments: the state-input record and the output from the implementation. The output alone does not provide enough context for the abstraction; for example, the free immersion `imm x` is needed to relate the timing of the two streams. With the ARM6 verification, the stream abstraction was not a simple function (i.e. sampling) of the output.

Example 3. All state-output functions are correct implementations of themselves; that is:

```
⊢ ∀f strm. ∃i. i ∈ strm ⇒ CORRECT f f (λx t. t) I SND state_inp_inp strm strm .
```

Here, `SND` gives the second value of a pair. For the `STREAM_ABSTRACTION` property to hold, there must be at least one member of the set `strm`.

The correctness definition has been proved to be transitive:

```
CORRECT_TRANS:
 ⊢ ∀f1 f2 f3 imm1 imm2 abs1 abs2 osmpl1 osmpl2 ismpl1 ismpl2 strm1 strm2 strm3.
       CORRECT f1 f2 imm1 abs1 osmpl1 ismpl1 strm1 strm2 ∧
       CORRECT f2 f3 imm2 abs2 osmpl2 ismpl2 strm2 strm3 ⇒
       CORRECT f1 f3
         (λx. imm2 x ∘ imm1 ⦇state := abs2 x.state; inp := ismpl2 x⦈) (abs1 ∘ abs2)
         (λ(x,stm). osmpl1
              (⦇state := abs2 x.state; inp := ismpl2 x⦈,osmpl2 (x,stm)))
         (λx. ismpl1 ⦇state := abs2 x.state; inp := ismpl2 x⦈) strm1 strm3
```

This theorem shows that if `f2` implements `f1`, and `f3` implements `f2`, then `f3` implements `f1`. Figure 3 gives a commuting diagram view for this theorem. The constraints places upon the abstractions (Section 2.2) ensure that they can be successfully composed. Therefore, it is possible to verify multiple levels of

[2] In HOL, ∘ is used for function composition; therefore, `state_out_state ∘ f 0` is the initialisation function for the state-output function `f`.

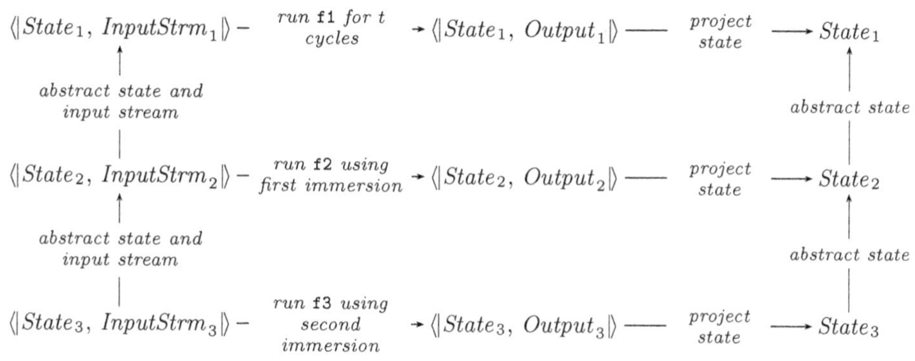

(a) State.

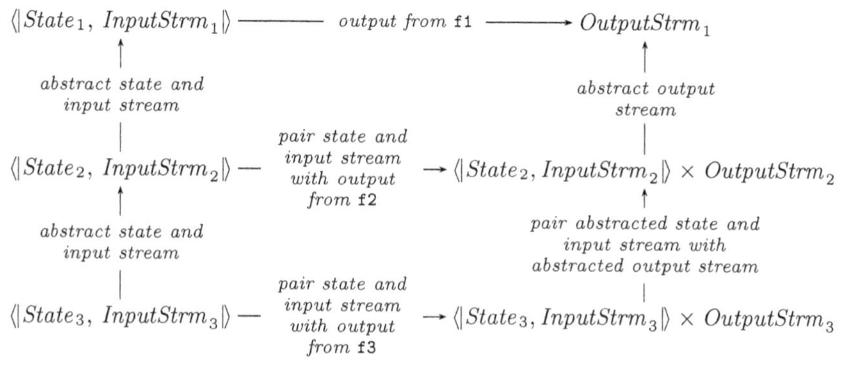

(b) Output.

Fig. 3. A commuting diagram view of correctness under transitivity. An informal presentation is shown i.e. the HOL types and abstraction maps are not used.

abstraction and prove that the most concrete level implements the most abstract level within a hierarchy. In computer architecture, this enables gate-level models to be related to high level languages, with micro-architecture and programmer's model levels in-between.

The correctness definition above differs from that in [5]; in particular, the treatment of output is different. In [5], correctness was expressed using a *single* equation of the form:

$$f(t, a, i) = \psi(g(s, b, j)) \ ,$$

where $\psi : A \times B \to C \times D$ abstracts the state and output of the implementation. This old definition is not suited to the verification of the ARM6 because the output from the implementation is checked only at the times given by the immersion i.e. at instruction boundaries. However, with ARM's block data transfer instructions, data and signals are sent to the memory unit over a number of cycles, so the entire output stream must be abstracted. The new correctness definition can be seen as a generalisation of the old definition i.e. it now covers a wider range of systems.

2.4 Stream Operations

Various stream operations have been defined in HOL. One basic operation is to advance a stream in time:

```
⊢def ADVANCE t1 s t2 = s (t1 + t2)
```

Example 4. If $\vdash_{def}$ `strm t = t + 4` then $\vdash$ `ADVANCE 1 strm = λt. t + 5`.

The function `PACK` takes a stream of type `num` $\rightarrow$ α and bundles values together to give a stream of type `num` $\rightarrow$ `(`α `list)`; this is done with respect to a free immersion:

```
⊢def PACK imm strm t = GENLIST (λs. strm (imm t + s)) (IMM_LEN imm t)
⊢def IMM_LEN imm t = imm (t + 1) - imm t
```

`GENLIST f n` creates the list `[f 0; f 1;...; f (n - 1)]`. Packing is a lossless stream sampling operation.

Example 5. Let $\vdash_{def}$ `strm t = t + 4` and let $\vdash_{def}$ `imm t = 4 * t` i.e. `imm` is a *linear* immersion in which each abstract cycle corresponds with four concrete cycles. Let $\vdash_{def}$ `pstrm = PACK imm strm`. This stream evaluates as follows:

`⊢ pstrm 0 = [4; 5; 6; 7]`   `⊢ pstrm 1 = [8; 9; 10; 11]`   `⊢ pstrm 2 = [12; 13; 14; 15]` .

The function `SERIALIZE` is the inverse of `PACK`:

```
⊢def SERIALIZE imm strm s = EL (s - IMM_START imm s) (strm (IMM_RET imm s))
⊢def IMM_RET imm s = LEAST t. s < imm (t + 1)
⊢def IMM_START imm = imm ∘ IMM_RET imm
```

Here, `IMM_RET imm` is the *retiming* corresponding with the free immersion `imm`; see [5]. The function `EL n` picks the nth element from a list.

Example 6. Given `strm`, `pstrm` and `imm` from Example 5, the following theorem holds: $\vdash$ `SERIALIZE pstrm imm = strm`.

The following theorem shows the relationship between `PACK` and `SERIALIZE`:

```
PACK_SERIALIZE:
 ⊢ ∀f. FREE_IMMERSION f ⇒
       (∀strm. PACKED_STRM f strm ⇒
          (PACK f (SERIALIZE f strm) = strm)) ∧
       ∀strm. SERIALIZE f (PACK f strm) = strm
⊢def ∀f strm. PACKED_STRM f strm = ∀t. LENGTH (strm t) = IMM_LEN f t
```

Serializing works when the length of the list at each time t is greater than or equal to the number of cycles in the interval t to $t + 1$ (as specified by the free immersion). When these lengths are equal, packing a serialized stream is an identity operation.

2.5 Algebraic Specification Framework

The work in this paper has its origins in algebraic specification [18,4]. Although HOL-4 is not based on equation or rewriting logic (cf. Maude [3]), it is useful to consider the modular structure of the specifications. For example, one can consider algebraic structures for:

1. Time:
$$(\mathbb{N} \mid 0, 1, \mathit{Suc} : \mathbb{N} \to \mathbb{N}, + : \mathbb{N} \times \mathbb{N} \to \mathbb{N})$$
2. Streams:
$$(\mathbb{N}, A, S = [\mathbb{N} \to A] \mid \mathit{Eval} : \mathbb{N} \times S \to A, \mathit{Advance} : \mathbb{N} \times S \to S, \ldots)$$
3. The top-level semantics:
$$(\mathbb{N}, A, [\mathbb{N} \to B], C \mid f : \mathbb{N} \times A \times [\mathbb{N} \to B] \to A \times C)$$
4. The next state semantics:
$$(A, B, C \mid \mathit{init} : A \to A, \mathit{next} : A \times B \to A, \mathit{out} : A \to C)$$
5. The machine semantics: this contains the operations used to define the next state semantics e.g. functions over vectors of n-bit words.

Each algebra has carrier sets and operations (constants are operations of arity zero). In HOL-4, specifications are split into *theory files*. For example, there are theories for natural number arithmetic, n-bit words and the top-level (state-output function) semantics of the ARM instruction set and the ARM6 processor.

3 One-Step Approach

Section 2.1 provides a definition of correctness for state systems with input and output. This section describes an approach to verifying correctness that is based on using *time-consistent* state-output functions and *uniform* abstractions. The *one-step theorems* of Section 3.4 show that, under certain circumstances, the correctness condition can be reformulated: producing a set of sub-goals in which the abstract clock is instantiated with values $t = 0$ and $t = 1$. With the ARM6, time $t = 1$ corresponds with the execution of a single machine code instruction: symbolic execution (term rewriting) is used to evaluate the state of the processor for each class of instruction and type of exception. However, the block data transfer and multiply instructions did require special treatment i.e. to verify these instructions, invariants were manually defined (at the processor level) and checked with an induction on time.

3.1 Uniform Immersions

An immersion is *uniform* if, and only if, it can be defined using a *duration* function, which computes the number of implementation cycles needed to complete one state transition at the specification level. The duration must be a function of the implementation's current state and of the stream of future input values; this is formalized in HOL as follows:

```
⊢def UIMMERSION imm f dur =
       (∀x. 0 < dur x) ∧
       (∀x. imm x 0 = 0) ∧
       ∀x t. imm x (t + 1) =
        dur (|state := (f (imm x t) x).state; inp := ADVANCE (imm x t) x.inp|) + imm x t
⊢def UNIFORM imm f = ∃dur. UIMMERSION imm f dur
```

For `imm` to be a free immersion, the duration must always be non-zero. The predicate `UNIFORM` asserts that a duration function exists that is consistent with the immersion and state-output function.

With the ARM6, the duration function examines the processor's state to determine which instruction is to be executed, and the input stream indicates whether an interrupt is going to occur. If the implementation is time-consistent (see below) then all possible timings (durations) can be deduced at cycle zero.

Example 7. Assume `f` satisfies ⊢ `IMAP f init next out` and `imm` is the uniform immersion satisfying ⊢ `UIMMERSION imm f dur`, where the initialisation, next state and duration functions are defined as follows:

```
⊢def dur x = x.state + x.inp 1 + 1
⊢def init a = a MOD 4
⊢def next a i = (a + i) MOD 4 .
```

Let `x = (|state:=7; inp := λt. t|)`. The uniform immersion `imm` evaluates as follows:

```
⊢ imm x 0 = 0
⊢ imm x 1 = (3 + 1 + 1) + 0 = 5
⊢ imm x 2 = (1 + 6 + 1) + 5 = 13
⊢ imm x 3 = (1 + 14 + 1) + 13 = 29 .
```

3.2 Time-Consistency

A state-output function is *time-consistent* if, and only if, it can be composed in time with respect to an immersion; this is defined in HOL as follows:

```
⊢def TCON_IMMERSION f imm strm =
       let g t x = (|state := (f t x).state; inp := ADVANCE t x.inp|) in
         ∀t1 t2 x. x.inp ∈ strm ⇒
           let s2 = imm x t2 in
           let s1 = imm (g s2 x) t1 in
             ADVANCE s2 x.inp ∈ strm ∧
             (g (s1 + s2) x = (g s1 ∘ g s2) x)
⊢def TCON f strm = TCON_IMMERSION f (λx t. t) strm
```

Advancing the input stream must be safe with respect to the set of valid streams `strm`. The function `g`, which computes the state and advances the input stream, must be composable i.e. it should not matter whether the state-output function is executed for `s1 + s2` cycles, or for `s2` cycles followed by `s1` cycles.

The input component is always time-consistent[3] because there is not an input stream initialisation function and

```
⊢ ∀t1 t2. ADVANCE (t1 + t2) = ADVANCE t1 ∘ ADVANCE t2 .
```

The property `TCON` asserts that `f` is time-consistent at all times i.e. the times given by the identity immersion. The one-step theorem in Section 3.4 will show that the time-consistency of an iterated map depends upon the definition of the initialisation function.

Example 8. If ⊢ `IMAP_INIT f I` then `f` is time-consistent with respect to all immersions.

Example 9. If ⊢ `IMAP f (λx. 0) (λx. x + 1) out` then the function `f` is not time-consistent with respect to *any* immersion because, if `g` is defined as above, then

```
⊢ ∀d x. ((g (d + d) x).state = d + d) ∧ (((g d ∘ g d) x).state = d) .
```

Example 10. The function $\vdash_{def}$ `f t x = x.inp` is time-consistent with respect to all immersions. However, `f` is not an iterated map because the initial state is not a function of the pre-initial state.

3.3 Time-Consistent and Uniform Sampling

An abstraction map for input streams is time-consistent if, and only if, advancing the abstract stream is the same as advancing the concrete stream by the time given by the immersion; this is defined in HOL as follows:

```
⊢def TCON_SMPL smpl imm f strm =
      ∀t x. x.inp ∈ strm ⇒
        (smpl ⟨|state := (f (imm x t) x).state;
                inp := ADVANCE (imm x t) x.inp|⟩ = ADVANCE t (smpl x))
```

In practice, this property is verified by using the fact that the state-output function is time-consistent and that the immersion is uniform. However, the exact proof obligations will vary depending on how one has defined the abstraction.

The function `OSMPL` constructs an output stream abstraction:

```
⊢def OSMPL f impl imm (x,strm) t =
      f ⟨|state := (impl (imm x t) x).state;
          inp := ADVANCE (imm x t) x.inp|⟩ (PACK (imm x) strm t)
```

[3] It is sufficient to have `(g (s1 + s2) x).state = ((g s1 ∘ g s2) x).state`.

The abstract output is some function of the implementation's state, input and packed output, at times given by the immersion. This form of abstraction has the advantage that, for correctness, one need only consider the output at time $t = 0$.

Example 11. If `imm x` is the free immersion from Figure 2, then the first three output values of `OSMPL f impl (x, strm)` are:

```
f ⟨|state := (impl 0 x).state; inp := x.inp|⟩            [strm 0; strm 1; strm 2]
f ⟨|state := (impl 3 x).state; inp := ADVANCE 3 x.inp|⟩ [strm 3]
f ⟨|state := (impl 4 x).state; inp := ADVANCE 4 x.inp|⟩ [strm 4; strm 5; strm 6] .
```

In correctness statements: ⊢ ∀`t. strm t = (impl t x).out` .

3.4 One-Step Theorems

The one-step theorem for time-consistency is as follows:

```
TCON_IMMERSION_ONE_STEP_THM:
 ⊢ ∀strm f init out imm.
       IS_IMAP_INIT f init ∧ UNIFORM imm f ⇒
       (TCON_IMMERSION f imm strm =
        ∀x. x.inp ∈ strm ⇒
          (init (f (imm x 0) x).state = (f (imm x 0) x).state) ∧
          (init (f (imm x 1) x).state = (f (imm x 1) x).state) ∧
          ADVANCE (imm x 1) x.inp ∈ strm)
```

If the state-output function `f` is an iterated map and the immersion is uniform then `f` is time-consistent if, and only if, the initialisation function `init` is an identity map for the states at time $t = 0$ and $t = 1$, and the stream starting at $t = 1$ is in `strm`. Note that, `init` need not be an identity map for the intermediate states i.e. at cycles greater than zero and less than `imm x 1`. This one-step theorem shows that time-consistency can be viewed as a closure property on the set of initial states. With the ARM6 model, the initialisation function constructs all processor states that correspond with an instruction boundary.

The one-step theorem for correctness is as follows:

```
ONE_STEP_THM:
 ⊢ ∀sstrm istrm spec impl imm abs osmpl ismpl f.
       IS_IMAP spec ∧ IS_IMAP impl ∧ UNIFORM imm impl ∧
       DATA_ABSTRACTION abs (state_out_state ∘ impl 0) (state_out_state ∘ spec 0) ∧
       STREAM_ABSTRACTION ismpl sstrm istrm ∧ TCON spec sstrm ∧
       TCON_IMMERSION impl imm istrm ∧ TCON_SMPL ismpl imm impl istrm ∧
       (osmpl = OSMPL f impl imm) ∧
       (∀x. x.inp ∈ istrm ⇒
          (let y = ⟨|state := abs x.state; inp := ismpl x|⟩ in
             ((spec 0 y).state = abs (impl (imm x 0) x).state) ∧
             ((spec 1 y).state = abs (impl (imm x 1) x).state) ∧
             (OUTPUT spec y 0 = osmpl (x,OUTPUT impl x) 0))) ⇒
       CORRECT spec impl imm abs osmpl ismpl sstrm istrm
```

This theorem can be used when the state-output functions are time-consistent iterated maps, the immersion is uniform, and when the input stream abstraction is time-consistent. It is assumed that the output stream abstraction is of the form given by `OSMPL`.

The two one-step theorems provide a mechanism through which correctness can be verified. Many of the properties are going to be straightforward to check. There are three main goals: time-consistency at cycle one; correctness at cycle one; and comparing the output at cycle zero. With the ARM6 verification, these goals are proved by case splitting (principally on the *instruction class*) followed by contextual term-rewriting with the *simplifier*. To verify the output, one must reason about lists of output values – some work is required to do this efficiently. Block data transfer and multiply instructions iterate an execute cycle until a termination condition is met – taking up to twenty and seventeen cycles to execute respectively. Here, brute force case splitting is not an option – invariants are defined and proved to hold by induction on time. In total, there is approximately eight thousand lines of HOL script: two thousand for formalizing the framework, four thousand for the verification, and the specifications are a thousand lines each. At present, the verification takes about fifteen minutes to run.

4 Formal Verification of the ARM6

This section provides an overview of the ARM6's input and output behaviour. The intent is to simply give a *flavour* of how the framework has been applied in practice.

A functional view of the ARM6 is shown in Figure 4. Some of the IO is not significant from a formal specification standpoint e.g. the clock, power, JTAG and chip test signals. Other signals have not been modelled for the formal verification, but may be added later e.g. LOCK, SEQ and the bus controls. The signals that are of significance are those for interrupts, resets and the memory unit (which includes the address bus, data bus and memory control signals). The correctness framework presented in this paper allows one to model these signals and show that they are correct with respect to an abstract, instruction set model.

The state-output function for the ISA (instruction set) model is:

```
ARM_SPEC:num →
       (state_arm_ex, interrupts option × word32 × word32 list) state_inp →
       (state_arm_ex, memop list) state_out
α option = NONE | SOME of α
interrupts = Reset of state_arm | Prefetch | Dabort of num | Fiq | Irq
memop = MemRead of word32 | MemWrite of bool⇒word32⇒word32
```

In HOL, one can use ML style *union* types. The state space `state_arm_ex` consists of the ISA registers, together with the current instruction's op-code and the exception status.

The input is a triple `(irpt,ireg,data)`. The word `ireg` is for an instruction fetch. Load instructions use the list of words, `data`. If `irpt` matches `SOME i` then an interrupt has occurred, with the type encoded by `i`. When a reset occurs the ARM6 processor *immediately* stops executing the current instruction, therefore, at the ISA level, a reset is parameterised by a value – this is used to set the

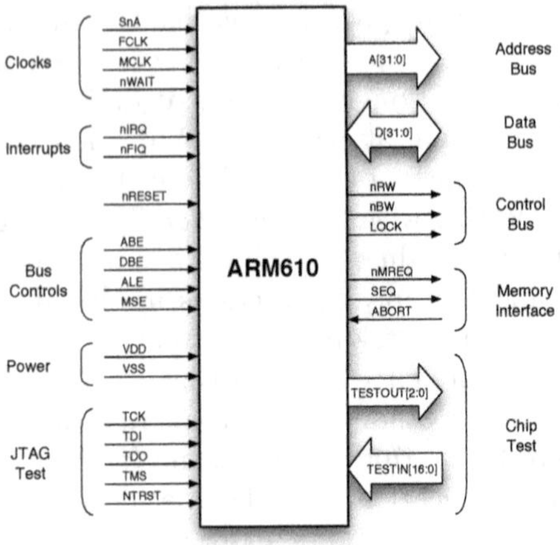

Fig. 4. Functional diagram for the ARM6 processor

registers after the reset. The data abort is parameterised by a number – this indicates how many loads were performed before there was an abort.

The output is a list of memory read/write requests: a memory read (load) is parameterised by an address; and a write (store) is parameterised by a byte/word flag, an address and a data word.

The state-output function for the ARM6 model is:

```
ARM6_SPEC:num →
      (state_arm6, bool × bool × bool × bool × word32) state_inp →
      (state_arm6, word32 × bool × bool × bool × bool × word32) state_out
```

The output is vector `(data,nMREQ,nOPC,nRW,nBW,areg)`. The flag `nMREQ` is set when there is not a memory request; `nRW` and `nBW` indicate whether a memory request is a read or write for a byte or a 32-bit word. An additional output has been added, `nOPC` – this value is part of the processor's control logic and indicates when the next memory operation is not an op-code fetch. This enables the output stream abstraction to hide all the instruction fetch memory operations: instruction fetching is very processor dependent, and it is best not to model this at the ISA level. The input vector is `(nRESET,ABORT,nFIQ,nIRQ,DATA)`. There are four interrupt signals and a word from the data bus.

Figure 5 shows how the ARM6's pipeline executes a sequence of instructions. At the ISA level, the output is: `[MemWrite F #512 #1; MemWrite F #516 #2]` and `[MemRead #512; MemRead #516]`. This output corresponds with the four cycles in which `nOPC` is true. The data input to the processor at cycles seven and eight should be `#1` and `#2`, but the actual values will depend on the memory unit. The *data* output for the store occurs at cycles five and six, this is one cycle after the address and control signals are sent – this complicates the output stream abstraction function – the last data output is not in the list of packed output values, and so it must be computed using the state-input record. The

```
MOV   r1, #1
MOV   r2, #2
MOV   r0, #512
STMIA r0!, {r1, r2}
LDMDB r0, {r3, r4}
```

(a) The program.

cycle	data	nMREQ	nOPC	nRW	nBW	areg
(STMIA 0) 3	-	F	F	F	T	pc
(1) 4	-	F	T	T	T	#512
(2) 5	#1	F	T	T	T	#516
(LDMDB 0) 6	#2	F	F	F	T	pc + #16
(1) 7	-	F	T	F	T	#512
(2) 8	-	F	T	F	T	#516
(3) 9	-	T	F	F	T	pc + #20

(b) The output.

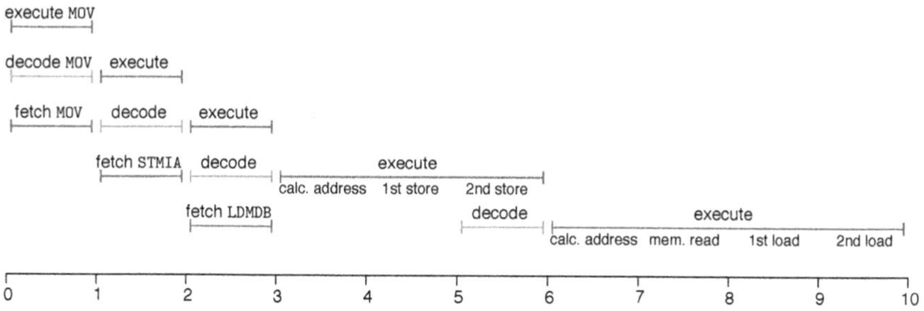

(c) ARM6's multi-cycle 3-stage pipeline.

Fig. 5. The execution of a simple ARM program

state-input record is also used to construct the output in the case when there is a reset signal.

5 Conclusions

This paper has demonstrated that the HOL-4 system is well-suited to reasoning about the correctness of hardware with input and output, one can:

- define an abstract model of correctness;
- reason about correctness for hierarchies of abstraction levels; and
- prove the one-step theorem, which enables one to verify designs with respect to the top-level correctness definition.

Section 2.4 shows that it is easy to reason about stream transforming maps using higher-order logic. On the downside, HOL-4 does not fully support the algebraic specification style e.g. there is no support for declaring algebraic structures in a modular style, with suitable interfaces. Instead, the logic is essentially flat, with the meta language ML simply providing a mechanism for organising *theories*.

The approach described in this paper could be implemented in other systems (including first-order tools). However, higher-order logics do provide a *natural* mechanism for reasoning about correctness *per se* e.g. verifying and applying the one-step theorems and proving that the correctness definition is transitive. Other tools could be used; for example, one could make use of *model checkers* in verifying certain lemmas. However, verifying an entire processor design with model checking is not yet tractable; for example, with the ARM6 verification, the proof that the processor performs 32-bit multiplication is non-trivial.

The original framework of Harman, Tucker and Fox has been generalised to deal with the IO behaviour of the ARM6 processor, but both the approach, and the HOL-4 system, have coped well when tested by the verification of a commercial processor design.

The ARM models will be used in future work. Konrad Slind, in Utah, has plans to use the ISA model as the target for a simple compiler, to be implemented in HOL. This would provide a means to verify software running on the ARM6. There are also plans to model and verify ARM's coprocessor instructions.

References

1. M. D. Aagaard, B. Cook, N. A. Day, and R. B. Jones. A framework for microprocessor correctness statements. In *CHARME 2001*, volume 2144 of *LNCS*, pages 433–448. Springer, 2001.
2. J. R. Burch and D. L. Dill. Automatic verification of pipelined microprocessor control. In D. L. Dill, editor, *Proceedings of the 6th International Conference, CAV '94: Computer Aided Verification*, volume 818 of *Lecture Notes in Computer Science*, pages 68–80, Berlin, 1994. Springer-Verlag.
3. M. Clavel, F. Durán, S. Eker, P. Lincoln, N. Martí-Oliet, J. Meseguer, and J. Quesada. Maude: Specification and programming in rewrite logic. Technical report, Computer Science Laboratory, SRI International, 1999.
4. H. Ehrig and B. Mahr. *Fundamentals of Algebraic Specification I: Equations and Initial Semantics.* EATCS Monograph vol. 6, Springer-Verlag, 1985.
5. A. C. J. Fox. *Algebraic Models for Advanced Microprocessors.* PhD thesis, University of Wales Swansea, 1998.
6. A. C. J. Fox. An algebraic framework for modelling and verifying microprocessors using HOL. Technical Report 512, University of Cambridge, Computer Laboratory, 2001.
7. S. Furber. *ARM: system-on-chip architecture.* Addison-Wesley, second edition, 2000.
8. M. J. C. Gordon. HOL: A machine oriented formulation of higher order logic. Technical Report 42, University of Cambridge Computer Laboratory, 1995.
9. M. J. C. Gordon, A. J. Milner, and C. P. Wadsworth. *Edinburgh* LCF*: A Mechanised Logic of Computation*, volume 78 of *Lecture Notes in Computer Science*. Springer-Verlag, 1979.
10. N. A. Harman and J. V. Tucker. Algebraic models and the correctness of microprocessors. In L. P. G Milne, editor, *Correct Hardware Design and Verification Methods.* Lecture Notes in Computer Science 683, Springer-Verlag, 1993.
11. N. A. Harman and J. V. Tucker. Algebraic models of microprocessors: Architecture and organisation. *Acta Informatica*, 33(5):421–456, 1996.
12. P. Manolios. Correctness of pipelined machines. In W. A. Hunt, Jr. and S. D. Johnson, editors, *Formal Methods in Computer-Aided Design, FMCAD 2000*, volume 1954 of *Lecture Notes in Computer Science*, pages 161–178. Springer-Verlag, 2000.
13. J. Sawada and W. A. Hunt, Jr. Verification of FM9801: An out-of-order model with speculative execution, exceptions, and program-modifying capability. *Formal Methods in System Design*, 20(2):187–222, 2002.

14. D. Seal, editor. *ARM Architectural Reference Manual.* Addison-Wesley, second edition, 2001.
15. R. Stephens. *Algebraic Stream Processing.* PhD thesis, Department of Computer Science, University College of Swansea, 1994.
16. B. C. Thompson. *A Mathematical Theory of Synchronous Concurrent Algorithms.* PhD thesis, Department of Computer Studies, University of Leeds, 1987.
17. J. V. Tucker and J. I. Zucker. Theory of computability over stream algebras and its application to the mathematical foundations of computer science. In I. M. Havel and V. Konbek, editors, *17th International Colloquium, Prague*, pages 62 – 80. Lecture Notes in Computer Science 629, Springer-Verlag, 1992.
18. M. Wirsing. Algebraic specification. In J. van Leeuwen, editor, *Handbook of Theoretical Computer Science, Volume B: Formal Models and Semantics*, pages 675 – 788. Elsevier, 1990.

Using Proofs by Coinduction to Find "Traditional" Proofs

Clemens Grabmayer

Department of Computer Science, Vrije Universiteit Amsterdam,
de Boelelaan 1081a, 1081 HV Amsterdam, The Netherlands
clemens@cs.vu.nl
http://www.cs.vu.nl/~clemens

Abstract. In the specific situation of formal reasoning concerned with "regular expression equivalence" we address instances of more general questions such as: how can coinductive argumentation be formalised logically and be applied effectively, as well as how is it linked to traditional forms of proof. For statements expressing that two regular expressions are language equivalent, we demonstrate that proofs by coinduction can be formulated in a proof system based on equational logic, where effective proof-search is possible. And we describe a proof-theoretic method for translating derivations in this proof system into a "traditional" axiom system: namely, into a "reverse form" of the axiomatisation of "regular expression equivalence" due to Salomaa. Hereby we obtain a coinductive completeness proof for the traditional proof system.

1 Introduction

Coalgebraic methods have been applied with much success in many areas of mathematics and computer science, contributing important new concepts as well as introducing fresh viewpoints at established theories. This has frequently led to the discovery of elegant new proofs of known results. Contrasting with the interest in applications, much less attention has been directed to formalising coalgebraic concepts, such as coinduction and corecursion, by using the tools of logic, and to relating these techniques with traditional methods, such as induction and recursion. This concerns also more specific questions such as whether proofs by coinduction that are formalised in an appropriate logical framework can be translated into formalised "conventional" proofs.

In this paper we consider the concrete example of a coinduction principle for proving that two regular expressions are language equivalent. We reformulate such a principle into one that can be used to decide equivalence of regular expressions effectively, and we give a logical formalisation. Furthermore, we describe a method that allows to translate proofs based on the coinduction principle into derivations in a "traditional" axiom system close to the well-known axiomatisation $\mathbf{F}_1$ of "the algebra of regular events" due to Salomaa in [7].

In [6] Rutten formulates a coinduction principle for showing equality of formal languages: to show that two languages L_1 and L_2 are equal, it suffices to prove

J.L. Fiadeiro et al. (Eds.): CALCO 2005, LNCS 3629, pp. 175–193, 2005.

that L_1 and L_2 are bisimilar in the "automaton of formal languages". Based on the differential calculus for regular expressions due to Brzozowski in [2], Rutten applies this principle to give coinductive demonstrations for a number of identities between regular expressions, and stresses the generality of this method. However, Rutten's proofs for exemplary identities use set-theoretical concepts in an essential way and do not lend themselves directly towards a formalisation in a proof system of equational logic comparable to Salomaa's axiomatisations. And to the author of the present paper some details have not been clear about why such a principle does in fact yield a generally applicable decision procedure.

Here we first introduce, by following and refining Rutten's approach, a "finitary" coinduction principle for "regular expression equivalence": to show that two regular expressions E and F are equivalent, prove that they are related by a *finite* bisimulation in a certain automaton on regular expressions whose transition function is based on the "Brzozowski derivative". We show that this principle is effective and lends itself to being mechanised. Next we introduce a natural-deduction style proof system $\mathbf{cREG_0}$ of equational logic with the property that derivations in $\mathbf{cREG_0}$ formalise, and correspond to, arguments by the finitary coinduction principle. It turns out that $\mathbf{cREG_0}$ is sound and complete with respect to regular expression equivalence. Finally, we describe an effective proof-theoretic transformation from derivations in $\mathbf{cREG_0}$ into derivations in a variant system $\mathbf{REG}$ of Salomaa's $\mathbf{F_1}$, where $\mathbf{REG}$ is the result of reversing all multiplicative parts of regular expressions in the axioms and in the rules of $\mathbf{F_1}$.

The proof system $\mathbf{cREG_0}$ we introduce is analogous in kind to an axiomatisation of "recursive type equality" introduced by Brandt and Henglein in [1] (together with its coinductive foundations) and to a system for "bisimilarity of normed recursive BPA-processes" due to Stirling given in [4] (without a coinductive motivation). All of these systems (and a number of similar, more recent ones) have in common the presence of inference rules that formalise "cyclic" forms of reasoning. Applications of such rules allow, roughly speaking, to detect that a bisimulation-building process that is formalised by a derivation has reached a subtask which it has already solved before. The transformation between $\mathbf{cREG_0}$-derivations and $\mathbf{REG}$-derivations that we develop here was inspired by a transformation given in [5, Ch.8, Sect.8.1], where proof-theoretic relations between proof systems for "recursive type equality" are investigated.

We give a short overview of the paper: In Section 2 we define basic notions concerning regular expressions and finite automata (such as the relation "regular expression equivalence" and the notion of bisimulation). Then in Section 3 we formulate the mentioned variant system $\mathbf{REG}$ of Salomaa's axiomatisation $\mathbf{F_1}$ and define three weaker systems. In Section 4, we review the most basic notions of the "differential calculus" for formal languages and of that for regular expressions; and we relate the coinduction principle due to Rutten. Subsequently in Section 5, we formulate and prove our "finitary" version of a conduction principle for regular expression equivalence, and argue that it can be used effectively. As a formalisation of this principle, we introduce the proof system $\mathbf{cREG_0}$ in Section 6 and show that it is sound and complete. Finally in Section 7 we describe an effective proof-theoretic transformation. from $\mathbf{cREG_0}$-derivations into **REG**-derivations. In the Conclusion, Section 8, we summarise our findings and explain how similar results can be obtained that apply directly to Salomaa's $\mathbf{F_1}$.

The proofs in this paper are generally only hinted or sketched, and the methods used are instantiated in supporting examples. However, the most important proofs can be found in a technical appendix that is contained in the electronic version of this paper which is available at `http://www.cs.vu.nl/~clemens/coind2tradproofs.pdf`.

2 Regular Expressions and Deterministic Automata

Let Σ be a finite nonempty set, called *alphabet*; elements of Σ are called *letters*. By Σ^* we denote the set of (finite) *words* over Σ. The *empty word* is designated by ϵ. Concatenation of words w and w' is denoted multiplicatively as $w.w'$. A *language* over Σ is any subset of Σ^*. By $\mathcal{L}(\Sigma)$ we denote the set of languages over Σ. On $\mathcal{L}(\Sigma)$ we define the *regular operators* $+$ (*sum*), . (*product*), and * (*star*), where $+$ and . are binary, and * is unary: for all $L_1, L_2 \in \mathcal{L}(\Sigma)$ we let

$$L_1 + L_2 =_{\text{def}} L_1 \cup L_2\,, \qquad L_1.L_2 =_{\text{def}} \{w_1.w_2 \mid w_1 \in L_1,\, w_2 \in L_2\}\,,$$

$$L^* =_{\text{def}} \bigcup_{n\in\omega} L^n\,, \quad \text{where} \quad \begin{array}{rl} L^0 =_{\text{def}} \{\epsilon\}\,, & \text{and} \\ L^{i+1} =_{\text{def}} L.L^i & (\text{for all } i \in \omega) \end{array}$$

(by ω we denote, here and below, the natural numbers including zero).

Let $\Sigma = \{a_1, \ldots, a_n\}$ be an alphabet (from now on, such a description is generally assumed for alphabets Σ). The set $\mathcal{R}(\Sigma)$ of *regular expressions* over Σ is defined as the set of those words over Σ that are generated by the grammar

$$E ::= 0 \mid a_1 \mid \ldots \mid a_n \mid E + E \mid E.E \mid E^*$$

We designate the regular expression 0^* by the symbol 1. By $\equiv$ we denote the binary relation "syntactical equality" between regular expressions. By $\sum_{i=1}^{n} E_i$ we denote, for all $n \in \omega\backslash\{0\}$ and $E_1, \ldots, E_n \in \mathcal{R}(\Sigma)$, the regular expression $E_1 + (E_2 + \ldots + (E_{n+1} + E_n))$. By a *context* C over $\mathcal{R}(\Sigma)$ we mean the result of replacing a single letter in a regular expression by a hole $[\,]$; by $C[E]$ we denote the result of hole-filling in C with the regular expression E. Every regular expression E denotes a language $L(E)$ via the function $L : \mathcal{R}(\Sigma) \to \mathcal{L}(\Sigma)$ that is inductively defined by

$$L(0) = \emptyset\,, \qquad L(a_i) = \{a_i\} \quad (1 \le i \le n)\,,$$

$$L(E + F) = L(E) \cup L(F)\,, \qquad L(E.F) = L(E).L(F) \qquad L(E^*) = L(E)^*\,.$$

(for all $E, F \in \mathcal{R}(\Sigma)$). Two regular expressions $E, F \in \mathcal{R}(\Sigma)$ are called *equivalent* (denoted by $E =_L F$) if and only if E and F denote the same formal language, i.e. iff $L(E) = L(F)$. In accordance with the notation just stipulated, we define a binary relation $=_L$ on $\mathcal{R}(\Sigma)$, called *regular expression equivalence*, by $=_L =_{\text{def}} \{\langle E, F\rangle \in \mathcal{R}(\Sigma) \times \mathcal{R}(\Sigma) \mid L(E) = L(F)\}$.

Let A be a (possibly) infinite set of *input symbols*. A *(deterministic) automaton* with input alphabet A is a triple $\mathcal{S} = \langle S, o, t \rangle$ consisting of a set S of *states*, an *output function* $o : S \to 2$, and a *transition function* $t : S \to S^A$, where S^A denotes the set of all functions from A to S, and $2 = \{0, 1\}$ (in this set 0 and 1 are usually numbers, but for convenience[1] we agree to consider them as regular expressions here). The output function o indicates whether a state s is $\mathcal{S}$ is *accepting* (if $o(s) = 1$) or not (if $o(s) = 0$). The transition function t assigns to a state s a function $t(s) : A \to S$ which defines the state $t(s)(a)$ that is reached by $\mathcal{S}$ after reading input symbol a. Sometimes we write $s\!\downarrow$ for $o(s) = 1$, $s\!\uparrow$ for $o(s) = 0$, and $s \xrightarrow{a} s'$ for $t(s)(a) = s'$.

Let $\mathcal{S} = \langle S, o, t \rangle$ and $\mathcal{S}' = \langle S, o', t' \rangle$ be automata. A *homomorphism* between $\mathcal{S}$ and $\mathcal{S}'$ is a function $f : S \to S'$ such that, for all $s \in S$ and $a \in A$, $o(s) = = o'(f(s))$ and $f(t(s)(a)) = t'(f(s))(a)$ holds. A *bisimulation* between $\mathcal{S}$ and $\mathcal{S}'$ is a nonempty relation $R \subseteq S \times S'$ such that for all $s \in S$, $s' \in S'$, and $a \in A$

$$s \, R \, s' \quad \Longrightarrow \quad o(s) = o'(s) \quad \text{and} \quad t(s)(a) \; R \; t'(s')(a)$$

holds. For $s \in S$ and $s' \in S'$ we write $s \sim s'$ if there exists a bisimulation R with sRs'; if there exists a finite bisimulation R with sRs', we write $s \sim_{\text{fin}} s'$.

3 The Axiom System REG

The first complete axiomatisations of regular expression equivalence were given by Salomaa in [7]. Here, our investigations will be based on Salomaa's first system $\mathbf{F_1}$. However, we introduce a variant system **REG** that arises from $\mathbf{F_1}$ essentially[2] by reversing all multiplicative expressions in axioms and rules. The reason is that, while having analogous properties as $\mathbf{F_1}$, the system **REG** will turn out to lend itself much better to establish a connection with the differential calculus for regular expressions in its usual form (as described in Section 4).

Let Σ be an alphabet. The axiom system $\mathbf{REG}(\Sigma)$ is defined as follows: its *formulas* are equations $E = F$ between regular expression E and F on Σ; its *axioms* are the formulas that belong to one of the schemes (B1)–(B11) listed in Figure 1; and its *inference rules* are the four rules SYMM, TRANS, CTXT, and FIX whose applications are schematically defined in Figure 1 (reflexivity axioms are not used in this definition as they can easily be recognised to be derivable).

Derivations in $\mathbf{REG}(\Sigma)$ are prooftrees, that is, finite upwards-growing labeled trees such that: all nodes are labeled by formulas of $\mathbf{REG}(\Sigma)$, the leaves at the top carry axioms of $\mathbf{REG}(\Sigma)$, and each internal node ν is labeled by a formula that is the conclusion of an application of a $\mathbf{REG}(\Sigma)$-rule with the formula(s) that label(s) the immediate successor(s) of ν as premises; the bottom-most formula of a prooftree is called its *conclusion*. For $E, F \in \mathcal{R}(\Sigma)$, we denote by $\vdash_{\mathbf{REG}(\Sigma)} E = F$ the statement that there exists a derivation in $\mathbf{REG}(\Sigma)$

[1] We want to be able to view outcomes of output functions as regular expressions.

[2] A less important change consists in dropping the substitution rule R2 specific to $\mathbf{F_1}$ in favour of the symmetry, transitivity, and context rules of equational logic.

The *axioms* of $\mathbf{REG}(\Sigma)$:

(B1)	$E + (F + G) = (E + F) + G$	(B7)	$E.1 = E$
(B2)	$(E.F).G = E.(F.G)$	(B8)	$E.0 = 0$
(B3)	$E + F = F + E$	(B9)	$E + 0 = E$
(B4)	$(E + F).G = E.G + F.G$	(B10)	$E^* = 1 + E.E^*$
(B5)	$E.(F + G) = E.F + E.G$	(B11)	$E^* = (1 + E)^*$
(B6)	$E + E = E$		

The *inference rules* of $\mathbf{REG}(\Sigma)$:

$$\frac{E = F}{F = E}\ \text{SYMM} \qquad \frac{E = G \qquad G = F}{E = F}\ \text{TRANS}$$

$$\frac{E = F}{C[E] = C[F]}\ \text{CTXT} \qquad \frac{E = F.E + G}{E = F^*.G}\ \text{FIX}\ (\text{if } o(F) = 0\ [\text{cf. Sect. 4}])$$

Fig. 1. The axiom system $\mathbf{REG}(\Sigma)$ for regular expression equivalence, which results from Salomaa's system $\mathbf{F_1}$ by reversing multiplicative expressions

with conclusion $E = F$. (We sometimes write **REG** in place of $\mathbf{REG}(\Sigma)$.)

The following theorem can be proved analogously to Salomaa's result for $\mathbf{F_1}$.

Theorem 1. *The system $\mathbf{REG}(\Sigma)$ is sound and complete with respect to regular expression equivalence. More formally, it holds:*

$$\textit{for all } E, F \in \mathcal{R}(\Sigma): \quad \left[\ \vdash_{\mathbf{REG}(\Sigma)} E = F \quad \Longleftrightarrow \quad E =_L F\ \right]. \tag{1}$$

For later use, we define three systems that are weaker than $\mathbf{REG}(\Sigma)$, but closely related: by $\mathbf{REG}^-(\Sigma)$ we designate the axiom system that results from $\mathbf{REG}(\Sigma)$ by excluding the rule FIX; by $\mathbf{ACI}(\Sigma)$ we denote the subsystem of $\mathbf{REG}^-(\Sigma)$ that contains only the axioms (B1), (B3), and (B6) for associativity, commutativity, and idempotency of $+$; and by $\mathbf{ACI}^+(\Sigma)$ we denote the extension of $\mathbf{ACI}(\Sigma)$ that contains of all the axioms (B1)–(B9) and furthermore

$$(\text{B7})^R \quad 1.E = E \qquad \text{and} \qquad (\text{B8})^R \quad 0.E = 0\,,$$

but that does not contain the rule FIX. For each of these three systems, we define binary relations on $\mathcal{R}(\Sigma)$ that denote "equality is derivable" in the respective system: for instance, we stipulate, for all $E, F \in \mathcal{R}(\Sigma)$,

$$E \equiv_{\mathbf{ACI}^+} F \quad \Longleftrightarrow_{\text{def}} \quad \vdash_{\mathbf{ACI}^+(\Sigma)} E = F\,; \tag{2}$$

the relations $\equiv_{\mathbf{ACI}}$ and $\equiv_{\mathbf{REG}^-}$ are defined analogously. It is easy to verify that all three relations are congruence relations on $\mathcal{R}(\Sigma)$. For all $E \in \mathcal{R}(\Sigma)$, we respectively denote by $[E]_{\mathbf{ACI}}$, $[E]_{\mathbf{ACI}^+}$, and $[E]_{\mathbf{REG}^-}$ the $\equiv_{\mathbf{ACI}}$-, $\equiv_{\mathbf{ACI}^+}$- and $\equiv_{\mathbf{REG}^-}$-equivalence classes of E. And by $\mathcal{R}(\Sigma)_{\mathbf{ACI}}$, $\mathcal{R}(\Sigma)_{\mathbf{ACI}^+}$, and $\mathcal{R}(\Sigma)_{\mathbf{REG}^-}$ we denote by factor sets of $\mathcal{R}(\Sigma)$ with respect to $\equiv_{\mathbf{ACI}}$, $\equiv_{\mathbf{ACI}^+}$, and $\equiv_{\mathbf{REG}^-}$.

4 The Differential Calculus for Regular Expressions

In this section we review the basic notions of a differential calculus for formal languages, as for example described by Conway [3, Ch.5], and for regular expressions, due to Brzozowski in [2]. We also state two coinduction principles.

Let Σ be an alphabet, and $L \in \mathcal{L}(\Sigma)$. For all words $w \in \Sigma^*$, the *w-derivative* of L is $L_w =_{\text{def}} \{v \in \Sigma^* \mid w.v \in L\}$. In the special case of letters $a \in \Sigma$ the *a-derivative* L_a can be used to turn the set $\mathcal{L}(\Sigma)$ of languages over Σ into an automaton $\langle \mathcal{L}(\Sigma), o_{\mathcal{L}}, t_{\mathcal{L}} \rangle$ by defining, for all $L \in \mathcal{L}(\Sigma)$ and $a \in \Sigma$,

$$o_{\mathcal{L}}(L) =_{\text{def}} \begin{cases} 1 \ldots \ \epsilon \in L \\ 0 \ldots \ \epsilon \notin L \end{cases} \qquad \text{and} \quad t_{\mathcal{L}}(L)(a) =_{\text{def}} L_a \,.$$

In [6, Section 4] Rutten shows the *coinduction principle* for proving equality of formal languages that is stated by the following proposition.

Proposition 1. *For all $L_1, L_2 \in \mathcal{L}(\Sigma)$ it holds:*

$$L_1 \sim L_2 \ \ in\ \mathcal{L}(\Sigma) \quad \Longrightarrow \quad L_1 = L_2 \,. \tag{3}$$

That is: to show $L_1 = L_2$ for two languages L_1 and L_2 over Σ, it suffices to demonstrate that L_1 and L_2 are bisimilar in the automaton $\mathcal{L}(\Sigma)$.

From Proposition 1 a similar proof principle for showing equivalence of regular expressions can be extracted by using the "Brzozowski derivative". This concept, here just called "derivative", allows to mimic language derivatives on regular expressions. Let again Σ be an alphabet. For all $a \in \Sigma$, and $G \in \mathcal{R}(\Sigma)$, the *a-derivative* G_a of a regular expression G over Σ is defined inductively by

$$0_a =_{\text{def}} 0\,, \qquad (E+F)_a =_{\text{def}} E_a + F_a\,, \qquad (E^*)_a =_{\text{def}} E_a.E^*\,,$$

$$b_a =_{\text{def}} \begin{cases} 1 & \ldots\ b = a \\ 0 & \ldots\ b \neq a \end{cases} \qquad (E.F)_a =_{\text{def}} \begin{cases} E_a.F + F_a & \ldots\ o(E) = 1 \\ E_a.F & \ldots\ o(E) = 0 \end{cases}$$

(for all $b \in \Sigma$ and $E, F \in \mathcal{R}(\Sigma)$). In a similar way, the function $o : \mathcal{R}(\Sigma) \to 2$ is inductively defined by (for all $b \in \Sigma$ and $E, F \in \mathcal{R}(\Sigma)$)

$$o(0) =_{\text{def}} 0\,, \qquad o(b) =_{\text{def}} 0\,, \qquad o(E+F) =_{\text{def}} \begin{cases} 0 & \ldots\ o(E) = o(F) = 0 \\ 1 & \ldots\ \text{else} \end{cases}$$

$$o(E.F) =_{\text{def}} \begin{cases} 1 & \ldots\ o(E) = o(F) = 1 \\ 0 & \ldots\ \text{else}\,, \end{cases} \qquad o(E^*) =_{\text{def}} 1\,.$$

We also define, for all $w \in \Sigma$ and $E \in \mathcal{R}(\Sigma)$, the *w-derivative* of E inductively: we let $E_\epsilon =_{\text{def}} E$, and, for all $w_0 \in \Sigma^*$ and $a \in \Sigma$, $E_{w_0.a} =_{\text{def}} (E_{w_0})_a$.

Now an automaton $\mathcal{R}(\Sigma) = \langle \mathcal{R}(\Sigma), o, t \rangle$ can be formed by letting o as above and $t : \mathcal{R}(\Sigma) \to \mathcal{R}(\Sigma)^\Sigma$ be defined by $t(E)(a) =_{\text{def}} E_a$ for all $a \in \Sigma$, $E \in \mathcal{R}(\Sigma)$. The function L is a homomorphism from $\mathcal{R}(\Sigma)$ to $\mathcal{L}(\Sigma)$ because, for $E \in \mathcal{R}(\Sigma)$ and $a \in \Sigma$, $L(E_a) = (L(E))_a$ and $o(E) = o_{\mathcal{L}}(L(E))$ hold (as is simple to prove). Due to this, the following statement is an easy consequence of Proposition 1.

Proposition 2. *The following* coinduction principle *holds for proving equivalence of regular expressions: for all* $E, F \in \mathcal{R}(\Sigma)$ *it holds*

$$E \sim F \;\; in\ \mathcal{R}(\Sigma) \quad \Longrightarrow \quad E =_L F \tag{4}$$

Although this principle can often be applied successfully in an informal manner (cf. the examples in [6, Section 6]), it does not itself define a general mechanisable method for deciding whether two regular expressions are equivalent. The reason is that the set of iterated derivatives of a regular expression is frequently infinite[3], and that therefore bisimulations in $\mathcal{R}(\Sigma)$ can be infinite.

5 A Finitary Coinduction Principle for $=_L$

One possible way of adopting the coinduction principle in Proposition 2 for deciding regular expression equivalence consists in refining it into a statement that only refers to finite bisimulations. As mentioned above, Proposition 2 relies on infinite bisimulations in an essential way since the number of derivatives of a regular expression may be infinite. However, it turns out that already "modulo" provability in the system **ACI** the number of derivatives of a regular expression is finite. This is stated by the second of the following two lemmas.

Lemma 1. *Let* Σ *be an alphabet, and let* $\equiv_{\mathcal{S}}$ *be one of the relations* $\equiv_{\mathbf{ACI}}$ *or* $\equiv_{\mathbf{ACI}^+}$ *on* $\mathcal{R}(\Sigma)$*. Then for all* $E, F \in \mathcal{R}(\Sigma)$ *and for all* $a \in \Sigma$ *it holds:*

$$E \equiv_{\mathcal{S}} F \quad \Longrightarrow \quad \big(o(E) = o(F) \;\;\&\;\; E_a \equiv_{\mathcal{S}} F_a \big)\,. \tag{5}$$

Proof (Sketch). In a first step, it can be verified in a straightforward way that (5) holds for all $a \in \Sigma$, and for all $E, F \in \mathcal{R}(\Sigma)$ such that $E = F$ is an axiom of **ACI** or **ACI**$^+$. The statement obtained hereby can then be "lifted" to apply to all $E, F \in \mathcal{R}(\Sigma)$ such that $E = F$ is a theorem of **ACI**, or of **ACI**$^+$, by using induction on the depth of derivations in **ACI**, or respectively, in **ACI**$^+$.

Lemma 2. *For all* $E \in \mathcal{R}(\Sigma)$*, the set* $\big\{ [E_w]_{\mathbf{ACI}} \;\big|\; w \in \Sigma^* \big\}$ *is finite. As a consequence, also* $\big\{ [E_w]_{\mathbf{ACI}^+} \;\big|\; w \in \Sigma^* \big\}$ *is finite for arbitrary* $E \in \mathcal{R}(\Sigma)$*.*[4]

Proof (Hint). The lemma can be shown by induction on the syntactical structure of regular expressions in $\mathcal{R}(\Sigma)$, using representation statements for w-derivatives of composite expressions like, in the case of an outermost product,

$$(\forall w \in \Sigma^*)\, (\exists V \subseteq \mathit{Suff}(w))\, \Big[(F.G)_w \;\equiv_{\mathbf{ACI}}\; F_w.G + \sum_{v \in V} G_v \Big]$$

(for all $F, G \in \mathcal{R}(\Sigma)$), where $\mathit{Suff}(w)$ means the set of all *suffixes* of w.

[3] For instance, by starting from a^* and computing the a-derivative repeatedly one is led to $1.a^*$, $0.a^* + 1.a^*$, …, $0.a^* + \ldots (0.a^* + 1.a^*)$, … .

[4] The part of Lemma 2 referring to **ACI** is comparable to Theorem 5.3 by Brzozowski in [2], which statement, however, is wrong (as Salomaa rightly points out in [7]). But the reason can easily be recognised in the fact that the derivative for multiplicative expressions is defined differently in [2] than in Section 4 here: there, for all $E, F \in \mathcal{R}(\Sigma)$ and $a \in \Sigma$, $(E.F)_a = E_a.F + o(E).F_a$ is stipulated.

We have formulated these lemmas also with respect to $\mathbf{ACI}^+$, on which system we base ourselves from now on, because it seems natural to apply also other identities than those of **ACI** to simplify derivatives.[5] Relying on Lemma 1, we can now define the "factor automaton" $\mathcal{R}(\Sigma)_{\mathbf{ACI}^+} = \langle \mathcal{R}(\Sigma)_{\mathbf{ACI}^+}, o_{\mathbf{ACI}^+}, t_{\mathbf{ACI}^+} \rangle$ of $\mathcal{R}(\Sigma)$ with respect to $\equiv_{\mathbf{ACI}^+}$ by letting

$$o_{\mathbf{ACI}^+} : \mathcal{R}(\Sigma)_{\mathbf{ACI}^+} \to 2, \quad [E]_{\mathbf{ACI}^+} \mapsto o(E), \qquad t_{\mathbf{ACI}^+} : \mathcal{R}(\Sigma)_{\mathbf{ACI}^+} \to (\mathcal{R}(\Sigma)_{\mathbf{ACI}^+})^{\Sigma}, \quad [E]_{\mathbf{ACI}^+} \mapsto (a \mapsto [E_a]_{\mathbf{ACI}^+}) .$$

And we are finally able to formulate the following *finitary coinduction principle* for proving or disproving that two given regular expressions are equivalent.

Theorem 2. *For all $E, F \in \mathcal{R}(\Sigma)$ it holds:*

$$[E]_{\mathbf{ACI}^+} \sim_{\mathit{fin}} [F]_{\mathbf{ACI}^+} \text{ in } \mathcal{R}(\Sigma)_{\mathbf{ACI}^+} \iff E =_L F . \tag{6}$$

Proof (Sketch). Let $E, F \in \mathcal{R}(\Sigma)$. The implication "$\Rightarrow$" in (6) is a consequence of Proposition 1 in view of the fact that the function $L^* : \mathcal{R}(\Sigma)_{\mathbf{ACI}^+} \to \mathcal{L}(\Sigma)$ which is defined by $L^*([G]_{\mathbf{ACI}^+}) \mapsto L(G)$ is a homomorphism. For the implication "$\Leftarrow$" in (6), assume $E =_L F$. Then $\{ \langle [E_w]_{\mathbf{ACI}^+}, [F_w]_{\mathbf{ACI}^+} \rangle \mid w \in \Sigma^* \}$ is a bisimulation between $[E]_{\mathbf{ACI}^+}$ and $[F]_{\mathbf{ACI}^+}$ in $\mathcal{R}(\Sigma)_{\mathbf{ACI}^+}$ (as is not difficult to verify); this bisimulation can easily be recognised to be finite by using Lemma 2.

As running example in this paper we consider $(a+b)^* = (a^*b)^*a^*$, a simple instance of the axiom scheme "sumstar" in a system due to Conway in [3, p.25]. We let $E^* \equiv (a+b)^*$, $F_1 \equiv (a^*b)^*a^*$, and $F_2 \equiv ((a^*b)(a^*b))^*a^* + a^*$. We find

$$(F_1)_a \equiv (((1.a^*).b + 0).(a.b)^*).a^* + 1.a^* \equiv_{\mathbf{ACI}^+} F_2 ,$$

and in a similar way, the other entries in the following tables can be verified:

	$[(\cdot)_a]_{\mathbf{ACI}^+}$	$[(\cdot)_b]_{\mathbf{ACI}^+}$	$o_{\mathbf{ACI}^+}(\cdot)$
E	$[E]_{\mathbf{ACI}^+}$	$[E]_{\mathbf{ACI}^+}$	$\downarrow$

	$[(\cdot)_a]_{\mathbf{ACI}^+}$	$[(\cdot)_b]_{\mathbf{ACI}^+}$	$o_{\mathbf{ACI}^+}(\cdot)$
F_1	$[F_2]_{\mathbf{ACI}^+}$	$[F_1]_{\mathbf{ACI}^+}$	$\downarrow$
F_2	$[F_2]_{\mathbf{ACI}^+}$	$[F_1]_{\mathbf{ACI}^+}$	$\downarrow$

From this it follows that $R = \{\langle [E]_{\mathbf{ACI}^+}, [F_1]_{\mathbf{ACI}^+} \rangle, \langle [E]_{\mathbf{ACI}^+}, [F_2]_{\mathbf{ACI}^+} \rangle\}$ is a finite bisimulation in $\mathcal{R}(\Sigma)_{\mathbf{ACI}^+}$ between $[E]_{\mathbf{ACI}^+}$ and $[F_1]_{\mathbf{ACI}^+}$. Using Theorem 2, this demonstrates $(a+b)^* =_L (a^*b)^*a^*$.

Based on the next lemma it is possible to extract an effective decision procedure for regular expression equivalence from our finitary coinduction principle.

Lemma 3. *For all alphabets Σ, the relation $\equiv_{\mathbf{ACI}}$ is decidable in $\mathcal{R}(\Sigma)$.*

---

[5] There is some arbitrariness in choosing a system of "basic" identities that one wants to have available for simplifying derivations. For instance, $\mathbf{REG}^-$ could be used as well. $\mathbf{ACI}^+$ has been chosen here partly because of the running example we employ.

Proof (Hint). Equations between $\equiv_{\mathbf{ACI}}$-equivalent sums of regular expressions can be decomposed into equations between $\equiv_{\mathbf{ACI}}$-equivalent parts of the sums. For example, for all $E_1, E_2, F_1, F_2, F_3 \in \mathcal{R}(\Sigma)$ that are not additive expressions,

$$\begin{aligned} & E_1 + E_2 \equiv_{\mathbf{ACI}} (F_1 + F_2) + F_3 \iff \\ & \iff (\exists f : \{1,2\} \to \{1,2,3\}) \, (\exists g : \{1,2,3\} \to \{1,2\}) \\ & \qquad (\forall i \in \{1,2\}) \, (\forall j \in \{1,2,3\}) \, [\, E_i \equiv_{\mathbf{ACI}} F_{f(i)} \;\&\; E_{g(j)} \equiv_{\mathbf{ACI}} F_j \,] \end{aligned}$$

holds. An obvious generalisation of this statement can be shown by structural induction on $\mathbf{ACI}(\Sigma)$-derivations, and it can be used to construct an effective (but clearly not efficient) search-algorithm that decides whether or not $E \equiv_{\mathbf{ACI}} F$ holds for given regular expressions $E, F \in \mathcal{R}(\Sigma)$.

Corollary 1. *Let Σ be an alphabet. Regular expression equivalence on $\mathcal{R}(\Sigma)$ can be decided by checking for the existence of finite bisimulations in $\mathcal{R}(\Sigma)_{\mathbf{ACI}^+}$.*

Proof (Sketch). Let $E, F \in \mathcal{R}(\Sigma)$ be arbitrary. It is an easy consequence of Proposition 2 that $E =_L F$ holds iff $R =_{\text{def}} \{ \langle [E_w]_{\mathbf{ACI}^+}, [F_w]_{\mathbf{ACI}^+} \rangle \mid w \in \Sigma^* \}$ is a bisimulation. Lemma 2 entails that R is always finite and that it can be determined effectively whether R is a bisimulation: for all pairs $\langle E_w, F_w \rangle$ with $w \in \Sigma^*$ and $w = b_1 \dots b_m$ such that the list $\langle [E]_{\mathbf{ACI}^+}, [F]_{\mathbf{ACI}^+} \rangle$, $\langle [E_{b_1}]_{\mathbf{ACI}^+}, [F_{b_1}]_{\mathbf{ACI}^+} \rangle$, $\langle [E_{b_1 b_2}]_{\mathbf{ACI}^+}, [F_{b_2 b_2}]_{\mathbf{ACI}^+} \rangle$, ..., $\langle [E_w]_{\mathbf{ACI}^+}, [F_w]_{\mathbf{ACI}^+} \rangle$ does not contain a loop (this can be decided due to Lemma 3) check whether $o_{\mathbf{ACI}^+}([E_w]_{\mathbf{ACI}^+}) = = o(E_w) = o(F_w) = o_{\mathbf{ACI}^+}([F_w]_{\mathbf{ACI}^+})$ holds. Because of Lemma 2 and König's lemma namely only finitely many such checks have to be performed. If one such check detects a mismatch, then R is not a bisimulation, and $E \neq_L F$ holds; if no mismatch is found, then R is a (finite) bisimulation and $E =_L F$ follows.

6 The Coinductively Motivated Proof System $\mathbf{cREG_0}$

Now we introduce a natural-deduction style proof system $\mathbf{cREG_0}(\Sigma)$ based on equational logic that allows to formalise arguments using the finitary coinduction principle for regular expression equivalence as finite derivations.

For the definition of $\mathbf{cREG_0}(\Sigma)$, we assume a countably infinite set Δ of assumption markers such that $\Sigma \cap \Delta = \emptyset$, and refer to the schemata listed in Figure 2: the *formulas* of $\mathbf{cREG_0}(\Sigma)$ are the equations between regular expressions over Σ; possible *assumptions* are formulas that have an assumption marker from the set Δ attached to them; and the *rules* of $\mathbf{cREG_0}(\Sigma)$ are the four rules $\text{App}_r\text{Ax}_{\mathbf{ACI}^+}$, $\text{App}_l\text{Ax}_{\mathbf{ACI}^+}$, COMP, and COMP/FIX that are schematically defined in Figure 2. Applications of the rule COMP/FIX have the special feature that at least one inhabited class of open assumptions is *discharged* (the marker of the assumptions belonging to this class is attached to the application).

Displaying this characteristic feature of proofs formalised in the format of natural-deduction systems (cf. the description of "N-systems" in [8]), namely the use of assumptions that may be "closed" (discharged) at a later stage in a deduction, *derivations* in $\mathbf{cREG_0}(\Sigma)$ are prooftrees such that: the leaves at the

Possible *assumptions* in $\mathbf{cREG_0}(\Sigma)$ and the *inference rules of* $\mathbf{cREG_0}(\Sigma)$:

$$\text{(Assm)} \quad (E = F)^d \quad (\text{with } d \in \Delta)$$

$$\frac{\begin{matrix}\mathcal{D}_1 \\ C[\tilde{E}] = F\end{matrix}}{C[\tilde{F}] = F}\ \text{App}_l\text{Ax}_{\mathbf{ACI^+}} \qquad \frac{\begin{matrix}\mathcal{D}_1 \\ E = C[\tilde{E}]\end{matrix}}{E = C[\tilde{F}]}\ \text{App}_r\text{Ax}_{\mathbf{ACI^+}}$$

$$(\text{if } \tilde{E} = \tilde{F} \text{ or } \tilde{F} = \tilde{E} \text{ is an axiom of } \mathbf{ACI^+})$$

$$\frac{\begin{matrix}\mathcal{D}_1 \\ E_{a_1} = F_{a_1}\end{matrix} \quad \dots \quad \begin{matrix}\mathcal{D}_n \\ E_{a_n} = F_{a_n}\end{matrix}}{E = F}\ \text{COMP} \quad (\text{if } o(E) = o(F))$$

$$\frac{\begin{matrix}[E = F]^d \\ \mathcal{D}_1 \\ E_{a_1} = F_{a_1}\end{matrix} \quad \dots \quad \begin{matrix}[E = F]^d \\ \mathcal{D}_n \\ E_{a_n} = F_{a_n}\end{matrix}}{E = F}\ \text{COMP/FIX}, d \quad (\text{if } o(E) = o(F))$$

Fig. 2. A coinductively motivated, natural-deduction style proof system $\mathbf{cREG_0}(\Sigma)$ for regular expression equivalence, given that $\Sigma = \{a_1, \dots, a_n\}$

top are labeled by assumptions such that different markers are attached to different formulas[6]; assumptions may be open (undischarged) or closed (discharged); formulas at an internal node ν arise through applications of $\mathbf{cREG_0}(\Sigma)$-rules from the formulas in the immediate successors of ν, whereby in the case of COMP/FIX-applications some open assumptions are discharged; the bottom-most formula is called the *conclusion*. Hereby an occurrence of an assumption $(E = F)^d$ at the top of a derivation $\mathcal{D}$ is called *open* iff on the path down to the conclusion of $\mathcal{D}$ there does not exist an application of COMP/FIX at which this assumption is discharged; otherwise the occurrence of $(E = F)^d$ is called *closed*. Assumptions in a derivation that are occurrences of the same formula with the same marker together form an *assumption class*.

For all $E, F \in \mathcal{R}(\Sigma)$, we denote by $\vdash_{\mathbf{cREG_0}(\Sigma)} E = F$ the statement that there exists a derivation $\mathcal{D}$ in $\mathbf{cREG_0}(\Sigma)$ without open assumptions such that $\mathcal{D}$ has conclusion $E = F$ (i.e. that $E = F$ is a *theorem* of $\mathbf{cREG_0}(\Sigma)$). (Sometimes we write $\mathbf{cREG_0}$ instead of $\mathbf{cREG_0}(\Sigma)$.)

Unlike as this is the case for the system **REG**, the basic axioms and rules of equational logic (the reflexivity axioms and the rules SYMM, TRANS, CTXT) are neither present nor in fact derivable in $\mathbf{cREG_0}$.[7] However, it turns out that these additional axioms rules are admissible in $\mathbf{cREG_0}$ in the following sense:

[6] The main reason for this proviso is that it is important for establishing a smooth proof-theoretical relationship (stated by Lemma 4 below) between $\mathbf{cREG_0}(\Sigma)$ and its annotated version $\mathbf{ann\text{-}cREG_0}(\Sigma, \Delta)$ defined in Section 7.

[7] In Remark 1, we comment on an extension of $\mathbf{cREG_0}$ with these axioms and rules.

if their use is limited to situations in which subderivations do not contain open assumptions, then no more theorems (than those of $\mathbf{cREG_0}$) become derivable.[8] This is an easy consequence of the fact, which is stated formally below, that the theorems of $\mathbf{cREG_0}$ are precisely the identities of regular expression equivalence.

Theorem 3. *The proof system* $\mathbf{cREG_0}(\Sigma)$ *is sound and complete with respect to regular expression equivalence; more formally, for all* $E, F \in \mathcal{R}(\Sigma)$ *it holds:*

$$\vdash_{\mathbf{cREG_0}(\Sigma)} E = F \quad\Longleftrightarrow\quad E =_L F\,. \tag{7}$$

Proof (Sketch). Let $E, F \in \mathcal{R}(\Sigma)$ be arbitrary. For the direction "$\Rightarrow$" in (7), let $\mathcal{D}$ be a derivation in $\mathbf{cREG_0}(\Sigma)$ without open assumptions and with conclusion $E = F$. Then $\left\{\langle[\tilde{E}]_{\mathbf{ACI^+}}, [\tilde{F}]_{\mathbf{ACI^+}}\rangle \mid \tilde{E} = \tilde{F} \text{ is formula in } \mathcal{D}\right\}$ is a finite bisimulation between E and F in $\mathcal{R}(\Sigma)_{\mathbf{ACI^+}}$. Hence Theorem 2 entails $E =_L F$. For the direction "$\Leftarrow$" in (7), suppose $E =_L F$. Then, again by Theorem 2, there exists a finite bisimulation between E and F in $\mathcal{R}(\Sigma)_{\mathbf{ACI^+}}$. From such a finite bisimulation a derivation in $\mathbf{cREG_0}(\Sigma)$ without open assumptions and with conclusion $E = F$ can be extracted in a rather straightforward way.

We consider again our running example $E \equiv (a+b)^* =_L (a^*b)^*a \equiv F_1$. From the finite bisimulation given in Section 5, it is easy to extract the following derivation in $\mathbf{cREG_0}(\{a, b\})$ that does not contain open assumptions (double lines indicate multiple successive applications of $\mathrm{App}_l\mathrm{Ax}_{\mathbf{ACI^+}}$ and/or $\mathrm{App}_r\mathrm{Ax}_{\mathbf{ACI^+}}$):

$$\text{COMP/FIX}, d\;\dfrac{\dfrac{\text{COMP/FIX}, e\;\dfrac{\dfrac{(E = F_2)^e}{E_a = (F_2)_a} \qquad \dfrac{(E = F_1)^d}{E_b = (F_2)_b}}{E = F_2}}{E_a = (F_1)_a} \qquad \dfrac{(E = F_1)^d}{E_b = (F_1)_b}}{E = F_1} \tag{8}$$

Remark 1. The fact that the system $\mathbf{cREG_0}$ does not contain the characteristic rules of equational logic is not absolutely necessary for showing a soundness and completeness theorem comparable to Theorem 3. In fact, for all alphabets Σ, the extension $\mathbf{cREG}(\Sigma)$ of $\mathbf{cREG_0}(\Sigma)$ by adding reflexivity axioms and the rules SYMM, TRANS, and CTXT is also sound and complete with respect to $=_L$ (but the soundness part requires a rather more involved proof, cf. the comparable situation treated in [1]). However, $\mathbf{cREG}(\Sigma)$ lacks a nice property of the system $\mathbf{cREG_0}(\Sigma)$: derivations in $\mathbf{cREG_0}(\Sigma)$ without open assumptions correspond, as reflected in the proof of Theorem 3, to finite bisimulations in $\mathcal{R}(\Sigma)_{\mathbf{ACI^+}}$; this is not the case for derivations in $\mathbf{cREG}(\Sigma)$ (due to, above all, the presence of the transitivity rule). Hence the system $\mathbf{cREG_0}$ is much more directly related to the finitary coinduction principle than the system $\mathbf{cREG}$. But there is a second

8 Note that admissibility in this sense does not demonstrate the soundness with respect to $=_L$ of the extension of $\mathbf{cREG_0}$ with the mentioned equational axioms and rules.

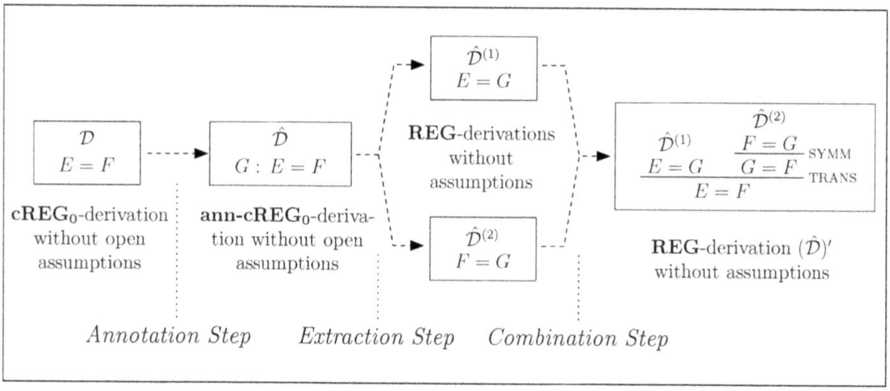

Fig. 3. Illustration of the three main steps in the transformation from an arbitrary derivation $\mathcal{D}$ in **cREG₀** without open assumptions into a derivation $(\hat{\mathcal{D}})'$ in **REG** with the same conclusion and without assumptions as $\mathcal{D}$

(although connected) reason for why we base ourselves on the system $\mathbf{cREG_0}$ here: it turns out that $\mathbf{cREG_0}$-derivations lend themselves much better to being transformed into **REG**-derivations than $\mathbf{cREG_0}$-derivations.[9]

7 A Transformation of $\mathbf{cREG_0}$- into REG-Derivations

In this section we sketch a proof-theoretic transformation of derivations in the coinductively motivated system $\mathbf{cREG_0}(\Sigma)$ into derivations in the variant system $\mathbf{REG}(\Sigma)$ of Salomaa's axiomatisation $\mathbf{F_1}$. The three steps of this transformation are the annotation step, the extraction step, and the combination step that are illustrated together in Figure 3 and that are described below separately.

7.1 The Annotation Step

In the annotation step, a given derivation $\mathcal{D}$ in $\mathbf{cREG_0}(\Sigma)$ is "analysed" by assigning to each formula a regular expression as an annotation. In this way a derivation $\hat{\mathcal{D}}$ in an annotated version $\mathbf{ann\text{-}cREG_0}(\Sigma, \Delta)$ of $\mathbf{cREG_0}(\Sigma)$ is built.

For an alphabet Σ and an infinite set Δ of assumption markers such that $\Sigma \cap \Delta = \emptyset$ holds, the system $\mathbf{ann\text{-}cREG_0}(\Sigma, \Delta)$ is defined as follows: the *formulas* of $\mathbf{ann\text{-}cREG_0}(\Sigma, \Delta)$ are expressions of the form $G : E = F$ with $E, F \in \mathcal{R}(\Sigma)$ and $G \in \mathcal{R}(\Sigma \cup \Delta)$; possible *assumptions* in $\mathbf{ann\text{-}cREG_0}(\Sigma, \Delta)$ are of the form $(d : E = F)^d$ with $E, F \in \mathcal{R}(\Sigma)$ and $d \in \Delta$; and the *rules* of $\mathbf{ann\text{-}cREG_0}(\Sigma, \Delta)$ are the four rules $\mathrm{App}_r\mathrm{Ax}_{\mathbf{ACI}^+}$, $\mathrm{App}_l\mathrm{Ax}_{\mathbf{ACI}^+}$, COMP, and COMP/FIX that are schematically defined in Figure 4; for both of the applications of COMP and COMP/FIX shown in Figure 4, $\bigcup_{i=1}^{n} J_i = \{1, \ldots, m\}$ is

[9] A possibility for extending the transformation from $\mathbf{cREG_0}$- into **REG**-derivations that is described in the next section to a transformation from **cREG**- into **REG**-derivations consists in the use of an elimination method for basic equational rules similar to one that is developed in [5, Ch.8].

Possible *assumptions* and the *inference rules in* **ann-cREG₀**(Σ, Δ):

$$\text{(Assm)}\quad (1.d:\ E = F)^d \qquad (\text{with } d \in \Delta;\ \text{ it is assumed: } \Sigma \cap \Delta = \emptyset)$$

$$\frac{\begin{array}{c}\mathcal{D}_1\\ G:\ C[\tilde{E}] = F\end{array}}{G:\ C[\tilde{F}] = E}\ \mathrm{App}_l\mathrm{Ax}_{\mathbf{ACI}^+} \qquad \frac{\begin{array}{c}\mathcal{D}_1\\ G:\ E = C[\tilde{E}]\end{array}}{G:\ E = C[\tilde{F}]}\ \mathrm{App}_r\mathrm{Ax}_{\mathbf{ACI}^+}$$

(given that $\tilde{E} = \tilde{F}$ or $\tilde{F} = \tilde{E}$ is an axiom of $\mathbf{ACI}^+$)

$$\frac{\begin{array}{c}\mathcal{D}_1\\ (G_{10}+)\sum_{j\in J_1} G_{1j}.d_j :\ E_{a_1} = F_{a_1}\end{array} \quad \cdots \quad \begin{array}{c}\mathcal{D}_n\\ (G_{n0}+)\sum_{j\in J_n} G_{nj}.d_j :\ E_{a_n} = F_{a_n}\end{array}}{\Big(o(E) + \sum_{i=1,\, G_{i0} \text{ occurs}}^{n} G_{i0}\Big) + \sum_{j=1}^{m}\Big(\sum_{i=1,\, j\in J_i}^{n} a_i.G_{ij}\Big).d_j\ :\ E = F}\ \begin{array}{l}\text{COMP}\\ (\text{if } o(E) = o(F))\end{array}$$

$$\frac{\begin{array}{c}[1.d_l:\ E = F]^{d_l}\\ \mathcal{D}_1\\ (G_{10}+)\sum_{j\in J_1} G_{1j}.d_j :\ E_{a_1} = F_{a_1}\end{array} \quad \cdots \quad \begin{array}{c}[1.d_l:\ E = F]^{d_l}\\ \mathcal{D}_n\\ (G_{n0}+)\sum_{j\in J_n} G_{nj}.d_j :\ E_{a_n} = F_{a_n}\end{array}}{\begin{array}{l}\Big(\sum_{i=1,\, l\in J_i}^{n} a_i.G_{il}\Big)^{*}.\Big(o(E) + \sum_{i=1,\, G_{i0} \text{ occurs}}^{n} a_i.G_{i0}\Big) + \\ \quad + \sum_{j=1,\, j\neq l}^{m}\Big(\Big(\sum_{i=1,\, l\in J_i}^{n} a_i.G_{il}\Big)^{*}.\Big(\sum_{i=1,\, j\in J_i}^{n} a_i.G_{ij}\Big)\Big).d_j\ :\ E = F\end{array}}\ \begin{array}{l}\text{COMP/FIX},\ d_l\\ (\text{if } o(E) = o(F))\\ \text{here: } 1 \le l \le m\end{array}$$

Fig. 4. The annotated version **ann-cREG₀**(Σ, Δ) of **cREG₀**(Σ)

assumed as well as that $G_{ij} \in \mathcal{R}(\Sigma)$ holds (i.e. that the G_{ij} do not contain letters from Δ), for all $i \in \{1, \ldots, n\}$ and $j \in J_i \cup \{0\}$. (We comment on the motivation for the specific way how the annotations have been chosen for the rules of the system **ann-cREG₀**(Σ, Δ) in Remark 2 at the end of this subsection.)

As in the system **cREG₀**(Σ), every application of the rule COMP/FIX discharges precisely one inhabited class of open assumptions (and the marker of this assumption class is attached to the application). *Derivations* in the system **ann-cREG₀**(Σ, Δ) are defined analogously as in **cREG₀**(Σ), and a similar proviso on the use of assumption markers is stipulated: in assumptions distinct equations must be annotated by distinct letters from Δ, i.e. if in a derivation the assumptions $(d_1 :\ E_1 = F_1)^{d_1}$ and $(d_2 :\ E_2 = F_2)^{d_2}$ occur, then $d_1 = d_2$ must entail $E_1 \equiv E_2$ and $F_1 \equiv F_2$.[10]

The following lemma states the basic proof-theoretic relationship between the systems **cREG₀**(Σ) and **ann-cREG₀**(Σ, Δ).

[10] This condition is necessary for the extraction step (i.p. for the proof of Lemma 6).

$$(\check{\hat{\mathcal{D}}} =) \quad \begin{array}{c} \{[E_i = F_i]^{d_i}\}_{i=1,\ldots,m} \\ \mathcal{D} \\ E = F \end{array} \quad \overset{(\hat{\cdot})}{\longmapsto} \; \underset{(\check{\cdot})}{\longleftarrow} \quad \begin{array}{c} \{[1.d_i : \; E_i = F_i]^{d_i}\}_{i=1,\ldots,m} \\ \tilde{\mathcal{D}} \\ (G_0 +) \sum_{j=1}^{m} G_j.d_j : E = F \end{array} \quad (= \hat{\mathcal{D}})$$

cREG₀(Σ)-derivation — **ann-cREG₀**(Σ, Δ)-derivation

Fig. 5. Proof-theoretic relation betw. **cREG₀**(Σ) and **ann-cREG₀**(Σ, Δ)

Lemma 4. *Every derivation $\mathcal{D}$ in* **ann-cREG₀**(Σ, Δ) *is of the form of the right derivation in Figure 5, for some $m \in \omega$, $E, F, G_0 \in \mathcal{R}(\Sigma)$, $E_i, F_i, G_i \in \mathcal{R}(\Sigma)$ for all $i \in \{1, \ldots, m\}$, and distinct $d_1, \ldots, d_n \in \Delta$; the expression at the top of this derivation denotes the family of all open assumption classes in $\mathcal{D}$.*

Every derivation in **cREG₀**(Σ) *that is of the form left in Figure 5 (for some $E, F, E_1, F_1, \ldots, E_m, F_m \in \mathcal{R}(\Sigma)$, $d_1, \ldots, d_m \in \Delta$) can effectively be transformed, by assigning an appropriate annotating regular expression in $\mathcal{R}(\Sigma \cup \Delta)$ to each formula in $\mathcal{D}$, into a derivation $\hat{\mathcal{D}}$ in* **ann-cREG₀**(Σ, Δ) *that is, for some $G_0, \ldots, G_m \in \mathcal{R}(\Sigma)$, of the form of the derivation $\tilde{\mathcal{D}}$ on the right in Figure 5.*

And vice versa, *every derivation $\tilde{\mathcal{D}}$ in* **ann-cREG₀**(Σ, Δ) *that is of the form on the right in Figure 5 can be transformed, by stripping annotated formulas in $\tilde{\mathcal{D}}$ of the annotating regular expressions, into a derivation $\check{\tilde{\mathcal{D}}}$ that is of the form of the left derivation in Figure 5.*

Proof (Hint). All three statements of the lemma can be shown by straightforward induction on the structure (or the depth) of derivations in **ann-cREG₀**(Σ, Δ), and respectively, by induction on the structure of derivations in **cREG₀**(Σ).

It is easy to verify that the result of annotating the **cREG₀**($\{a, b\}$)-derivation $\mathcal{D}$ in (8) for our running example is the following **ann-cREG₀**($\{a, b\}, \Delta$)-derivation $\hat{\mathcal{D}}$ without open assumptions (a number of annotations appear simplified):

$$\dfrac{\dfrac{\dfrac{(1.e: \; E = F_2)^e}{1.e: \; E_a = (F_2)_a} \qquad \dfrac{(1.d: \; E = F_1)^d}{1.d: \; E_b = (F_2)_b}}{\dfrac{a^* + a^*b.d: \; E = F_2}{a^* + a^*b.d: \; E_a = (F_1)_a}}\,\text{COMP/FIX}, e \qquad \dfrac{(1.d: \; E = F_1)^d}{1.d: \; E_b = (F_1)_b}}{(aa^*b + b)^*(1 + aa^*): \; E = F_1}\,\text{COMP/FIX}, d \tag{9}$$

Remark 2. Informally, the principal idea underlying the system **ann-cREG₀** and its relation with **cREG₀** is the following: annotating a **cREG₀**(Σ)-derivation $\mathcal{D}$ with conclusion $E = F$ and *without* open assumptions into a derivation $\hat{\mathcal{D}}$ in **ann-cREG₀**(Σ, Δ) with conclusion $G: \; E = F$ amounts to extracting from $\mathcal{D}$ a description as the regular expression G of the bisimulation between $[E]_{\mathbf{ACI}^+}$ and $[F]_{\mathbf{ACI}^+}$ in the automaton $\mathcal{R}(\Sigma)_{\mathbf{ACI}^+}$ that is formalised by $\mathcal{D}$ (cf. the proof of Theorem 3). For this regular expression $G \in \mathcal{R}(\Sigma)$, $[G]_{\mathbf{ACI}^+}$ is bisimilar in $\mathcal{R}(\Sigma)_{\mathbf{ACI}^+}$ to both $[E]_{\mathbf{ACI}^+}$ and $[F]_{\mathbf{ACI}^+}$; moreover, the "generated subautomaton" of $[G]_{\mathbf{ACI^+}}$ in $\mathcal{R}(\Sigma)_{\mathbf{ACI^+}}$ is a "common unfolding" of the subautomata in $\mathcal{R}(\Sigma)_{\mathbf{ACI^+}}$ that are generated by $[E]_{\mathbf{ACI^+}}$ and $[F]_{\mathbf{ACI^+}}$, respectively. These facts form the deeper reasons for why the extraction step (described in Subsection 7.2) of our transformation from $\mathbf{cREG_0}(\Sigma)$ to $\mathbf{REG}(\Sigma)$ is possible.

However, in the conclusion $G:\ E = F$ of an $\mathbf{ann\text{-}cREG_0}(\Sigma, \Delta)$-derivation $\tilde{\mathcal{D}}$ *with* open assumptions, the annotation $G \in \mathcal{R}(\Sigma \cup \Delta)$ only describes what could be called a "partial bisimulation" in $\mathcal{R}(\Sigma)_{\mathbf{ACI^+}}$ between $[E]_{\mathbf{ACI^+}}$ and $[F]_{\mathbf{ACI^+}}$. But nevertheless, and slightly apart from this, the annotation G in the conclusion of such a derivation $\tilde{\mathcal{D}}$ also specifies the common structure of a pair of "valid" equations that link the regular expressions on either side of "=" in the conclusion of $\tilde{\mathcal{D}}$ with the regular expressions on respectively the same side of "=" in the open assumptions of $\tilde{\mathcal{D}}$. More precisely, if $\tilde{\mathcal{D}}$ is a derivation in $\mathbf{ann\text{-}cREG_0}(\Sigma, \Delta)$ of the form

$$\begin{array}{c} \{[1.d_i\,:\ E_i = F_i]^{d_i}\}_{i=1,\ldots,m} \\ \tilde{\mathcal{D}} \\ (G_0+)\sum_{j=1}^{m} G_j.d_j : E = F \end{array} \tag{10}$$

for some $m \in \omega$, $E, F, E_1, F_1, \ldots, E_m, F_m, G_0, \ldots, G_m \in \mathcal{R}(\Sigma)$, $d_1, \ldots, d_m \in \Delta$, and with the expression at the top denoting the family of all inhabited open assumptions classes of $\tilde{\mathcal{D}}$ (due to Lemma 4 all $\mathbf{ann\text{-}cREG_0}(\Sigma, \Delta)$-derivations can be represented in this way), then the equations $E = (G_0+)\sum_{j=1}^{m} G_j.E_j$ and $F = (G_0+)\sum_{j=1}^{m} G_j.F_j$ are valid with respect to $=_L$, i.e. it holds:

$$E =_L (G_0+)\sum_{j=1}^{m} G_j.E_j \qquad \text{and} \qquad F =_L (G_0+)\sum_{j=1}^{m} G_j.F_j\,. \tag{11}$$

This property of $\mathbf{ann\text{-}cREG_0}$-derivations is essential for the extraction step. The annotations in the rules of $\mathbf{ann\text{-}cREG_0}(\Sigma, \Delta)$ have been chosen accordingly for this purpose, utilising the fundamental relation between regular expressions and their single-letter derivatives as formulated in Lemma 5 below.

7.2 The Extraction Step

In the extraction step, from a given derivation $\tilde{\mathcal{D}}$ in $\mathbf{ann\text{-}cREG_0}$ with conclusion $G : E = F$ two derivations $\tilde{\mathcal{D}}^{(1)}$ and $\tilde{\mathcal{D}}^{(2)}$ are constructed that, in case that $\tilde{\mathcal{D}}$ does not contain open assumptions, demonstrate respectively that E and F are equivalent with the annotating regular expression G. This is justified by Lemma 6 below; the proof of this lemma depends on Lemma 5, a version appropriate for regular expressions of the sometimes so called "fundamental theorem of formal languages" (due to the analogy with the "fundamental theorem of calculus").

Lemma 5. *For all* $E \in \mathcal{R}(\Sigma)$, $E \equiv_{\mathbf{REG^-}} o(E) + \sum_{i=1}^{n} a_i.E_{a_i}$ *holds. What is more, for every given* $E \in \mathcal{R}(\Sigma)$*, a derivation* $\mathcal{D}^{(E)}$ *in* $\mathbf{REG}^-(\Sigma)$ *with conclusion* $E = o(E) + \sum_{i=1}^{n} a_i.E_{a_i}$ *can effectively be constructed.*

Proof (Hint). The lemma can be shown by induction on the syntactical structure of regular expressions in $\mathcal{R}(\Sigma)$. For the treatment of the case $E = F^*$ in the induction step (for $E, F \in \mathcal{R}(\Sigma)$) the axioms (B10), (B11) of $\mathbf{REG}^-$ are needed.

Lemma 6. *From every derivation $\tilde{\mathcal{D}}$ in $\mathbf{ann\text{-}cREG_0}(\Sigma, \Delta)$ of the form (10), where $m \in \omega$, $d_1, \ldots, d_m \in \Delta$ distinct markers, $E, F, G_0 \in \mathcal{R}(\Sigma)$, and, for all $i \in \{1, \ldots, m\}$, $E_i, F_i, G_i \in \mathcal{R}(\Sigma)$, it is possible to construct effectively derivations in $\mathbf{REG}(\Sigma)$ of the respective forms*

$$\begin{array}{c}\tilde{\mathcal{D}}^{(1)}\\ E = (G_0+)\sum_{j=1}^{m} G_j.E_j\end{array} \quad , \qquad \textit{and} \qquad \begin{array}{c}\tilde{\mathcal{D}}^{(2)}\\ F = (G_0+)\sum_{j=1}^{m} G_j.F_j\end{array} \quad . \tag{12}$$

Proof (Hint). The lemma can be demonstrated by defining an effective extraction procedure of the two derivations $\tilde{\mathcal{D}}^{(1)}$ and $\tilde{\mathcal{D}}^{(2)}$ in $\mathbf{REG}(\Sigma)$ with the respective forms in (12) from an arbitrary derivation $\tilde{\mathcal{D}}$ in $\mathbf{ann\text{-}cREG_0}(\Sigma, \Delta)$ of the form (10), where $m \in \omega$, $E, F, G_0 \in \mathcal{R}(\Sigma)$, $E_i, F_i, G_i \in \mathcal{R}(\Sigma)$ for all $i \in \{1, \ldots, m\}$, and $d_1, \ldots, d_m \in \Delta$ are distinct. Such a procedure can be built by using induction on the structure (or on the depth) of the derivation $\tilde{\mathcal{D}}$.

Let us demonstrate the induction step for the extraction of the derivation $\hat{\mathcal{D}}^{(1)}$ from the annotated derivation $\hat{\mathcal{D}}$ in (9) relating to our running example. $\hat{\mathcal{D}}^{(1)}$ can be written as of the form

$$\frac{\begin{array}{c}[1.d:\ E = F_1]^d\\ \hat{\mathcal{D}}_1\\ a^* + a^*b.d:\ E_a = (F_1)_a\end{array} \qquad \begin{array}{c}[1.d:\ E = F_1]^d\\ \hat{\mathcal{D}}_2\\ 1.d:\ E_b = (F_1)_b\end{array}}{(aa^*b + b)^*(1 + aa^*):\ E = F_1}\ \text{COMP/FIX}$$

with $\hat{\mathcal{D}}_1$ and $\hat{\mathcal{D}}_2$ being the immediate left and right subderivations of $\hat{\mathcal{D}}$; to increase readability in this example, we suppress some "."-signs and brackets. We want to construct a derivation $\hat{\mathcal{D}}^{(1)}$ in $\mathbf{REG}(\{a, b\})$ with conclusion $E = (aa^* b + b)^*(1 + aa^*)$. By the induction hypothesis there exist derivations $\hat{\mathcal{D}}_1^{(1)}$ and $\hat{\mathcal{D}}_2^{(1)}$ in $\mathbf{REG}(\{a, b\})$ with the conclusions $E_a = a^* + a^*b.E$ and $E_b = E$. By temporarily using additional rules, from $\hat{\mathcal{D}}_1^{(1)}$ and $\hat{\mathcal{D}}_2^{(1)}$ the derivation $\hat{\mathcal{D}}_{0,\text{ar}}^{(1)}$

$$\frac{\dfrac{\dfrac{\text{CTXT}\ \dfrac{\begin{array}{c}\hat{\mathcal{D}}_1^{(1)}\\ E_a = a^* + a^*b.E\end{array}}{a.E_a = a.(a^* + a^*b.E)} \qquad \dfrac{\begin{array}{c}\hat{\mathcal{D}}_2^{(1)}\\ E_b = E\end{array}}{b.E_b = b.E}\ \text{CTXT}}{a.E_a + b.E_b = a.(a^* + a^*b.E) + b.E}\ +}{1 + a.E_a + b.E_b = 1 + a.(a^* + a^*b.E) + b.E}\ \text{CTXT}}{1 + a.E_a + b.E_b = (aa^*b + b).E + (1 + aa^*)}\ \text{App}_r\text{Ax}_{\text{ACI}+}$$

can be constructed. This derivation can be extended, by using the fixed-point rule FIX in $\mathbf{REG}(\{a, b\})$ in an essential way, into the derivation $\hat{\mathcal{D}}_{\text{ar}}^{(1)}$

$$\frac{\dfrac{\begin{array}{c}\mathcal{D}_{\text{ar}}^{(E)}\\ E = o(E) + a.E_a + b.E_b\end{array} \qquad \begin{array}{c}\hat{\mathcal{D}}_{0,\text{ar}}^{(1)}\\ 1 + a.E_a + b.E_b = (aa^*b + b).E + (1 + aa^*)\end{array}}{E = (aa^*b + b).E + (1 + aa^*)}\ \text{TRANS}}{E = (aa^*b + b)^*(1 + aa^*)}\ \text{FIX}$$

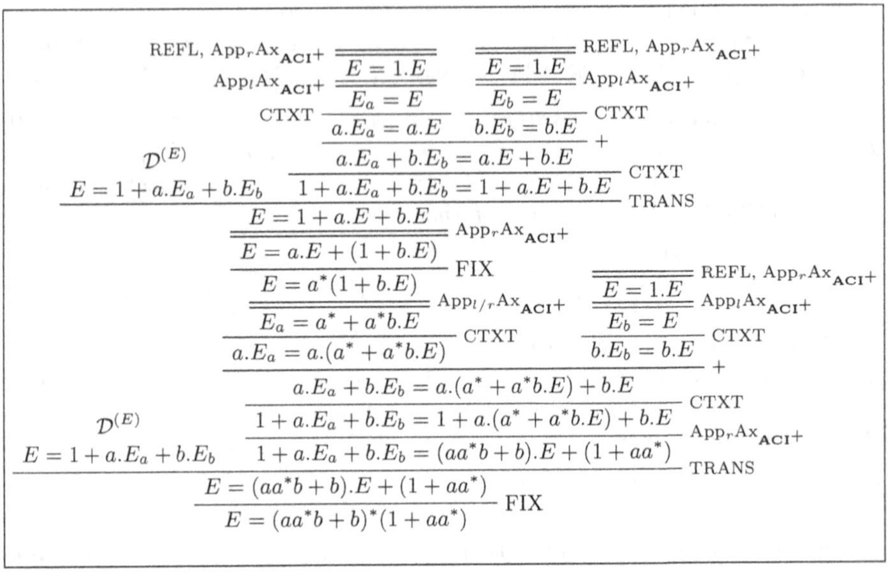

$$
\dfrac{\dfrac{\mathcal{D}^{(E)}}{E = 1 + a.E_a + b.E_b} \qquad \dfrac{\dfrac{\dfrac{\dfrac{\overline{\overline{E = 1.E}}\ \text{REFL, App}_r\text{Ax}_{\mathbf{ACI}+}}{E_a = E}\ \text{App}_l\text{Ax}_{\mathbf{ACI}+}}{a.E_a = a.E}\ \text{CTXT} \quad \dfrac{\dfrac{\overline{\overline{E = 1.E}}\ \text{REFL, App}_r\text{Ax}_{\mathbf{ACI}+}}{E_b = E}\ \text{App}_l\text{Ax}_{\mathbf{ACI}+}}{b.E_b = b.E}\ \text{CTXT}}{a.E_a + b.E_b = a.E + b.E}\ +}{1 + a.E_a + b.E_b = 1 + a.E + b.E}\ \text{CTXT}}{E = 1 + a.E + b.E}\ \text{TRANS}
$$

$$
\dfrac{\dfrac{\mathcal{D}^{(E)}}{E = 1 + a.E_a + b.E_b} \qquad \dfrac{\dfrac{\dfrac{\dfrac{\dfrac{\dfrac{\dfrac{E = 1 + a.E + b.E}{E = a.E + (1 + b.E)}\ \text{App}_r\text{Ax}_{\mathbf{ACI}+}}{E = a^*(1 + b.E)}\ \text{FIX}}{E_a = a^* + a^*b.E}\ \text{App}_{l/r}\text{Ax}_{\mathbf{ACI}+}}{a.E_a = a.(a^* + a^*b.E)}\ \text{CTXT} \quad \dfrac{\dfrac{\overline{\overline{E = 1.E}}\ \text{REFL, App}_r\text{Ax}_{\mathbf{ACI}+}}{E_b = E}\ \text{App}_l\text{Ax}_{\mathbf{ACI}+}}{b.E_b = b.E}\ \text{CTXT}}{a.E_a + b.E_b = a.(a^* + a^*b.E) + b.E}\ +}{1 + a.E_a + b.E_b = 1 + a.(a^* + a^*b.E) + b.E}\ \text{CTXT}}{1 + a.E_a + b.E_b = (aa^*b + b).E + (1 + aa^*)}\ \text{App}_r\text{Ax}_{\mathbf{ACI}+}}{\dfrac{E = (aa^*b + b).E + (1 + aa^*)}{E = (aa^*b + b)^*(1 + aa^*)}\ \text{FIX}}\ \text{TRANS}
$$

Fig. 6. Abbreviated result $\hat{\mathcal{D}}^{(1)}_{\text{ar}}$ of extracting the $\mathbf{REG}(\{a,b\})$-derivation $\hat{\mathcal{D}}^{(1)}$ from the **ann-cREG**$_\mathbf{0}$-deriv. $\hat{\mathcal{D}}$ in (9) (some **REG**-derivable rules are used)

where the derivation $\mathcal{D}^{(E)}_{\text{ar}}$ is guaranteed by Lemma 5 and can be chosen as

$$
\dfrac{\dfrac{\overset{(\text{B10})}{(a+b)^* = 1 + (a+b)(a+b)^*}}{(a+b)^* = 1 + a(a+b)^* + b(a+b)^*}\ \text{App}_l\text{Ax}_{\mathbf{ACI}+}}{E = 1 + a\underbrace{(1+0)(a+b)^*}_{E_a} + b\underbrace{(0+1)(a+b)^*}_{E_b}}\ \text{App}_l\text{Ax}_{\mathbf{ACI}+}
$$

The desired derivation $\hat{\mathcal{D}}^{(1)}$ in $\mathbf{REG}(\{a,b\})$ can then be found as the result of eliminating from $\hat{\mathcal{D}}^{(1)}_{\text{ar}}$ all applications of the additional rules "+", $\text{App}_l\text{Ax}_{\mathbf{ACI}+}$, and $\text{App}_r\text{Ax}_{\mathbf{ACI}+}$, which can easily be recognised to be derivable in $\mathbf{REG}(\{a,b\})$.

The result of the entire extraction process of $\hat{\mathcal{D}}^{(1)}$ from $\hat{\mathcal{D}}$ is displayed in Figure 6 as the derivation $\hat{\mathcal{D}}^{(1)}_{\text{ar}}$ in which applications of additional rules occur and the derivation $\mathcal{D}^{(E)}$ is abbreviated. In an analogous way, also the derivation $\hat{\mathcal{D}}^{(2)}$ in $\mathbf{REG}(\{a,b\})$ with conclusion $F = (aa^*b + b)^*(1 + aa^*)$ can be extracted from $\hat{\mathcal{D}}$; similar to $\hat{\mathcal{D}}^{(1)}_{\text{ar}}$, it is given as the abbreviated derivation $\hat{\mathcal{D}}^{(2)}_{\text{ar}}$ in Figure 7.

7.3 The Combination Step

The last step of the transformation is easy and consists in combining the two $\mathbf{REG}(\Sigma)$-derivations $\hat{\mathcal{D}}^{(1)}$ and $\hat{\mathcal{D}}^{(2)}$, which were extracted from the annotated version $\hat{\mathcal{D}}$ of a **cREG**$_\mathbf{0}$-derivation $\mathcal{D}$ on which the transformation was started. Building from $\hat{\mathcal{D}}^{(1)}$ and $\hat{\mathcal{D}}^{(2)}$ a $\mathbf{REG}(\Sigma)$-derivation $(\hat{\mathcal{D}})'$ with the same conclusion as $\mathcal{D}$ only requires the use of each an application of SYMM and TRANS, as

$$
\mathcal{D}^{(F_2)} \quad F_2 = 1 + a.(F_2)_a + b.(F_2)_b
$$

$$
\cfrac{
  \cfrac{
    \cfrac{
      \cfrac{\text{REFL, App}_r\text{Ax}_{\text{ACI}+} \;\; \overline{F_2 = 1.F_2}}{\text{App}_l\text{Ax}_{\text{ACI}+}\;\; \overline{(F_2)_a = F_2}}
    }{\text{CTXT}\;\; a.(F_2)_a = a.F_2}
    \quad
    \cfrac{
      \cfrac{\overline{F_1 = 1.F_1}\;\; \text{REFL, App}_r\text{Ax}_{\text{ACI}+}}{\overline{(F_1)_b = F_1}\;\; \text{App}_l\text{Ax}_{\text{ACI}+}}
    }{b.(F_1)_b = b.F_1\;\; \text{CTXT}}
  }{a.(F_2)_a + b.(F_2)_b = a.F_2 + b.F_1}\; +
}{
  \cfrac{1 + a.(F_2)_a + b.(F_2)_b = 1 + a.F_2 + b.F_1}{\cfrac{F_2 = 1 + a.F_2 + b.F_1}{\cfrac{F_2 = a.F_2 + (1 + b.F_1)}{F_2 = a^*(1 + b.F_1)}\;\text{FIX}}\;\text{App}_r\text{Ax}_{\text{ACI}+}}\;\text{TRANS}
}\;\text{CTXT}
$$

$$
\mathcal{D}^{(F_1)} \quad F_1 = 1 + a.(F_1)_a + b.(F_1)_b
$$

$$
\cfrac{
  \cfrac{
    \cfrac{\cfrac{(F_1)_a = a^* + a^*b.F_1}{a.(F_1)_a = a.(a^* + a^*b.F_1)}\;\text{CTXT}}{\text{App}_{l/r}\text{Ax}_{\text{ACI}+}}
    \quad
    \cfrac{\cfrac{\overline{F_1 = 1.F_1}\;\;\text{REFL, App}_r\text{Ax}_{\text{ACI}+}}{\overline{(F_1)_b = F_1}\;\;\text{App}_l\text{Ax}_{\text{ACI}+}}}{b.(F_1)_b = b.F_1}\;\text{CTXT}
  }{a.(F_1)_a + b.(F_1)_b = a.(a^* + a^*b.F_1) + b.F_1}\; +
}{
  \cfrac{1 + a.(F_1)_a + b.(F_1)_b = 1 + a.(a^* + a^*b.F_1) + b.F_1}{\cfrac{F_1 = 1 + a.(a^* + a^*b.F_1) + b.F_1}{\cfrac{F_1 = (aa^*b + b).F_1 + (1 + aa^*)}{F_1 = (aa^*b + b)^*(1 + aa^*)}\;\text{FIX}}\;\text{App}_r\text{Ax}_{\text{ACI}+}}\;\text{TRANS}
}\;\text{CTXT}
$$

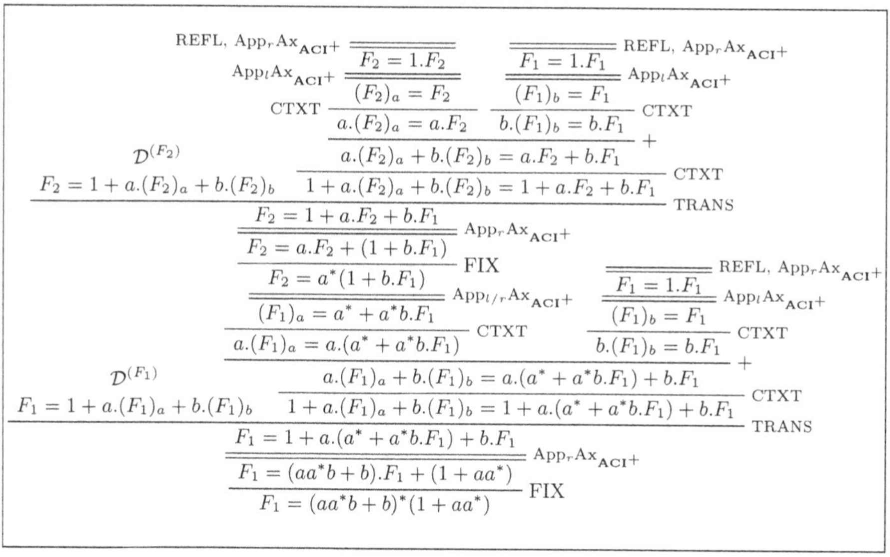

Fig. 7. Abbreviated result $\hat{\mathcal{D}}_{\text{ar}}^{(2)}$ of extracting the $\mathbf{REG}(\{a,b\})$-derivation $\hat{\mathcal{D}}^{(2)}$ from the $\mathbf{ann\text{-}cREG_0}$-deriv. $\hat{\mathcal{D}}$ in (9) (some $\mathbf{REG}$-derivable rules are used)

illustrated in Figure 3. Due to this, Lemma 4 and Lemma 6 together yield the transformation theorem below. In view of Theorem 3, this theorem facilitates an alternative completeness proof for $\mathbf{REG}$ and therefore entails the subsequent corollary, which is a restatement of the completeness part of Theorem 1.

Theorem 4. *Every derivation $\mathcal{D}$ in $\mathbf{cREG_0}(\Sigma)$ without open assumptions can effectively be transformed into a derivation $\mathcal{D}'$ in $\mathbf{REG}(\Sigma)$ with the same conclusion as $\mathcal{D}$.*

Corollary 2. $\mathbf{REG}(\Sigma)$ *is complete with respect to* $=_L$.

8   Conclusion

Using a coinduction principle for language equality given by Rutten in [6], we introduced a "finitary" coinduction principle for proving equivalence of regular expressions: for showing that two regular expressions E and F are equivalent, prove that, up to applying laws including associativity, commutativity and idempotency of $+$, E and F are bisimilar in an automaton of regular expressions whose transition function is based on the "Brzozowski derivative". We recognised that this principle can be used to decide regular expression equivalence in an effective way, and hence, that it can be implemented in principle (further considerations lead us to the belief that this is indeed a practical possibility).

Subsequently we introduced a proof system $\mathbf{cREG_0}$ of equational logic that formalises proofs using the finitary coinduction principle as finite derivations:

its soundness and completeness proof directly reflects the fact that a derivation in $\mathbf{cREG_0}$ (without open assumptions) corresponds to a finite bisimulation between the regular expressions in its conclusion. Finally, we showed that derivations in $\mathbf{cREG_0}$ can be transformed into derivations in a variant system **REG** of Salomaa's axiomatisation $\mathbf{F_1}$ in a very straightforward and "natural" way. Hereby we obtained a coinductive completeness proof for the system **REG**.

Our constructions, and in particular the transformation we sketched, can be adapted to yield also a coinductive completeness proof for Salomaa's $\mathbf{F_1}$. This is because an alternative differential calculus for formal languages and regular expressions can be introduced, in which derivatives take away letters from the end of words: one can define, for letters a, the language derivative $(\cdot)'_a : \mathcal{L}(\Sigma) \to \mathcal{L}(\Sigma)$ by $L \mapsto (L)'_a =_{\text{def}} \{v \mid v.a \in L\}$. Based on corresponding versions of derivatives for regular expressions, one can formulate an effective finitary coinduction principle analogous to Theorem 2, a sound and complete proof system $\mathbf{cREG'_0}$ for $=_L$ analogous to $\mathbf{cREG_0}$, and an effective transformation of $\mathbf{cREG'_0}$-derivations into $\mathbf{F_1}$-derivations analogous to the one described here. An effective completeness proof for $\mathbf{F_1}$ can directly be based on these elements.

Acknowledgement. The proof-transformation described in this paper was conceived after a talk by Jan Rutten in which $1 + a(a+b)^* + (a+b)^*aa(a+b)^* =_L$ $=_L ((b^*a)^*ab)^*$ was shown by coinduction and the "homework" was assigned of giving an alternative proof by a deduction in Salomaa's axiomatisation of $=_L$. Also, I want to thank Jan Rutten for calling it to my attention that Lemma 2 already holds w.r.t. **ACI** (and not only w.r.t. $\mathbf{REG}^-$ as I had used previously). Furthermore, I would like to convey my thanks to the anonymous referees for their remarks, observations, questions, and stimulating comments. Last, but not least thanks are due to Mihály Petreczky for a couple of useful discussions, and to Helle Hansen for suggesting a formulation concerning the name of Lemma 5.

References

1. Brandt, M., Henglein, F.: "Coinductive axiomatization of recursive type equality and subtyping", *Fundamenta Informaticae* **33** (1998) 1–30.
2. Brzozowski, J.A.: "Derivatives of regular expressions", *Journal of the ACM* **11** (1964) 481–494.
3. Conway, J.H.: *Regular Algebra and Finite Machines*, Chapman and Hall (1971).
4. Hüttel, H., Stirling, C.: "Actions Speak Louder Than Words: Proving Bisimilarity for Context-Free Processes", *Journ. of Logic and Computation* **8**:4 (1998) 485–509.
5. Grabmayer, C.: *Relating Proof Systems for Recursive Types*, PhD thesis, Vrije Universiteit Amsterdam (2005) `http://www.cs.vu.nl/~clemens/proefschrift.pdf` .
6. Rutten, J.J.M.M.: "Automata and Coinduction (an Exercise in Coinduction)", *Proceedings of CONCUR '98*, LNCS 1466, Springer (1998) 194–218.
7. Salomaa, A.: "Two complete axiom systems for the algebra of regular events", *Journal of the ACM* **13**:1 (1966) 158–169.
8. Troelstra, A.S., Schwichtenberg, H.: *Basic Proof Theory*, Cambridge University Press (1996, 2000).

From T-Coalgebras to Filter Structures and Transition Systems

H. Peter Gumm

Philipps-Universität Marburg,
35032 Marburg, Germany
gumm@mathematik.uni-marburg.de

Abstract. For any set-endofunctor $T : \mathcal{S}et \to \mathcal{S}et$ there exists a largest sub-cartesian transformation μ to the filter functor $\mathbb{F} : \mathcal{S}et \to \mathcal{S}et$. Thus we can associate with every T-coalgebra A a certain filter-coalgebra $A_{\mathbb{F}}$.

Precisely, when T (weakly) preserves preimages, μ is natural, and when T (weakly) preserves intersections, μ factors through the covariant powerset functor $\mathbb{P}$, thus providing for every T-coalgebra A a Kripke structure $A_{\mathbb{P}}$.

We characterize preservation of preimages, preservation of intersections, and preservation of both preimages and intersections via the existence of natural, sub-cartesian or cartesian transformations from T to either $\mathbb{F}$ or $\mathbb{P}$.

Moreover, we define for arbitrary T-coalgebras $\mathcal{A}$ a next-time operator $\bigcirc_{\mathcal{A}}$ with associated modal operators $\Box$ and $\Diamond$ and relate their properties to weak limit preservation properties of T. In particular, for any T-coalgebra $\mathcal{A}$ there is a transition system $\mathcal{K}$ with $\bigcirc_{\mathcal{A}} = \bigcirc_{\mathcal{K}}$ if and only if T preserves intersections.

1 Introduction

The importance of weak preservation properties of coalgebraic type functors has been clear since the seminal work of Rutten [Rut00]. Many of the results in the original 1996 preprint-version of his work assumed that the coalgebraic type functor weakly preserves pullbacks, or even arbitrary intersections.

In joint works with T. Schröder, we have subsequently shown that weak preservation of pullbacks decomposes into two more basic preservation properties, namely preservation of preimages and weak preservation of kernels. We have given numerous (co-)algebraic properties that correspond, in a one-to-one fashion, to these preservation properties of the type functor.

The current paper studies a transformation μ between an arbitrary $\mathcal{S}et$-endofunctor T and the *filter functor* that associates with a set X the set $\mathbb{F}(X)$ of all filters on a set X.

The basic idea is to capture the notion of *successors* of a point a, which plays a central role in Kripke Structures, and make it available for coalgebras of arbitrary type T.

J.L. Fiadeiro et al. (Eds.): CALCO 2005, LNCS 3629, pp. 194–212, 2005.

It turns out that, unless T preserves intersections, one cannot speak of a single set of successors, but must consider a family of successor sets. Fortunately, however, the successor sets form a filter. Therefore, one can construct a transformation μ between $T(X)$ and $\mathbb{F}(X)$, for arbitrary $\mathcal{S}et$-endofunctors T. Even though μ is not a natural transformation in general, it is enough to observe that it is always *sub-natural* and *sub-cartesian*, these terms are defined below. In fact, μ is the largest sub-cartesian transformation from T to $\mathbb{F}$. Now T (weakly) preserves preimages if and only if μ is a natural transformation.

Similarly, we always obtain a largest sub-natural transformation τ from T to the powerset functor $\mathbb{P}$. We show that τ is sub-cartesian iff T (weakly) preserves intersections and τ is cartesian iff T (weakly) preserves preimages and intersection.

For arbitrary T-coalgebras $\mathcal{A}$ this has the consequence that one always can define a filter-coalgebra on its base set which has the same subcoalgebras as $\mathcal{A}$. Closer connections between $\mathcal{A}$ and its associated filter-coalgebra or its associated Kripke structures correspond to the mentioned preservation properties of T.

Taking a logical viewpoint, one may generalize the nexttime operator $\bigcirc$ from Kripke structures, which associates to a subset S of a Kripke structure A the set of all points whose successors are all contained in S. We show how $\bigcirc$ can be defined for coalgebras of arbitrary type - even on base categories other than $\mathcal{S}et$. Preservation properties of the type functor T become very suggestive: T preserves preimages iff $\bigcirc$ commutes with homomorphic preimages, i.e. $\varphi^{-}\bigcirc Q = \bigcirc\varphi^{-}Q$, and T preserves intersections iff $\bigcirc$ commutes with forming intersections, i.e. $\bigcirc\bigcap_{i\in I} P_i = \bigcap_{i\in I}\bigcirc P_i$ for all subsets $P, P_i \subseteq A$ and homomorphisms φ.

2 Categorical Notions

We need only basic category theoretic notions and facts as found in the first few chapters of any textbook, such as e.g. [AHS90].

A functor $F : \mathcal{C} \to \mathcal{D}$ is said to *preserve monos*, if Ff is mono, whenever f was. When monos are left-invertible, as e.g. in the category of nonempty sets and mappings, they are, of course, automatically preserved.

Pullbacks are limits of two morphisms $f : A \to C$ and $g : B \to C$ with common codomain. Thus, the pullback of f and g is an object P with morphisms $p_1 : P \to A$ and $p_2 : P \to B$, so that $f \circ p_1 = g \circ p_2$, and for any other object Q with morphisms $p_1 : P \to A$ and $p_2 : Q \to B$ satisfying $f \circ q_1 = g \circ q_2$ there exists a unique "mediating" morphism $d : Q \to P$ with $p_i \circ d = q_i$, for $i = 1, 2$.

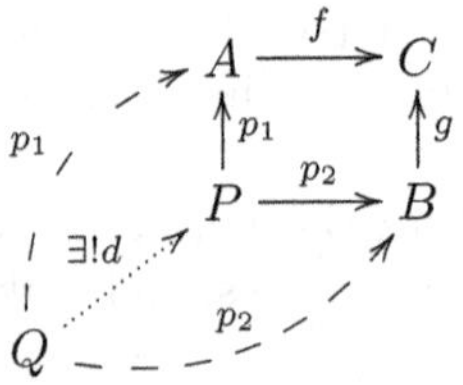

Weak pullbacks are the corresponding weak limits, i.e. where the uniqueness requirement for the mediating morphism is dropped.

Preimages are pullbacks where g is mono. Observe, that in this case, p_1 will automatically be mono, too. This is not necessarily the case for *weak preimages*. However, a weak pullback, in which one of p_1, p_2 is mono, is already a pullback.

Wide pullbacks are limits of infinite families $(f_i : A_i \to A)_{i \in I}$ with common codomain, and *intersections* (*weak intersections*) are limits (weak limits) of families of monomorphisms $(f_i : A_i \rightarrowtail A)_{i \in I}$ with common codomain.

A functor $F : \mathcal{C} \to \mathcal{D}$ is said to *weakly preserve pullbacks*, if it transforms every pullback diagram in $\mathcal{C}$ into a weak pullback diagram in $\mathcal{D}$. If pullbacks always exist in $\mathcal{C}$, then F weakly preserves pullbacks iff F *preserves weak pullbacks*, i.e. F transforms weak pullback diagrams into weak pullback diagrams.

Correspondingly, we say that F *weakly preserves preimages*, if it transforms each preimage diagram into a weak pullback diagram.

We say that F *weakly preserves intersections*, if F transforms every intersection diagram into a weak limit diagram.

If F preserves monos, as will often be the case, *weak preservation* of preimages (resp. intersections) is the same as *preservation* of preimages (resp. intersections).

3 Sub-natural and Sub-cartesian Transformations

Given categories $\mathcal{C}$, $\mathcal{D}$ and functors $F, G : \mathcal{C} \to \mathcal{D}$, a *transformation* $\nu : F \to G$ is just a family of $\mathcal{D}$-morphisms $\nu_A : FA \to GA$ for every object A in $\mathcal{C}$. It is called a *natural transformation*, if for every $\mathcal{C}$-morphism $f : A \to B$ the diagram

$$\begin{array}{ccc} FB & \xrightarrow{\nu_B} & GB \\ {\scriptstyle Ff}\uparrow & & \uparrow{\scriptstyle Gf} \\ FA & \xrightarrow{\nu_A} & GA \end{array}$$

commutes, and it is called *cartesian*, if the same diagram is a pullback.

We shall need to work with transformations, which are neither natural nor cartesian, but satisfy a weaker property:

Definition 1. *A transformation $\nu : F \to G$ will be called* sub-natural *if the above diagram commutes for every monomorphism f. It is called* sub-cartesian, *if the diagram is a weak pullback for every monomorphism f.*

The following observation will become important in later sections:

Theorem 1. *Assume that F preserves monos, and let $\nu : F \to G$ be a sub-cartesian transformation. Then*

(i) if G weakly preserves intersections then so does F,
(ii) if ν is natural and G weakly preserves preimages, then F preserves preimages.

Proof. To show (i), start with a family of monomorphisms $(e_i : A_i \hookrightarrow A)_{i \in I}$ and its limit M with morphisms $f_i : M \hookrightarrow A_i$, satisfying $e_i \circ f_i = e_k \circ f_k$ for all $i, k \in I$. Applying F and G and inserting the transformation morphisms, we obtain the following diagram, where the top row is a weak limit by the assumption on G. Since ν is subcartesian and the e_i and f_i are monos, the squares commute.

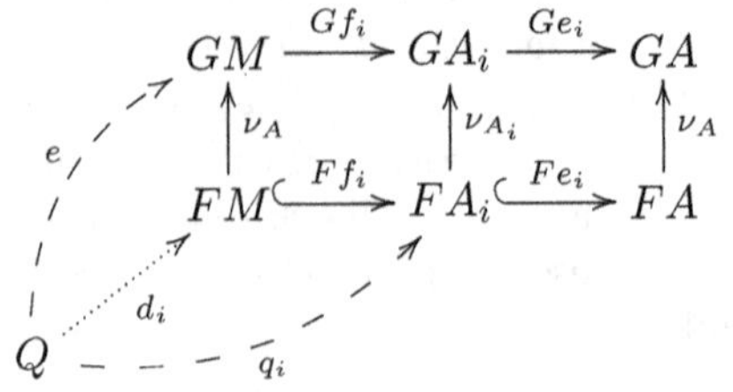

To show that the bottom row is a weak limit, too, let Q be a competitor with morphisms $q_i : Q \to FA_i$ satisfying $Fe_i \circ q_i = Fe_k \circ q_k$ for all $i, k \in I$. Then Q with morphisms $\nu_{A_i} \circ q_i$ becomes a competitor to the weak limit GM, yielding a morphism $e : Q \to GM$ with $\nu_{A_i} \circ q_i = Gf_i \circ e$ for all $i \in I$. Since ν is sub-cartesian, we obtain morphisms $d_i : Q \to FM$ for each $i \in I$ with $Ff_i \circ d_i = q_i$. We need to show that all the d_i are equal to a single morphism d. A diagram chase, utilizing $Fe_i \circ q_i = Fe_k \circ q_k$ and $e_k \circ f_k = e_i \circ f_i$, yields $Fe_i \circ Ff_i \circ d_i = Fe_i \circ Ff_i \circ d_k$. Since F preserves monos, we can cancel $Fe_i \circ Ff_i$ and obtain $d_i = d_k$.

To show (ii), let P with morphisms $p_1 : P \to A$ and $p_2 : P \to B$ be the preimage of $f : A \to C$ and monomorphism $g : B \hookrightarrow C$. It follows that p_1 is mono.

Applying F and G to this diagram and filling in the transformation ν, which we now assume to be natural, we obtain the following commutative cube:

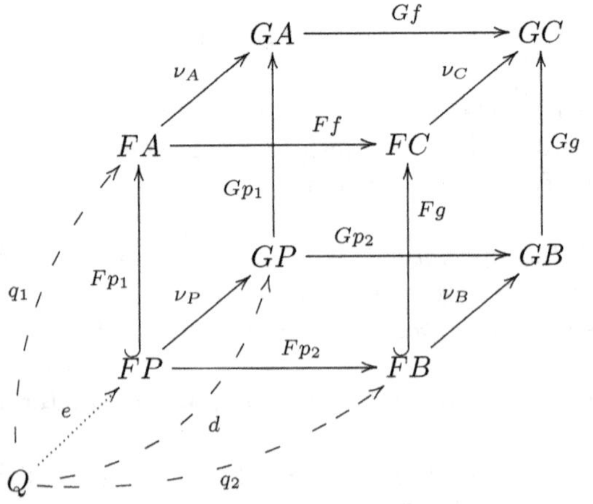

Since G weakly preserves preimages, the back face is a weak pullback diagram. Since F preserves monos, Fp_1 and Fg will be mono. We need to show that the front face is a weak pullback, too.

Given a competitor, i.e. an object Q with morphisms $q_1 : Q \to FA$ and $q_2 : Q \to FB$ satisfying $Ff \circ q_1 = Fg \circ q_2$, we extend q_1 and q_2 with the transformation morphisms ν_A and ν_B, to make Q into a competitor to the weak pullback of the back face. This yields a mediating morphism $d : Q \to GP$ with $\nu_A \circ q_1 = Gp_1 \circ d$. Since ν is sub-cartesian, the left face is a pullback, so we obtain a morphism $e : Q \to FP$ with $Fp_1 \circ e = q_1$. It follows that

$$\begin{aligned} Fg \circ Fp_2 \circ e &= Ff \circ Fp_1 \circ e \\ &= Ff \circ q_1 \\ &= Fg \circ q_2, \end{aligned}$$

so canceling the monomorphism Fg yields $Fp_2 \circ e = q_2$.

4 Functors on the Category $\mathcal{S}et$

Given a set X with subset $U \subseteq X$, we denote the canonical injection by $\subseteq_U^X : U \hookrightarrow X$. We sometimes drop the sub- and superscripts of $\subseteq$ when they are clear from the context.

Given a map $f : X \to Y$ with $U \subseteq X$ and $V \subseteq Y$, we denote by $f[U]$ the image of U and by $f^- V$ the preimage of V under f. Occasionally, we write $[f]$ for the image $f[X]$ of f.

4.1 $\mathcal{S}et$-Functors

For a $\mathcal{S}et$-functor $T : \mathcal{S}et \to \mathcal{S}et$, and for $U \subseteq X$, we write T_U^X for the application of T to the inclusion map $\subseteq_U^X$, and $[T_U^X]$ for the image of TU under the said map T_U^X, i.e.

$$[T_U^X] := T_U^X[TU].$$

We note two simple lemmas:

Lemma 1. *If $U \subseteq V \subseteq X$, then $[T_U^X] \subseteq [T_V^X]$.*

Proof. $[T_U^X] = T_U^X[TU] = (T_V^X \circ T_U^V)[TU] = T_V^X[T_U^V] \subseteq T_V^X[TV] = [T_V^X]$.

Lemma 2. *If $f : X \to Y$ and $U \subseteq X$, then $(Tf)[T_U^X] = [T_{f[U]}^Y]$.*

Proof. Let $f_|$ be the domain-codomain-restriction of f to U, then $f \circ \subseteq_U^X = \subseteq_{f[U]}^Y \circ f_|$, hence $Tf \circ T_U^X = T_{f[U]}^Y \circ Tf_|$. Applying the left side to TU yields $(Tf)[T_U^X]$. Since $f_|$ is surjective, it has a right inverse, hence $Tf_| : TU \to T(f[U])$ must be surjective, too. Therefore, $(T_{f[U]}^Y \circ Tf_|)[TU] = T_{f[U]}^Y[T(f[U])] = [T_{f[U]}^Y]$.

4.2 *Set*-Functors Preserve Finite Nonempty Intersections

For a *Set*-functor $T : Set \to Set$ we may assume $T(X) \neq \emptyset$, unless $X = \emptyset$, for otherwise T would have to be the trivial functor with $T(Y) = \emptyset$ for every set Y.

For nonempty sets X and Y, any injective $f : X \to Y$ has a left inverse. Hence Tf is invertible, too. As a consequence, every functor T on *Set* preserves monos with nonempty domain.

Rather surprisingly, every *Set*-endofunctor T also preserves nonempty intersections. This was proved by Trnková [Trn69]:

Lemma 3. *Whenever* $U \cap V \neq \emptyset$, *then* $[T_U^W] \cap [T_V^W] = [T_{U\cap V}^W]$.

A short proof of this result can be found in [GS02]. A corresponding theorem for infinite intersections is not valid in general.

4.3 Discharging Empty Sets and Mappings

The proviso about the empty set can be discarded by modifying the functor T on the empty set $\emptyset$ and on the empty mappings $\emptyset_A : \emptyset \to A$. To this end, consider the two-element set $\mathbf{2} = \{\mathbf{0}, \mathbf{1}\}$ with canonical injections $e_0, e_1 : \mathbf{1} \to \mathbf{2}$. Let $e : P \to T\mathbf{1}$ be the equalizer of Te_0 and Te_1:

$$P \xrightarrow{e} T\mathbf{1} \underset{Te_1}{\overset{Te_0}{\rightrightarrows}} T\mathbf{2}\,.$$

Then define a functor T^+ on objects X by

$$T^+(X) = \begin{cases} P, & \text{if } X = \emptyset, \\ T(X), & \text{otherwise.} \end{cases}$$

Identifying any $y \in Y$ with the map $\mathbf{1} \to Y$ with value y, we have $Ty : T\mathbf{1} \to TY$, and we can define for any $f : X \to Y$:

$$T^+ f := \begin{cases} Ty \circ e, & \text{if } X = \emptyset, y \in Y \\ T(f), & \text{otherwise.} \end{cases}$$

Due to the construction of e as an equalizer, one easily checks that the definition of $T^+ f$ does not depend on the choice of $y \in Y$. Then the following lemma can be verified:

Lemma 4. *([Trn69])* T^+ *is a Set-functor, preserving all monos and all finite intersections.* T^+ *agrees with* T *on all nonempty sets and on all mappings with nonempty domain.*

The above description of T^+ is from Barr([Bar93]) and it differs slightly from the original construction of Trnková, who defined $T^+(\emptyset)$ as the set of all natural transformations from $\hat{\mathbf{1}}$ to T, where $\hat{\mathbf{1}}$ is the functor with $\hat{\mathbf{1}}(\emptyset) = \emptyset$ and $\hat{\mathbf{1}}(X) = \mathbf{1}$ for $X \neq \emptyset$. Barr's description has the advantage that equalizers (in the category *Set*) are usually easier to calculate than natural transformations.

The following corollary will be needed later:

Corollary 1. *T^+ preserves preimages of injective maps.*

Since we are interested in coalgebras, nothing changes, when we replace T by T^+, so we will assume from now on that all $\mathcal{S}et$-functors under consideration satisfy the property of T^+ in the previous lemma. In particular, they preserve monos, finite intersections and preimages of injective maps.

4.4 Preservation Properties of $\mathcal{S}et$-Functors

Checking, whether a diagram is a pullback (weak pullback) is especially easy in the category $\mathcal{S}et$. Essentially, this is due to the fact that each set is a sum of one-element sets, so the pullback condition can be checked elementwise. This means that a set functor $T : \mathcal{S}et \to \mathcal{S}et$ weakly preserves the pullback $(P, (p_i)_{i\in I})$ of a family of maps $(f_i : X_i \to Z)_{i\in I}$ iff for each family $(u_i \in TX_i)_{i\in I}$ satisfying $(Tf_i)(u_i) = (Tf_k)(u_k)$ for all $i, k \in I$ there exists some $w \in TP$ with $(Tp_i)(w) = u_i$ for all $i \in I$.

In the following lemma we will use this criterion for checking preservation of infinite intersections, i.e. of wide pullbacks of monos:

Lemma 5. *A functor $T : \mathcal{S}et \to \mathcal{S}et$ preserves intersections iff for each family $(U_i \subseteq X)_{i\in I}$*

$$\bigcap_{i\in I} [T^X_{U_i}] = [T^X_{\bigcap_{i\in I} U_i}].$$

Proof. By Lemma 1, the inclusion "$\supseteq$" is always true.

Assume now, that T preserves intersections. For each $u \in \bigcap_{i\in I} [T^X_{U_i}]$ and each $i \in I$ there exists $u_i \in TU_i$ with $T^X_{U_i}(u_i) = u$. Abbreviate $W := \bigcap_{i\in I} U_i$, then TW with the maps $T^{U_i}_W$ is a limit of the sink $(T^X_{U_i})_{i\in I}$, so there exists some $w \in TW$ with $T^{U_i}_W(w) = u_i$, hence $T^X_W(w) = (T^X_{U_i} \circ T^{U_i}_W)(w) = T^X_{U_i}(u_i) = u$, so $u \in [T^X_W]$.

Conversely, assume that the formula is true and let P with maps $f_i : P \to X_i$ be the limit of the monomorphisms $e_i : X_i \rightarrowtail X$. Each e_i factors as $e_i = \subseteq \circ\, g_i$ with $g_i : X_i \to U_i \subseteq X$ bijective. It follows that there is a bijective map $g : P \to \bigcap_{i\in I} U_i$ with $g_i \circ f_i = \subseteq \circ\, g$. Applying T, we have the following commutative diagram:

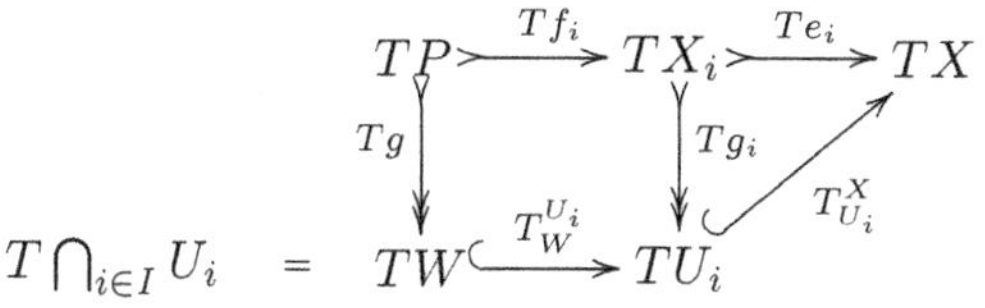

To see that TP with the maps Tf_i is the limit of the Te_i, assume a family of elements $u_i \in TX_i$ be given with $(Te_j)(u_j) = (Te_k)(u_k) =: u$ for all $j, k \in I$. We need to find an element $p \in TP$ with $Tf_i(p) = u_i$ for all $i \in I$.

Now, $T^X_{U_i}(Tg_i(u_i)) = u$ for all $i \in I$, so $u \in \bigcap_{i\in I}[T^X_{U_i}]$, hence by the assumption $u \in [T^X_{\bigcap_{i\in I} U_i}]$. Abbreviating $W := \bigcap_{i\in I} U_i$, then $u = T^X_W(v)$ for some $v \in TW$. We claim that $p := Tg^{-1}(v)$ is the sought element in TP. Indeed, $(Te_i \circ Tf_i)(p) = (T^X_{U_i} \circ T^{U_i}_W \circ Tg)(Tg^{-1}(v)) = T^X_W(v) = u = Te_i(u_i)$ for each $i \in I$. Since the Te_i are monos, it follows $Tf_i(p) = u_i$ for all $i \in I$.

The following is an easy but relevant corollary:

Corollary 2. *T preserves infinite intersections iff for any $u \in TX$ there is a smallest $U \subseteq X$ with $u \in [T^X_U]$.*

Proof. The smallest U with $u \in [T^X_U]$ must be $U := \bigcap_{i\in I}\{V \subseteq X \mid u \in [T^X_V]\}$. To check whether indeed $u \in [T^X_U]$, we apply the formula of the previous lemma and obtain the triviality $u \in \bigcap\{\, [T^X_V] \mid u \in [T^X_V] \,\}$.

For the other direction, choose any $x \in \bigcap_{i\in I}[T^X_{U_i}]$ and let W be the smallest $W \subseteq X$ with $x \in [T^X_W]$. Then $W \subseteq \bigcap_{i\in I} U_i$, so $x \in [T^X_{\bigcap_{i\in I} U_i}]$ by Lemma 1.

5 The Filter Functor

A *filter* $\mathcal{G}$ on a set X is a nonempty collection of subsets of X that is closed under finite intersections and supersets. In other words, $\emptyset \neq \mathcal{G} \subseteq \mathbb{P}(X)$ and

- $G_1, G_2 \in \mathcal{G} \Rightarrow G_1 \cap G_2 \in \mathcal{G}$, and
- $G \in \mathcal{G}$, and $G \subseteq H \subseteq X \Rightarrow H \in \mathcal{G}$.

On any set X, let $\mathbb{F}(X)$ be the set of all filters on X. $\mathbb{F}$ can be made into an endo-functor on $\mathcal{S}et$ by defining $\mathbb{F}f$ for any map $f : X \to Y$ as

$$(\mathbb{F}f)(\mathcal{G}) \;:=\; \uparrow \{f[G] \mid G \in \mathcal{G}\},$$

where $\mathcal{G}$ is an arbitrary filter on X. Here $\uparrow \mathcal{H}$, for any system of subsets $\mathcal{H} \subseteq \mathbb{P}(X)$, denotes the set of all supersets of sets in $\mathcal{H}$, i.e.

$$\uparrow \mathcal{H} := \{W \subseteq X \mid \exists H \in \mathcal{H}. H \subseteq W\}.$$

Here we shall work with the following equivalent definition for $\mathbb{F}f$:

Lemma 6. *For any map $f : X \to Y$, and any filter $\mathcal{G}$ on X, we have*

$$(\mathbb{F}f)(\mathcal{G}) \;=\; \{V \subseteq Y \mid f^{-}V \in \mathcal{G}\}.$$

It is shown in [Gum01] that $\mathbb{F}$ is a functor for which the following theorem holds:

Proposition 1. *$\mathbb{F}$ weakly preserves pullbacks, but not infinite intersections.*

Clearly, the covariant powerset functor $\mathbb{P}$ is a subfunctor of the filter functor $\mathbb{F}$. The natural embedding $\varepsilon : \mathbb{P} \to \mathbb{F}$ associates a set $U \subseteq X$ with the filter of all supersets of U in X. There is also an obvious transformation

$$\bigcap : \mathbb{F} \to \mathbb{P}$$

in the other direction, given by intersection. We have $\bigcap \circ \epsilon = id_F$, but $\bigcap$ is *not* a natural transformation. Instead we find:

Lemma 7. $\bigcap : \mathbb{F} \to \mathbb{P}$ *is sub-natural, but not sub-cartesian.*

Proof. For sub-naturality, it suffices to check that for every injective map $f : X \to Y$ and any $\mathcal{G} \in \mathbb{F}X$:

$$(\mathbb{P}f \circ \bigcap)(\mathcal{G}) = f[\bigcap \mathcal{G}] = \bigcap\{f[G] \mid G \in \mathcal{G}\} = \bigcap \uparrow \{f[G] \mid G \in \mathcal{G}\} = (\bigcap \circ \mathbb{F}f)(\mathcal{G}).$$

$\mathbb{P}$ preserves intersections, and $\mathbb{F}$ preserves monos. Therefore, we can invoke Theorem 1 to argue that if $\bigcap$ was sub-cartesian, then by (i), $\mathbb{F}$ would have to preserve intersections too, which is not the case, see e.g. [Gum01].

6 A Sub-cartesian Transformation to the Filter Functor

For an arbitrary functor $T : \mathcal{S}et \to \mathcal{S}et$, we now define a transformation $\mu : T \to \mathbb{F}$. When T preserves preimages, μ will be natural, otherwise it will be natural on injective maps only. Nevertheless, this property will suffice to prove our coalgebraic results of the following section.

Definition 2. *For any set X and any functor $T : \mathcal{S}et \to \mathcal{S}et$ define a map $\mu_X : T(X) \to \mathbb{F}(X)$ by*

$$\mu_X(u) := \{U \subseteq X \mid u \in [T_U^X]\}.$$

To see that μ_X is indeed a filter, we invoke Lemmas 1 and 3 in combination with 4. In general, μ is not a natural transformation, but we always have:

Lemma 8. *μ is a sub-cartesian transformation.*

Proof. For an injective map $f : X \to Y$, we first need to show commutativity of the following square:

$$\begin{array}{ccc} TY & \xrightarrow{\mu_Y} & \mathbb{F}Y \\ {\scriptstyle Tf}\uparrow & & \uparrow{\scriptstyle \mathbb{F}f} \\ TX & \xrightarrow[\mu_X]{} & \mathbb{F}X \end{array}$$

Given $u \in TX$, we have for any $V \subseteq Y$:

$$\begin{aligned} V \in \mu_Y((Tf)(u)) &\iff (Tf)(u) \in [T_V^Y] \\ &\iff \exists v \in TV.(Tf)(u) = T_V^Y(v) \\ &\overset{!}{\iff} \exists w \in T(f^- V).T_{f^- V}^X(w) = u \\ &\iff u \in [T_{f^- V}^X] \\ &\iff f^- V \in \mu_X(u) \\ &\iff V \in (\mathbb{F}f)(\mu_X(u)). \end{aligned}$$

The third (marked) equivalence is due to the fact that T preserves preimages with respect to injective maps, see Corollary 1.

To check that the diagram is a weak pullback, let $v \in TY$ and $\mathcal{G} \in \mathbb{F}X$ be given with $\mu_Y(v) = (\mathbb{F}f)(\mathcal{G})$. This implies that for every $V \subseteq Y$ we have:

$$v \in [T_V^Y] \iff f^- V \in \mathcal{G}.$$

Choosing $V := f[X]$, we have $f^- f[X] = X \in \mathcal{G}$, so we obtain $v \in [T_{f[X]}^Y]$. Hence, there exists $w \in T(f[X])$ with $T_{f[X]}^Y(w) = v$. Since $f = \subseteq_{f[X]}^Y \circ f'$ with f' bijective, we have that $Tf = T_{f[X]}^Y \circ Tf'$ with Tf' bijective. This yields an element $u \in TX$ with $(Tf')(u) = w$, i.e. $(Tf)(u) = v$. The condition $\mu_X(u) = \mathcal{G}$ follows from the commutativity of the diagram together with the fact that $\mathbb{F}f$ is mono.

Note that we did not claim that μ should be natural. In fact, we shall soon describe when this is the case. In the meantime, we can characterize μ amongst all sub-cartesian transformations:

Theorem 2. *μ is the largest sub-cartesian transformation from T to $\mathbb{F}$.*

Proof. Suppose $\nu : T \to \mathbb{F}$ is any sub-cartesian transformation. We need to prove $\nu_X(q) \subseteq \mu_X(q)$ for every set X and every $q \in TX$.

Put $\mathcal{G} := \nu_X(q)$ and assume $U \in \mathcal{G}$. For $\mathcal{G}_U := \{U \cap G \mid G \in \mathcal{G}\}$ we have $\mathcal{G} = \uparrow \mathcal{G}_U = (\mathbb{F}_U^X)\mathcal{G}_U$. We obtain the following situation:

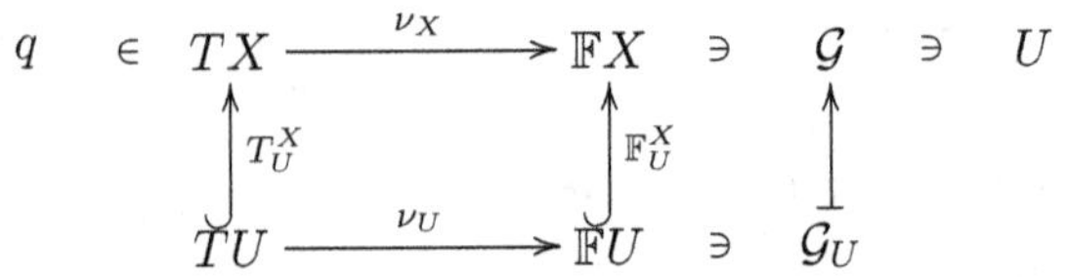

Since ν is sub-cartesian, there exists some $w \in TU$ with $(T_U^X)(w) = q$. Hence $q \in (T_U^X)[TU]$ which means $U \in \mu_X(q)$.

We can now formulate our first characterization theorem:

Theorem 3. *For a set functor $T : Set \to Set$ the following are equivalent:*

(i) T (weakly) preserves preimages
(ii) $\mu : T \to \mathbb{F}$ is a natural transformation
(iii) There exists a natural transformation $\nu : T \to \mathbb{F}$ which is sub-cartesian.

Proof. (i)⇒(ii): Let T weakly preserve preimages and let $f : X \to Y$ be any map. We need to show that the following diagram commutes:

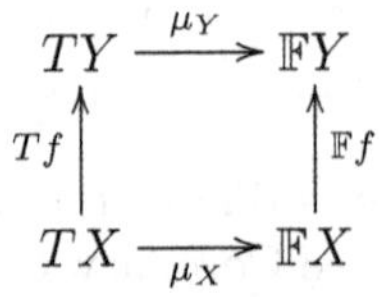

Given any $u \in TX$ we calculate

$$(\mathbb{F}f)(\mu_X(u)) = \{V \subseteq Y \mid u \in [T^X_{f^- V}\ ]\},$$
$$\mu_Y((Tf)(u)) = \{V \subseteq Y \mid (Tf)(u) \in [T^Y_V]\}.$$

The inclusion "$\subseteq$" between the above sets always holds, for given $u \in [T^X_{f^- V}]$, there is some $w \in T(f^- V)$ with $T^X_{f^- V}(w) = u$. Applying Tf, we find

$$\begin{aligned}(Tf)(u) &= (Tf)(T^X_{f^- V}(w)) \\ &= T^Y_V((Tf_|)(w))\end{aligned}$$

due to the commutativity of the following diagram which arises from applying T to the diagram describing the preimage of V under f:

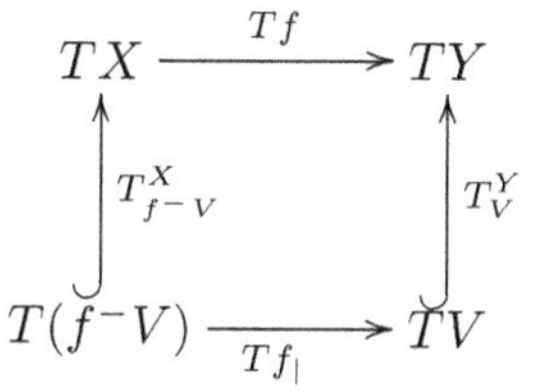

Hence $(Tf)(u) \in [T^Y_V]$.

For the other inclusion "$\supseteq$", we need to assume that T preserves preimages, which is to say that the above square is in fact a (weak) pullback. Given some $V \in \mu_Y((Tf)(u))$, i.e. $(Tf)(u) \in [T^Y_V]$, we have $v \in TV$ with $(Tf)(u) = T^Y_V(v)$. By the weak pullback property, there exists an element $w \in T(f^- V)$ with $T^X_{f^- V}(w) = u$, hence $f^- V \in \mu_X(u)$, i.e. $V \in (\mathbb{F}f)(\mu_X(u))$.

(ii)⇒(iii) follows from Lemma 8.

For (iii)⇒(i), we recall from [Gum01] that the filter functor weakly preserves pullbacks, in particular, preimages. Thus we are in a position to apply Theorem 1 to obtain the desired result.

6.1 Examples

It is instructive to calculate $\mu : T \to \mathbb{F}$ for some familiar functors on $\mathcal{S}et$. In the following table, the first column lists functors $T : \mathcal{S}et \to \mathcal{S}et$ and the second column gives, for an arbitrary $q \in TX$ the value of $\mu^T_X(q) \subseteq X$.

Functor T	$\mu_X(q)$
Id	$\uparrow \{q\}$
$\mathbb{P}(-)$	$\uparrow q$
$\mathbb{P}\mathbb{P}(-)$	$\uparrow \bigcup q$
$\mathbb{F}(-)$	q
$A \times -$	$\uparrow \{\pi_2(q)\}$
$(-)^I$	$\uparrow q[I]$
$\bar{\mathbb{P}}\bar{\mathbb{P}}(-)$	$\{U \subseteq X \mid \forall\, V \in q. \forall\, W \subseteq X. V \cap U = W \cap U \Rightarrow W \in q\}$

The last mentioned functor is the composition of the contravariant powerset functor $\bar{\mathbb{P}}$ with itself. Exemplary, we calculate $\mu_X(q)$ for this case:

On objects, $\bar{\mathbb{P}}(X) = \mathbb{P}(X)$, but for a map $f : X \to Y$, one has $\bar{\mathbb{P}}(f) = f^- : \mathbb{P}(Y) \to \mathbb{P}(X)$, hence $\bar{\mathbb{P}}(\subseteq^Y_X)(V) = X \cap V$ for $V \subseteq Y$. Next, for $\mathcal{S} \in \bar{\mathbb{P}}\bar{\mathbb{P}}(U)$ one gets $(\bar{\mathbb{P}}\bar{\mathbb{P}} \subseteq^X_U)(\mathcal{S}) = \{V \subseteq X \mid V \cap U \in \mathcal{S}\}$, hence

$$U \in \mu_X(q) \iff \exists\, \mathcal{S} \in \bar{\mathbb{P}}\bar{\mathbb{P}}(U).\ q = \{V \subseteq X \mid V \cap U \in \mathcal{S}\}.$$

The following formulas allow us to construct μ for functors that are combinations of simpler ones. Given T, T_1, and T_2 with associated sub-cartesian transformations μ^T, μ^{T_1}, and μ^{T_2} to $\mathbb{F}$, we get the following transformation for their sums, products and powers:

Functor	$\mu^T_X(q)$
$T_1 + T_2$	$\mathtt{if}\ q \in T_i(X)\ \mathtt{then}\ \mu^{T_i}_X(q)$
$T_1 \times T_2$	$\mu^{T_1}_X(\pi_1(q)) \cap \mu^{T_2}_X(\pi_2(q))$
T^I	$\{U \subseteq X \mid q[I] \subseteq [T^X_U]\}$

A system with state set X that takes an input from a set I and either produces an error $e \in E$ or moves to a new state, while producing an output $o \in O$, can be modeled by a coalgebra of type $T(X) = (E + O \times X)^I$. The above table tells us how to calculate μ^T. Using the fact that $[F^X_U] = FU$ for the standard functor $F(-) = E + O \times (-)$, we obtain: $\mu^T_X(q) = \uparrow q[I]$.

7 A Sub-natural Transformation to the Powerset Functor

The covariant powerset functor $\mathbb{P}$ which associates with a map $f : X \to Y$ the map $\mathbb{P}f : \mathbb{P}X \to \mathbb{P}Y$ with $(\mathbb{P}f)(U) := f[U]$ is obviously a subfunctor of the filter functor $\mathbb{F}$. The natural embedding is given by $\epsilon_X(U) := \uparrow \{U\}$ for any $U \in \mathbb{P}(X)$. When does the transformation $\mu : T \to \mathbb{F}$ factor through this embedding?

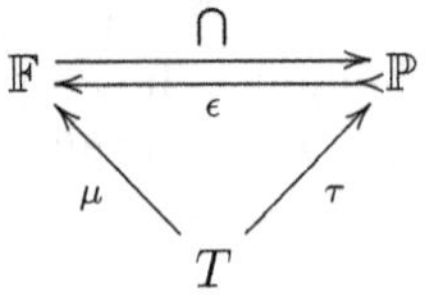

From $\bigcap \circ \epsilon = id_{\mathbb{P}}$, we obtain immediately:

Lemma 9. *$\tau := \bigcap \circ \mu$ is the only transformation $\tau : T \to \mathbb{P}$ with $\epsilon \circ \tau = \mu$.*

In other words, for an arbitrary functor $T : Set \to Set$ the following definition yields a transformation, which is sub-natural due to Lemmas 7 and 8:

Definition 3. *For an arbitrary set X, put $\tau_X(u) := \bigcap \mu_X(u)$.*

The just defined transformation τ is special amongst all sub-natural transformations $T \to \mathbb{P}$, for in analogy with Theorem 2, we obtain:

Theorem 4. *τ is the largest sub-natural transformation $T \to \mathbb{P}$.*

Proof. For any sub-natural $\nu : T \to \mathbb{P}$ and for any $q \in TX$, we need to show

$$\nu_X(q) \subseteq \bigcap \{U \subseteq X \mid q \in [T_U^X]\}.$$

Thus, for any $U \subseteq X$ with $q \in [T_U^X]$, we need to show $\nu_X(q) \subseteq U$.

By assumption, there is some $w \in TU$ with $q = (T_U^X)(w)$. This implies:

$$\begin{aligned} \nu_X(q) &= (\nu_X \circ T_U^X)(w) \\ &= (\mathbb{P}_U^X \circ \nu_U)(w) \\ &= \nu_U(w) \\ &\in \mathbb{P}(U). \end{aligned}$$

Further properties of τ will require conditions on the functor T. In particular, we are interested in preservation of intersections. Our aim is to characterize preservation of intersection by the existence of transformations to the powerset functor, in analogy to Theorem 3.

Theorem 5. *For a functor $T : \mathcal{S}et \to \mathcal{S}et$ the following are equivalent:*

(i) T (weakly) preserves intersections.
(ii) $\mu = \varepsilon \circ \tau$.
(iii) $\tau : T \to \mathbb{P}$ is sub-cartesian.
(iv) There exists a sub-cartesian transformation $\nu : T \to \mathbb{P}$.

Proof. (i)⇒ (ii): By Corollary 2, for every $u \in TX$, there is a smallest $U \subseteq X$ with $u \in [T_U^X]$. It follows that $U = \bigcap\{V \subseteq X \mid a \in [T_V^X]\}$ and $\mu_X(u) = \uparrow U = \varepsilon \circ \tau(u)$.

(ii)⇒ (iii): Since we know that $\mu = \varepsilon \circ \tau$ is sub-cartesian, the outer square of the following figure is a weak pullback. As ε_X is mono, one easily checks that the left square is a weak pullback, too, hence τ is sub-cartesian.

$$\begin{array}{ccccc} TY & \xrightarrow{\tau_Y} & \mathbb{P}Y & \xrightarrow{\varepsilon_Y} & \mathbb{F}Y \\ {\scriptstyle Tf}\uparrow & & {\scriptstyle \mathbb{P}f}\uparrow & & {\scriptstyle \mathbb{F}f}\uparrow \\ TX & \xrightarrow{\tau_X} & \mathbb{P}X & \xrightarrow{\varepsilon_X} & \mathbb{F}X \end{array}$$

(iv)⇒(i): This is exactly part (i) of Theorem 1, since $\mathbb{P}$ obviously preserves arbitrary intersections.

8 A Natural Transformation to $\mathbb{P}$

We now would like to characterize when τ is a natural transformation. This property is brought about jointly by the preservation of intersections and of preimages. Under the general assumption that T preserves preimages, the equivalence of (i) and (iii) is stated (without proof) in [Tay00]. In fact, this assumption is not necessary, as our proof will show:

Theorem 6. *For any functor $T : Set \to Set$ the following are equivalent:*

(i) T (weakly) preserves preimages and infinite intersections.
(ii) $\tau : T \to \mathbb{P}$ is natural and sub-cartesian.
(iii) There exists a natural transformation $\nu : T \to \mathbb{P}$ which is sub-cartesian.

Proof. (i)$\Rightarrow$ (ii): By Theorem 5, τ is sub-cartesian and $\mu = \varepsilon \circ \tau$, and by Theorem 3, μ is a natural transformation. So in the naturality diagram as in the previous proof, but this time for an arbitrary map $f : X \to Y$, we have the outer and the right square commuting. Since ε_Y is mono, the left square also commutes.

(iii)$\Rightarrow$ (i): Since $\mathbb{P}$ preserves preimages and intersections, this is once more a consequence of Theorem 1.

9 Modal Operators on Coalgebras

Coalgebras of the filter functor $\mathbb{F}$ have been described in [Gum01]. Given an $\mathbb{F}$-coalgebra $\mathcal{A} = (A, \alpha)$, i.e. a (structure)map $\alpha : A \to \mathbb{F}A$, one defines a relation $\twoheadrightarrow$ between A and $\mathbb{P}A$ by

$$a \twoheadrightarrow U :\iff U \in \alpha(a).$$

Then one has

(i) $a \twoheadrightarrow U$ and $a \twoheadrightarrow V \Rightarrow a \twoheadrightarrow U \cap V$,
(ii) $a \twoheadrightarrow U \subseteq V \Rightarrow a \twoheadrightarrow V$,

and conversely, a relation $\twoheadrightarrow$ between A and $\mathbb{P}A$ satisfying (i) and (ii) arises from a filter coalgebra on A.

Our sub-cartesian transformation μ can be used to associate to a coalgebra $\mathcal{A} = (A, \alpha)$ of arbitrary type T a filter coalgebra $\mathcal{A}_{\mathbb{F}} = (A, \alpha_{\mathbb{F}})$ on the same base set. The fact that μ is sub-cartesian has as consequence that the subcoalgebra structure is preserved and reflected:

Theorem 7. *To every coalgebra $\mathcal{A}_T = (A, \alpha)$, one can construct a filter-coalgebra $\mathcal{A}_{\mathbb{F}}$ on the same underlying set, so that $\mathcal{A}_T$ and $\mathcal{A}_{\mathbb{F}}$ have the same subcoalgebras.*

Proof. Define $\mathcal{A}_{\mathbb{F}} = (A, \mu \circ \alpha)$. From a T-coalgebra structure $\alpha : A \to TA$, we obtain the $\mathbb{F}$-coalgebra structure $\alpha_{\mathbb{F}} := \mu_A \circ \alpha$. Since μ is sub-natural, every subcoalgebra of $\mathcal{A}$ becomes a subcoalgebra of $\mathcal{A}_{\mathbb{F}}$, to. Since μ is sub-cartesian, every subcoalgebra of U of $\mathcal{A}_{\mathbb{F}}$ arises from a subcoalgebra of $\mathcal{A}$ on the same set. The required T-structure map on U arises as the mediating map for the weak pullback square in in the following figure:

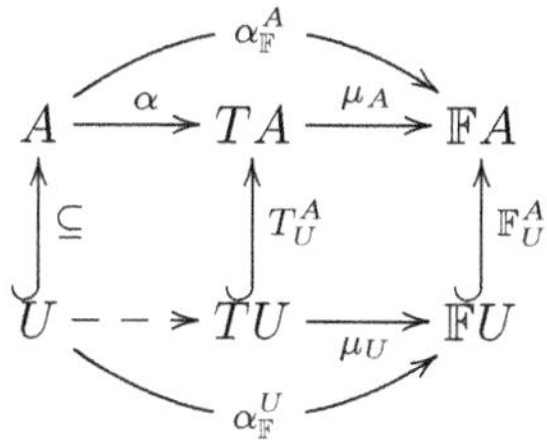

It would be nice, if in the previous theorem, we could replace the filter functor by the powerset functor. However, coalgebras of the powerset functors, i.e. Kripke-structures, have the property that the system of subcoalgebras is closed under arbitrary intersection. Therefore, we can hope for a similar theorem only when T preserves infinite intersections.

Nevertheless, without any assumptions on T, we can define an abstract next-time-operator $\bigcirc$, which recovers the vital properties of the next-time operator on Kripke-Structures. Indeed, if T preserves intersections, $\bigcirc P$ is the set of all states $s \in S$ whose immediate successors are all in P. The following lemmas, in particular Lemma 11, allow one to define $\bigcirc P$ for coalgebras in arbitrary categories. We begin with the following concrete definition:

Definition 4. *Let $\mathcal{A} = (A, \alpha)$ be a T-coalgebra on A. For any subset $P \subseteq A$ let*

$$\bigcirc_{\mathcal{A}} P := \{a \in A \mid \alpha(a) \in [T_P^A]\}.$$

We shall drop the subscript $_{\mathcal{A}}$, whenever the coalgebra structure is clear from context. When $\mathcal{A}$ is a Kripke structure, i.e. a $\mathbb{P}$-coalgebra, then $\bigcirc P$ is just the set of all states $a \in A$ such that all successors are in P. Guided by this intuition, the following properties are immediate:

Lemma 10. *(i) $\bigcirc : \mathbb{P}A \to \mathbb{P}A$ is monotone.*
(ii) $U \subseteq A$ is a subcoalgebra iff $\bigcirc U \subseteq U$.

Note that preimages are just pullbacks along injective maps, so the following lemma suggests a categorical definition of $\bigcirc P$:

Lemma 11. *$\bigcirc P$ is the preimage of $T_P^A : TP \hookrightarrow TA$ wrt. to $\alpha : A \to TA$.*

Proof. By definition, $\alpha[\bigcirc P] \subseteq [T_P^A]$, and since T_P^A is injective, there is (precisely) one map $\varphi : P \to TP$ making the square in the following figure commutative.

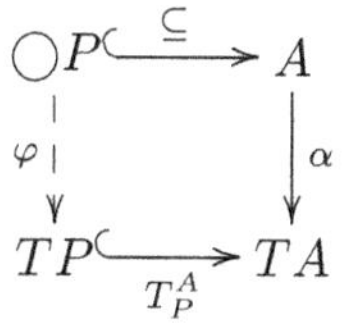

The square is in fact a preimage square, for given $a \in A$ and $q \in TP$ with $\alpha(a) = T_P^A(q)$, then $a \in \bigcirc P$ by definition and $(T_P^A \circ \varphi)(a) = (\alpha \circ \subseteq)(a) = \alpha(a) = T_P^A(q)$. Since T_P^A is injective, this means that $\varphi(a) = q$.

For arbitrary coalgebras $\mathcal{A} = (A, \alpha)$ we therefore can define modal operators $\Box_{\mathcal{A}}$ and $\Diamond_{\mathcal{A}}$ as largest and smallest fixed points:

$$\Box_{\mathcal{A}} S := \nu X. S \cap \bigcirc_{\mathcal{A}} X,$$

$$\Diamond_{\mathcal{A}} S := \mu X. S \cup \bigcirc_{\mathcal{A}} X.$$

Lemma 12. *Given a coalgebra $\mathcal{A} = (A, \alpha)$ and any subset $S \subseteq A$, then $\Box S$ is the largest subcoalgebra of $\mathcal{A}$ that is contained in S.*

Proof. We have $\Box S = S \cap \bigcirc \Box S$, so $\Box S \subseteq S$ and $\Box S$ is a subcoalgebra of $\mathcal{A}$ by Lemma 10. According to Tarski's description of the largest fixed point,

$$\Box S = \bigcup \{X \subseteq A \mid X \subseteq S \cap \bigcirc X\},$$

and using Lemma 10, $\Box S$ is the union of all subcoalgebras are contained in S.

Lemma 13. *Let $\mathcal{A} = (A, \alpha)$ and $\mathcal{B} = (B, \beta)$ be coalgebras, $\varphi : \mathcal{A} \to \mathcal{B}$ a homomorphism. Then for any subset $Q \subseteq B$ we have:*

$$\bigcirc \varphi^- Q \subseteq \varphi^- \bigcirc Q.$$

Proof. In the diagram below, the bottom face arises from applying the functor T to a preimage square. The front face commutes, since φ is a homomorphism, and the left and right faces are in fact pullbacks, due to Lemma 11.

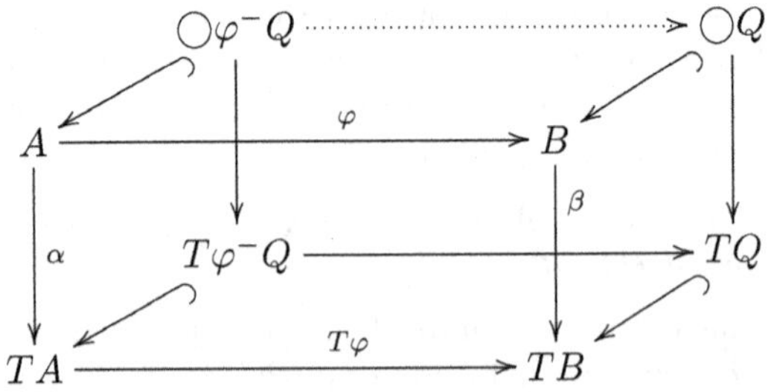

$\bigcirc \varphi^- Q$ with the obvious maps becomes a competitor to the pullback $\bigcirc Q$. This yields the dotted map, making the top face commutative. In particular, $\varphi[\bigcirc \varphi^- Q] \subseteq \bigcirc Q$, hence $\bigcirc \varphi^- Q \subseteq \varphi^- \bigcirc Q$.

Applying this lemma to a subset $Q = \varphi P$, we obtain the corollary:

Corollary 3. *For any homomorphism $\varphi : \mathcal{A} \to \mathcal{B}$ and any $P \subseteq A$ we have:*

$$\varphi[\bigcirc P] \subseteq \bigcirc \varphi[P].$$

Theorem 8. *T (weakly) preserves preimages if and only if for all homomorphisms $\varphi : \mathcal{A} \to \mathcal{B}$ we have:*

$$\varphi^- \bigcirc Q = \bigcirc \varphi^- Q.$$

Proof. We can reuse the figure from the previous proof. If T preserves preimages, then the bottom face is a weak pullback. We will show that the top face is a pullback, i.e. a preimage, too.

Given some $u \in \varphi^- \bigcirc Q$, i.e. $\varphi(u) \in \bigcirc Q$. Then

$$(T\varphi)(\alpha(u)) = \beta\varphi(u) \in [T_Q^B].$$

Thus there is a $w \in TQ$ with $T_Q^B(w) = (T\varphi)(\alpha(u))$ and the pullback $T\varphi^- Q$ contains an element $v \in T\varphi^- Q$ with $T^A_{\varphi^- Q}(v) = \alpha(u)$, i.e. $u \in \bigcirc\varphi^- Q$.

For the "only-if"-direction, it is enough, according to [GS03], to show that for any subcoalgebra $V \leq \mathcal{B}$ we have that $\varphi^- V \leq \mathcal{A}$.

Given $V \leq \mathcal{B}$, we have $V \subseteq \bigcirc V$, i.e. $\varphi^- V \subseteq \varphi^- \bigcirc V$. The hypothesis therefore yields $\varphi^- V \subseteq \bigcirc\varphi^- V$, meaning that $\varphi^- V$ is a subcoalgebra of $\mathcal{A}$.

Theorem 9. *T (weakly) preserves arbitrary intersections if and only if for every coalgebra $\mathcal{A}$ and each family $P_i \subseteq A$, $i \in I$, we have*

$$\bigcirc \bigcap_{i \in I} P_i = \bigcap_{i \in I} \bigcirc P_i.$$

Proof. Let T preserve intersections. In the following diagram the right squares are pullbacks by Lemma 11 and the bottom row is a weak intersection. This yields a map $f : \bigcap_{i \in I} \bigcirc P_i \to T \bigcap_{i \in I} P_i$, completing the outer square.

$$\begin{array}{ccccc}
\bigcap_{i\in I} \bigcirc P_i & \hookrightarrow & \bigcirc P_j & \hookrightarrow & A \\
\downarrow f & & \downarrow f_j & & \downarrow \alpha \\
T\bigcap_{i\in I} P_i & \hookrightarrow & TP_j & \hookrightarrow & TA
\end{array}$$

If we can show that this square is indeed a weak pullback square, then the result follows again from Lemma 11.

Any competitor Q for $\bigcap_{i \in I} P_i$ becomes a competitor to each of the $\bigcirc P_j$. The arising maps $\delta_j : Q \to \bigcirc P_j$ make Q into a copmpetitor to te intersection $\bigcap_{i \in I} \bigcirc P_i$. This yields the mediating map $\varepsilon : Q \to \bigcap_{i \in I} \bigcirc P_i$.

For the other direction, assume $\bigcap_{i \in I} \bigcirc_\alpha P_i \subseteq \bigcirc_\alpha \bigcap_{i \in I} P_i$ for every coalgebra $\mathcal{A} = (A, \alpha)$. This means

$$(\forall i \in I.\ \alpha(x) \in [T_{P_i}^A]) \Rightarrow \alpha(x) \in [T^A_{\bigcap_{i \in I} P_i}].$$

Given any element $u \in \bigcap [T_{P_i}^A]$, we consider the coalgebra on A with constant structure map $\alpha(a) = u$, in order to conclude that $u \in [T^A_{\bigcap_{i \in I} P_i}]$. Consequently,

$$\bigcap [T_{P_i}^A] \subseteq [T^A_{\bigcap_{i \in I} P_i}].$$

Lemma 1 provides the reverse inclusion, so by 5, T preserves intersections.

Now we can formulate the main theorem of this section:

Theorem 10. *For every T-coalgebra $\mathcal{A} = (A, \alpha)$ there is a Kripke structure $\mathcal{K} = (A, R)$ on the same set A with $\bigcirc_{\mathcal{A}} = \bigcirc_{\mathcal{K}}$ if and only if T (weakly) preserves intersections.*

The proof follows from the following lemma in combination with Theorem 9.

Lemma 14. *Let $\mathcal{A} = (A, \alpha)$ be a coalgebra. There exists a Kripke-Structure $\mathcal{K} = (A, \twoheadrightarrow)$ on the base set A with $\bigcirc_{\mathcal{A}} = \bigcirc_{K}$ if and only if for every family of subsets $P_i \subseteq A$, $i \in I$:*

$$\bigcap_{i\in I} \bigcirc_{\mathcal{A}} P_i = \bigcirc_{\mathcal{A}} \bigcap_{i\in I} P_i.$$

Proof. As Kripke-structures satisfy the formula, the necessity is clear. Conversely, assume the validity of the formula and define a Kripke structure $\mathcal{K} = (A, \twoheadrightarrow)$ by

$$a \twoheadrightarrow b \;\; :\Longleftrightarrow \;\; b \in \bigcap\{P \subseteq A \mid a \in \bigcirc_{\mathcal{A}} P\}$$

Given $a \in \bigcirc_{\mathcal{A}} Q$, then $a \twoheadrightarrow b$ implies that $b \in Q$ for every b, hence $a \in \bigcirc_{\mathcal{K}} Q$. For the other direction, compute:

$$\begin{aligned}
a \in \bigcirc_{\mathcal{K}} Q &\Longleftrightarrow \forall b \in A.\; a \twoheadrightarrow b \Rightarrow b \in Q \\
&\Longleftrightarrow \bigcap\{P \subseteq A \mid a \in \bigcirc_{\mathcal{A}} P\} \subseteq Q \\
&\Rightarrow \;\; \bigcirc_{\mathcal{A}} \bigcap\{P \subseteq A \mid a \in \bigcirc_{\mathcal{A}} P\} \subseteq \bigcirc_{\mathcal{A}} Q \\
&\Rightarrow \;\; \bigcap\{\bigcirc_{\mathcal{A}} P \subseteq A \mid a \in \bigcirc_{\mathcal{A}} P\} \subseteq \bigcirc_{\mathcal{A}} Q \\
&\Rightarrow \;\; a \in \bigcirc_{\mathcal{A}} Q
\end{aligned}$$

10 Discussion and Further Work

Starting with Rutten's seminal work [Rut00], weak preservation of pullbacks, resp. of intersections by the coalgebraic type functor has played an important role in the universal theory of coalgebra. Weak pullback preservation splits into weak preservation of kernels and (weak) preservation preimages [GS03].

Here we have seen that (weak) preimage preservation yields a natural and subcartesian transformation of T to the filter functor $\mathbb{F}$, thus establishing an intimate relationship between T-coalgebras and filter coalgebras. Similarly, preservation of intersections makes for a sub-cartesian transformation from T to the powerset functor $\mathbb{P}$, thus relating every T-coalgebra with a Kripke-Structure.

In the work of E.G. Manes [Man98] we can find the definition of sub-natural, resp. sub-cartesian, transformations under the names "mono-transformation", resp. "taut transformation". Manes proves that finitary collection monads are precisely the taut quotients of polynomial functors.

For coalgebras of polynomial type, B. Jacobs has introduced a nexttime operator in [Jac02]. Since polynomial functors are standard and preserve weak pullbacks and intersections, they are rather special. Our definition of $\bigcirc_{\mathcal{A}}$ does not rely on any assumption on the type functor, in fact, Lemma 11 suggests how to define a nexttime operator for coalgebras over base categories other than $\mathcal{S}et$.

We have seen that preservation of preimages, resp. of intersections, are reflected in very natural preservation properties of the $\bigcirc$ operator. It can now serve as a starting point for a CTL-like logic, for arbitray coalgebras. As an example, coalgebras of the 3-2-functor $F(X) = \{(x_1, x_2, x_3) \mid card(\{x_1, x_2, x_3\}) \leq 2\}$ (see [AM89]) are characterized in this logic by the following formulae:

1. $\bigcirc$ true,
2. $\bigcirc(\varphi \vee \psi \vee \theta) \Rightarrow \bigcirc(\phi \vee \psi) \vee \bigcirc(\phi \vee \theta) \vee \bigcirc(\psi \vee \theta)$.

References

AHS90. J. Adámek, H. Herrlich, and G.E. Strecker, *Abstract and concrete categories*, John Wiley & Sons, 1990.

AM89. P. Aczel and N. Mendler, *A final coalgebra theorem*, Proceedings category theory and computer science (D.H. Pitt et al., eds.), Lecture Notes in Computer Science, Springer, 1989, pp. 357–365.

Bar93. M. Barr, *Terminal coalgebras in well-founded set theory*, Theoretical Computer Science (1993), no. 144(2), 299–315.

GS02. H.P. Gumm and T. Schröder, *Coalgebras of bounded type*, Math. Struct. in Comp. Science (2002), no. 12, 565–578.

GS03. ______, *Types and coalgebraic structure*, Tech. report, Philipps-Universität Marburg (to appear in Algebra Universalis), 2003.

Gum01. H.P. Gumm, *Functors for coalgebras*, Algebra Universalis (2001), no. 45 (2-3), 135–147.

Jac02. B. Jacobs, *The temporal logic of coalgebras via Galois algebras*, Math. Struct. in Comp. Science (2002), no. 12, 875–903.

Man98. E.G. Manes, *Implementing collection classes with monads*, Math. Struct. in Comp. Science **8** (1998), 231–276.

Rut00. J.J.M.M. Rutten, *Universal coalgebra: a theory of systems*, Theoretical Computer Science (2000), no. 249, 3–80.

Tay00. P. Taylor, *Practical foundations of mathematics*, 2. ed., Cambridge University Press, 2000.

Trn69. V. Trnková, *Some properties of set functors*, Comm. Math. Univ. Carolinae (1969), no. 10,2, 323–352.

Context-Free Languages via Coalgebraic Trace Semantics

Ichiro Hasuo and Bart Jacobs

Institute for Computing and Information Sciences, Radboud University Nijmegen,
P.O. Box 9010, 6500 GL Nijmegen, The Netherlands
{ichiro, B.Jacobs}@cs.ru.nl
http://www.cs.ru.nl/{~ichiro, B.Jacobs}

Abstract. We show that, for functors with suitable mild restrictions, the initial algebra in the category of sets and functions gives rise to the final coalgebra in the (Kleisli) category of sets and relations. The finality principle thus obtained leads to the *finite trace semantics* of non-deterministic systems, which extends the *trace semantics* for coalgebras previously introduced by the second author. We demonstrate the use of our technical result by giving the first coalgebraic account on context-free grammars, where we obtain generated context-free languages via the finite trace semantics. Additionally, the constructions of both finite and possibly infinite parse trees are shown to be monads. Hence our extension of the application domain of coalgebras identifies several new mathematical constructions and structures.

1 Introduction

Context-free grammars and context-free languages are undoubtedly among the most fundamental notions in computer science. Introduced by Chomsky [Cho56], they have come to serve as a theoretical basis for formal (programming) languages [ASU86]. This paper presents the first steps in a coalgebraic analysis of those notions. In a sense it extends previous coalgebraic work [Jac05a, Rut03] on regular languages.

A context-free grammar is a clear example of a coalgebra: the state space consists of its non-terminal symbols and the coalgebraic structure is defined by its generation rules. Then the context-free language generated by the grammar should be the "behavior" of the coalgebra. Our motivation is to find a suitable setting which gives that behavior by *coinduction*, i.e. an argument using finality.

What is unusual here is that we are concerned only with the *finite* behavior (i.e. generated strings of only finite length). This suggests that the domain of the semantics might be the initial algebra, as the subcoalgebra of the final coalgebra consisting of all the finite behavior.

Interestingly, it turns out that for functors with mild restrictions, the initial algebra in **Sets** gives rise to the final coalgebra in the category **Rel** of sets and relations. The finality principle in **Rel** is called the *finite trace semantics* in this

J.L. Fiadeiro et al. (Eds.): CALCO 2005, LNCS 3629, pp. 213–231, 2005.

paper, in contrast to the (possibly infinite) *trace semantics* from [Jac04b] where the final coalgebra in **Sets** gives rise to a weakly final coalgebra in **Rel**.

A context-free grammar is identified as a coalgebra in **Rel** because of its non-deterministic nature. Now our technical result of finite trace semantics allows us to obtain the set of generated *finite skeletal parsed trees* (finite strings with additional tree structure) via finality. After applying the flattening function it yields the set of generated strings.

The category **Rel** can also be described as the Kleisli category $\mathbf{Sets}_{\mathcal{P}}$ of the powerset monad. This view is relevant in generalization of our work to other monads other than $\mathcal{P}$, such as the subdistribution functor $\mathcal{D}$.

The remainder of this paper is organized as follows. Later on this section gives a "sneak preview" of the technical result and its applications. Section 2 formulates context-free grammars as coalgebras, and introduces the notion of *skeletal parse trees (SPTs)* as strings with tree structure. It is shown in Section 3 that (finite) SPTs carry the initial algebra/final coalgebra for an appropriate functor, and that their formations have monad structures, related to one another via the "fundamental span" of monad maps. The details of our technical result of finite trace semantics for coalgebras are presented in Section 4. Section 5 puts the current work in the context of the previous work [Jac04b] of (possibly infinite) trace semantics for coalgebras. Section 6 is for conclusions and future work.

It is assumed that the reader is familiar with the basic categorical theory of algebras and coalgebras for both functors and monads. For these preliminaries see e.g. [Jac05b, Rut00, BW83].

1.1 Sneak Preview

As motivation we briefly present our main technical result and two illustrating examples of non-deterministic automata and context-free grammars. The details of constructions, definitions and proofs will follow later. For a functor $F : \mathbf{Sets} \to \mathbf{Sets}$ with mild restrictions, we have an initial algebra in **Sets** and a canonical lifting $F : \mathbf{Rel} \to \mathbf{Rel}$.

Theorem 1.1 (Finite trace semantics for coalgebras). *Let $\alpha : FA \stackrel{\cong}{\to} A$ be the initial F-algebra in* **Sets**. *The coalgebra $\mathsf{graph}(\alpha^{-1}) : A \stackrel{\cong}{\to} FA$ in* **Rel** *is final for the lifted functor F. Hence, given a coalgebra $c : X \to FX$ in* **Rel** *there exists a unique arrow $\mathsf{ft}_c : X \to A$ which makes the following diagram in* **Rel** *commute.*

$$\begin{array}{ccc} FX & \stackrel{F(\mathsf{ft}_c)}{\dashrightarrow} & FA \\ c\uparrow & & \uparrow\mathsf{graph}(\alpha^{-1}) \\ X & \underset{\mathsf{ft}_c}{\dashrightarrow} & A \end{array} \tag{1}$$

The relation ft_c thus obtained is called the ***finite trace*** *of c.*

Translating back to the category **Sets**, Theorem 1.1 assigns to each non-deterministic coalgebra $c : X \to \mathcal{P}FX$ its finite trace $\mathsf{ft}_c : X \to \mathcal{P}A$ into the powerset of the carrier of the initial F-algebra.

Example 1.2 (Non-deterministic automata). A non-deterministic automaton over alphabet Σ can be described as a coalgebra $X \to (\mathcal{P}X)^{\Sigma} \times 2$, or equivalently as $c : X \to \mathcal{P}(1 + \Sigma \times X)$ in **Sets**. The set $c(x)$ then contains the unique element $\checkmark$ of $1 = \{\checkmark\}$ if and only if x is an accepting state. For the functor $F = 1 + \Sigma \times -$ involved, the initial F-algebra (in **Sets**) consists of the *strings* (or *lists*) over Σ, as in: $[\mathsf{nil}, \mathsf{cons}] \colon 1 + \Sigma \times \Sigma^* \xrightarrow{\cong} \Sigma^*$. Hence, given a non-deterministic automaton $c : X \to \mathcal{P}FX$, Theorem 1.1 yields its finite trace $\mathsf{ft}_c : X \to \mathcal{P}\Sigma^*$. It is shown later in Example 4.13 that the set $\mathsf{ft}_c(x) \subseteq \Sigma^*$ is indeed the *language accepted by* the automaton when it starts in state x.

Example 1.3 (Context-free grammars). A context-free grammar (CFG) consists of a set Σ of terminal symbols, a set X of non-terminal symbols, and a relation $R \subseteq X \times (\Sigma + X)^*$ consisting of generation rules. It is described as a coalgebra $c : X \to \mathcal{P}\big((\Sigma + X)^*\big)$ in **Sets**.

For the functor $F = (\Sigma + -)^*$, the initial algebra $\Sigma^{\triangle}$ consists of *finite skeletal parse trees* (finite SPTs), which are strings with tree structure. An example of a finite SPT is given below on the left: it describes the formula $\mathsf{s}(\mathsf{x}) = \mathsf{0}$. The initial algebra structure on $\Sigma^{\triangle}$ is illustrated below on the right, where $\mathcal{S}_1$ and $\mathcal{S}_2$ are finite SPTs.

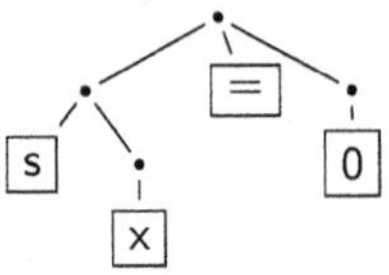

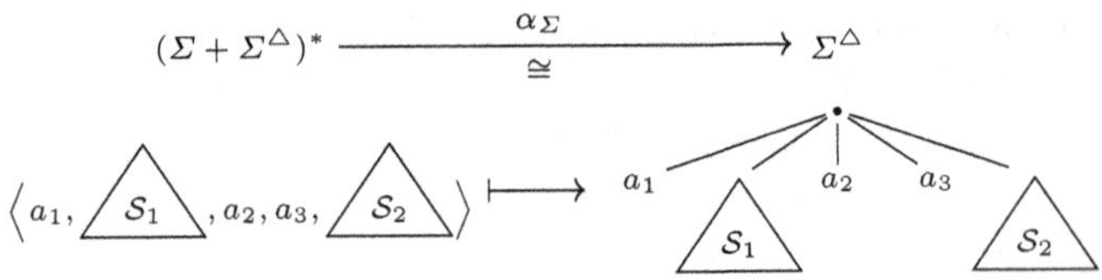

Given a CFG $c : X \to \mathcal{P}FX$, via the finite trace semantics we obtain a function $\mathsf{ft}_c : X \to \mathcal{P}\Sigma^{\triangle}$. Later in Example 4.14 it is shown that, for a non-terminal $x \in X$, the set $\mathsf{ft}_c(x)$ consists of all the finite SPTs that can be generated from x. By applying the flattening function $\Sigma^{\triangle} \to \Sigma^*$, which is defined via the initiality of $\Sigma^{\triangle}$, we obtain the set of strings generated from x.

1.2 Notations

The i-th coprojection into a coproduct $\coprod_{i \in I} X_i$ is denoted by κ_i. If the index set I is finite the coproduct will be written as $X_{i_1} + X_{i_2} + \cdots + X_{i_n}$.

The operator $(-)^*$ defined by $X^* = \coprod_{n < \omega} X^n$ is so-called the *Kleene star*. The set X^* consists of all the *strings* of finite length over X. It is standard that the Kleene star has a monad structure with unit $\overset{*}{\eta}$ creating a string of length one and multiplication $\overset{*}{\mu}$ "flattening" a string of strings into a string. We denote a string of length n by $\langle a_1, a_2, \ldots, a_n \rangle$; $\langle\rangle$ is then the empty string of length zero.

We write a symbol on top of the unit η and multiplication μ to indicate the relevant monad. *E.g.* $\overset{*}{\eta}_\Sigma$ is the Σ-component of the unit of Kleene monad $(-)^*$. Hence $(\overset{*}{\eta}_\Sigma)^*$ is the application of the functor $(-)^*$ to the arrow $\overset{*}{\eta}_\Sigma$.

We will heavily use two powerset functors: one is covariant $\mathcal{P}$ and the other is contravariant $\overline{\mathcal{P}}$. They act the same on objects. For a function $f : X \to Y$, $\mathcal{P}f$

maps a subset of X to its direct image under f, and $\overline{\mathcal{P}}f$ maps a subset $u \subseteq Y$ to its inverse image $f^{-1}(u)$.

It is standard that there is a bijective correspondence between a relation $R \subseteq X \times Y$ and a function $f : X \to \mathcal{P}Y$, given by $f_R(x) = \{y \in Y \mid (x, y) \in R\}$ (*relation-into-function*). In this paper we identify a relation with the corresponding function, and vice versa. Hopefully this will not cause any confusion.

2 Context-Free Grammars as Coalgebras

In this section we give a precise coalgebraic formulation of context-free grammars and context-free languages. For more about traditional treatment of those notions the reader is referred to [LP81].

A (traditional) context-free grammar, described as a triple of terminals, nonterminals and generation rules $R \subseteq X \times (\Sigma + X)^*$, is described as a coalgebra:

$$X \xrightarrow{c} \mathcal{P}\big((\Sigma + X)^*\big), \qquad \text{namely} \qquad x \longmapsto \{s \in (\Sigma + X)^* \mid \langle x, s\rangle \in R\}.$$

Definition 2.1 (Context-free grammar). In this paper a $\mathcal{P}\big((\Sigma + -)^*\big)$-coalgebra in **Sets** is called a *context-free grammar* (CFG in short) over Σ. Equivalently, via the relation-into-function, a CFG is a $(\Sigma + -)^*$-coalgebra in **Rel**.

Notice that not all $\mathcal{P}\big((\Sigma + -)^*\big)$-coalgebras are context-free grammars in the traditional sense, due to the lack of finiteness conditions on Σ, X and the generation rules. The above definition, which is more liberal and natural from the coalgebraic perspective, ignores algorithmic aspects of context-free grammars. They are not relevant in this paper.

Example 2.2. Consider the following CFG for the syntax of Peano arithmetic.

$$\begin{array}{l}
\Sigma = \{\mathsf{0}, \mathsf{s}, =, \wedge, \vee, \supset, \neg, \forall, \exists\} \cup \mathit{Var}, \qquad X = \{\mathbf{T}, \mathbf{Q}, \mathbf{F}\},\\
\mathbf{T} \to \mathsf{0}, \qquad \mathbf{T} \to \mathsf{x} \quad (\mathsf{x} \in \mathit{Var}), \qquad \mathbf{T} \to \mathsf{s}\mathbf{T},\\
\mathbf{Q} \to \forall\mathsf{x} \quad (\mathsf{x} \in \mathit{Var}), \qquad \mathbf{Q} \to \exists\mathsf{x} \quad (\mathsf{x} \in \mathit{Var}),\\
\mathbf{F} \to \mathbf{T}{=}\mathbf{T}, \quad \mathbf{F} \to \mathbf{F}{\wedge}\mathbf{F}, \quad \mathbf{F} \to \mathbf{F}{\vee}\mathbf{F}, \quad \mathbf{F} \to \mathbf{F}{\supset}\mathbf{F}, \quad \mathbf{F} \to \neg\mathbf{F}, \quad \mathbf{F} \to \mathbf{Q}\mathbf{F}.
\end{array}$$

The induced CFG $c : X \to \mathcal{P}\big((\Sigma + X)^*\big)$ is as follows.

$$\begin{array}{c}
c(\mathbf{T}) = \{\mathsf{0}\} \cup \mathit{Var} \cup \{\mathsf{s}\mathbf{T}\}, \qquad c(\mathbf{Q}) = \{\forall\mathsf{x} \mid \mathsf{x} \in \mathit{Var}\} \cup \{\exists\mathsf{x} \mid \mathsf{x} \in \mathit{Var}\},\\
c(\mathbf{F}) = \{\mathbf{T}{=}\mathbf{T}, \mathbf{F}{\wedge}\mathbf{F}, \mathbf{F}{\vee}\mathbf{F}, \mathbf{F}{\supset}\mathbf{F}, \neg\mathbf{F}, \mathbf{Q}\mathbf{F}\}.
\end{array}$$

Usually a context-free grammar over Σ is considered as a machine which generates strings over Σ, i.e. elements of Σ^*. However, from a coalgebraic perspective it is more natural to first obtain finite SPTs (i.e. strings with tree structure), and then by flattening obtain strings. In the following the precise definition of finite SPTs is presented, together with a few related notions.

The next definition is a bit complicated; the reader may find an alternative characterization (Proposition 3.1) in terms of initial algebra/final coalgebra.

Definition 2.3 ((Skeletal) parse trees). Let $c : X \to \mathcal{P}((\Sigma+X)^*)$ be a CFG over Σ. A *parse tree generated by c from $x \in X$* is a (possibly infinite-depth) tree which satisfy the following:

1. All leaf nodes are labelled from $\Sigma + X$;
2. All internal (i.e. non-leaf) nodes are labelled from X;
3. The root is labelled with x;
4. If a leaf node is labelled from X, say with y, then the empty string $\langle\rangle$ belongs to $c(y)$;
5. For each internal node let $y \in X$ be its label and let its immediate successors be labelled with $c_1, c_2, \dots, c_m$ ($c_i \in \Sigma+X$) from left to right. Then the string $\langle c_1, c_2, \dots, c_m\rangle$ is an element of $c(y)$.[1]

Condition 5 ensures that a parse tree is finitely-branching. A parse tree is *finite* if its depth is finite.

A *skeletal parse tree (SPT for short) generated by c from x* is a parse tree generated by c from x, with all of its labels from X deleted. It is *finite* if its depth is finite. A *skeletal parse tree (SPT) over Σ* is a skeletal parse tree generated by some CFG c over Σ. Equivalently, it is a finitely-branching, possibly infinite-depth tree with some of its leaves labelled from Σ and its internal nodes not labelled, and if it is trivial (i.e. root-only) then the sole node is not labelled.[2] An SPT is *finite* if its depth is finite.

The set of all the (possibly infinite) SPTs over Σ is denoted by $\Sigma^\wedge$, and the set of the finite SPTs is denoted by $\Sigma^\triangle$.

Example 2.4. Below are two parse trees generated by the context-free grammar in Example 2.2, from the non-terminal symbol **F**. The one on the left is finite.

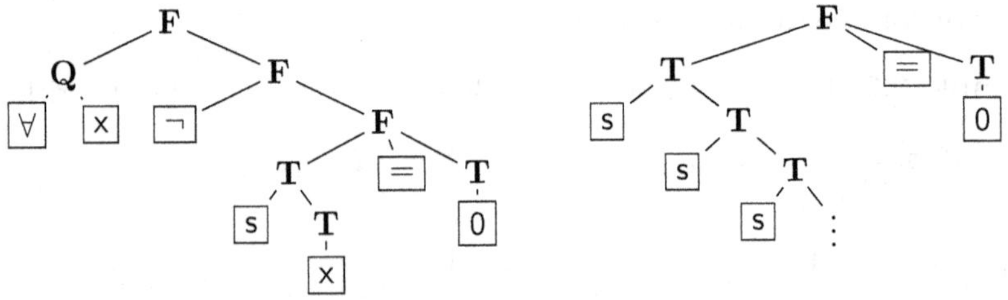

Forgetting about the non-terminal symbols from X, we obtain the following SPTs generated by the grammar from **F**.

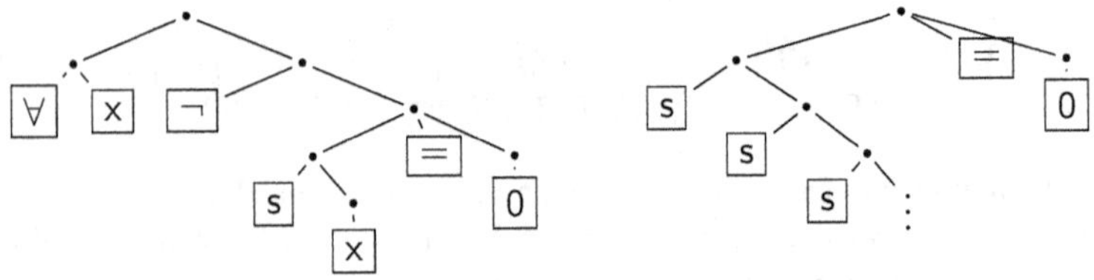

The infinite one on the right has no corresponding well-formed formula, while the other one can be read as ∀x.¬(s(x) = 0).

[1] Condition 4 may be considered as an instance of Condition 5 when $m = 0$.
[2] Since an SPT is generated from a non-terminal symbol.

3 Monad Structures on Languages

In this section we first investigate (co)algebraic structures on the set of finite/infinite SPTs. Then it turns out that the formation of finite SPTs $\Sigma^{\triangle}$ and SPTs $\Sigma^{\wedge}$, from Σ, are all monads just like that of strings Σ^*. Moreover, the embedding $\Sigma^{\triangle} \rightarrowtail \Sigma^{\wedge}$ and the flattening function $\Sigma^{\triangle} \twoheadrightarrow \Sigma^*$ are shown to be both *maps of monads* [BW83].

The following observation is the first step. It may also be read as a definition of $\Sigma^{\triangle}$ and $\Sigma^{\wedge}$. The proof is standard and left to the reader.

Proposition 3.1. *The set $\Sigma^{\triangle}$ of all finite SPTs over Σ carries the initial $(\Sigma + -)^*$-algebra. The algebraic structure α_Σ makes a sequence $\langle c_1, c_2, \ldots, c_n \rangle$ (where $c_i \in \Sigma + \Sigma^{\triangle}$) into a tree by adding a fresh root whose immediate successors are $c_1, c_2, \ldots, c_n$. An example is found in Example 1.3. The empty string $\langle\rangle$ is mapped by α_Σ to the trivial tree which is not labelled.*

The set $\Sigma^{\wedge}$ of (possibly infinite) SPTs over Σ carries the final $(\Sigma + -)^*$-coalgebra. The coalgebraic structure ζ_Σ removes the root and returns the sequence of its immediate successors. For example,[3]

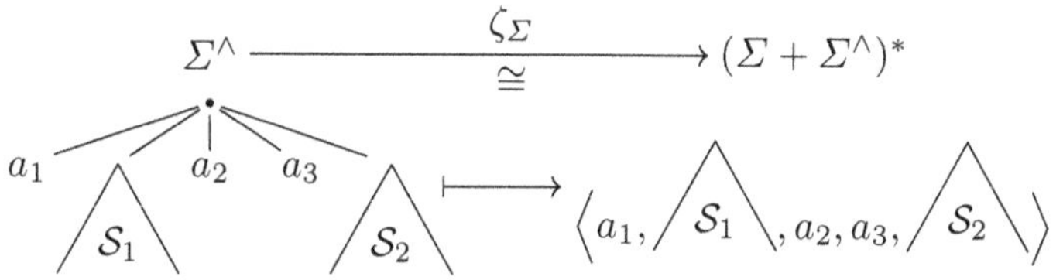

The trivial tree is mapped to the empty string $\langle\rangle$. □

Remark 3.2. We can think of the functor $(\Sigma + -)^*$ as a "signature" in the sense of traditional universal algebra. Let $\star$ be a fresh symbol, and for each $s \in (\Sigma + \{\star\})^*$ let $\|s\|$ denote the number of $\star$'s appearing in s. Then we have an obvious isomorphism

$$(\Sigma + X)^* \cong \textstyle\coprod_{s \in (\Sigma + \{\star\})^*} X^{\|s\|} \ .$$

Hence $(\Sigma + -)^*$ describes a signature such that $(\Sigma + \{\star\})^*$ is the set of operations and each operation $s \in (\Sigma + \{\star\})^*$ is $\|s\|$-ary.

The notation $(-)^{\triangle} : \Sigma \mapsto \Sigma^{\triangle}$ and $(-)^{\wedge} : \Sigma \mapsto \Sigma^{\wedge}$ is used to suggest an analogy with the Kleene (or list, string) monad $(-)^*$. Indeed, these constructions are closely related, as is shown in the sequel.

Each of the mappings $(-)^{\triangle}$ and $(-)^{\wedge}$ extends to a functor, using standard results about initial algebras and final coalgebras for a functor $F(\Sigma, -)$, where F is a bifunctor—in this case $F(X_1, X_2) = (X_1 + X_2)^*$. Moreover, it turns out that both functors $(-)^{\triangle}$ and $(-)^{\wedge}$ have a monad structure. The formation

[3] An open triangle designates a tree with a possibly infinite depth, while a closed one is a tree with a finite depth. This conforms to the notation $\Sigma^{\wedge}$ and $\Sigma^{\triangle}$.

of units $\overset{\vartriangle}{\eta}, \hat{\eta}$ and multiplications $\overset{\vartriangle}{\mu}, \hat{\mu}$ is much like for the *free monad* and *free iterative monad* generated by a functor [AAMV03, Jac04a]. The difference is that here the parameter set Σ is inside the Kleene monad $(-)^*$, which adds some complexity. The concrete constructions are described in Appendix A.1. It is straightforward, but laborious, to show that the constructions satisfy the requirements of a monad.

Proposition 3.3. *The triples* $\big((-)^{\vartriangle}, \overset{\vartriangle}{\eta}, \overset{\vartriangle}{\mu}\big)$ *and* $\big((-)^{\wedge}, \hat{\eta}, \hat{\mu}\big)$ *are monads.* □

Let $\iota_\Sigma : \Sigma^{\vartriangle} \rightarrowtail \Sigma^{\wedge}$ be the canonical embedding of the initial algebra into the final coalgebra. It is a mono by [Bar93, Theorem 3.2].

$$\begin{array}{ccc}
(\Sigma + \Sigma^{\vartriangle})^* & \overset{(\Sigma + \iota_\Sigma)^*}{\rightarrowtail} & (\Sigma + \Sigma^{\wedge})^* \\
\alpha_\Sigma \downarrow \cong & & \cong \uparrow \zeta_\Sigma \\
\Sigma^{\vartriangle} & \underset{\iota_\Sigma}{\rightarrowtail} & \Sigma^{\wedge}
\end{array}$$

It is straightforward to show that ι_Σ is natural in Σ, and is compatible with monad structures, i.e., is a map of monads.

The *flattening* function $\varphi_\Sigma : \Sigma^{\vartriangle} \to \Sigma^*$, which maps a finite SPT to a flat string demolishing the tree structure, is obtained via initiality of α_Σ.

$$\begin{array}{ccc}
(\Sigma + \Sigma^{\vartriangle})^* & \overset{(\Sigma + \varphi_\Sigma)^*}{\dashrightarrow} & (\Sigma + \Sigma^*)^* \\
\alpha_\Sigma \downarrow \cong & & \downarrow \overset{*}{\mu}_\Sigma \circ [\,\overset{*}{\eta}_\Sigma\,,\, \Sigma^*\,]^* \\
\Sigma^{\vartriangle} & \underset{\varphi_\Sigma}{\dashrightarrow} & \Sigma^*
\end{array}$$

It is easy to see that the flattening map is a map of monads. Moreover, it is obviously an epi: for a sequence $\langle a_1, a_2, \ldots, a_n \rangle \in \Sigma^*$ take the finite SPT of depth 2 that has leaves $a_1, a_2, \ldots, a_n$ from left to right.

Hence we have obtained the following result.

Proposition 3.4. *The embedding* ι_Σ *and the flattening map* φ_Σ *both form a map of monads. They yield the following "fundamental span of languages".*

$$\Sigma^* \overset{\varphi_\Sigma}{\twoheadleftarrow} \Sigma^{\vartriangle} \overset{\iota_\Sigma}{\rightarrowtail} \Sigma^{\wedge}$$

□

4 Finite Trace Semantics for Coalgebras

This section presents the main technical result (already previewed as Theorem 1.1) that an initial algebra in **Sets** (of a suitable functor) yields a final coalgebra in **Rel**. Examples 1.2 and 1.3 are also fully elaborated in greater detail.

4.1 Shapely Functors

The family of endofunctors F in **Sets** we are interested in is that of *shapely functors* [Jay95]. The following inductive definition is equivalent to the original one.

Definition 4.1 (Shapely functors). The family of *shapely functors* is defined inductively by the following BNF notation:

$$F, G, F_i ::= \mathrm{id} \mid \Sigma \mid F \times G \mid \textstyle\coprod_{i \in I} F_i \ ,$$

where Σ denotes the constant functor into Σ.

Notice that we can take the exponentiation $(-)^{\Sigma}$ to the power of a finite set Σ in building a shapely functor, because X^{Σ} is isomorphic to the $|\Sigma|$-fold product of X's. A shapely functor is different from a polynomial functor in the following points: we cannot take an exponentiation with an infinite set (because it makes Lemma 4.2.2 fail), but we can take an infinite coproduct—so that we can form the Kleene star $(-)^* = \coprod_{n<\omega} (-)^n$. A shapely functor has the following properties needed for our purpose.

Lemma 4.2. *Let F : **Sets** $\to$ **Sets** be a shapely functor.*

1. *F preserves weak pullbacks.*
2. *For an arrow $?_X : 0 \rightarrowtail X$ with domain 0, $F?_X : F0 \to FX$ is mono. Hence F preserves all monos in* **Sets**.
3. *F preserves ω-colimits and ω^{op}-limits. Hence F has both the initial algebra and the final coalgebra. They are, together with the canonical embedding, denoted as follows.*

$$\begin{array}{ccc} FA & \overset{F\iota}{\rightarrowtail} & FZ \\ {\scriptstyle\alpha}\downarrow{\scriptstyle\cong} & & {\scriptstyle\cong}\uparrow{\scriptstyle\zeta} \\ A & \underset{\iota}{\rightarrowtail} & Z \end{array}$$

Proof. The proofs are easy by induction on the construction of F. The preservation of ω-colimits (or ω^{op}-limits) allows us to obtain the initial F-algebra (or the final F-coalgebra) as the colimit (or limit) of the initial sequence of length ω (or final sequence, respectively): see e.g. [Bar93, AK95]. □

4.2 Relation Lifting, Distributive Law and Kleisli Category

An endofunctor F yields a *relation lifting*: given a relation $\langle r_1, r_2 \rangle : R \rightarrowtail X \times Y$, a lifted relation $\mathrm{Rel}_F(R) \rightarrowtail FX \times FY$ is defined by image factorization.

$$\begin{array}{ccc} FR & \twoheadrightarrow & \mathrm{Rel}_F(R) \\ & {\scriptstyle\langle Fr_1, Fr_2 \rangle}\searrow & \downarrow \\ & & FX \times FY \end{array}$$

The following compatibility results hold for a functor F which preserves weak pullbacks, hence in particular for a shapely F (Lemma 4.2).

Lemma 4.3. *Relation lifting is compatible with such operations on relations as:*

1. *Composition: for $R \rightarrowtail X \times Y$, $S \rightarrowtail Y \times Z$ and their composition $S \circ R = \{(x,z) \in X \times Z \mid \exists y \in Y.(x,y) \in R \text{ and } (y,z) \in Z\}$ we have $\mathrm{Rel}_F(S \circ R) = \mathrm{Rel}_F(S) \circ \mathrm{Rel}_F(R)$.*
2. *Graph of a function and functor application: for a function $f : X \to Y$ and its graph $\mathsf{graph}(f) = \{(x, f(x)) \mid x \in X\}$ we have $\mathrm{Rel}_F(\mathsf{graph}(f)) = \mathsf{graph}(Ff)$.*
3. *Inverse image and direct image: for functions $f_1 : X_1 \to Y_1$, $f_2 : X_2 \to Y_2$ and relations $R \rightarrowtail X_1 \times X_2$, $S \rightarrowtail Y_1 \times Y_2$, let us denote the inverse image and the direct image by $(f_1 \times f_2)^{-1}(S) = \{(x_1, x_2) \mid (f_1(x_1), f_2(x_2)) \in S\}$, and $\coprod_{f_1 \times f_2}(R) = \{\,(f_1(x_1), f_2(x_2)) \mid (x_1, x_2) \in R\}$. Then we have*

$$\mathrm{Rel}_F((f_1 \times f_2)^{-1}(S)) = (Ff_1 \times Ff_2)^{-1}(\mathrm{Rel}_F(S)) \ ,$$
$$\mathrm{Rel}_F(\textstyle\coprod_{f_1 \times f_2}(R)) = \coprod_{Ff_1 \times Ff_2}(\mathrm{Rel}_F(R)) \ . \qquad \square$$

The membership relation $\in_X \rightarrowtail X \times \mathcal{P}X$ on a set X is lifted to $\mathrm{Rel}_F(\in_X) \rightarrowtail FX \times F\mathcal{P}X$. By transposition we obtain the following function λ_X.

$$F\mathcal{P}X \xrightarrow{\lambda_X} \mathcal{P}FX \qquad u \longmapsto \{a \in FX \mid \langle a, u\rangle \in \mathrm{Rel}_F(\in_X)\}$$

Then the map λ_X is: 1) natural in X, and 2) compatible with the monad structure of $\mathcal{P}$: when we denote the unit (singleton map) by $\{-\}$ and the multiplication (union) by $\bigcup$, the following diagrams commute.

$$\begin{array}{ccc} FX & \xrightarrow{F\{-\}_X} & F\mathcal{P}X \\ & \searrow^{\{-\}_{FX}} & \downarrow \lambda_X \\ & & \mathcal{P}FX \end{array} \qquad \begin{array}{ccccc} F\mathcal{P}^2X & \xrightarrow{\lambda_{\mathcal{P}X}} & \mathcal{P}F\mathcal{P}X & \xrightarrow{\mathcal{P}\lambda_X} & \mathcal{P}^2FX \\ F\bigcup_X \downarrow & & & & \downarrow \bigcup_{FX} \\ F\mathcal{P}X & & \xrightarrow{\lambda_X} & & \mathcal{P}FX \end{array}$$

This says that the natural transformation $\lambda : F\mathcal{P} \Rightarrow \mathcal{P}F$ is a *distributive law.*[4]

Lemma 4.4 ([Jac04b]). *The maps λ_X thus defined form a distributive law of a functor F over a monad $\mathcal{P}$. It is called the "power law".* □

Example 4.5. For the functor $F = 1 + \Sigma \times -$, where $1 = \{\checkmark\}$, the lifted membership relation is as follows.

$$\mathrm{Rel}_{1+\Sigma\times-}(\in_X) = \{(\checkmark, \checkmark)\} \cup \{((a,x),(a,u)) \mid a \in \Sigma, x \in u\} \ .$$

For the functor $F = (\Sigma + -)^*$, the lifted membership relation $\mathrm{Rel}_{(\Sigma+-)^*}(\in_X)$ between $(\Sigma + X)^*$ and $(\Sigma + \mathcal{P}X)^*$ is described concretely as follows: a pair $\langle c_1c_2\ldots c_m, d_1d_2\ldots d_m\rangle$ belongs to $\mathrm{Rel}_{(\Sigma+-)^*}(\in_X)$ if and only if for each $i = 1, 2, \ldots, m$,

[4] The use of a distributive law in coalgebraic settings is investigated elaborately in [Bar04].

- if $c_i \in \Sigma$ then d_i is also from Σ and $c_i = d_i$;
- if $c_i \in X$ then d_i is in $\mathcal{P}X$ and $c_i \in d_i$.

The distributive law $\lambda : F\mathcal{P} \Rightarrow \mathcal{P}F$ gives rise to a lifting $F : \mathbf{Sets}_{\mathcal{P}} \to \mathbf{Sets}_{\mathcal{P}}$ of a functor F in the Kleisli category by

$$F : \quad \left(X \xrightarrow{\;f\;} Y \right) \quad \mapsto \quad \left(FX \xrightarrow{\;\lambda_Y \circ Ff\;} FY \right) .$$

In the sequel we identify the category **Rel** with the Kleisli category $\mathbf{Sets}_{\mathcal{P}}$ of the powerset monad. It is justified by the following straightforward observation.

Lemma 4.6. *The category* **Rel** *of sets and relations is isomorphic to the Kleisli category* $\mathbf{Sets}_{\mathcal{P}}$ *via the relation-into-function correspondence.*

Moreover, let F be an endofunctor in **Sets** *which preserves weak pullbacks. Then the canonical lifting of F in* **Rel** *in the sense of [CKW91], which maps an arrow $R : X \to Y$ to $\mathrm{Rel}_F(R) : FX \to FY$, coincides with the lifting of F in* $\mathbf{Sets}_{\mathcal{P}}$ *defined above via the distributive law.* □

Remark 4.7. As is already noted, working in the Kleisli category $\mathbf{Sets}_{\mathcal{P}}$ makes it easier to generalize to other monads than $\mathcal{P}$. A similar finality result holds for the *subdistribution monad* $\mathcal{D}$ such that $\mathcal{D}X = \{d : X \to [0,1] \mid \sum_{x \in X} d(x) \leq 1\}$. In that case we do not have the counterpart of the notion of relation lifting but start with a distributive law $F\mathcal{D} \Rightarrow \mathcal{D}F$. Details will be published later.

4.3 Contravariant Powerset Functor

We use the contravariant powerset functor $\overline{\mathcal{P}}$ in our construction. The following properties are used there.

Lemma 4.8. *1. For a mono $m : X \rightarrowtail Y$, $\mathcal{P}m$ is a split mono with its left inverse $\overline{\mathcal{P}}m$, i.e. $\overline{\mathcal{P}}m \circ \mathcal{P}m = \mathrm{id}_{\mathcal{P}X}$.*
2. For an iso $i : X \xrightarrow{\cong} Y$, $\mathcal{P}i$ is again an iso with inverse $\overline{\mathcal{P}}i$.
3. The union maps $\bigcup_X : \mathcal{P}^2X \to \mathcal{P}X$ form a natural transformation $\mathcal{P}\overline{\mathcal{P}} \Rightarrow \overline{\mathcal{P}}$.
4. For each n, the maps $\lambda^n_X : F^n\mathcal{P}X \to \mathcal{P}F^nX$ in Lemma 4.9 form a natural transformation $F^n\overline{\mathcal{P}} \Rightarrow \overline{\mathcal{P}}F^n$.

Proof. See Appendix A.2. □

4.4 Construction of Finite Trace via Composition of Coalgebra

In the construction of the finite trace, we use the *n-fold composition* $c^n : X \to \mathcal{P}F^nX$ of a coalgebra $c : X \to \mathcal{P}FX$ in **Sets**. Intuitively, one transition of c^n corresponds to n successive transitions of the original coalgebra c. It is defined inductively on n as follows.

$$c^0 = \{-\}_X \;, \qquad c^{n+1} = \left(X \xrightarrow{\;c\;} \mathcal{P}FX \xrightarrow{\;\mathcal{P}Fc^n\;} \mathcal{P}F\mathcal{P}F^nX \xrightarrow{\;\mathcal{P}\lambda_{F^nX}\;} \mathcal{P}^2F^{n+1}X \xrightarrow{\;\bigcup_{F^{n+1}X}\;} \mathcal{P}F^{n+1}X \right) .$$

The next observation is basic for the n-fold composition of a coalgebra.

Lemma 4.9 ([Wor]). *The distributive law $\lambda : F\mathcal{P} \Rightarrow \mathcal{P}F$ extends to n-fold distributive law $\lambda^n : F^n\mathcal{P} \Rightarrow \mathcal{P}F^n$ in the following way.*

$$\lambda^0_X = \mathrm{id}_{\mathcal{P}X}\ , \qquad \begin{array}{ccc} F^{n+1}\mathcal{P}X & \xrightarrow{F^n\lambda_X} & F^n\mathcal{P}FX \\ & \searrow^{\lambda^{n+1}_X} & \big\downarrow \lambda^n_{FX} \\ & & \mathcal{P}F^{n+1}X \end{array} .$$

Let $c : X \to \mathcal{P}FX$ be a coalgebra in **Sets**. *For each n, m the following diagrams commute.*

$$\begin{array}{ccc} F^{n+m}\mathcal{P}X & \xrightarrow{F^n\lambda^m_X} & F^n\mathcal{P}F^mX \\ {\scriptstyle F^m\lambda^n_X}\big\downarrow & \searrow^{\lambda^{n+m}_X} & \big\downarrow{\scriptstyle \lambda^n_{F^mX}} \\ F^m\mathcal{P}F^nX & \xrightarrow[\lambda^m_{F^nX}]{} & \mathcal{P}F^{n+m}X \end{array} \qquad \begin{array}{ccccc} X & \xrightarrow{c^n} & \mathcal{P}F^nX & \xrightarrow{\mathcal{P}F^nc^m} & \mathcal{P}F^n\mathcal{P}F^mX \\ & & & & \big\downarrow \mathcal{P}\lambda^n_{F^mX} \\ & & & & \mathcal{P}^2F^{n+m}X \\ & & \searrow^{c^{n+m}} & & \big\downarrow \bigcup_{F^{n+m}X} \\ & & & & \mathcal{P}F^{n+m}X \end{array}$$

Proof. By induction. □

Now we are ready to prove our main technical result.

Theorem 4.10 (Finite trace semantics for coalgebras, Theorem 1.1). *Let F be a shapely functor, and $\alpha : FA \xrightarrow{\cong} A$ be the initial F-algebra in* **Sets**. *The coalgebra $\{-\}_{FA} \circ \alpha^{-1} : A \xrightarrow{\cong} FA$ in* $\mathbf{Sets}_{\mathcal{P}}$ *is final for the lifted functor F.*

Proof. Given a coalgebra $c : X \to FX$ in $\mathbf{Sets}_{\mathcal{P}}$, we construct an arrow $\mathsf{ft}_c : X \to A$, and show that it is the unique arrow which makes the diagram in $\mathbf{Sets}_{\mathcal{P}}$ on the left (equivalently, the diagram in **Sets** on the right) commute.

$$\begin{array}{ccc} FX & \overset{F\mathsf{ft}_c}{\dashrightarrow} & FA \\ {\scriptstyle c}\big\uparrow & & \big\uparrow{\scriptstyle \{-\}_{FA}\circ\alpha^{-1}} \\ X & \underset{\mathsf{ft}_c}{\dashrightarrow} & A \end{array} \ , \qquad \begin{array}{ccc} \mathcal{P}FX & \overset{\mathcal{P}F\mathsf{ft}_c}{\dashrightarrow} & \mathcal{P}F\mathcal{P}A \\ & & \big\downarrow \mathcal{P}\lambda_A \\ & & \mathcal{P}^2FA \\ {\scriptstyle c}\big\uparrow & & \big\downarrow \bigcup_{FA} \\ & & \mathcal{P}FA \\ & & \cong\big\downarrow \mathcal{P}\alpha \\ X & \underset{\mathsf{ft}_c}{\dashrightarrow} & \mathcal{P}A \end{array} . \qquad (2)$$

In the rest of the proof we work in the category **Sets**.

As is stated in Lemma 4.2, the initial F-algebra in **Sets** for shapely F is obtained via the initial sequence $0 \to F0 \to F^20 \to \cdots$ as follows.

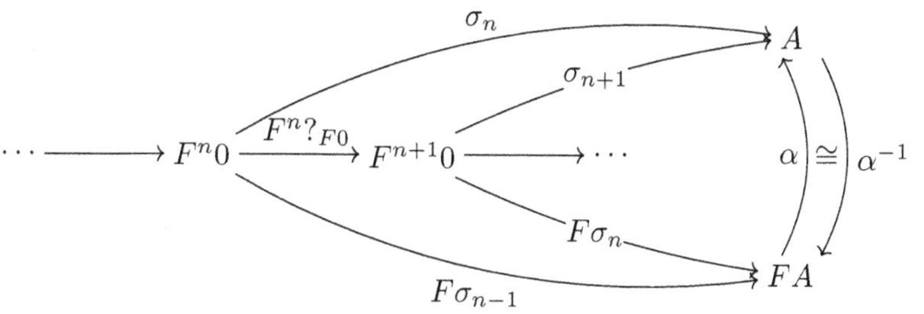

(3)

The cocone $\{\sigma_n : F^n 0 \to A\}_{n<\omega}$ is by construction the colimit of the initial sequence. Since a shapely F preserves ω-colimits the cocone $\{?_{FA} : 0 \to FA\} \cup \{F\sigma_n : F^{n+1}0 \to FA\}_{n<\omega}$ is again a colimit, yielding the initial algebra α as the mediating iso arrow. Lemma 4.2.2 shows that each σ_n is mono.

We define the *n-th trace* $\mathsf{trace}_c^n : X \to \mathcal{P}A$ of c by the following composite. The n-th trace $\mathsf{trace}_c^n(x) \subseteq A$ is understood as the set of behavior of x which terminates within n steps.

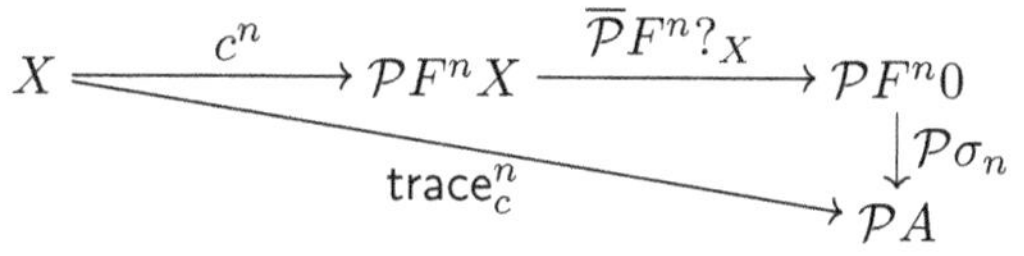

For n-th traces the following equality holds, which says that all behavior within n steps are already included in trace^n. For $n \le m$,

$$\mathrm{Im}\,\sigma_n \cap \mathsf{trace}_c^m(x) = \mathsf{trace}_c^n(x) \ , \tag{4}$$

where $\mathrm{Im}\,\sigma_n$ is the direct image $\sigma_n[F^n 0]$. The proof is given in Appendix A.3.

Finally, we define the finite trace $\mathsf{ft}_c : X \to \mathcal{P}A$ of c as the union of n-th traces: for each $x \in X$,

$$\mathsf{ft}_c(x) \stackrel{\text{def}}{=} \bigcup_{n<\omega} \mathsf{trace}_c^n(x) \ .$$

By the equality (4) we have another characterization of $\mathsf{ft}_c(x)$: for each n and $t_n \in F^n 0$, $\sigma_n(t_n) \in \mathsf{ft}_c(x)$ if and only if $\sigma_n(t_n) \in \mathsf{trace}_c^n(x)$. Hence, by Lemma 4.8.1, for each n we have the following equality of functions $X \to F^n 0$.

$$\overline{\mathcal{P}}\sigma_n \circ \mathsf{ft}_c = \overline{\mathcal{P}}\sigma_n \circ \mathsf{trace}_c^n = \overline{\mathcal{P}}F^n ?_X \circ c^n \ . \tag{5}$$

In the following Lemmas 4.11 and 4.12 we show that the arrow ft_c thus constructed is indeed the unique arrow that makes the diagram (2) commute. □

Lemma 4.11. *The arrow* $\mathsf{ft}_c : X \to \mathcal{P}A$ *in* **Sets**, *as defined in the proof of Theorem 4.10, makes the diagram (2) commute.*

Proof. By the construction of the initial algebra as the colimit (i.e. coequalizer of coproduct), it suffices to prove that: for each $n < \omega$ and $t_n \in F^n 0$,

$$\sigma_n(t_n) \in \mathsf{ft}_c(x) \iff \sigma_n(t_n) \in (\mathcal{P}\alpha \circ \bigcup_{FA} \circ \mathcal{P}\lambda_A \circ \mathcal{P}F\mathsf{ft}_c \circ c)(x) \ .$$

When $n = 0$, we have $F^n 0 = 0$ hence the equivalence trivially holds. When $n > 0$, we proceed as follows.

$$\begin{aligned}
&\sigma_n(t_n) \in (\mathcal{P}\alpha \circ \bigcup\nolimits_{FA} \circ \mathcal{P}\lambda_A \circ \mathcal{P}F\mathsf{ft}_c \circ c)(x) \\
&\iff t_n \in (\overline{\mathcal{P}}\sigma_n \circ \mathcal{P}\alpha \circ \bigcup\nolimits_{FA} \circ \mathcal{P}\lambda_A \circ \mathcal{P}F\mathsf{ft}_c \circ c)(x) \\
&\iff t_n \in (\overline{\mathcal{P}}\sigma_n \circ \overline{\mathcal{P}}\alpha^{-1} \circ \bigcup\nolimits_{FA} \circ \mathcal{P}\lambda_A \circ \mathcal{P}F\mathsf{ft}_c \circ c)(x) \\
&\qquad\qquad (\mathcal{P}\alpha = (\mathcal{P}\alpha^{-1})^{-1} = \overline{\mathcal{P}}\alpha^{-1} \text{ by Lemma 4.8.2}) \\
&\iff t_n \in (\overline{\mathcal{P}}F\sigma_{n-1} \circ \bigcup\nolimits_{FA} \circ \mathcal{P}\lambda_A \circ \mathcal{P}F\mathsf{ft}_c \circ c)(x) \qquad (\alpha^{-1} \circ \sigma_n = F\sigma_{n-1}) \\
&\iff t_n \in (\bigcup\nolimits_{F^n 0} \circ \mathcal{P}\lambda_{F^{n-1}0} \circ \mathcal{P}F\overline{\mathcal{P}}\sigma_{n-1} \circ \mathcal{P}F\mathsf{ft}_c \circ c)(x) \qquad \text{(Lemma 4.8.3,4)} \\
&\iff t_n \in (\bigcup\nolimits_{F^n 0} \circ \mathcal{P}\lambda_{F^{n-1}0} \circ \mathcal{P}F\overline{\mathcal{P}}F^{n-1}?_X \circ \mathcal{P}Fc^{n-1} \circ c)(x) \qquad \text{(By (5))} \\
&\iff t_n \in (\overline{\mathcal{P}}F^n?_X \circ \bigcup\nolimits_{F^n X} \circ \mathcal{P}\lambda_{F^{n-1}X} \circ \mathcal{P}Fc^{n-1} \circ c) \qquad \text{(Lemma 4.8.4,3)} \\
&\iff t_n \in (\overline{\mathcal{P}}F^n?_X \circ c^n) \qquad \text{(Definition of } c^n) \\
&\iff \sigma_n(t_n) \in \mathsf{ft}_c(x) \ . \qquad \text{(By (5))}
\end{aligned}$$

This concludes the proof. □

Lemma 4.12. *If an arrow* $f : X \to \mathcal{P}A$ *in* **Sets** *makes the diagram (2) commute in place of* ft_c*, then* f *is equal to* ft_c *as defined in the proof of Theorem 4.10.*

Proof. It suffices to show that

$$\overline{\mathcal{P}}\sigma_n \circ f = \overline{\mathcal{P}}\sigma_n \circ \mathsf{ft}_c \ , \qquad (6)$$

since, if it holds, for each $x \in X$, $n < \omega$ and $t_n \in F^n 0$ we have

$$\sigma_n(t_n) \in f(x) \iff \sigma_n(t_n) \in \mathsf{ft}_c(x)$$

which yields the lemma. We show (6) by induction on n.

When $n = 0$ the claim trivially holds. For $n + 1$,

$$\begin{aligned}
\overline{\mathcal{P}}\sigma_{n+1} \circ f &= \overline{\mathcal{P}}\sigma_{n+1} \circ \mathcal{P}\alpha \circ \bigcup\nolimits_{FA} \circ \mathcal{P}\lambda_A \circ \mathcal{P}Ff \circ c \\
&\qquad\qquad (f \text{ makes the diagram (2) commute}) \\
&= \bigcup\nolimits_{F^{n+1}0} \circ \mathcal{P}\lambda_{F^n 0} \circ \mathcal{P}F\overline{\mathcal{P}}\sigma_n \circ \mathcal{P}Ff \circ c \\
&\qquad\qquad \text{(As in the proof of Lemma 4.11)} \\
&= \bigcup\nolimits_{F^{n+1}0} \circ \mathcal{P}\lambda_{F^n 0} \circ \mathcal{P}F\overline{\mathcal{P}}\sigma_n \circ \mathcal{P}F\mathsf{ft}_c \circ c \\
&\qquad\qquad (\overline{\mathcal{P}}\sigma_n \circ f = \overline{\mathcal{P}}\sigma_n \circ \mathsf{ft}_c \text{ by induction hypothesis}) \\
&= \overline{\mathcal{P}}\sigma_{n+1} \circ \mathsf{ft}_c \ . \quad \text{(Same calculation as above, but now backwards)}
\end{aligned}$$

This concludes the proof. □

Example 4.13 (Non-deterministic automata). We continue from Example 1.2. For the functor $F = 1 + \Sigma \times -$, the commutation of the diagram (2) amounts to the following conditions.

$$\begin{aligned} \langle\rangle \in \mathsf{ft}_c(x) &\iff \checkmark \in c(x) \ , \\ \mathsf{cons}(a,s) \in \mathsf{ft}_c(x) &\iff \exists x' \in X. \quad (a,x') \in c(x) \ \wedge \ s \in \mathsf{ft}_c(x') \ . \end{aligned}$$

These conditions indeed (corecursively) characterize the language $\mathsf{ft}_c(x)$ accepted by the non-deterministic automaton c when we start from x.

Bartels [Bar04] gives an alternative characterization of the accepted language, using a different distributive law. The precise relationship with our work is yet to be determined.

Example 4.14 (Context-free grammar). We continue from Example 1.3. For the functor $F = (\Sigma + -)^*$, the commutation of the diagram (2) amounts to the following conditions. For each element

of $\mathsf{ft}_c(x)$ (here $c_1, c_2, \ldots, c_n \in \Sigma + \Sigma^{\triangle}$), there exists a string $\langle d_1, d_2, \ldots, d_n \rangle \in c(x)$ such that for each i:

- if $c_i \in \Sigma$ then d_i is also in Σ and $c_i = d_i$;
- if $c_i \in \Sigma^{\triangle}$ then d_i is in X and $c_i \in \mathsf{ft}_c(d_i)$.

Hence we obtain the set of finite SPTs generated by c from x as $\mathsf{ft}_c(x)$ via finality in $\mathbf{Sets}_{\mathcal{P}}$.

5 (Possibly Infinite) Trace Semantics

In this section we relate our current work to earlier work [Jac04b], where the final coalgebra in **Sets** gives rise to a weakly final coalgebra in **Rel**.

Theorem 5.1 (Main result of [Jac04b]). *Let F be a shapely functor, and $\zeta : Z \overset{\cong}{\to} FZ$ be the final coalgebra in* **Sets**.

1. *The coalgebra $\mathsf{graph}(\zeta) : Z \to FZ$ is weakly final for the lifted functor F in* **Rel**. *That is, given a coalgebra $c : X \to FX$, there exists a (not necessarily unique) relation $t : X \to Z$ that makes the following diagrams commute.*

$$\begin{array}{ccc} FX & \overset{Ft}{\rightsquigarrow} & FZ \\ {\scriptstyle c}\uparrow & & \uparrow{\scriptstyle \mathsf{graph}(\zeta)} \\ X & \underset{t}{\rightsquigarrow} & Z \end{array} \qquad (7)$$

2. *There is a canonical choice* mt_c *(maximum trace) of a trace of* c*, namely the maximum one with respect to the inclusion order.* □

It turns out that the finite trace of a coalgebra gives rise to the smallest trace via canonical embedding $\iota : A \rightarrowtail Z$.

Corollary 5.2. *Let* F *be a shapely functor, and* $c : X \to \mathcal{P}FX$ *be a coalgebra in* **Sets**.

1. *Each trace* t *of* c *gives rise to the finite trace of* c *by* $X \xrightarrow{t} \mathcal{P}Z \xrightarrow{\overline{\mathcal{P}}\iota} \mathcal{P}A$ *in* **Sets**.
2. *The finite trace* ft_c *gives rise to a trace of* c *by* $X \xrightarrow{\mathsf{ft}_c} \mathcal{P}A \xrightarrow{\mathcal{P}\iota} \mathcal{P}Z$ *in* **Sets**. *Moreover, this trace is the smallest among the traces of* c.

Proof. A trace induces the finite trace, and vice versa, since the following diagram in **Sets** commutes. For the former take the three squares on the left and put them on the right of the definition of a trace, and for the latter take those on the right.

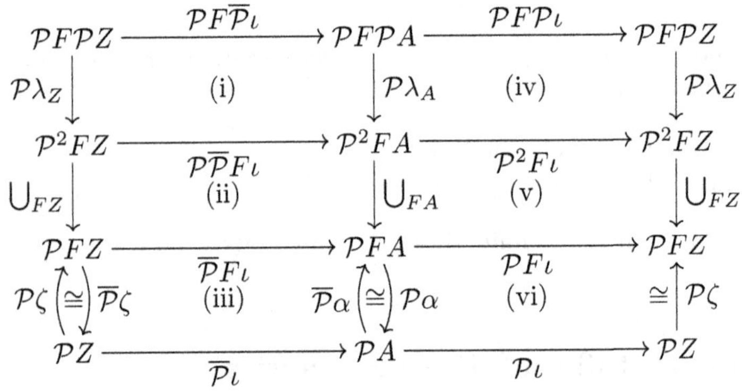

Square (i) commutes by Lemma 4.8.4, (ii) by Lemma 4.8.3, (iii) is the definition of ι mapped by $\overline{\mathcal{P}}$, (iv) commutes by naturality of λ, (v) by naturality of $\bigcup$, and (vi) is the definition of ι.

It remains to be shown that the trace $\mathcal{P}\iota \circ \mathsf{ft}_c$ is the smallest trace. Take an arbitrary trace $t : X \to \mathcal{P}Z$ of c. It induces the finite trace by $\overline{\mathcal{P}}\iota \circ t$, and by Theorem 4.10 (uniqueness of the finite trace) we have $\mathsf{ft}_c = \overline{\mathcal{P}}\iota \circ t$. Since in general $(\mathcal{P}f \circ \overline{\mathcal{P}}f)(u) \subseteq u$ holds, we have $\mathcal{P}\iota \circ \mathsf{ft}_c = \mathcal{P}\iota \circ \overline{\mathcal{P}}\iota \circ t \subseteq t$. □

6 Conclusions and Future Work

We have presented that under suitable mild restrictions the initial algebra in **Sets** gives rise to the final coalgebra in **Rel**. The relation induced by the finality in **Rel** extracts the set of finite behavior of non-deterministic systems. The technical result is applied to non-deterministic automata and the first coalgebraic account of context-free grammars/languages. The (co)algebraic and monadic structures on strings and skeletal parse trees have been also elaborated.

The well-known relationship between context-free languages and *pushdown automata* (see e.g.[LP81]) would be an interesting topic to consider from a coalgebraic perspective. So is the problem of *parsing*, which is a partial inverse of the flattening function φ_Σ in Section 3.

As mentioned in Remark 4.7 we are now applying the current approach to another monad than $\mathcal{P}$, namely the subdistribution monad.

Acknowledgements

Thanks are due to Ana (Sokolova) Woracek and anonymous referees for valuable comments and discussion.

References

[AAMV03] P. Aczel, J. Adámek, S. Milius, and J. Velebil. Infinite trees and completely iterative theories: a coalgebraic view. *Theor. Comp. Sci.*, 300:1–45, 2003.

[AK95] J. Adámek and V. Koubek. On the greatest fixed point of a set functor. *Theor. Comp. Sci.*, 150(1):57–75, 1995.

[ASU86] A.V. Aho, R. Sethi, and J.D. Ullman. *Compilers: Principles, Techniques, and Tools*. Addison-Wesley series in Computer Science. Addison-Wesley, 1986.

[Bar93] M. Barr. Terminal coalgebras in well-founded set theory. *Theor. Comp. Sci.*, 114:299–315, 1993.

[Bar04] F. Bartels. *On Generalized Coinduction and Probabilistic Specification Formats: Distributive Laws in Coalgebraic Modelling*. PhD thesis, Free Univ. Amsterdam, 2004.

[BW83] M. Barr and C. Wells. *Toposes, Triples and Theories*. Springer-Verlag, 1983. Available free for downloading at `http://www.cwru.edu/artsci/math/wells/pub/ttt.html`.

[Cho56] N. Chomsky. Three models for the description of language. *IRE Transactions on Information Theory*, 2:113–124, 1956.

[CKW91] A. Carboni, G. Kelly, and R. Wood. A 2-categorical approach to change of base and geometric morphisms I. *Cah. de Top. et Géom. Diff.*, 32(1):47–95, 1991.

[Jac04a] B. Jacobs. Relating two approaches to coinductive solution of recurisve equations. In *Coalgebraic Methods in Computer Science (CMCS 2004)*, volume 106 of *Elect. Notes in Theor. Comp. Sci.* Elsevier, Amsterdam, 2004.

[Jac04b] B. Jacobs. Trace semantics for coalgebras. In *Coalgebraic Methods in Computer Science (CMCS 2004)*, volume 106 of *Elect. Notes in Theor. Comp. Sci.* Elsevier, Amsterdam, 2004.

[Jac05a] B. Jacobs. A bialgebraic review of regular expressions, deterministic automata and languages. Techn. Rep. NIII-R05003, Inst. for Computing and Information Sciences, Radboud Univ. Nijmegen, 2005.

[Jac05b] B. Jacobs. Introduction to coalgebra. Towards mathematics of states and observations. Draft of a book, `www.cs.ru.nl/B.Jacobs/PAPERS/index.html`, 2005.

[Jay95] C. Jay. A semantics for shape. *Science of Comput. Progr.*, 25:251–283, 1995.

[LP81] H.R. Lewis and C.H. Papadimitriou. *Elements of the Theory of Computation.* Prentice-Hall, 1981.

[Rut00] J. Rutten. Universal coalgebra: a theory of systems. *Theor. Comp. Sci.*, 249:3–80, 2000.

[Rut03] J. Rutten. Behavioural differential equations: a coinductive calculus of streams, automata, and power series. *Theor. Comp. Sci.*, 308:1–53, 2003.

[Wor] A. (Sokolova) Woracek. Personal communication.

A Appendix

A.1 Functor/Monad Structure of $(-)^{\triangle}$ and $(-)^{\wedge}$

On a function $f : \Sigma \to \Phi$, the action of $(-)^{\triangle}$ and $(-)^{\wedge}$ is obtained as follows.

$$\begin{array}{ccc}
(\Sigma+\Sigma^{\triangle})^* & \xrightarrow{(\Sigma+f^{\triangle})^*} & (\Sigma+\Phi^{\triangle})^* \\
 & & \downarrow (f+\Phi^{\triangle}) \\
\alpha_\Sigma \downarrow \cong & & (\Phi+\Phi^{\triangle})^* \\
 & & \cong \downarrow \alpha_\Phi \\
\Sigma^{\triangle} & \xrightarrow[f^{\triangle}]{} & \Phi^{\triangle}
\end{array}
\quad\text{and}\quad
\begin{array}{ccc}
(\Phi+\Sigma^{\wedge})^* & \xrightarrow{(\Phi+f^{\wedge})^*} & (\Phi+\Phi^{\wedge})^* \\
(f+\Sigma^{\wedge})^* \uparrow & & \\
(\Sigma+\Sigma^{\wedge})^* & & \cong \uparrow \zeta_\Phi \\
\zeta_\Sigma \uparrow \cong & & \\
\Sigma^{\wedge} & \xrightarrow[f^{\wedge}]{} & \Phi^{\wedge}
\end{array}$$

The monad structure is constructed as follows.

$$\begin{array}{ccc}
\Sigma & \xrightarrow{\kappa_1} & \Sigma+\Sigma^{\triangle} \\
 & & \downarrow \overset{*}{\eta}_{\Sigma+\Sigma^{\triangle}} \\
 & & (\Sigma+\Sigma^{\triangle})^* \\
 & & \cong \downarrow \alpha_\Sigma \\
 & \overset{\triangle}{\eta}_\Sigma \searrow & \Sigma^{\triangle}
\end{array}
\qquad
\begin{array}{ccc}
(\Sigma+\Sigma)^* & \xrightarrow{(\Sigma+\hat{\eta}_\Sigma)^*} & (\Sigma+\Sigma^{\wedge})^* \\
\overset{*}{\eta}_{\Sigma+\Sigma} \uparrow & & \\
\Sigma+\Sigma & & \cong \uparrow \zeta_\Sigma \\
\kappa_1 \uparrow & & \\
\Sigma & \xrightarrow[\hat{\eta}_\Sigma]{} & \Sigma^{\wedge}
\end{array}$$

$$\begin{array}{ccc}
(\Sigma^{\triangle}+\Sigma^{\triangle\triangle})^* & \xrightarrow{(\Sigma^{\triangle}+\overset{\triangle}{\mu}_\Sigma)^*} & (\Sigma^{\triangle}+\Sigma^{\triangle})^* \\
 & & \downarrow [\alpha_\Sigma^{-1}, \overset{*}{\eta}_{\Sigma+\Sigma^{\triangle}} \circ \kappa_2]^* \\
 & & (\Sigma+\Sigma^{\triangle})^{**} \\
\cong \downarrow \alpha_{\Sigma^{\triangle}} & & \downarrow \overset{*}{\mu}_{\Sigma+\Sigma^{\triangle}} \\
 & & (\Sigma+\Sigma^{\triangle})^* \\
 & & \cong \downarrow \alpha_\Sigma \\
\Sigma^{\triangle\triangle} & \xrightarrow[\overset{\triangle}{\mu}_\Sigma]{} & \Sigma^{\triangle}
\end{array}$$

The definition of $\hat{\mu}$ is rather complicated. Let a_Σ be the following composite on the left.

$$
\begin{array}{ccccc}
\Sigma^{\wedge\wedge} & \xrightarrow[\cong]{\zeta_{\Sigma^{\wedge}}} & (\Sigma^{\wedge}+\Sigma^{\wedge\wedge})^* & \xrightarrow[\cong]{(\zeta_\Sigma+\Sigma^{\wedge\wedge})^*} & ((\Sigma+\Sigma^{\wedge})^*+\Sigma^{\wedge\wedge})^* \\
{\scriptstyle a_\Sigma}\downarrow & & & & \downarrow{\scriptstyle [\kappa_1^*,\, \overset{*}{\eta}_{(\Sigma+\Sigma^{\wedge})+\Sigma^{\wedge\wedge}} \circ \kappa_2]^*} \\
(\Sigma+(\Sigma^{\wedge}+\Sigma^{\wedge\wedge}))^* & \xleftarrow[{[\Sigma+\kappa_1,\kappa_2\circ\kappa_2]^*}]{} & ((\Sigma+\Sigma^{\wedge})+\Sigma^{\wedge\wedge})^* & \xleftarrow[\overset{*}{\mu}_{(\Sigma+\Sigma^{\wedge})+\Sigma^{\wedge\wedge}}]{} & ((\Sigma+\Sigma^{\wedge})+\Sigma^{\wedge\wedge})^{**}
\end{array}
$$

This map a_Σ is used in the coalgebraic structure map on the left in:

$$
\begin{array}{ccc}
(\Sigma+(\Sigma^{\wedge}+\Sigma^{\wedge\wedge}))^* & \overset{(\Sigma+b_\Sigma)^*}{\dashrightarrow} & (\Sigma+\Sigma^{\wedge})^* \\
{\scriptstyle [(\Sigma+\kappa_1)^*\circ\zeta_\Sigma,\, a_\Sigma]}\uparrow & & \cong\uparrow{\scriptstyle \zeta_\Sigma} \\
\Sigma^{\wedge}+\Sigma^{\wedge\wedge} & \underset{b_\Sigma}{\dashrightarrow} & \Sigma^{\wedge} \\
{\scriptstyle \kappa_2}\uparrow & \nearrow{\scriptstyle \hat{\mu}_\Sigma} & \\
\Sigma^{\wedge\wedge} & &
\end{array}
\quad .
$$

By finality one easily obtains $b_\Sigma \circ \kappa_1 = \Sigma^{\wedge}$.

A.2 Proof of Lemma 4.8

Points 1 and 2 are straightforward. Point 3 is equivalent to saying that union is preserved by taking an inverse image. For Point 4, we show the proof for $n = 1$. The case for general n is easy by induction. Let $s \in FX$ and $r \in F\overline{\mathcal{P}}Y$. Then

$$
\begin{aligned}
s \in (\lambda_X \circ F\overline{\mathcal{P}}f)(r) &\iff (\, s,\, (F\overline{\mathcal{P}}f)(r)\,) \in \mathrm{Rel}_F(\in_X) && \text{(Definition of } \lambda) \\
&\iff (s,r) \in (\mathrm{id}\times F\overline{\mathcal{P}}f)^{-1}\,\mathrm{Rel}_F(\in_X) && \\
&\iff (s,r) \in \mathrm{Rel}_F\big((\mathrm{id}\times\overline{\mathcal{P}}f)^{-1}(\in_X)\big) && \text{(Lemma 4.3.3)} \\
&\iff (s,r) \in \mathrm{Rel}_F\big((f\times\mathrm{id})^{-1}(\in_Y)\big) && (\dagger\text{, see below}) \\
&\iff (\,(Ff)(s), r\,) \in \mathrm{Rel}_F\,(\in_Y) && \text{(Lemma 4.3.3)} \\
&\iff (Ff)(s) \in \lambda_Y(r) && \text{(Definition of } \lambda) \\
&\iff s \in (\overline{\mathcal{P}}Ff \circ \lambda_Y)(r)\ ,
\end{aligned}
$$

where (†) holds because

$$
\begin{aligned}
(x,u) \in (\mathrm{id}\times\overline{\mathcal{P}}f)^{-1}(\in_X) &\iff x \in (\overline{\mathcal{P}}f)(u) \\
&\iff f(x) \in u \\
&\iff (x,u) \in (f\times\mathrm{id})^{-1}(\in_Y)\ . \qquad \square
\end{aligned}
$$

A.3 Proof of Theorem 4.10

First we show that, for each n,

$$
\mathrm{Im}\,\sigma_n \cap \mathsf{trace}_c^{n+1}(x) = \mathsf{trace}_c^{n}(x)\ . \tag{8}
$$

It is proved as follows.

$$
\begin{aligned}
&\mathrm{Im}\,\sigma_n \cap \mathsf{trace}_c^{n+1}(x) \\
&= (\mathcal{P}\sigma_n \circ \overline{\mathcal{P}}\sigma_n \circ \mathsf{trace}_c^{n+1})(x) \\
&= (\mathcal{P}\sigma_n \circ \overline{\mathcal{P}}F^n?_{F0} \circ \overline{\mathcal{P}}\sigma_{n+1} \circ \mathsf{trace}_c^{n+1})(x) \qquad (\sigma_n = \sigma_{n+1} \circ F^n?_{F0} \text{ by (3)}) \\
&= (\mathcal{P}\sigma_n \circ \overline{\mathcal{P}}F^n?_{F0} \circ \overline{\mathcal{P}}F^{n+1}?_X \circ c^{n+1})(x) \\
&\qquad \text{(Definition of } \mathsf{trace}_c^{n+1}\text{, and } \overline{\mathcal{P}}\sigma_n \circ \mathcal{P}\sigma_n = \mathrm{id} \text{ by Lemma 4.8.1)} \\
&= (\mathcal{P}\sigma_n \circ \overline{\mathcal{P}}F^n?_{FX} \circ c^{n+1})(x) \qquad (F?_X \circ ?_{F0} = ?_{FX}) \\
&= (\mathcal{P}\sigma_n \circ \overline{\mathcal{P}}F^n?_{FX} \circ \textstyle\bigcup_{F^{n+1}X} \circ \mathcal{P}\lambda^n_{FX} \circ \mathcal{P}F^n c \circ c^n)(x) \\
&= (\mathcal{P}\sigma_n \circ \textstyle\bigcup_{F^n 0} \circ \mathcal{P}\lambda^n_0 \circ \mathcal{P}F^n\overline{\mathcal{P}}?_{FX} \circ \mathcal{P}F^n c \circ c^n)(x) \qquad \text{(Lemma 4.8.3,4)} \\
&= (\mathcal{P}\sigma_n \circ \textstyle\bigcup_{F^n 0} \circ \mathcal{P}\lambda^n_0 \circ \mathcal{P}F^n\overline{\mathcal{P}}?_X \circ \mathcal{P}F^n\{-\}_X \circ c^n)(x) \\
&\qquad (\overline{\mathcal{P}}?_{FX} \circ c = \overline{\mathcal{P}}?_X \circ \{-\}_X : X \to \mathcal{P}0\text{, with terminal codomain } 1 = \mathcal{P}0) \\
&= (\mathcal{P}\sigma_n \circ \overline{\mathcal{P}}F^n?_X \circ \textstyle\bigcup_{F^n X} \circ \mathcal{P}\lambda^n_X \circ \mathcal{P}F^n\{-\}_X \circ c^n)(x) \qquad \text{(Lemma 4.8.4,3)} \\
&= (\mathcal{P}\sigma_n \circ \overline{\mathcal{P}}F^n?_X \circ \textstyle\bigcup_{F^n X} \circ \mathcal{P}\{-\}_{F^n X} \circ c^n)(x) \\
&\qquad (\lambda^n \text{ is compatible with the unit of } \mathcal{P}) \\
&= (\mathcal{P}\sigma_n \circ \overline{\mathcal{P}}F^n?_X \circ c^n)(x) \qquad \text{(Unit law of the monad } \mathcal{P}) \\
&= \mathsf{trace}_c^n(x) \ .
\end{aligned}
$$

Obviously the sequence $\{\mathrm{Im}\,\sigma_n\}_{n<\omega}$ of subsets of A is increasing, since $\sigma_n = \sigma_{n+1} \circ F^n?_{F0}$. Now, for arbitrary $n \leq m$,

$$
\begin{aligned}
\mathrm{Im}\,\sigma_n \cap \mathsf{trace}_c^m(x) &= \mathrm{Im}\,\sigma_n \cap \mathrm{Im}\,\sigma_{n+1} \cap \cdots \cap \mathrm{Im}\,\sigma_{m-1} \cap \mathsf{trace}_c^m(x) \\
&\qquad (\text{Since } \mathrm{Im}\,\sigma_n \subseteq \mathrm{Im}\,\sigma_{n+1} \subseteq \cdots \subseteq \mathrm{Im}\,\sigma_{m-1}) \\
&= \mathrm{Im}\,\sigma_n \cap \mathrm{Im}\,\sigma_{n+1} \cap \cdots \cap \mathsf{trace}_c^{m-1}(x) \qquad \text{(By (8))} \\
&= \cdots \\
&= \mathsf{trace}_c^n(x) \ .
\end{aligned}
$$

□

Towards a Coalgebraic Semantics of the Ambient Calculus

Daniel Hausmann, Till Mossakowski, and Lutz Schröder

BISS, Dept. of Computer Science, University of Bremen

Abstract. Recently, various process calculi have been introduced which are suited for the modelling of mobile computation and in particular the mobility of program code; a prominent example is the ambient calculus. Due to the complexity of the involved spatial reduction, there is — in contrast to the situation in standard process algebra — up to now no satisfying coalgebraic representation of a mobile process calculus. Here, we discuss work towards a unifying coalgebraic framework for the denotational semantics of mobile systems. The connection between the ambient calculus and a coalgebraic approach which uses an extension of labelled transition systems in the representation of the reduction relation is analyzed in more detail. The formal representation of this framework is cast in the algebraic-coalgebraic specification language CoCasl.

1 Introduction

Coalgebra has in recent years gained importance as a framework which allows the modelling of reactive systems at an appropriate level of generality [14]. Here, coalgebra serves as a basis for the semantics of processes, giving rise to generic notions in particular of bisimilarity, coinduction, corecursion, and modal logic [12]. In analogy to the (largely algebraic and order theoretic) denotational semantics of programming languages and logics, it is desirable to find a coalgebraic denotational semantics for process calculi, which then profit from the above-mentioned generic semantic notions and results. Such a denotational semantics also adds clarity to the calculi themselves and facilitates the comparison and, possibly, unification of process calculi.

For (the finitely branching fragment of) the classical process calculus CCS, a coalgebraic semantics has been defined in [10]; further work in similar directions is found e.g. in [6,7]. Here, we present work leading towards a coalgebraic denotational semantics for mobile process calculi, in particular the ambient calculus [4]. This poses rather more involved problems than in the case of classical calculi, since the spatial structure interacts with the dynamic structure of processes in a complex way.

An LTS semantics for the ambient calculus has been defined in [5]. This semantics involves the use of a *hardening* relation, which singles out active top-level processes to be involved in spatial reductions. The hardening relation inspires our design of a coalgebraic functorial signature for the ambient calculus. We

J.L. Fiadeiro et al. (Eds.): CALCO 2005, LNCS 3629, pp. 232–246, 2005.

then give corecursive definitions of the process-building operations of a subset of the ambient calculus in the final coalgebra for this functor, and we show that the arising coalgebraic notion of bisimulation coincides with the one induced by the LTS semantics of [5]. As a framework supporting the formal specification of these concepts, we use the algebraic-coalgebraic language CoCasl [10].

The material is organized as follows. Section 2 provides a brief overview of CoCasl. An introduction to the ambient calculus is given in Section 3. Finally, our coalgebraic semantics for the ambient calculus is presented in Section 4.

2 CoCasl

The algebraic-coalgebraic specification language CoCasl has been introduced in [10] as an extension of the standard algebraic specification language Casl. For the basic Casl syntax, the reader is referred to [2,11]. We briefly explain the CoCasl features relevant for the understanding of the present work.

A simple but typical CoCasl specification is shown in Fig. 1. This specification defines the final finitely branching labelled transition system (LTS) over a given set of labels, exploiting both algebraic and coalgebraic aspects of CoCasl.

```
spec FinalLTS =
  sort Label
  then cofree {
       sort State
       then free {
            type Set ::= {} | {__}(State) | __ ∪ __(Set; Set)
            op __ ∪ __ : Set × Set → Set,
                      assoc, comm, idem, unit {}  }
       then cotype State ::= (next : Label → Set)  }
end
```

Fig. 1. CoCasl specification of the final LTS

Several CoCasl features are nicely illustrated here. To begin, CoCasl offers a **cotype** construct which defines coalgebraic process types, dually to Casl's datatype construct **type**. Without further qualifications, type or cotype declarations essentially amount to just operator declarations; e.g., the type declaration in Fig. 1 gives rise to operators $\{\_\} : State \to Set$ etc. called *constructors*, while the cotype declaration produces an operator $next : State \times Label \to Set$, called a *selector* or *observer*. Like type declarations, cotype declarations may have several alternatives separated by |; while for types, this is just an enumeration of constructors, the effect of alternatives in a cotype declaration is the generation of axioms emulating sum types, i.e. guaranteeing that the cotype is disjointly decomposed into the domains of the (partial) observers. E.g. writing

$$\textbf{cotype}\ \mathit{Process} ::= \mathit{cont}(\mathit{hd1} :?\mathit{Elem}; \mathit{next} :?\mathit{Process}) \mid \mathit{fork}(\mathit{hd2} :?\mathit{Elem}; \mathit{left} :?\mathit{Process}; \mathit{right} :?\mathit{Process})$$

produces a process type that can in each step either just advance one step (*next*) or fork (*left*/*right*). It is shown in [10] that one can indeed define for each cotype signature a functor T such that models of the cotype correspond to T-coalgebras. E.g., the cotype *State* of Fig. 1 corresponds to coalgebras for $TX = \mathit{Label} \to \mathcal{P}_\omega(X)$, where $\mathcal{P}_\omega$ denotes the finite powerset functor, and the cotype *Process* above to coalgebras for $TX = \mathit{Elem} \times X + \mathit{Elem} \times X \times X$.

Cotypes can be qualified by keywords expressing further constraints. In particular, the keyword **cofree**, placed directly before the keyword **cotype**, restricts the models of a simple cotype such as *Process* to the final coalgebra (uniquely up to isomorphism), which in the case of *Process* consists of infinite *Elem*-labelled trees with branching degree either 1 or 2 at each node. In the context of this work, a more powerful mechanism is more important, which applies to complex cotypes such as *State* in Fig. 1: The keyword **cofree** may also be used to restrict the models of an entire specification, delimited as in Fig. 1 by curly brackets, to final models over a given model of the preceding specification — in the case of Fig. 1 over a given set of labels. This concept is dual to the CASL construct **free**, also appearing in Fig. 1, which restricts models of the following specification to be initial over a given model of the preceding specification, in the case of the type *Set* in Fig. 1 over a given set of states (and, irrelevantly, a given set of labels). (Subtle differences between **cofree** and **free** are discussed in [10]; this is not relevant for the understanding of the present work.)

Explicitly, this means that the type *Set* indeed consists of the set of constructor terms modulo associativity, commutativity, idempotence, and neutrality of $\{\}$, i.e. essentially of all finite subsets of *State*. The cotype *State* is thus really the final coalgebra for the functor $TX = \mathit{Label} \to \mathcal{P}_\omega(X)$, i.e. the final finitely branching LTS. This cotype, or process type, has been used in [10] in order to define a coalgebraic denotational semantics for CCS, exploiting the fact that final coalgebras admit corecursive definitions; e.g. the parallel operator may be defined in CoCASL by the corecursive equation

- $\mathit{next}(l, s1 \parallel s2) = \mathit{power}[\_\_ \parallel \_\_](\mathit{next}(l, s1) * s2 \cup s1 * \mathit{next}(l, s1))$

(omitting the silent action), where $\mathit{power}[\_\_ \parallel \_\_]$ has previously been defined as the image function of $\_\_ \parallel \_\_ : \mathit{State} \times \mathit{State} \to \mathit{State}$, and $A * s$ denotes the cartesian product $A \times \{s\}$. In this work, we pursue similar goals for the ambient calculus.

The form of corecursion used above, also called coiteration or coinductive definition, is a very simple one which is based directly on the definition of the final coalgebra: the corecursive equation essentially expresses that *next* is the unique morphism from a coalgebra determined by the right hand side of the equation into the final coalgebra. In the definition of our coalgebraic semantics of the ambient calculus, we will need a more complex form of corecursion to be explained in Section 4.

3 The Ambient Calculus

The ambient calculus [4] models mobile computing (i.e. in mobile computing devices, like latops or mobile phones) as well as mobile computations (i.e. processes that move among devices, like applets). A central issue is the handling of administrative domains and their boundaries (e.g. protected by firewalls). The ambient calculus hence comprises agents, their ambients, and mobility of these ambients.

Space is understood to be hierarchical in the ambient calculus, and the hierarchical fragmentation of space is represented using the notion of ambient: An ambient is a single entity with a clear separation from its environment. It may contain processes or further ambients. Thus ambients may be nested or they may be residing in parallel on the same level. The dynamic change of the position of ambients in space over time is represented in the ambient calculus by several reduction rules which utilize so called *capabilities*. The capabilities model the opportunity for processes to enter, leave, or open ambients.

The syntax of the ambient calculus is defined as follows: For a set $\mathcal{N}$ of names (m, n will range over names in the sequel), the set of processes for the ambient calculus $\mathcal{AC}$ is defined inductively as the least set which is closed under

- the nil process $\mathbf{0}$,
- parallel composition of processes $P|Q$,
- capability prefixing $M.P$, where $M \in \{\mathbf{in}\, n, \mathbf{out}\, n, \mathbf{open}\, n\}$
- the ambient operator $n[P]$,
- name restriction $(\nu n)P$,
- replication $!P$.

The set of free names $fn(P)$ of an ambient calculus process P is, roughly speaking, the set of names that appear in the process either in ambient operators or in the prefixing of capabilities, minus the set of all names that appear in name restrictions.

$$\frac{}{m[\mathbf{in}\, n.P|Q] \mid n[R] \longrightarrow n[m[P|Q]|R]} \qquad \frac{}{n[m[\mathbf{out}\, n.P|Q] \mid R] \longrightarrow m[P|Q] \mid n[R]}$$

$$\frac{}{\mathbf{open}\, n.Q \mid n[R] \longrightarrow Q|R} \quad \frac{P \longrightarrow Q}{P|R \longrightarrow Q|R} \quad \frac{P \longrightarrow Q}{n[P] \longrightarrow n[Q]} \quad \frac{P \longrightarrow Q}{(\nu n)P \longrightarrow (\nu n)Q}$$

$$\frac{P \equiv P' \qquad P' \longrightarrow Q' \qquad Q' \equiv Q}{P \longrightarrow Q}$$

Fig. 2. The reduction relation of the ambient calculus

$$P|\mathbf{0} \equiv P \qquad P|Q \equiv Q|P \qquad (\nu m)(\nu n)P \equiv (\nu n)(\nu m)P$$

$$(\nu n)\mathbf{0} \equiv \mathbf{0} \qquad (P|Q)|R \equiv P|(Q|R) \qquad (\nu n)(P|Q) \equiv P|(\nu n)Q \text{ if } n \notin fn(P)$$

$$!\mathbf{0} \equiv \mathbf{0} \qquad !P \equiv P|!P \qquad (\nu m)n[P] \equiv n[(\nu m)P] \text{ if } n \neq m$$

Fig. 3. The structural congruence of the ambient calculus

The semantics of the ambient calculus is defined by the *reduction relation* $\longrightarrow \subset \mathcal{AC} \times \mathcal{AC}$, which is the least relation that satisfies the rules displayed in Fig. 2, where $\equiv \subset \mathcal{AC} \times \mathcal{AC}$ denotes *structural congruence*. The latter is defined as the smallest congruence relation satisfying the rules of Fig. 3.

Example 1. The following typical example shows two possible chains of reductions of the process $m[\,\mathbf{in}\,n.\,\mathbf{out}\,n.\,\mathbf{0}\,] \mid n[\,\mathbf{open}\,m.\,\mathbf{0}\,]$:

$$\begin{aligned} m[\,\mathbf{in}\,n.\,\mathbf{out}\,n.\,\mathbf{0}\,] \mid n[\,\mathbf{open}\,m.\,\mathbf{0}\,] &\longrightarrow n[\,m[\,\mathbf{out}\,n.\,\mathbf{0}\,] \mid \mathbf{open}\,m.\,\mathbf{0}\,] \\ &\longrightarrow m[\,\mathbf{0}\,] \mid n[\,\mathbf{open}\,m.\,\mathbf{0}\,] \end{aligned}$$

$$\begin{aligned} m[\,\mathbf{in}\,n.\,\mathbf{out}\,n.\,\mathbf{0}\,] \mid n[\,\mathbf{open}\,m.\,\mathbf{0}\,] &\longrightarrow n[\,m[\,\mathbf{out}\,n.\,\mathbf{0}\,] \mid \mathbf{open}\,m.\,\mathbf{0}\,] \\ &\longrightarrow n[\,\mathbf{out}\,n.\,\mathbf{0} \mid \mathbf{0}\,] \end{aligned}$$

It is easy to see that the ambient calculus is non-deterministic and non-confluent (and due to the replication operation also potentially non-terminating).

4 A Coalgebraic Semantics for the Ambient Calculus

Following [4], we can equivalently define the reduction relation of the ambient calculus in terms of a labelled transition system together with a hardening relation; the role of the latter is to single out active top-level processes, in order to prevent processes 'tagging along' in spatial reductions performed by parallel processes. Below, we present such a system in a somewhat modified form which will allow us to embed the ambient calculus semantically into a coalgebraic framework; in particular, we give a suitable behaviour functor for the ambient calculus and define the process building operations of the ambient calculus as corecursive operations on the final coalgebra of this functor. Besides making standard coalgebraic machinery available for the ambient calculus, this clarifies the observational aspects of the calculus. The corecursive definitions will be presented in CoCasl.

4.1 The LTS

We recall the definition of labelled transition systems:

Definition 2. A *labelled transition system* (LTS) is a triple (S, A, T) where S is a set of *states*, A is a set of *actions* and $T \subset S \times A \times S$ is the *transition relation*. For $P, Q \in S$ and $a \in A$, $(P, a, Q) \in T$ means that the system evolves by the action a from source P to target Q; this is denoted in the form $P \xrightarrow{a} Q$.

In order to allow for a coalgebraic specification of the semantics, we design a variant of the LTS given in [4] in such a way that the conclusion of each inference rule of the LTS has the application of exactly one process-building operation of the ambient calculus on the left hand side of the conclusion (and no process building operations appear in the premises). This facilitates the subsequent corecursive definition of the operations. As indicated above, we need a hardening relation as in [5], extended by intermediate capabilities in the spirit of [8]:

As in [5], a *concretion* is an expression of the form: $(\nu \overrightarrow{p})\langle P\rangle Q$, where $P, Q \in \mathcal{AC}$ and $\overrightarrow{p} = \{p_1, \ldots, p_n\}$. The process P is called the *prime*, the process Q is called the *residue*. The intuition is that a process, which may have many top-level processes, may harden to a concretion that singles out an active subprocess P, leaving behind the residue Q, where $\overrightarrow{p}$ is the set of private names shared by P and Q.

$$\frac{P \xrightarrow{\mathbf{in}\,n} Q}{m[P] \overset{\mathbf{enter}\,n}{\succ} (\nu)\langle m[Q]\rangle \mathbf{0}} \qquad \frac{}{n[P] \overset{\overline{\mathbf{enter}}\,n}{\succ} (\nu)\langle P\rangle \mathbf{0}}$$

$$\frac{P \xrightarrow{\mathbf{out}\,n} Q}{m[P] \overset{\mathbf{exit}\,n}{\succ} (\nu)\langle m[Q]\rangle \mathbf{0}} \qquad \frac{}{n[P] \overset{\overline{\mathbf{open}}\,n}{\succ} (\nu)\langle P\rangle \mathbf{0}}$$

$$\frac{\overrightarrow{p} \cap fn(Q) = \emptyset \quad P \overset{\alpha}{\succ} (\nu \overrightarrow{p})\langle P'\rangle P''}{P|Q \overset{\alpha}{\succ} (\nu \overrightarrow{p})\langle P'\rangle P''|Q} \qquad \frac{\overrightarrow{p} \cap fn(Q) = \emptyset \quad P \overset{\alpha}{\succ} (\nu \overrightarrow{p})\langle P'\rangle P''}{Q|P \overset{\alpha}{\succ} (\nu \overrightarrow{p})\langle P'\rangle Q|P''}$$

$$\frac{P \overset{\alpha}{\succ} (\nu \overrightarrow{p})\langle P'\rangle P''}{!P \overset{\alpha}{\succ} (\nu \overrightarrow{p})\langle P'\rangle P''|!P} \qquad \frac{P \overset{\alpha}{\succ} C \quad n \notin fn(\alpha)}{(\nu n)P \overset{\alpha}{\succ} (\overline{\nu n})C}$$

Fig. 4. The hardening relation

In order to keep track of the structure of an ambient calculus process over several inference steps, we use a labelled version of the hardening relation. So-called *intermediate capabilities* are used to store the information that a process is of a specific shape, and this information then appears as the premise of an inference rule which is used to derive a transition of the process. Thus, the LTS itself works entirely on processes, rather than also on concretions as in [8].

The hardening relation and the LTS are defined by mutual recursion; i.e. LTS relations may appear as assumptions in the rules for the hardening relation, and vice versa. The hardening relation has the format $\overset{\alpha}{\succ} \subset \mathcal{AC} \times \mathit{HAction} \times \Upsilon$ for the set $\mathit{HAction} = \{\mathbf{enter}, \overline{\mathbf{enter}}, \mathbf{exit}, \overline{\mathbf{open}}\} \times \mathcal{N}$. The rules are given in Fig. 4, where $(\overline{\nu n})C$ for $C = (\nu \overrightarrow{p})\langle P' \rangle P''$ is defined as

$$(\overline{\nu n})C = \begin{cases} (\nu \overrightarrow{p})\langle P' \rangle (\nu n) P'' & \text{if } n \notin fn(P') \\ (\nu \overrightarrow{p} \langle m[(\nu n)P''']\rangle) P'' & \text{if } n \notin fn(P'') \text{ and } P' = m[P'''] \ (n \neq m) \\ (\nu n, \overrightarrow{p})\langle P' \rangle P'' & \text{otherwise.} \end{cases}$$

The labelled transition system has the format $\xrightarrow{\alpha} \subset \mathcal{AC} \times \mathit{Action} \times \mathcal{AC}$ for the set $\mathit{Action} = \{\tau\} \cup \{\mathbf{in}, \mathbf{out}, \mathbf{open}\} \times \mathcal{N}$ of transition labels. Its rules are displayed in Fig. 5. The rules imply that concretions in hardenings labelled **enter** or **exit** are always of the form $(\nu \overrightarrow{p})\langle n[P] \rangle Q$. Note that while we do not impose the full structural congruence on terms for purposes of the hardening and transition relations, we do assume that bound names are given only up to α-equivalence. Thus the hardening rule for parallel composition and the transition rules for opening and entering ambients can always be made applicable by suitably renaming bound names. Concerning the transition rule for exiting ambients, one can show that the premise implies $n \in fn(P)$, in particular $n \notin \overrightarrow{p}$.

$$\frac{}{M.P \xrightarrow{M} P} \qquad \frac{P \xrightarrow{\alpha} P'}{P|Q \xrightarrow{\alpha} P'|Q} \qquad \frac{P \xrightarrow{\alpha} P'}{Q|P \xrightarrow{\alpha} Q|P'} \qquad \frac{P \xrightarrow{\alpha} P'}{P! \xrightarrow{\alpha} P'|P!}$$

$$\frac{P \overset{\mathbf{exit}\, n}{\succ} (\nu \overrightarrow{p})\langle P' \rangle P''}{n[P] \xrightarrow{\tau} (\nu \overrightarrow{p})(n[P''] \mid P')} \qquad \frac{P \xrightarrow{\tau} Q}{n[P] \xrightarrow{\tau} n[Q]} \qquad \frac{P \xrightarrow{\alpha} Q \quad n \notin fn(\alpha)}{(\nu n)P \xrightarrow{\alpha} (\nu n)Q}$$

$$\frac{P \xrightarrow{\mathbf{open}\, n} P' \quad Q \overset{\overline{\mathbf{open}}\, n}{\succ} (\nu \overrightarrow{q})\langle Q' \rangle Q'' \quad \overrightarrow{q} \cap fn(P) = \emptyset}{P|Q \xrightarrow{\tau} (\nu \overrightarrow{q})(P'|Q'|Q'')}$$

$$\frac{Q \xrightarrow{\mathbf{open}\, n} Q' \quad P \overset{\overline{\mathbf{open}}\, n}{\succ} (\nu \overrightarrow{p})\langle P' \rangle P'' \quad \overrightarrow{p} \cap fn(Q) = \emptyset}{P|Q \xrightarrow{\tau} (\nu \overrightarrow{p})(P'|P''|Q')}$$

$$\frac{P \overset{\mathbf{enter}\, n}{\succ} (\nu \overrightarrow{p})\langle P' \rangle P'' \quad Q \overset{\overline{\mathbf{enter}}\, n}{\succ} (\nu \overrightarrow{q})\langle Q' \rangle Q'' \quad \overrightarrow{p} \cap \overrightarrow{q} = \emptyset}{P|Q \xrightarrow{\tau} (\nu \overrightarrow{p})(\nu \overrightarrow{q})(n[P'|Q'] \mid P''|Q'')}$$

$$\frac{Q \overset{\mathbf{enter}\, n}{\succ} (\nu \overrightarrow{q})\langle Q' \rangle Q'' \quad P \overset{\overline{\mathbf{enter}}\, n}{\succ} (\nu \overrightarrow{p})\langle P' \rangle P'' \quad \overrightarrow{p} \cap \overrightarrow{q} = \emptyset}{P|Q \xrightarrow{\tau} (\nu \overrightarrow{p})(\nu \overrightarrow{q})(n[P'|Q'] \mid P''|Q'')}$$

Fig. 5. The transition relation

Theorem 3. *The labelled transition system defined above is, up to structural congruence, sound and complete with respect to the reduction relation of the* *ambient calculus as recalled in Section 3. Formally:* $P \xrightarrow{\tau} Q$ *implies* $P \longrightarrow Q$ *(soundness) and* $P \longrightarrow Q$ *implies* $\exists P'.P \xrightarrow{\tau} P' \wedge P' \equiv Q$ *(completeness).*

Cardelli and Gordon present a different labelled transition system for the ambient calculus in [5]. We write $P \xrightarrow{\alpha}_{CG} Q$ to indicate that the process P can reduce to the process Q by a transition with label α which is justified by a rule of the labelled transition system from [5]. (Unlike the system in [5], our system does not take input and output primitives into account; however, these features can easily be added to the theory.)

Theorem 4. *The labelled transition system defined above is sound and complete with respect to the labelled transition system of [5]:* $P \xrightarrow{\alpha} Q$ *implies* $P \xrightarrow{\alpha}_{CG} Q$ *(soundness) and* $P \xrightarrow{\alpha}_{CG} Q$ *implies* $P \xrightarrow{\alpha} Q$ *(completeness).*

4.2 Coalgebraic Semantics

The mobility aspects of the labelled transition system defined above can be modelled in a coalgebraic manner. This amounts to designing a behaviour functor which captures the possible observations on an ambient calculus process. These observations apparently include not only the reductions in the LTS, but also the concretions to which a process hardens. An further important aspect is the handling of name creation, which however poses certain technical problems in the coalgebraic setting. For the time being, we therefore give a coalgebraic semantics of a reduced calculus without name restriction (comparable in this respect e.g. to the Basic Sail calculus [13]). The combination of reduction and hardening suggests using the functor

$$AC = \lambda X.(Action \to \mathcal{P}_\omega(X)) \times (HAction \to \mathcal{P}_\omega(X \times X)).$$

Coalgebras $(\mathcal{A}, \psi : \mathcal{A} \to AC(\mathcal{A}))$ for this functor have a function $\psi = \langle next, harden \rangle$ as structure. The first part of this function $(next : \mathcal{A} \to (Action \to \mathcal{P}_\omega(\mathcal{A})))$ assigns to each state of the coalgebra a function which maps each transition label to the set of the corresponding successor states. The second function $(harden : \mathcal{A} \to (HAction \to \mathcal{P}_\omega(\mathcal{A} \times \mathcal{A})))$ assigns to each state a function which maps hardening labels to the set of the corresponding concretions of the state.

It should be noted that the crucial difference between AC and the standard functor $\mathcal{P}_\omega(A \times \_)$ for LTS lies in the fact that the hardening part is essentially of the type $\lambda X. \mathcal{P}_\omega(X \times X)$; here, the peculiarity is captured that a process splits into two parts for purposes of further reduction. There is good indication that this feature is indeed the essence of 'mobility', since it is instrumental in the modelling of 'moving and leaving others behind'.

A CoCasl specification of the final coalgebra of the functor AC is shown in Fig. 6. The specification is based on a specification of finite sets, which is parametric in the type of elements and which needs to be instantiated with states and concretions. The cotype *State* is the mentioned final AC-coalgebra; it serves as a semantic domain for the interpretation of the ambient calculus operations. Since CoCasl does not have product and sum types, the presentation of AC

```
spec SET [sort Elem] =
free {
   type Set[Elem] ::= {} | {__}(Elem) | __ ∪ __(Set[Elem]; Set[Elem])
   op __ ∪ __ : Set[Elem] × Set[Elem] → Set[Elem],
            assoc, comm, idem, unit {}   }
   then
   pred __ ∈ __ : Elem × Set[Elem]
   op __ − __ : Set[Elem] × Elem → Set[Elem]
   ... %% recursive Definitions of ∈, −

spec ACDomain[sort Name] = SET [sort Name] then
cofree {
   sort State
   free type Cap ::= in | out | open
   free type HCap ::= enter | coenter | exit | coopen
   free type Action ::= tau | action(Cap; Name)
   free type HAction ::= haction(HCap; Name)
   free type Concretion ::= conc(State; State)
   then SET [sort State] and SET [sort Concretion]
   then cotype State ::= (next :   Action → Set[State];
                          harden : HAction → Set[Concretion])
   }
```

Fig. 6. CoCasl specification of the semantic domain for the ambient calculus

needs to be split up into various datatype definitions; this makes for a somewhat more verbose, but also clearer and more readable specification style.

The process building operations of the ambient calculus can then be defined as functions into the semantic domain *State* by mutual corecursion. The corecursive definitions are clear from the corresponding inference rules for the hardening relation and the transition relation as shown in Figs. 4–5. The CoCasl specifications of the operations are shown in Figs. 7–11. (The operator declarations should be considered redeclarations, following an initial declaration of all operators which is necessary for mutually corecursive definitions due to Casl's linear visibility principle. Moreover, the label *in* would have to be introduced via a display annotation, since **in** is a reserved word.)

In the description of the successor and concretion sets, the definitions below use explicit equivalences for elementhood in the successor sets rather than equations involving application of the powerset functor to the defined functions. One can also apply the latter style in CoCasl, making extensive use of parametrized specifications as e.g. in the CCS semantics of [10]. This style is actually preferable, but would require rather more infrastructure than the available space permits to present here.

Remark 5. A word of explanation is in order as to why the equations in Figs. 7–11 actually constitute good corecursive definitions. The format of these equations

$$
\begin{array}{ll}
\textbf{op} & zero\ :\ State \\
\textbf{vars} & a : Action;\ b : HAction \\
\bullet & next(a, zero) = \{\} \\
\bullet & harden(b, zero) = \{\}
\end{array}
$$

Fig. 7. CoCasl specification of the nil process

$$
\begin{array}{ll}
\textbf{op} & cap\ :\ Cap \times Name \times State \rightarrow State \\
\textbf{vars} & a : Action;\ b : HAction;\ c : Cap;\ n : Name;\ p : State \\
\bullet & next(a, cap(c, n, p)) = \{p\}\ when\ a = action(c, n)\ else\ \{\} \\
\bullet & harden(b, cap(c, n, p)) = \{\}
\end{array}
$$

Fig. 8. CoCasl specification of capability prefixing

$$
\begin{array}{ll}
\textbf{op} & \_\_ \parallel \_\_\ :\ State \times State \rightarrow State \\
\textbf{vars} & a : Action;\ b : HAction;\ n : Name;\ p, q, r : State;\ c : Concretion \\
\bullet & r \in next(a, p \parallel q) \Leftrightarrow \\
& \quad (a = \tau \wedge \exists p', q', q'' : State\ \bullet \\
& \qquad\qquad r = p' \parallel q' \parallel q'' \wedge p' \in next(action(open, n), p) \wedge \\
& \qquad\qquad conc(q', q'') \in harden(haction(coopen, n), q)) \vee \\
& \quad (a = \tau \wedge \exists q', p', p'' : State\ \bullet \\
& \qquad\qquad r = p' \parallel p'' \parallel q' \wedge q' \in next(action(open, n), q) \wedge \\
& \qquad\qquad conc(p', p'') \in harden(haction(coopen, n), p)) \vee \\
& \quad (a = \tau \wedge \exists p', p'', q', q'' : State\ \bullet \\
& \qquad\qquad r = amb(n, p' \parallel q') \parallel p'' \parallel q'' \wedge \\
& \qquad\qquad conc(p', p'') \in harden(haction(enter, n), p) \wedge \\
& \qquad\qquad conc(q', q'') \in harden(haction(coenter, n), q)) \vee \\
& \quad (a = \tau \wedge \exists p', p'', q', q'' : State\ \bullet \\
& \qquad\qquad r = amb(n, p' \parallel q') \parallel p'' \parallel q'' \wedge \\
& \qquad\qquad conc(p', p'') \in harden(haction(coenter, n), p) \wedge \\
& \qquad\qquad conc(q', q'') \in harden(haction(enter, n), q)) \vee \\
& \quad (\exists p' : State\ \bullet\ r = p' \parallel q \wedge p' \in next(a, p)) \vee \\
& \quad (\exists q' : State\ \bullet\ r = p \parallel q' \wedge q' \in next(a, q)) \\
\bullet & c \in harden(b, (p \parallel q)) \Leftrightarrow \\
& \quad (\exists p', p'' : State\ \bullet\ c = conc(p', p'' \parallel q) \wedge conc(p', p'') \in harden(b, p)) \vee \\
& \quad (\exists q', q'' : State\ \bullet\ c = conc(q', p \parallel q'') \wedge conc(q', q'') \in harden(b, q))
\end{array}
$$

Fig. 9. CoCasl specification of parallel composition

$$
\begin{array}{ll}
\textbf{op} & amb : Name \times State \rightarrow State \\
\textbf{vars} & a : Action; b : HAction; cap : Cap; m, n : Name; p, q : State; c : Concretion \\
\bullet & q \in next(\tau, amb(n, p)) \Leftrightarrow \\
& \quad (\exists p', p'' : State \bullet \\
& \qquad q = amb(n, p'') \parallel p' \wedge \\
& \qquad conc(p', p'') \in harden(haction(exit, n), p)) \vee \\
& \quad (\exists p' : State \bullet q = amb(n, p') \wedge p' \in next(\tau, p)) \\
\bullet & next(action(cap, n), amb(m, p)) = \{\} \\
\bullet & c \in harden(haction(enter, n), amb(m, p)) \Leftrightarrow \\
& \quad (\exists p' : State \bullet c = conc(amb(m, p'), zero) \wedge \\
& \qquad p' \in next(action(in, n), p)) \\
\bullet & harden(haction(coenter, n), (amb(n, p))) = \{conc(p, zero)\} \\
\bullet & c \in harden(haction(exit, n), amb(m, p)) \Leftrightarrow \\
& \quad (\exists p' : State \bullet c = conc(amb(m, p'), zero) \wedge \\
& \qquad p' \in next(action(out, n), p)) \\
\bullet & harden(haction(coopen, n), (amb(n, p))) = \{conc(p, zero)\}
\end{array}
$$

Fig. 10. CoCasl specification of the ambient operator

$$
\begin{array}{ll}
\textbf{op} & rep : State \rightarrow State \\
\textbf{vars} & a : Action; b : HAction; p, q : State; c : Concretion \\
\bullet & q \in next(a, rep(p)) \Leftrightarrow \\
& \quad (\exists p' : State \bullet q = p' \parallel rep(p) \wedge p' \in next(a, p)) \\
\bullet & c \in harden(b, rep(p)) \Leftrightarrow \\
& \quad (\exists p', p'' : State \bullet c = conc(p', p'' \parallel rep(p)) \wedge \\
& \qquad conc(p', p'') \in harden(b, p))
\end{array}
$$

Fig. 11. CoCasl specification of replication

deviates from the standard coiteration format in that the right hand sides contain composite expressions of the language being interpreted, rather than just one application of the single operation being defined. According to the results of [16], a semantics for a process calculus with signature functor Σ in coalgebras for a behaviour functor B can be defined by exhibiting an *abstract GSOS law*, i.e. a natural transformation

$$\rho : \Sigma(Id \times B) \rightarrow BT,$$

where T is the free monad (i.e. the term algebra functor) over Σ (cf. also [1]). The appearance of T in the target offers the possibility of using composite terms, as required. The semantic function $h : \Sigma S \rightarrow S$, where (S, ζ) is the final B-coalgebra, is the unique so-called ρ-model over (S, ζ), i.e. uniquely determined by the equation

$$\zeta \circ h = Bh^* \circ \rho_S \circ \Sigma\langle id, \zeta \rangle, \qquad (*)$$

where $h^* : TS \rightarrow S$ is the T-algebra determined by h. If the semantic function h is omitted from the notation, as done above, then equation $(*)$ becomes precisely the format of our corecursive definitions. This shows that our corecursive equations have a unique solution provided that ρ_S is part of a natural transformation ρ. If, as is the case here, Σ and B are κ-accessible for some regular cardinal κ and $|S| \geq \kappa$, then it suffices to check that ρ_S is natural for self-maps of S: we then obtain the components ρ_X for $|X| < \kappa$, natural in X, by restriction of ρ_Z, and from these ρ_X we can assemble all of ρ by taking κ-directed unions.

Verification of the naturality condition for ρ_S as given in Figs. 4–5 is tedious but straightforward. The point is essentially that the rules adhere to similar restrictions as standard GSOS rules for the definition of labelled transition systems, in particular do not depend on equality of states, do not introduce new state variables in the conclusion, and never look ahead more than one step (the latter could in fact also be handled by means of so-called tree rules [16]).

Generally, GSOS semantics is *compositional*, i.e. the interpretation of composite terms is recursively derived from that of single operations [16]. This does not, incidentally, contradict the previously diagnosed impossibility of a compositional LTS semantics for the ambient calculus [17], since our semantic domain is more than just an LTS.

Remark 6. The reason that restriction is currently omitted in the specification is that it is as yet unclear how to model the sharing of private names in concretions: a concretion needs to contain information about the potential future interaction between the prime and the residue, but on the other hand should not leak information about the shared bound names.

We explicitly record the agreement between the coalgebraic specification and the LTS semantics given above:

Lemma 7. *The* CoCasl *specification is sound and complete with respect to the LTS defined in Section 4.1: for ambient calculus terms P and P' not involving restriction, $P \xrightarrow{\alpha} P'$ iff $P' \in next(\alpha, P)$ follows from the corecursive equations (omitting an obvious translation between ambient calculus terms and their representation in* CoCasl*). Furthermore, the LTS and the* CoCasl *specification agree w.r.t. hardening: $P \overset{\beta}{\succ} (\nu)\langle P_1'\rangle P_2'$ iff $conc(P_1', P_2') \in harden(\beta, P)$ follows from the corecursive equations (again for terms without restriction, and omitting an obvious translation of fixed finite sets).*

For the reduced calculus, the notion of bisimulation arising from the coalgebraic modelling can be brought into agreement with a natural notion of behavioral indistinguishability:

Definition 8 (Ambient bisimulation). A symmetric relation $\simeq$ on ambient calculus processes without name restriction is called an *ambient bisimulation* if for any two processes P, Q such that $P \simeq Q$ the following hold:

1. $P \xrightarrow{\alpha} P' \Rightarrow (\exists Q'. Q \xrightarrow{\alpha} Q' \wedge P' \simeq Q')$ for any $\alpha \in Act$.
2. $P \overset{\beta}{\succ} (\nu)\langle P_1'\rangle P_2' \Rightarrow (\exists Q_1', Q_2'. Q \overset{\beta}{\succ} (\nu)\langle Q_1'\rangle Q_2' \wedge P_1' \simeq Q_1' \wedge P_2' \simeq Q_2')$ for each $\beta \in HAction$.

If $P \simeq Q$ for some ambient bisimulation $\simeq$, then P and Q are called *ambient bisimilar.*

Recall that for any endofunctor G, a binary relation R between two G-coalgebras E and F is called a *bisimulation* if there exists a G-coalgebra structure on R that makes the projection functions $\pi_1 : R \to E$ and $\pi_2 : R \to F$ into coalgebra homomorphisms. If for two elements $e \in E$ and $f \in F$ and a bisimulation R it holds that $e\,R\,f$, then e and f are said to be *bisimilar*. In the final coalgebra, bisimilarity is equality.

From the preceding lemma, the following is easily shown:

Theorem 9. *Ambient bisimulation and coalgebraic bisimulation on AC-coalgebras coincide; formally: for $P, Q \in \mathcal{AC}$ not involving restriction, $P \simeq Q$ iff P and Q denote the same element of the final AC-coalgebra, i.e. iff $P = Q$ follows from the above* CoCasl *specification.*

5 Conclusion

We have described a transition semantics for the ambient calculus that correctly captures the ambient calculus in the sense that its reductions coincide with the reduction of the ambient calculus. Similar results have been obtained in [8] for the safe ambient calculus with passwords, and in [9] for the ambient calculus itself. In both cases, however, labelled transition systems are used which mix processes and concretions, while our system separates the two types of entities by keeping the reduction relation and the hardening relation apart. A consequence is that our semantics fits into a coalgebraic framework.

The coalgebraic treatment of the transition semantics (using the specification language CoCasl, or a corresponding functor) exhibits more structure than labelled transition systems. The coalgebraic structure is based on two kinds of observations: one can observe firstly the successor states in the sense of process algebra, and secondly the set of ways ('concretions') in which a top-level process can be singled out for interaction with the ambient structure. Here, the set of concretions is particularly noteworthy; we expect that this part of the functor, being essentially the composite of the powerset functor and the squaring functor, points to a fundamental aspect of mobile calculi — processes split up into parts that move and others that remain behind. The coalgebraic treatment of private names, i.e. the security aspect of the ambient calculus, had to be left open for the time being; this problem is hoped to be resolved in future work. One should note that the first step in this direction has already been taken in the shape of our transition semantics, which does include name restriction: the transition rules adhere to a generalized GSOS format, which is an important precondition for a corecursive formulation.

The corecursive definition of process building operations implies the possibility to prove algebraic laws about the ambient calculus using coinduction, based on a notion of bisumulation arising from the coalgebraic semantics. An open problems that remains in this respect is the relation of this bisimilarity to the contextual equivalence of ambients [4] and to the notion of reduction barbed congruence [9].

We expect that the program of characterizing process and mobile calculi using coalgebras over certain functors will eventually lead to a systematic understanding of the nature of these calculi. The calculi are usually presented using some concrete syntax plus some transition rules; and there are many variations of the syntax and the rules whose impact on the nature of the respective calculus is not clear from the outset. By contrast, the representation using operations on a coalgebra for a certain functor immediately determines (through the functor) the fundamental observations that can be made, while the operations (that correspond to the syntax of the respective calculus) may vary without changing the fundamental nature of the meaning of processes. Hence, it is also expected that different calculi can be related and combined much more easily using the coalgebraic representation. Future work will substantiate this point by considering further mobile calculi. Moreover, the behaviour functor more or less automatically comes with an expressive modal logic (cf. [15] and references therein); further work will include the investigation of this logic in particular in relation to ambient logic [3]. In advance, we note that the hardening part of our behaviour functor will naturally give rise to a binary modality which appears also in ambient logic.

References

1. F. Bartels, *Generalised coinduction*, Math. Struct. Comput. Sci. **13** (2003), 321–348.
2. M. Bidoit and P. D. Mosses, CASL *user manual*, LNCS, vol. 2900, Springer, 2004.
3. L. Cardelli and A. Gordon, *Ambient logic*, Math. Struct. Comput. Sci., to appear.
4. ______, *Mobile ambients*, Theoret. Comput. Sci. **240** (2000), 177–213.
5. A. Gordon and L. Cardelli, *Equational properties of mobile ambients*, Math. Struct. Comput. Sci. **13** (2003), 371–408.
6. F. Honsell, M. Lenisa, U. Montanari, and M. Pistore, *Final semantics for the π-calculus*, Programming Concepts and Methods, Chapman & Hall, 1998, pp. 225–243.
7. B. Klin, *A coalgebraic approach to process equivalence and a coinduction principle for traces*, Coalgebraic Methods in Computer Science, ENTCS, vol. 106, Elsevier, 2004, pp. 201–218.
8. M. Merro and M. Hennessy, *Bisimulation congruences in safe ambients*, ACM SIGPLAN Notices **37** (2002), 71–80.
9. M. Merro and F. Zappa Nardelli, *Behavioural theory for mobile ambients*, Tech. Report RR-5375, INRIA, 2004.
10. T. Mossakowski, L. Schröder, M. Roggenbach, and H. Reichel, *Algebraic-co-algebraic specification in* CoCASL, J. Logic Algebraic Programming, to appear.
11. P. D. Mosses (ed.), CASL *reference manual*, LNCS, vol. 2960, Springer, 2004.
12. D. Pattinson, *Expressive logics for coalgebras via terminal sequence induction*, Notre Dame J. Formal Logic **45** (2004), 19–33.
13. D. Pattinson and M. Wirsing, *Making components move: a separation of concerns approach*, Formal Methods for Components and Objects (FMCO 02), LNCS, vol. 2852, Springer, 2003, pp. 487–507.
14. J. Rutten, *Universal coalgebra: a theory of systems*, Theoret. Comput. Sci. **249** (2000), 3–80.

15. L. Schröder, *Expressivity of coalgebraic modal logic: The limits and beyond*, Foundations of Software Science And Computation Structures, LNCS, vol. 3441, Springer, 2005, pp. 440–454.
16. D. Turi and G. Plotkin, *Towards a mathematical operational semantics*, Logic in Computer Science, IEEE Computer Society Press, 1997, pp. 280–291.
17. M. Vigliotti, *Reduction semantics for ambient calculi*, Ph.D. thesis, Imperial College, London, 2004.

The Least Fibred Lifting and the Expressivity of Coalgebraic Modal Logic

Bartek Klin

University of Sussex, Warsaw University
klin@brics.dk

Abstract. Every endofunctor B on the category **Set** can be lifted to a fibred functor on the category (fibred over **Set**) of equivalence relations and relation-preserving functions. In this paper, the least (fibre-wise) of such liftings, $L(B)$, is characterized for essentially any B. The lifting has all the useful properties of the relation lifting due to Jacobs, without the usual assumption of weak pullback preservation; if B preserves weak pullbacks, the two liftings coincide. Equivalence relations can be viewed as Boolean algebras of subsets (predicates, tests). This correspondence relates $L(B)$ to the least test suite lifting $T(B)$, which is defined in the spirit of predicate lifting as used in coalgebraic modal logic. Properties of $T(B)$ translate to a general expressivity result for a modal logic for B-coalgebras. In the resulting logic, modal operators of any arity can appear.

1 Introduction

Coalgebras are used as models for various kinds of transition systems, offering a general view on the notions of coinduction, bisimulation, and on logics used to reason about systems. For example, a finitely branching labelled transition system with carrier X is a coalgebra $h : X \to \mathcal{P}_f(A \times X)$, where A is a set of labels. Replacing the behaviour functor $\mathcal{P}_f(A \times -)$ with various functors on the category **Set** of sets and functions, one models other kinds of systems [1].

Final coalgebras are abstract models of the behaviour of systems. For a coalgebra $h : X \to BX$, two processes in X are considered behaviourally equivalent if they are identified by the unique morphism from h to a final B-coalgebra. If $B = \mathcal{P}_f(A \times -)$, behavioural equivalence coincides with bisimilarity.

Usually, bisimilarity is defined as the greatest bisimulation on a transition system. A natural coalgebraic generalization of the classical notion of bisimulation [2,3] is based on spans of coalgebras [1]. There, a bisimulation is a relation that lifts to a span of coalgebra morphisms. If the behaviour functor B preserves weak pullbacks, the greatest bisimulation exists and coincides with behavioural equivalence. Another, closely related approach was presented by Jacobs et al. [4,5,6]. There, the functor B is provided with a relation lifting $J(B)$, which is a functor on the category of binary relations that behaves as B on the underlying sets. A bisimulation is then a $J(B)$-coalgebra. Provided B preserves weak pullbacks, bisimulations satisfy many useful properties, e.g., they

J.L. Fiadeiro et al. (Eds.): CALCO 2005, LNCS 3629, pp. 247–262, 2005.

are closed under unions and preserved by coalgebra morphisms; also, as in the span approach, the greatest bisimulation coincides with behavioural equivalence.

In the case of labelled transition systems, bisimilarity is characterized by Hennessy-Milner logic [7]. The coalgebraic framework provides a more general perspective, where a modal logic can be derived for coalgebras for any functor. It is expected that coalgebraic modal logic is invariant under behavioural equivalence. Moreover, it should be expressive: logically indistinguishable states should be behaviourally equivalent. Several approaches to deriving coalgebraic modal logics have been proposed [8,9,10,11,12,13], based on different abstract notions of modal operators (modalities).

In [10], a single modal operator is associated with every functor B. The resulting logic is expressive for essentially all functors. However, the modal operator involved is rather complex and difficult to relate to modalities usually considered in particular cases (e.g., the box and diamond modalities of Hennessy-Milner logic). In [11], modalities are defined to be predicate liftings, i.e., natural transformations $\lambda : 2^{-} \to 2^{B}$, which transform predicates on any set X to predicates on BX. These correspond more closely to the modal operators usually considered in modal logics. However, the logic thus obtained fails to be expressive for many functors. In [9], modalities are defined to be test constructors, i.e. functions $w : B2 \to 2$.

Very recently, Schröder [13] provided a characterization of functors which admit an expressive logic based on predicate liftings. He also observed that predicate liftings and test constructors are in a one-to-one correspondence. To enhance expressivity, he then introduced polyadic predicate liftings, corresponding to functions $w : B(2^{\kappa}) \to 2$, and proved that a modal logic based on those is expressive for all accessible functors.

In this paper, we treat both relation lifting and coalgebraic modal logics in a fibrational setting. The relation lifting $J(B)$ is viewed as a lifting of B to a fibred functor on the category **Rel** of binary relations and relation-preserving functions. Similarly, any family of predicate liftings (test constructors) induces a fibred lifting of B to the category **TS** of test suites, i.e., families of predicates.

Our first observation is that, when one wants to lift a functor B to a fibred functor on **Rel**, the choice of $J(B)$ among many other liftings is not entirely clear. Despite its many useful properties, it does not really seem canonical. We therefore study the *least* fibred liftings of functors. The least fibred liftings to **Rel** turn out to be trivial, but when restricted to the category **ERel** of equivalence relations, the least fibred lifting $L(B)$ is nontrivial, and it enjoys all the useful properties of $J(B)$, without the usual assumption of weak pullback preservation. Indeed, if B preserves weak pullbacks, both liftings coincide. It seems, therefore, that $L(B)$ is the canonical choice of relation lifting, at least to equivalence relations, and that $J(B)$ inherits its properties from it. Our lifting $L(B)$ is defined only for equivalence relations, but this does not seem a serious limitation in this context, since behavioural equivalence, which one aims to model by bisimulations, is always an equivalence relation. (Note, however, that the lifting $J(B)$ has been used in other contexts [14], where arbitrary relations are essential.)

The category **ERel** is isomorphic to the category **BTS** of test suites closed under union, intersection and complementation. This isomorphism relates $L(B)$ corresponds to the least fibred lifting $T(B)$ of B to **TS**. Lifting functors to test suites is in the spirit of coalgebraic modal logic, where families of predicate liftings induce such constructions. From a characterization of $T(B)$ we derive a modal logic for any functor on **Set**, and prove an expressivity result for a wide class of functors (we give proofs for ω-continuous functors here, but all results generalize to all accessible functors, as will be shown in the full version of this paper). The logic we obtain is exactly the same as the logic of polyadic modalities of [13]. This provides further insight into the canonicity and importance of Schröder's results.

The paper is organized as follows. Section 2 treats preliminaries. In Section 3, basic definitions concerning fibrations are given together with two examples used throughout the paper. In Section 4, the least relation lifting $L(B)$ is characterized and compared with Jacobs's lifting $J(B)$. In Section 5, the least test suite lifting $T(B)$ is characterized and studied, and in Section 6 we define the modal logic related to it. We also give some examples suggesting that polyadic modalities can be useful to describe practically important cases.

I would like to thank Gordon Plotkin for inspiring discussions, and Lutz Schröder for letting me know of his recent related work. I am grateful to Alexander Kurz and to an anonymous referee for pointing out several inaccuracies.

2 Preliminaries

Let **Set** denote the category of sets and functions, and **Pos** the category of partial orders and monotonic functions. For an endofunctor B on any category, a B-coalgebra is a morphism $h : X \to BX$. A coalgebra morphism from $g : X \to BX$ to $h : Y \to BY$ is a morphism $f : X \to Y$ such that $Bf \circ g = h \circ f$. For any B, all B-coalgebras and their morphisms form a category. For any $h : X \to BX$, the unique morphism to the final B-coalgebra (if it exists) is called the coinductive extension of h.

Many functors on **Set** admit final coalgebras; in particular, all ω-continuous ones do. A functor B is called ω-continuous if it preserves limits of diagrams of the shape

$$X_0 \longleftarrow X_1 \longleftarrow X_2 \longleftarrow X_3 \cdots\cdots$$

The carrier Z of a final B-coalgebra arises from the limiting cone

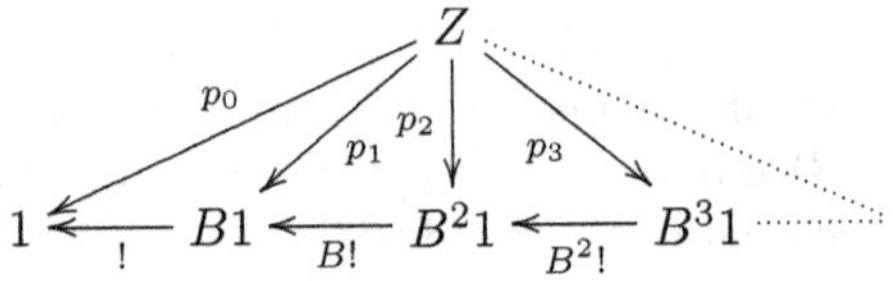

(1)

where 1 is the final object in the underlying category, and $! : B1 \to 1$ is the unique map from $B1$.

Many kinds of state-based systems can be modelled as coalgebras. For example, finitely branching labelled transition systems are coalgebras $h : X \to \mathcal{P}_f(A \times X)$, where $\mathcal{P}_f$ is the covariant finite powerset functor, and A is a fixed set of labels. Coalgebras for this functor are described by Hennessy-Milner logic [7], which is defined by the grammar:

$$\phi ::= \mathtt{tt} \mid \phi \wedge \phi \mid \neg\phi \mid \langle a\rangle\, \phi$$

Given any coalgebra $h : X \to \mathcal{P}_f(A\times X)$, every formula of this logic is interpreted as a predicate on X or, equivalently, as a function $[\![\phi]\!] : X \to 2$ (here and in the following, 2 denotes the two-element set $\{\mathtt{tt}, \mathtt{ff}\}$), defined inductively in the natural way, with the only interesting clause:

$$[\![\langle a\rangle\, \phi]\!](x) = \mathtt{tt} \iff \exists\, \langle a, y\rangle \in hx.\ [\![\phi]\!](y) = \mathtt{tt}$$

This logic is invariant under behavioural equivalence. Moreover, it is expressive: if two states are not distinguished by Hennessy-Milner formulae, they are identified by the coinductive extension of h.

3 Fibrations and Lifting

Any functor $(-)^* : \mathbf{Set}^{op} \to \mathbf{Pos}$ gives rise to a *total category* $\mathbf{Set}^*$, via the so-called Grothendieck construction: objects of $\mathbf{Set}^*$ are pairs $\langle X, R\rangle$ such that $R \in X^*$, and morphisms $f : \langle X, R\rangle \to \langle Y, S\rangle$ are functions $f : X \to Y$ such that $R \le f^*S$, where $\le$ is the ordering relation in X^*.

The obvious forgetful functor $p : \mathbf{Set}^* \to \mathbf{Set}$, mapping a pair $\langle X, R\rangle$ to its underlying set X, is then a *fibration*. This is only a special case of a more general definition used in fibration theory (see [15] for a detailed presentation), but only examples of this kind are considered in this paper. Slightly abusing the terminology, we will simply call functors $(-)^* : \mathbf{Set}^{op} \to \mathbf{Pos}$ fibrations. The partial order X^* is called the *fibre* over X. The map $f^* : Y^* \to X^*$ is called the *reindexing* along f.

Example 1. Define $(-)^\dagger : \mathbf{Set}^{op} \to \mathbf{Pos}$ by

$$X^\dagger = \langle \mathcal{P}(X \times X), \subseteq\rangle$$
$$f^\dagger(S) = \{\, \langle x, y\rangle \in X \times X \ :\ (fx)S(fy)\,\}$$

for any set X, function $f : X \to Y$ and relation $S \subseteq Y \times Y$. The proof of functoriality is very easy and omitted here. In the total category arising from this fibration, objects are pairs $\langle X, R\rangle$ with R a binary relation on X, and a function $f : X \to Y$ is a morphism between $f : \langle X, R\rangle \to \langle Y, S\rangle$ if and only if xRy implies $(fx)S(fy)$. The total category is therefore the category $\mathbf{Rel}$ of binary relations and relation-preserving functions. The fibre over a set X is the set of all binary relations on X, partially ordered by inclusion.

Note that for any $f : X \to Y$, the reindexing $f^\dagger : Y^\dagger \to X^\dagger$ maps equivalence relations to equivalence relations. It is therefore possible to restrict the functor

$(-)^\dagger$ to yield equivalence relations only; the total category arising from this fibration is the category **ERel** of equivalence relations and relation-preserving functions.

Example 2. Denote $2 = \{\texttt{tt}, \texttt{ff}\}$. A *test* on a set X is a function $V : X \to 2$. (A test can be also seen as a subset of its domain, but the functional representation will make some definitions and results in Section 5 look more natural.) A *test suite* on X is simply a set of tests on X. To say that θ is a test suite on a set X, we write $\theta : X \rightrightarrows 2$. Define $(-)^\ddagger : \mathbf{Set}^{op} \to \mathbf{Pos}$ by

$$X^\ddagger = \langle\{\theta : X \rightrightarrows 2\}, \supseteq\rangle$$
$$f^\ddagger(\vartheta) = \{\, V \circ f \,:\, V \in \vartheta \,\}$$

for any set X, function $f : X \to Y$ and test suite $\vartheta : Y \rightrightarrows 2$. Note that fibres $X^\ddagger$ are ordered by reverse inclusion. The proof of functoriality is again very easy and omitted here. The total category arising from this fibration is denoted by **TS**. As object in **TS** is a pair $\langle X, \theta\rangle$ where $\theta : X \rightrightarrows 2$, and a morphism $f : \langle X, \theta\rangle \to \langle Y, \vartheta\rangle$ is a function $f : X \to Y$ such that for any test $V \in \vartheta$, one has $V \circ f \in \theta$.

To get some intuition, note that any topology can be seen as a test suite subject to additional closure conditions. Morphisms in **TS** between such test suites are then exactly continuous functions between topological spaces.

Intersections, unions and complements of tests are defined in the obvious way. It is easy to see that if a test suite $\vartheta : Y \rightrightarrows 2$ is a complete field of sets (i.e., it is closed under arbitrary intersection, union and complementation), then for any $f : X \to Y$, the test suite $f^\ddagger\vartheta$ is also a complete field of sets. It is therefore possible to restrict $(-)^\ddagger$ to complete fields of sets only; the total category arising from this fibration will be denoted **BTS**.

Proposition 3. The categories **ERel** and **BTS** are isomorphic; the two maps defined on objects as

$$\begin{aligned} \langle X, \theta\rangle &\mapsto \langle X, \{\, \langle x, y\rangle \in X \times X \,:\, \forall V \in \theta.\ Vx = Vy \,\}\rangle \\ \langle X, R\rangle &\mapsto \langle X, \{\, V : X \to 2 \,:\, \forall x, y \in X.\ xRy \Rightarrow Vx = Vy \,\}\rangle \end{aligned} \qquad (2)$$

and as identities on morphisms are mutually inverse functors. □

For any fibration $(-)^* : \mathbf{Set}^{op} \to \mathbf{Pos}$, say that a functor $B^* : \mathbf{Set}^* \to \mathbf{Set}^*$ *lifts* a functor $B : \mathbf{Set} \to \mathbf{Set}$ if

$$p \circ B^* = B \circ p$$

where $p : \mathbf{Set}^* \to \mathbf{Set}$ is the forgetful functor associated to $(-)^*$. Given a functor $B : \mathbf{Set} \to \mathbf{Set}$, a lifting of B to B^* is uniquely determined by its action on the elements of fibres, i.e., by a family (indexed by sets X) of functions

$$B_X : X^* \to (BX)^*$$

For any such family, the map defined by

$$\begin{aligned} B^* \langle X, R \rangle &= \langle BX, B_X R \rangle \\ B^* f &= Bf \end{aligned} \tag{3}$$

is a functor on $\mathbf{Set}^*$ if and only if all functions B_X are monotonic and, for any function $f : X \to Y$ and any $S \in Y^*$,

$$B_X(f^* S) \leq (Bf)^*(B_Y S)$$

If, moreover, the above inequality holds as equality:

$$B_X(f^* S) = (Bf)^*(B_Y S) \tag{4}$$

we say that B^* is *fibred.*

In the following, an iterated version of B_X will be useful. For any $R \in X^*$, define $B^n R \in (B^n X)^*$ by induction:

$$\begin{aligned} B_X^0 R &= R \\ B_X^{n+1} R &= B_{B^n X} B_X^n R \end{aligned}$$

The following result, used in Section 5, is a characterization of final coalgebras for fibred liftings of ω-continuous functors.

Proposition 4. Assume a fibration $(-)^* : \mathbf{Set}^{op} \to \mathbf{Pos}$ such that

- every fibre has arbitrary intersections $\bigwedge$, and
- reindexing functions preserve intersections.

Every fibred functor $B^* : \mathbf{Set}^* \to \mathbf{Set}^*$ lifting an ω-continuous functor $B : \mathbf{Set} \to \mathbf{Set}$ admits a final coalgebra

$$\phi : \langle Z, \zeta \rangle \to \langle BZ, B_Z \zeta \rangle$$

where $\phi : Z \to BZ$ is the final B-coalgebra obtained as in (1), and

$$\zeta = \bigwedge_{n \in \mathbb{N}} p_n^*(B_1^n T)$$

where T is the largest element (i.e., the empty intersection) in 1^*.

4 The Least Relation Lifting

In [4,5,6], Jacobs et al. showed how to lift any functor $B : \mathbf{Set} \to \mathbf{Set}$ to a functor $J(B) : \mathbf{Rel} \to \mathbf{Rel}$ (in fact they work in a slightly more general setting, with relations of the type $X \times Y$ rather than $X \times X$; here their definition is simplified to fit into the setting of this paper), defined by the following action:

$$J(B)_X(R) = \{ \langle \alpha, \beta \rangle \in BX \times BX \ : \ \exists w \in BR.(B\pi_1)w = \alpha, B(\pi_2)w = \beta \}$$

where $\pi_1, \pi_2 : R \to X$ are projections. They then define a bisimulation (on the underlying B-coalgebra) to be a $J(B)$-coalgebra. If B preserves weak pullbacks, $J(B)$ has many useful properties. For example, it is fibred, bisimulations are closed under unions, and (the relation component of the carrier of) the greatest bisimulation on the final B-coalgebra is the equality relation.

The definition of $J(B)$ is elegant in that it does not depend on the structure of B. It is also closely related to the coalgebra span approach to bisimulation [16,1]. However, speaking in terms of fibrations, it does not really seem canonical: it is not immediately clear why one should choose this particular lifting from many available ones, and only the numerous useful properties of $J(B)$ convince one that the lifting is "right".

In search for a canonical lifting of a functor B to **Rel** it is natural to look at the least and the greatest (fibre-wise) fibred liftings. These turn out to be trivial: as is easy to check, the least (greatest) fibred lifting of any B to **Rel** is defined by the action mapping any relation on X to the empty (resp. full) relation on BX. However, when one restricts to the category **ERel** of equivalence relations, the least fibred lifting becomes more interesting. In the remainder of this section we shall see the construction of this lifting, denoted $L(B)$, and some of its properties. The construction works for any functor that preserves monos. This is a much weaker assumption than that of weak pullback preservation. Practically all functors on **Set** preserve monos, and every functor on **Set** preserves all monos with nonempty domains.

Observe that $L(B)$ is fully determined by its action on equality relations. Indeed, take any set X and an equivalence R on X, and consider the abstraction function

$$[-]_R : X \to X/_R$$

Note that $[-]_R^\dagger \Delta_{X/_R} = R$ (here and in the following, Δ_X denotes the equality relation on X.) Pictorially:

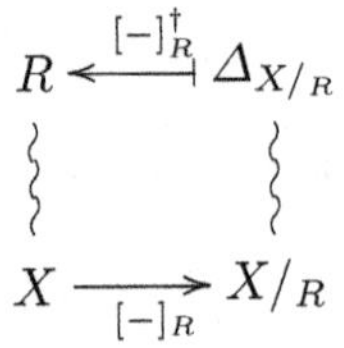

Then

$$L(B)_X R = L(B)_X([-]_R^\dagger(\Delta_{X/_R})) = (B[-]_R)^\dagger(L(B)_{X/_R}(\Delta_{X/_R}))$$

The second equality holds by the fibredness condition (4) on $L(B)_X$. Note that $(B[-]_R)^\dagger$ in the rightmost expression does not depend on the lifting $L(B)_X$, so the left hand side is fully determined by $L(B)_{X/_R}(\Delta_{X/_R})$.

The least possible value of the latter is $\Delta_{B(X/_R)}$, as this is the smallest equivalence relation on $B(X/_R)$. Then one has

$$\begin{aligned} L(B)_X R &= (B[-]_R)^\dagger(\Delta_{B(X/_R)}) = \\ &= \{ \langle \alpha, \beta \rangle \subseteq BX \times BX \,:\, (B[-]_R)\alpha = (B[-]_R)\beta \} \end{aligned} \tag{5}$$

From the above reasoning it is clear that every fibred lifting of B is fibre-wise larger than $L(B)$ defined as in (5). It is also obvious that for any equivalence relation R on X, $L(B)_X R$ is also an equivalence relation. It must yet be checked that $L(B)_X$ defines a fibred functor. This follows from a collection of properties analogous to those of $J(B)$ as listed in [6]:

Proposition 5. If B preserves monos, then the maps $L(B)_X$:

(i) preserve equality relations: $L(B)(\Delta_X) = \Delta_{BX}$;
(ii) half-preserve the transitive closure of relational composition: for R, S equivalence relations on X, the relational composition $S \circ R = \{\langle x, z\rangle : \exists y.\, xRy, ySz\}$ satisfies: $L(B)_X(\overline{S \circ R}) \supseteq \overline{L(B)_X R \circ L(B)_X S}$;
(iii) are monotonic: if $R \subseteq S$ then $L(B)_X R \subseteq L(B)_X S$;
(iv) preserve reversals: $L(B)_X(R^{op}) = (L(B)_X R)^{op}$;
(v) preserve reindexing: for any $f : X \to Y$ and any relation S on Y, the condition (4) from Section 3 holds.

Note that the property (iv) above is trivially satisfied, and is included here for a comparison with an analogous result in [6].

Corollary 6. If B preserves monos, then $L(B)$ defined as in (3) from maps $L(B)_X$ defined in (5) is a fibred functor on **ERel**.

Proof. : Immediate from properties (iii) and (v) in Proposition 5. □

By a *bisimulation* on a B-coalgebra $h : X \to BX$ we will mean simply an $L(B)$-coalgebra

$$h : \langle X, R\rangle \to \langle BX, L(B)_X R\rangle$$

In other words, it is an equivalence relation R on X such that

$$xRy \Longrightarrow (B[-]_R)(hx) = (B[-]_R)(hy)$$

The next theorem states some properties of bisimulations, analogous to those mentioned in [6] (see also [1]).

Proposition 7. If B preserves monos, then for any coalgebras $g : X \to BX$, $h : Y \to BY$:

(i) Bisimulations are closed under transitive closures of arbitrary unions, hence there exists a greatest bisimulation, called *bisimilarity.*
(ii) Equality relations are bisimulations.
(iii) If $f : X \to Y$ is a B-coalgebra homomorphism from g to h, then any x is bisimilar to fx in the coalgebra

$$[B\iota_1 \circ g, B\iota_2 \circ h] : X + Y \to B(X + Y)$$

(iv) If $f : X \to Y$ is a coalgebra homomorphism from g to h then x is bisimilar to x' in X if and only if fx is bisimilar to fx' in Y.

(v) If B is ω-continuous, then (a final B-coalgebra exists and) bisimilarity on the final B-coalgebra is equality.

Propositions 5 and 7 show that $L(B)$ satisfies all the useful properties of $J(B)$, without the assumption that B preserves weak pullbacks. Moreover:

Proposition 8. If B preserves weak pullbacks, then $L(B) = J(B)$.

Therefore $J(B)$, when restricted to equivalence relations, is essentially a special case of the canonical lifting $L(B)$.

5 The Least Test Suite Lifting

In Section 3, it was noted that the category **ERel** of equivalence relations and relation-preserving functions is isomorphic to the category **BTS** of test suites that are complete fields of sets. This means that for any $B : \mathbf{Set} \to \mathbf{Set}$, the least fibred lifting $L(B)$ to **ERel** corresponds to the least fibred lifting to **BTS**; its concrete description can be obtained from (5) and the isomorphisms (2). However, the resulting definition is rather unwieldy. Since least fibred liftings of functors to test suites will be of use in Section 6, instead we use techniques as in Section 4 to derive a characterization of the least fibred lifting to **TS**, denoted $T(B)$ in the following. It turns out that the new characterization does not require even the very mild condition of B preserving monos.

For any set X, denote the test suite of all tests on X by $\mathbb{T}_X$. Begin by observing that any fibred lifting of $B : \mathbf{Set} \to \mathbf{Set}$ to **TS** is fully determined by its values on the $\mathbb{T}_X$. Indeed, consider any set X and any test suite $\theta : X \rightrightarrows 2$. Define a function $e_\theta : X \to 2^\theta$ by

$$e_\theta(x)(V) = Vx \qquad \text{for } V \in \theta$$

Lemma 9. For any $\theta : X \rightrightarrows 2$, $\theta = e_\theta^\ddagger \mathbb{T}_{2^\theta}$; pictorially,

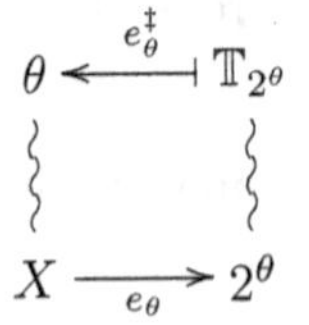

□

Then, by the fibredness condition (4), for any $\theta : X \rightrightarrows 2$ we have

$$T(B)_X\theta = T(B)_X(e_\theta^\ddagger \mathbb{T}_{2^\theta}) = (Be_\theta)^\ddagger T(B)_{2^\theta}\mathbb{T}_{2^\theta}$$

The least (wrt. the fibre ordering) possible value for $T(B)_{2^\theta}\mathbb{T}_{2^\theta}$ is $\mathbb{T}_{B(2^\theta)}$, hence the least candidate for a fibred lifting $T(B)_X$ of B is defined by

$$T(B)_X\theta = (Be_\theta)^\ddagger \mathbb{T}_{B(2^\theta)} = \{\, W \circ Be_\theta \;:\; W : B(2^\theta) \to 2 \,\} \qquad (6)$$

Theorem 10. For any functor B on **Set**, the above action $T(B)_X$ defines a fibred functor on **TS** as in (3).

Proof. For any X, the action $T(B)_X$ is clearly monotonic, therefore the only condition to check is that for any $f : X \to Y$ and any $\theta : Y \rightrightarrows 2$ the equality

$$(Bf)^{\ddagger}T(B)_Y\theta = T(B)_X(f^{\ddagger}\theta) \tag{7}$$

holds. To prove this, we begin with two easy lemmas, assuming arbitrary $f : X \to Y$ and $\theta : Y \rightrightarrows 2$.

Lemma 11. For any $Z : 2^{f^{\ddagger}\theta} \to 2$ there exists $W : 2^{\theta} \to 2$ such that

$$Z = W \circ 2^{(-\circ f)}$$

Proof. The function $2^{(-\circ f)} : 2^{f^{\ddagger}\theta} \to 2^{\theta}$ is always a mono, so (since its domain is nonempty) it is a section. Take $W = Z \circ u$, where $u : 2^{\theta} \to 2^{f^{\ddagger}\theta}$ is the corresponding retraction. □

Lemma 12. The following diagram commutes:

$$\begin{array}{ccc} X & \xrightarrow{e_{f^{\ddagger}\theta}} & 2^{f^{\ddagger}\theta} \\ {\scriptstyle f}\downarrow & & \downarrow{\scriptstyle 2^{(-\circ f)}} \\ Y & \xrightarrow[e_{\theta}]{} & 2^{\theta} \end{array}$$

Proof. Calculate, for any $x \in X$ and $V \in \theta$,

$$\begin{aligned}(2^{(-\circ f)}(e_{f^{\ddagger}\theta}(x)))(V) &= ((e_{f^{\ddagger}\theta}(x)) \circ (- \circ f))(V) \\ &= e_{f^{\ddagger}\theta}(x)(V \circ f) \\ &= (V \circ f)(x) = V(f(x)) = e_{\theta}(f(x))(V)\end{aligned}$$

□

We are now ready to prove (7). Calculate

$$\begin{aligned}(Bf)^{\ddagger}T(B)_Y\theta &= \{ W \circ Be_{\theta} \circ Bf \ : \ W : B(2^{\theta}) \to 2 \} \\ &= \left\{ W \circ B(2^{(-\circ f)}) \circ Be_{f^{\ddagger}\theta} \ : \ W : B(2^{\theta}) \to 2 \right\} \\ &= \left\{ Z \circ Be_{f^{\ddagger}\theta} \ : \ Z : B(2^{f^{\ddagger}\theta}) \to 2 \right\} = T(B)_X(f^{\ddagger}\theta)\end{aligned}$$

using two lemmas above. □

Note that the characterization of $T(B)$ does not require B to preserve monos. From properties of $L(B)$ and $T(B)$ it follows that:

Corollary 13. If B preserves monos then $T(B)$, restricted to **BTS**, coincides with $L(B)$ along the isomorphisms between **BTS** and **ERel**.

Moreover, the properties of $L(B)$ as stated in Propositions 5 and 7 translate to analogous properties of $T(B)$. Two of these properties (corresponding to properties (iv) and (v) from Proposition 7) will be useful in Section 6, so we restate them here, and we provide independent proofs.

In the following theorem, given any $h : X \to BX$, elements $x, y \in X$ are called *bisimilar* if there exists a $T(B)$-coalgebra $h : \langle X, \theta \rangle \to \langle BX, T(B)_X\theta \rangle$ such that x, y are not distinguishable by tests from θ, which is then called a bisimulation suite.

Theorem 14. If $f : X \to Y$ is a B-coalgebra homomorphism from g to h then x is bisimilar to x' in X if and only if fx is bisimilar to fx' in Y.

Proof (sketch). For any bisimulation suite θ on h such that $V(fx) = V(fx')$ for all $V \in \theta$, $f^\ddagger\theta$ is a bisimulation suite on g. Moreover, $Vx = Vx'$ for all $V \in f^\ddagger\theta$. Similarly, for any bisimulation suite θ on g such that $Vx = Vx'$ for all $V \in \theta$,

$$f_\ddagger = \{ V : Y \to 2 : V \circ f \in \theta \}$$

is a bisimulation suite on h, and $V(fx) = V(fx')$ for all $V \in f_\ddagger\theta$. □

Theorem 15. (Expressivity) Assume $B : \mathbf{Set} \to \mathbf{Set}$ is ω-continuous and preserves monos. Then a final $T(B)$-coalgebra

$$\phi : \langle Z, \zeta \rangle \to \langle BZ, T(B)_Z\zeta \rangle$$

exists, and the test suite $\zeta : Z \rightrightarrows 2$ is jointly monic, i.e., any two elements of Z are distinguished by a test from ζ.

Proof. Since B is ω-continuous, a final B-coalgebra

$$\phi : Z \to BZ$$

exists and is obtained as in (1). Since $T(B)$ is fibred, by Proposition 4 a final $T(B)$-coalgebra as above exists and

$$\zeta = \bigcup_{n \in \mathbb{N}} p_n^\ddagger (T(B)_1^n \emptyset)$$

where $\emptyset : 1 \rightrightarrows 2$ is the final test suite, i.e., the empty test suite on 1. Pictorially:

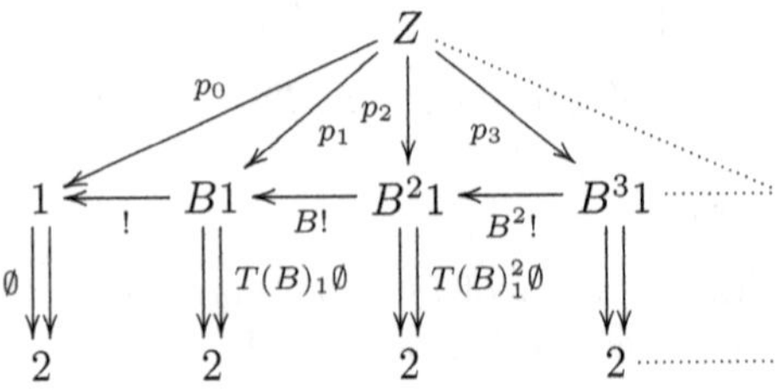

The family of functions $\{ p_n : Z \to B^n 1 : n \in \mathbb{N} \}$, being a limiting cone, is jointly monic. Since compositions of jointly monic families is always jointly monic, it is enough to ensure that all test suites $T(B)_1^n\emptyset$ are jointly monic. The empty test suite on 1 is jointly monic, so it is enough to show $T(B)$ preserves joint monicity of test suites: whenever $\theta : X \rightrightarrows 2$ is jointly monic, then $T(B)_X\theta : BX \rightrightarrows 2$ is jointly monic.

To show that, recall that

$$T(B)_X\theta = \left\{ W \circ Be_\theta \; : \; W : B(2^\theta) \to 2 \right\}$$

First, if θ is jointly monic then e_θ is a mono. Indeed, assume any $x, y \in X$ such that $e_\theta(x) = e_\theta(y)$. This, by definition of e_θ, means that for all $V \in \theta$, $Vx = Vy$. Since θ is jointly monic, $x = y$.

Since B preserves monos, also Be_θ is a mono. Moreover, the family of all functions from $B(2^\theta)$ to 2 is jointly monic, as 2 is a cogenerator in **Set**. Precomposing a jointly monic family with a mono yields a jointly monic family, therefore $T(B)_X\theta$ is jointly monic. This completes the proof. s □

Results and examples of Section 6 should give some intuition on how the final $T(B)$-coalgebra looks in concrete cases.

6 Application to Coalgebraic Modal Logic

We proceed to show how the least fibred lifting $T(B)$ of an ω-continuous functor B to **TS** gives rise to an expressive modal logic for B-coalgebras.

Given a functor $B : \mathbf{Set} \to \mathbf{Set}$, define the set $\mathcal{L}$ of *formulae* inductively: for any $k \in \mathbb{N}$, and for any function $U : B(2^k) \to 2$, if $\phi_1, \ldots, \phi_k \in \mathcal{L}$ then

$$[U](\phi_1, \ldots, \phi_k) \in \mathcal{L}$$

The interpretation of $\mathcal{L}$ in a coalgebra $h : X \to BX$ is a function

$$[\![-]\!] : \mathcal{L} \to (X \to 2)$$

defined inductively by

$$[\![[U](\phi_1, \ldots, \phi_k)]\!] = U \circ B([\![\phi_1]\!] \times \cdots \times [\![\phi_k]\!]) \circ B\Delta_k \circ h$$

where $\Delta_k : X \to X^k$ is the diagonal map, defined by $\Delta_k(i)(x) = x$.

The relation of this logic to the lifting $T(B)$ from Section 5 is made apparent in the following theorem:

Theorem 16. If B is ω-continuous and preserves monos, then for any coalgebra $h : X \to BX$ with the coinductive extension $f : X \to Z$,

$$\{ [\![\phi]\!] \; : \; \phi \in \mathcal{L} \} = f^\ddagger \zeta$$

where $\zeta : Z \rightrightarrows 2$ comes from the final $T(B)$-coalgebra as in Proposition 4.

Proof. (Sketch). To prove the left-to right inclusion, proceed by structural induction on $\mathcal{L}$. Consider any $\phi = [U](\phi_1, \ldots, \phi_k)$, for some $U : B(2^k) \to 2$. By the inductive assumption, $V_1, \ldots, V_k \in f^\ddagger\zeta$, hence there exists a function $\rho : k \to f^\ddagger\zeta$. Define $W : B(2^{f^\ddagger\zeta}) \to 2$ by

$$W = U \circ B(2^\rho)$$

Note that for any $x \in X$, $i \leq k$,

$$(2^\rho(e_{f^\ddagger\zeta}(x))(i) = e_{f^\ddagger\zeta}(x)(\rho(i)) = V_i x$$

therefore

$$2^\rho \circ e_{f^\ddagger\zeta} = (V_1 \times \cdots \times V_k) \circ \Delta_k$$

Then calculate

$$\begin{aligned} [\![\phi]\!] &= U \circ B(V_1 \times \cdots \times V_k) \circ B\Delta_k \circ h = \\ &= U \circ B2^\rho \circ Be_{f^\ddagger\zeta} \circ h = \\ &= W \circ Be_{f^\ddagger\zeta} \circ h \in h^\ddagger T(B)_X(f^\ddagger\zeta) \subseteq f^\ddagger\zeta \end{aligned}$$

To prove the right-to-left inclusion, define $\theta_n = f^\ddagger(p_n^\ddagger(T(B)_1^n\emptyset))$ (see the proof of Theorem 15). Then $f^\dagger\zeta = \bigcup_{n\in\mathbb{N}} \theta_n$. One shows by induction on n that every test from θ_n is an interpretation of a formula from $\mathcal{L}$. To this end, prove from fibredness of $T(B)$ that

$$\theta_n = h^\ddagger(T(B)_X(\theta_{n-1}))$$

therefore each test from θ_n is of the form

$$V = W \circ B(e_{\theta_{n-1}}) \circ h$$

for some $W : B(2^{\theta_{n-1}}) \to 2$. One proves by induction that all θ_n are finite. Take $k = |\theta_{n-1}|$, fix a bijection $\sigma : k \to \theta_{n-1}$, and define

$$U = W \circ B(2^\sigma)$$

Consider the formula

$$\phi = [U](\phi_1, \ldots, \phi_k)$$

such that $[\![\phi_i]\!] = \sigma(i)$ (such ϕ_i exist by the inductive assumption). Then calculate, similarly as above,

$$\begin{aligned} [\![\phi]\!] &= U \circ B(\sigma(1) \times \cdots \times \sigma(k)) \circ B\Delta_k \circ h = \\ &= W \circ B2^\sigma \circ B(\sigma(1) \times \cdots \times \sigma(k)) \circ B\Delta_k \circ h = \\ &= W \circ Be_{\theta_{n-1}} \circ h = V \end{aligned}$$

□

Corollary 17. If B is ω-continuous and preserves monos, the logic $\mathcal{L}$ is adequate and expressive.

Proof. Combine Theorems 14, 15 and 16. □

The logic $\mathcal{L}$ and its interpretation is essentially the expressive logic of polyadic predicate liftings as defined in [13], so the above expressivity result should come as no surprise. The purpose of this section was to show a strong link of the logic to the least fibred lifting of B, thus providing further insight into the canonicity and importance of results from [13].

Example 18. Consider $B = \mathcal{P}_f(A \times X)$, where A is a fixed set of labels. Coalgebras for B are finitely branching labelled transition systems. The finitary Hennessy-Milner logic, adequate and expressive for such coalgebras, can be equivalently defined inductively as follows: for any logical operator $b : 2^k \to 2$, and for any labels $a_1, \ldots, a_k \in A$ and for any formulae $\phi_1, \ldots, \phi_k$,

$$b[\langle a_1\rangle \phi_1, \ldots, \langle a_k\rangle \phi_k]$$

is a formula. The interpretation of formulae in a coalgebra $h : X \to BX$ is a function from formulae to $X \to 2$ defined inductively:

$$[\![b[\langle a_1\rangle \phi_1, \ldots, \langle a_k\rangle \phi_k]]\!](x) = b(v_1, \ldots, v_k)$$

where $v_i = \mathtt{tt}$ iff $\langle a_i, y\rangle \in hx$ such that $[\![\phi_i]\!] y = \mathtt{tt}$. An interpretation-preserving translation between this and the usual representation of Hennessy-Milner logic is straightforward. The translation is not bijective; for example, Hennessy-Milner formulae $\mathtt{tt} \wedge \langle a\rangle (\mathtt{tt} \vee \langle a\rangle \mathtt{tt})$ and $\langle a\rangle \mathtt{tt}$ are mapped to the same formula. However, all identified formulae are logically equivalent.

This version of Hennessy-Milner logic can be translated bijectively to the logic $\mathcal{L}$ defined above. The translation, denoted γ, is defined by induction:

$$\gamma(b[\langle a_1\rangle \phi_1, \ldots, \langle a_k\rangle \phi_k]) = [U](\gamma(\phi_1), \ldots, \gamma(\phi_k))$$

where $U : \mathcal{P}_f(A \times 2^k) \to 2$ is defined by

$$U(\beta) = b(u_1, \ldots, u_k)$$

where $u_i = \mathtt{tt}$ iff $\langle a_i, \mathtt{tt}\rangle \in \beta$. It is straightforward to check that this translation is bijective and preserves interpretation of formulae: $[\![\phi]\!] = [\![\gamma(\phi)]\!]$.

Example 19. In [13], Schröder proves that if, for a functor B, the family of functions

$$(Bf : BX \to B2)_{f:X\to 2}$$

is jointly monic for any set X, then B admits an expressive modal logic with unary modalities. He also shows a few functors for which this condition fails. Here is another example, maybe simpler and more natural: the functor $BX = \mathcal{P}_f(A \times X \times X)$, for A a nonempty fixed set of labels. Indeed, for $X = \{x, y, z\}$, the set

$$\{\langle a, x, x\rangle, \langle a, x, y\rangle, \langle a, y, z\rangle, \langle a, x, z\rangle\}$$

is identified with

$$\{\langle a,x,x\rangle, \langle a,x,y\rangle, \langle a,y,z\rangle\}$$

under Bf for any $f : X \to 2$. This means that the expressive logic $\mathcal{L}$ for this functor makes essential use of modalities of multiple arity. Proceeding as in Example 18 above, one shows a correspondence between $\mathcal{L}$ and a logic similar to Hennessy-Milner logic and defined by:

$$\phi ::= \mathtt{tt} \mid \phi \wedge \phi \mid \neg\phi \mid \langle a\rangle(\phi,\phi)$$

(where a ranges over A), with an interpretation (for a given $h : X \to BX$) as for Hennessy-Milner logic, except:

$$[\![\langle a\rangle(\phi_1,\phi_2)]\!](\beta) = \mathtt{tt} \text{ iff } \langle a,x,y\rangle \in \beta \text{ s.t. } [\![\phi_1]\!]x = \mathtt{tt}, [\![\phi_2]\!]x = \mathtt{tt}$$

This logic has actually been used in the literature, to describe a simple calculus with process passing [17]. This example shows that the extension of coalgebraic modal logic to modalities of arbitrary arity, as done first in [13], is of practical importance. Note, however, that allowing multiple sorts of logical formulae one can define an expressive logic for this functor with only unary modalities, along the lines of [18].

References

1. Rutten, J.J.M.M.: Universal coalgebra: a theory of systems. Theoretical Computer Science **249** (2000) 3–80
2. Park, D.M.: Concurrency and automata on infinite sequences. Lecture Notes in Computer Science **140** (1981) 195–219
3. Milner, R.: Communication and Concurrency. Prentice Hall (1988)
4. Hermida, C., Jacobs, B.: Structural induction and coinduction in a fibrational setting. Information and Computation **145** (1998) 107–152
5. Jacobs, B.: Exercises in coalgebraic specification. In Crole, R., Backhouse, R., Gibbons, J., eds.: Algebraic and Coalgebraic Metods in the Mathematics of Program C onstruction. Volume 2297 of Lecture Notes in Computer Science. Springer (2002)
6. Jacobs, B., Hughes, J.: Simulations in coalgebra. Electronic Notes in Theoretical Computer Science **82** (2003)
7. Hennessy, M., Milner, R.: Algebraic laws for nondeterminism and concurrency. Journal of the ACM **32** (1985) 137–161
8. Jacobs, B.: Towards a duality result in the modal logic for coalgebras. In: Proc. CMCS 2000. Volume 33 of Electronic Notes in Theoretical Computer Science. (2000)
9. Klin, B.: Abstract Coalgebraic Approach to Process Equivalence for Well-Behaved Operational Semantics. PhD thesis, BRICS, Aarhus University (2004)
10. Moss, L.: Coalgebraic logic. Annals of Pure and Applied Logic **96** (1999) 177–317
11. Pattinson, D.: Expressivity Results in the Modal Logic of Coalgebras. PhD thesis, Universität München (2001)
12. Pattinson, D.: Semantical principles in the modal logic of coalgebras. In: Proc. STACS 2001. Volume 2010 of Lecture Notes in Computer Science., Springer Verlag (2001)

13. Schröder, L.: Expressivity of coalgebraic modal logic: the limits and beyond. In: Procs. FOSSACS'05. (to appear)
14. Jacobs, B.: Trace semantics for coalgebras. In: Proc. CMCS 2004. Volume 106 of Electronic Notes in Theoretical Computer Science. (2004)
15. Jacobs, B.: Categorical Logic and Type Theory. Number 141 in Studies in Logic and the Foundations of Mathematics. North Holland (1999)
16. Aczel, P., Mendler, N.: A final coalgebra theorem. In: Proc. CTCS'89. Volume 389 of Lecture Notes in Computer Science. (1989) 357–365
17. Thomsen, B.: A theory of higher-order communicating systems. Information and Computation **116** (1995)
18. Cirstea, C., Pattinson, D.: Modular construction of modal logic. In: Proc. CONCUR'04. Volume 3170 of Lecture Notes in Computer Science. (2004)

Ultrafilter Extensions for Coalgebras

C. Kupke[1], A. Kurz[2,⋆], and D. Pattinson[3]

[1] CWI and Universiteit van Amsterdam, Amsterdam, The Netherlands
[2] Department of Computer Science, University of Leicester, UK
[3] Imperial College, London, UK

Abstract. This paper studies finitary modal logics as specification languages for Set-coalgebras (coalgebras on the category of sets) using Stone duality. It is well-known that Set-coalgebras are not semantically adequate for finitary modal logics in the sense that bisimilarity does not in general coincide with logical equivalence. Stone-coalgebras (coalgebras over the category of Stone spaces), on the other hand, do provide an adequate semantics for finitary modal logics. This leads us to study the relationship of finitary modal logics and Set-coalgebras by uncovering the relationship between Set-coalgebras and Stone-coalgebras. This builds on a long tradition in modal logic, where one studies canonical extensions of modal algebras and ultrafilter extensions of Kripke frames to account for finitary logics. Our main contributions are the generalisations of two classical theorems in modal logic to coalgebras, namely the Jónsson-Tarski theorem giving a set-theoretic representation for each modal algebra and the bisimulation-somewhere-else theorem stating that two states of a coalgebra have the same (finitary modal) theory iff they are bisimilar (or behaviourally equivalent) in the ultrafilter extension of the coalgebra.

1 Introduction

To formalise transition systems as coalgebras for a functor $T : \mathsf{Set} \to \mathsf{Set}$ has many advantages. In particular, the theory of transition systems can be set up parametric in the 'type' T of the transition system and a number of techniques for coalgebras (e.g. final semantics, isomorphism theorems, final sequence, co-Birkhoff theorems) can be obtained by dualising the corresponding concepts for algebras (Rutten [18]). Unfortunately, when it comes to *specification languages for coalgebras*, it is more difficult to achieve results parametric in the functor T.

The idea that (variants of) modal logics are the natural logics for coalgebras goes back to Moss seminal paper [14]. Applying to modal logic dualised algebraic methods, leads to the insight that modal logic for coalgebras is dual to equational logic for algebras [11,13]. But the methods derived from this approach are adequate only for *infinitary* logics. This can be seen as a consequence of the fact that $\mathsf{Set}^{\mathrm{op}}$ is equivalent to the category of *complete* atomic Boolean algebras which correspond to *infinitary* propositional logic in the same way as Boolean algebras capture finitary propositional logic.

⋆ Partially support by the Nuffield Foundation Grant NUF-NAL04.

J.L. Fiadeiro et al. (Eds.): CALCO 2005, LNCS 3629, pp. 263–277, 2005.

Maybe for this reason, the approach towards (more realistic) finitary logics for coalgebras has been somewhat ad hoc. It essentially consisted in giving up parametricity in T and restricting attention to particular classes of functors [12,17,6]. More recently, Pattinson [15,16] has shown how these logic arise uniformly as *logics given by predicate liftings*. It is one of the aims of this paper to further develop this approach towards a theory of logics for coalgebras that is fully parametric in the functor T.

Another approach to finitary logics for coalgebras is to change the model theory, that is, to replace coalgebras over Set (Set-coalgebras) by coalgebras over Stone spaces (Stone-coalgebras) [10]. Stone-coalgebras generalise the so-called descriptive general frames which are known in modal logic as the standard adequate semantics for finitary modal logics. Here *adequate* means that the logic is sound and complete and that two states are bisimilar iff they have the same theory. The deeper reason for the adequateness of finitary modal logics and Stone-coalgebras is the duality of Boolean algebras and Stone spaces, see Johnstone [7].

In [9], we have shown that every sound logic $\mathcal{L}$ given by predicate liftings induces a functor L on the category BA of Boolean algebras. Using the dual equivalence of BA and the category Stone of Stone spaces, it follows that L has a 'dual' L^∂ on Stone and that L^∂-coalgebras provide an adequate semantics for $\mathcal{L}$.

The main issue of this paper can now be explained as follows: If a finitary modal logic for T-coalgebras is given by a functor L on BA, then an adequate semantics for this logic is provided by the Stone-coalgebras for the dual functor L^∂. The quest for a model theory of finitary modal logics for coalgebras now boils down to a comparison of T-coalgebras over Set and Stone-coalgebras for L^∂. This is the main theme of this paper. By building on the well-developed model theory of modal logics, where this question has been studied for the special case of Kripke frames and Kripke models, our main contribution is the generalisation of two important theorems of modal logic: The Jónsson-Tarski theorem and bisimulation-somewhere-else. The former result provides us with an completeness theorem, and the latter with a model-theoretic characterisation of logical equivalence.

Summary of Techniques: The main ingredients of our approach are depicted in the following *non-commuting* diagram

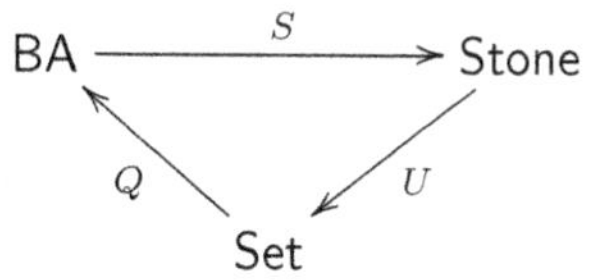

(1)

The category BA of Boolean algebras is the main building block of our logics, which are obtained by 'adding modal operators' to BA. The category Stone of Stone spaces is our main technical tool. Stone is 'categorically the same' as BA in the sense that Stone is dually equivalent to BA. But, as a category of topological spaces, Stone is sufficiently Set-like to be useful in the study of Set-based coalgebras.

The functor Q : Set $\to$ BA is the contravariant powerset functor mapping a set X to the algebra of predicates over X. The functor S is one part of the dual equivalence between Stone and BA and maps a Boolean algebra A to its space SA of ultrafilters giving a topological representation[1] of A. Finally, U is the forgetful functor that maps a space to its carrier set. Note that the one traversal of this diagram, starting at BA, produces the perfect [8] or canonical extension $QUS(A)$ for any Boolean algebra A. The traversal starting at Set produces the set of ultrafilters $USQ(X)$ over a set X (see e.g. [3] for more information).

One of our aims is to lift these constructions to T-coalgebras, where $T : \mathsf{Set} \to \mathsf{Set}$. This will be achieved by first translating a T-coalgebra to an L-algebra, for a suitable $L : \mathsf{BA} \to \mathsf{BA}$, then to transport this algebra by duality to an L^∂-coalgebra over Stone and finally back to a T-coalgebra where we use Q, S, U to map the carriers of the respective structures.

It has been shown in [9] that any logic $\mathcal{L}$ for T-coalgebras (as e.g. the logics in [15,16,6,17,12]) given by predicate liftings can be described by a functor L on BA (capturing syntax and proof rules) and a natural transformation $\delta : LQ \to QT$ (giving the coalgebraic semantics).

$\delta : LQ \to QT$

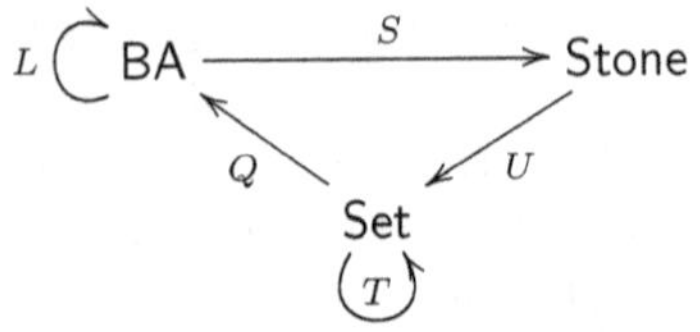

(2)

The transformation δ allows to lift Q to a functor $\tilde{Q} : \mathsf{Coalg}(T) \to \mathsf{Alg}(L)$. The semantics of formulas w.r.t. to a coalgebra $\xi : X \to TX$ is given by by the unique morphism from the initial L-algebra to $\tilde{Q}\xi$. The initial L-algebra is commonly known as the Lindenbaum algebra of the logic $\mathcal{L}$.

Summary of Results: We will show how to generalise two classic results from modal logic to coalgebras, namely the Jónsson-Tarski theorem and the bisimulation-somewhere-else result for ultrafilter extensions.

Jónsson-Tarski Theorem (Completeness). Given a modal logic described by L and δ, we extend $US : \mathsf{BA} \to \mathsf{Set}$ to a map $\tilde{U}\tilde{S} : \mathsf{Alg}(L) \to \mathsf{Coalg}(T)$. Applying

$$\tilde{Q}\tilde{U}\tilde{S} : \mathsf{Alg}(L) \to \mathsf{Coalg}(T) \to \mathsf{Alg}(L) \tag{3}$$

to an algebra $LA \to A$, there will be an injective L-algebra morphism

$$j_A : A \to QUSA.$$

This is known in modal logic, in the case of Kripke frames, as the Jónsson-Tarski theorem. As a corollary, completeness of the logic w.r.t. T-coalgebras then follows because the T-coalgebra corresponding to the initial L-algebra provides a counter-model for any non-derivable formula.

[1] The elements of A are represented by the clopen (closed and open) subsets of the topological space SA. $\wedge, \vee, \neg$ in A become intersection, union and complement.

Lifting Functors from Set *to* Stone. We will lift a functor T : Set $\to$ Set to a functor $\hat{T}$: Stone $\to$ Stone in such a way that SQ extends to a functor

$$\tilde{S}\tilde{Q} : \mathsf{Coalg}(T) \to \mathsf{Coalg}(\hat{T}).$$

$\hat{T}$ will depend on a choice of logic for T, but there is a canonical such, namely the logic given by *all* predicate liftings for T. We show that two states in a T-coalgebra have the same theory if and only if they are bisimilar in the corresponding $\hat{T}$-coalgebra.

Ultrafilter Extensions. Ultrafilter extensions are one of the central notions in the model theory of modal logics. In order to define ultrafilter extensions we need to find, for each coalgebra $X \to TX$ a suitable coalgebra $USQ(X) \to T(USQ(X))$ where USQ : Set $\to$ Set maps a set X to the set of ultrafilters on X. We determine conditions that allow us to obtain a transformation $t : U\hat{T} \to TU$, thus completing Diagram (2) to

$\delta : LQ \to QT$

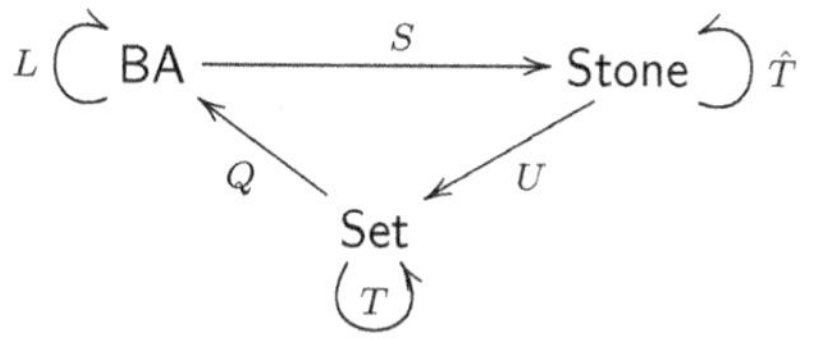

$t : U\hat{T} \to TU$ (4)

The transformation t allows to lift U to $\tilde{U}$: $\mathsf{Coalg}(\hat{T}) \to \mathsf{Coalg}(T)$. The ultrafilter extension of a coalgebra is then given by the composition

$$\tilde{U}\tilde{S}\tilde{Q} : \mathsf{Coalg}(T) \to \mathsf{Alg}(L) \to \mathsf{Coalg}(\hat{T}) \to \mathsf{Coalg}(T). \quad (5)$$

Under the assumption that the transformation t above is natural, we show that two states in a T-coalgebra (X, ξ) have the same theory if and only if they are bisimilar in the ultrafilter extension $\tilde{U}\tilde{S}\tilde{Q}(X, \xi)$. This provides a model-theoretic characterisation of logical equivalence for finitary logics.

Related Work. The first attempt of formulating a duality which accounts for an algebraic semantics of modal logic, for the special class of Kripke-polynomial functors, goes back to Jacobs [6]. Moreover, Section 5 of *loc.cit.* contains some material on ultrafilter extensions of coalgebras but fails to give an account of bisimilarity somewhere else, as there the function embedding a coalgebra into its ultrafilter extension is a morphism of coalgebras.

2 Preliminaries and Notation

Stone Duality. Unfortunately we have space only to indicate the most important notions. For a general introduction we refer to [7,2]. We write Set for the category of sets and functions, BA for the category of Boolean algebras and their morphisms and Stone for the category of Stone-spaces and continuous maps. The *contravariant* functors witnessing the dual equivalence between Set and Stone are denoted by

$$P : \mathsf{Stone} \to \mathsf{BA} \text{ and } S : \mathsf{BA} \to \mathsf{Stone}$$

where $P\mathbb{X}$ is the Boolean algebra of clopen (closed and open) subsets of $\mathbb{X}$ and SA is the space consisting of ultrafilters over A; on arrows, these functors act as inverse image; for more on this duality see [7]. The forgetful functors are denoted by U : Stone $\rightarrow$ Set and V : BA $\rightarrow$ Set throughout, and Q : Set $\rightarrow$ BA is the contravariant powerset functor, which is assumed to take values in BA. The composition QUS constructs the perfect [8] or canonical extension of a Boolean algebra, and we write

$$j_A : A \rightarrow QUSA, \quad a \mapsto \{u \in USA \mid a \in u\}$$

for the canonical embedding. The fact that $j_A : A \rightarrow QUSA$ is an injective Boolean algebra morphism is known as *Stone's representation theorem for Boolean algebras*: j_A represents A as an algebra of subsets where $\wedge, \vee, \neg$ in A become intersection, union and complement. Another map which we will need throughout the paper is the map

$$\eta_X : X \rightarrow USQX, \quad x \mapsto \{Y \subseteq X \mid x \in Y\}$$

embedding a set X into the set of ultrafilters of QX. (In fact, but we will not use this, Q and US are adjoint on the right and j and η are the (co)units of the adjunction.)

The category Stone allows familiar type constructions. For example, whereas *Kripke polynomial functors (KPF)* [6] on Set are given by the left-hand side below, *Vietoris polynomial functors (VPF)* [10] on Stone are given by the right-hand side.

$$T ::= \mathrm{Id} \mid K \mid T^I \mid T{+}T \mid T{\times}T \mid \mathcal{P}{\circ}T \qquad T ::= \mathrm{Id} \mid \mathbb{K} \mid T^I \mid T{+}T \mid T{\times}T \mid \mathcal{V}{\circ}T$$

$K, \mathbb{K}$ denote constant functors, I denotes a set. $\mathcal{P}$ is covariant powerset and $\mathcal{V}$ the Stone space analogue: $\mathcal{V}\mathbb{X}$ is the Stone space of closed subsets of $\mathbb{X}$; the topology is generated by $\{\{b \subseteq U\mathbb{X} \mid b \text{ closed and } b \subseteq a\} \mid a \text{ clopen}\}$.

Coalgebraic Modal Logic. (See [9] for more details). Our treatment of coalgebras and modal logic is parametric in an endofunctor Set $\rightarrow$ Set, which is denoted by T throughout. By an n-ary predicate lifting for T we mean a natural transformation $\lambda : (2^{\cdot})^n \rightarrow 2^{T\cdot}$ where $2^{\cdot}$: Set $\rightarrow$ Set is contravariant powerset (note that $2^{\cdot} = VQ$). A set Λ of predicate liftings and associated arities gives rise to a functor L_0 : Set $\rightarrow$ BA by mapping $A \mapsto F\{[\lambda](a_1, \dots, a_n) \mid \lambda\ n\text{-ary}, a_1, \dots, a_n \in A\}$; here F : Set $\rightarrow$ BA is the functor that constructs free Boolean algebras and expressions of the form $[\lambda](a_1, \dots, a_n)$ are understood purely syntactically. To every set of predicate liftings we associate a logic $\mathcal{L}(\Lambda)$ given by

$$\mathcal{L}(\Lambda) \ni \varphi ::= \mathrm{ff} \mid \varphi \rightarrow \varphi \mid [\lambda](\varphi_1, \dots, \varphi_n) \qquad (\lambda \in \Lambda \quad n\text{-ary})$$

It follows by induction that $\mathcal{L}(\Lambda) = \bigcup_{n \geq 0} (UL_0)^n (VF\{\mathrm{tt}, \mathrm{ff}\})$ where V : BA $\rightarrow$ Set is the forgetful functor.

A *modal axiom* is an expression $\varphi \leftrightarrow \psi$ where $\varphi, \psi \in L_0(FX)$ for a denumerable set X of variables. We write $\mathcal{A} \vdash \varphi$ if φ is derivable using propositional reasoning, congruence (if $\varphi_1 \leftrightarrow \psi_1, \dots, \varphi_n \leftrightarrow \psi_n$ then $[\lambda](\varphi_1, \dots, \varphi_n) \leftrightarrow [\lambda](\psi_1, \dots, \psi_n)$) and substitution instances of axioms in $\mathcal{A}$.

Given a set $\mathcal{A}$ of modal axioms, we define a functor L : BA $\rightarrow$ BA by $LA = L_0UA/\!\sim$ where $\sim$ is the least equivalence relation on UL_0A that contains all substitution instances of axioms $\varphi \leftrightarrow \psi \in \mathcal{A}$. This allows us to view syntax and proof calculus of a logic given by a set of predicate liftings and modal axioms as endofunctor $L : \mathsf{BA} \to \mathsf{BA}$. Note that the n-fold application of L to the initial Boolean algebra $\mathbb{2}$ yields the set

$$\mathcal{L}^n(\Lambda, \mathcal{A}) = \{\varphi \in \mathcal{L}(\Lambda) \mid \text{rank}(\varphi) \leq n\}/\!\sim$$

where $\sim$ is the inter-derivability relation given by $\mathcal{A}$.

For a T-coalgebra (C, γ), the semantics $[\![\varphi]\!]_\gamma \subseteq C$ of a formula is given by the inductive extension of the assignment

$$[\![[\lambda](\varphi_1, \ldots, \varphi_n)]\!]_\gamma = \gamma^{-1} \circ \lambda(C)([\![\varphi_1]\!]_\gamma, \ldots, [\![\varphi_n]\!]_\gamma)$$

to the whole of $\mathcal{L}(\Lambda)$. Assuming soundness of the semantics, that is $\mathcal{A} \vdash \varphi \leftrightarrow \psi$ implies $[\![\varphi]\!]_\gamma = [\![\psi]\!]_\gamma$ for all T-coalgebras (C, γ), we can define a natural transformation

$$\delta_X : LQ(X) \to QT(X)$$

by the inductive extension of the assignment $([\lambda](\varphi_1, \ldots, \varphi_n))_\sim \mapsto \lambda(X)(\varphi_1, \ldots, \varphi_n)$ where $(\cdot)_\sim$ is the equivalence class of $\cdot$ by $\sim$.

This allows us to recast the coalgebraic semantics of $\mathcal{L}(\Lambda)$ as follows: For $\varphi \in VF(\{\mathrm{tt}, \mathrm{ff}\})$, $[\![\varphi]\!]_\gamma$ is given canonically; if $\varphi \in (UL_0)^{n+1}(VF(\{\mathrm{tt}, \mathrm{ff}\}))$ we obtain $[\![\varphi]\!]_\gamma = \gamma^{-1} \circ \delta(\pi(\varphi))$ where $\pi : L_0 U \to L$ takes equivalence classes. Assuming that the initial L-algebra exists, we arrive at the following compact characterisation of the coalgebraic semantics. The semantics of formulas w.r.t. to a coalgebra $\xi : X \to TX$ is given by by the unique morphism from the initial algebra $LI \to I$

$$\begin{array}{ccccc} I & \longleftarrow & & & LI \\ \downarrow{\scriptstyle [\![\cdot]\!]} & & & & \downarrow{\scriptstyle L[\![\cdot]\!]} \\ QX & \xleftarrow{Q\xi} & QTX & \xleftarrow{\delta_X} & LQX \end{array} \tag{6}$$

We say that two states x, y in two coalgebras are **behaviourally equivalent** or **bisimilar** if they can be identified by some coalgebra morphism. If two states are bisimilar, then they satisfy the same formulae. The converse is not true in general. This failure plays an important role in this paper.

3 Jónsson-Tarski Theorem (Completeness)

Given an algebra $\alpha : LA \to A$, we want to transform it to the Set-coalgebra

$$\tilde{U}\tilde{S}(\alpha) \;=\; USA \overset{US\alpha}{\to} USLA \overset{h_A}{\to} TUSA.$$

Thinking of the elements of $USLA$ as ultrafilters over LA, we define

$$h_A : USLA \longrightarrow TUSA \tag{7}$$

$$u \;\mapsto\; h_A(u) \in \bigcap\{\delta(Lj_A(a)) \mid a \in u\} \tag{8}$$

that is, h_A chooses an element in $\bigcap\{\delta(Lj_A(a)) \mid a \in u\}$ for each ultrafilter u on LA. This definition is constructed in such a way that $\tilde{U}\tilde{S}$ preserves the semantics (compare Diagram(9) below with Diagram (6)). The notation $\tilde{U}\tilde{S}$ suggests that both U and S can be lifted seperately, see Section 5. Here we neither require h_A to be natural nor $\tilde{U}\tilde{S}$ to be a functor.

Definition 1. We say that h is definable if for all algebras A and all ultrafilters u on LA we have that $\bigcap\{\delta(Lj_A(a)) \mid a \in u\}$ is non-empty.

Remark 2. A necessary condition for h to be definable is that δ is injective. For suppose otherwise. Then there will be an $a \in LA$ such that $a \neq \perp$ and $\delta(Lj_A(a)) = \emptyset$. As $a \neq \perp$ we find an ultrafilter $u \in USLA$ s.t. $a \in u$. But then $\bigcap\{\delta(Lj_A(a)) \mid a \in u\} = \emptyset$.

The essence of completeness w.r.t. to the coalgebraic semantics is that

$$j_A : A \to QUSA$$

is an injective $\mathsf{Alg}(L)$-morphism. This is known as the Jónsson-Tarski theorem. It is an extension of Stone's representation theorem from Boolean algebras to modal algebras (ie L-algebras).

To see how completeness follows, assume that φ is not derivable and $\alpha : LA \to A$ is the initial algebra. We have $\alpha \models \varphi \neq \top$, hence $\tilde{Q}\tilde{U}\tilde{S}(\alpha) \models \varphi \neq \top$ by j_A being an injective morphism, hence $\tilde{U}\tilde{S}(\alpha) \not\models \varphi$ by definition of the coalgebraic semantics (see Diagram (6)), thus providing the countermodel for φ.

From Stone's theorem, we know that j_A is an injective BA-morphism. To see what is needed to make j_A an L-algebra morphism we take a look at the following diagram.

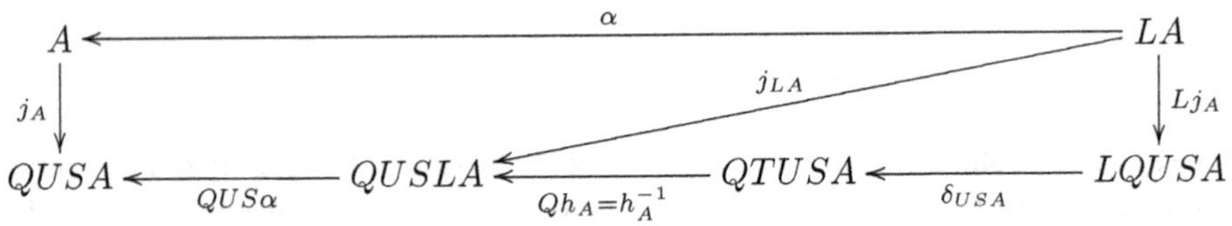

(9)

The lower part, which is an L-algebra on $QUSA$, is obtained by transforming (A, α) into a T-coalgebra and back to an L-algebra. From the naturality of j, it follows that j_A is an L-algebra morphism if the triangle commutes. This leads us to

Theorem 3. *Assuming that h is definable, the logic given by δ is complete w.r.t. the coalgebraic semantics.*

Proof. We show that the triangle in the diagram above commutes. For $b \in LA$, let us write $\hat{b}$ for $j_{LA}(b) = \{u \in USLA \mid b \in u\}$. Eliding subscripts, we have to show $h^{-1}(\delta(Lj(b))) = \hat{b}$, that is,

$$h(u) \in \delta(Lj(b)) \Leftrightarrow b \in u.$$

'$\Leftarrow$' holds by definition of h. For '$\Rightarrow$' assume $b \notin u$. It follows $\neg b \in u$, hence $h(u) \in \delta(Lj(\neg b))$, hence $h(u) \in \neg\delta(Lj(b))$, ie $h(u) \notin \delta(Lj(b))$.

Remark 4. The completeness proof of Jacobs [6] works essentially this way (his r is our h). Compared to the completeness proof of [9] (which mimicked the induction along the final coalgebra sequence of [15]), the Jónsson-Tarski approach to completeness is simpler as it avoids an induction along the final sequence. On the other hand not all logics admit such a completeness proof: If we take the finite powerset functor together with the standard modal logic, then h is not definable, see Example 23.

4 Lifting Functors from Set to Stone

In this section we are going to use predicate liftings to lift a functor $T : \mathsf{Set} \to \mathsf{Set}$ to a functor $\hat{T} : \mathsf{Stone} \to \mathsf{Stone}$. We will give two descriptions of $\hat{T}$. First, $\hat{T}\mathbb{X}$ is the dual of the Boolean algebra generated by the images of the predicate liftings $QU\mathbb{X} \to QTU\mathbb{X}$ (Definition 7). Second, $\hat{T}$ is the dual of the functor L on BA that describes the complete logic corresponding to the given predicate liftings (Remark 16).

Given a collection S of subsets of X we denote by $\langle S \rangle_{\mathsf{BA}}$ the subalgebra of the Boolean algebra $\mathcal{P}(X)$ generated by S, i.e. by closing S under taking finite unions, intersections and under complementation. We will use the following technical lemma.

Definition and Lemma 5. Given a functor $F : \mathsf{C} \to \mathsf{Set}$ and a functor $G : \mathsf{C}^{\mathrm{op}} \to \mathsf{Set}$ such that there is a natural transformation $j : G \to VQF^{\mathrm{op}}$. Then we can define a functor $\langle G \rangle_{\mathsf{BA}} : \mathsf{C}^{\mathrm{op}} \to \mathsf{BA}$ by letting $\langle G \rangle_{\mathsf{BA}} X := \langle j_X[GX] \rangle_{\mathsf{BA}}$ and $\langle G \rangle_{\mathsf{BA}} f := VQF^{\mathrm{op}} f \restriction_{\langle G \rangle_{\mathsf{BA}} Y}$ for arbitrary X, Y and $f : X \to Y \in \mathsf{C}$.

Proof. Using the naturality of j it is easy to show that $\langle G \rangle_{\mathsf{BA}}$ is well defined on objects and morphisms. Functoriality of $\langle G \rangle_{\mathsf{BA}}$ then follows from the functoriality of VQF^{op}.

Definition 6. Let $F, G : \mathsf{C} \to \mathsf{Set}$ be functors and $\tau : F \to G$ a natural transformation. Then we define a functor $\Im(\tau) : \mathsf{C} \to \mathsf{Set}$ by $\Im(\tau)(X) := \tau_X[FX]$ for $X \in \mathsf{C}$ and by letting $\Im(\tau)(f)$ to be the unique map such that the following diagram commutes

$$\begin{array}{ccccc} FX & \twoheadrightarrow & \Im(\tau)(X) & \hookrightarrow & GX \\ \downarrow{\scriptstyle Ff} & & \downarrow{\scriptstyle \Im(\tau)(f)} & & \downarrow{\scriptstyle Gf} \\ FY & \twoheadrightarrow & \Im(\tau)(Y) & \hookrightarrow & GY \end{array}$$

where $f : X \to Y \in \mathsf{C}$ was arbitrary.

We are now ready for the definition of a lifting of a Set-endofunctor to Stone.

Definition 7. Given $T : \mathsf{Set} \to \mathsf{Set}$ and a set Λ of predicate liftings $\lambda : VQ^{n_\lambda} \to VQT$ define

$$\hat{T} := S(\langle \Im(\tau^\Lambda) \rangle_{\mathsf{BA}})$$

where $\tau^\Lambda := [(\lambda_{U\_} \circ i_{\_}^{n_\lambda})_{\lambda \in \Lambda}] : \coprod_{\lambda \in \Lambda} VP^{n_\lambda} \to VQT$ denotes the natural transformation obtained by cotupling of all the transformations $\lambda_{U\_} \circ i_{\_}^{n_\lambda}$ and the maps $i_{\mathbb{X}}^{n_\lambda}$ are the embeddings $VP^{n_\lambda}\mathbb{X} \to VQ^{n_\lambda}U\mathbb{X}$.

Proposition 8. $\hat{T}$ is a functor.

Proof. Clearly $\tau^{\Lambda} = [(\lambda_{U\_} \circ i_{\_}^{n_\lambda})_{\lambda\in\Lambda}]$ is a natural transformation from $\coprod VP^{n_\lambda}$ to $VQTU$. Therefore $\Im(\tau)$ is a functor from $\mathsf{Stone}^{\mathrm{op}} \to \mathsf{Set}$ and there is a natural transformation $j : \Im(\tau) \to VQTU$. But then by Lemma 5 $\langle\Im(\tau)\rangle_{\mathsf{BA}}$ is a functor from $\mathsf{Stone}^{\mathrm{op}}$ to BA. Therefore $\hat{T}$ is a functor from $\mathsf{Stone}^{\mathrm{op}}$ to $\mathsf{Stone}^{\mathrm{op}}$ or, equivalently, $\hat{T} : \mathsf{Stone} \to \mathsf{Stone}$.

The previous definition pre-supposes a set Λ of predicate liftings to define the lifted functor $\hat{T} : \mathsf{Stone} \to \mathsf{Stone}$. The next proposition, which was stated in [19] and which is an instance of the Yoneda lemma, shows that there is a canonical choice for the set of liftings.

Proposition 9. There is a 1-1 correspondence

$$\{n\text{-ary predicate liftings } \lambda_X : (2^n)^X \to 2^{TX}\} \;\cong\; \{\text{ subsets of } T(2^n)\}$$

given by $S \subseteq T(2^n) \mapsto \lambda$ where

$$\lambda(C) : (P_1, \ldots, P_n) \in \mathcal{P}(C)^n \mapsto \{t \in TC \mid \mathbb{1}_S \circ T\langle \mathbb{1}_{P_1}, \ldots, \mathbb{1}_{P_n}\rangle(t) = 1\}$$

where, for $Y \subseteq X$, $\mathbb{1}_Y : X \to 2$ is the characteristic function of Y.

Given this canonical choice of liftings, it is instructive to look at some concrete examples.

Example 10. 1. Suppose $TX = K$ is constant with some finite set K as its value. Then $\hat{T} \cong \mathbb{K}$ where $\mathbb{K}$ is the set K with the discrete topology. To see that, note that every lifting is determined by a subset $k \subseteq K$, which gives rise to the algebra QK of all subsets of K, which in turn induces the lifted functor $\hat{T}\mathbb{X} = SQK \cong \mathbb{K}$.

2. For $TX = X$, i.e. $T = \mathrm{Id}$, we get $\hat{T} \cong \mathrm{Id}$. For $n = 1$, we obtain a unary lifting λ_S for every $S \subseteq 2$; this gives rise to the liftings

$$\lambda_1 = id \quad \lambda_2 = \neg \quad \lambda_3 = \mathrm{tt} \quad \lambda_4 = \mathrm{ff}$$

where $\lambda_i(C) : \mathcal{P}(C) \to \mathcal{P}(C)$. One can show, that all n-ary liftings can be obtained as Boolean combinations of λ_1. Hence the generated Boolean algebra $\langle\Im(\tau^{\Lambda})\rangle_{\mathsf{BA}}\mathbb{X}$ is isomorphic to $P\mathbb{X}$, whence $\widehat{\mathrm{Id}} \cong \mathrm{Id}$.

3. For $TX = \mathcal{P}(X)$, we obtain $\hat{T} \cong \mathcal{V}$ where $\mathcal{V} : \mathsf{Stone} \to \mathsf{Stone}$ denotes the Vietoris functor. Invoking Proposition 9, we obtain 8 unary liftings of type $VQC \to VQTC$, which are generated by Boolean combinations of $\Box$ and $\Diamond$, where $\Box(C) : 2^C \to 2^{TC}$ is given by $c \mapsto \{d \subseteq C \mid d \subseteq c\}$ and $\Diamond = \neg\circ\Box\circ\neg$. Similarly, all n-ary liftings can be defined, and one obtains that for the case $TX = \mathcal{P}X$, $\langle\Im(\tau^{\Lambda})\rangle_{\mathsf{BA}}\mathbb{X}$ is the Boolean algebra generated by $\{\Box a \mid a \in P\mathbb{X}\} \cup \{\Diamond a \mid a \in P\mathbb{X}\}$ quotiented by the axioms of standard modal logic, i.e. $\Box\varphi \leftrightarrow \neg\Diamond\neg\varphi$ and $\Box(\varphi_1, \ldots, \varphi_n) \leftrightarrow (\Box\varphi_1 \wedge \cdots \wedge \Box\varphi_n)$. From this it follows that $\hat{T} \cong \mathcal{V}$, see [10] for details.

Remark 11. It is possible to prove that $\widehat{\cdot}$ commutes with the formation of products, coproducts and the composition of functors, i.e.

$$\widehat{T_1 \times T_2} \cong \widehat{T_1} \times \widehat{T_2}, \quad \widehat{T_1 + T_2} \cong \widehat{T_1} + \widehat{T_2} \quad \text{and} \quad \widehat{T_1 \circ T_2} \cong \widehat{T_1} \circ \widehat{T_2}.$$

Combining this fact with the above mentioned examples one can show that for every Kripke polynomial functor T the corresponding Vietoris polynomial functor is isomorphic to the functor $\hat{T}$.

We will now show that we can extend the functor $SQ : \mathsf{Set} \to \mathsf{Stone}$ to a functor $\mathsf{Coalg}(T) \to \mathsf{Coalg}(\hat{T})$. As a first step of this construction let us see how we can transform ultrafilter of $QTU\mathbb{X}$ naturally into ultrafilter of $\langle\Im(\tau^{\Lambda})\rangle_{\mathsf{BA}}\mathbb{X}$ by simply forgetting the sets in $QTU\mathbb{X} \setminus \langle\Im(\tau^{\Lambda})\rangle_{\mathsf{BA}}\mathbb{X}$.

Definition and Lemma 12. The function $\hat{\pi}_{\mathbb{X}}$ defined by

$$\begin{aligned} \hat{\pi}_{\mathbb{X}} : SQTU\mathbb{X} &\to \hat{T}\mathbb{X} \\ u &\mapsto u \cap \left(\langle\Im(\tau^{\Lambda})\rangle_{\mathsf{BA}}\mathbb{X}\right) \end{aligned}$$

is well-defined and continuous. The family of functions $\hat{\pi}_{\mathbb{X}\in\mathsf{Stone}}$ gives rise to a natural transformation $\hat{\pi} : SQTU \to \hat{T}$.

Proof. Let j be the natural embedding of $\langle\Im(\tau^{\Lambda})\rangle_{\mathsf{BA}}\mathbb{X}$ into $QTU\mathbb{X}$. Then it is easy to see that $Sj = \hat{\pi}_{\mathbb{X}}$. Hence $\hat{\pi}_{\mathbb{X}}$ is well defined and continuous. Naturality of $\hat{\pi}$ then follows from the naturality of j.

With the help of $\hat{\pi}$ we can turn T-coalgebras into $\hat{T}$-coalgebras.

Definition 13. Let $(X, \gamma) \in \mathsf{Coalg}(T)$. Then we define a function $\hat{\gamma} : SQX \to \hat{T}SQX$ by letting $\hat{\gamma} := \hat{\pi}_{SQX} \circ SQT\eta_X \circ SQ\gamma$.

$$\overbrace{SQX \xrightarrow{SQ\gamma} SQTX \xrightarrow{SQT\eta_X} SQTUSQX \xrightarrow{\hat{\pi}_{SQX}} \hat{T}SQX}^{\hat{\gamma}}$$

The operation of turning a T-coalgebra into a $\hat{T}$-coalgebra is functorial.

Proposition 14. The mapping

$$\begin{aligned} (X, \gamma) \in \mathsf{Coalg}(T) &\mapsto (SQX, \hat{\gamma}) \in \mathsf{Coalg}(\hat{T}) \\ f \in \mathsf{Coalg}(T) &\mapsto SQf \in \mathsf{Coalg}(\hat{T}) \end{aligned}$$

defines a functor $\tilde{S}\tilde{Q} : \mathsf{Coalg}(T) \to \mathsf{Coalg}(\hat{T})$.

Proof. The claim follows from the fact that η and $\hat{\pi}$ are both natural.

The semantics of the logic w.r.t. $\hat{T}$-coalgebras is given by the following predicate liftings.

Definition 15. A predicate lifting $\lambda : (VQ)^n \to VQT$ for T induces a predicate lifting $\hat{\lambda} : (VP)^n \to VP\hat{T}$ for $\hat{T}$ via

$$\hat{\lambda}_{\mathbb{X}} = Vk_{\langle\Im(\tau^{\Lambda})\rangle_{\mathsf{BA}}\mathbb{X}} \circ \lambda_{U\mathbb{X}} \circ i^n_{\mathbb{X}}$$

where $i^n_{\mathbb{X}} : (VP\mathbb{X})^n \to (VQU\mathbb{X})^n$ and $k_{\langle\Im(\tau^{\Lambda})\rangle_{\mathsf{BA}}\mathbb{X}} : \langle\Im(\tau^{\Lambda})\rangle_{\mathsf{BA}}\mathbb{X} \to PS\langle\Im(\tau^{\Lambda})\rangle_{\mathsf{BA}}\mathbb{X}$ is the isomorphism given by Stone duality.

Remark 16. $\hat{T}$ can be described more abstractly. Let $\delta' : L'Q \to QT$ describe the semantics of the logic $\mathcal{L}$ given as above by predicate liftings (and no axioms). We can define 'an improved version' L of L' 'with axioms' by factoring $L'A \to LQUSA \to QTUSA$ through its image as $L'A \twoheadrightarrow LA \hookrightarrow QTUSA$. One then shows the following.

1. L is a functor.
2. LQX is obtained by factoring $\delta' : L'QX \to QTX$ through its image. The image $\delta_X : LQX \to TQX$ gives the interpretation of L w.r.t. T-coalgebras whereas, intuitively, the quotient $L'QX \to LQX$ describes the axioms added to $\mathcal{L}$. That δ is injective corresponds to the completeness of the logic described by L, see [9].
3. L is dual to $\hat{T}$, that is, there is an isomorphism $SL \to \hat{T}S$, or, equivalently, $\hat{\delta} : LP \to P\hat{T}$. The iso $\hat{\delta}$ gives a $\hat{T}$-coalgebra semantics to the logic $\mathcal{L}$ which agrees with the one from Definition 15.
4. The functor $\tilde{S}\tilde{Q} : \mathsf{Coalg}(T) \to \mathsf{Coalg}(\hat{T})$ can now be described as mapping $X \to TX$ to $SQX \to SQTX \xrightarrow{S\delta_X} SLQX \overset{\cong}{\to} \hat{T}SQX$.

Proposition 17. 1. Consider a state x of a T-coalgebra and the state $\eta_X(x)$ in the corresponding $\hat{T}$-coalgebra. x and $\eta_X(x)$ have the same theory.
2. Two states of a $\hat{T}$-coalgebra are bisimilar iff they have the same theory.

Proof. 1. Let $\iota : LI \to L$ be the initial L-algebra and $\varphi \in I$. The semantics of φ w.r.t. a coalgebra $X \to TX$ and its ultrafilter extension $SQX \to \hat{T}SQX$ is given by the initial algebra maps as in the following diagram (see Remark 16).

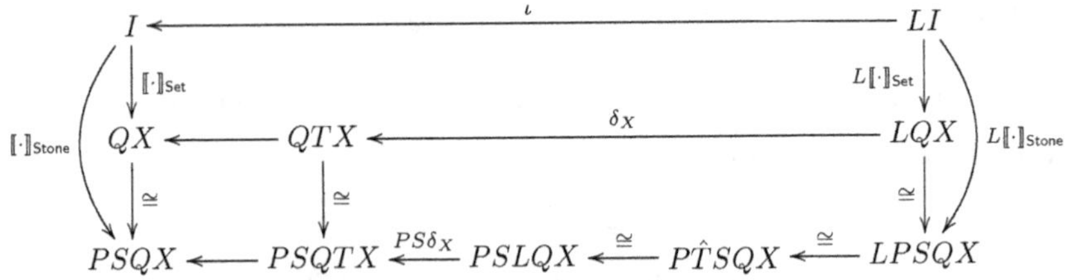

The left column means that $u \in [\![\varphi]\!]_{\mathsf{Stone}}$ iff $[\![\varphi]\!]_{\mathsf{Set}} \in u$ (note the similarity with the truth lemma of the canonical model known in modal logic). This implies the claim.
2. This follows from $\hat{T}$ being dual to L.

The following corollary reconciles logical equivalence and bisimilarity. Although two logically equivalent states in a Set-coalgebra may fail to be bisimilar, they will be bisimilar in the corresponding Stone-coalgebra:

Theorem 18. *Given $T : \mathsf{Set} \to \mathsf{Set}$ and a logic $\mathcal{L}$ for T-coalgebras, let $\hat{T} : \mathsf{Stone} \to \mathsf{Stone}$ be the lifted functor. Then, given $(X, \gamma) \in \mathsf{Coalg}(T)$ and $x, y \in X$, we have that x and y have the same theory iff $\eta_X(x)$ is bisimilar to $\eta_X(y)$ in $\tilde{S}\tilde{Q}(X, \gamma)$.*

5 The Ultrafilter Extension of a Coalgebra

In this section we define $\tilde{U}$, thus lifting Diagram (1) to algebras and coalgebras[2]

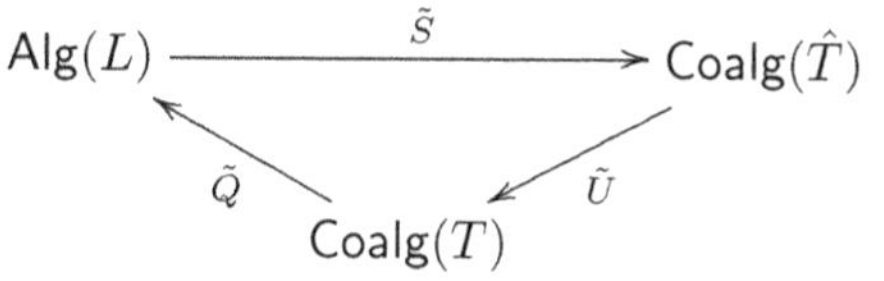

(10)

$\tilde{U}\tilde{S}\tilde{Q}(X \xrightarrow{\xi} TX)$ will be the ultrafilter extension of ξ. Although SQ is left-adjoint to U, this will not hold in general for the lifted functors. The reason is that the unit $\eta_X : X \to USQX$ may fail to be a coalgebra morphism. This is the observation that gives rise to Theorem 27.

We need a transformation $t : U\hat{T} \to TU$. This can be done if **ultrafilters in $\hat{T}$ have non-empty intersection**, that is, if for all Stone spaces $\mathbb{X}$ and all ultrafilters $u \in U\hat{T}\mathbb{X}$ we have $\bigcap u \neq \emptyset$. We then define

$$t_{\mathbb{X}} : U\hat{T}\mathbb{X} \to TU\mathbb{X}$$
$$u \mapsto t_{\mathbb{X}}(u) \in \bigcap u$$

Remark 19. Using $\hat{T}S \cong SL$, we see that $t_{\mathbb{X}}$ appeared already as $h_{P\mathbb{X}}$ in (7). Similarly, h_A is t_{SA}. Note that naturality was not required in Section 3.

Under the assumption that t is natural, we can now lift $U : \mathsf{Stone} \to \mathsf{Set}$ to a functor

$$\tilde{U} : \mathsf{Coalg}(\hat{T}) \to \mathsf{Coalg}(T)$$

mapping $\xi : \mathbb{X} \to \hat{T}\mathbb{X}$ to $U\mathbb{X} \xrightarrow{U\xi} \hat{U}T\mathbb{X} \xrightarrow{t_{\mathbb{X}}} TU\mathbb{X}$. In the following proposition we prove two useful properties of t.

Proposition 20. For all $\mathbb{X} \in \mathsf{Stone}$ let $t_{\mathbb{X}}$ be defined as above. Then

1. $t_{\mathbb{X}}$ is injective for all $\mathbb{X}$.
2. If for all $\mathbb{X}$ and for all $u \in U\hat{T}\mathbb{X}$ we have that $\bigcap u$ is a singleton set, then t is a natural transformation.

Proof. The first item follows from the fact that for two ultrafilters $u \neq u'$ we always have $\bigcap u \cap \bigcap u' = \emptyset$. To prove that t is natural we have to show that $TUf \circ t_{\mathbb{X}} = t_{\mathbb{Y}} \circ U\hat{T}f$ for some arbitrary $f : \mathbb{X} \to \mathbb{Y}$. Let $u \in U\hat{T}X$. Then

$$\begin{aligned} t_{\mathbb{Y}}(U\hat{T}f(u)) &= t_{\mathbb{Y}}((TUf^{-1})^{-1}(u)) = F \\ &\Leftrightarrow F \in \bigcap (TUf^{-1})^{-1}(u) = (TUf^{-1})^{-1}(F') \\ &\qquad \text{for the } F' \text{such that} \quad (\bigcap u) = \{F'\} \\ &\Leftrightarrow F = TUf[F'] \Leftrightarrow TUf(t_{\mathbb{X}}(u)) = F. \end{aligned}$$

[2] $\tilde{S}(LA \to A) = SA \to SLA \cong \hat{T}SA$, see Remark 16.3.

Kripke polynomial functors fulfill this criterion, for example:

Example 21. Let $T = \mathcal{P}$ and Λ the canonical set of liftings. Then it is easy to see that $\langle \Im(\tau^\Lambda)\rangle_{\mathsf{BA}}\mathbb{X} = P\mathcal{V}\mathbb{X}$ and therefore we have for all $u \in U\hat{T}\mathbb{X} = S(\langle \Im(\tau^\Lambda)\rangle_{\mathsf{BA}}\mathbb{X})$ that $\bigcap u = \{F\}$ for some $F \in \mathcal{V}\mathbb{X}$ by Stone duality. Therefore t is natural according to Proposition 20. The reader is invited to check that in fact $\mathcal{V}\mathbb{X} = (\Im(t)\mathbb{X}, \tau_t)$ where $\Im(t)$ is defined as in 6 and τ is the quotient topology induced by $t_\mathbb{X}$. Therefore our definition of an ultrafilter extension for $\mathcal{P}$-coalgebras coincides with the one used in modal logic.

Remark 22. The construction sketched in the example works also for other functors: If t is natural, the mapping $\bar{T} : \mathsf{Stone} \to \mathsf{Stone}, \quad \mathbb{X} \mapsto (\Im(t)\mathbb{X}, \tau_t)$, can be extended to a functor with the property that $\hat{T} \cong \bar{T}$ and that $U\bar{T}\mathbb{X} \subseteq TU\mathbb{X}$ for all $\mathbb{X}$. We can then use the inclusion $U\bar{T}\mathbb{X} \subseteq TU\mathbb{X}$ which simply forgets the topology in place of the t-map to define the ultrafilter extension. This works in particular for a KPF T where we get that $\bar{T}$ is equal to the corresponding VPF.

There are also functors for which we cannot define an ultrafilter extension.

Example 23. Let $T = \mathcal{P}_\omega$ and $\Lambda = \{\Diamond\}$ where $\mathcal{P}_\omega$ denotes the finite power set functor and $\Diamond(Y) := \{Y' \mid Y' \text{ is finite and } Y' \cap Y \neq \emptyset\}$. Then t cannot be defined in general. For a counterexample consider $\mathbb{X} = (\omega \cup \{*\}, \tau)$ where τ is generated by the Boolean set algebra of all finite subsets of ω and all cofinite subsets of $\omega \cup \{*\}$ that contain $*$. Then $\mathbb{X}$ is a Stone space. If we define $U := \{\Diamond(\{n\}) \mid n \in \omega\} \subseteq \langle \Im(\tau^\Lambda)\rangle_{\mathsf{BA}}\mathbb{X}$ one can easily check that U has the finite intersection property. Therefore we can extend U to an ultrafilter $u \in U\hat{\mathcal{P}}_\omega\mathbb{X}$. But obviously $\bigcap U = \emptyset$ and hence also $\bigcap u = \emptyset$.

Of course, finitely branching Kripke frames, ie coalgebras for $\mathcal{P}_\omega$, do have ultrafilter extensions. The point of the example above is that these ultrafilter extensions are $\mathcal{P}$-coalgebras but not $\mathcal{P}_\omega$-coalgebras.

The important property we need is that t preserves the semantics. The semantics of the logic w.r.t. $\hat{T}$-coalgebras was given in Definition 15 and Remark 16.3.

Proposition 24. $t : U\hat{T} \to TU$ preserves the semantics. That is, the subsets of $U\mathbb{X}$ determined by interpreting a formula on $\xi : \mathbb{X} \to \hat{T}\mathbb{X}$ and on $t_\mathbb{X} \circ U\xi : U\mathbb{X} \to TU\mathbb{X}$ are identical.

Proof. The claim is proven by induction on the structure of formulas. We only provide the inductive step for formulas of the form $[\lambda]\varphi$. Let $x \in X$ and $\psi = [\lambda]\varphi$, then

$$\begin{aligned} x \in [\![\psi]\!]_{t_\mathbb{X}\circ U\xi} &\Leftrightarrow x \in (t_\mathbb{X} \circ U\xi)^{-1}(\lambda_{U\mathbb{X}}([\![\varphi]\!]_{t_\mathbb{X}\circ U\xi})) \overset{\text{I.H.}}{\Leftrightarrow} x \in (t_\mathbb{X} \circ U\xi)^{-1}(\lambda_{U\mathbb{X}}([\![\varphi]\!]_\xi)) \\ &\overset{(*)}{\Leftrightarrow} x \in U\xi^{-1}\left(\{u \in U\hat{T}\mathbb{X} \mid \bigcap u \subseteq \lambda_{U\mathbb{X}}([\![\varphi]\!]_\xi)\}\right) \\ &\Leftrightarrow x \in U\xi^{-1}\left(\{u \in U\hat{T}\mathbb{X} \mid \lambda_{U\mathbb{X}}([\![\varphi]\!]_\xi) \in u\}\right) = U\xi^{-1}\left(\hat{\lambda}_\mathbb{X}([\![\varphi]\!]_\xi)\right) \\ &\Leftrightarrow x \in [\![\psi]\!]_\xi, \end{aligned}$$

where the $\Rightarrow$-part of $(*)$ is true because

$$\bigcap u \not\subseteq \lambda_{U\mathbb{X}}([\![\varphi]\!]_\xi) \Rightarrow \lambda_{U\mathbb{X}}([\![\varphi]\!]_\xi) \notin u \Rightarrow -\lambda_{U\mathbb{X}}([\![\varphi]\!]_\xi) \in u \Rightarrow \bigcap u \subseteq -\lambda_{U\mathbb{X}}([\![\varphi]\!]_\xi).$$

Remark 25. That t preserves the semantics means that the left-hand column of the diagram

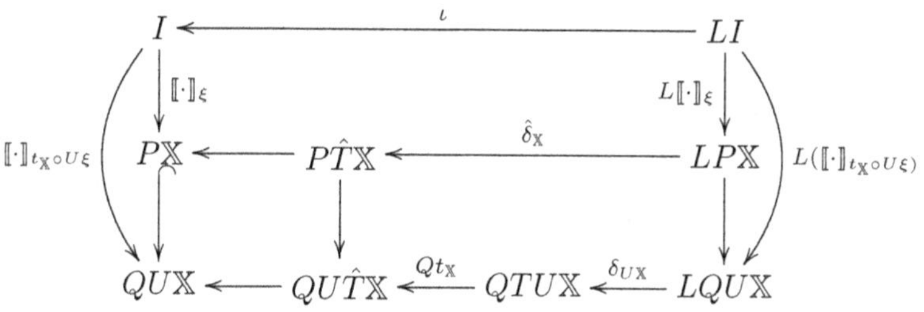

commutes. We can therefore allow as transformation $t : U\hat{T} \to TU$ any transformation making the lower right square commute, or, redrawing it a bit, making the following commute.

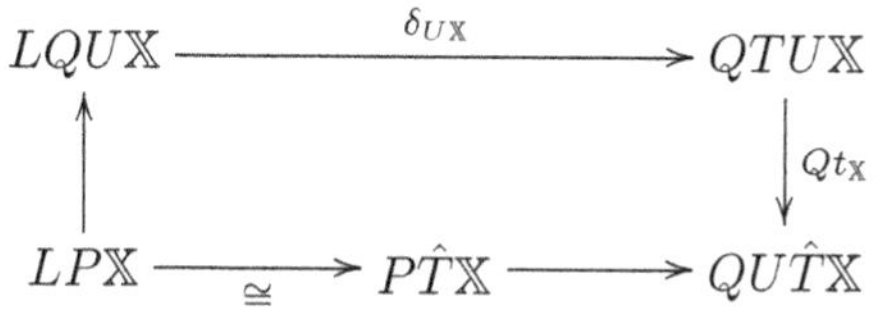

$$
\begin{array}{ccc}
LQ\mathbb{X} & \xrightarrow{\ \delta_{U\mathbb{X}}\ } & QTU\mathbb{X} \\
\uparrow & & \downarrow{\scriptstyle Qt_\mathbb{X}} \\
LP\mathbb{X} & \xrightarrow[\cong]{} P\hat{T}\mathbb{X} \longrightarrow QU\hat{T}\mathbb{X}
\end{array}
\tag{11}
$$

This diagram appeared already as the upper square of Diagram (9), compare Remark 19.

Proposition 26. Assume that t is natural. Then Stone-bisimilarity equals Set-bisimilarity. That is, two states in $\xi : \mathbb{X} \to \hat{T}\mathbb{X}$ are bisimilar iff they are bisimilar in $\tilde{U}\xi$.

Proof. $\subseteq$ follows from t being natural. $\supseteq$: If two states in $\tilde{U}\xi$ are bisimilar than they have the same theory. Now apply Propositions 24 and 17.2.

We can now improve on the bisimulation-somewhere-else result of Theorem 18. Together with the proposition above, it implies that two states in $X \to TX$ that have the same theory are in fact bisimilar in some other *Set*-coalgebra, namely the ultrafilter extension of $X \to TX$.

Theorem 27. *Given* $T :$ Set $\to$ Set *and a logic* $\mathcal{L}$ *for T-coalgebras, let* $\hat{T} :$ Stone $\to$ Stone *be the lifted functor. Assume that ultrafilters in* $\hat{T}$ *have non-empty intersection and that* $t : U\hat{T} \to TU$ *is natural. Then, given* $(X, \gamma) \in$ Coalg(T) *and* $x, y \in X$, *we have that* x *and* y *have the same theory iff* $\eta_X(x)$ *is bisimilar to* $\eta_X(y)$ *in* $\tilde{U}\tilde{S}\tilde{Q}(X, \gamma)$.

Remark 28. The result holds, in particular, for all Kripke polynomial functors.

6   Conclusion and Future Work

The focus of this paper was on the relationship between Stone-coalgebras and Set-coalgebras. This is a special instance of a more general phenomenon in computer science where topology-based structures and set-based structures interact. This was observed already in Abramsky [1] where powerdomain-coalgebras and powerset-coalgebras were compared. We believe that the methods used here will generalise to other such situations.

First, we can treat other logics than classical ones by replacing the duality between BA and Stone by one for, e.g. Heyting algebras or distributive lattices. Infinitary logics can be treated as well, see e.g. [4]. Second, we can replace Set by other categories of interest in semantics. Third, we can make algebraic tools available by upgrading the triangle of Diagram (1) to a square where Set is now accompanied by its dual category of complete atomic Boolean algebras. This will enable the use of methods developed in the study of perfect or canonical extensions of Boolean algebras (see e.g. [20, Section 7]).

There are also a number of more immediate open questions. Formulate a finitary definability result for classes of coalgebras in the style of Goldblatt-Thomason [5], based on ultrafilter extensions. If T preserves finite sets than it has a canonical lifting to from Set to Stone; show that then this lifting agrees with $\hat{T}$. Find nice conditions guaranteeing that ultrafilters in $\hat{T}$ have non-empty intersection.

References

1. S. Abramsky. A domain equation for bisimulation. *Information and Computation*, 92, 1991.
2. S. Abramsky and A. Jung. Domain theory. In *Handbook of Logic in Computer Science*. OUP, 1994.
3. P. Blackburn, M. de Rijke, and Y. Venema. *Modal Logic*. CSLI, 2001.
4. M. Bonsangue and A. Kurz. Duality for logics of transition systems. In V. Sassone, editor, *FoSSaCS'05*, volume 3441 of *LNCS*, 2005.
5. R. Goldblatt. Metamathematics of modal logic I. *Reports on Mathematical Logic*, 6, 1976.
6. B. Jacobs. Many-sorted coalgebraic modal logic: a model-theoretic study. *Theor. Inform. Appl.*, 35, 2001.
7. P. Johnstone. *Stone Spaces*. Cambridge University Press, 1982.
8. B. Jónsson and A. Tarski. Boolean algebras with operators, part 1. *Amer. J. Math.*, 73, 1951.
9. C. Kupke, A. Kurz, and D. Pattinson. Algebraic semantics for coalgebraic logics. In *CMCS'04*, ENTCS, 2004.
10. C. Kupke, A. Kurz, and Y. Venema. Stone coalgebras. *Theoret. Comput. Sci.*, 327, 2004.
11. A. Kurz. A co-variety-theorem for modal logic. In *Advances in Modal Logic 2*. CSLI, 2001. Selected Papers from AiML 2, Uppsala, 1998.
12. A. Kurz. Specifying coalgebras with modal logic. *Theoret. Comput. Sci.*, 260, 2001. Earlier version appeared in the Proceedings of CMCS'98, ENTCS, Vol. 11, 1998.
13. A. Kurz and J. Rosický. Operations and equations for coalgebras. *Math. Structures Comput. Sci.*, 15, 2005.
14. L. Moss. Coalgebraic logic. *Annals of Pure and Applied Logic*, 96, 1999.
15. D. Pattinson. Coalgebraic modal logic: Soundness, completeness and decidability of local consequence. *Theoret. Comput. Sci.*, 309, 2003.
16. D. Pattinson. Expressive logics for coalgebras via terminal sequence induction. *Notre Dame Journal of Formal Logic*, 45, 2004.
17. M. Rößiger. From modal logic to terminal coalgebras. *Theoret. Comput. Sci.*, 260, 2001.
18. J. Rutten. Universal coalgebra: A theory of systems. *Theoret. Comput. Sci.*, 249, 2000.
19. L. Schroeder. Expressivity of Coalgebraic Modal Logic: The Limits and Beyond. In V. Sassone, editor, *FoSSaCS'05*, volume 3441 of *LNCS*, 2005.
20. Y. Venema. *Handbook of Modal Logic*, chapter Algebras and Coalgebras. To appear. Electronically available.

Equational Logic of Recursive Program Schemes

John G. Mersch

Department of Mathematics, Northwestern State University of Louisiana, Natchitoches LA 71497, USA
merschj@nsula.edu

Abstract. In this paper we present FLRS, a sound and complete equational logic for proving the equivalence of recursive program schemes. We use extended versions of the Solution Theorems from [1] and [2] to provide coalgebraic semantics to recursive program schemes. We end the paper with a proof that FLRS is complete with respect to the coalgebraic semantics.

This paper studies the equational logic of recursive program schemes. For example, one takes a system of equations such as the following:

$$\begin{aligned} \mathsf{f}(x) &= \mathsf{G}(x, \mathsf{f}(\mathsf{g}(x))) \\ \mathsf{g}(x) &= \mathsf{G}(\mathsf{f}(x), \mathsf{g}(\mathsf{g}(x))) \end{aligned} \tag{1}$$

where G is a *given* or *base* function and where f and g are defined in terms of G, f, and g by (1). One could also consider the following recursive scheme:

$$\begin{aligned} \mathsf{f}(x) &= \mathsf{G}(x, \mathsf{f}(\mathsf{g}(x))) \\ \mathsf{g}(x) &= \mathsf{G}(\mathsf{G}(x, \mathsf{f}(\mathsf{g}(x))), \mathsf{g}(\mathsf{g}(x))) . \end{aligned} \tag{2}$$

Since (2) is obtained from (1) by replacing $\mathsf{f}(x)$ in the second equation with its definition, one would expect (1) and (2) to be provably equivalent in some formal system. The goal of this paper is to provide such a system which is both sound and complete. Along with the syntax , one needs to provide a semantics to recursive program schemes. Traditional approaches to semantics involve putting a partial order on the set of Σ-trees and then defining the semantics in terms of least fixed-points (see [3]) or by putting a metric on the set of trees and then defining semantics in terms of convergence of Cauchy sequences (see [4]). More recent work use coalgebraic methods to develop semantics. Moreover, in [5] Moss shows how coalgebraic semantics can be used to prove soundness and completeness of the FLR_0 fragment of the formal language of recursion (see [6] and [7]). The logic presented in this paper extends FLR_0 to include the operation of taking second-order fixed-points and uses the coalgebraic methods developed in [2] and [5] to provide soundness and completeness results.

1 Syntax

The Formal Logic of Recursive Schemes, FLRS, provides a framework for reasoning about the equivalence of uninterpreted recursive program schemes. As such,

J.L. Fiadeiro et al. (Eds.): CALCO 2005, LNCS 3629, pp. 278–292, 2005.

the logic will contain constructors for fixed-point definitions of functions. Since FLRS is also intended to be an extension of FLR_0, the language will contain constructors for recursive definitions of first-order variables as well. Formally, the syntax of FLRS is obtained by starting with a denumerable set of (first-order) variables $\mathsf{Var} = \{\mathsf{v}_1, \mathsf{v}_2, \mathsf{v}_3, \ldots\}$ and, for each arity n, a denumerable set of function variables $\mathsf{fVar}_n = \{\mathsf{f}_1^n, \mathsf{f}_2^n, \mathsf{f}_3^n, \ldots\}$ of arity n. We denote by fVar the set $\bigcup_{n\in\omega} \mathsf{fVar}_n$. When the arity is clear from context we write f_i instead of f_i^k. We use lower case roman letters, x, y, z, and so on to denote first-order variables and sans serif letters, f, g, h, G, and so on to denote function variables.

The terms of FLRS are defined recursively as follows:

1. Each variable v_i is a term.
2. If $E_1, \ldots, E_n$ are terms then $\mathsf{f}_i^n(E_1, \ldots, E_n)$ is a term.
3. If $A_0, A_1, \ldots, A_n$ are terms, then $A_0 \ \mathtt{where}\,\{x_1 = A_1, \ldots, x_n = A_n\}$ is also a term.
4. Suppose that $F_0, F_1, \ldots, F_n$ are terms and that $\mathsf{f}_1, \ldots, \mathsf{f}_n$ are function symbols. Suppose further that, for $i = 1, \ldots, n$, F_i is not a variable and that the free variables of F_i are contained in $\mathsf{v}_1, \ldots, \mathsf{v}_{\mathrm{arity}(\mathsf{f}_i)}$. Then the following is also a term: $F_0 \llbracket \mathsf{f}_1 = F_1, \ldots, \mathsf{f}_i = F_i \rrbracket$.

Clauses 3 and 4 in the above definition correspond, respectively, to first-order and second-order fixed-point operations. In FLRS we use $\llbracket\rrbracket$-terms to represent second-order function definitions. In particular

$$\mathsf{f}(\mathsf{v}_1) \llbracket \mathsf{f} = \mathsf{G}(\mathsf{v}_1, \mathsf{f}(\mathsf{g}(\mathsf{v}_1))),\ \mathsf{g} = \mathsf{G}(\mathsf{f}(\mathsf{v}_1), \mathsf{g}(\mathsf{g}(\mathsf{v}_1))) \rrbracket \tag{3}$$

is the representation of the solution of f according to (1).

Syntactic notions such as *free* and *bound* variables are defined in the usual way: $\mathtt{where}\,\{x_1 = A_1, \ldots, x_n = A_n\}$ binds the first-order variables $x_1, \ldots, x_n$ and $\llbracket \mathsf{f}_1 = F_1, \ldots, \mathsf{f}_n = F_n \rrbracket$ binds the function variables $\mathsf{f}_1, \ldots, \mathsf{f}_n$. There are two syntactic substitutions: one each for the first- and second-order variables. For the first-order substitution we start with a map $s : \mathsf{Var} \to \mathsf{Term}$. This extends to a map $[s] : \mathsf{Term} \to \mathsf{Term}$, which is defined recursively as follows:

1. $[s](\mathsf{v}_i) = s(\mathsf{v}_i)$.
2. $[s](\mathsf{f}(E_1, \ldots, E_n)) = \mathsf{f}([s](E_1), \ldots, [s](E_n))$.
3. $[s](A \ \mathtt{where}\,\{x_1 = A_1, \ldots, x_n = A_n\}) =$
   $[t](A) \ \mathtt{where}\,\{x_1 = [t](A_1), \ldots, x_n = [t](A_n)\}$ where t is the substitution $t(x_i) = x_i$ and $t(y) = s(y)$ for all other variables.
4. $[s](F \llbracket \mathsf{f}_1 = F_1, \ldots, \mathsf{f}_n = F_n \rrbracket) = [s](F) \llbracket \mathsf{f}_1 = F_1, \ldots, \mathsf{f}_n = F_n \rrbracket$.

It is important to note that in the fourth clause of the above definition the first-order substitution is not performed on the terms inside the recursive program scheme; the variables in the F_i terms serve as place holders for arguments of the unknown functions rather than as syntactic variables.

For the second-order substitution, we start with a map $\sigma : \mathsf{fVar} \to \mathsf{Term}$ with the property that the only free variables in $\sigma(\mathsf{f}_i^n)$ are $\mathsf{v}_1, \ldots, \mathsf{v}_n$ and define a map $[\sigma] : \mathsf{Term} \to \mathsf{Term}$ which extends σ. The recursive definition of $[\sigma]$ is as follows.

1. $[\sigma]\,\mathsf{v}_i = \mathsf{v}_i$.
2. $[\sigma]\,\mathsf{f}\,(E_1,\ldots,E_k) = [s]\,\sigma\,(\mathsf{f})$ where s is the first-order substitution $s\,(\mathsf{v}_i) = [\sigma]\,E_i$.
3. $[\sigma]\left(A \texttt{ where}\left\{\overrightarrow{x} = \overrightarrow{A}\right\}\right) = [\sigma]\,(A) \texttt{ where}\left\{\overrightarrow{x} = \overrightarrow{[\sigma]\,(A)}\right\}$.
4. $[\sigma]\,F\,[\![\mathsf{f}_1 = F_1,\ldots,\mathsf{f}_k = F_k]\!] = ([\gamma]\,F)\left[\!\!\left[\overrightarrow{\mathsf{f}} = \overrightarrow{[\gamma]\,F}\right]\!\!\right]$ where γ is the second order substitution $\gamma\,(\mathsf{f}_i) = \mathsf{f}_i\,(\mathsf{v}_1,\ldots,\mathsf{v}_k)$ and $\gamma\,(\mathsf{g}) = \sigma\,(\mathsf{g})$ for $\mathsf{g} \neq \mathsf{f}_i$.

2 Semantics

In [8] and in this paper we use completely iterative monads to provide semantics for recursive program schemes. In this section we recall the basic definitions and present the major theorems of this theory.

Definition 1. *Let $(C,+)$ be a category with a coproduct. An endofunctor $H : C \to C$ is said to be* iteratable *if the functor $H\,(-) + a$ has a final coalgebra for all objects a of C.*

In this case we write $T_H\,(a)$ for the final coalgebra, and $\alpha_a : T_H\,(a) \to HT_H\,(a) + a$ for the final coalgebra map. By Lambek's lemma, α_a is an isomorphism, and we write $[\tau_a, \eta_a] : HT_H\,(a) + a \to T_H\,(a)$ for the inverse of α_a. In the case where no confusion is likely to occur, we will drop the subscript from T_H and simply write $T\,(a)$ for the final $H\,(-) + a$-coalgebra.

Example 1. Given a signature, that is, given a set Σ and a map arity : $\Sigma \to \mathbb{N}$, the signature functor for Σ is the functor $H_\Sigma : \mathsf{Set} \to \mathsf{Set}$ defined by

$$H_\Sigma\,(X) = \{\langle \mathsf{f}, s\rangle : \mathsf{f} \in \Sigma \text{ and } s : \text{arity}\,(\mathsf{f}) \to X\}\,.$$

For a morphism $k : X \to Z$, $H_\Sigma\,(f)$ is the morphism which takes $\langle \mathsf{f}, s\rangle$ to $\langle \mathsf{f}, k \circ s\rangle$. Since $H_\Sigma\,(-) + a$ is a bounded functor for any set a, $H_\Sigma\,(-) + a$ has a final coalgebra. Therefore, signature functors are iteratable. We write $T_\Sigma\,(a)$ for the final $H_\Sigma\,(-) + a$-coalgebra. In the category Set, $T_\Sigma\,(a)$ is the set of all finite and infinite Σ-trees with leaves labeled in a. In this case the map $\eta_a^\Sigma : a \to T_\Sigma\,(a)$ is the map which takes $x \in a$ to the (degenerate) tree consisting of one leaf labeled x. The map $\tau_a^\Sigma : H_\Sigma T_\Sigma\,(a) \to T_\Sigma\,(a)$ is the map that takes $\langle \mathsf{f}, t_1, \ldots, t_k\rangle$ to its tree representation.

Theorem 1 (First-Order Substitution[1]). *Let $f : a \to T\,(b)$. Then there is a unique $f^\star : T\,(a) \to T\,(b)$ such that*

1. *$f^\star$ is a morphism of H-algebras. That is, $f^\star \circ \tau_a = \tau_b \circ H\,(f^\star)$.*
2. *$f = f^\star \circ \eta_a$*

It is not without reason that Theorem 1 is given the name First-Order Substitution. Let $f : a \to T_\Sigma\,(b)$. Intuitively, a and b are interpreted as sets of variables and the elements of $T_\Sigma\,(b)$ are interpreted as `where`$\{\}$- and $[\![]\!]$-free terms of

FLRS[1], and the map f can be viewed as defining a first-order substitution on the variables in a. Theorem 1 gives us the existence of a map $f^\star : T_\Sigma(a) \to T_\Sigma(b)$. Given $t \in T_\Sigma(a)$, that is, given a `where`{}- and [[]]-free term with variables contained in the set a, the element $f^\star(t)$ is the result of performing the first-order substitution f on t.

In the special case of signature functors, the map $f^\star$ allows us to model first-order substitutions. In a general category the $\star$ operation allows us to put a monad structure on the functor T. This fact is contained in the following lemma.

Lemma 1 ([9]). *$(T, \eta, \star)$ is a Kleisli triple.*

Definition 2. *Given an iteratable functor H, the Kleisli triple from Lemma 1 is called the* Kleisli triple associated with H. *Standard results from the literature on monads give us that (T, η, μ), where μ_a is defined to be $(\mathsf{id}_{T(a)})^\star$, is a monad. This monad is called the* free completely iterative monad on H.

Definition 3. *Let $(C, +)$ be a category with a coproduct and let $H : C \to C$ be iteratable. The* ideal monad generated by H *is the tuple $(T, \eta, \mu, HT, \tau, H(\mu))$ where (T, η, μ) is the free completely iterative monad on H and $[\tau, \eta]$ is the pointwise inverse of the final coalgebra map.*

We also define a natural transformation $\kappa : H \to T$ by $\kappa = \tau \circ H(\eta)$.

Example 2. In the case of signature functors, $H_\Sigma(\eta_a^\Sigma)$ is the map which takes $\langle \mathsf{f}, x_1, \ldots, x_n \rangle$, where the x_i are considered as variables, to $\langle \mathsf{f}, x_1, \ldots, x_n \rangle$, where the x_i are considered to be trees and τ_a^Σ takes $\langle \mathsf{f}, x_1, \ldots, x_k \rangle$ to itself considered as a tree. Combining these two facts we see that κ^Σ is the *canonical map*, that is, the natural transformation which takes function symbols to the trees representing the function symbols.

Theorem 1 gave us an operation on morphisms which corresponded to the operation of extending a first-order substitution defined on variables to a first-order substitution defined on terms. The following theorem gives us an operation on natural transformations which corresponds to extending a second-order substitution.

Theorem 2 (Second-Order Substitution[2]). *Given two iteratable functors, H and H' and the ideal monads they generate, let $\nu : H \to T'$ be a natural transformation which factors as*

$$H \xrightarrow{\nu_0} H'T' \xrightarrow{\tau'} T'$$

Then there is a unique natural transformation $\nu^* : T \to T'$ such that

1. *$\nu = \nu^* \circ \kappa$; and*
2. *ν^* is a morphism of monads.*

---
[1] More precisely, only the finite elements of $T_\Sigma(b)$ can be interpreted as `where`{}- and [[]]-free terms. However, this distinction is not important for the present discussion.

The previous discussion suggests that ideal monads generated by signature functors can be used to model the syntax of FLRS. In particular, the natural transformations η and κ allow us to model v_i and $\mathsf{f}(t_1, \ldots, t_n)$. The map $f^\star$ models the syntax of first-order substitution and the natural transformation ν^* models second-order substitution. We now show how ideal monads can be used to model the `where` $\{\}$- and $[\![ ]\!]$-terms of FLRS.

Theorem 3 (Solution Theorem [1]). *Let H be an iteratable monad and let $(T, \eta, \mu, HT, \tau, H(\mu))$ be the ideal monad generated by H. Suppose $f : a \to T(a+b)$ factors as*

$$a \xrightarrow{f_0} HT(a+b) + b \xrightarrow{[\tau_{a+b}, \eta_{a+b} \circ \mathsf{inr}]} T(a+b).$$

Then there is a unique $f^\dagger : a \to T(b)$ such that $f^\dagger = [f^\dagger, \eta_b]^\star \circ f$.

Example 3. Let Σ be a signature which contains binary connectives $\times$ and $*$, let a be the set $\{x, y\}$, let b be the set $\{p, q\}$, and consider the following set of recursive equations:

$$x = y \times p \qquad y = q * x \tag{4}$$

We now define a map $f : a \to T_\Sigma(a+b)$ by

$f(x) =$ 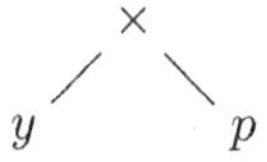

$f(y) =$ 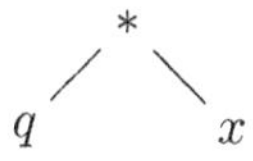

Clearly f factors through $H_\Sigma T_\Sigma(a+b) + b$. Thus we let $f^\dagger$ be the map given by Theorem 3. The goal of this example is to compute the value of $f^\dagger(x)$. In order to do so, we use the fact that $f^\dagger = [f^\dagger, \eta_b]^\star \circ f$. In particular, $f^\dagger(x) = [f^\dagger, \eta_b]^\star (y \times p)$. We can now use the fact that $[f^\dagger, \eta_b]^\star$ is an H_Σ-algebra homomorphism to conclude $f^\dagger(x) = [f^\dagger, \eta_b]^\star(y) \times [f^\dagger, \eta_b]^\star(p) = f^\dagger(y) \times p$. Using a similar argument we see that $f^\dagger(y) = q * f^\dagger(x)$. Therefore, we conclude that

$f^\dagger(x) =$ 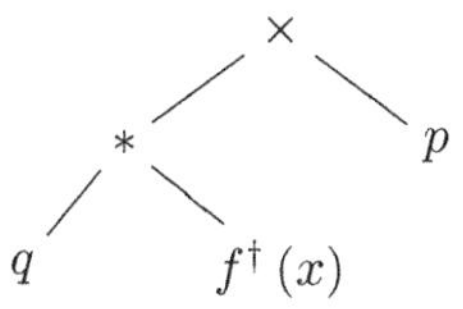

However, unwinding x according to (4) yields the following equation:

$x =$ 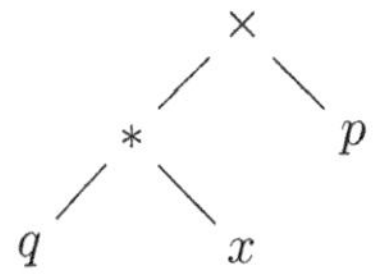

This shows that $f^\dagger(x)$ is a solution of x according to the recursive definition (4). Moreover, the uniqueness condition of Theorem 3 guarantees that the solution to (4) is unique.

In the statement of Theorem 3, we require that the map f factor through $HT(a+b)+b$. This is a guardedness requirement. In terms of signature functors, this requirement states that the right-hand side of each first-order recursive definition must be either a compound term, that is, not a variable, or a variable that is distinct from the variables being defined. In particular, the following set of recursive equations is not guarded:

$$x = y \qquad y = \mathsf{f}(x, w) \qquad z = w \tag{5}$$

The first equation of (5) violates the guardedness requirement. The third equation, on the other hand, does not violate the guardedness requirement as w is not one of the variables being recursively defined. In Theorem 5 we will remove the guardedness requirement for systems of recursive equations.

Now that we have an operation which models the `where`{}-terms of FLRS using ideal monads, we turn our attention to modeling the ⟦⟧-terms. The following theorem is the second-order analog of Theorem 3 and gives us the solution to a recursive program scheme provided that the scheme is in Greibach normal form, that is, provided that the right-hand sides of the recursive definitions are complex terms where the head of the term is a base function symbol.

Theorem 4 (Second-Order Solution Theorem [2]). *Let H and H' be iteratable functors such that $H+H'$ is also iteratable. Write $\widehat{H}$ for the functor $H+H'$ and consider the ideal monads generated by H, H', and $\widehat{H}$. Let $\nu : H \to \widehat{T}$ be a natural transformation which factors as*

$$H \xrightarrow{\nu_0} H'\widehat{T} \xrightarrow{\mathsf{inr}_{\widehat{T}}} \widehat{H}\widehat{T} \xrightarrow{\widehat{\tau}} \widehat{T}\,.$$

Then there is a unique natural transformation $\nu^\ddagger : H \to T'$ which factors through $H'T'$ such that $\nu^\ddagger = \left[\nu^\ddagger, \kappa'\right]^* \circ \nu$.

Again, the factorization requirement is the guardedness requirement: all second-order systems of recursive equations must be in Greibach normal form.

3 Extended Solution Theorems

Our goal in this section is to provide, in the category Set, extended versions of the solution theorems of the previous section. To see why these extensions are necessary, consider the following recursive program scheme:

$$\begin{aligned}\mathsf{f}(x) &= \mathsf{g}(x)\\ \mathsf{g}(x) &= \mathsf{H}(x, \mathsf{h}(x))\\ \mathsf{h}(x) &= \mathsf{G}(\mathsf{h}(x), \mathsf{h}(\mathsf{g}(x)))\end{aligned} \tag{6}$$

Clearly (6) should have a solution since we could replace the equation $\mathsf{f}(x) = \mathsf{g}(x)$ with $\mathsf{f}(x) = \mathsf{H}(x, \mathsf{h}(x))$ and obtain an equivalent scheme that is in Greibach normal form. However, we cannot use Theorem 4 to provide this solution. Now consider a slight modification to (6):

$$\begin{aligned} \mathsf{f}(x) &= \mathsf{f}(x) \\ \mathsf{g}(x) &= \mathsf{H}(x, \mathsf{h}(x)) \\ \mathsf{h}(x) &= \mathsf{G}(\mathsf{h}(x), \mathsf{h}(\mathsf{g}(x))) \end{aligned} \tag{7}$$

Clearly (7) should have a solution. In fact, there are infinitely many solutions to the equation $\mathsf{f}(x) = \mathsf{f}(x)$. However, given a choice of a canonical solution, called the ungroundedness scapegoat, for the equation $\mathsf{f}(x) = \mathsf{f}(x)$, (7) should have a unique solution. Again, Theorem 4 does not provide this solution.

Our other goal is to establish an extended version of Theorem 3 which, again provided we have fixed an ungroundedness scapegoat, provides a solution to the following set of equations.

$$x = x \qquad y = p \times z \qquad z = x * q \tag{8}$$

Before we are able to provide solutions to systems such as (8), we need to discuss ungrounded equations and to define the ungroundedness scapegoat.

Definition 4. *Let $f : a \to T(a+b)$ and let $x \in a$. Write f_0 for the composition $\alpha_{a+b} \circ f$. The f-*sequence *of x is defined as follows:*

1. $x_0 = x$.
2. *x_{n+1} is defined if x_n is defined and $f_0(x_n) \in a \subseteq HT(a+b) + (a+b)$. In this case $x_{n+1} = f_0(x_n)$.*
3. *x_{n+1} is undefined in all other cases.*

An element x is said to be grounded *if the f-sequence of x is finite. x is* ungrounded *otherwise.*

Definition 5 (Ungroundedness Scapegoat). *Let H be an iteratable functor such that $T(0)$ is non-empty. Pick some element $\bot_0 \in T(0)$, and for each object a of C define $\bot_a = T(!_a)(\bot_0)$ where $!_a$ is the unique morphism from 0 to a. $\bot_a$ is called the* ungroundedness scapegoat *for a.*

Theorem 5 (Extended Solution Theorem [8]). *Let a be a finite set and let $f : a \to T(a+b)$. Then there exists a unique $f^\dagger : a \to T(b)$ such that*

1. *$f^\dagger = [f^\dagger, \eta_b]^\star \circ f$; and*
2. *$f^\dagger(x) = \bot_b$ if x is ungrounded for f.*

Having removed the factorization requirement of Theorem 3, Theorem 5 is able to provide a solution to (8). In particular, the solution is

$$x = \bot_b \qquad y = p \times (\bot_b * q) \qquad z = \bot_b * q,$$

where a is the set $\{x, y, z\}$, b is the set $\{p, q\}$ and the signature contains some 0-ary function symbol which was chosen as the ungroundedness scapegoat.

We also have and extended version of Theorem 4. Again, we begin with a discussion of ungroundedness. Let H and H' be two iteratable functors such that $\widehat{H} = H + H'$ is also iteratable. Assume that $T'(0)$ is non-empty and pick some $\bot_0'' \in T'(0)$. We define the ungroundedness scapegoat for second-order systems as follows: $\bot_a'' \in T'(a)$ is $T'(!_a)(\bot_0'')$.

Let $\nu : H \to \widehat{T}$ be a natural transformation which factors as $\nu = \widehat{\tau} \circ \gamma$ and define a natural transformation $\sigma : H\widehat{T} + H'\widehat{T} \to \widehat{H}\widehat{T}$ by $\sigma = \left[\widehat{H}(\widehat{\mu}) \circ \gamma_{\widehat{T}}, \mathsf{inr}_{\widehat{T}}\right]$. Let $w \in H(\{*\})$. We say that w is grounded with respect to ν if there is some $n \geq 0$ such that $\sigma^n_{\{*\}} \circ \gamma_{\{*\}}(w) \in H'\widehat{T}(\{*\}) \subseteq \widehat{H}\widehat{T}(\{*\})$. Otherwise, we say that w is ungrounded. For $w \in H(a)$, we say that w is grounded with respect to ν if $H(\downarrow_a)(w)$ is grounded, where $\downarrow_a$ is the unique map from a to $\{*\}$.

Theorem 6 (Extended Second-Order Solution Theorem [8]). *Given two iteratable functors, H and H', such that $\widehat{H} = H + H'$ is also iteratable and the ideal monads they generate, let $\nu : H \to \widehat{T}$ be a natural transformation which factors as*

$$H \xrightarrow{\gamma} \widehat{H}\widehat{T} \xrightarrow{\widehat{\tau}} \widehat{T}$$

Then there is a unique natural transformation $\nu^\ddagger : H \to T'$ which factors through $H'T'$ such that

1. *$\nu^\ddagger = \left[\nu^\ddagger, \kappa'\right]^* \circ \nu$; and*
2. *$\nu^\ddagger_a(x) = \bot_a''$ if x is ungrounded with respect to ν.*

4 The Denotation Map Λ

In this section we define a denotation map, Λ, for the terms of FLRS. This denotation map is parameterized by the free variables and free function variables of the term A. To wit: for every non-repeating list of variables $x_1, \ldots, x_n$ which contains all the free variables of A and for every signature Σ containing all the free function symbols of A, $\Lambda(x_1, \ldots, x_n; \Sigma)(A)$ will be defined to be a function from 1 to $T_\Sigma(n)$. The map Λ is defined recursively on the construction on A.

1. $\Lambda(\overrightarrow{x}; \Sigma)(x_i)$ is the function $\eta_n \circ n_i$, where n_i is the ith coproduct injection which makes n a coproduct of 1.
2. $\Lambda(\overrightarrow{x}; \Sigma)(\mathsf{f}^k(E_1, \ldots, E_k)) = [\ldots, \Lambda(\overrightarrow{x}; \Sigma)(E_i), \ldots]^\star \circ \kappa_k(\langle \mathsf{f}, \mathsf{id}_k \rangle)$. In this clause, $\langle \mathsf{f}, \mathsf{id}_k \rangle$ is taken to be a map from 1 to $H_\Sigma(k)$ in the obvious way.
3. $\Lambda(\overrightarrow{x}; \Sigma)(A\ \mathtt{where}\{y_1 = A_1, \ldots, y_k = A_k\}) = \left[s^\dagger, \eta_n\right]^\star \circ \Lambda(\overrightarrow{y}, \overrightarrow{x}; \Sigma)(A)$ where $s : k \to T_\Sigma(k+n)$ is the map $s(i) = \Lambda(y_1, \ldots, y_k, x_1, \ldots, x_n; \Sigma)(A_i)$ and $s^\dagger$ is the map given by Theorem 5. In the above definition we must assume that the y_i's are distinct from the x_j's.
4. $\Lambda(\overrightarrow{x}; \Sigma)\left(A\left[\!\left[\overrightarrow{\mathsf{f}} = \overrightarrow{F}\right]\!\right]\right) = \left[\nu^\ddagger_n, \kappa_n\right]^* \circ \Lambda(x_1, \ldots, x_n; \Delta + \Sigma)(A)$ where Δ is the signature $\{\mathsf{f}_1, \ldots, \mathsf{f}_m\}$, $\nu : H_\Delta \to T_{\Delta+\Sigma}$ is the natural transformation whose component at n takes $\langle \mathsf{f}^k_i, s \rangle$ to $T_{\Delta+\Sigma}(s) \circ \Lambda(\mathsf{v}_1, \ldots, \mathsf{v}_k; \Delta + \Sigma)(F_i)$,

and $\nu^{\ddagger}$ is the natural transformation given by Theorem 6. The natural transformation ν defined in this clause is referred to as the *natural transformation induced by* $\left[\!\left[\overrightarrow{\mathsf{f}} = \overrightarrow{F}\right]\!\right]$.

Let ϕ and ψ be terms and let Γ be a set of equations of terms. We define the semantic relation $\Gamma \models \phi = \psi$ to hold if, for any substitutions σ and s, for any list of variables $\overrightarrow{x}$ and any signature Σ for which the following denotations are defined, we have $\Lambda(\overrightarrow{x}; \Sigma)([\sigma][s]\phi) = \Lambda(\overrightarrow{x}; \Sigma)([\sigma][s]\psi)$ whenever $\Lambda(\overrightarrow{x}; \Sigma)([\sigma][s]\chi_1) = \Lambda(\overrightarrow{x}; \Sigma)([\sigma][s]\chi_2)$ for all $\chi_1 = \chi_2 \in \Gamma$.

5 The Logic FLRS

In this section we present the axioms of FLRS. Throughout this section we use ϕ, ψ, and χ as variables standing for terms in Term. We use Γ and Δ for sets of equations.

Equational Axioms

$$\vdash \phi = \phi \qquad \phi = \psi \vdash \psi = \phi \qquad \phi = \psi, \psi = \chi \vdash \phi = \chi$$

Logical Inference Rules

Weakening. If $\Gamma \vdash \phi = \psi$, then $\Gamma \cup \Delta \vdash \phi = \psi$.
Substitution. If $\Gamma \vdash \phi = \psi$ and s is a first-order substitution, then $[s](\Gamma) \vdash [s](\phi) = [s](\psi)$. If σ is a second-order substitution, then $[\sigma](\Gamma) \vdash [\sigma](\phi) = [\sigma](\psi)$.

Recursion Axioms

Head. $\vdash \mathsf{f}_i(\mathsf{v}_1, \ldots, \mathsf{v}_n)\left[\!\left[\overrightarrow{\mathsf{f}} = \overrightarrow{F}\right]\!\right] = F_i\left[\!\left[\overrightarrow{\mathsf{f}} = \overrightarrow{F}\right]\!\right]$.
Fixpoint. Consider the term $F\left[\!\left[\overrightarrow{f} = \overrightarrow{F}\right]\!\right]$ and define a second-order substitution by $\sigma(\mathsf{f}_i) = \mathsf{f}_i(\overrightarrow{\mathsf{v}})\left[\!\left[\overrightarrow{\mathsf{f}} = \overrightarrow{F}\right]\!\right]$. Given this second-order substitution we have $\vdash F\left[\!\left[\overrightarrow{\mathsf{f}} = \overrightarrow{F}\right]\!\right] = [\sigma](F)$.
Bekič-Scott. $\vdash F\left[\!\left[\overrightarrow{\mathsf{f}} = \overrightarrow{F}, \overrightarrow{\mathsf{g}} = \overrightarrow{G}\right]\!\right] = \left(F\left[\!\left[\overrightarrow{\mathsf{g}} = \overrightarrow{G}\right]\!\right]\right)\left[\!\left[\overrightarrow{\mathsf{f}} = \overrightarrow{F\left[\!\left[\overrightarrow{\mathsf{g}} = \overrightarrow{G}\right]\!\right]}\right]\!\right]$.
Bottom. In order to apply this axiom, a 0-ary function symbol must have be designated as the ungroundedness scapegoat. Suppose that $\bot$ is the function symbol so designated. If $E_1, \ldots, E_k$ are any terms, then we have the following equality.

$$\vdash A\left[\!\left[\overrightarrow{\mathsf{f}} = \overrightarrow{F}, \mathsf{g} = \mathsf{g}(E_1, \ldots, E_k)\right]\!\right] = A\left[\!\left[\overrightarrow{\mathsf{f}} = \overrightarrow{F}, \mathsf{g} = \bot\right]\!\right].$$

Mix. Consider a term of the form $A\ \mathtt{where}\{x_1 = A_1, \ldots, x_n = A_n\}$ with free variables $z_1, \ldots, z_m$. Let $\mathsf{h}_1, \ldots \mathsf{h}_n$ be n new function variables of arity $n+m$ and define a first-order substitution by $s(x_i) = \mathsf{h}_i(x_1, \ldots, x_n, z_1, \ldots, z_n)$. Then

$$\vdash A\ \mathtt{where}\{x_1 = A_1, \ldots, x_n = A_n\} = [s](A)\left[\!\left[\ldots, \mathsf{h}_i = [s](A_i), \ldots\right]\!\right].$$

Recursion Inference Rule

Strong Recursion Inference. In order to apply this inference rule, we start with two terms of the form $\mathsf{f}_i(\overrightarrow{\mathsf{v}})\left[\!\left[\overrightarrow{\mathsf{f}} = \overrightarrow{F}\right]\!\right]$ and $\mathsf{g}_j(\overrightarrow{\mathsf{v}})\left[\!\left[\overrightarrow{\mathsf{g}} = \overrightarrow{G}\right]\!\right]$. Let Σ' be the signature $\left\{\overrightarrow{\mathsf{f}}\right\}$, let Σ'' be the signature $\left\{\overrightarrow{\mathsf{g}}\right\}$, let Σ be the signature of functions free in either term, and let n be such that all the free variables in either term are contained in $\{\mathsf{v}_1, \ldots, \mathsf{v}_n\}$. We assume that Σ and Σ' are disjoint signatures. By a *basic term* over Σ' we mean a `where`$\{\}$- and $[\![]\!]$-free term over Σ', that is, we mean a finite element of $T_{\Sigma'}(n)$. Suppose that each F_i is in the form $\mathsf{G}(t_1, \ldots, t_k)$ where $\mathsf{G} \in \Sigma$ and each t_i is a basic term over Σ' and that each G_i is in a similar form. For each basic term, t, over Σ' with $t = \langle \mathsf{f}_i, s\rangle$, let $unf(t) = [s]F_i$. $unf(w)$ is similarly defined for basic terms over Σ''. Now let Δ be a set of equations of the form $t = w$ where t and w are basic terms over Σ' and Σ'', respectively. Suppose $\Gamma, \Delta \vdash \mathsf{f}_i(\overrightarrow{\mathsf{v}}) = \mathsf{g}_j(\overrightarrow{\mathsf{v}})$, and for each equation $t = w \in \Delta$ we have $\Gamma, \Delta \vdash unf(t) = unf(w)$. Under all these assumptions, the Strong Recursion Inference Rule allows us to conclude $\Gamma \vdash \mathsf{f}_i(\overrightarrow{\mathsf{v}})\left[\!\left[\overrightarrow{\mathsf{f}} = \overrightarrow{F}\right]\!\right] = \mathsf{g}_j(\overrightarrow{\mathsf{v}})\left[\!\left[\overrightarrow{\mathsf{g}} = \overrightarrow{G}\right]\!\right]$.

We also have first order versions of the Head, Fixpoint, and Bekič-Scott axioms. These axioms are not formally member of the system as they are derivable from their second-order counterparts via the Mix Axiom.

Theorem 7 (Soundness [8]). *The logic* FLRS *is sound with respect to the semantics presented in Section 4.*

6 Normal Form

We are now in a position to prove the completeness of FLRS with respect to the coalgebraic semantics. As is customary, completeness is proved in two steps. First we prove that every term can be written in a normal form and then we use the normal form to prove completeness. We begin with a definition and a "flattening" type lemma.

Definition 6. *A term is said to be in* semi-normal form *if it is either a variable or written in the form* $\mathsf{h}_i(\overrightarrow{\mathsf{v}})\left[\!\left[\overrightarrow{\mathsf{h}} = \overrightarrow{H}\right]\!\right]$ *where each of the* H_i *is of the form* $\mathsf{G}(t_1, \ldots, t_k)$ *where all the* t_j *are* $[\![]\!]$*- and* `where`$\{\}$*-free terms all of whose function symbols are one of the* h_i*. The function symbol* G *may be, but is not required to be, one of the* h_j*.*

Lemma 2.

$$\vdash \mathsf{f}_i(\overrightarrow{\mathsf{v}})\left[\!\left[\overrightarrow{\mathsf{f}} = \overrightarrow{F}\right]\!\right]\left[\!\left[\ldots, \mathsf{h}_j = \mathsf{g}^j_{n_j}(\overrightarrow{\mathsf{v}})\left[\!\left[\overrightarrow{\mathsf{g}^j} = \overrightarrow{G^j}\right]\!\right], \ldots\right]\!\right] =$$

$$\mathsf{f}_i(\overrightarrow{\mathsf{v}})\left[\!\left[\overrightarrow{\mathsf{f}} = \overrightarrow{F}, \ldots, \mathsf{h}_j = G^j_{n_j}, \ldots, \overrightarrow{\mathsf{g}^j} = \overrightarrow{G^j}, \ldots\right]\!\right]$$

Proof. We may assume that all the g_k^j, h_k, and f_k are distinct. We may also assume that no f_j appears in the recursive program scheme

$$\left[\!\left[\ldots, \mathsf{h}_j = \mathsf{g}_{n_j}^j\left(\vec{\mathsf{v}}\right)\left[\!\left[\vec{\mathsf{g}^j} = \vec{G^j}\right]\!\right], \ldots\right]\!\right]$$

and that no g_k^j appears in the recursive program scheme $\left[\!\left[\vec{\mathsf{f}} = \vec{F}\right]\!\right]$.

We start the proof with a derivation. In lines 5 and 6 of the derivation we let A be the term $\mathsf{f}_i\left(\vec{\mathsf{v}}\right)\left[\!\left[\vec{\mathsf{f}} = \vec{f}\right]\!\right]\left[\!\left[\vec{\mathsf{g}} = \vec{G}\right]\!\right]$.

1) $\vdash \mathsf{g}_{n_j}^j\left(\vec{\mathsf{v}}\right)\left[\!\left[\vec{\mathsf{g}^j} = \vec{G^j}\right]\!\right] = G_{n_j}^j\left[\!\left[\vec{\mathsf{g}^j} = \vec{G^j}\right]\!\right]$

Head

2) $\vdash \mathsf{f}_i\left(\vec{\mathsf{v}}\right)\left[\!\left[\vec{\mathsf{f}} = \vec{F}\right]\!\right]\left[\!\left[\ldots, \mathsf{h}_j = \mathsf{g}_{n_j}^j\left(\vec{\mathsf{v}}\right)\left[\!\left[\vec{\mathsf{g}^j} = \vec{G^j}\right]\!\right], \ldots\right]\!\right] =$
$\mathsf{f}_i\left(\vec{\mathsf{v}}\right)\left[\!\left[\vec{\mathsf{f}} = \vec{F}\right]\!\right]\left[\!\left[\ldots, \mathsf{h}_j = G_{n_j}^j\left[\!\left[\vec{\mathsf{g}^j} = \vec{G^j}\right]\!\right], \ldots\right]\!\right]$

1, Substitution

3) $\vdash \mathsf{f}_i\left(\vec{\mathsf{v}}\right)\left[\!\left[\vec{\mathsf{f}} = \vec{F}\right]\!\right]\left[\!\left[\vec{\mathsf{g}^j} = \vec{G^j}\right]\!\right] = \mathsf{f}_i\left(\vec{\mathsf{v}}\right)\left[\!\left[\vec{\mathsf{f}} = \vec{F}\right]\!\right]$

Fixpoint

4) $\vdash \mathsf{f}_i\left(\vec{\mathsf{v}}\right)\left[\!\left[\vec{\mathsf{f}} = \vec{F}\right]\!\right]\left[\!\left[\ldots, \mathsf{h}_j = G_{n_j}^j\left[\!\left[\vec{\mathsf{g}^j} = \vec{G^j}\right]\!\right], \ldots\right]\!\right] =$
$\mathsf{f}_i\left(\vec{\mathsf{v}}\right)\left[\!\left[\vec{\mathsf{f}} = \vec{F}\right]\!\right]\left[\!\left[\vec{\mathsf{g}^j} = \vec{G^j}\right]\!\right]\left[\!\left[\ldots, \mathsf{h}_j = G_{n_j}^j\left[\!\left[\vec{\mathsf{g}^j} = \vec{G^j}\right]\!\right], \ldots\right]\!\right]$

3, Symmetry, Substitution

5) $\vdash A\left[\!\left[\ldots, \mathsf{h}_i = \mathsf{g}_{n_i}^i\left(\vec{\mathsf{v}}\right)\left[\!\left[\vec{\mathsf{g}^i} = \vec{G^i}\right]\!\right], \ldots, \mathsf{h}_j = G_{n_j}^j\left[\!\left[\vec{\mathsf{g}^j} = \vec{G^j}\right]\!\right], \ldots\right]\!\right]$
$= A\left[\!\left[\ldots, \mathsf{h}_i = \mathsf{g}_{n_i}^i\left(\vec{\mathsf{v}}\right)\left[\!\left[\vec{\mathsf{g}^i} = \vec{G^i}\right]\!\right]\left[\!\left[\vec{\mathsf{g}^j} = \vec{G^j}\right]\!\right], \ldots, \mathsf{h}_j = G_{n_j}^j\left[\!\left[\vec{\mathsf{g}^j} = \vec{G^j}\right]\!\right], \ldots\right]\!\right]$

Lemma 5.2 of [8]

6) $\vdash A\left[\!\left[\ldots, \mathsf{h}_i = \mathsf{g}_{n_i}^i\left(\vec{\mathsf{v}}\right)\left[\!\left[\vec{\mathsf{g}^i} = \vec{G^i}\right]\!\right]\left[\!\left[\vec{\mathsf{g}^j} = \vec{G^j}\right]\!\right], \ldots, \mathsf{h}_j = G_{n_j}^j\left[\!\left[\vec{\mathsf{g}^j} = \vec{G^j}\right]\!\right], \ldots\right]\!\right]$
$= \mathsf{f}_i\left(\vec{\mathsf{v}}\right)\left[\!\left[\vec{\mathsf{f}} = \vec{F}\right]\!\right]\left[\!\left[.., \mathsf{h}_i = \mathsf{g}_{n_i}^i\left(\vec{\mathsf{v}}\right)\left[\!\left[\vec{\mathsf{g}^i} = \vec{G^i}\right]\!\right], \ldots, \mathsf{h}_j = G_{n_j}^j, .., \vec{\mathsf{g}^j} = \vec{G^j}\right]\!\right]$

Beckič-Scott

If we continue in the same manner, we see that

$$\vdash \mathsf{f}_i\left(\vec{\mathsf{v}}\right)\left[\!\left[\vec{\mathsf{f}} = \vec{F}\right]\!\right]\left[\!\left[\ldots, \mathsf{h}_j = \mathsf{g}_{n_j}^j\left(\vec{\mathsf{v}}\right)\left[\!\left[\vec{\mathsf{g}^j} = \vec{G^j}\right]\!\right], \ldots\right]\!\right] =$$

$$\mathsf{f}_i\left(\vec{\mathsf{v}}\right)\left[\!\left[\vec{\mathsf{f}} = \vec{F}\right]\!\right]\left[\!\left[\ldots, \mathsf{h}_j = G_{n_j}^j, \ldots, \vec{\mathsf{g}^j} = \vec{G^j}, \ldots\right]\!\right]$$

Since none of the f_i appear in $\left[\!\left[\ldots, \mathsf{h}_j = \mathsf{g}_{n_j}^j\left(\vec{\mathsf{v}}\right)\left[\!\left[\vec{\mathsf{g}^j} = \vec{G^j}\right]\!\right], \ldots\right]\!\right]$, we can use the Fixpoint Axiom and Substitution to conclude that

$$\vdash \mathsf{f}_i\left(\overrightarrow{\mathsf{v}}\right)\left[\!\left[\overrightarrow{\mathsf{f}} = \overrightarrow{F}\right]\!\right]\left[\!\left[\ldots, \mathsf{h}_j = G^j_{n_j}, \ldots, \overrightarrow{\mathsf{g}^j} = \overrightarrow{G^j}, \ldots\right]\!\right] =$$

$$\mathsf{f}_i\left(\overrightarrow{\mathsf{v}}\right)\left[\!\left[\overrightarrow{\mathsf{f}} = \overrightarrow{F}\right]\!\right]\left[\!\left[\ldots, \mathsf{h}_j = G^j_{n_j}\left[\!\left[\overrightarrow{\mathsf{f}} = \overrightarrow{F}\right]\!\right], \ldots, \overrightarrow{\mathsf{g}^j} = \overrightarrow{G^j\left[\!\left[\overrightarrow{\mathsf{f}} = \overrightarrow{F}\right]\!\right]}, \ldots\right]\!\right]$$

The desired result then follows from an application of the Bekič-Scott Axiom. □

Theorem 8. *Any term with the property that all the* `where{}`*-fragments do not contain any equations of the form* $\mathsf{v}_i = \mathsf{v}_j$ *can be written in semi-normal form.*

Proof. This theorem is proved by induction on the construction of A. The base case is trivial and the inductive steps, while not trivial, are consequences of Lemma 2. □

The requirement that no `where{}`-fragment contains an equation of the form $\mathsf{v}_i = \mathsf{v}_j$ is a result of our syntactical restriction on the $[\![]\!]$-terms. Recall that none of the F_i in the term $F\left[\!\left[\overrightarrow{\mathsf{f}} = \overrightarrow{F}\right]\!\right]$ are allowed to be variables. If we were to allow equations of the form $\mathsf{v}_i = \mathsf{v}_j$ in the `where{}`-term, the Mix Axiom would yield a $[\![]\!]$-term which does not satisfy the syntactical restriction.

Unfortunately, the semi-normal form is not sufficient for proving completeness. Thus, we turn our attention to defining and finding a normal form for the terms of FLRS.

Definition 7. *A term is said to be in* normal form *if it is either a variable or written in the form* $\mathsf{h}_i\left(\overrightarrow{\mathsf{v}}\right)\left[\!\left[\overrightarrow{\mathsf{h}} = \overrightarrow{H}\right]\!\right]$ *with each of the* H_i *is of the form* $\mathsf{G}(t_1, \ldots, t_k)$ *where* G *is not one of the* h_i *and all the* t_j *are* $[\![]\!]$*- and* `where{}`*-free terms all of whose function symbols are one of the* h_i*.*

The only difference between the definition of normal form and the definition of semi-normal form is that the function symbol G is not allowed to be one of the h_i in the definition of normal form. This restriction does not apply to terms written in semi-normal form.

Lemma 3 ([8]). *Consider the term* $A\left[\!\left[\overrightarrow{\mathsf{f}} = \overrightarrow{F}, \mathsf{g} = \mathsf{f}_i(t_1, \ldots, t_k)\right]\!\right]$. *If* s *is the first-order substitution* $s(\mathsf{v}_i) = t_i$, *then*

$$\vdash A\left[\!\left[\overrightarrow{\mathsf{f}} = \overrightarrow{F}, \mathsf{g} = \mathsf{f}_i(t_1, \ldots, t_k)\right]\!\right] = A\left[\!\left[\overrightarrow{\mathsf{f}} = \overrightarrow{F}, \mathsf{g} = [s](F_i)\right]\!\right].$$

In order to write terms in normal form we need to assume that we have a signature which contains an ungroundedness scapegoat. Thus, for the remainder of this section we fix a 0-ary function symbol, written $\perp$, to serve as such a scapegoat. We assume that $\perp$ does not appear on the left-hand side of any equation in the recursive program schemes that follow.

Suppose that $\mathsf{f}_i\left(\overrightarrow{\mathsf{v}}\right)\left[\!\left[\overrightarrow{\mathsf{f}} = \overrightarrow{F}\right]\!\right]$ is in semi-normal form. If all the head symbols of the F_i are distinct from the f_is, then the term is already in normal form. If not, there are three possible cases.

Case I In this case we consider a recursive program scheme in the following form:

$$\mathsf{h}\left(\overrightarrow{\mathsf{v}}\right)\left[\!\left[\overrightarrow{\mathsf{f}} = \overrightarrow{F}, \mathsf{g}_1 = \mathsf{g}_2\left(\overrightarrow{t}\right), \mathsf{g}_2 = \mathsf{g}_3\left(\overrightarrow{w}\right), \ldots, \mathsf{g}_l = \mathsf{G}\left(\overrightarrow{u}\right)\right]\!\right]$$

where $\mathsf{G} \notin \left\{\overrightarrow{\mathsf{f}}, \overrightarrow{\mathsf{g}}\right\}$ and $\mathsf{h} \in \left\{\overrightarrow{\mathsf{f}}, \overrightarrow{\mathsf{g}}\right\}$. In this case we apply Lemma 3 to rewrite our term in the form

$$\mathsf{h}\left(\overrightarrow{\mathsf{v}}\right)\left[\!\left[\overrightarrow{\mathsf{f}} = \overrightarrow{F}, \mathsf{g}_1 = \mathsf{G}\left(\overrightarrow{r}\right), \mathsf{g}_2 = \mathsf{g}_3\left(\overrightarrow{w}\right), \ldots, \mathsf{g}_l = \mathsf{G}\left(\overrightarrow{u}\right)\right]\!\right]$$

where all of the r_i are $[\![]\!]$- and `where`{}-free terms over $\left\{\overrightarrow{\mathsf{f}}, \overrightarrow{\mathsf{g}}\right\}$. Using the same procedure, we can write the recursive definitions of $\mathsf{g}_2, \ldots, \mathsf{g}_{l-1}$ in the appropriate form.

Case II In this case we consider a recursive program scheme in the following form:

$$\mathsf{h}\left(\overrightarrow{\mathsf{v}}\right)\left[\!\left[\overrightarrow{\mathsf{f}} = \overrightarrow{F}, \mathsf{g}_1 = \mathsf{g}_2\left(\overrightarrow{t}\right), \mathsf{g}_2 = \mathsf{g}_3\left(\overrightarrow{w}\right), \ldots, \mathsf{g}_l = \mathsf{g}_1\left(\overrightarrow{u}\right)\right]\!\right]$$

where $\mathsf{h} \in \left\{\overrightarrow{\mathsf{f}}, \overrightarrow{\mathsf{g}}\right\}$. In this case we apply Lemma 3 to rewrite our term in the form

$$\mathsf{h}\left(\overrightarrow{\mathsf{v}}\right)\left[\!\left[\overrightarrow{\mathsf{f}} = \overrightarrow{F}, \mathsf{g}_1 = \mathsf{g}_1\left(\overrightarrow{r}\right), \mathsf{g}_2 = \mathsf{g}_3\left(\overrightarrow{w}\right), \ldots, \mathsf{g}_l = \mathsf{g}_1\left(\overrightarrow{u}\right)\right]\!\right].$$

We can then use the Bottom Axiom to rewrite this as

$$\mathsf{h}\left(\overrightarrow{\mathsf{v}}\right)\left[\!\left[\overrightarrow{\mathsf{f}} = \overrightarrow{F}, \mathsf{g}_1 = \bot, \mathsf{g}_2 = \mathsf{g}_3\left(\overrightarrow{w}\right), \ldots, \mathsf{g}_l = \mathsf{g}_1\left(\overrightarrow{u}\right)\right]\!\right].$$

Since $\bot$ is not in $\left\{\overrightarrow{\mathsf{f}}, \overrightarrow{\mathsf{g}}\right\}$, the recursive definition of g_1 is now in the appropriate form. The next case shows how to write the definitions of $\mathsf{g}_2, \ldots, \mathsf{g}_l$ in normal form.

Case III In this case we consider a recursive program scheme in the following form:

$$\mathsf{h}\left(\overrightarrow{\mathsf{v}}\right)\left[\!\left[\overrightarrow{\mathsf{f}} = \overrightarrow{F}, \mathsf{g}_1 = \mathsf{g}_2\left(\overrightarrow{t}\right), \mathsf{g}_2 = \mathsf{g}_3\left(\overrightarrow{w}\right), \ldots, \mathsf{g}_l = \bot\right]\!\right]$$

where $\mathsf{h} \in \left\{\overrightarrow{\mathsf{f}}, \overrightarrow{\mathsf{g}}\right\}$. In this case we can apply Lemma 3 $l-1$ times to rewrite our term in the form

$$\mathsf{h}\left(\overrightarrow{\mathsf{v}}\right)\left[\!\left[\overrightarrow{\mathsf{f}} = \overrightarrow{F}, \mathsf{g}_1 = \bot, \mathsf{g}_2 = \bot, \ldots, \mathsf{g}_l = \bot\right]\!\right].$$

Since the three cases above are the only cases in which a term in semi-normal form is not in normal form we have established the following theorem.

Theorem 9. *Any term with the property that none of its* `where`{}*-fragments contain equations of the form* $\mathsf{v}_i = \mathsf{v}_j$ *can be written in normal form.*

7 Completeness

In this section we prove completeness of the system FLRS with respect to the final coalgebra semantics. Towards that end we start with two terms, A and B with the property that none of their `where{}`-fragments contain equations of the form $\mathsf{v}_i = \mathsf{v}_j$. If A is a variable and $\models A = B$, then clearly B must be the same variable. Therefore we focus on the case where neither A nor B is a variable. By Theorem 9 we may assume that A and B are both written in normal form and are not variables. That is we may assume that

$$A = \mathsf{f}_i\left(\overrightarrow{\mathsf{v}}\right)\left[\!\left[\overrightarrow{\mathsf{f}} = \overrightarrow{F}\right]\!\right] \quad \text{and} \quad B = \mathsf{g}_j\left(\overrightarrow{\mathsf{v}}\right)\left[\!\left[\overrightarrow{\mathsf{g}} = \overrightarrow{G}\right]\!\right]$$

where all the F_i are of the form $\langle \mathsf{G}, t_1, \ldots, t_k\rangle$ with each $t_i \in T_{\Sigma'}(k)$ and all the G_i are of the form $\langle G, w_1, \ldots, w_k\rangle$ with each $w_i \in T_{\Sigma''}(k)$. Further, we may assume that no f_i appears in the normal form of B and that no g_j appears in the normal form of A. We let Σ be the signature of functions which appear free in either A or B, Δ' be the signature $\left\{\overrightarrow{\mathsf{f}}\right\}$, Δ'' be the signature $\left\{\overrightarrow{\mathsf{g}}\right\}$, and Δ be the signature $\Delta' + \Delta''$. We write H for the functor H_Σ, H' for the functor H_Δ, and $\widehat{H}$ for the functor $H' + H$. Finally, we let $\nu : H' \to \widehat{T}$ be the natural transformation induced by the recursive program scheme $\left[\!\left[\overrightarrow{\mathsf{f}} = \overrightarrow{F}, \overrightarrow{\mathsf{g}} = \overrightarrow{G}\right]\!\right]$.

Since all the F_i and G_j are in normal form, the natural transformation ν factors as follows

$$H_\Delta \xrightarrow{\nu_0} H_\Sigma T_\Delta \xrightarrow{H_\Sigma((\kappa \circ \mathsf{inl})^*)} H_\Sigma T_{\Sigma+\Delta} \xrightarrow{\mathsf{inr}_{T_{\Sigma+\Delta}}} \widehat{H}\widehat{T} \xrightarrow{\widehat{\tau}} \widehat{T}$$

We now define an H_Σ-coalgebra as follows

$$T_\Delta \xrightarrow{\alpha_\Delta} H_\Delta T_\Delta + 1 \xrightarrow{\nu_{0 T_\Delta} + \mathsf{id}} H_\Sigma T_\Delta T_\Delta + 1 \xrightarrow{H_\Sigma(\mu) + \mathsf{id}} H_\Sigma T_\Delta + 1$$

Now we let β be the unique natural transformation such that the following diagram commutes

$$\begin{array}{ccccccc}
T_\Delta & \xrightarrow{\alpha_\Delta} & H_\Delta T_\Delta + 1 & \xrightarrow{\nu_{0 T_\Delta} + \mathsf{id}} & H_\Sigma T_\Delta T_\Delta + 1 & \xrightarrow{H_\Sigma(\mu) + \mathsf{id}} & H_\Sigma T_\Delta + 1 \\
{\scriptstyle \beta}\downarrow & & & & & & \downarrow{\scriptstyle H_\Sigma(\beta) + \mathsf{id}} \\
T_\Sigma & & & \xrightarrow{\alpha_\Sigma} & & & H_\Sigma T_\Sigma + 1
\end{array}$$

Lemma 4. $\beta = \left(\nu^\ddagger\right)^*$

Proof. We show that $\beta = \left(\nu^\ddagger\right)^*$ by showing that the relation $\beta_a(t)\, R \left(\nu^\ddagger\right)^*_a(t)$ is a bisimulation for any a. First, if $t = \mathsf{v}_i$, then we have $\beta_a(t) = \left(\nu^\ddagger\right)^*_a(t) = \mathsf{v}_i$. Otherwise we have $t = \langle \mathsf{f}, s : k \to T_\Delta(a)\rangle$. Suppose $\nu_a(f) = \langle \mathsf{G}, s' : k \to T_\Delta(k)\rangle$. Using the definition of β above we see that $\alpha_{\Sigma a} \circ \beta_a(t) = \langle G, \beta_a \circ s^\star \circ s'\rangle$. Using the standard properties of second-order substitution we see that $\alpha_{\Sigma a} \circ \left(\nu^\ddagger\right)^*_a(t) = \left\langle \mathsf{G}, \left(\nu^\ddagger\right)^*_a \circ s^\star \circ s'\right\rangle$. This shows that $\beta_a(t)$ and $\left(\nu^\ddagger\right)^*_a(t)$ have the same head and that their children stand in the same relation. Since equality is the largest bisimulation on a final coalgebra, we conclude that $\beta = \left(\nu^\ddagger\right)^*$. □

We now show $\vdash A = B$ via an application of the Strong Recursion Inference Rule. Let l be such that all the free variables in A and B appear in $\{\mathsf{v}_1, \ldots, \mathsf{v}_l\}$. Now let Γ be the set of equations $\{t = w : t, w \in T_\Delta(l) \text{ and } \beta_l(t) = \beta_l(w)\}$. Since we assume that $\models \mathsf{f}_i(\overrightarrow{\mathsf{v}}) \left[\!\left[\overrightarrow{\mathsf{f}} = \overrightarrow{F}\right]\!\right] = \mathsf{g}_j(\overrightarrow{\mathsf{v}}) \left[\!\left[\overrightarrow{\mathsf{g}} = \overrightarrow{G}\right]\!\right]$, we can use the soundness of the Fixpoint Axiom and the fact that $\left[\!\left[\overrightarrow{\mathsf{f}} = \overrightarrow{F}\right]\!\right]$ and $\left[\!\left[\overrightarrow{\mathsf{g}} = \overrightarrow{G}\right]\!\right]$ are disjoint recursive program schemes to conclude

$$\models \mathsf{f}_i(\overrightarrow{\mathsf{v}}) \left[\!\left[\overrightarrow{\mathsf{f}} = \overrightarrow{F}, \overrightarrow{\mathsf{g}} = \overrightarrow{G}\right]\!\right] = \mathsf{g}_j(\overrightarrow{\mathsf{v}}) \left[\!\left[\overrightarrow{\mathsf{f}} = \overrightarrow{F}, \overrightarrow{\mathsf{g}} = \overrightarrow{G}\right]\!\right].$$

From this we conclude that

$$\left[\nu^\ddagger, \kappa^\Sigma\right]_l^* \circ \kappa_l^{\Delta+\Sigma}(\langle \mathsf{f}_i, \mathsf{id}_{k_1}\rangle) = \left[\nu^\ddagger, \kappa^\Sigma\right]_l^* \circ \kappa_l^{\Delta+\Sigma}(\langle \mathsf{g}_j, \mathsf{id}_{k_2}\rangle).$$

Thus we have $\nu_l^\ddagger(\langle \mathsf{f}_i, \mathsf{id}_k\rangle) = \nu_l^\ddagger(\langle \mathsf{g}_j, \mathsf{id}_k\rangle)$. Finally, since $\beta \circ \kappa_\Delta = \nu^\ddagger$, we have $\mathsf{f}_i(\overrightarrow{\mathsf{v}}) = \mathsf{g}_j(\overrightarrow{\mathsf{v}}) \in \Gamma$. This shows $\Gamma \vdash \mathsf{f}_i(\overrightarrow{\mathsf{v}}) = \mathsf{g}_j(\overrightarrow{\mathsf{v}})$.

Now suppose that $t = w \in \Gamma$ with $t = \langle \mathsf{f}, s\rangle$ and $w = \langle \mathsf{g}, p\rangle$. Since $\beta_l(t) = \beta_l(w)$ and since the recursive program scheme $\left[\!\left[\overrightarrow{\mathsf{f}} = \overrightarrow{F}, \overrightarrow{\mathsf{g}} = \overrightarrow{G}\right]\!\right]$ is in normal form, we may assume that $\nu_l(\langle \mathsf{f}, \mathsf{id}_{k_3}\rangle) = \langle \mathsf{G}, s'\rangle$ and $\nu_l(\langle \mathsf{g}, \mathsf{id}_{k_4}\rangle) = \langle \mathsf{g}, p'\rangle$. We need to show $\Gamma \vdash \mathit{unf}(t) = \mathit{unf}(w)$, that is, we need to show $\Gamma \vdash \langle \mathsf{G}, s^\star \circ s'\rangle = \langle \mathsf{G}, p^\star \circ p'\rangle$. However, as in the proof of the lemma above, we may use the definition of β to conclude that $\langle \mathsf{G}, \beta_l \circ s^\star \circ s'\rangle = \langle \mathsf{G}, \beta_l \circ p^\star \circ p'\rangle$. In particular, for $i = 0, \ldots, \text{arity}(G) - 1$ we have $\beta_l \circ s^\star \circ s'(i) = \beta_l \circ p^\star \circ p'(i)$, and hence we have $s^\star \circ s'(i) = p^\star \circ p'(i) \in \Gamma$. This shows that $\Gamma \vdash \mathit{unf}(t) = \mathit{unf}(w)$.

This shows that the hypotheses of the Strong Recursion Inference Rule are satisfied. Therefore, we show $\vdash A = B$ with an application of this rule.

References

1. Aczel, P., Adámek, J., Milius, S., Velebil, J.: Infinite trees and completely iterative theories: A coalgebraic view. Theoretical Computer Science **300** (2003) 1–45
2. Milius, S., Moss, L.S.: The category theoretic solution of recursive program schemes. preprint available at http://www.iti.cs.tu-bs.de/~milius/ (2004)
3. Guessarian, I.: Algebraic Semantics. Volume 99 of Lecture Notes in Computer Science. Springer-Verlag, Berlin (1981)
4. Courcelle, B.: Fundamental properties of infinite trees. Theoretical Computer Science **25** (1983) 95–169
5. Moss, L.S.: Recursion and corecursion have the same equational logic. Theoretical Computer Science **294** (2003) 233–267
6. Hurkens, A.J.C., McArthur, M., Moschovakis, Y.N., Moss, L.S., Whitney, G.T.: The logic of recursive equations. The Journal of Symbolic Logic **63** (1998) 451–478
7. Moschovakis, Y.N.: The formal language of recursion. The Journal of Symbolic Logic **54** (1989) 1216–1252
8. Mersch, J.G.: Equational Logic of Recursive Program Schemes. PhD thesis, Indiana University (2004)
9. Moss, L.S.: Parametric corecursion. Theoretical Computer Science **260** (2001) 139–163

The Category Theoretic Solution of Recursive Program Schemes

Stefan Milius[1] and Lawrence S. Moss[2]

[1] Institute of Theoretical Computer Science,
Technical University, Braunschweig, Germany
milius@iti.cs.tu-bs.de

[2] Department of Mathematics, Indiana University, Bloomington, IN, USA
lsm@cs.indiana.edu

Abstract. This paper provides a general account of the notion of recursive program schemes, their uninterpreted and interpreted solutions, and related concepts. It can be regarded as the category-theoretic version of the classical area of algebraic semantics. The overall assumptions needed are small indeed: working only in categories with "enough final coalgebras" we show how to formulate, solve, and study recursive program schemes. Our general theory is algebraic and so avoids using ordered, or metric structures. Our work generalizes the previous approaches which do use this extra structure by isolating the key concepts needed to study recursion, e. g., substitution in infinite trees, including second-order substitution. As special cases of our interpreted solutions we obtain the usual denotational semantics using complete partial orders, and the one using complete metric spaces. Our theory also encompasses implicitly defined objects which are not usually taken to be related to recursive program schemes at all. For example, the classical Cantor two-thirds set falls out as an interpreted solution (in our sense) of a recursive program scheme. In this short version of our paper we can only sketch some proofs.

1 Introduction

The theory of *recursive program schemes* is a topic at the heart of semantics. One takes a system of equations such as

$$\varphi(x) \approx F(x, \varphi(Gx)) \qquad \psi(x) \approx F(\varphi(Gx), GGx)) \tag{1.1}$$

where F and G are *given* functions and where φ and ψ are defined in terms of them by (1.1). The problems are: to give some sort of semantics to schemes, and to say what it means to *solve* a scheme. Actually, we should distinguish between *interpreted* schemes, where one also has an algebra A with operations for all the given operation symbols, and *uninterpreted* schemes.

This paper presents a generalization of the classical theory based on *Elgot algebras* and *coalgebras*. The point in a nutshell is that knowing that the infinite trees are the final coalgebra of a functor on sets leads to a purely algebraic

J.L. Fiadeiro et al. (Eds.): CALCO 2005, LNCS 3629, pp. 293–312, 2005.

account of first-order substitution and (co-)recursion, as shown in [1,28]. One does not need to assume any metric or order to study infinite trees: the finality principle is sufficient. However, to extend the result to second-order substitution, we need additional ideas. In this paper we show that corecursion allows us to give an uninterpreted semantics to a scheme; i. e., we show how to solve a scheme in final coalgebras.

For our interpreted semantics we work with Elgot algebras, a simple and fundamental notion introduced in [5]. We show how to give an interpreted solution to recursive program schemes in arbitrary Elgot algebras. We believe that our results in this area generalize and extend the previous work on this topic. Our method for obtaining interpreted solutions easily specializes to the usual denotational semantics using complete partial orders. As a second application we show how to solve recursive program schemes in complete metric spaces. Finally, we also provide examples of recursive program schemes and their solutions which cannot be treated with the classical theory: recursive definitions of operations satisfying equations like commutativity, examples in non-well founded sets (solving $x = \{\mathcal{P}_{\mathsf{f}}(x)\}$), and fractal self-similarity (the Cantor set c solves $c = \frac{1}{3}c \cup (\frac{2}{3} + \frac{1}{3}c)$). In some cases, one could use the classical theory (but even in those our application appears new). Our overall claim is that we have a unified view of solution principles for a large class of implicit definitions including recursive program schemes.

Our theory is based on notions from category theory (monads, Eilenberg-Moore algebras, Elgot algebras) and coalgebra (finality, solution theorems, completely iterative algebras). Our overall assumptions are weak: there must be finite coproducts, and all functors we deal with must have "enough final coalgebras". More precisely, we work in a category $\mathcal{A}$ with finite coproducts and with functors $H : \mathcal{A} \to \mathcal{A}$ such that for all objects X a final coalgebra TX of $H(\_) + X$ exists. The price we pay for working in such a general setting is that our theory takes somewhat more effort to build. But this is not excessive, and perhaps our categorical proofs reveal more conceptual clarity than the classical ones.

Related Work. The classical theory of recursive program schemes is compactly presented by Guessarian [20]. There one finds results on uninterpreted solutions of program schemes and interpreted ones in continuous algebras.

The realization that solutions of formal recursive equations can be studied with coalgebraic methods goes back to the second author [28], and appears also in the work of Aczel et al. [1], which generalizes and extends results of Elgot et al. [17,18], see also [26] of the first author. From [1,26] it also follows how to generalize *second order substitution* of infinite trees (see Courcelle [16]) to final coalgebras. The types of recursive equations studied in [28,1] did not go as far as program schemes. It is thus an important test problem for coalgebra to see if work on solving systems of equations can extend to (un)interpreted recursive program schemes. We are pleased that this paper reports a success in this matter.

Ghani et al. [19] obtained a general solution theorem with the aim of providing a categorical treatment of uninterpreted program scheme solutions. Part of our proof for the solution theorem for uninterpreted schemes is inspired by their proof of the same fact. However, the connection to (generalized) second order substitution as presented in [1] is new here.

Complete metric spaces as a basis for the semantics of recursive program schemes have been studied by Arnold and Nivat, see e. g. [10]. Bloom [14] studied interpreted solutions of recursive program schemes in so-called contraction theories. The semantics of recursively defined data types as fixed points of functors on the category of complete metric spaces has been investigated in [9,7]. We build on this with our treatment of self-similar objects. These have also recently been studied in a categorical framework by Leinster, see [22,23,24]. The example in this paper uses standard results on complete metric spaces, see e.g. [11].

In this short abstract we can only present sketches of proofs and we are forced to omit most of the technical detail. We refer the interested reader to the full version [27] of this paper.

2 Preliminaries

Assumption 2.1. Throughout this paper we assume that a category $\mathcal{A}$ with finite coproducts (having monomorphic injections) is given. In addition, all endofunctors H on $\mathcal{A}$ we consider are assumed to be *iteratable* [sic]: for each object X, the functor $H(\_) + X$ has a final coalgebra.

These are fairly mild conditions. However, we admit that iteratability is not a very nice notion with respect to closure properties of functors—for example, iteratable functors need not be closed under coproducts or composition. In the concrete categories we consider here there are stronger yet much nicer conditions that ensure iteratability:

Examples 2.2. (i) In the category Set of sets and maps any accessible (equivalently, bounded, see [6]) endofunctor is iteratable, see [12].

(ii) Consider the category CPO of complete partial orders; i. e., posets (not necessarily with a least element) in which every ascending chain has a join, and continuous maps. Notice that coproducts in CPO are disjoint unions with elements of different summands incompatible. Usually, one considers *locally continuous* endofunctors on categories of cpos, i. e., endofunctors H where each derived function $\mathsf{CPO}(X, Y) \to \mathsf{CPO}(HX, HY)$ is continuous. Unfortunately, on CPO not every locally continuous functor has a final coalgebra. For a counterexample consider the endofunctor assigning to a cpo X the powerset of the set of order components of X. This is a locally continuous endofunctor but it does not have a final coalgebra. However, any accessible endofunctor H on CPO has a final coalgebra, see [12], and moreover, H is iteratable.

(iii) Let CMS be the category of complete metric spaces with distances measured in the interval $[0, 1]$ together with non-expanding maps $f : X \to Y$; i. e., $d_Y(fx, fy) \leq d_X(x, y)$ for all $x, y \in X$. A functor H on CMS is called *contracting* if there exists a constant $\varepsilon < 1$ such that each derived function $\mathsf{CMS}(X, Y) \to$

$\mathsf{CMS}(HX, HY)$ is a contracting map; i.e., $d_{HX,HY}(Hf, Hg) \leq \varepsilon \cdot d_{X,Y}(f, g)$ for all non-expanding maps $f, g : X \to Y$. Contracting functors are iteratable, see [7].

Remark 2.3. We denote for an endofunctor H on $\mathcal{A}$ by $\mathcal{T}(H)X$ the final coalgebra, of $H(\_)+X$. Whenever confusion is unlikely we will drop the parenthetical (H) and simply write T for $\mathcal{T}(H)$. By the Lambek Lemma [21], the structure map of the final coalgebra is an isomorphism, and consequently, TX is a coproduct of HTX and X with injections

$$\eta_X^H : X \to TX \qquad \text{and} \qquad \tau_X^H : HTX \to TX\,.$$

Again, the superscripts will be dropped if confusion is unlikely.

It has been shown in the previous work [1,26] that the object assignment T gives rise to a monad on $\mathcal{A}$. And this monad is characterized by a universal property—it is the *free completely iterative monad* on H.

Examples 2.4. (i) Given any signature $\Sigma = (\Sigma_n)_{n<\omega}$ there is an associated polynomial endofunctor $H_\Sigma X = \Sigma_0 + \Sigma_1 \times X + \Sigma_2 \times X^2 + \cdots$ on Set, which is iteratable. Consider the algebra $T_\Sigma X$ of finite and infinite Σ-trees over the set X. That is, (ordered and rooted) trees labelled so that a node with n children, $n > 0$, is labelled by an n-ary operation symbol, and leaves are labelled by constant symbols or elements of X.

Notice that $T_\Sigma X$ is the final coalgebra of $H_\Sigma(\_)+X$. The coalgebra structure is given by the inverse of tree tupling. Finally, notice that T_Σ is a monad on Set.

(ii) A functor $H : \mathsf{Set} \to \mathsf{Set}$ is finitary (i.e., it preserves filtered colimits) iff it is a quotient of some polynomial functor H_Σ, see [8], III.4.3. The latter means that we have a natural transformation $\varepsilon : H_\Sigma \to H$ with epimorphic components ε_X, which are fully described by their kernel equivalence whose pairs can be presented in the form of so-called basic equations

$$\sigma(x_1, \ldots, x_n) = \rho(y_1, \ldots, y_m)$$

for $\sigma \in \Sigma_n$, $\rho \in \Sigma_m$ and $\sigma(x_1, \ldots, x_n), \rho(y_1, \ldots, y_m) \in H_\Sigma X$ for some set X including all x_i and y_j. Adámek [2] has proved that the final coalgebra TX of $H(\_) + X$ is given by the quotient $T_\Sigma X/{\sim_X}$ where $\sim_X$ is the following congruence: for every Σ-tree t denote by $\partial_n t$ the finite tree obtained by cutting t at level n and labelling all leaves at that level by some symbol $\perp$ not from Σ. Then we have $s \sim_X t$ for two Σ-trees s and t iff for all $n < \omega$, $\partial_n s$ can be obtained from $\partial_n t$ by finitely many applications of the basic equations describing the kernel of ε_X. Example: The functor H which assigns to a set X the set of unordered pairs of X is a quotient of $H_\Sigma X = X \times X$ expressing one binary operation b where ε_X is presented by commutativity of b; i.e., by the basic equation $b(x, y) = b(y, x)$. And TX is the coalgebra of all unordered binary trees with leaves labelled in the set X.

(iii) Consider the finite power set functor $\mathcal{P}_\mathrm{f} : \mathsf{Set} \to \mathsf{Set}$. Under the Anti-Foundation Axiom (AFA), its final coalgebra is the set HF_1 of hereditarily finite sets; see [13]. Analogously, the final coalgebra of $\mathcal{P}_f(\_) + X$ is the set $HF_1(X)$ of hereditarily finite sets generated from the set X. Even without AFA, the final coalgebra of $\mathcal{P}_f$ may be described as in Worrell [30]; it is the coalgebra formed by all strongly extensional trees; i. e., unordered trees so that for every node the subtrees defined by any two different children are not bisimilar. Analogously, the final coalgebra of $\mathcal{P}_f(\_) + X$ is the coalgebra of all strongly extensional trees where some leaves have a label from the set X.

(iv) For our later applications we shall find it convenient to work with an iteratable endofunctor $H : \mathsf{Set} \to \mathsf{Set}$ that has a locally continuous lifting H' on CPO; i. e., $U \cdot H' = H \cdot U$ where $U : \mathsf{CPO} \to \mathsf{Set}$ is the forgetful functor. It can be proved analogously as in [8], Theorem IV.5.3, that H' is iteratable. Moreover, for every CPO X, there is a cpo structure on the final coalgebra $\mathcal{T}(H)X$ of $H(\_)+X$ on Set; and this yields the final coalgebra of $H'(\_) + X$ on CPO. It follows that $U \cdot \mathcal{T}(H') = \mathcal{T}(H) \cdot U$. For example, every polynomial functor H_Σ has a locally continuous lifting H', and $\mathcal{T}(H')X$ is the Σ-tree algebra $T_\Sigma X$ with the order induced by the order of the cpo X.

(v) For a set endofunctor H with a contracting lifting H' on CMS; i. e., $U \cdot H' = H \cdot U$ for $U : \mathsf{CMS} \to \mathsf{Set}$ the forgetful functor, we have that H is iteratable and $U \cdot \mathcal{T}(H') = \mathcal{T}(H) \cdot U$. In fact, this follows from the results of [7] since U preserves limits. Any polynomial functor H on Set has a contracting lifting to CMS. For $HX = X^n$, define $H'(X, d) = (X^n, \frac{1}{2}d_{\max})$ (where $d_{\max}$ is the maximum metric) which is a contracting functor with $\varepsilon = \frac{1}{2}$. And coproducts of $\frac{1}{2}$-contracting liftings are $\frac{1}{2}$-contracting liftings of coproducts. The final coalgebra $\mathcal{T}(H')X$ is the Σ-tree algebra $T_\Sigma X$ equipped with a suitable metric, see [27] for details.

Example 2.5. The classical treatment of recursive program schemes fits into our work in the following way: Suppose we have a signature Σ of *given* operation symbols. Let Φ be a (finite) signature of new operation symbols. Classically a *recursive program scheme* (or shortly, *RPS*) gives for each operation symbol $f \in \Phi_n$ a term t^f over $\Sigma + \Phi$ in n variables. That is, a scheme is a set of formal equations

$$f(x_1, \dots, x_n) \approx t^f(x_1, \dots, x_n), \quad f \in \Phi_n, \quad n \in \mathbb{N}\,. \tag{2.2}$$

Now a signature is the same as a functor $\mathbb{N} \to \mathsf{Set}$, where $\mathbb{N}$ is understood as the discrete category with natural numbers as objects. Now observe that the names of the variables in (2.2) do not matter. More precisely, any RPS as in (2.2) gives rise to a natural transformation $\Phi \to T_{\Sigma+\Phi} \cdot J$, where $J : \mathbb{N} \to \mathsf{Set}$ is the inclusion functor mapping any number n to the set $\{0, \dots, n-1\}$. Notice that there is no need to consider only finite signatures Φ here. Moreover, this natural transformation extends the classical notion of RPS in the sense that by taking $T_{\Sigma+\Phi}$ we allow infinite trees on the right-hand sides of systems like (2.2). Here is how we render this example in our approach: Observe that $T_{\Sigma+\Phi} = \mathcal{T}(H_\Sigma + H_\Phi)$ where H_Σ and H_Φ denote the polynomial functors associated to Σ and Φ, respectively. Thus, to give the above natural transformation is the same as to give a natural transformation $H_\Phi \to \mathcal{T}(H_\Sigma + H_\Phi)$.

Example 2.6. Let Σ contain a unary operation symbol G and a binary one F. The signature Φ of recursively defined operations contains two unary symbols φ and ψ. Consider the recursive program scheme (1.1). The polynomial functor expressing the givens is $H_\Sigma = X + (X \times X)$ and the recursively defined operations Φ are expressed by $H_\Phi X = X + X$. Thus, the scheme (1.1) gives a natural transformation $H_\Phi \to \mathcal{T}(H_\Sigma + H_\Phi)$.

In this paper we will abstract away from signatures and sets and study the uninterpreted and the interpreted semantics of recursive program schemes considered as natural transformations of the form $V \to \mathcal{T}(H + V)$ where H, V, and $H + V$ are iteratable endofunctors of the category $\mathcal{A}$.

3 Completely Iterative Algebras and Complete Elgot Algebras

For interpreted solutions of recursive program schemes we need a suitable notion of algebras which can serve as interpretation of the givens. In the classical theory one works with continuous algebras; i. e., algebras carried by a cpo such that all operations are continuous maps. Here we work with *completely iterative algebras*, see [26], and *complete Elgot algebras*, see [5].

Definition 3.1. *Let $H : \mathcal{A} \to \mathcal{A}$ be an endofunctor. By a* flat equation morphism *in an object A (of parameters) we mean a morphism $e : X \to HX + A$. If $a : HA \to A$ is an H-algebra, a* solution *of e in A is a morphism $e^\dagger : X \to A$ such that $e^\dagger = [a, A] \cdot (He^\dagger + A) \cdot e$. We call A a* completely iterative algebra *(CIA) if every flat equation morphism has a unique solution.*

We explain this notion in the classical setting. Let Σ be a signature of givens. A flat recursive system of equations is given by a set X of variables and for each variable $x \in X$ a formal equation

$$x \approx t \quad \text{where } t = \sigma(x_1, \dots, x_n), \sigma \in \Sigma_n, x_i \in X \text{ or } t \in A.$$

The system corresponds to $e : X \to H_\Sigma X + A$, and a solution $e^\dagger$ assigns to every variable an element of A such that the formal equations become identities in A.

Examples 3.2. (i) Final coalgebras. In [26] it is shown that for a final H-coalgebra $\alpha : T \to HT$ the inverse $\tau : HT \to T$ of the structure map yields a CIA. Analogously, for every object X of $\mathcal{A}$ a final coalgebra TX of $H(\_) + X$ yields a CIA, see Theorem 3.9 below. Furthermore, the algebras $T_\Sigma X$ of all Σ-trees over a set X are CIAs, see Example 2.4(i).

(ii) Let $\mathcal{P}_\mathsf{f}$ be the finite power set functor on Set, and assume the Anti-Foundation Axiom. Let HF_1 be the set of hereditarily finite sets. Let τ be the inclusion of $\mathcal{P}_\mathsf{f}(HF_1)$ into HF_1. This map τ turns HF_1 into a CIA with respect to $\mathcal{P}_\mathsf{f}$. This is a special case of the first example.

(iii) Let H be a contracting endofunctor of the category CMS of complete metric spaces, see Example 2.2(iii). Then any non-empty H-algebra is completely iterative; see [5] for this and further examples.

Remark 3.3. In order to define Elgot algebras below we will need two operations. The first one formalizes the renaming of parameters in a flat equation morphism. For a flat equation morphism $e : X \to HX + A$ and a morphism $h : A \to B$ we define

$$h \bullet e \equiv X \xrightarrow{e} HX + A \xrightarrow{HX+h} HX + B\,.$$

The second operation allows us to combine two flat equations morphisms $e : X \to HX + Y$ and $f : Y \to HY + A$ where the parameters of the first are the variables of the second into one "simultaneous" flat equation morphism:

$$f \blacksquare e \equiv X + Y \xrightarrow{[e,\mathsf{inr}]} HX + Y \xrightarrow{HX+f} HX + HY + A \xrightarrow{\mathsf{can}+A} H(X+Y) + A\,.$$

Definition 3.4. *A (complete)* Elgot H-algebra *is a triple* $(A, a, (\_)^\dagger)$*, where* (A, a) *is an* H*-algebra, and* $(\_)^\dagger$ *assigns to every flat equation morphism* $e : X \to HX + A$ *a solution* $e^\dagger : X \to A$ *such that the following two laws hold:*

Functoriality: *Solution respects renaming of variables. Given two flat equation morphisms* $e : X \to HX + A$ *and* $f : Y \to HY + A$ *and a morphism* $h : X \to Y$ *of equations between them; i. e.,* $f \cdot h = (Hh + A) \cdot e$*, we then have* $e^\dagger = f^\dagger \cdot h$.

CIA-identity. *Simultaneous recursion may be performed sequentially. For every flat equation morphisms* $e : X \to HX + Y$ *and* $f : Y \to HY + A$*, the solution of the combined equation morphism* $f \blacksquare e$ *may be obtained by first solving* f *and then solving* e*, "plugging in"* $f^\dagger$ *for the parameters:* $(f^\dagger \bullet e)^\dagger = (f \blacksquare e)^\dagger \cdot \mathsf{inl}$.

Remark 3.5. (i) Notice that there is a notion of (non-complete) Elgot algebra, see [5]. However, since we are only concerned with complete Elgot algebras here, we will henceforth abuse terminology and just speak of Elgot algebras in lieu of complete ones.

(ii) The axioms of Elgot algebras are inspired by the axioms of iteration theories of Bloom and Ésik [15]. In fact, the two laws above are essentially "flat" versions of the functorial dagger implication and the left pairing identity (also known as Bekić-Scott identity) from [15].

One justification for the above axioms is that Elgot algebras turn out to be the Eilenberg-Moore category of the monad T, see Remark 2.3. We shall mention this result at the end of this section. Applied to a polynomial functor H_Σ on Set that means a Σ-algebra A is an Elgot algebra precisely if all Σ-trees over A can be canonically interpreted in A.

Examples 3.6. (i) Completely iterative algebras are Elgot algebras. Cf. [5].

(ii) Continuous algebras. Let H be a locally continuous endofunctor on CPO, see Example 2.2(ii). It was shown in [5] that any H-algebra (A, a) with a least element $\bot$ is an Elgot algebra when to a flat equation morphism $e : X \to HX + A$ the least solution $e^\dagger$ is assigned.

(iii) Suppose that $H : \mathsf{Set} \to \mathsf{Set}$ is a functor with a locally continuous lifting $H' : \mathsf{CPO} \to \mathsf{CPO}$, see Example 2.4(iv). We call an H-algebra $\alpha : HA \to A$ CPO-*enrichable* if there exists a complete partial order $\sqsubseteq$ on A such that A

becomes a continuous algebra $\alpha : H'(A, \sqsubseteq) \to (A, \sqsubseteq)$ with a least element. Any CPO-enrichable algebra A is an Elgot algebra: to every equation morphism $e : X \to HX + A$ assign the least solution of $e : (X, \leq) \to H'(X, \leq) + (A, \sqsubseteq)$, where $\leq$ is the discrete order on X; i. e., $x \leq y$ iff $x = y$.

Definition 3.7. *A homomorphism h from an Elgot algebra $(A, a, (\_)^\dagger)$ to an Elgot algebra $(B, b, (\_)^\ddagger)$ is a morphism $h : A \to B$ preserving solutions: for each $e : X \to HX + A$ we have $(h \bullet e)^\ddagger = h \cdot e^\dagger$.*

Proposition 3.8. *Every homomorphism $h : (A, a, (\_)^\dagger) \to (B, b, (\_)^\ddagger)$ of Elgot algebras is a homomorphism of H-algebras; i. e., $h \cdot a = b \cdot Hh$. The converse is false in general. If, however, A and B are CIAs then any H-algebra morphism is a homomorphism of Elgot algebras.*

The classical theory of recursive program schemes rests on the fact that in any continuous algebra A all Σ-trees over A can be interpreted; i. e., there is a canonical map $T_\Sigma A \to A$. In a suitable category of cpos the structures $T_\Sigma X$ play the rôle of *free algebras*. The freeness is used to define maps *out* of those algebras. In our setting, the Σ-trees are the final coalgebra. So in order to generalize the classical theory, we need a setting in which the final coalgebras TY are free algebras. The following result gives such a setting. It is fundamental for the rest of the paper and collects the results of Theorems 2.8 and 2.10 of [26] and Theorem 5.6 of [5].

Theorem 3.9. *The following are equivalent:*
(i) TY is a final coalgebra of $H(\_) + Y$,
(ii) TY is a free completely iterative H-algebra on Y, and
(iii) TY is a free (complete) Elgot H-algebra on Y.

In more detail: if (TY, α_Y) is a final coalgebra of $H(\_)+Y$, the inverse $[\tau_Y, \eta_Y] : HTY + Y \to Y$ of α_Y gives a CIA, which as an Elgot algebra is free on Y. Conversely, given a free Elgot H-algebra $(TY, \tau_Y, (\_)^\dagger)$ with a universal arrow $\eta_Y : Y \to TY$, then this is a CIA, whence a free CIA on Y, and $[\tau_Y, \eta_Y]$ is an isomorphism whose inverse is the structure map of a final coalgebra of $H(\_)+Y$.

Recall that we assume H is iteratable, so $H(\_)+Y$ *does* have a final coalgebra for all Y. The next result gives the *dramatis personae* for the rest of the paper.

Theorem 3.10. *There is a left adjoint to the forgetful functor from $\mathsf{Alg}^\dagger H$, the category of Elgot algebras and their homomorphisms, to the base category $\mathcal{A}$; in symbols, $L \dashv U : \mathsf{Alg}^\dagger H \to \mathcal{A}$. The left-adjoint L assigns to each object Y of $\mathcal{A}$ a free Elgot algebra $(TY, \tau_Y, (\_)^\dagger)$ on Y. The components of the unit η are the universal arrows $\eta_Y : Y \to TY$ of the free Elgot algebras. The counit ε gives for each Elgot algebra $(A, a, (\_)^\ddagger)$ the unique homomorphism $\widetilde{a} : TA \to A$ such that $\widetilde{a} \cdot \eta_A = id$, and we also have $\widetilde{a} = (\alpha_A)^\ddagger$ where the coalgebra structure $\alpha_A : TA \to HTA + A$ is considered as a flat equation morphism.*

Moreover, we obtain additional structure:

(i) A monad $(\mathcal{T}(H), \eta^H, \mu^H)$ on $\mathcal{A}$ such that for all objects Y of $\mathcal{A}$,
(a) $\mathcal{T}(H)Y = TY$ is the carrier of a final coalgebra of $H(\_) + Y$;

(b) μ_Y *is the (unique) solution of* α_{TY}, *considered as a flat equation morphism with parameters in* TY.

(ii) *A natural transformation* $\alpha^H : T \to HT + Id$ *expressing the coalgebra structures of* TY.

(iii) *A natural transformation* $\tau^H : HT \to T$ *such that* $[\tau^H, \eta^H]$ *is the pointwise inverse of* α^H.

(iv) *A "canonical embedding"* κ^H *of* H *into* T: $\kappa^H \equiv H \xrightarrow{H\eta^H} HT \xrightarrow{\tau^H} T$.

As always, we just write $\mathcal{T}(H)$, or even just T, to denote the monad of 3.10(i) above, and we shall frequently drop the superscripts when dealing with the structure coming from a single endofunctor H.

Theorem 3.11. [5] *The category* $\mathsf{Alg}^\dagger\, H$ *of Elgot algebras is isomorphic to the Eilenberg-Moore category* $\mathcal{A}^T$ *of monadic* T*-algebras. The isomorphism assigns to an Elgot algebra* $(A, a, (\_)^\ddagger)$ *the Eilenberg-Moore algebra* $\widetilde{a} : TA \to A$.

4 Second Order Substitution

In [1,26] it is proved that the monad $\mathcal{T}(H)$ of Theorem 3.10 is characterized by an important universal property—it is the free *completely iterative monad* on H. In this short abstract we will not need the full strength of this result in order to present our results on recursive program schemes. However, the freeness of $\mathcal{T}(H)$ specializes to second order substitution. We will first recall this concept for Σ-trees, see e.g. [16], and then present a generalization to the final coalgebras $\mathcal{T}(H)$.

Example 4.1. Let Σ and Γ be signatures. Each symbol $\sigma \in \Sigma(n)$ is considered as a flat tree in n variables. A second order substitution gives an "implementation" to each such σ as a Γ-tree in the same n variables. We model this by a natural transformation $\Sigma \to T_\Gamma \cdot J$ between signatures (considered as functors $\mathbb{N} \to \mathsf{Set}$), and this gives rise to a natural transformation $\lambda : H_\Sigma \to T_\Gamma$. Now for any set X of variables the action of the second order substitution $\overline{\lambda}_X : T_\Sigma X \to T_\Gamma X$ is to replace every Σ-symbol in a tree t from $T_\Sigma X$ by its implementation according to λ; i. e., if $t = \sigma(t_1, \ldots, t_n)$ with $\sigma \in \Sigma(n)$ and if $t'(x_1, \ldots x_n) \in T_\Gamma X$ is the implementation of σ, then $\overline{\lambda}_X(t) = t'(\overline{\lambda}_X(t_1), \ldots, \overline{\lambda}_X(t_n))$. Example: Suppose that Σ consists of two binary symbols $+$ and $*$ and a constant 1, and Γ consists of a binary symbol b, a unary one u and a constant c. Furthermore, let λ be given by the assignment $x + y \mapsto b(x, u(y))$, $x * y \mapsto b(u(x), y)$, and $1 \mapsto u(c)$. For the set $Z = \{z, z'\}$, the second order substitution morphism $\overline{\lambda}_Z$ acts for example as follows: $(z + z') * 1 \mapsto b(u(b(z, u(z'))), u(c))$. Notice that although in this example we assigned finite trees to all Σ-symbols in general we may assign also infinite Γ-trees. When infinite trees are involved there is usually the restriction to so-called non-erasing substitutions; i. e., all Σ-symbols are assigned to trees which are not just single node trees labelled by a variable. Finally, the reader may check that $\overline{\lambda}$ is again natural in X and that it is in fact a monad morphism from T_Σ to T_Γ.

Theorem 4.2. [1] *Let H and K be iteratable functors. Suppose that $\lambda : H \to \mathcal{T}(K)$ is an* ideal *natural transformation; i. e., there exists some natural transformation $\lambda' : H \to K\mathcal{T}(K)$ with $\tau^K \cdot \lambda' = \lambda$. Then there exists a unique monad morphism $\overline{\lambda} : \mathcal{T}(H) \to \mathcal{T}(K)$ such that $\overline{\lambda} \cdot \kappa^H = \lambda$.*

Second-order substitution is not trivial to define in the classical setting, and so it is significant that it generalizes to our setting.

5 Uninterpreted Recursive Program Schemes

In the classical treatment of recursive program schemes one gives an uninterpreted semantics to systems as in (2.2) which are in Greibach normal form; i. e., every term on the right-hand side of the system has as its head symbol a symbol from the given signature Σ. The semantics assigns to each of the new operation symbols a tree over Σ. These trees are obtained as the result of unfolding the recursive specification of the RPS.

We have seen in Example 2.4(i) that Σ-trees can be characterized as the final coalgebra of the polynomial endofunctor associated to Σ. It is the universal property of this final coalgebra which allows one to give a semantics to the given RPS. We will in this section provide a conceptually easy and general way to give an uninterpreted semantics to recursive program schemes considered more abstractly as natural transformations, see our discussion in Example 2.5.

Definition 5.1. *Let V and H be endofunctors on $\mathcal{A}$. A* recursive program scheme *(or* RPS, *for short) is a natural transformation $e : V \to \mathcal{T}(H + V)$.*

The RPS e is called guarded *if it factors as follows:*

$$e \equiv V \xrightarrow{f} H\mathcal{T}(H+V) \xrightarrow{\mathsf{inl} * \mathcal{T}(H+V)} (H+V)\mathcal{T}(H+V) \xrightarrow{\tau^{H+V}} \mathcal{T}(H+V)\,,$$

for some natural transformation $f : V \to H\mathcal{T}(H+V)$.

A solution *of e is an ideal transformation $e^\dagger : V \to \mathcal{T}(H)$ such that the following equation holds:*

$$e^\dagger \equiv V \xrightarrow{e} \mathcal{T}(H+V) \xrightarrow{\overline{[\kappa^H, e^\dagger]}} \mathcal{T}(H)\,.$$

Remark 5.2. Recall that $\overline{[\kappa^H, e^\dagger]}$ is the unique monad morphism extending $\sigma = [\kappa^H, e^\dagger] : H+V \to \mathcal{T}(H)$, see Theorem 4.2. Observe that therefore it is important to require that $e^\dagger$ be an ideal transformation since otherwise $\overline{\sigma}$ is not defined.

Remark 5.3. From Example 2.5, our definition is a generalization of the classical notion of RPS (to the category-theoretic setting), and it extends the classical work as well by allowing infinite trees on the right-hand sides of equations. Furthermore, any recursive program scheme as in (2.2) which is in Greibach normal form yields a guarded recursive program scheme in the sense of Definition 5.1.

Theorem 5.4. *Any guarded recursive program scheme has a unique solution.*

Sketch of Proof. Let $H : \mathcal{A} \to \mathcal{A}$ be any iteratable functor. Then $T = \mathcal{T}(H)$ is a final coalgebra of the functor $H \cdot \_ + Id$ on the endofunctor category $[\mathcal{A}, \mathcal{A}]$. This result extends to the comma-category $H/\mathbf{Mon}(\mathcal{A})$ whose objects are pairs $(S, \sigma : H \to S)$ where S is a monad on $\mathcal{A}$ and σ is a natural transformation, and whose morphisms $h : (S_1, \sigma_1) \to (S_2, \sigma_2)$ are monad morphisms $h : S_1 \to S_2$ with $h \cdot \sigma_1 = \sigma_2$. In fact, we obtain a functor $\mathcal{H}$ on $H/\mathbf{Mon}(\mathcal{A})$ with $\mathcal{H}(S, \sigma) = (HS + Id, \mathsf{inl} \cdot H\eta)$, where $\eta : Id \to S$ is the unit of the monad S. Furthermore, T together with $\kappa^H : H \to T$, see Theorem 3.10(iv), is the final $\mathcal{H}$-coalgebra.

Now suppose we are given a guarded RPS $e : V \to \mathcal{T}(H+V)$. Then we have a natural transformation $\sigma = \mathsf{inl} \cdot [H\eta^{H+V}, f] : H + V \to H\mathcal{T}(H+V) + Id$ which is ideal in the sense that it factors through $H\mathcal{T}(H+V)$. Notice that $H\mathcal{T}(H+V) + Id$ is obtained by applying the functor $\mathcal{H}$ to $(\mathcal{T}(H+V), \kappa^{H+V} \cdot \mathsf{inl})$. Thus it is a monad; in fact, it is a completely iterative monad in the sense of [1]. Use the full universal property of the free completely iterative monad $\mathcal{T}(H+V)$ to obtain a monad morphism $\overline{e} : \mathcal{T}(H+V) \to H\mathcal{T}(H+V) + Id$, see [1], Theorem 4.14. It is easy to see that this gives rise to a $\mathcal{H}$-coalgebra, and so there exists a unique coalgebra homomorphism h from this coalgebra to the final one carried by $(\mathcal{T}(H), \kappa^H)$, which gives a monad morphism. Now define

$$e^\dagger \equiv V \xrightarrow{\mathsf{inr}} H + V \xrightarrow{\kappa^{H+V}} \mathcal{T}(H+V) \xrightarrow{h} \mathcal{T}(H) .$$

A non-trivial proof shows that this is indeed the desired unique solution of e, see the full version [27] for details.

Remark 5.5. (i) The first part of the proof of Theorem 5.4 showing the finality of $\mathcal{T}(H)$ and defining the monad morphism h uses similar ideas than the proof of the main result of [19]. However, the second part in which it is proved that $e^\dagger$ is a unique solution of e is new. It connects solutions to the (generalized) second order substitution as presented in Theorem 4.2.

(ii) Notice that in the classical setting not every recursive program scheme which has a solution needs to be in Greibach normal form. For example, consider the system formed by the first equation in (1.1) and by the equation $\psi(x) \approx \varphi(\psi(x))$. This system gives rise to an unguarded RPS. Thus, Theorem 5.4 does not provide a solution of this RPS. However, the system is easily rewritten to an equivalent one in Greibach normal form which gives a guarded RPS that we can uniquely solve using our Theorem 5.4.

Example 5.6. Let us now present an example of RPSs which are *not* captured in the classical setting. Sometimes one might wish to recursively define new operations from old ones where the new operations should satisfy certain extra properties automatically. We demonstrate this with an RPS defining recursively a new operation which is commutative. Suppose the signature Σ of givens consists of a ternary symbol F and a unary one G. Let us assume that we want to require that F satisfies the equation $F(x,y,z) = F(y,x,z)$ in any interpretation. Then Σ is modelled by the endofunctor $HX = X^3/{\sim} + X$ where $\sim$ is the smallest equivalence on X^3 with $(x,y,z) \sim (y,x,z)$. To be an H-algebra is equivalent to being an algebra A with a unary operation G_A and a ternary one F_A satisfying $F_A(x, y, z) = F_A(y, x, z)$. Suppose that one wants to define a commutative binary operation φ by the formal equation

$$\varphi(x, y) \approx F(x, y, \varphi(Gx, Gy)) . \tag{5.3}$$

To do it we express φ by the endofunctor V assigning to a set X the set of unordered pairs of X. It is not difficult to see that the formal equation (5.3) gives rise to a guarded RPS $e : V \to \mathcal{T}(H + V)$. In fact, to see the naturality use the description of the terminal coalgebra $\mathcal{T}(H + V)Y$ given in [2], see Example 2.4(ii). Notice that in the classical setting we are unable to demand that (the solution of) φ is a commutative operation at this stage: this fact would be proved separately once a solution has been obtained. Here we have encoded this additional requirement into our RPS—any solution will be commutative. In fact, the components of the uninterpreted solution $e^\dagger_X : VX \to \mathcal{T}(H)X$ assign to an unordered pair $\{ x, y \}$ in VX the tree

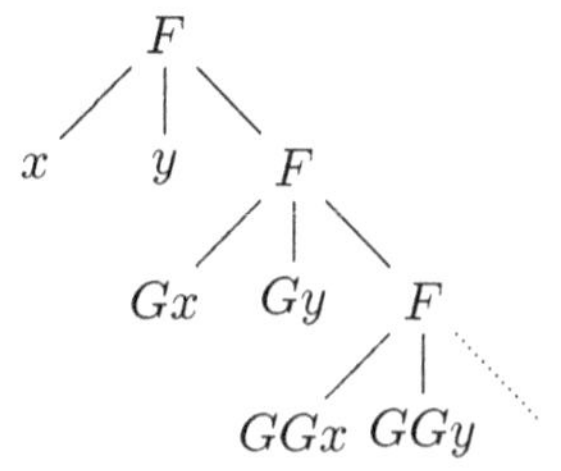

where for every node labelled by F the first two children are unordered.

6 Interpreted Recursive Program Schemes

We have seen in the previous section that for any guarded recursive program scheme we can find a unique uninterpreted solution. In practice, however, one is more interested in finding *interpreted* solutions. In the classical treatment of RPS this means that a recursive program scheme defining new operation symbols of a signature Φ from given ones in a signature Σ comes together with some Σ-algebra A. An interpreted solution of the recursive program scheme in question is, then, an operation on A for each operation symbol in Φ such that the formal equations of the RPS become valid equations in A.

Of course, in general an algebra A will not admit interpreted solutions. We shall show in this section that any Elgot algebra $(A, a, (\_)^*)$ admits an interpreted solution of any guarded recursive program scheme. Moreover, if A is a CIA, interpreted solutions are unique. We also state that uninterpreted solutions and interpreted ones correspond to one another. This is a fundamental result for algebraic semantics.

Definition 6.1. *Let $(A, a, (\_)^*)$ be an Elgot algebra w.r.t. H and let $e : V \to \mathcal{T}(H + V)$ be an RPS. An* interpreted solution *of e in A is a structure of a*

V-algebra $e_A^\ddagger : VA \to A$, such that the $(H+V)$-algebra $[a, e_A^\ddagger] : (H+V)A \to A$ is an Elgot algebra and such that the equation

$$e_A^\ddagger \equiv VA \xrightarrow{e_A} \mathcal{T}(H+V)A \xrightarrow{\widetilde{[a,e_A^\ddagger]}} A \tag{6.4}$$

holds, where the second arrow denotes the Eilenberg-Moore algebra structure associated to the Elgot algebra $[a, e_A^\ddagger]$, see Theorem 3.11.

Theorem 6.2. *Let $(A, a, (\_)^*)$ be an Elgot algebra w. r. t. H and let $e : V \to \mathcal{T}(H+V)$ be a guarded RPS. Then the following hold:*

(i) there exists an interpreted solution $e_A^\ddagger$ of e in A,

(ii) if A is a completely iterative algebra, then $e_A^\ddagger$ is the unique interpreted solution of e in A.

Sketch of Proof. Recall the $\mathcal{H}$-coalgebra structure $\overline{e}$ from the proof of Theorem 5.4. Let us write T for $\mathcal{T}(H+V)$. Then the component at A of $\overline{e}$ yields a flat equation morphism $\overline{e}_A : TA \to HTA + A$ w.r.t. the given Elgot algebra. Denote its solution $(\overline{e}_A)^*$ by β, and define

$$e_A^\ddagger \equiv VA \xrightarrow{\mathsf{inr}} (H+V)A \xrightarrow{\kappa_A^{H+V}} \mathcal{T}(H+V)A \xrightarrow{\beta} A\,.$$

A non-trivial proof shows that this morphism is an interpreted solution of e, and that it is the unique solution if A is a CIA. For details see [27].

Finally, we state the "Fundamental Theorem of Algebraic Semantics", which establishes that uninterpreted and interpreted solutions are connected in the "proper way".

Theorem 6.3. *Let $(A, a, (\_)^*)$ be an Elgot algebra considered as an Eilenberg-Moore algebra $\widetilde{a} : \mathcal{T}(H)A \to A$, and let e be any guarded recursive program scheme. If $e_A^\ddagger : VA \to A$ is the interpreted solution of e in A of Theorem 6.2 and $e^\dagger : V \to \mathcal{T}(H)$ is the (uninterpreted) solution of Theorem 5.4, then the equation $\widetilde{a} \cdot (e^\dagger)_A = e_A^\ddagger$ holds.*

Remark. Notice that $(e^\dagger)_A$ is the component at A of the natural transformation $e^\dagger$ whereas $e_A^\ddagger$ is not a component of some natural transformation but merely a morphism from VA to A.

Sketch of Proof. Recall the morphisms $h : \mathcal{T}(H+V) \to \mathcal{T}(H)$ and $\beta : \mathcal{T}(H+V)A \to A$ from the proofs of Theorems 5.4 and 6.2. It is not difficult to show that the equation $\beta = \widetilde{a} \cdot h_A$ holds. Precompose both sides with $\kappa_A^{H+V} \cdot \mathsf{inr} : VA \to \mathcal{T}(H+V)A$ to obtain the desired result.

6.1 Interpreted Solutions in CPOs

We shall show in this subsection that if we have $\mathcal{A} = \mathsf{CPO}$ as our base category, then interpreted solutions of a guarded RPS e in an Elgot algebra $(A, a, (\_)^*)$ are given as least fixed points of a continuous function on a function space. In this way we recover denotational semantics from our categorical interpreted semantics of recursive program schemes.

Example 6.4. The standard example in classical algebraic semantics is the factorial function. So let Σ be a signature containing a constant zero, two unary symbols succ and pred, a binary symbol $*$ and a ternary one if. The interpretation we have in mind is the natural numbers. The signature Φ of the recursively defined operations consists just of one unary symbol f. Consider the formal recursive equation

$$f(n) \approx \mathsf{if}(n, \mathsf{succ}(\mathsf{zero}), f(\mathsf{pred}(n)) * n)) \tag{6.5}$$

defining the factorial function. It gives rise to a guarded recursive program scheme $e : H_\Phi \to \mathcal{T}(H_\Sigma + H_\Phi)$ as demonstrated in Example 2.5.

To obtain a suitable Elgot algebra in which we can find an interpreted solution of e, we turn the natural numbers into a CPO. Let $\mathbb{N}_\perp$ be the flat CPO obtained from the discretely ordered set of natural numbers by adding a least element $\perp$. We equip $\mathbb{N}_\perp$ with the obvious continuous operations corresponding to the names of the operation symbols of Σ. Hence we have a continuous Σ-algebra, and therefore $\mathbb{N}_\perp$ is an Elgot H_Σ-algebra, as in Example 3.6(iii).

The interpreted solution $e^\ddagger_{\mathbb{N}_\perp} : H_\Phi \mathbb{N}_\perp \to \mathbb{N}_\perp$ is given by a function on $\mathbb{N}_\perp$. But how do we know that this function is the desired factorial function? Usually one would simply regard the RPS (6.5) itself as a continuous function R which maps every continuous operation f on $\mathbb{N}_\perp$ to $\mathsf{if}_{\mathbb{N}_\perp}(\_, 1, f(\mathsf{pred}_{\mathbb{N}_\perp}(\_) *_{\mathbb{N}_\perp} \_)$; i. e., we interpret all the operation symbols of Σ in the algebra $\mathbb{N}_\perp$. It is clear that the least fixed point of R is the desired factorial function. And we show that this least fixed point coincides with the interpreted solution obtained from Theorem 6.2.

In general any recursive program scheme can be turned into a continuous function R on the function space $\mathsf{CPO}(VA, A)$ and the least fixed point of R is the same as the interpreted solution obtained from Theorem 6.2.

We assume throughout this subsection that H and V are locally continuous (and, as always, iteratable) endofunctors of CPO. We also consider a fixed guarded RPS $e : V \to \mathcal{T}(H+V)$, and an H-algebra (A, a) with a least element $\perp$. By Example 3.6(ii), we know that this carries the structure of an Elgot algebra $(A, a, (\_)^*)$, where $(\_)^*$ assigns to every flat equation morphism a least solution. Furthermore, for any continuous map $f : VA \to A$ we have an Elgot algebra on A with structure $[a, f] : (H + V)A \to A$. Its associated Eilenberg-Moore algebra structure is denoted by $\widetilde{[a, f]}$, see Theorem 3.11.

Theorem 6.5. *The following function R on $\mathsf{CPO}(VA, A)$*

$$f \mapsto VA \xrightarrow{e_A} \mathcal{T}(H + V)A \xrightarrow{\widetilde{[a,f]}} A \tag{6.6}$$

is continuous. Its least fixed point is the interpreted solution $e^\ddagger_A : VA \to A$ of Theorem 6.2.

Sketch of Proof. To see the continuity of R is suffices to prove that $\widetilde{(\_)} : \mathsf{CPO}(HA, A) \to \mathsf{CPO}(\mathcal{T}(H)A, A)$ is continuous. Let us write T for $\mathcal{T}(H)$. Recall from Theorem 3.10 that for any continuous map $a : HA \to A$ the Eilenberg-Moore algebra structure $\widetilde{a}$ is the least solution of the flat equation morphism $\alpha_A : TA \to HTA + A$, i.e., $\widetilde{a}$ is the least fixed point of the continuous function $F(a, -) : \mathsf{CPO}(TA, A) \to \mathsf{CPO}(TA, A)$ with $F(a, f) = [a, A] \cdot (Hf + A) \cdot \alpha_A$. Observe that F is continuous in the first argument a, and so F is a continuous function on the product $\mathsf{CPO}(HA, A) \times \mathsf{CPO}(TA, A)$. It follows from standard arguments that taking the least fixed point in the second argument yields a continuous map $\mathsf{CPO}(HA, A) \to \mathsf{CPO}(TA, A)$. But this is precisely the map $\widetilde{(\_)}$.

We prove that $e_A^\ddagger$ is the least fixed point of R. Notice that the least fixed point of R is the join t of the increasing chain in $\mathsf{CPO}(VA, A)$ given by $t_0 = \mathsf{const}_\perp$ and $t_{i+1} = \widetilde{[a, t_i]} \cdot e_A$, for $i \in \mathbb{N}$.

Furthermore, recall that the interpreted solution $e_A^\ddagger$ is defined by $\beta \cdot \kappa_A^{H+V} \cdot \mathsf{inr}$, where $\beta = g^*$ is the least solution of the flat equation morphism g which is obtained from the component at A of the $\mathcal{H}$-coalgebra $\overline{e}$, see Theorem 6.2. By Example 3.6(ii), the solution β of g is the join of the chain given by $\beta_0 = \mathsf{const}_\perp$ and $\beta_{i+1} = [a, A] \cdot H(\beta_i + A) \cdot g$, for $i \in \mathbb{N}$.

Observe that $e_A^\ddagger$ is a fixed point of R, see (6.4). Thus, we have $t \sqsubseteq e_A^\ddagger$. To show the reverse inequality one proves by induction on i the inequalities $\beta_i \sqsubseteq \widetilde{[a, t]}$, for $i \in \mathbb{N}$, see [27]. This implies that $\beta \sqsubseteq \widetilde{[a, t]}$ and therefore $e_A^\ddagger = \beta \cdot \kappa_A^{H+V} \cdot \mathsf{inr} \sqsubseteq \widetilde{[a, t]} \cdot \kappa_A^{H+V} \cdot \mathsf{inr} = t$.

Remark 6.6. Suppose that H, V and $H + V$ are iteratable endofunctors of Set, which have locally continuous liftings H', V' and $H' + V'$ to CPO. Then we have $\mathcal{T}(H + V) \cdot U = U \cdot \mathcal{T}(H' + V')$, see Example 2.4(iv). Furthermore, assume that the guarded RPS $e : V \to \mathcal{T}(H + V)$ has a lifting $e' : V' \to \mathcal{T}(H' + V')$; i.e., a natural transformation e' such that $U * e' = e * U$. Now consider any CPO-enrichable H-algebra (A, a) as an Elgot algebra, see Example 3.6(iii). Then we can apply Theorem 6.5 to obtain an interpreted solution $e_A^\ddagger$ of e in the algebra A as a least fixed point of the above function R of (6.6).

Example 6.7. (i) Suppose we have signatures Σ and Φ. Then the polynomial functors H_Σ and H_Φ satisfy the requirements of Remark 6.6. Consider any system as in (2.2) in Greibach normal form, and form the associated guarded RPS $e : H_\Phi \to \mathcal{T}(H_\Sigma + H_\Phi)$. Then e has a lifting e'. Let (A, a) be a CPO-enrichable H_Σ-algebra; i.e., a continuous Σ-algebra with a least element $\perp$. We wish to consider the continuous function R which assigns to any continuous algebra structure $\varphi : H_\Phi A \to A$ the algebra structure $\widetilde{[a, \varphi]} \cdot e'_A$. It maps for a given continuous Φ-algebra structure φ on A any operation $f_A : A^n \to A$, $f \in \Phi_n$, to the operation that computes the right-hand side t_A^f, where operation symbols of Σ are interpreted according to a and symbols of Φ according to the given φ.

Theorem 6.5 states that an interpreted solution $e_A^\ddagger$ of e in the algebra A is obtained by taking the least fixed point of R; in other words the interpreted solution $e_A^\ddagger$ gives the usual denotational semantics.

(ii) Apply the previous example to the RPS of Example 6.4. Then Theorem 6.5 states that the interpreted solution of the RPS (6.5) in the Elgot algebra

$\mathbb{N}_\perp$ is obtained as the least fixed point of the function R of Example 6.4. That is, the interpreted solution gives the desired factorial function.

(iii) Recall the guarded RPS e from Example 5.6 and its uninterpreted solution. Consider again the algebra $\mathbb{N}_\perp$ together with the following two operations:

$$F_{\mathbb{N}_\perp}(x, y, z) = \begin{cases} x & \text{if } x = y \\ z & \text{else} \end{cases} \qquad G_{\mathbb{N}_\perp}(x) = \begin{cases} \lfloor \frac{x}{2} \rfloor & \text{if } x \in \mathbb{N} \\ \perp & x = \perp \end{cases}$$

Since the first operation obviously satisfies $F_{\mathbb{N}_\perp}(x, y, z) = F_{\mathbb{N}_\perp}(y, x, z)$ we have defined an H-algebra. It is not difficult to check that the set functor H has a locally continuous lifting H' to CPO and that $\mathbb{N}_\perp$ is a continuous H'-algebra. In fact, the existence of the lifting H' follows from the fact that the unordered pair functor $V : \mathsf{Set} \to \mathsf{Set}$ can be lifted to CPO; the lifting assigns to a cpo $(X, \leq)$ the set of unordered pairs with the following order: $\{x, y\} \sqsubseteq \{x', y'\}$ iff either $x \leq x'$ and $y \leq y'$ or $x \leq y'$ and $y \leq x'$. Thus, we have defined an Elgot algebra w.r.t. $H : \mathsf{Set} \to \mathsf{Set}$, see Example 3.6(iii). The interpreted solution $e^\ddagger_{\mathbb{N}_\perp} : V\mathbb{N}_\perp \to \mathbb{N}_\perp$ is given by one commutative binary operation $\varphi_{\mathbb{N}_\perp}$ on $\mathbb{N}_\perp$. We leave it to the reader to verify that for natural numbers x and y, $\varphi_{\mathbb{N}_\perp}(x, y)$ is the natural number represented by the greatest common prefix in the binary representation of x and y, e.g., $\varphi_{\mathbb{N}_\perp}(12, 13) = 6$. Notice that we do not have to prove separately that $\varphi_{\mathbb{N}_\perp}$ is commutative. The way we have formed the RPS e in Example 5.6 ensures that the interpreted solution will be given by a commutative operation.

(iv) Least fixed points are RPS solutions. Let A be a poset with joins of all subsets which are at most countable, and let $f : A \to A$ be a function preserving joins of ascending chains. Take f and binary joins to obtain an algebra structure on A of the polynomial set functor $H_\Sigma X = X + X \times X$ expressing a binary operation symbol F and a unary one G. Obviously, this functor has a lifting $H' : \mathsf{CPO} \to \mathsf{CPO}$ and A is a CPO-enrichable algebra; i.e., A is an Elgot algebra. Turn the formal equations (1.1) into a recursive program scheme $e : H_\Phi \to \mathcal{T}(H_\Sigma + H_\Phi)$, see Example 2.6. The RPS e has a lifting $e' : V' \to \mathcal{T}(H' + V')$, where V' denotes the lifting of H_Φ. The interpreted solution $e^\ddagger_A : V'A \to A$ gives two continuous functions $\varphi^\dagger, \psi^\dagger : A \to A$. Clearly, we have $\varphi^\dagger(a) = \bigvee_{n \in \mathbb{N}} f^n(a)$, and in particular $\varphi^\dagger(\perp)$ is the least fixed point of f.

6.2 Interpreted Solutions in CMS

In [14] the interpreted semantics of classical recursive program schemes is studied in certain interpretations arising from the use of complete metric spaces. It is proved there that recursive program schemes which are in Greibach normal form admit a unique interpreted solution in any contraction theory. We shall prove a similar result in our categorical setting now.

Recall the category CMS of complete metric spaces from Example 3.2(iii), and let H and V be contracting endofunctors. We shall show in this subsection that for any guarded RPS e we can find a unique interpreted solution in any non-empty H-algebra A. More precisely, assume that we have a guarded RPS e :

$V \to \mathcal{T}(H+V)$, and let (A, a) be a non-empty H-algebra. Then A is a CIA, and therefore it has the structure of an Elgot algebra, see Examples 3.2(iii) and 3.6(i). Notice that for any non-expanding map $f : VA \to A$ we obtain an algebra structure $[a, f] : (H+V)A \to A$, thus we have an Eilenberg–Moore algebra $\widetilde{[a, f]} : \mathcal{T}(H+V)A \to A$. As in CPO, the RPS e induces a function R on $\mathsf{CMS}(VA, A)$. The standard procedure for obtaining an interpreted solution would be to prove that R is a contracting map, and then invoke Banach's Fixed Point theorem to obtain a unique fixed point of R. Here we simply apply Theorem 6.2. Notice, however, that we cannot completely avoid Banach's Fixed Point theorem: it is used in the proof that final coalgebras exist for contracting functors, see [7].

Corollary 6.8. *The interpreted solution* $e_A^\ddagger : VA \to A$ *of* e *in* A *as obtained in Theorem 6.2 is the unique fixed point of the function* R *defined as in (6.6).*

Proof. In fact, being a fixed point of R is equivalent to being an interpreted solution of e in the CIA A, whose unique existence we have by Theorem 6.2.

Example 6.9. (i) Self similar sets are solutions of interpreted program schemes. Let (X, d) be a metric space. Then $(C(X), h)$ is the complete metric space formed by all non-empty compact subsets of X with the Hausdorff metric; i. e., for two compact subspaces A and B of X we have $h(A, B) = \max\{\, d(A \to B), d(B \to A)\,\}$, where $d(A \to B) = \max_{a \in A} \min_{b \in B} d(a, b)$. It is well-known that any contracting map f on X gives rise to a contracting map $f' : A \mapsto f[A]$ on $C(X)$, see [11]. Now consider the functor H' on CMS with $H'(X, d) = (X^3, \frac{1}{3} d_{\max})$, where $d_{\max}$ is the maximum metric. It is a lifting of the polynomial functor H_Σ on Set expressing one ternary operation α. Let $A = [0, 1]^2$ be equipped with the usual Euclidean metric. Consider the contracting maps $f(x, y) = (\frac{1}{3}x, \frac{1}{3}y)$, $g(x, y) = (\frac{1}{3}x + \frac{1}{3}, \frac{1}{3}y)$, and $h(x, y) = (\frac{1}{3}x + \frac{2}{3}, \frac{1}{3}y)$ of A. Then it follows that $\alpha_A : C(A)^3 \to C(A)$ with $\alpha_A(U, W, Z) = f[U] \cup g[W] \cup h[Z]$ is a $\frac{1}{3}$-contracting map, whence a structure of an H'-algebra. The formal equation $\varphi(x) \approx \alpha(\varphi(x), x, \varphi(x))$ gives rise to a guarded RPS on Set, viz. $e : \mathit{Id} \to \mathcal{T}(H_\Sigma + \mathit{Id})$, where the identity functor expresses the operation φ. Consider the lifting of the identity to CMS given by $V'(X, d) = (X, \frac{1}{3}d)$. Using this lift, e gives rise to a natural transformation $e' : V' \to \mathcal{T}(H' + V')$. In fact, it is easy to see that all components e'_X are non-expanding maps once the metric on $\mathcal{T}(H' + V')X$ is understood, see the full version [27] or [7] for more details on this. The interpreted solution of e' in the algebra $(C(A), \alpha_A)$ is given by a map $\varphi^\dagger : C(A) \to C(A)$ which maps a non-empty compact subspace U of A to a space of the following form: $\varphi^\dagger(U)$ has three parts, the middle one is a copy of U scaled by $\frac{1}{3}$, and the left-hand and right-hand one look like copies of the whole space $\varphi^\dagger(U)$ scaled by $\frac{1}{3}$. For example we have the assignment

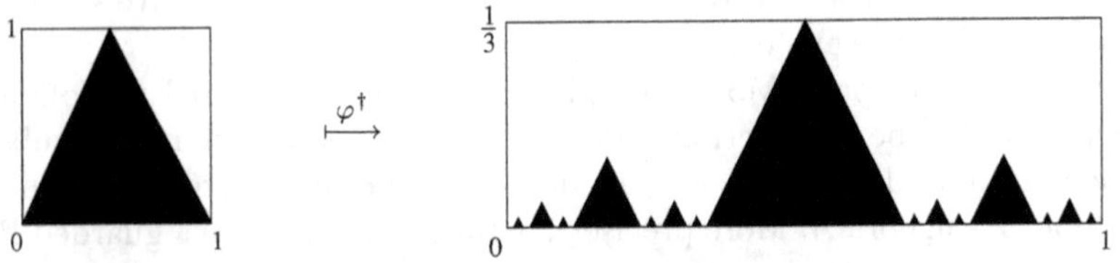

(ii) The Cantor "middle third" set c is the unique nonempty subset of the interval $[0,1]$ which satisfies $c = \frac{1}{3}c \cup (\frac{2}{3} + \frac{1}{3}c)$. So as in (i), there is an RPS e on CMS whose interpreted solution in the algebra $C(A)$, $A = [0,1]$, is the Cantor set. More detailed, the functions $f(x) = \frac{1}{3}x$ and $g(x) = \frac{2}{3} + \frac{1}{3}x$ yield a $\frac{1}{3}$-contraction $\alpha_A : C(A)^2 \to C(A)$ with $\alpha(U,W) = f[U] \cup g[W]$, whence an algebra of the functor given by $H'(X,d) = (X^2, \frac{1}{3}d_{\max})$. The formal equation $c \approx \alpha(c,c)$ gives rise to a guarded RPS $e' : V' \to \mathcal{T}(H' + V')$ where V' is the constant functor whose value is the trivial one point space 1, a lifting of the set functor expressing the constant c. Its interpreted solution is given by the function $1 \to C(A)$ choosing the Cantor space.

6.3 Interpreted Solutions in Non-well Founded Sets

In this subsection we show that in hypersets unique solutions of recursive equations as discussed in [13] can be obtained as special instances of our Theorem 6.2. For example, assuming the Anti-Foundation-Axiom there exists a unique set c such that $c = \{\, p(c)\,\}$ where $p(c)$ denotes the finite power set of c, i. e., c uniquely solves the formal equation $x \approx \{\, p(x)\,\}$, where we understand p and the set brackets as formal operation symbols.

Now in order to see how to obtain this and many other examples from our results we shall work with the finite power set functor $\mathcal{P}_\mathrm{f} : \mathsf{Set} \to \mathsf{Set}$ whose initial CIA (equivalently, final coalgebra) is HF_1 with a structure map $\tau : \mathcal{P}_\mathrm{f}(HF_1) \to HF_1$, see Example 3.2(ii). In the first step we extend this CIA by additional operations needed for our examples. So consider the unary operation $a \mapsto p(a)$ and the binary ones $(a,b) \mapsto a \cup b$ and $(a,b) \mapsto p(a) \times b$ on HF_1. If we describe the type of these operations by the set functor $KX = X + X \times X + X \times X$, then HF_1 is a K-algebra with a structure $b : K(HF_1) \to HF_1$.

Lemma 6.10. *The algebra HF_1 with the structure $k = \tau \cdot \mathcal{P}_\mathrm{f} b : \mathcal{P}_\mathrm{f} K(HF_1) \to HF_1$ is a CIA of the functor $\mathcal{P}_\mathrm{f} K$.*

In fact, this follows from a general result from the theory of CIAs, see the full version [27]. Here we do not have the space to present this in full generality. However, let us point out that from the general result it follows that formal equations with additional operations can be uniquely solved in HF_1 as soon as these additional operations "are guarded" by a set bracket and they "commute with substitutions".

Example 6.11. (i) The formal equation $x \approx \{\, p(x)\,\}$ gives rise to a flat equation morphism $X \to \mathcal{P}_\mathrm{f} KX + HF_1$, where X is a singleton set. Its unique solution $e^\dagger : X \to HF_1$ is essentially a set c with the desired property. Notice that for this example we do not need our machinery for recursive program schemes since no recursive function definitions are made.

(ii) Consider the formal equation $f(x) \approx \{\, f(x) \cup x\,\}$. Take the functor $V = \mathit{Id}$ on Set describing the unary operation f. Then the above formal equation gives rise to a recursive program scheme $e : V \to \mathcal{T}(\mathcal{P}_\mathrm{f} K + V)$ analogously as in

Example 2.5. The unique interpreted solution $e^{\ddagger}_{HF_1}$ obtained by Theorem 6.2 is given by a function $f^{\dagger}$ on HF_1 which satisfies $f^{\dagger}(a) = \{ f^{\dagger}(a) \cup a \}$.

(iii) Consider the system of formal equations $f(x, y) \approx \{ p(x) \times f(gx, gy) \}, g(x) \approx \{ x \cup f(x, x) \}$. Describe the operations f and g by the set functor $VX = X \times X + X$. Then as before the given system gives rise to a guarded RPS $e : V \to \mathcal{T}(\mathcal{P}_f K + V)$. And the interpreted solutions $e^{\ddagger}_{HF_1}$ yields two functions $f^{\dagger} : HF_1^2 \to HF_1$ and $g^{\dagger} : HF_1 \to HF_1$ turning the formal equations into valid identities as desired.

7 Conclusions and Future Research

We have presented a general and conceptually clear way of treating the uninterpreted and the interpreted semantics of recursive program schemes. For this we have used recent results on Elgot algebras and from the theory of coalgebras. Now one must go forward in reinventing algebraic semantics with category theoretic methods. We strongly suspect that there is much to be said about the relation of our work to operational semantics. We have not investigated higher-order recursive program schemes using our tools, and it would be good to know whether our approach applies in that area as well. The paper [25] addresses variable binding and infinite terms coalgebraically, and this may well be relevant. Back to the classical theory, one of the main goals of the original theory is to serve as a foundation for program equivalence. It is not difficult to prove the soundness of fold/unfold transformations in an algebraic way using our semantics; this was done in [29] for uninterpreted schemes. One would like more results of this type. The equivalence of interpreted schemes in the natural numbers is undecidable, and so one naturally wants to study the equivalence of interpreted schemes in *classes of interpretations*. The classical theory proposes classes of interpretations, many of which are defined on ordered algebras (see [20]). It would be good to revisit this part of the classical theory to see whether Elgot algebras suggest tractable classes of interpretations.

References

1. P. Aczel, J. Adámek, S. Milius and J. Velebil, Infinite Trees and Completely Iterative Theories: A Coalgebraic View, *Theoret. Comput. Sci.*, 300 (2003), 1–45.
2. J. Adámek, On a Description of Terminal Coalgebras and Iterative Theories, *Electron. Notes Theor. Comput. Sci.*, 82.1 (2003).
3. J. Adámek, S. Milius and J. Velebil, Free Iterative Theories: A Coalgebraic View, *Math. Structures Comput. Sci.*, 13 (2003), 259–320.
4. J. Adámek, S. Milius and J. Velebil, From Iterative Algebras to Iterative Theories, *Electron. Notes Theor. Comput. Sci.*, 106 (2004), 3–24, full version submitted and available at the URL `http://www.iti.cs.tu-bs.de/~milius`.
5. J. Adámek, S. Milius and J. Velebil, Elgot Algebras, to appear in *Electron. Notes Theor. Comput. Sci.*, available at the URL `http://www.iti.cs.tu-bs.de/~milius`.

6. J. Adámek and H. E. Porst, On tree coalgebras and coalgebra presentations, *Theoret. Comput. Sci.*, 311 (2004), 257–283.
7. J. Adámek and J. Reitermann, Banach's Fixed-Point Theorem as a Base for Data-Type Equations, *Appl. Categ. Structures*, 2 (1994), 77–90.
8. J. Adámek and V. Trnková, *Automata and Algebras in Categories.* Kluwer Academic Publishers, 1990.
9. P. America and J. Rutten, Solving Reflexive Domain Equations in a Category of Complete Metric Spaces, *J. Comput. System Sci.*, 39 (1989), 343–375.
10. A. Arnold and M. Nivat, The metric space of infinite trees. Algebraic and topological properties, *Fund. Inform.* III, no. 4 (1980), 445–476.
11. M. F. Barnsley, *Fractals everywhere*, Academic Press 1988.
12. M. Barr, Terminal coalgebras in well-founded set theory, *Theoret. Comput. Sci.*, 114 (1993), 299–315.
13. J. Barwise and L. S. Moss, *Vicious Circles*, CSLI Publications, Stanford, 1996.
14. S. L. Bloom, All Solutions of a System of Recursion Equations in Infinite Trees and Other Contraction Theories, *J. Comput. System Sci.* 27 (1983), 225–255.
15. S. L. Bloom and Z. Ésik, *Iteration Theories: The equational logic of iterative processes,* EATCS Monographs on Theoretical Computer Science, Berlin: Springer-Verlag (1993).
16. B. Courcelle, Fundamental properties of infinite trees, *Theoret. Comput. Sci.*, 25 (1983), no. 2, 95–169.
17. C. C. Elgot, Monadic Computation and Iterative Algebraic Theories, in: *Logic Colloquium '73* (eds: H. E. Rose and J. C. Shepherdson), North-Holland Publishers, Amsterdam, 1975.
18. C. C. Elgot, S. L. Bloom and R. Tindell, On the Algebraic Structure of Rooted Trees, *J. Comput. System Sci.*, 16 (1978), 361–399.
19. N. Ghani, C. Lüth and F. De Marchi, Solving Algebraic Equations using Coalgebra, *Theor. Inform. Appl.*, 37 (2003), 301–314.
20. I. Guessarian, *Algebraic Semantics.* Lecture Notes in Comput. Sci., 99, Springer, 1981.
21. J. Lambek, A Fixpoint Theorem for Complete Categories, *Math. Z.*, 103 (1968), 151–161.
22. T. Leinster, General self-similarity: an overview, e-print math.DS/0411343 v1.
23. T. Leinster, A general theory of self-similarity I, e-print math.DS/041344.
24. T. Leinster, A general theory of self-similarity II, e-print math.DS/0411345.
25. R. Matthes and T. Uustalu, Substitution in Non-Wellfounded Syntax with Variable Binding. In H. P. Gumm, (ed.), *Electron. Notes Theor. Comput. Sci.*, 82 (2003).
26. S. Milius, Completely Iterative Algebras and Completely Iterative Monads, *Inform. and Comput.*, 196 (2005), 1–41.
27. S. Milius and L. S. Moss, The Category Theoretic Solution of Recursive Program Schemes, full version, available at the URL `http://www.iti.cs.tu-bs.de/~milius`.
28. L. S. Moss, Parametric Corecursion, *Theoret. Comput. Sci.*, 260 (2001), no. 1–2, 139–163.
29. L. S. Moss, The Coalgebraic Treatment of Second-Order Substitution and Uninterpreted Recursive Program Schemes, preprint, 2002.
30. J. Worrell, On the Final Sequence of a Finitary Set Functor, accepted for publication in *Theoret. Comput. Sci.*

A Categorical Approach to Simulations*

Miguel Palomino[1], José Meseguer[2], and Narciso Martí-Oliet[1]

[1] Departamento de Sistemas Informáticos, Universidad Complutense de Madrid
[2] Computer Science Department, University of Illinois at Urbana-Champaign

Abstract. Simulations are a very natural way of relating concurrent systems, which are mathematically modeled by Kripke structures. The range of available notions of simulations makes it very natural to adopt a categorical viewpoint in which Kripke structures become the objects of several categories while the morphisms are obtained from the corresponding notion of simulation. Here we define in detail several of those categories, collect them together in various institutions, and study their most interesting properties.

1 Introduction

Simulations are a very natural way of relating concurrent systems. They are particularly useful for proving temporal logic properties, because we can use simulations to *shift our ground*; that is, to prove that a system $\mathscr{A}$ satisfies a property φ by considering a perhaps much simpler system (for example a finite-state abstraction) $\mathscr{B}$, proving that $\mathscr{B}$ satisfies φ, and showing that $\mathscr{B}$ *simulates* $\mathscr{A}$. This transfer result holds for the universal fragment ACTL* of CTL*, and in particular for all linear temporal logic formulas. Similarly, we may want to prove that a possibly more complex but more efficient concurrent system $\mathscr{C}$ is a *correct implementation* of another system $\mathscr{A}$; again this amounts to showing that $\mathscr{A}$ simulates $\mathscr{C}$, and will then allow transferring all ACTL* properties already established for $\mathscr{A}$ to its implementation $\mathscr{C}$. Obviously, the more flexibly we can shift our ground by means of suitable simulations, the more easily we can reason about concurrent systems, their abstractions, and their implementations. There are therefore good practical reasons to look for the *most general* notions of simulation possible, as a way to support very general and flexible reasoning methods.

The point of this paper is to systematically exploit a *categorical* point of view in the quest for general notions of simulation. That is, we consider increasingly more general categories whose objects are Kripke structures, and whose morphisms are adequate simulations between them. There are several orthogonal dimensions along which simulations can be generalized as discussed in detail in [13,11]. We can extend them: (1) from functions to relations; (2) from strictly preserving state predicates to only doing so in a looser way; (3) from simulations in which one step is simulated by another to "stuttering" simulations in which several steps in one system can correspond to several steps in the other; and (4) from the case in which all systems we relate share the same

* Research supported by ONR Grant N00014-02-1-0715, NSF Grant CCR-0234524, by DARPA through Air Force Research Laboratory Contract F30602-02-C-0130, and by the Spanish CICYT projects MELODIAS TIC 2002-01167 and MIDAS TIC 2003–0100.

J.L. Fiadeiro et al. (Eds.): CALCO 2005, LNCS 3629, pp. 313–330, 2005.

set AP of atomic predicates to one in which systems with different atomic predicates can be related among each other. All these extensions (1)–(4), and their possible combinations are mathematically characterized in this paper by increasingly more general categories.

A theme running in parallel with such generalizations is characterizing corresponding sets of *temporal logic formulas* that can be "reflected" by (that is, lifted along) increasingly more general simulation maps. This is closely related to another theme also developed in detail, namely the different *temporal logic institutions* involved. Indeed, Kripke structures are the most frequently used models for temporal logic. From an institutional viewpoint we will expect, for a given signature, a corresponding *category* of Kripke structures, which is precisely what we are investigating. The point then is that different choices of increasingly more general categories give rise to a corresponding family of temporal logic institutions, for which we study under what conditions the amalgamation property (semi-exactness) holds.

Another theme also studied in detail is the issue of *categorical constructions*, including limits, colimits, and epi-mono factorizations. As far as we know, most of the constructions we give are new. They shed further light on Kripke structures and the morphisms that we have available for relating them.

An extended version of this paper can be found at `http://maude.sip.ucm.es/~miguelpt/papers/cap.pdf`.

2 Kripke Structures and Simulations

When reasoning about computational systems, it is convenient to abstract from as many details as possible by means of simple mathematical models that can be used to reason about them. For a state-based system we can represent its behavior by means of a *transition system*, which is a pair $\mathscr{A} = (A, \to_{\mathscr{A}})$ with A a set of states and $\to_{\mathscr{A}} \subseteq A \times A$ a binary relation called the transition relation. A transition system, however, does not include any information about the relevant properties of the system. In order to reason about such properties it is necessary to add information about the atomic properties that hold in each state. In what follows, we assume a fixed set AP of atomic propositions and define a *Kripke structure* as a triple $\mathscr{A} = (A, \to_{\mathscr{A}}, L_{\mathscr{A}})$, where $(A, \to_{\mathscr{A}})$ is a transition system with $\to_{\mathscr{A}}$ a *total* relation (this is a customary requirement, which simplifies the semantics of the temporal logic), and $L_{\mathscr{A}} : A \to \mathscr{P}(AP)$ is a labeling function associating to each state the set of atomic propositions that hold in it.

We use the notation $a \to_{\mathscr{A}} b$ to state that $(a,b) \in \to_{\mathscr{A}}$. A *path* in $\mathscr{A}$ is a function $\pi : \mathbb{N} \longrightarrow A$ such that, for each $n \in \mathbb{N}$, $\pi(n) \to_{\mathscr{A}} \pi(n+1)$.

To specify system properties we use the logic ACTL$^*(AP)$, which is a sublogic of the branching-time temporal logic CTL$^*(AP)$ (see for example [5, Sect. 3.1]). There are two types of formulas in CTL$^*(AP)$: state formulas, denoted by State(AP), and path formulas, denoted by Path(AP). The semantics of the logic, specifying the satisfaction relations $\mathscr{A}, a \models \varphi$ and $\mathscr{A}, \pi \models \psi$ for a Kripke structure $\mathscr{A}$, an initial state $a \in A$, a state formula φ, a path π, and a path formula ψ, is defined as usual [5]. ACTL$^*(AP)$ is the restriction of CTL$^*(AP)$ to those formulas such that their negation-normal forms (with negations pushed to atoms) do not contain any existential path quantifiers. To avoid introducing existential quantifiers implicitly, it is more convenient to restrict ourselves to the negation-free fragment $\mathrm{ACTL}^*\backslash\neg(AP)$ of $\mathrm{ACTL}^*(AP)$, defined as follows:[1]

$$\begin{array}{ll}\text{state formulas:} & \varphi = p \in AP \mid \top \mid \bot \mid \varphi \vee \varphi \mid \varphi \wedge \varphi \mid \mathbf{A}\psi \\ \text{path formulas:} & \psi = \varphi \mid \psi \vee \psi \mid \psi \wedge \psi \mid \mathbf{X}\psi \mid \psi\mathbf{U}\psi \mid \psi\mathbf{R}\psi \mid \mathbf{G}\psi \mid \mathbf{F}\psi .\end{array}$$

We write $\mathrm{State}\backslash\neg(AP)$ and $\mathrm{Path}\backslash\neg(AP)$ for the sets of state and path formulas in $\mathrm{ACTL}^*\backslash\neg(AP)$, respectively. When working with stuttering simulations, we also use $\mathrm{ACTL}^* \setminus \mathbf{X}$, respectively $\mathrm{ACTL}^*\backslash\{\neg,\mathbf{X}\}$, for the fragment of the logic without the operator $\mathbf{X}$, respectively $\mathbf{X}$ and $\neg$.

2.1 Generalized Stuttering Simulations

In general, we are not only interested in the study of isolated systems, but would also like to be able to study their interrelationships. To do that we introduce a very general notion of simulation in increasingly more general steps; in a first step, we slightly extend the simulations in [5] (which essentially correspond to our *strict* simulations). Examples of simulations can be found in [13,12].

Given transition systems $\mathscr{A} = (\mathscr{A}, \rightarrow_{\mathscr{A}})$ and $\mathscr{B} = (\mathscr{B}, \rightarrow_{\mathscr{B}})$, a *simulation of transition systems* $H : \mathscr{A} \longrightarrow \mathscr{B}$ is a binary relation $H \subseteq A \times B$ such that if $a \rightarrow_{\mathscr{A}} a'$ and aHb then there is b' such that $b \rightarrow_{\mathscr{B}} b'$ and $a'Hb'$. A *map of transition systems* H is a simulation such that H is a function. If both H and H^{-1} are simulations, then we call H a *bisimulation*. We can extend a simulation of transition systems H to paths by defining $\pi H \rho$ if $\pi(n) H \rho(n)$ for each $n \in \mathbb{N}$.

Similarly, for Kripke structures $\mathscr{A} = (A, \rightarrow_{\mathscr{A}}, L_{\mathscr{A}})$ and $\mathscr{B} = (B, \rightarrow_{\mathscr{B}}, L_{\mathscr{B}})$ over the same set AP of atomic propositions, an *AP-simulation* $H : \mathscr{A} \longrightarrow \mathscr{B}$ of $\mathscr{A}$ by $\mathscr{B}$ is given by a simulation $H : (A, \rightarrow_{\mathscr{A}}) \longrightarrow (B, \rightarrow_{\mathscr{B}})$ between the underlying transition systems such that if aHb, then $L_{\mathscr{B}}(b) \subseteq L_{\mathscr{A}}(a)$. We say that H is an *AP-map* if its underlying simulation of transition systems is a map. We call H *strict* if aHb implies $L_{\mathscr{B}}(b) = L_{\mathscr{A}}(a)$. Also, we call H an *AP-bisimulation* if H and H^{-1} are AP-simulations.

Simulations of transition systems and of Kripke structures *compose* and the identity function $1_{\mathscr{A}} : \mathscr{A} \longrightarrow \mathscr{A}$ is trivially a simulation of transition systems and of Kripke structures. Therefore, transition systems together with their simulations define a category **TSys** with corresponding subcategories for maps and bisimulations. Similarly, Kripke structures together with AP-simulations define a category $\mathbf{KSim}_{AP}$, with two corresponding subcategories $\mathbf{KMap}_{AP}$ and $\mathbf{KBSim}_{AP}$ whose morphisms are, respectively, AP-maps and AP-bisimulations. Of course, there is also a subcategory $\mathbf{KSim}^{\mathrm{str}}_{AP}$ of strict AP-simulations, and corresponding subcategories $\mathbf{KMap}^{\mathrm{str}}_{AP}$ and $\mathbf{KBSim}^{\mathrm{str}}_{AP} = \mathbf{KBSim}_{AP}$. Note that if H is an isomorphism in $\mathbf{KSim}_{AP}$ then it must be a map and a bisimulation. Note, finally, that the mapping $(A, \rightarrow_{\mathscr{A}}, L_{\mathscr{A}}) \mapsto (A, \rightarrow_{\mathscr{A}})$ extends to a forgetful functor $\mathbf{TS} : \mathbf{KSim}_{AP} \longrightarrow \mathbf{TSys}$, with corresponding restrictions to the appropriate subcategories.

The definition of simulation can be extended by allowing the presence of stuttering [3,14,10]. For $\mathscr{A} = (A, \rightarrow_{\mathscr{A}})$ and $\mathscr{B} = (B, \rightarrow_{\mathscr{B}})$ transition systems and $H \subseteq A \times B$

[1] **X**, **G**, and **F** stand for the classic *next* (◯), *henceforth* (□), and *eventually* (◇) LTL operators.

a relation, we say that a path ρ in $\mathscr{B}$ *H-matches* a path π in $\mathscr{A}$ if there are strictly increasing functions $\alpha, \beta : \mathbb{N} \longrightarrow \mathbb{N}$ with $\alpha(0) = \beta(0) = 0$ such that, for all $i, j, k \in \mathbb{N}$, if $\alpha(i) \leq j < \alpha(i+1)$ and $\beta(i) \leq k < \beta(i+1)$, it holds that $\pi(j) H \rho(k)$. For example, the following diagram shows the beginning of two matching paths, where related elements are joined by dashed lines and $\alpha(0) = \beta(0) = 0$, $\alpha(1) = 2$, $\beta(1) = 3$, $\alpha(2) = 5$.

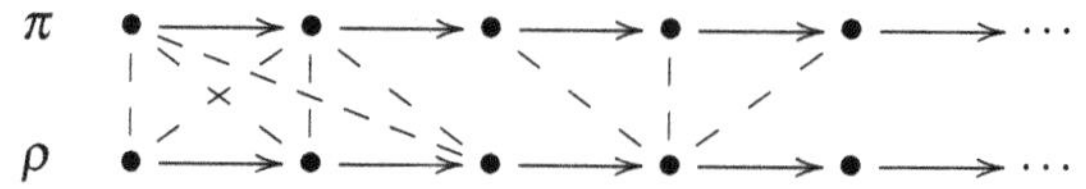

Then, a *stuttering simulation of transition systems* $H : \mathscr{A} \longrightarrow \mathscr{B}$ is a binary relation $H \subseteq A \times B$ such that if aHb, then for each path π in $\mathscr{A}$ starting at a there is a path ρ in $\mathscr{B}$ starting at b that H-matches π. If H is a function we say that H is a *stuttering map of transition systems*. If both H and H^{-1} are stuttering simulations, then we call H a *stuttering bisimulation*. Stuttering simulations of transition systems compose [10] and together with transition systems define a category that we denote **STSys** and which contains **TSys** as subcategory.

Given Kripke structures $\mathscr{A} = (A, \rightarrow_{\mathscr{A}}, L_{\mathscr{A}})$ and $\mathscr{B} = (B, \rightarrow_{\mathscr{B}}, L_{\mathscr{B}})$ over AP, a *stuttering AP-simulation* $H : \mathscr{A} \longrightarrow \mathscr{B}$ is a stuttering simulation of transition systems $H : (A, \rightarrow_{\mathscr{A}}) \longrightarrow (B, \rightarrow_{\mathscr{B}})$ such that if aHb then $L_{\mathscr{B}}(b) \subseteq L_{\mathscr{A}}(a)$. We call it *strict* if aHb implies $L_{\mathscr{B}}(b) = L_{\mathscr{A}}(a)$. Again, stuttering AP-simulations compose and define a category $\mathbf{KSSim}_{AP}$ with corresponding subcategories of strict and stuttering AP-maps.

We can generalize simulations even further by allowing them to relate Kripke structures over different sets of atomic propositions. This provides a very flexible way of relating Kripke structures and will allow us to gather all the previous categories $\mathbf{KSSim}_{AP}$, for different choices of AP, into a single one.

Given a function $\alpha : AP \longrightarrow \text{State}(AP')$ and a Kripke structure $\mathscr{A} = (A, \rightarrow_{\mathscr{A}}, L_{\mathscr{A}})$ over AP', we define the *reduct Kripke structure* $\mathscr{A}|_{\alpha} = (A, \rightarrow_{\mathscr{A}}, L_{\mathscr{A}|_{\alpha}})$ over AP, with labeling function $L_{\mathscr{A}|_{\alpha}}(a) = \{p \in AP \mid \mathscr{A}, a \models \alpha(p)\}$. α is extended in the expected, homomorphic way to formulas $\varphi \in \text{CTL}^*(AP)$, replacing each atomic proposition p occurring in φ by $\alpha(p)$; we denote this extension by $\overline{\alpha}(\varphi)$.

Proposition 1. *Let* $\alpha : AP \rightarrow \text{State}(AP')$ *be a function and* φ *be a formula in* $\text{CTL}^*(AP)$. *Then, for all Kripke structures* $\mathscr{A} = (A, \rightarrow_{\mathscr{A}}, L_{\mathscr{A}})$ *over* AP', *states* $a \in A$, *and paths* π:

- *if* φ *is a state formula,* $\mathscr{A}, a \models \overline{\alpha}(\varphi) \iff \mathscr{A}|_{\alpha}, a \models \varphi$, *and*
- *if* φ *is a path formula,* $\mathscr{A}, \pi \models \overline{\alpha}(\varphi) \iff \mathscr{A}|_{\alpha}, \pi \models \varphi$.

The definition of generalized stuttering simulations is now immediate. For Kripke structures $\mathscr{A}$ over a set AP of atomic propositions and $\mathscr{B}$ over a set AP', a *stuttering simulation* (resp. *strict stuttering simulation*) $(\alpha, H) : (AP, \mathscr{A}) \longrightarrow (AP', \mathscr{B})$ consists of a function $\alpha : AP \longrightarrow \text{State}\backslash\{\neg, \mathbf{X}\}(AP')$ (resp. $\alpha : AP \longrightarrow \text{State} \setminus \mathbf{X}(AP')$) and a stuttering AP-simulation (resp. strict stuttering AP-simulation) $H : \mathscr{A} \longrightarrow \mathscr{B}|_{\alpha}$.

To simplify notation, from now on we will write $(\alpha, H) : \mathscr{A} \longrightarrow \mathscr{B}$ instead of $(\alpha, H) : (AP, \mathscr{A}) \longrightarrow (AP', \mathscr{B})$, except in those cases where it could lead to confusion.

Composition of generalized stuttering simulations can be defined by $(\beta, G) \circ (\alpha, F) = (\overline{\beta} \circ \alpha, G \circ F)$. Using as objects pairs $(AP, \mathscr{M})$ with AP a set of atomic propositions and

$\mathcal{M}$ a Kripke structure over AP, this immediately gives rise to categories **KSSim** and **KSMap** of Kripke structures and stuttering simulations and simulation maps, respectively. However, generalized *strict* simulations between Kripke structures over different sets of atomic propositions *do not* compose, unless we only use functions of the form $\alpha : AP \longrightarrow \text{Bool}(AP')$, where $\text{Bool}(AP')$ is the set of Boolean formulas over AP'. (This situation will recur in Sections 5 and 6.)

The relationships between some of the different categories of Kripke structures introduced can be summarized in the following diagram, where the horizontal arrows are inclusions while the vertical ones are the expected forgetful functors.

$$\begin{array}{ccccccc}
\mathbf{KMap}_{AP} & \longrightarrow & \mathbf{KSim}_{AP} & \longrightarrow & \mathbf{KSSim}_{AP} & \longrightarrow & \mathbf{KSSim} \\
\downarrow & & \downarrow & & \downarrow & & \downarrow \\
\mathbf{TSys} & \longrightarrow & \mathbf{TSys} & \longrightarrow & \mathbf{STSys} & \longrightarrow & \mathbf{STSys}
\end{array}$$

The important fact about stuttering simulations is that they reflect satisfaction of appropriate classes of formulas. Given Kripke structures $\mathcal{A}$ over AP and $\mathcal{B}$ over AP', we say that a stuttering simulation $(\alpha, H) : \mathcal{A} \longrightarrow \mathcal{B}$ *reflects* the satisfaction of a formula $\varphi \in \text{CTL}^*(AP)$ if either:

- φ is a state formula, and $\mathcal{B}, b \models \overline{\alpha}(\varphi)$ and aHb imply that $\mathcal{A}, a \models \varphi$; or
- φ is a path formula, and $\mathcal{B}, \rho \models \overline{\alpha}(\varphi)$ and ρ H-matches π imply that $\mathcal{A}, \pi \models \varphi$.

Theorem 1 ([11]). *Stuttering simulations always reflect satisfaction of* $\text{ACTL}^* \setminus \{\neg, \mathbf{X}\}$ *formulas. Strict stuttering simulations also reflect satisfaction of* $\text{ACTL}^* \setminus \mathbf{X}$ *formulas.*

Appendix B contains a summary of the categories presented in this section. The "best" one is **KSSim**, the most general one, in that it provides the greater flexibility for relating arbitrary Kripke structures which otherwise could not be related; on the other hand, as we will see in Section 7, we know less about its categorical properties than for most of the others.

3 Some Categorical Concepts

Almost all the notions from category theory [9,2] that we use are rather basic and we only review those concepts that may not be so familiar. To try to avoid confusions with simulation morphisms, we refer to the morphisms in a category simply as arrows.

Opfibrations. What determines an *opfibration* [8] is the capacity of "lifting" an arrow in a base category to another category in an "initial" (and hence minimal) manner in an appropriate sense.

Let $F : \mathcal{C} \longrightarrow \mathcal{D}$ be a functor. An arrow $f : X \longrightarrow Y$ in $\mathcal{C}$ is *opcartesian* over u if $F(f) = u$ and for every arrow $g : X \longrightarrow Z$ in $\mathcal{C}$ such that $F(g) = v \circ u$ for some $v : F(Y) \longrightarrow F(Z)$ there exists a unique arrow $h : Y \longrightarrow Z$ such that $g = h \circ f$ and $F(h) = v$. The functor F is an *opfibration* if there exists an opcartesian morphism over every arrow $u : F(X) \longrightarrow J$. The dual notions are those of *cartesian morphism* and *fibration*.

Institutions. The notion of institution is due to Goguen and Burstall's seminal work [6]; their goal was to capture the notion of model in a formalism independent way. An *institution* is a 4-tuple $\mathcal{I} = (\mathbf{Sign}, sen, \mathbf{Mod}, \models)$ such that:

- **Sign** is a category whose objects are called *signatures*,
- sen : **Sign** $\longrightarrow$ **Set** is a functor that associates to each signature Σ a set of Σ-*sentences*,
- **Mod** : $\mathbf{Sign}^{\mathrm{op}} \longrightarrow$ **Cat** is a functor that associates to each signature Σ a category whose objects are called Σ-*models*, and
- $\models$ is a function that associates to each $\Sigma \in |\mathbf{Sign}|$ a binary relation $\models_\Sigma \subseteq |\mathbf{Mod}(\Sigma)| \times sen(\Sigma)$ called Σ-*satisfaction*, in such a way that the following property holds for every $H : \Sigma \longrightarrow \Sigma'$, $M' \in |\mathbf{Mod}(\Sigma')|$, and every $\varphi \in sen(\Sigma)$: $M' \models_{\Sigma'} sen(H)(\varphi) \iff \mathbf{Mod}(H)(M') \models_\Sigma \varphi$.

A *theory morphism* $H : (\Sigma, \Gamma) \longrightarrow (\Sigma', \Gamma')$ is a signature morphism $H : \Sigma \longrightarrow \Sigma'$ such that every model in $\mathbf{Mod}(\Sigma')$ that satisfies Γ' also satisfies $sen(H)(\varphi)$, for all $\varphi \in \Gamma$. By defining $\mathbf{Mod}(\Sigma, \Gamma)$ as the full subcategory of $\mathbf{Mod}(\Sigma)$ determined by those models that satisfy Γ, we can extend **Mod** to a functor **Mod** : $\mathbf{Th}^{\mathrm{op}} \longrightarrow$ **Cat**, where **Th** is the category of theories and theory morphisms.

A property expressing the possibility of "putting theories together" by colimits is the exactness of an institution. An institution is *exact* if its category of signatures is cocomplete and the model functor **Mod** preserves limits, and is *semiexact* if **Sign** has pushouts and **Mod** sends pushouts in **Sign** to pullbacks in **Cat**.

Monads and Kleisli categories. A *monad* (called a *triple* in [2]) is a tuple (T, η, μ), where $T : \mathscr{C} \longrightarrow \mathscr{C}$ is a functor, and $\eta : 1_\mathscr{C} \longrightarrow T$ and $\mu : T \circ T \longrightarrow T$ are natural transformations satisfying $\mu \circ \eta T = \mu \circ T \eta = 1_T$ and $\mu \circ \mu T = \mu \circ T \mu$.

All monads can be obtained from adjunctions. One possible construction makes use of the so-called *Kleisli category*. The Kleisli category $\mathscr{C}_T$ of a monad (T, η, μ) has as objects those of $\mathscr{C}$. If X and Y are objects of $\mathscr{C}$, an arrow $X \longrightarrow Y$ in the Kleisli category is an arrow $X \longrightarrow T(Y)$ in $\mathscr{C}$. Composition of two arrows $f : X \longrightarrow T(Y)$ and $g : Y \longrightarrow T(Z)$ is defined as $\mu_Z \circ Tg \circ f$.

Grothendieck construction. Often we are interested in considering all the components of an *indexed category* together in a "flat" category obtained by taking the disjoint union of the components and adding some new arrows. This is called, for example in [15], the *Grothendieck construction*. Given a functor $\mathbf{C} : \mathscr{I}^{\mathrm{op}} \longrightarrow$ **Cat**, the associated *Grothendieck construction* is defined by:

- its objects are pairs (I, X), where I is an object of $\mathscr{I}$ and X is an object of $\mathbf{C}(I)$;
- an arrow $(I, X) \longrightarrow (J, Y)$ is a pair (u, f) with $u : I \longrightarrow J$ in $\mathscr{I}$ and $f : X \longrightarrow \mathbf{C}(u)(Y)$ in $\mathbf{C}(I)$;
- the composition of arrows $(u, f) : (I, X) \longrightarrow (J, Y)$ and $(v, g) : (J, Y) \longrightarrow (K, Z)$ is given by $(v, g) \circ (u, f) = (v \circ u, \mathbf{C}(u)(g) \circ f)$.

Regular epis and monos. As defined in [7], an arrow $m : X \longrightarrow Y$ is a *regular monomorphism* if there exist arrows f and g such that m is the equalizer of f and g. Dually, $e : X \longrightarrow Y$ is a *regular epimorphism* if it is the coequalizer of two arrows.

Given two classes $\mathscr{E}$ and $\mathscr{M}$ of epimorphisms and monomorphisms respectively, closed under composition with isomorphisms, a $(\mathscr{E}, \mathscr{M})$-*factorization* of an arrow f is a factorization $f = m \circ e$ with e in $\mathscr{E}$ and m in $\mathscr{M}$. A category is (univocally) $(\mathscr{E}, \mathscr{M})$-*factorizable* if every arrow has a (unique up to isomorphism) $(\mathscr{E}, \mathscr{M})$-factorization.

A category is a $(\mathscr{E},\mathscr{M})$-*category* if it is univocally factorizable and both $\mathscr{E}$ and $\mathscr{M}$ are closed under composition.

4 Minimal Kripke Structures

Theorem 1 is the basis of the method of model checking by abstraction: given an infinite (or too large) system $\mathscr{M}$, one tries to find a system $\mathscr{A}$ with a finite set of reachable states that simulates it and uses a model checker to prove properties of $\mathscr{M}$ by means of $\mathscr{A}$. But usually, one only has the concrete system $\mathscr{M}$ and a *surjective* function $h : M \longrightarrow A$ mapping concrete states to a simplified abstract domain A. In this situation, we are interested in using h to find a Kripke structure that best simulates $\mathscr{M}$ under certain conditions. In [4] the *minimal transition system* associated to a transition system $\mathscr{M}$ and a surjective function $h : M \to A$ was defined; using our notion of simulation this can be extended to the level of Kripke structures.

Definition 1. *The* minimal *Kripke structure $\mathscr{M}^h_{\min}$ corresponding to a Kripke structure $\mathscr{M}$ and the surjective function $h : M \longrightarrow A$ is given by the triple $(A, (h \times h)(\to_{\mathscr{M}}), L_{\mathscr{M}^h_{\min}})$, where $L_{\mathscr{M}^h_{\min}}(a) = \bigcap_{x \in h^{-1}(a)} L_{\mathscr{M}}(x)$.*

The following proposition is an immediate consequence of the definitions.

Proposition 2 ([13]). *For any Kripke structure $\mathscr{M}$ and any surjective function h, $h : \mathscr{M} \longrightarrow \mathscr{M}^h_{\min}$ is an AP-map.*

The use of the adjective "minimal" is appropriate since, as pointed out in [4], $\mathscr{M}^h_{\min}$ is the most accurate approximation to $\mathscr{M}$ that is consistent with h. Within our framework, the notion of minimality can be expressed in a precise categorical sense by means of an opcartesian morphism.

Proposition 3. *For a Kripke structure $\mathscr{M}$ and surjective function $h : M \longrightarrow A$, the AP-map $h : \mathscr{M} \longrightarrow \mathscr{M}^h_{\min}$ is an opcartesian morphism in the context of the forgetful functor $U : \mathbf{KMap}_{AP} \longrightarrow \mathbf{Set}$ mapping a Kripke structure $\mathscr{M} = (M, \to_{\mathscr{M}}, L_{\mathscr{M}})$ to its underlying set M and an AP-map to itself.*

Proof. Given $f : \mathscr{M} \longrightarrow \mathscr{N}$ in $\mathbf{KMap}_{AP}$ such that it can be factorized in **Set** as $f = g \circ h$ for some function $g : A \longrightarrow N$, we have to find a unique g' in $\mathbf{KMap}_{AP}$ such that $g' : \mathscr{M}^h_{\min} \longrightarrow \mathscr{N}$, $f = g' \circ h$, and $U(g') = g$. By definition of U, it must be $g' = g$; we have to check that g is actually an *AP*-map.

By definition of $\mathscr{M}^h_{\min}$, if $a \to_{\mathscr{M}^h_{\min}} b$ there exist x and y in M such that $h(x) = a$, $h(y) = b$, and $x \to_{\mathscr{M}} y$. Hence, since f is an *AP*-map, $g(a) = g(h(x)) = f(x) \to_{\mathscr{N}} f(y) = g(h(y)) = g(b)$. In addition, using again the fact that f is an *AP*-map, if $p \in L_{\mathscr{N}}(s)$ then $p \in L_{\mathscr{M}}(x)$ for all x in M such that $f(x) = s$. Let then $a \in A$ such that $g(a) = s$: for all y in M such that $h(y) = a$, since $f(y) = g(h(y)) = s$, it is the case that $p \in L_{\mathscr{M}}(y)$. Therefore, $p \in L_{\mathscr{M}^h_{\min}}(a)$, and for all a with $g(a) = s$ we have $L_{\mathscr{N}}(s) \subseteq L_{\mathscr{M}^h_{\min}}(a)$. □

Note that this result can be extended to the category $\mathbf{KSMap}_{AP}$: then, whenever f is a stuttering *AP*-map, so will be g. The result also holds for generalized simulations in which the set of atomic propositions may vary.

Proposition 4. *The simulation* $(\eta_{AP}, h) : \mathscr{M} \longrightarrow \mathscr{M}^h_{\min}$*, where* $h : M \longrightarrow A$ *is a surjective function and* η_{AP} *is the inclusion* $AP \hookrightarrow \mathrm{State}\backslash\neg(AP)$*, is an opcartesian morphism for the forgetful functor* $U : \mathbf{KMap} \longrightarrow \mathbf{Set}$ *mapping a pair* $(AP, \mathscr{M})$ *to the underlying set* M *and a simulation map* (α, h) *to the corresponding function* h.

Proof. The proof follows the same steps as the one for Proposition 3, despite the fact that the set of atomic propositions now may vary from one Kripke structure to another. □

5 Borrowing

Simulations, in all their different variants, require suitable preservation of transitions and of atomic propositions. Sometimes, however, it is more natural and easier to think just in terms of the underlying transition systems; in those cases we can still recover a full-fledged simulation by *borrowing* the Kripke structure of the domain using the labeling function of the codomain.

Definition 2. *Let* $\mathscr{A} = (A, \rightarrow_{\mathscr{A}})$ *be a transition system and let* $\mathscr{B} = (B, \rightarrow_{\mathscr{B}}, L_{\mathscr{B}})$ *be a Kripke structure over a set* AP *of atomic propositions. If* $h : A \longrightarrow B$ *is a stuttering map between the underlying transition systems, then* $\mathscr{A}$ *can be extended to a Kripke structure over* AP *with* $L_{\mathscr{A}} = L_{\mathscr{B}} \circ h$*. We say that* $\mathscr{A}$ borrows *its properties from* $\mathscr{B}$.

Proposition 5. *If* $\mathscr{A} = (A, \rightarrow_{\mathscr{A}})$ *borrows its properties from a Kripke structure* $\mathscr{B} = (B, \rightarrow_{\mathscr{B}}, L_{\mathscr{B}})$ *over a set* AP *of atomic propositions through a stuttering map of transition systems* $h : (A, \rightarrow_{\mathscr{A}}) \longrightarrow (B, \rightarrow_{\mathscr{B}})$*, then* h *becomes a strict stuttering* AP*-map. Furthermore,* h *is a cartesian morphism for the forgetful functor* $U : \mathbf{KSMap}_{AP} \longrightarrow \mathbf{STSys}$ *mapping a Kripke structure to its underlying transition system.*

Proof. h is clearly a strict stuttering AP-map because, by definition of $L_{\mathscr{A}}$, atomic propositions are preserved. To show that it is a cartesian morphism, assume a stuttering AP-map $f : \mathscr{C} \longrightarrow \mathscr{B}$ and a stuttering map of transition systems $g : U(\mathscr{C}) \longrightarrow (A, \rightarrow_{\mathscr{A}})$ such that $f = h \circ g$: we have to show that there is a unique stuttering AP-map g' such that $h \circ g' = f$ and $U(g') = g$. The only possible candidate is g, and we have to check that $g : \mathscr{C} \longrightarrow \mathscr{A}$ is indeed a stuttering AP-map. By hypothesis, g is a map of transition systems. Now, assume that $g(c) = a$ and $p \in L_{\mathscr{A}}(a)$. It follows that $p \in L_{\mathscr{B}}(h(a))$, and since $f(c) = (h \circ g)(c) = h(a)$ and f is a stuttering AP-map, $p \in L_{\mathscr{C}}(c)$ as required. □

It is interesting to note that this proposition also holds even if h is not a function (but the resulting AP-simulation may not be strict).

One could ask whether this result can be extended to the Grothendieck category **KSMap** so that (η_{AP}, h) becomes a cartesian morphism for the forgetful functor $U : \mathbf{KSMap} \longrightarrow \mathbf{STSys}$. The answer is *no* and the reason lies in the generality of the functions $\alpha : AP \longrightarrow \mathrm{State}(AP')$ used to relate Kripke structures over different sets of atomic propositions. However, the result can be recovered by working in the subcategory $\mathbf{KSMap}^{bool}$ of **KSMap** in which the codomain of the functions α is restricted to $\mathrm{Bool}(AP')$. That is the content of the following proposition.

Proposition 6. *If $\mathscr{A} = (A, \rightarrow_{\mathscr{A}})$ borrows its properties from $\mathscr{B} = (B, \rightarrow_{\mathscr{B}}, L_{\mathscr{B}})$ through a stuttering map of transition systems $h : (A, \rightarrow_{\mathscr{A}}) \longrightarrow (B, \rightarrow_{\mathscr{B}})$, then (η_{AP}, h) becomes a strict stuttering map. Furthermore, (η_{AP}, h) is a cartesian morphism for the forgetful functor $U : \mathbf{KSMap}^{bool} \longrightarrow \mathbf{STSys}$ mapping a Kripke structure to its underlying transition system.*

Proof. (η_{AP}, h) is clearly a strict stuttering map because, by definition of $L_{\mathscr{A}}$, atomic propositions are preserved. To show that it is a cartesian morphism, assume a stuttering map $(\alpha, f) : \mathscr{C} \longrightarrow \mathscr{B}$ and a stuttering map of transition systems $g : U(\mathscr{C}) \longrightarrow (A, \rightarrow_{\mathscr{A}})$ such that $f = h \circ g$. We have to show that there is a unique stuttering map (α', g') such that $(\eta_{AP}, h) \circ (\alpha', g') = (\alpha, f)$ and $U(\alpha', g') = g$. The only possible candidate is (α, g), and therefore we have to check that $g : \mathscr{C} \longrightarrow \mathscr{A}|_{\alpha}$ is indeed a stuttering AP'-map, where AP' is the set of atomic propositions of $\mathscr{C}$. By hypothesis, g is a map of transition systems. Now, assume that $g(c) = a$ and $p \in L_{\mathscr{A}|_{\alpha}}(a)$; it follows that $\mathscr{A}, a \models \alpha(p)$. Since $L_{\mathscr{A}}(a) = L_{\mathscr{B}}(h(a))$, it is immediate to show that for all $\varphi \in \mathrm{Bool}(AP)$, $\mathscr{A}, a \models \varphi$ iff $\mathscr{B}, h(a) \models \varphi$. Therefore, $\mathscr{B}, h(a) \models \alpha(p)$, and since $f(c) = (h \circ g)(c) = h(a)$ and (α, f) is a stuttering map, $\mathscr{C}, c \models p$ by Theorem 1, that is, $p \in L_{\mathscr{C}}(c)$ as required. □

6 Temporal Logic Institutions

It is not hard to notice that the result in Proposition 1 has a distinct institutional flavor. Indeed, Kripke structures can be organized as the models of a temporal logic institution [6] in which Proposition 1 corresponds to the property required of the satisfaction relation. Other institutions for temporal logics are discussed in [1], but their notions of signature morphism and of simulation are more restricted. Some of the ideas in this section were presented in [11]: we also include them here for the sake of completeness.

Let us first define the category of signatures. For that, let $\mathrm{State}\backslash\neg : \mathbf{Set} \longrightarrow \mathbf{Set}$ be the functor mapping a set AP to the set of state formulas $\mathrm{State}\backslash\neg(AP)$, and a function $\alpha : AP \longrightarrow AP'$ to its homomorphic extension $\overline{\alpha} : \mathrm{State}\backslash\neg(AP) \longrightarrow \mathrm{State}\backslash\neg(AP')$. Then, the triple $\langle \mathrm{State}\backslash\neg, \eta, \mu \rangle$ is a monad (Section 3), where $\eta : Id_{\mathbf{Set}} \Rightarrow \mathrm{State}\backslash\neg$ and $\mu : \mathrm{State}\backslash\neg \circ \mathrm{State}\backslash\neg \Rightarrow \mathrm{State}\backslash\neg$ are natural transformations such that $\eta_{AP}(p) = p$ and μ "unnests" a formula into its basic atomic propositions. Our category of signatures will be $\mathbf{Set}_{\mathrm{State}\backslash\neg}$, the Kleisli category of the monad; its objects are just sets, and the morphisms $AP \longrightarrow AP'$ are functions $\alpha : AP \longrightarrow \mathrm{State}\backslash\neg(AP')$.

Definition 3. *The institution of Kripke structures is given by:*

- $\mathbf{Sign_K} = \mathbf{Set}_{\mathrm{State}\backslash\neg}$.
- $sen_{\mathbf{K}}$ *is the functor mapping a set AP to* $\mathrm{State}\backslash\neg(AP)$*, and a function* $\alpha : AP \longrightarrow \mathrm{State}\backslash\neg(AP')$ *to its homomorphic extension* $\overline{\alpha} : \mathrm{State}\backslash\neg(AP) \longrightarrow \mathrm{State}\backslash\neg(AP')$.
- $\mathbf{Mod_K} : \mathbf{Set}_{\mathrm{State}\backslash\neg} \longrightarrow \mathbf{Cat}^{\mathrm{op}}$ *is given by* $\mathbf{Mod_K}(AP) = \mathbf{KSim}_{AP}$ *and, for* $\alpha : AP \longrightarrow AP'$ *in* $\mathbf{Set}_{\mathrm{State}\backslash\neg}$, $\mathbf{Mod_K}(\alpha)(\mathscr{A}) = \mathscr{A}|_{\alpha}$ *and* $\mathbf{Mod_K}(\alpha)(H) = H$.
- *The satisfaction relation is defined as* $\mathscr{A} \models \varphi$ *iff* $\mathscr{A}, a \models \varphi$ *for all* $a \in A$.

Proposition 7. $\mathscr{I}_{\mathbf{K}} = (\mathbf{Sign_K}, sen_{\mathbf{K}}, \mathbf{Mod_K}, \models)$ *is an institution.*

Analogously, we could think of defining an institution for Kripke structures and strict morphisms. However, the fact that α can map an atomic proposition to an arbitrary formula makes it impossible. The problem is that the putative model functor is not such: the reduct of a strict simulation may not itself be strict. As happened in Sections 2.1 and 5, to solve this problem and get an institution for strict simulations it is enough to restrict the signature morphisms to be functions of the form $\alpha : AP \longrightarrow \mathrm{Bool}(AP)$.

Notice also that the category **KSim** can be obtained by means of the Grothendieck construction [15]. Indeed, **KSim** is just the Grothendieck category corresponding to the indexed category $\mathbf{Mod_K}$. (The same would happen for strict simulations if we were to work with the restricted α functions.) Similarly, **KMap** and **KBSim** can be obtained by modifying $\mathbf{Mod_K}$ so that *AP* is mapped to $\mathbf{KMap}_{AP}$ and $\mathbf{KBSim}_{AP}$, respectively.

Of course, analogous results exist for the general case of stuttering simulations. Now the functor used to define the Kleisli category of signatures is $\mathrm{State}\backslash\{\neg,\mathbf{X}\}$, mapping *AP* to the set of state formulas $\mathrm{State}\backslash\{\neg,\mathbf{X}\}(AP)$ (and to $\mathrm{Bool}(AP)$ for the strict case). Similarly, the model functor maps the set of atomic propositions *AP* to the corresponding category of sttutering *AP*-simulations, $\mathbf{KSSim}_{AP}$. Actually, as the proof reveals, the construction also applies to any temporal logic whose formulae are reflected by simulations; in particular, we could restrict the institutions to the LTL sublogic of ACTL*.

The institutions just introduced use the most general notion of signature morphism compatible with the reflection of suitable temporal formulas. But precisely because of this generality, they do not have the *exactness* property. To see this, it is enough to consider the set of atomic propositions $AP = \{p,q\}$ and signature morphisms $\alpha_1, \alpha_2 : AP \longrightarrow \mathrm{State}\backslash\neg(AP)$ such that $\alpha_1(p) = p \wedge q$ and $\alpha_2(p) = p \vee q$. Then, for any signature morphisms $\beta_1, \beta_2 : AP \longrightarrow \mathrm{State}\backslash\neg(AP')$, $(\beta_1 \circ \alpha_1)(p) = \beta_1(p) \wedge \beta_1(q)$, which is different from $(\beta_2 \circ \alpha_2)(p) = \beta_2(p) \vee \beta_2(q)$. This shows that $\mathbf{Sign_K}$ does not have pushouts. The situation, however, changes when the signature morphisms are restricted to mapping atomic propositions to atomic propositions. Note that the counterexample shows also that this time it is not enough to consider $\mathrm{Bool}(AP)$: we have to map *atomic* propositions to *atomic* propositions.

Proposition 8. *Let* $\mathscr{I}'_{\mathbf{K}}$ *be obtained from the institution* $\mathscr{I}_{\mathbf{K}}$ *by replacing* $\mathbf{Set}_{\mathrm{State}\backslash\neg}$ *by* **Set** *as the category of signatures. Then* $\mathscr{I}'_{\mathbf{K}}$ *is a semiexact institution.*

The same result also applies to the institutions of strict and stuttering simulations described above.

7 Limits and Colimits in Categories of Simulations

We collect in this section categorical properties about existence of limits and colimits in some of the categories of Kripke structures that have been presented in Section 2. We focus on the categories over a fixed set of atomic propositions. For the Grothendieck categories, colimits can be obtained by mimicking the constructions presented below; however, we conjecture that in such Grothendieck categories limits do not exist in general.

7.1 Products and Coproducts

Proposition 9. *For all sets of atomic propositions AP, the category* $\mathbf{KMap}_{AP}$ *has finite products.*

Proof. Given Kripke structures $\mathscr{A}$ and $\mathscr{B}$, define $\mathscr{A} \times \mathscr{B} = (A \times B, \to_{\mathscr{A}\times\mathscr{B}}, L_{\mathscr{A}\times\mathscr{B}})$, where $(a,b) \to_{\mathscr{A}\times\mathscr{B}} (a',b')$ iff $a \to_{\mathscr{A}} a'$ and $b \to_{\mathscr{B}} b'$, and $L_{\mathscr{A}\times\mathscr{B}}(a,b) = L_{\mathscr{A}}(a) \cup L_{\mathscr{B}}(b)$, with the usual projections $\pi_{\mathscr{A}} : \mathscr{A} \times \mathscr{B} \longrightarrow \mathscr{A}$ and $\pi_{\mathscr{B}} : \mathscr{A} \times \mathscr{B} \longrightarrow \mathscr{B}$. The relation $\to_{\mathscr{A}\times\mathscr{B}}$ is total and thus $\mathscr{A} \times \mathscr{B}$ is well-defined, and it is immediate to check that $\pi_{\mathscr{A}}$ and $\pi_{\mathscr{B}}$ are AP-maps.

Now, if $f : \mathscr{C} \longrightarrow \mathscr{A}$ and $g : \mathscr{C} \longrightarrow \mathscr{B}$ are AP-maps, the unique arrow $\langle f,g \rangle : \mathscr{C} \longrightarrow \mathscr{A} \times \mathscr{B}$ such that $\pi_{\mathscr{A}} \circ \langle f,g \rangle = f$ and $\pi_{\mathscr{B}} \circ \langle f,g \rangle = g$ is given by $\langle f,g \rangle(c) = (f(c), g(c))$. Uniqueness is clear: we have to check that $\langle f,g \rangle$ is indeed an AP-map. If $c \to_{\mathscr{C}} c'$ then $f(c) \to_{\mathscr{A}} f(c')$ and $g(c) \to_{\mathscr{B}} g(c')$, and therefore $\langle f,g \rangle(c) \to_{\mathscr{A}\times\mathscr{B}} \langle f,g \rangle(c')$. And if $p \in L_{\mathscr{A}\times\mathscr{B}}(\langle f,g \rangle(c))$ then $p \in L_{\mathscr{A}}(f(c))$ or $p \in L_{\mathscr{B}}(g(c))$: either way, $p \in L_{\mathscr{C}}(c)$. □

Note that this construction can be extended to infinite products in the expected way and, since the Kripke structure with a single state $*$, single transition $* \to *$, and $L(*) = \emptyset$ is a final object, $\mathbf{KMap}_{AP}$ has arbitrary products.

This result is also true for the category of strict AP-maps, but the constructions are slightly more involved. The final object in $\mathbf{KMap}^{\mathrm{str}}_{AP}$ is $(\mathscr{P}(AP), \mathscr{P}(AP) \times \mathscr{P}(AP), id_{\mathscr{P}(AP)})$. The construction of finite products is shown in the proof of the next result; this is sometimes called the *synchronous product* of Kripke structures in the literature.

Proposition 10. *For all sets of atomic propositions AP, the category* $\mathbf{KMap}^{\mathrm{str}}_{AP}$ *has finite products.*

Proof. Given Kripke structures $\mathscr{A}$ and $\mathscr{B}$, let $\mathscr{A} \times \mathscr{B}$ be the Kripke structure defined in the proof of Proposition 9. Let us define

$$\mathrm{Path}(\mathscr{A} \times \mathscr{B})^{=} = \{\pi \in \mathrm{Path}(\mathscr{A} \times \mathscr{B}) \mid \text{for all } i,\, L_{\mathscr{A}}(\pi_{\mathscr{A}}(\pi(i))) = L_{\mathscr{B}}(\pi_{\mathscr{B}}(\pi(i)))\},$$
$$D = \{(a,b) \mid \text{there exists } \pi \in \mathrm{Path}(\mathscr{A} \times \mathscr{B})^{=} \text{ and } i \in \mathbb{N} \text{ such that } (a,b) = \pi(i)\}.$$

The product of $\mathscr{A}$ and $\mathscr{B}$ in $\mathbf{KMap}^{\mathrm{str}}_{AP}$ is given by $\mathscr{A} \times^{\mathrm{st}} \mathscr{B} = (D, \to_{\mathscr{A}\times\mathscr{B}}|_{D^2}, L_{\mathscr{A}\times\mathscr{B}}|_D)$ with the expected projections. By construction, $\to_{\mathscr{A}\times\mathscr{B}}|_{D^2}$ is total. Note that for arbitrary strict AP-maps $f : \mathscr{C} \longrightarrow \mathscr{A}$ and $g : \mathscr{C} \longrightarrow \mathscr{B}$, the function $\langle f,g \rangle : \mathscr{C} \longrightarrow \mathscr{A} \times^{\mathrm{st}} \mathscr{B}$ is well-defined. For each $c \in C$, let π be a path with $\pi(0) = c$ (it must exist because $\to_{\mathscr{C}}$ is total). Since both f and g are strict, $L_{\mathscr{A}}(f(\pi(i))) = L_{\mathscr{B}}(g(\pi(i)))$ and the path

$$(f(\pi(0)), g(\pi(0))) \to_{\mathscr{A}\times\mathscr{B}} (f(\pi(1)), g(\pi(1))) \to_{\mathscr{A}\times\mathscr{B}} \cdots$$

belongs to $\mathrm{Path}(\mathscr{A} \times \mathscr{B})^{=}$; thus, $(f(c), g(c)) \in D$. And $\langle f,g \rangle$ is clearly strict. □

Note that in some cases we may have $\mathscr{A} \times^{\mathrm{str}} \mathscr{B} = \emptyset$ even though neither $\mathscr{A}$ nor $\mathscr{B}$ are empty. This simply means that the *only* Kripke structure $\mathscr{C}$ from which there exist strict AP-simulations to both $\mathscr{A}$ and $\mathscr{B}$ is the empty one. Note also that the construction can be extended in the expected way to infinite products.

If we take a look at what happens when considering AP-simulations instead of just maps, it turns out that finite producs also exist in $\mathbf{KSim}_{AP}$ although its definition is quite different from the previous ones.

Proposition 11. *For all sets of atomic propositions AP, the category* $\mathbf{KSim}_{AP}$ *has finite products.*

Proof. Define the product of $\mathscr{A}$ and $\mathscr{B}$ to be $\mathscr{A} \times \mathscr{B} = (A \uplus B, \rightarrow_{\mathscr{A}} \uplus \rightarrow_{\mathscr{B}}, L_{\mathscr{A}\times\mathscr{B}})$, where $L_{\mathscr{A}\times\mathscr{B}}(x)$ is $L_{\mathscr{A}}(x)$ if $x \in A$ or $L_{\mathscr{B}}(x)$ if $x \in B$, with projections $\Pi_{\mathscr{A}}$ and $\Pi_{\mathscr{B}}$ defined by $a\Pi_{\mathscr{A}}a$ for all $a \in A$ and $b\Pi_{\mathscr{B}}b$ for all $b \in B$. Then, for AP-simulations $F : \mathscr{C} \longrightarrow \mathscr{A}$ and $G : \mathscr{C} \longrightarrow \mathscr{B}$, the unique $\langle F,G\rangle$ is defined by $c\langle F,G\rangle a$ iff cFa, and $c\langle F,G\rangle b$ iff cGb. □

Again, the above construction can be extended to arbitrary families $\{\mathscr{A}_i\}_{i\in I}$ of Kripke structures, and since the empty Kripke structure is trivially a final object, the category $\mathbf{KSim}_{AP}$ has arbitrary products.

By contrast, there are no products in the category $\mathbf{KSMap}_{AP}$ of stuttering AP-simulations. To see this, consider $\mathscr{A}$ given by $a_1 \rightarrow_{\mathscr{A}} a_2 \rightarrow_{\mathscr{A}} \dots$ and $\mathscr{B}$ by $b_1 \rightarrow_{\mathscr{B}} b_2 \rightarrow_{\mathscr{B}} \dots$, where both labeling functions are empty. Now, assume $\mathscr{C}$ is given by $c_1 \rightarrow_{\mathscr{C}} c_2 \rightarrow_{\mathscr{C}} \dots$. Consider now stuttering AP-simulations $f : \mathscr{C} \longrightarrow \mathscr{A}$ with $f(c_1) = a_1$, $f(c_{2*i}) = a_{i+1}$, and $f(c_{2*i+1}) = a_{i+1}$ for $i \geq 1$, and $g : \mathscr{C} \longrightarrow \mathscr{B}$ with $g(c_{2*i+1}) = a_{i+1}$ and $g(c_{2*i+2}) = a_{i+1}$ for $i \geq 0$. Assume that $\mathscr{D}$ is the product of $\mathscr{A}$ and $\mathscr{B}$ with projections $\pi_{\mathscr{A}}$ and $\pi_{\mathscr{B}}$, and let $d_1 \rightarrow_{\mathscr{D}} d_2 \rightarrow_{\mathscr{D}} \dots$ be the path in $\mathscr{D}$ $\langle f,g\rangle$-matching the path in $\mathscr{C}$ that starts at c_1. We have $\langle f,g\rangle(c_1) = d_1$, $\pi_{\mathscr{A}}(d_1) = a_1$, and $\pi_{\mathscr{B}}(d_1) = b_1$. Now, $\langle f,g\rangle(c_2)$ must be equal to d_1 or to d_2. But the first alternative cannot hold because then $\pi_{\mathscr{A}}(\langle f,g\rangle(c_2)) \neq f(c_2)$; therefore $\langle f,g\rangle(c_2) = d_2$, and $\pi_{\mathscr{A}}(d_2)$ and $\pi_{\mathscr{B}}(d_2)$ have to be a_2 and b_1, respectively. And we are done, because if we swap the definitions of f and g the same argument leads to $\pi_{\mathscr{A}}(d_1) = \pi_{\mathscr{B}}(d_2) = a_1$: a contradiction.

Coproducts exist in all the categories mentioned in the previous section and their definition is the same in all cases. Here we present the details for $\mathbf{KSim}_{AP}$.

Proposition 12. *For all sets of atomic propositions AP, the category* $\mathbf{KSim}_{AP}$ *has finite coproducts.*

Proof. Given Kripke structures $\mathscr{A}$ and $\mathscr{B}$, define $\mathscr{A} + \mathscr{B}$ as $(A \uplus B, \rightarrow_{\mathscr{A}} \uplus \rightarrow_{\mathscr{B}}, L_{\mathscr{A}+\mathscr{B}})$, where $L_{\mathscr{A}+\mathscr{B}}(x)$ is $L_{\mathscr{A}}(x)$ if $x \in A$ or $L_{\mathscr{B}}(x)$ if $x \in B$, and with inclusions $I_{\mathscr{A}}$ and $I_{\mathscr{B}}$ defined by $aI_{\mathscr{A}}a$ for all $a \in A$ and $bI_{\mathscr{B}}b$ for all $b \in B$. $\mathscr{A} + \mathscr{B}$ is clearly well-defined and it is trivial to check that $I_{\mathscr{A}}$ and $I_{\mathscr{B}}$ are AP-simulations. Now, for $F : \mathscr{A} \longrightarrow \mathscr{C}$ and $G : \mathscr{B} \longrightarrow \mathscr{C}$ arbitrary AP-simulations, define $[F,G] : \mathscr{A} + \mathscr{B}$ by $a[F,G]c$ iff aFc, and $b[F,G]c$ iff bGc. It is easy to check that $[F,G]$ so defined is the only AP-simulation that satisfies $I_{\mathscr{A}} \circ [F,G]$ and $I_{\mathscr{B}} \circ [F,G]$. □

Note that the Kripke structure $\mathscr{A} + \mathscr{B}$ is the same as the Kripke structure $\mathscr{A} \times \mathscr{B}$ of Proposition 11, and that the construction also applies to infinite families. The initial object corresponds to the empty Kripke structure.

7.2 Equalizers and Coequalizers

Proposition 13. *For all sets of atomic propositions AP, the category* $\mathbf{KMap}_{AP}$ *has equalizers.*

Proof. Let $f, g : \mathscr{A} \longrightarrow \mathscr{B}$ be AP-maps. Let us define

$$\mathrm{Path}(\mathscr{A})_{f,g} = \{\pi \in \mathrm{Path}(\mathscr{A}) \mid f \circ \pi = g \circ \pi\},$$
$$E = \{a \in A \mid \text{there exists } \pi \in \mathrm{Path}(\mathscr{A})_{f,g} \text{ and } i \in \mathbb{N} \text{ such that } a = \pi(i)\}.$$

The equalizer of f and g is given by the Kripke structure $\mathscr{E} = (E, \rightarrow_{\mathscr{A}}|_{E^2}, L_{\mathscr{A}}|_E)$ and the inclusion $e : \mathscr{E} \longrightarrow \mathscr{A}$. By definition, $\rightarrow_{\mathscr{E}}$ is total and thus $\mathscr{E}$ is a well-defined Kripke structure; e is trivially a (strict) AP-map. Now, suppose that $h : \mathscr{D} \longrightarrow \mathscr{A}$ is an AP-map such that $f \circ h = g \circ h$. Define $m : \mathscr{D} \longrightarrow \mathscr{E}$ by $m(d) = h(d)$. Obviously $f(h(d)) = g(h(d))$ and, since $\rightarrow_{\mathscr{D}}$ is total, there is a path π in $\mathscr{D}$ such that $\pi(0) = d$: its image by h belongs to $\mathrm{Path}(\mathscr{A})_{f,g}$ and therefore $h(d) \in E$ and m is well-defined. It is clear that m is unique and that $h = e \circ m$. Finally, m is an AP-map: if $d \rightarrow_{\mathscr{D}} d'$ then $h(d) \rightarrow_{\mathscr{A}} h(d')$ and by definition of m and $\rightarrow_{\mathscr{E}}$ it is $m(d) \rightarrow_{\mathscr{E}} m(d')$; and if $p \in L_{\mathscr{E}}(m(d))$ then $p \in L_{\mathscr{A}}(h(d))$ and hence $p \in L_{\mathscr{D}}(d)$. □

It is easy to check that the same construction gives equalizers in the categories $\mathbf{KMap}^{\mathrm{str}}_{AP}$ and $\mathbf{KSMap}_{AP}$. As for $\mathbf{KSim}_{AP}$, we have not been able to prove or disprove the existence of equalizers.

Proposition 14. *For all sets of atomic propositions AP, the category* $\mathbf{KMap}_{AP}$ *has coequalizers.*

Proof. Assume that $f, g : \mathscr{A} \longrightarrow \mathscr{B}$ are AP-simulations, and define $\equiv$ to be the least equivalence relation over B containing $\{(f(a), g(a)) \mid a \in A\}$. Then the coequalizer of f and g is given by the quotient Kripke structure $\mathscr{B}/{\equiv}$ and the projection $c : \mathscr{B} \longrightarrow \mathscr{B}/{\equiv}$. For assume that $h : \mathscr{B} \longrightarrow \mathscr{D}$ is an AP-map such that $h \circ f = h \circ g$; we can define $m : \mathscr{B}/{\equiv} \longrightarrow \mathscr{D}$ by $m([b]) = h(b)$ with $h = m \circ c$. We have to check that m is well-defined and that it is an AP-map. The first part is proved by showing that if $b_1 \equiv b_2$ then $h(b_1) = h(b_2)$, by induction on the definition of $\equiv$. The base case corresponds to $f(a) \equiv g(a)$, and by hypothesis it is $h(f(a)) = h(g(a))$. And it is immediate that the result also holds for $b \equiv b$, for $b_2 \equiv b_1$ if it holds for $b_1 \equiv b_2$, and for $b_1 \equiv b_3$ if it holds for $b_1 \equiv b_2$ and for $b_2 \equiv b_3$. For the second part, if $[b_1] \rightarrow_{\mathscr{B}/\equiv} [b_2]$ it must be $b_1' \rightarrow_{\mathscr{B}} b_2'$ for some $b_1' \equiv b_1$ and $b_2' \equiv b_2$ and hence $h(b_1') \rightarrow_{\mathscr{D}} h(b_2')$. And if $p \in L_{\mathscr{D}}(m([b]))$ then $p \in L_{\mathscr{B}}(b')$ for all $b' \in [b]$ and therefore $p \in L_{\mathscr{B}/\equiv}([b])$. □

Again, this construction also applies to the category $\mathbf{KMap}^{\mathrm{str}}_{AP}$ but we do not know what happens in $\mathbf{KSim}_{AP}$ or $\mathbf{KSMap}_{AP}$.

7.3 Satisfaction for Products and Coproducts

At this point, it is interesting to ask ourselves whether there is any relation between the formulas satisfied by two Kripke structures $\mathscr{A}$ and $\mathscr{B}$ and those satisfied by their product $\mathscr{A} \times \mathscr{B}$, or more generally, whether there is any relation between the properties satisfied by a family of Kripke structures and those of their corresponding limits and colimits. Unfortunately, there is no general pattern.

Let us consider products. In one direction, the relation is immediate: there exist simulations from $\mathscr{A} \times \mathscr{B}$ to both $\mathscr{A}$ and $\mathscr{B}$ (the projections) and therefore any property that holds in any of the latter will also be true of $\mathscr{A} \times \mathscr{B}$. In the other direction, since products and coproducts coincide in $\mathbf{KSim}_{AP}$ there are also simulations from $\mathscr{A}$ and $\mathscr{B}$ to $\mathscr{A} \times \mathscr{B}$ (the inclusions) and thus properties of $\mathscr{A} \times \mathscr{B}$ can be transferred to both $\mathscr{A}$ and $\mathscr{B}$. This relation however does not hold when simulations are restricted to be maps. For example, in the category $\mathbf{KMap}_{AP}$ for $AP = \{p, q\}$, if $\mathscr{A} = (\{a\}, a \rightarrow_{\mathscr{A}}$

$a, L_{\mathscr{A}})$ with $L_{\mathscr{A}}(a) = \{p\}$, and $\mathscr{B} = (\{b\}, b \rightarrow_{\mathscr{B}} b, L_{\mathscr{B}})$ with $L_{\mathscr{B}}(b) = \{q\}$, we have $\mathscr{A} \times \mathscr{B}, (a,b) \models \mathbf{G}(p \wedge q)$ but $\mathscr{A}, a \not\models \mathbf{G}(p \wedge q)$ and $\mathscr{B}, b \not\models \mathbf{G}(p \wedge q)$.

The same reasoning applies in general and hence it follows that limits inherit the properties of their objects while these satisfy those of their colimits, but the converse implications do not always hold.

8 Factorizations in Categories of Simulations

First we characterize the classes of *AP*-simulations that correspond to the (regular) epimorphisms and monomorphisms.

Proposition 15. *A morphism in* $\mathbf{KMap}_{AP}$, $\mathbf{KMap}^{\mathrm{str}}_{AP}$, *or* $\mathbf{KSMap}_{AP}$ *is an epimorphism if and only if it is a surjective function.*

Proof. Assume that $f : \mathscr{A} \longrightarrow \mathscr{B}$ is surjective. Then, if $g \circ f = h \circ f$ it must be the case that $g = h$, and hence f is epi, because the range of f is B.

Conversely, assume now that f is an epimorphism. If f were not surjective there would be an element $b \in B$ not in the image of f. Define a Kripke structure $\mathscr{B}'$ which is like $\mathscr{B}$ but with b replaced by b_1 and b_2 with the same labeling as b and such that $b_i \rightarrow_{\mathscr{B}'} b'$ iff $b \rightarrow_{\mathscr{B}} b'$ and $b' \rightarrow_{\mathscr{B}'} b_i$ iff $b' \rightarrow_{\mathscr{B}} b$. Now, if $g : \mathscr{B} \longrightarrow \mathscr{B}'$ maps b to b_1 and the other elements to themselves, and $h : \mathscr{B} \longrightarrow \mathscr{B}'$ maps b_2 to b and is the identity elsewhere, we have that g and h are well-defined (strict/stuttering) *AP*-maps, $g \circ f = h \circ f$, but $g \neq h$: a contradiction with the assumption that f was an epimorphism. □

Proposition 16. *A morphism in* $\mathbf{KMap}_{AP}$, $\mathbf{KMap}^{\mathrm{str}}_{AP}$, *or* $\mathbf{KSMap}_{AP}$ *is a monomorphism if and only if it is injective over paths (which is weaker than just injectivity).*

Proof. Assume that $f : \mathscr{A} \longrightarrow \mathscr{B}$ is injective over paths, that is, that the function f from $\mathrm{Path}(\mathscr{A})$ to $\mathrm{Path}(\mathscr{B})$ defined by $f(\pi) = f \circ \pi$ is injective. Let $g, h : \mathscr{C} \longrightarrow \mathscr{A}$ be morphisms such that $f \circ g = f \circ h$. Let $c \in C$ and π be a path starting at c: then it is $f(g(\pi)) = f(h(\pi))$ from where it follows $g(\pi) = h(\pi)$ and therefore $g(c) = h(c)$.

Conversely, assume that f is mono but there are paths π and π' in $\mathscr{A}$ such that $f(\pi) = f(\pi')$ and $\pi \neq \pi'$. Let us define a Kripke structure $\mathscr{C}$ with a single path $c_1 \rightarrow_{\mathscr{C}} c_2 \rightarrow_{\mathscr{C}} \dots$ and with $L_{\mathscr{C}}(c_i) = L_{\mathscr{A}}(\pi(i)) \cup L_{\mathscr{A}}(\pi'(i))$. Then, if $g, h : \mathscr{C} \longrightarrow \mathscr{A}$ are defined by $f(c_i) = \pi(i)$ and $g(c_i) = \pi'(i)$, g and h are *AP*-maps by construction (and strict, if f is so), and it is $f \circ g = f \circ h$ and $g \neq h$: a contradiction. □

The characterization of regular monomorphisms is now immediate.

Proposition 17. *A morphism* $f : \mathscr{A} \longrightarrow \mathscr{B}$ *is a regular mono in* $\mathbf{KMap}_{AP}$, $\mathbf{KMap}^{\mathrm{str}}_{AP}$, *or* $\mathbf{KSMap}_{AP}$ *if and only if* f *is injective,* $L_{\mathscr{A}}(a) = L_{\mathscr{B}}(f(a))$ *for all* $a \in A$, *and* $a \rightarrow_{\mathscr{A}} a'$ *iff* $f(a) \rightarrow_{\mathscr{B}} f(a')$.

Proof. The implication from left to right follows from the construction in the proof of Proposition 13. In the other direction, let $\mathscr{C}$ be the Kripke structure obtained from $\mathscr{B}$ by splitting each state b into b_1 and b_2, with $L_{\mathscr{C}}(b_1) = L_{\mathscr{C}}(b_2) = L_{\mathscr{B}}(b)$, and with

$b_i \rightarrow_{\mathscr{C}} b'_j$ iff $b \rightarrow_{\mathscr{B}} b'$. Now, define $f, g : \mathscr{B} \longrightarrow \mathscr{B}'$ by $g(b) = b_1$ and $h(b) = b_1$ if $b \in f(A)$ and $h(b) = b_2$ otherwise: f and g so defined are (strict) AP-maps and, since $f(\mathscr{A})$ is a Kripke substructure of $\mathscr{B}$ and it is isomorphic to $\mathscr{A}$ due to the assumptions, it is easy to check that the result of the construction of the equalizer in Proposition 13 is (isomorphic to) f. □

There is also a rather more involved characterization of regular epis which is described in the next proposition.

Proposition 18. *A morphism $f : \mathscr{A} \longrightarrow \mathscr{B}$ is a regular epi in $\mathbf{KMap}_{AP}$ or $\mathbf{KMap}^{\mathrm{str}}_{AP}$ if and only if: (1) $f(a) = f(a')$ implies that there are paths π and π' such that $\pi(0) = a$, $\pi'(0) = a'$, and $f(\pi) = f(\pi')$; (2) if $b \rightarrow_{\mathscr{B}} b'$ there exist $a, a' \in A$ with $f(a) = b$, $f(a') = b'$, and $a \rightarrow_{\mathscr{A}} a'$; (3) for all b, $L_{\mathscr{B}}(b) = \bigcap_{f(a)=b} L_{\mathscr{A}}(a)$.*

Proof. The implication from left to right follows from Proposition 14; item (2) is proved by induction over $\equiv$. In the other direction, we define a Kripke structure $\mathscr{C}$ and two AP-maps $g, h : \mathscr{C} \longrightarrow \mathscr{A}$ as follows. For each pair of states a, a' such that $f(a) = f(a')$ let π and π' be paths as in (1). Now, add to $\mathscr{C}$ a fresh path ρ in such a way that $g(\rho(i)) = \pi(i)$ and $h(\rho(i)) = \pi'(i)$. Then, if we apply the construction in Proposition 14 that returns the coequalizer of g and h through a quotient Kripke structure $\mathscr{A}/{\equiv}$, the result is, by items (2) and (3), isomorphic to $f : \mathscr{A} \longrightarrow \mathscr{B}$. □

Proposition 19. $\mathbf{KMap}_{AP}$, $\mathbf{KMap}^{\mathrm{str}}_{AP}$, *and* $\mathbf{KSMap}_{AP}$ *are (epi, regular mono)-categories.*

Proof. Let $f : \mathscr{A} \longrightarrow \mathscr{B}$ be a morphism in any of these categories, and let us write $f(\mathscr{A})$ for $(f(A), \rightarrow_{\mathscr{B}}|_{f(A)}, L_{\mathscr{B}}|_{f(A)})$. Then, define $e : \mathscr{A} \longrightarrow f(\mathscr{A})$ to be like f and $m : f(\mathscr{A}) \longrightarrow \mathscr{B}$ to be the obvious inclusion: (e, m) is the unique (epi, regular mono)-factorization of f (up to isomorphism).

By Propositions 15 and 17, e and m are indeed epi and regular mono respectively. Now, assume that $e' : \mathscr{A} \longrightarrow \mathscr{C}, m' : \mathscr{C} \longrightarrow \mathscr{B}$ is another (epi, regular mono)-factorization of f. Define $g : f(\mathscr{A}) \longrightarrow \mathscr{C}$ by $g(b) = e'(a)$ where $e(a) = b$ (recall that e is surjective), and $h : \mathscr{C} \longrightarrow f(\mathscr{A})$ by $h(c) = e(a)$ where $e'(a) = c$. Let us check that they are well-defined. If $e(a) = e(a')$ then $m(e(a)) = m(e(a'))$ and therefore $m'(e'(a)) = m'(e'(a'))$; now, since m' is regular mono, $e'(a) = e'(a')$ and g is well-defined, and analogously for h. It is also clear that they are inverses of each other and that $g \circ e = e'$ and $m' \circ g$, so we are only left with checking that they are simulations; we present the arguments for g: those for h are symmetric. If $b \rightarrow_{f(\mathscr{A})} b'$, where $e(a) = b$ and $e(a') = b'$, then $m(e(a)) \rightarrow_{\mathscr{B}} m(e(a'))$ or, equivalently, $m'(e'(a)) \rightarrow_{\mathscr{B}} m'(e'(a'))$ and thus, since m' is a regular mono, $e'(a) \rightarrow_{\mathscr{B}} e'(a')$ and hence $g(a) \longrightarrow_{f(\mathscr{A})} g(a')$. In the case of stuttering simulations, a path π in $f(\mathscr{A})$ translates to a path $m(\pi)$ in $\mathscr{B}$ which is m'-matched by ρ in $\mathscr{C}$; this same ρ also h-matches π. Finally, $L_{f(\mathscr{A})}(b) = L_{\mathscr{B}}(e(a)) = L_{\mathscr{B}}(m(e(a))) = L_{\mathscr{B}}(m'(e'(a))) = L_{\mathscr{C}}(e'(a)) = L_{\mathscr{C}}(g(b))$. The first equality assumes that $b = e(a)$, and the second and the fourth one hold because m and m' are regular monos. □

Although we do not have a counterexample we believe that (regular epi, mono) factorizations do not exist in general. When they do, however, they are unique: the argument is similar to that for (epi, regular mono)-factorizations.

9 Conclusions

In previous papers [13,11] we have studied the suitability of different kinds of simulations between transition systems and Kripke structures for the study of the relationships between formal models of concurrent systems. The range of available notions of simulations makes it very natural to adopt a categorical viewpoint in which Kripke structures become the objects of several categories while the morphisms are obtained from the corresponding notion of simulation. In this paper we have defined in detail several of those categories and studied their most interesting properties: minimal Kripke structures as opcartesian morphisms, borrowing of properties as cartesian morphisms, temporal logic institutions, constructions of limits and colimits, and factorizations.

There are two main directions left open for future work. On the one hand, we would like to finally prove or disprove the existence of limits in the Grothendieck categories. On the other hand, as briefly discussed in [11], rewriting logic theories representing Kripke structures can also be organized in categories: we plan to organize them in an institution and to study its relationship with $\mathscr{I}_{\mathbf{K}}$.

Acknowledgments. We would like to thank the anonymous referees for very interesting and useful comments.

References

1. M. Arrais and J. L. Fiadeiro. Unifying theories in different institutions. In M. Haveraaen, O. Owe, and O.-J. Dahl, editors, *Recent Trends in Data Type Specification. 11th Workshop on Specification of Abstract Data Types*, volume 1130 of *LNCS*, pages 81–101. Springer, 1996.
2. M. Barr and C. Wells. *Category Theory for Computing Science. Third Edition.* Centre de Recherches Mathématiques, 1999.
3. M. C. Browne, E. M. Clarke, and O. Grümberg. Characterizing finite Kripke structures in propositional temporal logic. *Theoretical Computer Science*, 59:115–131, 1988.
4. E. M. Clarke, O. Grumberg, and D. E. Long. Model checking and abstraction. *ACM Transactions on Programming Languages and Systems*, 16(5):1512–1542, Sept. 1994.
5. E. M. Clarke, O. Grumberg, and D. A. Peled. *Model Checking*. MIT Press, 1999.
6. J. Goguen and R. Burstall. Institutions: Abstract model theory for specification and programming. *Journal of the Association for Computing Machinery*, 39(1):95–146, 1992.
7. H. Herrlich and G. E. Strecker. *Category Theory: An Introduction.* Advanced Mathematics. Allyn and Bacon, Boston, 1973.
8. B. Jacobs. *Categorical Logic and Type Theory*, volume 141 of *Studies in Logic and the Foundations of Mathematics*. North-Holland, 1999.
9. S. Mac Lane. *Categories for the Working Mathematician. Second Edition*. Springer, 1998.
10. P. Manolios. *Mechanical Verification of Reactive Systems*. PhD thesis, University of Texas at Austin, Aug. 2001.
11. N. Martí-Oliet, J. Meseguer, and M. Palomino. Theoroidal maps as algebraic simulations. In J. L. Fiadeiro and P. Mosses and F. Orejas, editors, *Recent Trends in Algebraic Development Techniques, WADT 2004*, volume 3423 of *LNCS*, pages 126–143. Springer, 2005.
12. N. Martí-Oliet, J. Meseguer, and M. Palomino. Algebraic simulations. Submitted. `http://maude.sip.ucm.es/~miguelpt/bibliography`, 2005.
13. J. Meseguer, N. Martí-Oliet, and M. Palomino. Equational abstractions. In F. Baader, editors, *Automated Deduction - CADE-19*, volume 2741 of *LNCS*, pages 2–16. Springer, 2003.

14. K. S. Namjoshi. A simple characterization of stuttering bisimulation. In S. Ramesh and G. Sivakumar, editors, *Foundations of Software Technology and Theoretical Computer Science, 17th Conference*, volume 1346 of *LNCS*, pages 284–296. Springer, 1997.
15. A. Tarlecki, R. M. Burstall, and J. A. Goguen. Some fundamental algebraic tools for the semantics of computation. Part 3: Indexed categories. *Theoretical Computer Science*, 91(2):239–264, 1991.

A Proofs of Some of the Results

Proposition 7. *$\mathscr{I}_{\mathbf{K}} = (\mathbf{Sign}_{\mathbf{K}}, sen_{\mathbf{K}}, \mathbf{Mod}_{\mathbf{K}}, \models)$ is an institution.*

Proof. It is a routine exercise to check that the purported functors are actually so. For example, let us check that $\mathbf{Mod}_{\mathbf{K}}$ is well-defined. Given $\alpha : AP \longrightarrow AP'$, $\mathbf{Mod}_{\mathbf{K}}(\alpha)$ is a functor. It is well-defined over objects and preserves identities and composition: we only need to check that $\mathbf{Mod}_{\mathbf{K}}(\alpha)(H) : \mathscr{A}|_\alpha \longrightarrow \mathscr{B}|_\alpha$ is an AP-simulation whenever $H : \mathscr{A} \longrightarrow \mathscr{B}$ is an AP'-simulation. Since the transition systems do not change, $\mathbf{Mod}_{\mathbf{K}}(\alpha)(H)$ preserves the transition relation. Now, if aHb and $p \in L_{\mathscr{B}|_\alpha}(b)$, then by definition we have $\mathscr{B}, b \models \alpha(p)$ and, by Theorem 1, this yields $\mathscr{A}, a \models \alpha(p)$, which again by definition implies that $p \in L_{\mathscr{A}|_\alpha}(a)$, as required. Thus, $\mathbf{Mod}_{\mathbf{K}}$ is well-defined over both objects and morphisms. It clearly preserves identities, so we are only left with showing that it preserves composition, for which it is enough to show that, given arrows $\alpha : AP \longrightarrow AP'$ and $\beta : AP' \longrightarrow AP''$, and a Kripke structure $\mathscr{A}$ over AP'', $\mathscr{A}_{\beta\circ\alpha} = (\mathscr{A}|_\beta)|_\alpha$. The equality at the level of transition systems is immediate. For the labeling function, $p \in L_{\mathscr{A}|_{\beta\circ\alpha}}(a)$ iff $\mathscr{A}, a \models \overline{\beta}(\alpha(p))$ (by definition) iff $\mathscr{A}|_\beta \models \alpha(p)$ (by Proposition 1) iff $p \in L_{(\mathscr{A}|_\beta)|_\alpha}(a)$ (by definition). Finally the property required of the satisfaction relation follows from Proposition 1. □

Proposition 8. *Let $\mathscr{I}'_{\mathbf{K}}$ be obtained from the institution $\mathscr{I}_{\mathbf{K}}$ by replacing $\mathbf{Set}_{\mathsf{State}\backslash\neg}$ by* **Set** *as the category of signatures. Then $\mathscr{I}'_{\mathbf{K}}$ is a semiexact institution.*

Proof. That $\mathscr{I}'_{\mathbf{K}}$ is an institution is immediate, and since the category of signatures is **Set** we know that it has pushouts. Therefore, we are left with checking that pushouts are transformed into pullbacks by the model functor. Consider then a pushout

$$\begin{array}{ccc} AP_0 & \xrightarrow{\quad\alpha_2\quad} & AP_2 \\ {\scriptstyle\alpha_1}\downarrow & & \downarrow{\scriptstyle\beta_2} \\ AP_1 & \xrightarrow{\quad\beta_1\quad} & AP_3 = (AP_1 \uplus AP_2)/\equiv \end{array}$$

where $\equiv$ is the least equivalence relation on $AP_1 \uplus AP_2$ verifying $\alpha_1(p) \equiv \alpha_2(p)$, and β_1 and β_2 take each element to its quotient class. To see that it is mapped to a pullback, let $F_1 : C \longrightarrow \mathbf{KSim}_{AP_1}$ and $F_2 : C \longrightarrow \mathbf{KSim}_{AP_2}$ be functors such that $\_|_{\alpha_1} \circ F_1 = \_|_{\alpha_2} \circ F_2$; we have to find a unique functor $F : C \longrightarrow \mathbf{KSim}_{AP_3}$ such that $\_|_{\beta_1} \circ F = F_1$ and $\_|_{\beta_2} \circ F = F_2$.

Let c be an object in C and $f : c \longrightarrow c'$ an arrow, with $F_1(c) = \mathscr{A}$ and $F_2(c) = \mathscr{B}$. It follows from the hypothesis that A is equal to B, $\rightarrow_{\mathscr{A}}$ equal to $\rightarrow_{\mathscr{B}}$, and $F_1(f)$ equal to $F_2(f)$. This leads us to define $F(c) = (A, \to_{\mathscr{A}}, L_{F(c)})$ and $F(f) = F_1(f)$, where we choose to define the labeling function as $L_{F(c)} = \beta_1(L_{\mathscr{A}}) \cup \beta_2(L_{\mathscr{B}})$. Since it is straightforward to check that $F(f)$ is an AP_3-simulation, F is well-defined, and it is a functor because F_1 (or F_2) is so.

We are left with checking that F satisfies the commutativity condition and proving that it is the only one that does it. For the first part, note that by the definition of the pushout it is not possible for any two p and p' in AP_1 to be such that $\beta_1(p) = \beta_1(p')$ and $p \in L_{\mathscr{A}}(a)$ but $p' \notin L_{\mathscr{A}}(a)$ (a detailed proof proceeds by induction on the definition of $\equiv$). We need to use this property to show that $F(c)|_{\beta_1} = F_1(c)$. We already know that their objects and transition relations are the same; as for the atomic predicates:

$$p \in L_{F(c)|_{\beta_1}}(a) \iff F(c), a \models \beta_1(p) \iff \beta_1(p) \in L_{F(c)}(a) \iff p \in L_{F_1(c)}(a),$$

where the property is required for the last implication to the right. The result for F_2 is symmetric. Uniqueness follows from the definitions of the functors and the previous equivalences. □

B Summary of Categories

The following table summarizes most of the categories introduced in this paper; for each of them, the third column contains the (co)limits for which explicit constructions have been given.

Categories	Objects	Arrows	(Co)limits
TSys	transition systems	simulations of transition systems	(Co)products
$\mathbf{KSim}_{AP}$	Kripke str. over AP	AP-simulations	(Co)products
$\mathbf{KMap}_{AP}$	Kripke str. over AP	AP-simulation maps	(Co)products, (co)equalizers
$\mathbf{KMap}^{\text{str}}_{AP}$	Kripke str. over AP	strict AP-simulations	(Co)products, (co)equalizers
$\mathbf{KSSim}_{AP}$	Kripke str. over AP	stuttering AP-simulations	?
$\mathbf{KSMap}_{AP}$	Kripke str. over AP	stuttering AP-simulation maps	Equalizers
KSim	arbitrary Kripke str.	simulations	?/colimits as above
KSSim	arbitrary Kripke str.	stuttering simulations	?/colimits as above

Behavioral Extensions of Institutions*

Andrei Popescu and Grigore Roşu

Department of Computer Science,
University of Illinois at Urbana-Champaign
{popescu2, grosu}@cs.uiuc.edu

Abstract. We show that any institution $\mathcal{I}$ satisfying some reasonable conditions can be transformed into another institution, $\mathcal{I}_{beh}$, which captures formally and abstractly the intuitions of adding support for behavioral equivalence and reasoning to an existing, particular algebraic framework. We call our transformation an "extension" because $\mathcal{I}_{beh}$ has the same sentences as $\mathcal{I}$ and because its entailment relation includes that of $\mathcal{I}$. Many properties of behavioral equivalence in concrete hidden logics follow as special cases of corresponding institutional results. As expected, the presented constructions and results can be instantiated to other logics satisfying our requirements as well, thus leading to novel behavioral logics, such as partial or infinitary ones, that have the desired properties.

1 Introduction

Many approaches to behavioral equivalence are defined as extensions of more standard algebraic frameworks, following relatively well understood methodologies. For example, hidden algebra is defined as an extension of algebraic specification: it adds appropriate machinery for experiments and then uses it to define behavioral equivalence as "indistinguishability under experiments", also known to be the largest behavioral congruence consistent with the visible data.

Here we explore this problem from an abstract model theoretical perspective. We investigate conditions under which an institution admits behavioral extensions. The intuition of a behavioral signature extending an algebraic signature is captured categorically in a general way covering all cases of operations in current use, including the ones that tend to be problematic: constants of hidden sorts and operations with multiple arguments of hidden sort. Let the original institution be $\mathcal{I} = (Sign, Sen, Mod, \models)$, let Ψ be a fixed signature in $Sign$ called the *visible signature*, and let D be a Ψ-model called the *data model*. Then we build the *behavioral extension of* $\mathcal{I}$ *over* (Ψ, D), say $\mathcal{I}_{beh} = (Sign_{beh}, Sen_{beh}, Mod_{beh}, \models)$, as follows. The objects in $Sign_{beh}$ are those in the comma category $\Psi/Sign$; the $(\varphi : \Psi \to \Sigma, \Sigma)$-sentences in $\mathcal{I}_{beh}$ are exactly the Σ-sentences in $\mathcal{I}$, while the $(\varphi : \Psi \to \Sigma, \Sigma)$-models in $\mathcal{I}_{beh}$ are the *data-consistent* Σ-models in $\mathcal{I}$; finally,

* Supported in part by joint NSF/NASA grant CCF-0234524, by NSF CAREER grant CCF-0448501, and by NSF grant CNS-0509321.

J.L. Fiadeiro et al. (Eds.): CALCO 2005, LNCS 3629, pp. 331–347, 2005.

satisfaction $A \models_{(\varphi,\Sigma)} \rho$ in $\mathcal{I}_{beh}$ is defined as $A_\varphi \models_\Sigma \rho$ in $\mathcal{I}$, for a carefully chosen model A_φ that symbolizes the "quotient" of A by its behavioral equivalence. An appropriate novel notion of *quotient system* is introduced for this purpose.

The abstract relationship between behavioral and normal satisfactions is studied via a model-theoretic notion of "visibility", and some structural properties preserved by the behavioral extension are pointed out. We show that many of the relevant properties of particular hidden logics can be proved at institutional level. The motivation for such a generalization is, as usual, its logic-independent status: a plethora of concrete algebraic logics formalizable as institutions satisfy our mild restrictions, so they all admit behavioral extensions.

Notice that from the way we define the concepts, we restrict ourselves to the *fixed-data* approach. An adaptation of our construction to the *loose-data* setting seems possible, and we shall sketch it in Section 7. Due to space limitations, proofs of our results are omitted, but they can all be found in [24].

Preliminaries. We assume the reader familiar with basic categorical notions: functor, colimit, etc. We use the terminology and notation from [23], with the following exceptions: we let ";" denote the morphisms' composition, which is considered in diagrammatic order; by colimit and limit we mean small colimit and small limit; by a *filtered (chain) colimit* we mean a colimit of a functor defined on a *non-empty* filtered (total respectively) ordered set. We use the following *comma category* notations: if $A \in |\mathcal{C}|$, $A/\mathcal{C}$ denotes the category whose objects are pairs (h, B), where $h : A \to B$ is a morphism in $\mathcal{C}$, and whose morphisms $u : (h, B) \to (g, C)$ are such that $u : B \to C$ is a morphism in $\mathcal{C}$ with $h; u = g$; there is a canonical forgetful functor U from $A/\mathcal{C}$ to $\mathcal{C}$, which maps each object (h, B) to B and each morphism $u : (h, B) \to (g, C)$ to $u : B \to C$; when $u : A \to A'$ is a morphism in $\mathcal{C}$, there is a canonical comma functor $u/\mathcal{C}$ between $A'/\mathcal{C}$ and $A/\mathcal{C}$, mapping each object (h, B) to $(u; h, B)$ and each morphism to itself; to each functor $F : \mathcal{C} \to \mathcal{D}$ and object A in $\mathcal{C}$, one can associate a functor between comma categories $F_A : A/\mathcal{C} \to F(A)/\mathcal{D}$, which maps each object (h, B) to $(F(h), F(B))$ and each morphism g to $F(g)$.

Since we need a special notion of quotient object, we define a parameterized notion of co-well-powered-ness: let $\mathcal{C}$ be a category and $\mathcal{E}$ be a class of morphisms in $\mathcal{C}$. $|\mathcal{C}|$ is said to be *$\mathcal{E}$-co-well-powered* if for each $A \in |\mathcal{C}|$ there is some *set* $\mathcal{D}$ of morphisms in $\mathcal{E}$ of source A, such that any morphism of source A in $\mathcal{E}$ is isomorphic in $A/\mathcal{C}$ to some morphism in $\mathcal{D}$. If $\mathcal{E}$ is taken to be the class of all epimorphisms, we get the usual notion of co-well-powered-ness. If $\mathcal{C}$ is a category, $\mathcal{C}^{op}$ denotes its dual. We let *Set* denote the category of sets and functions and *Cat* the category of categories and functors.

2 Institutions

In this section, we discuss several institutional concepts, many already known.

An *institution* [17] consists of: a category *Sign*, whose objects are called *signatures*; a functor $Sen : Sign \to Set$, giving for each signature Σ a set whose elements are called *Σ-sentences*; a functor $Mod : Sign \to Cat^{op}$ giving for each signature Σ a category whose objects are called *Σ-models* and whose arrows are called *Σ-morphisms*; a *Σ-satisfaction* relation $\models_\Sigma \subseteq |Mod(\Sigma)| \times Sen(\Sigma)$ for each $\Sigma \in |Sign|$, such that for each morphism $\varphi : \Sigma \to \Sigma'$ in *Sign*, the *satisfaction condition* "$M' \models_{\Sigma'} Sen(\varphi)(e)$ iff $Mod(\varphi)(M') \models_\Sigma e$" holds for all $M' \in |Mod(\Sigma')|$ and $e \in Sen(\Sigma)$. As usual, we may let $\_\restriction_\varphi$ denote the reduct functor $Mod(\varphi)$ and φ denote $Sen(\varphi)$. When $M = M'\restriction_\varphi$ we say that M' is a *φ-expansion of M* and M is the *φ-reduct of M'*.

The satisfaction relation is extended to sets of Σ-sentences and classes of Σ-models: if $E \subseteq Sen(\Sigma)$ and $\mathcal{M} \subseteq |Mod(\Sigma)|$, then we write $\mathcal{M} \models_\Sigma E$ whenever $M \models_\Sigma e$ for each $e \in E$ and $M \in \mathcal{M}$. We let E^* denote the class $\{M \mid M \models_\Sigma E\}$ and dually, $\mathcal{M}^*$ the set of Σ-sentences $\{e \mid \mathcal{M} \models_\Sigma e\}$. The two "$*$" operators form a Galois connection [17]; we let "$\bullet$" denote the two corresponding closure operators. The satisfaction relation is also extended to a (semantic) consequence relation, for which we use the same symbol, following classical logic tradition: if $E, E' \subseteq Sen(\Sigma)$, we write $E \models_\Sigma E'$ whenever $E^* \subseteq E'^*$. To simplify notation, we may write $\models$ instead of $\models_\Sigma$. A *presentation* [17] is a pair (Σ, E), where $E \subseteq Sen(\Sigma)$. A *theory* [17] is a presentation (Σ, E) with E with $E^\bullet = E$. A *presentation morphism* $\varphi : (\Sigma, E) \to (\Sigma', E')$ is a signature morphism $\varphi : \Sigma \to \Sigma'$ with $\varphi(E) \subseteq E'^\bullet$. A presentation morphism between theories is called a *theory morphism.* We let $Mod(\Sigma, E)$ denote the full sub-category of $Mod(\Sigma)$ having as objects all the Σ-models which satisfy E. An institution is *ω-exact* if *Mod* preserves colimits of functors defined on the ordered set of natural numbers.

A signature morphism $\varphi : \Sigma \to \Sigma'$ is *representable* [10] if there exists a Σ-model $T_{[\varphi]}$ (called the *representation* of φ) and an isomorphism of categories $I_\varphi : Mod(\Sigma') \to T_{[\varphi]}/Mod(\Sigma)$ such that $I_\varphi ; U = Mod(\varphi)$, where $U : T_{[\varphi]}/Mod(\Sigma) \to Mod(\Sigma)$ is the usual forgetful functor. Representable signature morphisms capture the idea of first-order variable. For instance, in the institution of first-order predicate logic with equality ($FOPL_=$; see Example 1.(1)), given a set of constant symbols X, the inclusion of $\Sigma = (S, F, P)$ into $\Sigma' = (S, F \cup X, P)$ is represented by $T_\Sigma(X)$, the term algebra over variables X and operations in F, with all the relations in P empty.

The sentences of an institution $\mathcal{I}$ can be naturally extended with first-order-like constructions [29]: if $\varphi : \Sigma \to \Sigma'$, $\rho, \delta \in Sen(\Sigma)$, $\rho' \in Sen(\Sigma')$, and $E \subseteq Sen(\Sigma)$, one can build the sentences $\bigwedge E$, $\bigvee E$, $\neg\rho$, $\delta \Rightarrow \rho$, $(\forall\varphi)\rho'$, $(\exists\varphi)\rho'$, with the following semantics, for each Σ-model M: $M \models \bigwedge E$ iff $M \models E$; $M \models \bigvee E$ iff $M \models e$ for some $e \in E$; $M \models \neg\rho$ iff $M \not\models \rho$; $M \models \delta \Rightarrow \rho$ iff $M \models \delta$ implies $M \models \rho$; $M \models (\forall\varphi)\rho'$ iff $M' \models \rho'$ for all φ-expansions M' of M; $M \models (\exists\varphi)\rho'$ iff there exists some φ-expansion M' of M such that $M' \models \rho'$. It might be the case that the newly constructed sentences are equivalent to some existing sentences in $\mathcal{I}$ - we take the convention that whenever we mention such a sentence, say $(\forall\varphi)\rho'$, we tacitly assume that it is equivalent to an existing one in $\mathcal{I}$ and we simply identify them, i.e., consider that $(\forall\varphi)\rho' \in Sen(\Sigma)$.

Given a signature Σ, a Σ-sentence ρ is called: *basic* [10] if there exits a Σ-model T_ρ such that for each Σ-model M, $M \models \rho$ iff there exists some morphism $T_\rho \to M$; *universal* if there exists a signature morphism $\varphi : \Sigma \to \Sigma'$ and a basic sentence $\rho' \in Sen(\Sigma')$ such that ρ is of the form $(\forall\varphi)\rho'$; *positive* if it is either basic or is obtained from basic sentences by a finite number of conjunctions ($\bigwedge E$), disjunctions ($\bigvee E$), universal quantification ($(\forall\varphi)\rho'$), and existential quantification ($(\exists\varphi)\rho'$). The notion of basic sentence is an institutional generalization for ground atom (equation, predicate etc.) - in our examples of institutions, the basic sentences are the primary bricks used to construct the more complicated sentences. For instance, in FOPL$_=$, the basic sentences are just finite conjunctions of ground term equalities $t_1 = t_2$ and/or of relational statements over ground terms $R(t_1, \ldots, t_n)$; in the institution of equational logic (EQL - see Example 1.(2)), the basic sentences are just ground term equalities. Universal sentences capture institutionally the universally quantified atoms. Universal sentences contain basic sentences: any basic sentence $\rho \in Sen(\Sigma)$ is equivalent to $(\forall 1_\Sigma)\rho$. The institution $\mathcal{I}$ is said to: *have basic Horn implications* iff for each signature Σ, each set of basic sentences $E \subseteq Sen(\Sigma)$, and each basic sentence $\rho \in Sen(\Sigma)$, the sentence $(\bigwedge E) \Rightarrow e$ is in $Sen(\Sigma)$; *have finitary basic Horn implications* if the above condition is satisfied for E finite.

A signature morphism $\varphi : \Sigma \to \Sigma'$ is called *liberal* [17] iff $Mod(\varphi)$ has a left adjoint. An institution is called *liberal* iff each of its signature morphisms is liberal. Let $\mathcal{I}$ be an institution, $\mathcal{U}$ be a $|Sign|$-indexed class of model morphisms closed under composition and images by reduct functors, and $\varphi : \Sigma \to \Sigma'$ be a morphism in $Sign$. We say that: φ *creates $\mathcal{U}$-morphisms* iff for any $A' \in |Mod(\Sigma')|$ and any $h : A'\!\upharpoonright_\varphi \to B$ in $\mathcal{U}_\Sigma$, there exists $f : A' \to B'$ in $\mathcal{U}_{\Sigma'}$ such that $f\!\upharpoonright_\varphi = h$; also, φ *weakly creates* $\mathcal{U}$-morphisms iff for any $A' \in |Mod(\Sigma')|$ and any $h : A'\!\upharpoonright_\varphi \to B$ in $\mathcal{U}_\Sigma$, there exist $g : B \to C$ in $\mathcal{U}_\Sigma$ and $f : A' \to B'$ in $\mathcal{U}_{\Sigma'}$ such that $f\!\upharpoonright_\varphi = h; g$. Morphism creation condition is used in [12] and [10] (under the name *lifting*) for institution-independent interpolation and ultraproducts results. We shall use weak creation at the bare definition of hidden signature morphisms.

Example 1. We briefly discuss two important institutions that will be used as working examples. Their detailed descriptions, as well as several other examples of institutions on which our results apply, are discussed in AppendixC of [24].

(1) FOPL$_=$ [17] - the institution of (many-sorted) first order predicate logic with equality. The signatures are triples (S, F, P), where S is a set of *sorts*, $F = \bigcup\{F_{w,s} | w \in S^*, s \in S\}$ is a set of (S-sorted) *operation symbols*, and $P = \bigcup\{P_w | w \in S^*\}$ is a set of (S-sorted) *relation symbols*. A signature morphism is a triple $\varphi = (\varphi^{sort}, \varphi^{op}, \varphi^{rel}) : (S, F, P) \to (S', F', P')$, where $\varphi^{sort} : S \to S'$, $\varphi^{op} : F \to F'$, and $\varphi^{rel} : P \to P'$ are mappings such that $\varphi^{op}(F_{w,s}) \subseteq F'_{\varphi^{sort}(w), \varphi^{sort}(s)}$ and $\varphi^{rel}(P_w) \subseteq P'_{\varphi^{sort}(w)}$ for each $w \in S^*$ and $s \in S$. (We may write φ instead of φ^{sort}, φ^{rel} and φ^{op}.) Given a signature $\Sigma = (S, F, P)$, a Σ-model is a triple $M = (\{M_s\}_{s\in S}, \{M_{w,s}(\sigma)\}_{(w,s)\in S^*\times S}, \{M_w(\sigma)\}_{w\in S^*})$ interpreting each sort as a set, each operation symbol as a function, and each relation symbol as a relation, with appropriate arities. (We may write M_σ and M_π instead of $M_{w,s}(\sigma)$ and $M_w(\pi)$.) The model morphisms are S-sorted functions which preserve operations and relations. The set of Σ-sentences and the satisfaction relation are the usual first-order ones. Each $Sen(\varphi)$ translates sentences symbol-wise, and $Mod(\varphi)$ is the usual forgetful functor.

(2) EQL, the institution of equational logic [17], is a restriction of $\mathrm{FOPL}_=$, with no relation symbols (its signatures are pairs (S, F)), and with only conditional equations $(\forall X)t_1 = t'_1 \wedge \ldots t_n = t'_n \Rightarrow t = t')$ as sentences.

3 Hidden Algebra Logic and Behavioral Satisfaction

Hidden algebra extends algebraic specification to handle states naturally, using behavioral equivalence. Systems need only satisfy their requirements behaviorally, in the sense of *appearing* to satisfy them under all possible experiments. Hidden algebra was introduced in [16] and developed further in [18,19,20,27] among many other places. CafeOBJ [14] and BOBJ [20], are executable specification languages that support behavioral specification and reasoning. One distinctive feature of hidden algebra logics is to split sorts into *visible* for data and *hidden* for states. A model, or *hidden algebra*, is an abstract implementation of a system, consisting of its possible states, with functions for operations. The restriction of a model to the visible subsignature is called *data*. *Hidden logics* refer to close relatives of hidden algebra, including both *fixed-data* and *loose-data* variants. This paper is concerned with the fixed-data approach. Hidden algebra is constructed on top of many-sorted algebra and equational logic - we shall use the notations of EQL (see Example 1).

Given a set V of *visible sorts*, a V-sorted signature Ψ called the *data signature*, and a Ψ-algebra D called the *data algebra*, then a *fixed-data hidden* (Ψ, D)-*signature* is a $(V \cup H)$-sorted signature Σ with $\Sigma\lceil_V = \Psi$, where H is a set disjoint from V of *hidden sorts*. Hereafter we write "hidden signature" instead of "fixed-data hidden (Ψ, D)-signature". The operations in Σ with one hidden argument and visible result are called *attributes*, those with one hidden argument and hidden result are called *methods*, those with two hidden arguments and hidden result are called *binary methods*, and so on; those with only visible arguments and hidden result are called *hidden constants*. Let $\Sigma = (S, F)$ be a hidden signature, where $S = V \cup H$. A *hidden Σ-algebra* is a Σ-algebra A with $A\lceil_\Psi = D$; it can be regarded as a universe of possible states of a system. A system can be seen as a "black-box," the inside of which is not seen, one being only concerned with its behavior under "experiments". A *hidden Σ-morphism* between two hidden Σ-algebras A and B is a usual Σ-homomorphism $h : A \to B$ such that $h\lceil_\Psi = 1_D$.

An *experiment* is an observation of a system after it has been perturbed; the $\bullet$ below is a placeholder for the state being experimented upon. A *context for sort* s is a term in $T_\Sigma(\{\bullet : s\} \cup Z)$ having exactly one occurrence of a special variable $\bullet$ of sort s, where Z is an S-indexed componentwise infinite set of special variables. Let $\mathbb{C}[\bullet : s]$ denote the S-indexed set of all contexts for sort s, and $var(c)$ the finite set of variables in a context c except $\bullet$. A context with visible result sort is called an *experiment*; let $\mathbb{E}[\bullet : s]$ denote the V-indexed set of all experiments for sort s. The interesting experiments are those for hidden sorts $s \in H$. We sometimes say that an experiment or a context for sort s is *appropriate* for terms or equations of sort s. Contexts can be "applied" as follows. If $c \in \mathbb{C}_{s'}[\bullet : s]$ and $t \in T_{\Sigma,s}(X)$, then $c[t]$ denotes the term in $T_{\Sigma,s'}(var(c) \cup X)$ obtained from c by substituting t for $\bullet$. Further, c generates a map $A_c\colon A_s \to [A^{var(c)} \to A_{s'}]$ on each Σ-algebra A, defined by $A_c(a)(\theta) = a^*_\theta(c)$, where a^*_θ is the unique extension of the map (denoted a_θ) that takes $\bullet$ to a and each $z \in var(c)$ to $\theta(z)$.

We recall the important notion of behavioral equivalence. Given a hidden Σ-algebra A, the equivalence $a \equiv_\Sigma a'$ iff $A_\gamma(a)(\theta) = A_\gamma(a')(\theta)$ for all experiments γ and all maps $\theta\colon var(\gamma) \to A$ is called *behavioral equivalence on A*. A *hidden congruence* is a congruence which is the identity on visible sorts. The following supports several important results in hidden logics. Since final models may not exist when operations of zero or more than one hidden argument are allowed, the existence of a largest hidden congruence does not depend on them.

Theorem 1. *Given a hidden Σ-algebra A, the behavioral equivalence is the largest hidden congruence on A* (see [26] for a proof).

Given a hidden Σ-algebra A and a Σ-equation $(\forall X)\ t = t'$, say ρ, then A *behaviorally satisfies* ρ, written $A \models\!\!\!\equiv_\Sigma \rho$, iff $\theta(t) \equiv_\Sigma \theta(t')$ for all $\theta\colon X \to A$. Let $\mathbb{E}[\rho]$ be either the set $\{(\forall X, var(\gamma))\ \gamma[t] = \gamma[t'] \mid \gamma \in \mathbb{E}[\bullet : h]\}$ when the sort h of t, t' is hidden, or the set $\{\rho\}$ when the sort of t, t' is visible. $\mathbb{E}[E]$ is the set $\bigcup_{e\in E} \mathbb{E}[\rho]$. Behavioral satisfaction of an equation can be reduced to strict satisfaction of a potentially infinite set of equations:

Proposition 1. *If A is a hidden Σ-algebra then $A \models\!\!\!\equiv_\Sigma E$ iff $A \models_\Sigma \mathbb{E}[E]$.*

Behavioral satisfaction is "reflected" by hidden morphisms [19]:

Proposition 2. *If $h : A \to B$ is a hidden Σ-morphism and ρ a Σ-equation, then $B \models\!\!\!\equiv \rho$ implies $A \models\!\!\!\equiv \rho$.*

The notion of morphism of hidden signatures [16] reflects at a *syntactic level* the object-oriented principles of data encapsulation. A morphism of (Ψ, D)-hidden signatures $\chi : (V \cup H, F) \to (V \cup H', F')$ of (Ψ, D)-hidden signatures is a many sorted signature morphism such that: (C1) χ is an identity on Ψ; (C2) $\chi^{sort}(H) \subseteq H'$; (C3) for each operation $\sigma' \in F'$ having an argument sort in $\chi^{sort}(H)$, it is the case that $\sigma' \in \chi^{op}(F)$. These conditions have natural interpretations in terms of information encapsulation: visible data remains unchanged (C1); hidden states are not unhidden by imports (C2); and no new methods or attributes are added on imported states (C3). Condition (C3), although has a rather restrictive character, is quite faithful to the principle of "behavior-protecting" inheritance mechanism. The above conditions ensure that behavioral equivalence and satisfaction are preserved by the reduct functor:

Proposition 3. *If $\chi\colon\Sigma\to\Sigma'$ is a hidden signature morphism with $\Sigma = (V \cup H, F)$ and A' is a hidden Σ'-algebra, then: (1) for all $h \in H$ and $a, b \in A'_{\chi^{sort}(h)}$, $a \equiv_{\Sigma'} b$ iff $a \equiv_\Sigma b$; (2) $(A'\restriction_\chi)/_{\equiv_\Sigma} = (A'/_{\equiv_{\Sigma'}})\restriction_\chi$; (3) $A' \models\!\!\!\equiv \chi(\rho)$ iff $A'\restriction_\chi \models\!\!\!\equiv \rho$, for each Σ-equation ρ.*

4 Quotient Systems

Image factorization systems [1] are a categorical generalization of the system of injections and surjections from set theory. Unlike bare monics and epics, the morphisms of a factorization system work together to provide, up to an isomorphism, a unique factorization for each morphism. *Inclusion systems* [15] and *weak inclusion systems* [8], modifications of factorization systems by dropping the "up to an isomorphism" relaxation, turn out to be more suitable for the categorical study of algebraic specification concepts. In this paper, because of the coalgebraic nature of the involved notions, we introduce a variant of a factorization system that is dual to the weak inclusion system:

Definition 1. *A* ***quotient system*** *for a category $\mathcal{C}$ is a pair $(\mathcal{E}, \mathcal{M})$, where $\mathcal{E}$ and $\mathcal{M}$ are subcategories of $\mathcal{C}$ such that: (1) $\mathcal{E}$ is a* partial order, *in the sense that $\mathcal{E}(A, B)$ contains at most one morphism for any $A, B \in |\mathcal{C}|$, and $A = B$ whenever $\mathcal{E}(A, B) \neq \emptyset$ and $\mathcal{E}(B, A) \neq \emptyset$; (2) Morphisms in $\mathcal{C}$ can be factored uniquely as $e; m$, with $e \in \mathcal{E}$, $m \in \mathcal{M}$. The elements of $\mathcal{E}$ are called* ***quotients*** *and those of $\mathcal{M}$* ***injections****. B is called a* ***quotient object*** *of A when $\mathcal{E}(A, B) \neq \emptyset$.*

Note that $(\mathcal{E}, \mathcal{M})$ is a quotient system for $\mathcal{C}$ iff $(\mathcal{M}, \mathcal{E})$ is a weak inclusion system for $\mathcal{C}^{op}$. Thus, w.r.t. category theory, quotient systems bring nothing essentially new. However, they model properly the important notion of *congruence*, which is not to be considered, like in the case of factorization systems, *up to an isomorphism*, but chosen in a *unique*, canonical way. This will have important semantical and technical consequences when we define behavioral satisfaction: first, we can model faithfully in an institutional framework the process of constructing the behavioral equivalence, originally defined in an *internal* fashion within the set-theoretical structure of the algebras (see Section 3); second, by regarding models as *universes for congruences*, we do not need to postulate the existence of final objects; finally, delicate technical issues regarding lifting and preserving properties can be elegantly treated using quotient systems.

The category of sets, as well as that of algebras, have natural quotient systems if we allow a slight and non-problematic *foundational modification:* we assume that all elements in the considered sets or carriers are sets themselves and in addition they are mutually disjoint. That anything is a set is a harmless principle of the Zermelo-Fraenkel Set Theory,[1] but note that we only take this assumption about algebras (models), and not about sentences. Moreover, any algebra can be isomorphically and uniformly transformed into one satisfying the above condition by simply replacing its elements x with singletons $\{x\}$. Now, we can take $\mathcal{M}$ as the category of all injective morphisms and $\mathcal{E}$ as that of those surjective morphisms $f : A \to B$ such that, for each element $b \in B$, the elements $a \in A$ with $f(a) = b$ form a partition of b. Therefore, $\mathcal{E}$ provides canonical ways to factor algebras by refining their carrier sets, viewed as partitions, in a dual manner to inclusions that give a canonical way to embed an algebra into another. We next list some properties of quotient systems, some of them dual to ones for weak inclusion systems [8]. Let $(\mathcal{E}, \mathcal{M})$ be a quotient system for $\mathcal{C}$.

[1] This set-theoretical assumption that we take should be regarded as a meta-level setting, having nothing to do with the duality algebra-coalgebra. In particular, it does not imply that we are planning to treat the coalgebraic phenomena with algebraic methods; at least not to a greater extent than any other "mathematical" approach.

Proposition 4. (see Fact 5 in [8]) *(1) Any $e \in \mathcal{E}$ in an epic; (2) $\mathcal{M}$ contains all the isomorphisms in $\mathcal{C}$; and (3) all isomorphisms in $\mathcal{E}$ are identities.*

Proposition 5. (see also Corollary 26 in [8]) *If $e, e' \in \mathcal{E}$ of same source admit pushout in $\mathcal{C}$, then they have a* unique *pushout whose morphisms are in $\mathcal{E}$. If $(I,\leq)$ is a filtered set and $c = (e_{i,j} : A_i \to A_j)_{i,j \in I, i \leq j}$ an I-diagram in $\mathcal{E}$ admitting a colimit in $\mathcal{C}$, then there is a unique colimit of c in $\mathcal{C}$ whose morphisms are in $\mathcal{E}$. In particular, if $\mathcal{C}$ is {pushout and filtered}-cocomplete, then so is $\mathcal{E}$.*

Example 2. For each signature (S, F) in EQL, $\mathcal{E}_{(S,F)}$ consists of all surjective morphisms $h : A \to B$ such that $b = \bigcup_{a \in A, h_s(a)=b} a$ for each sort $s \in S$ and $b \in B_s$, and $\mathcal{M}_{(S,F)}$ consists of all injective morphisms. In the case of $\mathrm{FOPL}_=$, we can consider two canonical ways to provide quotient systems, following the idea of inclusion systems for $\mathrm{FOPL}_=$ [13]. Let (S, F, P) be a signature. An (S, F, P)-morphism $f : A \to B$ is called *strong* if, for each (n-ary) relation symbol $R \in P$ and each $(a_1, \ldots, a_n)$, it holds that $(a_1, \ldots, a_n) \in A_R$ iff $(f(a_1), \ldots, f(a_n)) \in B_R$. (1) The quotients are morphisms $h : A \to B$ such that h is a (S, F)-quotient in EQL; the injections are the strong injective morphisms; (2) The quotients are morphisms $h : A \to B$ such that h is a strong (S, F)-quotient in EQL; the injections are the injective morphisms.

All the institutions that use some form of set-theoretical notion of model tend to have quotient systems on models, although the choice is not always unique.

5 The Behavioral Extension of an Institution

Next we provide an institutional generalization of fixed-data hidden logic.

Definition 2. *An* ***institution with quotients*** *is an institution equipped with quotient systems $(\mathcal{E}_\Sigma, \mathcal{M}_\Sigma)$ on each category of models $Mod(\Sigma)$, such that all reducts $Mod(\varphi)$ along signature morphisms $\varphi : \Sigma \to \Sigma'$ preserve quotients and injections. (That is, for each e in $\mathcal{E}_{\Sigma'}$ and m in $\mathcal{M}_{\Sigma'}$, it holds that $e{\upharpoonright_\varphi}$ is in $\mathcal{E}_\Sigma$ and $m{\upharpoonright_\varphi}$ is in $\mathcal{M}_\Sigma$.) An institution with quotients is* **co-well-powered** *if each $Mod(\Sigma)$ is $\mathcal{E}_\Sigma$-co-well-powered.*

Notice that the notion of $\mathcal{E}_\Sigma$-co-well-powered-ness becomes particularly simple thanks to Proposition 4.(3): one only asks that, for each $A \in |Mod(\Sigma)|$, the class of morphisms in $\mathcal{E}_\Sigma$ of source A is a set. All throughout this section, we shall work inside the following framework:

> **Framework 1:** A co-well-powered institution with quotients $\mathcal{I}$, having filtered colimits and pushouts of models, such that all reducts $Mod(\varphi)$ along signature morphisms $\varphi : \Sigma \to \Sigma'$ preserve filtered colimits and pushouts of quotient diagrams (i.e., diagrams consisting of morphisms in $\mathcal{E}$).

Our examples of institutions with quotients all satisfy the above conditions. While these institutions have not only filtered colimits and pushouts, but also arbitrary colimits on models, the arbitrary colimits are usually not preserved by reduct functors. The only property that needs explanation is the preservation of pushouts of quotients. In EQL, this follows from the fact that the supremum of two congruences of a model does not depend on the signature where the supremum is taken - see Appendix D of [24]. As for the case of the two possible families of quotient systems in $FOPL_=$, the quotient preservation property follows from the equational case, using the fact that the forgetful functor $Mod(S, F, P) \to Mod(S, F, \emptyset)$ creates colimits (and pushouts in particular).

Let Ψ be a fixed signature of $\mathcal{I} = (Sign, Mod, Sen, \models)$, that we call the *visible signature*, and D be a fixed Ψ-model, that we call the *data model*. We define an institution $\mathcal{I}_{beh}(\Psi, D)$, the *behavioral extension of* $\mathcal{I}$ *over* (Ψ, D). We let $\mathcal{I}_{beh} = (Sign_{beh}, Mod_{beh}, Sen_{beh}, \models)$ denote $\mathcal{I}_{beh}(\Psi, D)$ without forgetting though that our construction is parameterized by Ψ and D.

Signatures. The signatures of $\mathcal{I}_{beh}$ are pairs $(\varphi : \Psi \to \Sigma, \Sigma)$, where Σ is a signature in $\mathcal{I}$. (Instead of the entire class of objects of $\Psi/Sign$, one could also consider, without adding any technical difficulties, only a subclass, like the class of inclusions [20].) We postpone the definition of signature morphisms.

Sentences. For a signature (φ, Σ) in $\mathcal{I}_{beh}$, let $Sen_{beh}(\varphi, \Sigma)$ be precisely $Sen(\Sigma)$. However, the sentences will get in $\mathcal{I}_{beh}$ a different meaning than in $\mathcal{I}$.

Models. For a signature (φ, Σ) in $\mathcal{I}_{beh}$, let $Mod_{beh}(\varphi, \Sigma)$ be the *fiber category* [2] $D\lceil_\varphi^{-1}$ of the functor $\_\lceil_\varphi : Mod(\Sigma) \to Mod(\Psi)$ over D: its objects are those $A \in |Mod(\Sigma)|$ with $A\lceil_\varphi = D$ and its morphisms are those $h : A \to B$ in $Mod(\Sigma)$ with $h\lceil_\varphi = 1_D$. Interestingly, this fiber category captures precisely the intuition of hidden algebra: models protect data and morphisms are data-consistent.

We are next going to define behavioral satisfaction (in $\mathcal{I}_{beh}$) as satisfaction in $\mathcal{I}$ on smallest data-consistent quotient objects. We first need to introduce some notation and show that such objects indeed exist.

Definition 3. *For a signature (φ, Σ) and a (φ, Σ)-model A in $\mathcal{I}_{beh}$, let $A/_D\mathcal{E}_\Sigma$ be the category of* ***data-consistent quotients*** *of A: its objects are morphisms $e : A \to B$ in $\mathcal{E}_\Sigma$ with $e\lceil_\varphi = 1_D$ and its morphisms $h : (e : A \to B) \to (e' : A \to B')$ are morphisms $h : B \to B'$ with $h\lceil_D = 1_D$ and $e; h = e'$.*

It follows from the above definition that all the mentioned morphisms $h : B \to B'$ are actually in $\mathcal{E}_\Sigma$ (one can see that by decomposing h as $e_h; i_h$ and using the unique factorization property for $e; e_h; i_h = e'$). Moreover, the category $A/_D\mathcal{E}_\Sigma$ is isomorphic to the full subcategory of $\mathcal{E}_\Sigma$ having the class of objects restricted to quotient objects of A.

Proposition 6. *The category $A/_D\mathcal{E}_\Sigma$ has a unique final object, $e_{A,\varphi} : A \to A_\varphi$.*

The morphism $e_{A,\varphi}$ can be intuitively regarded as the "largest congruence on A that is data-consistent", or the "behavioral equivalence" on A. Note that the construction of A_φ follows a final approach, without assuming the existence of *globally final models* - rather, we get a final model, i.e., a greatest congruence, starting from any given model. This allows our formalization to capture non-coalgebraic variants of hidden algebra at no additional cost.

Satisfaction relation. We can now define satisfaction in $\mathcal{I}_{beh}$, called *behavioral satisfaction* and written $\models$, as follows: for a signature (φ, Σ), a (φ, Σ)-model A and a (φ, Σ)-sentence ρ, let $A \models_{(\varphi,\Sigma)} \rho$ in $\mathcal{I}_{beh}$ iff $A_\varphi \models_\Sigma \rho$ in $\mathcal{I}$.

The only thing left to define in $\mathcal{I}_{beh}$ is the morphism of signatures. As discussed in Section 3, this is a delicate concept to define even in the concrete framework of hidden algebra, because it needs to imply the property that its semantic counterpart, the reduct, preserves behavioral equivalences on models. Whether the morphisms in $Sign_{beh}$ can be defined categorically in some "syntactic" way capturing the conditions (C1), (C2), (C3) from Section 3 seems to be a difficult problem and perhaps not worthwhile the effort. Our approach, instead, is to define morphisms of signatures by capturing precisely the above crucial property.

Proposition 7. *Let $\varphi : \Psi \to \Sigma$, $\varphi' : \Psi \to \Sigma'$ and $\chi : \Sigma \to \Sigma'$ be three signature morphisms in $\mathcal{I}$ such that $\varphi; \chi = \varphi'$. Then the following are equivalent: (a) χ weakly creates data-consistent quotients; and (b) for each Σ'-model A' with $A'\lceil_\varphi = D$, it is the case that $(e_{A',\varphi'})\lceil_\chi = e_{(A'\lceil_\chi),\varphi}$.*

Signature morphisms. The morphisms $\chi : (\varphi, \Sigma) \to (\varphi', \Sigma')$ in $Sign_{beh}$ are now defined to be morphisms $\chi : \Sigma \to \Sigma'$ in $Sign$ such that $\varphi; \chi = \varphi'$ and the equivalent conditions in Proposition 7 hold. It is not hard to see that $Sign_{beh}$ is now a (broad) subcategory of $\Psi/Sign$. Sen_{beh} and Mod_{beh} can be defined on signature morphisms $\chi : (\varphi, \Sigma) \to (\varphi', \Sigma')$ as expected, that is, exactly as the functors Sen and Mod are defined on $\chi : \Sigma \to \Sigma'$, but using the appropriate restricted classes of models and model morphisms.

Condition (b) in Proposition 7 provides the motivation for the definition of signature morphisms: one wants the "behavioral equivalence", i.e. the largest hidden quotient, to be preserved by reduct functors - this is in fact the main reason for the conditions (C2) and (C3) in the definition of hidden signature morphisms (see Section 3). As for condition (a), one can use the following intuition for the weak creation property stated there. Let $\chi : \Sigma \to \Sigma'$ be a morphism in $\Psi/Sign$. Also, let $A \in Mod_{beh}(\varphi, \Sigma)$ and $A' \in Mod_{beh}(\varphi', \Sigma')$ such that $A = A'\lceil_\chi$. The existence of a quotient $e : A \to B$ with $e\lceil_\varphi = 1_D$ means that the hidden structure of A can be flattened in a behaviorally consistent way, i.e., not affecting the data. This situation should not depend on notation, so one should be able to alternatively perform this flattening on A'. Yet, because of the larger number of expressible entities in Σ', here consistent flattening might cause more effects-hence the "weak" nature of creation.

Theorem 2. *$\mathcal{I}_{beh}$ is an institution with quotients, where, for each $(\varphi, \Sigma) \in |Sign|$, $\mathcal{E}_{(\varphi,\Sigma)}$ and $\mathcal{M}_{(\varphi,\Sigma)}$ are the restrictions of $\mathcal{E}_\Sigma$ and $\mathcal{M}_\Sigma$ to $Mod_{beh}(\Sigma, \varphi)$, respectively. Moreover, there exists a canonical morphism of institutions (in the sense of [17]) between $\mathcal{I}_{beh}$ and $\mathcal{I}$, projecting each $\mathcal{I}_{beh}$ signature (φ, Σ) into Σ, not changing the sentences, and mapping each (φ, Σ)-model A to A_φ.*

The institution $\mathcal{I}_{beh}$ above generalizes the institutions of variants of fixed-data hidden algebra [16,20,26], constructed in a similar fashion on top of many-sorted equational logic. Theorem 2 tells us that similar behavioral extensions of many other logics are possible, in for particular those in Appendix C of [24], including partial and infinitary ones. A first important property of behavioral satisfaction is that entailment in $\mathcal{I}$ is "sound" in $\mathcal{I}_{beh}$. The next proposition generalizes former results on "behavioral soundness of equational deduction" [27], with syntactic proofs in the concrete hidden algebraic framework.

Proposition 8. *If* $(\varphi, \Sigma) \in |Sign_{beh}|$, $\rho \in Sen(\Sigma)$ *and* $E \subseteq Sen(\Sigma)$, *then* $E \models_{\Sigma} \rho$ *implies* $E \models\!\!\!\equiv_{(\varphi,\Sigma)} \rho$.

The following proposition generalizes another standard result in hidden algebra, namely that behavioral satisfaction coincides with usual satisfaction on sentences over the visible syntax.

Proposition 9. *Let* $(\varphi, \Sigma) \in |Sign_{beh}|$, $\rho \in Sen_{\mathcal{I}}(\Psi)$ *and* $A \in |Mod_{beh}(\varphi, \Sigma)|$. *Then* $A \models\!\!\!\equiv_{(\varphi,\Sigma)} \varphi(\rho)$ *iff* $A \models_{\Sigma} \varphi(\rho)$ *iff* $D \models_{\Psi} \rho$.

In hidden algebra, "visibility" does not concern only sentences over the visible signature. The sentences of visible sort need not contain only data constructs; indeed, sentences of visible sort may involve several attributes and methods. There is no notion of "visible sort" in our abstract framework. However, we can still define an institutional generalization of "sentences of visible sorts", that we call "visible sentences", by model-theoretic means; the visible sentences will be those preserved back and forth by data-consistent flattening, following the intuition that these sentences should sense only modifications in the visible part of a system. We also introduce "quasi-visible sentence", for which the preservation property holds only backwards. But let us set some terminology first:

Definition 4. *Let* $(\varphi, \Sigma) \in |Sign_{beh}|$, $\rho \in Sen(\Sigma)$, *and* $\mathcal{K}$ *a subcategory of* $Mod_{beh}(\varphi, \Sigma)$. *Then* ρ *is* ***closed (behaviorally closed) under*** $\mathcal{K}$ *if, for each* $A \to B$ *in* $\mathcal{K}$, $A \models \rho$ *implies* $B \models \rho$ ($A \models\!\!\!\equiv \rho$ *implies* $B \models\!\!\!\equiv \rho$, *respectively).*

Definition 5. *Let* (φ, Σ) *be a signature in* $\mathcal{I}_{beh}$. *Then* $\rho \in Sen_{beh}(\varphi, \Sigma)$ *is* ***φ-visible*** *if it is closed under both* $\mathcal{E}_{(\Sigma,\varphi)}$ *and* $\mathcal{E}^{op}_{(\Sigma,\varphi)}$ *and* ***φ-quasi-visible*** *if it is closed under* $\mathcal{E}^{op}_{(\Sigma,\varphi)}$. *If the signature* φ *is clear, we shall say "visible" ("quasi-visible") instead of "φ-visible" ("φ-quasi-visible").*

Proposition 10. *Let* $(\varphi, \Sigma) \in |Sign_{beh}|$ *and* $\rho \in Sen_{beh}(\varphi, \Sigma)$. *Then: (1)* ρ *is visible iff, for each* $A \in |Mod_{beh}(\varphi, \Sigma)|$, *[*$A \models \rho$ *iff* $A \models\!\!\!\equiv \rho$*]; (2) if* ρ *is quasi-visible then, for each* $A \in |Mod_{beh}(\varphi, \Sigma)|$, *[*$A \models\!\!\!\equiv \rho$ *implies* $A \models \rho$*]; (3) if* ρ *is closed under* $\mathcal{M}^{op}_{(\varphi,\Sigma)}$ *and under* $\mathcal{E}_{(\varphi,\Sigma)}$, *then it is behaviorally closed under* $Mod_{beh}(\varphi, \Sigma)^{op}$.

Thus, according to Proposition 10, the visible sentences are precisely those for which behavioral satisfaction coincides with usual satisfaction. On the other hand, the quasi-visible sentences have the property that, in order to satisfy them behaviorally, one has to satisfy them strictly. Moreover, (3) in Proposition 10 is the abstract version of the hidden algebraic result (Proposition 2) saying that equational behavioral satisfaction is preserved by reflexions of arbitrary hidden morphisms. (Recall that in the usual algebraic settings, equations are closed under arbitrary quotients and reflexions of embedding.)

Proposition 11. *Visible and quasi-visible sentences are preserved by signature morphisms and closed under conjunctions, disjunctions, universal and existential quantifications. In addition, visible sentences are also closed under negation.*

An immediate consequence of the above proposition is that both visible and quasi-visible sentences provide subinstitutions of $\mathcal{I}_{beh}$. Also, in the case of positive sentences (a very wide class, containing the basic and the universal sentences), the notions of visibility and quasi-visibility coincide:

Corollary 1. *Let (φ, Σ) be a signature in $\mathcal{I}_{beh}$ and ρ be a positive Σ-sentence in $\mathcal{I}$. Then ρ is φ-visible iff it is φ-quasi-visible.*

The next proposition deals with some structural properties inherited from $\mathcal{I}$ to $\mathcal{I}_{beh}$: filtered colimits of models and signatures. The former are usually important for Birkhoff-like axiomatizability results, while the latter, which also bring filtered colimits of theories [17], can be used for approximating finite refinements towards a fixed point. The comma nature of the signatures in $\mathcal{I}_{beh}$ "invite" us to construct filtered colimits, starting from those of $\mathcal{I}$.

Proposition 12. *(1) If (φ, Σ) is a signature in $\mathcal{I}_{beh}$ such that φ creates isomorphisms in $\mathcal{I}$, then $Mod_{beh}(\varphi, \Sigma)$ has filtered colimits; (2) If $\mathcal{I}$ has countable filtered colimits of signatures and is ω-exact, then $\mathcal{I}_{beh}$ also has countable filtered colimits of signatures.*

In the case of many-sorted algebraic signatures, the signature morphisms that create model isomorphisms are precisely those that are injective on sorts. In particular, Proposition 12.(1) holds for the case, usually considered for hidden algebra, of φ being an inclusion.

6 Behavioral Satisfaction of Universal Sentences

We next focus our study on basic and universal sentences. As already mentioned, these are institutional generalizations of ground equations and arbitrary equations, respectively. Some important properties of hidden logics depend on the equational character of these special sentences.

Before we define our next framework, let us first recall that, in $\mathrm{FOPL}_=$ or EQL, if ρ is some ground Σ-equation, then T_ρ is the quotient by ρ of the ground Σ-term model; then because of the special way to construct direct sums in these logics, it follows that for any Σ-model A, the direct sum $A \amalg T_\rho$ is actually isomorphic to A "factored" by ρ, i.e., the least restrictive "flattening" of A that satisfies ρ (this property is actually institution-independent). Following this intuition, from here on we assume:

> **Framework 2:** An institution $\mathcal{I}$ satisfying Framework 1, such that for any Σ, any $A \in |Mod(\Sigma)|$, and any basic $\rho \in Sen(\Sigma)$, the coproduct $(\amalg_A : A \to A \amalg T_\rho, \amalg_{T_\rho} : T_\rho \to A \amalg T_\rho)$ exists and can be taken such that $\amalg_A \in \mathcal{E}_\Sigma$. Then $A \amalg T_\rho$ is unique with this property and we denote it $A/_\rho$.

The following says that behavioral satisfaction of basic sentences can be equivalently regarded as data-consistent factoring:

Proposition 13. *If* (φ, Σ) *is a signature,* A *is a* (φ, Σ)*-model in* $\mathcal{I}_{beh}$*, and* ρ *is a basic* Σ*-sentence (in* $\mathcal{I}$*), then* $A \models\!\!\!\equiv \rho$ *iff* $(\amalg_A)\restriction_\varphi = 1_D$.

In what follows, we shall place the discussion in the context of elementary diagrams. Diagrams are a main concept in classical model theory [7]. The diagram of a model M consists of a set of sentences in its parameterized language which describe its structure well enough in order to axiomatize the class of morphisms of source M. A first institutional definition of diagrams was given in [29]. We shall make use of a more recent definition in [11], which has the advantage that asks the morphisms between models and signatures to yield smooth translations of the diagram sentences. An institution $\mathcal{I} = (Sign, Sen, Mod, \models)$ is said to have *elementary diagrams* [11] if: (1) for each signature Σ and each Σ-model M there exists a signature morphism $\iota_\Sigma(M) : \Sigma \to \Sigma_M$ (called the *elementary extension of* Σ *via* M) and a set E_M of Σ_M-sentences (called *the elementary diagram* of the model M) such that $Mod(\Sigma_M, E_M)$ and $M/Mod(\Sigma)$ are isomorphic by an isomorphism $i_{\Sigma,M}$ such that $i_{\Sigma,M}; U = Mod(\iota_\Sigma(M))^r$, where $U : M/Mod(\Sigma) \to Mod(\Sigma)$ is the usual forgetful functor from the comma category and $Mod(\iota_\Sigma(M))^r : Mod(\Sigma_M, E_M) \to Mod(\Sigma)$ is the restriction of $Mod(\iota_\Sigma(M)) : Mod(\Sigma_M) \to Mod(\Sigma)$; (2) ι is *functorial*, i.e., for each signature morphism $\varphi : \Sigma \to \Sigma'$, each $M \in |Mod(\Sigma)|$, $M' \in |Mod(\Sigma')|$ and $h : M \to M'\restriction_\varphi$, there exists a presentation morphism $\iota_\varphi(h) : (\Sigma_M, E_M) \to (\Sigma'_{M'}, E_{M'})$ such that $\iota_\Sigma(M); \iota_\varphi(h) = \varphi; \iota_{\Sigma'}(M')$; (3) i is *natural*, i.e., for each signature morphism $\varphi : \Sigma \to \Sigma'$, each $M \in |Mod(\Sigma)|$, $M' \in |Mod(\Sigma')|$ and $h : M \to M'\restriction_\varphi$ in $Mod(\Sigma)$, $i_{\Sigma',M'}; Mod(\varphi)_{M'}; (h/Mod(\varphi)) = Mod(\iota_\varphi(h))^{rcr}; i_{\Sigma,M}$, where $h/Mod(\varphi) : M/Mod(\Sigma) \to (M'\restriction_\varphi)/Mod(\Sigma')$ and $Mod(\varphi)_{M'} : (M'\restriction_\varphi)/Mod(\Sigma') \to M'/Mod(\Sigma')$ are the usual functors between comma categories (see the end of Section 1), and $Mod(\iota_\varphi(h))^{rcr} : Mod(\Sigma_M, E_M) \to Mod(\Sigma'_{M'}, E_{M'})$ is the restriction and corestriction of $Mod(\iota_\varphi(h)) : Mod(\Sigma_M) \to Mod(\Sigma'_{M'})$.

For each $h : A \to B$ in $Mod(\Sigma)$, we shall write $\iota_\Sigma(h)$ instead of $\iota_{1_\Sigma}(h)$.

An important result in hidden algebra is that behavioral satisfaction of unconditional equational sentences can be reduced to usual satisfaction *in the same model* of a set of visible sentences (see Proposition 1). We shall provide an institutional version of this result. For this, we further assume that the institution $\mathcal{I}$ is liberal and either has basic Horn implications, or {is compact and has finitary basic Horn implications}. Regarding the elementary diagrams, we assume that they are: *basic*, in the sense that, for each signature Σ and Σ-model A, each $\rho \in E_A$ is basic and $(E_A)^\bullet \cap Basic(\Sigma) = (A_A)^* \cap Basic(\Sigma)$;[2] *D-representable*, i.e., $\iota_\Sigma(D)$ is representable; *basic-sensitive*, i.e., for each signature Σ, Σ-model A and basic Σ-sentence ρ, $\iota_\Sigma(i_A)^{-1}((E_{A\amalg T_\rho})^\bullet) = (E_A \cup \iota_\Sigma(A)(\rho))^\bullet$ (thus, if a model is factored by a basic sentence, its diagram gains precisely that sentence); *quotient-sensitive*, i.e., for each Σ-quotient $e : A \to B$, if $A \neq B$, there exists a

[2] $Basic(\Sigma)$ denotes the set of basic Σ-sentences.

basic Σ_A-sentence α such that $A_A \not\models \alpha$ and $B_e \models \alpha$ (so the fact that B is smaller than A by a quotient is expressible in the language of A as a simple sentence).

For each $(\varphi, \Sigma) \in |Sen_{beh}|$ and $\rho \in Sen_{beh}(\varphi, \Sigma)$, define $\mathcal{QV}_\rho = \{(\forall\phi)\alpha \mid \phi$ signature morphism of source Σ, α quasi-visible sentence, $\rho \models (\forall\phi)\alpha\}$.

Proposition 14. *Let $(\varphi, \Sigma) \in |Sen_{beh}|$, let ρ be a universal Σ-sentence, and let $A \in |Mod_{beh}(\varphi, \Sigma)|$. Then $A \models_{(\varphi,\Sigma)} \rho$ iff $A \models_\Sigma \mathcal{QV}_\rho$.*

Our two working examples of institutions, as well as the others listed in Appendix C in [24], satisfy the hypotheses from our Frameworks 1 and 2, as well as those needed for Proposition 14. Let us take $\text{FOPL}_=$ for instance. The only properties which might not be clear (like the existence of basic Horn implications) or well-known (like liberality or semi-exact-ness), are some of those regarding diagrams: $(E_A)^\bullet \cap Basic(\Sigma) = (A_A)^* \cap Basic(\Sigma)$ simply because the first-order entailment system extends conservatively the ground equational entailment system; each $\iota_\Sigma(A)$ is representable: it only adds some constants to the source signature; basic-sensitivity asks that, if A is a model factored by a ground equation or atomic relation ρ becoming $A/_\rho$, all that one can infer from E_{A_ρ}, can be equivalently inferred from E_A together with ρ, which is obviously true; quotient-sensitivity is fulfilled as follows: if B is a quotient object of A (by $h : A \to B$), different from A, then there exists a sort s and $a, b \in A_s$ such that $a \neq b$ and $h_s(a) \neq h_s(b)$ - then $a = b$ is the desired sentence α from E_A.

In the case of EQL, it happens that the quasi-visible sentences α can be taken to be basic, hence visible (since "quasi-visible" plus "basic" implies "visible"), so the concrete equational result actually says more than we were able to prove at our institutional level. Yet, it is not clear that a similar neater result as the equational one holds for our other examples of institutions (like $\text{FOPL}_=$). Another question would be whether Proposition 14 holds for other types of sentences besides universal ones - one could easily find examples of conditional equations and existentially quantified sentences for which the property of reducing behavioral satisfaction to normal satisfaction in the same model does not hold; thus the class of universal sentences of an institution might be close to maximality w.r.t. this property, if one wants to cover the classical relevant cases. Note that universal sentences cover the cases when second-order quantification, i.e., over relation and function symbols, are considered (see also [22] for a higher-order result related to our Proposition 14).

7 Related Work and Concluding Remarks

The paper [25] was, at our knowledge, the first to introduce the notion of behavioral, or observational equivalence as we interpret it in this paper, and [28] was the first to sketch a treatment of observational equivalence in arbitrary institutions, where it is defined as existential elementary equivalence w.r.t. some signature morphism. Then [6] considered the notions of hiding and behavior in institutions; since this paper was an important source of inspiration for us, we shall discuss it below. The framework there was inspired by the following situation from "monadic" hidden algebra: the hidden models can be seen as *behavior algebras*, some forms of Lawvere-like algebras, equipped with a distinguished terminal object, having a fixed interpretation; moreover, the category of behavior algebras has a final object constructed using the sets of all possible behaviors of the (hidden) states; hence, thanks to a smooth back and forth communication between the categories of hidden algebras and behavioral algebras, a final semantics can be given for behavioral satisfaction of a sentence by a hidden model. This situation is generalized in [6] to the institutional level, where the notion of behavior algebra is provided as an extra data: a functor from a subcategory, of *hidden signatures*, to Cat^{op}, for which the relevant properties (finality, communication to the hidden models, etc.) are postulated. Our approach shares with [6] the idea of defining behavioral satisfaction as (normal) satisfaction inside a quotient. However, our approach is not tributary to the monadic framework, which only considers hidden operations with *precisely* one hidden argument, framework which loses two important cases: that of hidden constants (in particular, that of different cases of classical automata used in formal languages), and that of operations having multiple hidden-sort arguments; also we do not use data provided "from outside" the institution (as is the case of abstract behavior algebras in [6]), but construct the behavioral extension only by *internal* means of the considered institution. A quasi-abstract treatment of behavioral equivalence can also be found in [5], where a setting similar to the institutional one is used, but localized to a fixed *satisfaction frame*; the behavioral satisfaction (in one of the proposed variants) is also defined as usual satisfaction in a quotient, but in order for the quotient to enjoy good set-theoretical properties, a concrete many-sorted "carrier" set is considered attached to each model, through a *concretization functor*. Another paper in the vicinity of our work, but more concerned with *hiding* than with *behavior*, is [21], discussing compositional operations on modules that can hide some of the information.

We believe that our results can be adapted to also cover loose-data behavioral approach, such as *observational logic* [3,4]. The main point towards such an adaptation is that the loose-data setting is still based on a notion of behavioral equivalence, called *observational equality* in [3,4], hence it can still be formalized by our final construction in a fiber category. The main difference is that loose-data behavioral logics allows arrows between algebras that do not have the same data reduct. However, roughly speaking, if we express the concepts in [4] using our notations, we find that the arrows between two (φ, Σ)-models A and B are the usual morphisms between their quotients A_φ and B_φ, quotients which can be constructed independently, taking the data model D to be first $A\restriction_\varphi$ and then $B\restriction_\varphi$. One can show that this construction yields yet another institution, which takes only the data signature Ψ as a parameter this time. The latter institution could be seen as a form of Grothendieck construction (in the style of [9]) obtained by flattening the "indexed" institution $\{\mathcal{I}_{beh}(\Psi, D)\}_{D\in|Mod(\Psi)|}$.

Acknowledgments. We warmly thank the assigned reviewers for their very detailed and meaningful reports.

References

1. J. Adamek, H. Herrlich, and G. Strecker. *Abstract and Concrete Categories.* John Wiley & Sons, 1990.
2. J. Benabou. Fibred categories and the foundations of naive category theory. *Journal of Symbolic Logic*, 50:10–37, 1985.
3. M. Bidoit and R. Hennicker. On the integration of observability and reachability concepts. In *FOSSACS'02*, volume 2303 of *LNCS*, pages 21–36, 2002.
4. M. Bidoit, R. Hennicker, and A. Kurz. Observational logic, constructor-based logic, and their duality. *Theoretical Computer Science*, 3(298):471–510, 2003.
5. M. Bidoit and A. Tarlecki. Behavioural satisfaction and equivalence in concrete model categories. In *Trees in Algebra and Programming (CAAP'96)*, volume 1059 of *LNCS*, pages 241–256, 1996.
6. R. Burstall and R. Diaconescu. Hiding and behaviour: an institutional approach. In *A Classical Mind: Essays in Honour of C.A.R. Hoare*, pages 75–92. Prentice Hall, 1994.
7. C.C.Chang and H.J.Keisler. *Model Theory.* North Holland, Amsterdam, 1973.
8. V. E. Căzănescu and G. Roşu. Weak inclusion systems. *Mathematical Structures in Computer Science*, 7(2):195–206, 1997.
9. R. Diaconescu. Grothendieck institutions. *Applied Categorical Structures*, 10(4):383–402, 2002.
10. R. Diaconescu. Institution-independent ultraproducts. *Fundamenta Informaticae*, 55(3-4):321–348, 2003.
11. R. Diaconescu. Elementary diagrams in institutions. *Logic and Computation*, 14(5):651–674, 2004.
12. R. Diaconescu. An institution-independent proof of Craig interpolation theorem. *Studia Logica*, 77:59–79, 2004.
13. R. Diaconescu. *Institution-independent Model Theory.* To appear. Book draft. (Ask author for current draft at `Razvan.Diaconescu@imar.ro`).
14. R. Diaconescu and K. Futatsugi. *CafeOBJ Report.* World Scientific, 1998. AMAST Series in Computing, volume 6.
15. R. Diaconescu, J. Goguen, and P. Stefaneas. Logical support for modularization. In *Logical Environments*, pages 83–130. Cambridge, 1993.
16. J. Goguen. Types as theories. In *Topology and Category Theory in Computer Science*, pages 357–390. Oxford, 1991.
17. J. Goguen and R. Burstall. Institutions: Abstract model theory for specification and programming. *Journal of the ACM*, 39(1):95–146, January 1992.
18. J. Goguen and R. Diaconescu. Towards an algebraic semantics for the object paradigm. In *Proceedings of WADT*, volume 785 of *LNCS*. Springer, 1994.
19. J. Goguen and G. Malcolm. A hidden agenda. *J. of TCS*, 245(1):55–101, 2000.
20. J. Goguen and G. Roşu. Hiding more of hidden algebra. In *Proceeding of FM'99*, volume 1709 of *LNCS*, pages 1704–1719. Springer, 1999.
21. J. Goguen and G. Roşu. Composing hidden information modules over inclusive institutions. In *From Object Orientation to Formal Methods: Dedicated to the memory of Ole-Johan Dahl*, volume 2635 of *LNCS*, pages 96–123. Springer, 2004.
22. M. Hofmann and D. Sanella. On behavioral abstraction and behavioral satisfaction in higher-order logic. *Theoretical Computer Science*, pages 167:3–45, 1996.
23. S. M. Lane. *Categories for the Working Mathematician.* Springer, 1971.
24. A. Popescu and G. Roşu. Behavioral extensions of institutions. Technical Report UIUCDCS-R-2005-2582 and UILU-ENG-2005-1778, Department of Computer Science, University of Illinois at Champaign-Urbana, May 2005.

25. H. Reichel. Behavioural equivalence – a unifying concept for initial and final specifications. In *Proceedings of the 3rd Hungarian Computer Science Conference*. Akademiai Kiado, 1981.
26. G. Roşu. *Hidden Logic*. PhD thesis, University of California at San Diego, 2000.
27. G. Roşu and J. Goguen. Hidden congruent deduction. In *Automated Deduction in Classical and Non-Classical Logics*, volume 1761 of *LNAI*. Springer, 2000.
28. D. Sannella and A. Tarlecki. On observational equivalence and algebraic specification. *Journal of Computer and System Science*, 34:150–178, 1987.
29. A. Tarlecki. Bits and pieces of the theory of institutions. In *Proceedings, Summer Workshop on Category Theory and Computer Programming*, volume 240 of *LNCS*, pages 334–360. Springer, 1986.

Discrete Lawvere Theories

John Power*

Laboratory for the Foundations of Computer Science, University of Edinburgh,
King's Buildings, Edinburgh EH9 3JZ, Scotland
Tel: +44 131 650 5159, Fax: +44 131 667 7209
ajp@inf.ed.ac.uk

Abstract. We introduce the notion of discrete countable Lawvere V-theory and study constructions that may be made on it. The notion of discrete countable Lawvere V-theory extends that of ordinary countable Lawvere theory by allowing the homsets of an ordinary countable Lawvere theory to become homobjects of a well-behaved axiomatically defined category such as that of ω-cpo's. Every discrete countable Lawvere V-theory induces a V-enriched monad, equivalently a strong monad, on V. We show that discrete countable Lawvere V-theories allow us to model all the leading examples of computational effects other than continuations, and that they are closed under constructions of sum, tensor and distributive tensor, which are the fundamental ways in which one combines such effects. We also show that discrete countable Lawvere V-theories are closed under taking an image, allowing one to treat observation as a mathematical primitive in modelling effects.

1 Introduction

Lawvere theories are a category-theoretic formulation of universal algebra for which the notion of operation is primitive. Unlike universal algebra, the notion of Lawvere theory is presentation-independent, i.e., the category of models determines the theory uniquely up to coherent isomorphism. The concept has proved to be particularly fruitful, generalising to the study of finite limit theories and beyond [2,3]. It corresponds to the study of finitary monads on the category *Set*, but its definition is more in the spirit of universal algebra, the notion of Lawvere theory being essentially an axiomatisation of the notion of a clone of an equational theory.

Denotational semantics is less concerned with sets with structure than it is with ω-cpo's with structure, as the latter allow for an account of partiality and recursion. So one would like to extend the notion of, and results about, Lawvere theories to include ω-cpo's. Many results, several of them explained herein, can be extended elegantly and axiomatically by reference to enriched Lawvere theories [25] with enrichment in the category $V = \omega Cpo$: to allow for recursion, one needs to replace finitariness assumptions by countability assumptions, but that amounts to a minor technical adjustment.

* This work has been done with the support of EPSRC grant GR/586372/01.

J.L. Fiadeiro et al. (Eds.): CALCO 2005, LNCS 3629, pp. 348–363, 2005.

As we shall explain in Section 3, if V is locally countably presentable as a symmetric monoidal closed category, e.g., V is ωCpo or Set or $Poset$ or Cat, a countable Lawvere V-theory is a small V-category L with countable cotensors, which we shall define, together with an identity-on-objects strict countable cotensor preserving V-functor from $V_{\aleph_1}^{op}$ to L, where $V_{\aleph_1}$ is a skeleton of the full sub-V-category of V determined by the countably presentable objects of V, to L. A model of L in a V-category C with countable cotensors is a countable cotensor preserving V-functor from L to C. Extending the result for ordinary Lawvere theories, countable Lawvere V-theories correspond to V-monads on V with countable rank.

Taking V to be Set in the definition of countable Lawvere V-theory, we recover a countable version of the usual notion of Lawvere theory, together with its associated body of theory. Taking V to be Cat, enrichment yields an elegant body of theory for categories with equational structure such as finite or countable product structure, finite or countable limit structure, finite or countable coproduct structure, and various forms of monoidal structure. It is similarly fruitful where V is $Poset$ or ωCpo or any number of other naturally arising base categories. In this paper, which is oriented towards computer science, we focus on the example of V being ωCpo.

Here, we refine the notion of countable Lawvere V-theory to something we call a *discrete* countable Lawvere V-theory: L is still a small V-category, but it comes equipped with an identity-on-objects functor from a skeleton of $\aleph_1^{op}$ rather than from $V_{\aleph_1}^{op}$, so an object of L is either a natural number or is $\aleph_0$ rather than being an arbitrary countably presentable object of V such as Sierpinski space when V is $Poset$. Again, taking V to be Set, we recover the notion of countable Lawvere V-theory. In contrast, when V is Cat, the restriction from V-theories to discrete countable Lawvere V-theories is substantial, cutting out examples such as those involving limits and colimits. But when V is ωCpo, discrete countable Lawvere V-theories still include the bulk of the structures that appear in practice, in particular all those of primary interest in analysing computational effects [10,11].

The reason for the restriction is that the category of discrete countable Lawvere V-theories is closed under constructions that are important in the analysis of computational effects but that one cannot make of arbitrary countable Lawvere V-theories. In particular, one can make a distributive tensor of discrete countable Lawvere V-theories, which one cannot do of arbitrary countable Lawvere V-theories. And discrete countable Lawvere V-theories, unlike arbitrary countable Lawvere V-theories, are closed under the construction of an image determined by a model. The distributive tensor appears in both concurrency [7] and in combining probabilistic and ordinary nondeterminism [20]; and one wants the image in order to take operations and observations as primitive notions in analysing computational effects [26]. The notion of discrete countable Lawvere V-theory is also closed under the two operations on countable Lawvere V-theories that have been investigated to date, namely the sum and the tensor [10,11]. There is a sense in which the correspondence between Lawvere theories and monads extends to one between discrete Lawvere theories and a kind of monads, but that occurs more by fiat than by a natural condition on the definition of monad [16].

Space precludes us from generalising these ideas from discreteness. But the definition of discrete Lawvere theory enriches without fuss: one can start with categories V and W that are locally countably presentable as symmetric monoidal closed categories and a symmetric monoidal adjunction $U : V \longrightarrow W$ of countable rank, then systematically replace the implicit use of Set in the definition by W. That does not allow for the distributive tensor or for the image, but it may shed light on the subtle relationship between the categories ωCpo and $Poset$ for example. In particular, the various freeness results relating ordinary Lawvere theories, discrete countable Lawvere V-theories, and countable Lawvere V-theories all extend without fuss.

The paper is organised as follows. In Section 2, we recall the definitions and some leading examples of countable Lawvere V-theories. In Section 3, we give our definition of discrete countable Lawvere V-theory together with a preliminary analysis of it. In Sections 4, 5, 6, and 7 respectively, we analyse sum, tensor, distributive tensor, and image. We use Kelly's book [14] as the source book for all definitions and notation for enriched categories, with [15] being the basic text for locally presentable V-categories.

2 Enriched Lawvere Theories

In this section, we recall the notions of Lawvere theory and enriched Lawvere theory and at least one strand of thought that motivates their use in computer science. The work in this section is adapted from [10,11], which in turn was motivated by the desire for a more profound formulation of Moggi's unification of computational effects as monads in [21,22].

Definition 1. *A* Lawvere theory *consists of a small category L with finite products together with a strict finite product preserving identity-on-objects functor $I : Nat^{op} \longrightarrow L$, where Nat is the category of all natural numbers and maps between them [2,3]. A* model *of a Lawvere theory L in a category C with finite products is a finite-product preserving functor from L to C.*

Implicit in the definition is the fact that the objects of L are exactly the natural numbers. The definition provides a category theoretic formulation of universal algebra, with the notion of operation taken as primitive: a map in L from n to m is to be understood as being given by m operations of arity n. Unlike the notion of equational theory, the concept of Lawvere theory is presentation-independent, i.e., if a pair of Lawvere theories have equivalent categories of models, the two theories are isomorphic.

The definition of model extends to the definition of the category $Mod(L, C)$ of models of L in any category C with finite products: maps of models are defined to be natural transformations. Note that naturality forces maps of models to respect the product structure in the definition of model. Observe also that we do not demand strict preservation in the definition of model: to do so would eliminate many of the leading examples [24]!

For any Lawvere theory L and any locally finitely presentable category C, the functor $ev_1 : Mod(L, C) \longrightarrow C$ has a left adjoint, inducing a monad T_L on C: we shall return to this later in the section.

The usual way in which one obtains Lawvere theories is by means of sketches, with the Lawvere theory given freely on the sketch: Barr and Wells' book [3] treats sketches in loving detail and gives a range of examples of both sketches and Lawvere theories. To give a sketch amounts to giving operations and universally defined equations, i.e., an equational theory. The Lawvere theory is an axiomatisation of the notion of the clone of an equational theory, equivalently of a sketch.

Example 1. The Lawvere theory L_E for exceptions is the free Lawvere theory generated by an E-indexed family of nullary operations with no equations. The monad on Set induced by L_E is $T_E = - + E$. More generally, if C is any category with finite powers and sums then $Mod(L_E, C)$ is equivalent to the category of algebras for the monad $- + \underline{E}$ where $\underline{E}$ is the E-fold copower of 1, i.e., $\coprod_E 1$.

Interactive input/output works similarly to exceptions [11], so we omit details. For the next example, we use the evident generalisation of the notion of Lawvere theory to countable Lawvere theory as used in [10,11] and as we shall make precise shortly: it allows us to use countable arities.

Example 2. The countable Lawvere theory L_S for side-effects, where $S = V^L$, is the free countable Lawvere theory generated by the operations $lookup : V \longrightarrow L$ and $update : 1 \longrightarrow L \times V$ subject to the seven natural equations listed in [23], four of them specifying interaction equations for *lookup* and *update* and three of them specifying commutation equations. Our presentation of the operations here is in terms of generic effects, corresponding to the evident functions of the form $L \longrightarrow (S \times V)^S$ and $L \times V \longrightarrow S^S$ respectively [11]. It is shown in [23] that L_S induces the side-effects monad. More generally, if C is any category with countable powers and copowers then, slightly generalising the result in [23], $Mod(L_S, C)$ is equivalent to the category of algebras for the monad $(S \times -)^S$ where we write $(S \times -)$ for the S-fold copower $\coprod_S -$, and $(-)^S$ for the S-fold power $\prod_S -$.

Example 3. The Lawvere theory L_N for (binary) nondeterminism is the Lawvere theory freely generated by a binary operation $\vee : 2 \longrightarrow 1$ subject to equations for associativity, commutativity and idempotence, i.e., the Lawvere theory for a semilattice. The induced monad on Set is the finite non-empty subset monad, $\mathcal{F}^+$.

Example 4. The Lawvere theory L_P for probabilistic nondeterminism is that freely generated by $[0, 1]$-many binary operations $+_r : 2 \longrightarrow 1$ subject to the equations for associativity, commutativity and idempotence in [6]. The induced monad on Set is the distributions with finite support monad, $\mathcal{D}_f$.

In denotational semantics, one is not primarily interested in sets but rather in ω-cpo's, as the latter may be used to account for recursion. We therefore seek to generalise the study of Lawvere theories from sets to ω-cpo's. This may be done elegantly and axiomatically by recourse to the notion of a countable *enriched* Lawvere theory and considering the example of enrichment in the category of ω-cpo's [10,11,25]. Countability is essential to account for recursion, as seen for instance in studying side-effects.

Axiomatically, the details are as follows. We first assume our base category V is locally countably presentable as a symmetric monoidal closed category [1,14,15]. For the purposes of this paper, we do not require a definition of that: we simply need to know that the categories *Set*, *Poset*, ω*Cpo* and *Cat* are all examples. In all those examples, the relevant symmetric monoidal closed structure is, in fact, cartesian closed structure. So we shall assume that when convenient.

The least obvious point to note when enriching Lawvere theories is that the notion of countable product of a single generator does not generalise most naturally to a notion of countable product but rather to a notion of countable *cotensor* [14]. The notion of cotensor is the most natural enrichment of the notion of a power-object. Given an object x of a category C and given a set A, the A-fold power x^A satisfies the defining condition that there is a bijection of sets

$$C(y, x^A) \cong C(y, x)^A$$

natural in y. This enriches to the notion of cotensor as follows.

Definition 2. *Given an object x of a V-category C and given an object a of V, the* cotensor x^a *satisfies the defining condition that there is an isomorphism in V*

$$C(y, x^a) \cong C(y, x)^a$$

V-natural in y. The cotensor x^a is called countable *if a is a countably presentable object of V.*

There is an evident dual notion of *tensor* $a \otimes x$. When $C = V$, x^a is the exponential and $a \otimes x$ agrees with the monoidal structure of V.

Example 5. Taking V to be *Poset*, the notion of cotensor allows us not only to consider objects such as $x \times x$ $(= x^2)$ in a locally ordered category, but also to consider objects such as $x^{\leq}$, where $\leq$ is Sierpinski space, the two point partial order $\bot \leq \top$. This possibility allows us, in describing *Poset*-theories, not only to retain countable products but also to consider a greater range of arities and, in particular, to incorporate inequations. For the latter, suppose one wishes to say that $f \leq g$ for two morphisms $f, g : x \to y$; this is accomplished by introducing a third morphism $h : x \to y^{\leq}$ and asserting the equations $f = y^{\bot} \circ h$ and $g = y^{\top} \circ h$, where $\bot$ and $\top$ are the two evident maps from 1 to $\leq$. Observe that $y^{\leq}$ only appears here as a codomain of an operation, not as a domain of one. *Poset*-enriched Lawvere theories are at the heart of Ghani and Lüth's work on term rewriting systems in [5].

Define $V_{\aleph_1}$ to be a skeleton of the full sub-V-category of V determined by the countably presentable objects of V. It is equivalent to the free V-category with countable tensors on 1 [14,25].

Definition 3. *A* countable Lawvere V-theory *is a small V-category L with countable cotensors together with a strict countable-cotensor preserving identity-on-objects V-functor $I : V_{\aleph_1}^{op} \longrightarrow L$. A map of countable Lawvere V-theories from L to L' is a strict countable-cotensor preserving V-functor from L to L' that commutes with I and I'. A* model *of L in a V-category C with countable cotensors is a countable-cotensor preserving V-functor $M : L \longrightarrow C$.*

Routinely generalising the unenriched case, for any countable Lawvere V-theory L and any V-category with countable cotensors C, we have a V-category of models of L in C, $Mod(L, C)$; the homobjects are given by homobjects of all V-natural transformations [14], and the V-naturality condition implies they respect countable cotensors. There is a canonical forgetful V-functor U_L from $Mod(L, C)$ to C. If it has a left V-adjoint, as it does whenever C is a locally presentable V-category, this forgetful V-functor exhibits $Mod(L, C)$ as equivalent to the V-category T_L-*Alg* for the induced V-monad T_L on C. We denote the category of countable Lawvere V-theories by Law_V.

To give a V-enriched V-monad is equivalent to giving a strong monad on V [19]. So, in order to make the comparison with Moggi's unified account of computational effects as modelled by strong monads a little more direct, we express the main abstract result of [25] in terms of strong monads.

Theorem 1. *If V is locally countably presentable as a symmetric monoidal closed category, the construction of T_L from L induces an equivalence of categories between the category of countable Lawvere V-theories on V and the category of strong monads on V with countable rank. Moreover, the comparison V-functor exhibits an equivalence between the V-categories $Mod(L, V)$ and T_L-Alg.*

A common and important way to generate countable Lawvere V-theories is by taking the free countable Lawvere V-theory on an unenriched countable Lawvere theory. For instance, let V be ωCpo. Given an unenriched countable Lawvere theory L, the free countable Lawvere ωCpo-theory L_ω on L is generated by the operations and equations of L, but it has more objects as there are countably presentable ω-cpo's other than flat ones, and these additional objects generate additional maps. See [11] for details, but suffice it for here to note that Examples 1, 2, 3 and 4 all thereby freely yield countable Lawvere V-theories where $V = \omega Cpo$.

For the leading example of a countable Lawvere V-theory that does not arise freely from an unenriched countable Lawvere theory, let V be ωCpo and consider nontermination.

Example 6. The countable Lawvere ωCpo-theory L_Ω for nontermination is the theory freely generated by a nullary operation $\Omega : 0 \longrightarrow 1$ subject to the condition that there is an inequality

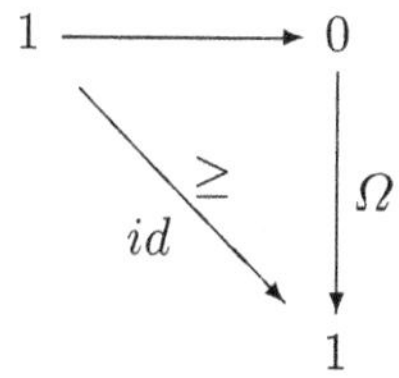

where the unlabelled map is the unique map determined because 0 is the initial object of $V_{\aleph_1}$ and therefore the terminal object of $V_{\aleph_1}^{op}$. The models of L_Ω in ωCpo are the ω-cpo's with a least element. The corresponding strong monad T_Ω is the lifting construction $(-)_\perp$ which adds a new least element. Observe that there is at most one morphism from L_Ω to any other countable ωCpo-theory L, reflecting the fact that a least element is unique if it exists.

Adding a nontermination effect allows us to model recursion in the context of ωCpo. If we want to model other effects in addition to recursion, we have to combine them with nontermination: switching to L_ω from L does not suffice (see [11] for details).

3 Discrete Lawvere Theories

As we have seen, countable Lawvere V-theories, for $V = \omega Cpo$, allow us to account for all the leading examples of computational effects other than continuations. Moreover, they are closed under two constructions that model natural ways in which computational effects are combined: sum and tensor [10,11]. But there are other constructions that may be made of countable Lawvere theories in the original Set-based case, that do not exist for arbitrary countable Lawvere V-theories where $V = \omega Cpo$, but that do appear in combining computational effects in practice. So we seek a refinement of the notion of countable Lawvere V-theory to account for them.

In particular, as we shall discuss in Section 6, consider the operations of one theory L distributing over those of another theory L', as appear in concurrency [7] and in the combination of nondeterminism and probabilistic nondeterminism [20]. The construction does not enrich. The problem lies with the arities: an arbitrary countable Lawvere V-theory has arities that may be any countably presentable objects of V, but the notion of distributivity a priori only makes sense for arities that are sets (see Section 6 for details).

Lawvere V-theories also are not closed under taking the image of a model, as one wants when taking observations as a primitive notion, as we shall outline in Section 7, cf [26]. Again, the problem lies with the arbitrariness of the arities.

So in this section, we refine the notion of enriched Lawvere theory by restricting the arities to be sets, but still including all our examples, still allowing us to take sum and tensor, but also allowing us to take a distributive tensor and an image. This yields our notion of discrete Lawvere theory. Motivated by recursion, we describe a countable rather than a finite version here.

The notion of discrete countable Lawvere theory lies between that of ordinary countable Lawvere theory and that of countable Lawvere V-theory: in the former, the arities are objects of $\aleph_1$ and the homs are sets; in the latter, arities are countably presentable objects of V and the homs are objects of V; but for discrete countable Lawvere V-theories, we demand that the arities lie in $\aleph_1$ but allow the homs to be objects of V. The definition, using the notation we have already established, is as follows. The underlying ordinary category of a V-category C is denoted by C_0 [14].

Definition 4. *A* discrete countable Lawvere V-theory *is a small V-category L with countable products and a strict countable-product preserving identity-on-objects functor $I : \aleph_1^{op} \longrightarrow L_0$. A map of discrete countable Lawvere V-theories from L to L' is a (necessarily strict countable-product preserving) V-functor from L to L' that commutes with I and I'. A* model *of L in a V-category C with countable products is a countable product preserving V-functor $M : L \longrightarrow C$.*

We denote the category of discrete countable Lawvere V-theories by $DLaw_V$.

The definition of model extends, as before, to yield a natural V-category $Mod(L, C)$ of models of L in any V-category C with countable products: the homobjects are given by the usual internalisation to V of the set of all V-natural transformations between models [14].

Let us consider, in concrete terms, what a model M of a discrete countable Lawvere V-theory L in a V-category C with countable products is. First, one must send the object 1 to an object X of C. Since M preserves countable products and since all objects of L are countable products of copies of 1, this completely determines the behaviour of M on objects, indeed also on all maps in $\aleph_1^{op}$, up to coherent isomorphism. It remains to give the behaviour of M on homs. That amounts to giving, for each a and b in $\aleph_1$, a map in V of the form

$$M_{a,b} : L(a, b) \longrightarrow C(X^a, X^b)$$

subject to preservation of identities and composition. So, in particular, if L is freely generated by some sort of signature and equations, equivalently by some generalised notion of sketch (see [18] for a general definition and treatment of the sketch idea), it simply amounts to giving a model of each constructor of the generalised signature subject to the generalised equations, exactly as for Lawvere V-theories as in Section 2 but with arities restricted to be objects of $\aleph_1$ rather than arbitrary countably presentable objects of V.

One can construct a forgetful functor from the category Law_V of countable Lawvere V-theories to $DLaw_V$ as follows. Observe that $\aleph_1^{op}$ being the free category with countable products on 1 and $V_{\aleph_1}^{op}$ including the object $1 \otimes I$ (where I is the unit of V, and so $1 \otimes I$ is I) and having countable products determine a canonical functor $J : \aleph_1^{op} \longrightarrow V_{\aleph_1}^{op}$.

Definition 5. *The* forgetful functor $U : Law_V \longrightarrow DLaw_V$ *sends a countable Lawvere V-theory $I : V_{\aleph_1}^{op} \longrightarrow L$ to the V-category and functor determined by factorising the composite*

$$\aleph_1^{op} \xrightarrow{\;J\;} V_{\aleph_1}^{op} \xrightarrow{\;I\;} L$$

as an identity-on-objects functor followed by full faithful one, and using the latter functor to induce V-structure on the factorisation from that on L.

So the objects of the factorisation are exactly the objects of $\aleph_1 op$, with the hom from a to b given by $L(IJa, IJb)$, and with composition determined by that of L.

Theorem 2. *The forgetful functor $U : Law_V \longrightarrow DLaw_V$ has a left adjoint F. Moreover, given any V-category C with countable cotensors, the unit of the adjunction determines a canonical equivalence*

$$Mod(L, C) \simeq Mod(F(L), C)$$

where the first occurrence of Mod refers to models of a discrete theory, thus V-functors that preserve countable products, while the second refers to models of a V-theory, thus V-functors preserving countable V-cotensors [4].

Proof. The first statement may be seen as an instance of the theory of (countable) essentially algebraic theories [1,3] or alternatively via the study of V-categories with equational structure [4,17]. The second follows from the work relating strict maps and pseudo-maps of categories with structure in [4]. Alternatively, one can check it by direct calculation.

There is also a forgetful functor U' from $DLaw_V$ to the category of ordinary countable Lawvere theories Law_c: it simply forgets the V-structure of L by taking its underlying ordinary category L_0.

Theorem 3. *The forgetful functor $U' : DLaw_V \longrightarrow Law_c$ has a left adjoint F' with the additional property that for any V-category C with countable products, the unit of the adjunction determines a canonical equivalence*

$$Mod(L, C_0) \simeq Mod(F'(L), C)$$

where here, the first occurrence of Mod refers to models of an ordinary theory, thus functors that preserve countable products, while the second refers to models of a discrete V-theory, thus V-functors preserving countable V-products [4].

Proof. The free V-category (without insisting upon a countable product condition) on L in fact has countable products and acts as $F'(L)$: one must observe that countable products of the former freely yield countable products of the latter.

By Theorem 3, Examples 1, 2, 3 and 4 may all be seen as discrete countable Lawvere V-theories: it is safe for us to identify these ordinary countable Lawvere theories with the discrete countable Lawvere V-theories they freely generate. A non-free example is as follows.

Example 7. Consider Example 6, the Lawvere ωCpo-theory for nontermination. As presented, it gives a sketch from which the Lawvere ωCpo-theory is given freely. But one can see by inspection that the sketch may equally be seen as a sketch for a discrete ωCpo-theory, in that one could equally consider the free discrete ωCpo-theory generated by it and consider its models: the arities appearing in Example 6 are all discrete sets, in this case finite sets, and the homs of a discrete Lawvere V-theory may be arbitrary objects of V, in this case allowing us to express the inequality in the example, i.e., we can replace the use of $y^{\leq}$ as the codomain of an operation in Example 6 by two operations into y with an inequality between them. But the example does not, in any reasonable sense, freely generate an ordinary Lawvere theory as ordinary Lawvere theories do not allow us to treat inequality nontrivially.

There are four constructions on Lawvere theories that are of primary interest to us: the sum, the tensor, the distributive tensor, and the image. These all arise in the study of computational effects. One considers the sum in combining exceptions with any other computational effect [10,11]; one considers the tensor in combining side-effects with all other effects other than exceptions [10,11]; the distributive tensor or a two-sided version of it appears in combining two sorts of nondeterminism, for instance internal and external nondeterminism as used by Hennessy in modelling concurrency [7], or in combining ordinary nondeterminism with probabilistic nondeterminism [20]; and one considers the image in deriving equations from a theory of observations. So, in forthcoming sections we consider these constructions in turn.

4 Sum

In this section, we consider the sum of discrete Lawvere V-theories. We only do so briefly as one can already consider sums of arbitrary Lawvere V-theories [10,11] and we mainly just need to check that the construction of a sum restricts from arbitrary Lawvere V-theories to discrete ones.

Theorem 4. *The category of discrete countable Lawvere V-theories is cocomplete.*

One way to prove this is by observing that the category of discrete countable Lawvere V-theories is locally countably presentable [1]. An explicit construction of the sum is complicated as a general construction involves a transfinite induction, with inductive steps being given by a complicated coequaliser, cf [13]. But all our examples of discrete countable Lawvere theories are given freely by sketches. And in those terms, the sum is easy to describe: one takes all operations of both equational theories, including information such as partial order information, subject to all axioms of both. The complication arises in passing from the induced sketch to the Lawvere theory freely generated by it, as, in doing so, one may apply the operations of one theory to the operations of the other, hence the transfinite induction.

Even in terms of sketches, care is required. For instance, given Lawvere theories L and L', there are always maps of Lawvere theories given by coprojections

$L \longrightarrow L + L'$ and $L' \longrightarrow L + L'$. But these coprojection functors need not be faithful. For instance, L might be the trivially collapsing theory, i.e., its equations may force L to be equivalent to 1. In that case, $L + L'$ is also equivalent to 1, so the coprojection from L' is trivial.

It is a simple observation, comparing the above with [11], that the sum of discrete countable Lawvere V-theories qua countable Lawvere V-theories is discrete. One would certainly hope so as left adjoints preserve sums! For the same reason, a sum of countable Lawvere theories qua discrete countable Lawvere V-theories is free on the sum of ordinary countable Lawvere theories. For computational effects, the leading examples are given by the combination of exceptions with all other effects and the combination of interactive I/O with most other effects [11].

5 Tensor

The tensor of theories arises when one wants to combine side-effects with most other computational effects. Again, we already know that a tensor product of countable Lawvere V-theories exists [11], and it is just a matter of observing that we can adapt the construction and characterising theorem to discrete Lawvere V-theories.

Definition 6. *[11] Given discrete countable Lawvere V-theories L and L', the discrete countable Lawvere V-theory $L \otimes L'$, which we call the tensor product of L and L', is defined by the universal property of having maps of discrete countable Lawvere V-theories from L and L' to $L \otimes L'$, subject, suppressing canonical isomorphisms, to commutativity of*

$$\begin{array}{ccc} L(a,b) \otimes L'(a',b') & \longrightarrow & L(a \times b', b \times b') \otimes L'(a \times a', a \times b') \\ \downarrow & & \downarrow \scriptstyle{comp} \\ L(a \times a', b \times a') \otimes L'(b \times a', b \times b') & \xrightarrow[comp]{} & (L \otimes L')(a \times a', b \times b') \end{array}$$

The tensor product exists because it is determined by the free theory on an enriched sketch [18]. But it may equally, indeed more elegantly, be proved to exist by appeal to an enrichment of the delicate 2-categorical analysis of the form appearing in [9], from which the following result also follows:

Theorem 5. – *The construction $L \otimes L'$ on discrete countable Lawvere V-theories extends to a symmetric monoidal structure on $DLaw_V$, and*
– *for any small V-category C with countable products, there is a coherent equivalence of V-categories between $Mod(L \otimes L', C)$ and $Mod(L, Mod(L', C))$.*

The unit for the tensor product is the initial discrete countable Lawvere V-theory, i.e, the theory generated by no operations and no equations. This is the initial object of the category of discrete countable Lawvere V-theories, so is also the unit for the sum; it corresponds to the identity monad.

As we have already observed, Examples 1, 2, 3, 4 and 6 may all be seen as discrete countable Lawvere V-theories, and so we may simply translate our study of tensor of countable Lawvere V-theories in [11] to that of discrete countable Lawvere V-theories here, while retaining all our examples. Its importance, as studied in detail in [11], lies in combining side-effects with almost all other effects, the main counter-example to that being in the combination with exceptions, which is covered by the previous section. As was the case for sum, the left adjoints F and F' preserve the tensor product, making the adjunctions $U : Law_V \longrightarrow DLaw_V$ and $U' : DLaw_V \longrightarrow Law_c$ into symmetric monoidal adjunctions.

6 Distributive Tensor

We now turn to the distributive tensor. This is where the concept of discrete countable Lawvere V-theory starts to yield its value relative to arbitrary countable Lawvere V-theories: one can speak of a distributive tensor product of ordinary Lawvere theories without difficulty, but there does not seem to be any natural way to speak of a distributive tensor of arbitrary countable Lawvere V-theories, although it appears in computational practice.

For ordinary Lawvere theories, the distributive tensor is defined similarly to the tensor except that the two sets of operations are not required to commute, but rather the first are required to distribute over the second. This yields the following.

Definition 7. *Given Lawvere theories L and L', the Lawvere theory $L \rhd L'$, called the* distributive tensor *of L over L', is defined by the universal property of having maps of Lawvere theories from L and L' to $L \rhd L'$, with all operations of L distributing over all operations of L', i.e., given $f : (n+1) \longrightarrow 1$ in L and $f' : n' \longrightarrow 1$ in L', we demand commutativity of the diagram*

$$\begin{array}{ccc} n + n' & \xrightarrow{(f.(n \times \pi_{i'}))_{i' \in n'}} & n' \\ {\scriptstyle id_n \times f'} \downarrow & & \downarrow {\scriptstyle f'} \\ n + 1 & \xrightarrow[f]{} & 1 \end{array}$$

together with commutativity of all other n variants of the diagram given by varying the choice of an element of $n+1$ in the bottom left-hand corner from being the last element to being any of its n predecessors.

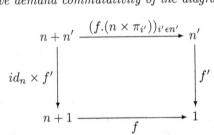

It is possible to characterise the distributive tensor in terms of models of L in $Mod(L', C)$ for any category with finite products C, along the lines of Theorem 5. But it is a little complicated, involving the use of operads and symmetric monoidal structure or more generally multicategory structure on $Mod(L', C)$. So we shall not do that here. We just remark that the distributive tensor does arise naturally in computation, sometimes in a two-sided framework such as in Hennessy's modelling of internal and external choice in [7] for modelling concurrency, and sometimes for modelling the combination of probabilistic nondeterminism and ordinary nondeterminism [12,20].

But now try extending this to arbitrary countable Lawvere V-theories: one would replace the $n+1$ in the bottom left-hand corner of the diagram by an arbitrary finitely presentable object v of V that, in some coherent and elegant sense, can be subdivided into an element together with the rest of v. But I cannot see any natural way to do that in general in the spirit of enriched categories.

The simple elegant solution that includes all our examples seems to be to restrict to discrete countable Lawvere V-theories: so the arities are all discrete sets, from which we can choose an element as a possible codomain and naturally consider the rest of the set; but we are still able to consider inequalities as in Example 6 seen as a discrete theory in Example 7. So that is what we do.

Definition 8. *Given discrete countable Lawvere V-theories L and L', the discrete countable Lawvere V-theory $L \rhd L'$, called the* distributive tensor *of L over L', is defined by the universal property of having maps of discrete countable Lawvere V-theories from L and L' to $L \rhd L'$, with all operations of L distributing over all operations of L'. I.e., for all objects a and a' of $\aleph_1$ and for all elements x of a (so $a \geq 1$), suppressing canonical isomorphisms, we demand commutativity of the diagram*

$$\begin{array}{ccc}
L(a,1) \times L'(a',1) & \xrightarrow{L(\pi_{i'},1)_{i' \epsilon a'} \times id} & L(a-x+a',a') \times L'(a',1) \\
{\scriptstyle id \times (a-x+(-))} \Big\downarrow & & \Big\downarrow {\scriptstyle comp} \\
L(a,1) \times L'(a-x+a',a) & \xrightarrow[comp]{} & (L \rhd L')(a-x+a',1)
\end{array}$$

It requires a little thought, but given that thought, this definition internalises a countable version of the notion of distributivity, in particular allowing us to include Example 7, the example of nontermination, without running into an arity problem. Thus we can extend our list of constructions on discrete countable Lawvere V-theories to include combinations involving various forms of nondeterminism.

7 Image

The image construction is of a somewhat different nature to those we have studied so far, and it is used for a somewhat different purpose. Here, rather than starting with a pair of Lawvere theories, one starts with a single Lawvere theory and a model of it. We first consider a construction that is not quite what we want but which does exist for arbitrary countable Lawvere V-theories.

Definition 9. *Given a countable Lawvere V-theory L and a model $M : L \longrightarrow C$, the* full image $L_f M$ *of M is the (bijective-on-objects,fully faithful) factorisation of M, i.e., up to isomorphism of V-categories, it is determined by putting*

$$(L_f M)(m,n) = C(Mm, Mn)$$

A variant of this, where one removes the size limitation at the cost of considerably less elegance, appears in the study of continuations, where M is typically taken to be the free model on a test set R [8]. But what we want to do here is a little more subtle. First we consider the situation for ordinary Lawvere theories.

The image, as opposed to the full image, appears when one wants to take observations rather than equations as primitive [26]. The idea is that one only considers the signature with which one starts, and then puts equations or inequations between derived terms depending upon what an observational model demands. One cannot just take the full image, i.e., the relatively familiar (bijective-on-objects,fully faithful) factorisation, as that allows additional maps that are not generated by a signature, reflecting the fact that continuations allow one to write additional programs. So we need a more subtle factorisation, one which, in a precise sense, moves the fullness from the right to the left of the factorisation system. The factorisation system we need is defined as follows:

Definition 10. *Given a Lawvere theory L and a model $M : L \longrightarrow C$, the* image L_M *of M is determined by the (bijective-on-objects and full,faithful) factorisation of M, i.e., up to isomorphism of categories, it is determined by putting $L_M(m,n)$ equal to the image of the function*

$$M_{m,n} : L(m,n) \longrightarrow C(Mm, Mn)$$

Example 8. Following Example 2, let $S = V^L$. The standard semantics of a command is generally understood to be a state-changing function, i.e., a function of the form

$$S \longrightarrow S$$

So the operations *lookup* and *update* should act on powers of this set. They are generally deemed to act as follows: the operation *lookup* is modelled by the function

$$(S \to S)^V \longrightarrow (S \to S)^L$$

determined by composition with the function from $L \times S$ to $V \times S$ that, given (loc, σ), "looks up" loc in $\sigma : L \to V$ to determine its value, and is given by the projection to S; and the operation *update* is modelled by the function

$$(S \to S) \longrightarrow (S \to S)^{L \times V}$$

determined by composition with the function from $L \times V \times S$ to S that, given (loc, v, σ), "updates" $\sigma : L \to V$ by replacing the value at loc by v. We wish to set a pair of operations generated by *lookup* and *update* equal precisely when they yield the same functions on powers of $S \to S$. And that is given as follows:

first take the free countable Lawvere theory generated by operations *lookup* and *update*. Then take the model of this countable Lawvere theory determined by $S \to S$ together with the functions defined above. Now take the image determined by this model. The result is exactly L_S.

The above example appears, along with several others, in [26]. The difficulty with enriching the idea arises even when V is *Poset*. If one replaces categories by V-categories with finite cotensors in the definition of image, the factorisation need not have finite cotensors: the problem arises when one considers cotensors with non-discrete posets such as Sierpinski space. Looking harder at why that is not a problem for ordinary Lawvere theories, one notes that the homs are sets, and in *Set*, all epimorphisms are retracts, so are preserved by all functors, whereas that is not the case in *Poset*. In contrast, products often do preserve epimorphisms or at least strong epimorphisms. For instance, in *Poset*, an epimorphism is pointwise a surjective function, and the product of surjective functions is again surjective. The central result we need is as follows.

Theorem 6. *Given a factorisation system (E, M') on V for which E is closed under countable product, and given a discrete countable Lawvere theory L and a model $M : L \longrightarrow C$, the (bijective-on-objects and locally E,M') factorisation of M, i.e, determined by putting $L_{M(a,b)}$ equal to the (E,M')-factorisation of the map in V*

$$M_{a,b} : L(a, b) \longrightarrow C(Ma, Mb)$$

lifts to a factorisation of V-categories with countable products.

This result applies to *Poset* if we take E to be the class of epimorphisms. But ωCpo requires more care, as countable products there involve more subtlety than finite products do: a finite product of epimorphisms is always an epimorphism by cartesian closedness. The solution to that difficulty seems likely to involve making more subtle use of the relationships developed in this paper between discrete countable Lawvere V-theories and arbitrary countable Lawvere V-theories, and their relationships with finitary versions.

References

1. J. Adámek and J. Rosický, *Locally Presentable and Accessible Categories*, London Mathematical Society Lecture Note Series, Vol. 189, Cambridge University Press, 1994.
2. M. Barr and C. Wells, *Toposes, Triples and Theories*, Springer-Verlag, 1985.
3. M. Barr and C. Wells, *Category Theory for Computing Science*, Prentice-Hall, 1990.
4. R. Blackwell, G. M. Kelly, and A. J. Power, Two-Dimensional Monad Theory, *J. Pure Appl. Algebra* Vol. 59, pp. 1–41, 1989.
5. N. Ghani and C. Lüth, Monads and Modular Term Rewriting, *Proc. CTCS '97*, LNCS, Vol. 1290, pp. 69–86, Springer-Verlag, 1997.
6. R. Heckmann, Probabilistic Domains, *Proc. CAAP '94*, LNCS, Vol. 136, pp. 21-56, Springer-Verlag, 1994.

7. M. C. B. Hennessy, *Algebraic Theory of Processes*, MIT Press, 1988.
8. M. Hyland, P. Levy, G. D. Plotkin, and A. J. Power, Combining Continuations with Other Effects, *Proc. Continuations Workshop 2004*, Birmingham Technical Report, No. CSR-04-1, refereed presentation, 2004.
9. J. M. E. Hyland and A. J. Power, Pseudo-Commutative Monads and Pseudo-Closed 2-Categories, *J. Pure Appl. Algebra* Vol. 175, pp. 141–185, 2002.
10. J. M. E. Hyland, A. J. Power and G. D. Plotkin, Combining Computational Effects: Commutativity and Sum, *Proc. 2nd IFIP Conf on Theoretical Computer Science* (eds. Ricardo A. Baeza-Yates, Ugo Montanari and Nicola Santoro), pp. 474–484, Kluwer, 2002.
11. J. M. E. Hyland, A. J. Power and G. D. Plotkin, Combining Computational Effects: Sum and Tensor, *Theoretical Computer Science*, to appear.
12. C. Jones and G. D. Plotkin, A Probabilistic Powerdomain of Evaluations, *Proc. LICS '89*, pp. 186–195, IEEE Press, 1989.
13. G. M. Kelly, A Unified Treatment of Transfinite Constructions for Free Algebras, Free Monoids, Colimits, Associated Sheaves, and so on, *Bull. Austral. Math. Soc.*, Vol. 22, pp. 1–83, 1980.
14. G. M. Kelly, *Basic Concepts of Enriched Category Theory*, Cambridge University Press, 1982.
15. G. M. Kelly, Structures Defined by Finite Limits in the Enriched Context I, *Cahiers de Topologie et Géométrie Différentielle*, Vol. 23, No. 1, pp. 3–42, 1982.
16. G. M. Kelly and Stephen Lack, Finite-product-preserving Functors, Kan Extensions, and Strongly-finitary Monads, *Applied Categorical Structures*, 1993.
17. G. M. Kelly and A. J. Power, Adjunctions whose Counits are Coequalizers, and Presentations of Finitary Enriched Monads, *J. Pure Appl. Algebra*, Vol. 89, pp. 163–179, 1993.
18. Y. Kinoshita, A. J. Power, and M. Takeyama, Sketches, *J. Pure Appl. Algebra*, Vol. 143, pp. 275–291, 1999.
19. A. Kock, Monads on Symmetric Monoidal Closed Categories, *Arch. Math.*, Vol. 21, pp. 1–10, 1970.
20. M. W. Mislove, Nondeterminism and Probabilistic Choice: Obeying the Laws, *Proc. CONCUR 2000* (ed. C. Palamidessi), LNCS, Vol. 1877, pp. 350–364, Springer-Verlag, 2000.
21. E. Moggi, Computational Lambda-Calculus and Monads, *Proc. LICS '89*, pp. 14–23, IEEE Press, 1989.
22. E. Moggi, Notions of Computation and Monads, *Inf. and Comp.*, Vol. 93, No. 1, pp. 55–92, 1991.
23. G. D. Plotkin and A. J. Power, Notions of Computation Determine Monads, *Proc. FOSSACS '02*, (eds. M. Nielsen and U. Engberg), LNCS, Vol. 2303, pp. 342–356, Springer-Verlag, 2002.
24. A. J. Power, Why Tricategories? *Information and Computation*, Vol. 120, pp. 251–262, 1995.
25. A. J. Power, Enriched Lawvere Theories, *Theory and Applications of Categories*, Vol. 6, pp. 83–93, 2000.
26. A. J. Power, Canonical models for computational effects, *Proc. FOSSACS 2004*, LNCS Vol. 2987, pp. 438–452, Springer-Verlag, 2004.

Final Semantics for Event-Pattern Reactive Programs

César Sánchez, Henny B. Sipma, Matteo Slanina, and Zohar Manna*

Computer Science Department,
Stanford University, Stanford, CA 94305-9025
{cesar, sipma, matteo, zm}@CS.Stanford.EDU

Abstract. Event-pattern reactive programs are front-end programs for distributed reactive components that preprocess an incoming stream of event stimuli. Their purpose is to recognize temporal patterns of events that are relevant to the serviced program and ignore all other events, outsourcing some of the component's complexity and shielding it from event overload. Correctness of event-pattern reactive programs is essential, because bugs may result in loss of relevant events and hence failure to react appropriately.

We introduce PAR, a specification language for event-pattern reactive programs. We propose a new approach for defining such languages in terms of observations and actions. This approach applies standard techniques from coalgebra to obtain instances of the corecursion and coinduction principles. Corecursion is used to formally define the operational semantics of PAR, and coinduction allows to prove general equivalences between (ground and parameterized) PAR programs.

This is the first of a series of papers in which we study questions of expressive completeness, complexity, and formal verification techniques for specification languages of event-pattern reactive programs.

1 Introduction

Reactive programs are software components that maintain an ongoing interaction with their environment. With the introduction of middleware technologies and the emphasis on component-based systems, this interaction is increasingly performed through *events*. Reactive components, which can range from simple sensors to sophisticated monitors or controllers, operate relatively autonomously and communicate using events. This gives rise to *publish-subscribe* architectures, in which producer components publish events to the middleware and consumer components subscribe with it to receive relevant events.

Different subscription policies are possible. The simplest uses a list of event types and/or senders that can be syntactically matched by an attribute in the event. This is known as *event filtering* and is available in most popular platforms,

* This research was supported in part by NSF grants CCR-01-21403, CCR-02-20134, CCR-02-09237, CNS-0411363, and CCF-0430102, by ARO grant DAAD19-01-1-0723, and by NAVY/ONR contract N00014-03-1-0939.

J.L. Fiadeiro et al. (Eds.): CALCO 2005, LNCS 3629, pp. 364–378, 2005.

including GRYPHON [1], ACE-TAO [18], SIENA [3], and ELVIN [19]. A more sophisticated policy is *content filtering*, in which the subscription contains a list of predicates on the data of the event. This approach is especially popular in active databases and stock market applications. With these policies every single event is either discarded or dispatched, independently of the event history. Another extension of event filtering, orthogonal to content filtering, is *event correlation*, the approach studied in this paper. Here, subscriptions may contain temporal patterns of either attributes or content predicates on events.

Event correlation is attractive for several reasons: it may substantially reduce unnecessary component activations, thereby improving the performance; it separates event-pattern recognition from event processing, thus increasing analyzability of component interactions; it allows automatic synthesis of the pattern recognition code, thus reducing ad-hoc implementations and improving reusability. At present, some middleware platforms provide limited forms of event correlation services. Unfortunately, formal semantics are not given, making their use risky: unclear semantics or incorrect implementations may result in the loss of important events, potentially causing failure to respond to critical situations.

In a previous paper [16] we introduced ECL, a language to specify event correlation patterns, developed under the DARPA PCES program for the Boeing Boldstroke [20] platform to support mission-critical avionics applications. We gave a formal semantics in terms of correlation machines, an extension of finite-state transducers that enabled direct translation into event-processing code. Prototype implementations were integrated in ACE-TAO [18] and FACET [9].

In this paper we shift focus from implementation to analysis, in particular program equivalences. Practical applications need to determine whether a given pattern expression can be replaced by a simpler one, or merged with that of another component, without affecting its behavior. Correlation machines, however, are not well suited to answer these questions, since they explicitly model operational details, such as parallelism, into the semantics. Instead, we are interested in behavioral equivalences, in which two programs are considered equivalent if they exhibit the same notification behavior.

We intend to study languages for event-pattern reactive programming algebraically, influenced by the pioneering work on languages for the study of concurrency, mostly process algebras [13,8,2] and Hennessy-Milner logic [7]. The main difference is that we specifically design our languages to be deterministic, because we want to synthesize executable behaviors from the expressions, while every reasonable concurrency model is intrinsically nondeterministic.

Coalgebra is a convenient framework to study dynamic systems, and, in general, systems with hidden state spaces, where only the observable behavior is of interest [10]. For example, in [15] Rutten shows how equivalence of regular expressions can be analyzed in this framework. He constructs an automaton, whose states correspond to languages, that is final with respect to all other automata; then he shows that language equivalence can be reduced to proving bisimilarity of their corresponding states in this final automaton. In this paper we develop a similar theory for the behaviors of event-pattern machines, and proof techniques to decide equivalence of classes of event-pattern expressions.

We develop a framework for specification formalisms for event-pattern reactive programs, based on standard coalgebraic techniques. We chose to develop our techniques directly from the basic definitions, rather than treat them as special cases of general results about the existence of final semantics and coinduction principles (for example, from Hidden Algebras [5,14,4]). Since our expressions do not describe experiments and observations, developing all the necessary machinery to use one such general result would not significantly simplify the presentation.

We introduce PAR—a subset of ECL—a declarative language for the specification of event-pattern reactive programs. We illustrate the application of our coalgebraic framework by defining the formal semantics of PAR and studying some of its properties.

The paper is organized as follows. Section 2 presents PAR and informally describes the intended semantics. Section 3 presents the notions of event-pattern machines and behaviors, and proves that the so-called "machine of all behaviors" is final among all machines, from which we obtain the principles of coinduction and corecursion for machines. The use of corecursion is illustrated in Sect. 4 to obtain the formal (behavioral operational) semantics of PAR; the use of coinduction is shown in Sect. 5 and 6, where we discuss some equivalences between PAR programs. Section 7 contains a final discussion and sketches some future work. Because of space limitations, we only include the most relevant proofs. The omitted ones can be found in the online version of this paper, available from the authors' web page.

2 The Language PAR

Event-pattern reactive programs are components that recognize temporal patterns of events. In this section we introduce the PAR expression language[1] which enables a declarative specification of these patterns. PAR is a subset of ECL, which we proposed in [16], but is equally expressive. In fact, every finitely expressible event pattern can be described in PAR [17].

Syntax. We assume that the input event stream consists of input symbols taken from a finite set Σ. A PAR expression can describe multiple patterns to be searched in parallel in the input stream. To handle multiple notifications, the output $\mathcal{O}$ consists of sets of symbols from a finite set Γ. The simplest notification is a singleton, notifications are combined by set union, and the absence of output is denoted by $\varnothing$.

A *simple* PAR expression is an equality test for an input symbol, that is, for each $a \in \Sigma$ there is an expression $\mathbf{a}$. If $A \in \mathcal{O}$ and x and y are PAR expressions, then so are

$x \mid y$	$\overline{x}$	**repeat** x	**silent**
$x \,;\, y$	$x[A]$	**try** x **unless** y	

[1] The name PAR stands for event-**PA**ttern **R**eactions.

Informal Semantics. A PAR program processes input events, one at a time, and produces a (possibly empty) output after each event is processed. The semantics of PAR expressions can be defined by their behavior in response to all prefixes of input streams. This behavior is characterized by two aspects: the output and the completion status. The output is the information transmitted to the served reactive component, where a nonempty output usually causes a component activation. The completion status is introduced to assist in the compositional definition of the semantics. We distinguish three completion statuses: **success**, to represent that the pattern has *just* been observed; **failure**, to indicate that the pattern cannot be observed in any stream that extends the current prefix; **incomplete**, which represents that more input is needed or the input symbol is not relevant. We use the symbols $\top$ to represent success, ι for incomplete, and $\bot$ to represent failure. All PAR behaviors have the property that, once success or failure is declared, the output will be empty and the completion status will become and remain incomplete for all subsequent inputs.

Informally, the PAR constructs behave as follows:

Simple: The expression **a** ignores every event that does not match a, and declares success as soon as the first a event is received.

Negation: $\overline{x}$ behaves as x except it reverses success with failure (and vice-versa).

Selection: The expression $x \mid y$ evaluates x and y in parallel, offering each the same events, and generating as output the combination of the subexpressions' outputs. Selection succeeds as soon as one of the branches succeeds and only fails when both branches have failed.

Sequential: Sequential composition, $x; y$, evaluates the first subexpression, and upon successful completion starts the evaluation of the second. If one of them fails, sequential immediately fails.

Repetition: The expression **repeat** x starts by evaluating x, called the *body*. If the evaluation of the body completes with success, it continues evaluating **repeat** x again, called the *continuation*. If the body fails, repetition declares failure.

Output: $x[A]$ evaluates x. Upon successful completion, the output A is generated and combined with simultaneous outputs of x's subexpressions. The completion status of $x[A]$ is the same as that of x.

Preemption: **try** x **unless** y evaluates x and y in parallel. It succeeds when x does. It fails if x fails or if y succeeds before x does.

Silent: It does not generate any output and always declares incomplete.

Example 1. The expression (**try a unless** (**b**|**c**))$[A]$ waits to receive an a without receiving a b or a c. If the evaluation succeeds, then it notifies the component with an A and terminates.

Example 2. The expression **repeat** (**a** ; **try b**$[A]$ **unless** (**c** ; **c**)) represents the behavior that notifies the component as soon as b occurs after the first a (subsequent occurrences of a are ignored) without two c events in between (in which case the pattern reactive program stops). If the pattern is successfully observed, then the component is notified with an A and the expression restarts.

3 Event-Pattern Machines and Coinduction

In this section we develop the abstract theory of event-pattern machines, following the approach of [15]. We first define the notion of machines and behaviors and then we construct the final machine, whose states correspond to the behaviors they represent.

A machine is a black-box device whose behavior can be studied by means of observations and experiments. In our context, an observation consists of the output and completion status generated in response to an input symbol. An experiment is a sequence of observations.

To model completion statuses we define the following three element lattice $\mathcal{C} = \{\top, \iota, \bot\}$, where $\bot$ (failure) $<$ ι (incomplete) $<$ $\top$ (success). Apart from the usual lattice operations ($\wedge$, $\vee$), $\mathcal{C}$ is equipped with a unary "opposite" operation $\widehat{\cdot}$, defined as: $\widehat{\bot} = \top \quad \widehat{\iota} = \iota \quad \widehat{\top} = \bot$.

Definition 1 (Machine). *An event-pattern machine $M : \langle M, o, \alpha, \partial \rangle$ consists of the following components:*
- *M: a (possibly infinite) set of states;*
- *o: $M \to \mathcal{O}^{\Sigma}$, an output function, mapping states to functions from input symbols to output values;*
- *α: $M \to \mathcal{C}^{\Sigma}$, a completion function, mapping states to functions from input symbols to completion statuses.*
- *∂: $M \to M^{\Sigma}$, a transition function, mapping states to functions from input symbols to states;*

and satisfies the following "adequacy" condition

$$\textit{for every } m \in M, a \in \Sigma, \quad \textit{if } \alpha(m)(a) \neq \iota \textit{ then } \partial(m)(a) \textit{ is silent,} \qquad (S1)$$

where a set of states $S \subseteq M$ is defined to be silent if, for every $s \in S$ and $a \in \Sigma$, $\alpha(s)(a) = \iota$, $o(s)(a) = \varnothing$, and $\partial(s)(a) \in S$. A silent state is any state that belongs to some silent set.

The adequacy condition reflects the property about behaviors, stated earlier, that success or failure is always followed by silence, that is, empty output and incomplete status. When a machine enters a silent state it will remain in some silent state thereafter. In Sect. 4 we will assign a meaning to a PAR expression by mapping it to a machine state; when a PAR expression signals success (or failure) its enclosing expression can use that information, for example in the case of **repeat** to reset the machine. However, the intended behavior of a PAR expression by itself after it has failed (or succeeded) is to remain silent.

We refer to event-pattern machines simply as machines. We use the same symbol for the name of the machine and its state space because usually the state space is equipped with the appropriate functions to become a machine.

Example 3. Figure 1(a) shows an example of a machine with three states, where s is a silent state. In order to simplify this graphical representation, incomplete completion statuses and empty outputs are omitted in the arrows' decorations.

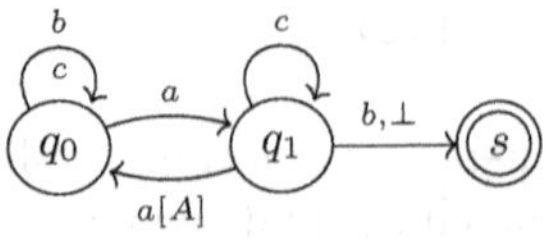

(a) Machine M.

Σ	c	a	c	c	c	a	b	b	c	b	c	a	c	b	...
$\mathcal{O}$	$\varnothing$	$\varnothing$	$\varnothing$	$\varnothing$	$\varnothing$	A	$\varnothing$	$\varnothing$	$\varnothing$	$\varnothing$	$\varnothing$	$\varnothing$	$\varnothing$	$\varnothing$	...
$\mathcal{C}$	ι	ι	ι	ι	ι	ι	ι	ι	ι	ι	ι	ι	ι	$\bot$	...
M	q_0	q_1	q_1	q_1	q_1	q_0	q_0	q_0	q_0	q_0	q_0	q_1	q_1	s	...

(b) Sample run from initial node q_0.

Fig. 1. Example machine with a behavior evaluation

Transitions outgoing from the silent state are also omitted. Figure 1(b) shows a sample run for input "*cacccabbcbcacb*", starting from initial state q_0. The rows labeled $\mathcal{O}$, $\mathcal{C}$ and M contain the output generated, the completion status declared and the state reached (resp.) after processing the corresponding input symbol.

The notation for o, α and ∂ emphasizes the coalgebraic nature of the definition of machine, since we can easily compose these functions into a functor $\Gamma_1(X) = (\Sigma \to (\mathcal{O} \times \mathcal{C} \times X))$. Machines are "adequate" coalgebras of the functor Γ_1. It is often more convenient, however, to represent these functions as maps from input symbols into functions from states to output, completion, and next state (resp.) Abusing notation, we write $\alpha_a(m)$ instead of $\alpha(m)(a)$; the distinction should be clear from the context.[2] Using this notation we can extend these operators from single input symbols a to strings va as follows:

$$\alpha_{va}(m) \stackrel{\text{def}}{=} \alpha_a \partial_v(m) \qquad o_{va}(m) \stackrel{\text{def}}{=} o_a \partial_v(m) \qquad \partial_{va}(m) \stackrel{\text{def}}{=} \partial_a \partial_v(m),$$

with also $\partial_\epsilon(m) \stackrel{\text{def}}{=} m$. *Behaviors* play the same rôle in the theory of event-pattern machines and expressions, that languages (subsets of Σ^*) play in the theory of automata and regular expressions.

Definition 2 (Behaviors). *Let B be a map from stream prefixes Σ^+ into $\mathcal{O} \times \mathcal{C}$. We say that B is a behavior if*

$$\textit{for every } w, v \in \Sigma^+, \quad \textit{if } \pi_2 B(w) \neq \iota \textit{ then } B(wv) \textit{ is silent,} \tag{S2}$$

where a behavior is silent if it returns $\langle \varnothing, \iota \rangle$ for every input prefix. The set of all behaviors is denoted by $\mathcal{B}$.

Here, π_1 represents the projection function from the pair $\mathcal{O} \times \mathcal{C}$ into the first component $\mathcal{O}$. Similarly π_2 projects into $\mathcal{C}$. Abusing notation,[3] we also use $\pi_1 B$ to represent the map from input prefixes w into the corresponding output value $\pi_1(Bw)$.

Condition ($S2$), similar to the adequacy condition ($S1$) for machines, establishes that once a behavior declares a pattern successfully found (or impossible to find) it should subsequently exhibit no other action. It is easy to see that there is a unique silent behavior, namely the function that for every input returns $\langle \varnothing, \iota \rangle$.

[2] Technically, the overloaded notation $\alpha_a(m)$ is defined as $\lambda a{:}\Sigma.m{:}M.\alpha(m)(a)$.
[3] This overloaded use of the projection function $\pi_1 B$ is defined as $\lambda s{:}\Sigma^+.\pi_1(Bs)$.

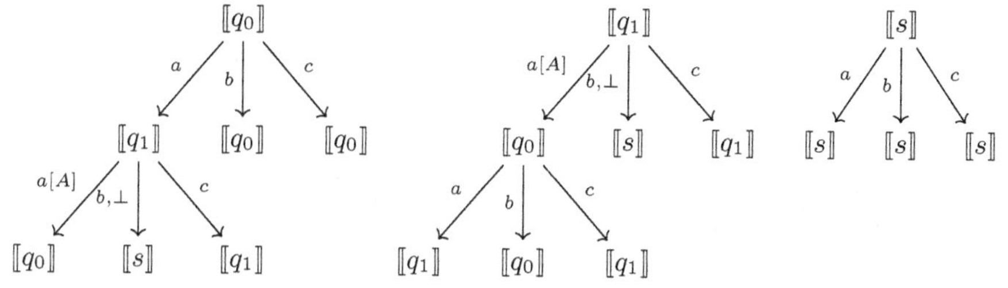

(a) Behavior $[\![q_0]\!]$. (b) Behavior $[\![q_1]\!]$. (c) Behavior $[\![s]\!]$.

Fig. 2. Behaviors associated to states q_0, q_1 and s of Fig. 1

Every state of a machine can be naturally associated with a behavior:

Definition 3 (Associated behavior). *Given a state* $m \in M$, *the function* $[\![m]\!]$ *describes its associated behavior:*

$$\begin{aligned} [\![m]\!] : \Sigma^+ &\to \mathcal{O} \times \mathcal{C} \\ wa &\mapsto \langle\, o_a(\partial_w m),\ \alpha_a(\partial_w m)\, \rangle. \end{aligned}$$

It is easy to see that the adequacy condition ($S2$) holds for $[\![m]\!]$: Consider a word wa such that $\pi_2([\![m]\!]wa) \neq \iota$. Since $\pi_2([\![m]\!]wa) = \alpha_a(\partial_w m)$, the state $\partial_{wa} m$ is silent by condition ($S1$); consequently—by definition of silent state—every subsequent output is $\varnothing$ and every subsequent status is ι.

Example 4. Figure 2 depicts the behaviors associated to the states q_0, q_1 and s in Fig. 1. These behaviors are described by infinite trees but, to simplify the graphical representation, subtrees labeled with a behavior already represented are not further expanded. To see in this representation the value of a behavior for an input prefix, simply traverse the tree following the appropriate edges. The value returned corresponds to the label of the last edge traversed. In Fig. 2(c) the behavior $[\![s]\!]$ corresponds to the *unique* silent behavior.

Our next goal is to construct a machine from the set of all behaviors. The transition function for this machine uses *behavior derivatives*:

Definition 4 (Behavior derivative). *Given an input string* w, *the* w*-derivative of a behavior* B *is the behavior* B_w *which, given an input* v, *returns* $B_w(v) = B(wv)$.

To see that B_w is indeed a behavior, assume that its completion status is not ι for some input v; the completion status of B is also not incomplete for input wv. Consequently, every extension of $B(wv)$ is the silent behavior, so B_w is also the silent behavior.

Example 5. We can use Fig. 2 to illustrate some behavior derivatives: to calculate the w-derivative of a behavior simply traverse the edges corresponding to w. For example: $[\![q_0]\!]_{aaa} = [\![q_1]\!]$ and $[\![s]\!]_{aaa} = [\![s]\!]$.

The *machine of all behaviors* can now be defined as $\mathcal{B} : \langle \mathcal{B}, o, \alpha, \partial \rangle$ with the set of all behaviors $\mathcal{B}$ as set of states, and for each input a and behavior B:

$$o_a B = \pi_1(Ba) \qquad \alpha_a B = \pi_2(Ba) \qquad \partial_a B = B_a.$$

Since the adequacy condition holds, $\mathcal{B}$ is indeed a machine: observe that if $\alpha_a B \neq \iota$ then, by Condition ($S2$), B_a is the silent behavior. Moreover, since $B_{ab} = B_a$ is also the silent behavior, B_a is a silent state in the machine $\mathcal{B}$.

We will show, in Theorem 2, that $\mathcal{B}$ is final among all machines, that is, for all machines M there exists a unique homomorphism from M into $\mathcal{B}$.

Definition 5 (Homomorphism). *A machine homomorphism from M to M' is a function $f : M \to M'$ such a that, for all $m \in M$ and $a \in \Sigma$:*

$$\begin{aligned} o(m) &= o'(f(m)), \\ \alpha(m) &= \alpha'(f(m)) \text{ and} \\ f(\partial_a m) &= \partial'_a f(m). \end{aligned}$$

Homomorphisms are functions preserving observations and experiments. The notion of *bisimulation relation* captures whether two states are indistinguishable by experiments:

Definition 6 (Bisimulation). *A bisimulation between machines M and M' is a binary relation $\#$ such that for all $m \in M$, $m' \in M'$ and input symbol a:*

$$\text{if } m\#m' \text{ then } \begin{cases} o(m) = o'(m'), \\ \alpha(m) = \alpha'(m') \text{ and} \\ \partial_a m \# \partial'_a m'. \end{cases}$$

Two states m and m' are called *bisimilar*, written $m \approx m'$, if there exists a bisimulation that relates them. The relation $\approx$ is the largest bisimulation relation between two machines, and is called *bisimilarity*.

Theorem 1 (Coinduction). *For all behaviors A and B, if $A \approx B$ then $A = B$.*

Proof. We proceed by showing a stronger result, by induction on the length of input prefixes: for all $w \in \Sigma^+$ and for all behaviors A and B, if $A \approx B$ then $A(w) = B(w)$ and $\partial_w A \approx \partial_w B$:

- *Base* ($w = a$): First, $\pi_1 A(a) = o_a A = o_a B = \pi_1 B(a)$, by $A \approx B$, and $\pi_2 A(a) = \alpha_a A = \alpha_a B = \pi_2 B(a)$, by $A \approx B$.

  Also, $\partial_a A \approx \partial_a B$ holds immediately from the definition of bisimulation.
- *Inductive step* ($w = va$). Here,

$$\begin{aligned} \pi_1 A(va) &= o_a(\partial_v A) = o_a(\partial_v B) = \pi_1 B(va), && \text{by } A \approx B \text{ and IH,} \\ \pi_2 A(va) &= \alpha_a(\partial_v A) = \alpha_a(\partial_v B) = \pi_2 B(va), && \text{by } A \approx B \text{ and IH, and} \\ \partial_w A &= \partial_a \partial_v A \approx \partial_a \partial_v B = \partial_w B, && \text{by } A \approx B \text{ and IH.} \end{aligned}$$

□

Coinduction justifies the following proof principle: to show the equality between behaviors A and B it is sufficient to establish the existence of a bisimulation relation on $\mathcal{B}$ that contains $\langle A, B\rangle$. We can use coinduction to show that $\mathcal{B}$ is final among all machines:

Theorem 2 (Finality of $\mathcal{B}$). *For every machine M, there is a unique homomorphism from M to $\mathcal{B}$.*

Proof. Existence is guaranteed since the behavior function $[\![\cdot]\!] : M \to \mathcal{B}$ that maps every m to $[\![m]\!]$ (see Definition 3) is a homomorphism. For uniqueness, suppose that f and g are two homomorphisms. It is enough to show that the relation $\# = \{\langle f(m), g(m)\rangle \mid m \in M\}$ is a bisimulation, since in that case—by coinduction—$f(m) = g(m)$ for all m, and consequently $f = g$. Let m be an arbitrary state; since f and g are homomorphisms:

$$\begin{aligned} o_a f(m) &= o_a m &&= o_a g(m), \\ \alpha_a f(m) &= \alpha_a m &&= \alpha_a g(m), \quad \text{and} \\ \partial_a f(m) &= f(\partial_a m) \;\#\; g(\partial_a m) &&= \partial_a g(m). \end{aligned}$$

Therefore, # is a bisimulation. □

The unique homomorphism $[\![\cdot]\!]$ identifies the behaviors of two states precisely when they are bisimilar. Moreover, homomorphisms preserve bisimulation:

Theorem 3. *Let R be a bisimulation, and f and g homomorphisms. Then, $\{\langle f(m), g(n)\rangle \mid \langle m, n\rangle \in R\}$ is also a bisimulation.*

The previous theorem, together with coinduction, allows to prove whether two states of *arbitrary* machines define the same behavior, simply by showing the existence of a bisimulation that relates them.

Corecursion. The finality of $\mathcal{B}$ justifies the following principle of definition by *corecursion*: To associate behaviors to the elements of a set M, turn M into a machine by defining an output function o, a completion function α and a transition function ∂, such that the adequacy condition for machines ($S1$) is satisfied. The desired semantics is then obtained as the unique homomorphism $[\![\cdot]\!]$ from M to $\mathcal{B}$, which assigns to each element m the behavior it describes.

4 Formal Semantics of PAR

In this section we illustrate the use of corecursion by defining the operational semantics of PAR. We build a machine whose states are all PAR expressions and whose functions α, o and ∂ are described by rules. By showing (see the formal proof in the longer version) that this is indeed a machine, we guarantee that each PAR expression defines a unique behavior.

The rules describing the functions, shown in Fig. 3, 4, and 5 use the following notation: $x \overset{a}{\rightsquigarrow} c$ stands for $\alpha_a x = c$, $x \overset{a}{\rightarrow} y$ stands for $\partial_a x = y$ (with $x \overset{a}{\rightarrow}_\iota y$ as an abbreviation for both $x \overset{a}{\rightsquigarrow} \iota$ and $x \overset{a}{\rightarrow} y$), and $x \overset{a}{\Rightarrow} u$ stands for $o_a x = u$. Below we briefly explain some of the rules.

$$\alpha\mathbf{Ev_1}:\ \mathbf{a} \overset{a}{\rightsquigarrow} \top \qquad \alpha\mathbf{Ev_2}:\ \mathbf{a} \overset{b}{\rightsquigarrow} \iota \ (\text{if } b \neq a)$$

$$\alpha\mathbf{Seq}\ \frac{x \overset{a}{\rightsquigarrow} c}{x\,;y \overset{a}{\rightsquigarrow} c \wedge \iota} \qquad \alpha\mathbf{Sel}\ \frac{x \overset{a}{\rightsquigarrow} c \qquad y \overset{a}{\rightsquigarrow} d}{x \mid y \overset{a}{\rightsquigarrow} c \vee d} \qquad \alpha\mathbf{Rep}\ \frac{x \overset{a}{\rightsquigarrow} c}{\mathbf{repeat}\, x \overset{a}{\rightsquigarrow} c \wedge \iota}$$

$$\alpha\mathbf{Push}\ \frac{x \overset{a}{\rightsquigarrow} c}{x[A] \overset{a}{\rightsquigarrow} c} \qquad \alpha\mathbf{Neg}\ \frac{x \overset{a}{\rightsquigarrow} c}{\overline{x} \overset{a}{\rightsquigarrow} \widehat{c}}$$

$$\alpha\mathbf{Try_1}\ \frac{x \overset{a}{\rightsquigarrow} c}{\mathbf{try}\ x\ \mathbf{unless}\ y \overset{a}{\rightsquigarrow} c}\ c \neq \iota \qquad \alpha\mathbf{Try_2}\ \frac{x \overset{a}{\rightsquigarrow} \iota \qquad y \overset{a}{\rightsquigarrow} d}{\mathbf{try}\ x\ \mathbf{unless}\ y \overset{a}{\rightsquigarrow} \widehat{d} \wedge \iota}$$

Fig. 3. Rules for the completion function α of the machine of PAR expressions

$$\mathbf{Ev}:\ \mathbf{a} \xrightarrow{b} \mathbf{a}\ (b \neq a) \qquad \mathbf{Neg}\ \frac{x \xrightarrow{a}_{\iota} x'}{\overline{x} \xrightarrow{a} \overline{x'}} \qquad \mathbf{Push}\ \frac{x \xrightarrow{a}_{\iota} x'}{x[A] \xrightarrow{a} x'[A]}$$

$$\mathbf{Seq_1}\ \frac{x \xrightarrow{a}_{\iota} x'}{x\,;y \xrightarrow{a} x'\,;y} \qquad \mathbf{Seq_2}\ \frac{x \overset{a}{\rightsquigarrow} \top}{x\,;y \xrightarrow{a} y}$$

$$\mathbf{Sel_1}\ \frac{x \xrightarrow{a}_{\iota} x' \qquad y \xrightarrow{a}_{\iota} y'}{x \mid y \xrightarrow{a} x' \mid y'} \qquad \mathbf{Sel_2}\ \frac{x \overset{a}{\rightsquigarrow} \bot \qquad y \xrightarrow{a}_{\iota} y'}{x \mid y \xrightarrow{a} y'} \qquad \mathbf{Sel_3}\ \frac{x \xrightarrow{a}_{\iota} x' \qquad y \overset{a}{\rightsquigarrow} \bot}{x \mid y \xrightarrow{a} x'}$$

$$\mathbf{Rep_1}\ \frac{x \xrightarrow{a}_{\iota} x'}{\mathbf{repeat}\, x \xrightarrow{a} x'\,;\mathbf{repeat}\, x} \qquad \mathbf{Rep_2}\ \frac{x \overset{a}{\rightsquigarrow} \top}{\mathbf{repeat}\, x \xrightarrow{a} \mathbf{repeat}\, x}$$

$$\mathbf{Try_1}\ \frac{x \xrightarrow{a}_{\iota} x' \qquad y \xrightarrow{a}_{\iota} y'}{\mathbf{try}\ x\ \mathbf{unless}\ y \xrightarrow{a} \mathbf{try}\ x'\ \mathbf{unless}\ y'} \qquad \mathbf{Try_2}\ \frac{x \xrightarrow{a}_{\iota} x' \qquad y \overset{a}{\rightsquigarrow} \bot}{\mathbf{try}\ x\ \mathbf{unless}\ y \xrightarrow{a} x'}$$

Fig. 4. Rules for the step function ∂ of the machine of PAR expressions

Completion Function (Fig. 3). Rule ($\alpha\mathbf{Ev_1}$) says that the simple expression **a** declares success upon receiving an a event, while rule ($\alpha\mathbf{Ev_2}$) states that **a** is incomplete otherwise. More interesting is rule ($\alpha\mathbf{Seq}$): the completion status of $x\,;y$ is that of x, but no higher than ι (i.e., either $\bot$ or ι). Rule ($\alpha\mathbf{Try_1}$) says that if the try part completes in $\top$ or $\bot$, then so does the **try-unless** expression. Rule ($\alpha\mathbf{Try_2}$) says that if the try part is incomplete and the unless part succeeds then **try-unless** fails, and that it remains incomplete otherwise.

Transition Function (Fig. 4). Rule ($\mathbf{Rep_1}$) says that if, after an event is processed, the body x is still incomplete, with x' as derivative, then the successor expression is $x'\,;\mathbf{repeat}\, x$. If, on the other hand, x declares success, rule ($\mathbf{Rep_2}$) states that the successor expression is the continuation **repeat** x. The last case, when x declares failure, the successor expression of **repeat** x is **silent**. This is handled by a global rule (**Silent**), which complements all rules in Fig. 4:

$$\textbf{oEv}: \ \mathbf{a} \overset{b}{\Rrightarrow} \varnothing \qquad \textbf{oNeg}\ \frac{x \overset{a}{\Rrightarrow} o}{\overline{x} \overset{a}{\Rrightarrow} o} \qquad \textbf{oSeq}\ \frac{x \overset{a}{\Rrightarrow} o}{x\,;y \overset{a}{\Rrightarrow} o}$$

$$\textbf{oSel}\ \frac{x \overset{a}{\Rrightarrow} o \qquad y \overset{a}{\Rrightarrow} u}{x \mid y \overset{a}{\Rrightarrow} o \cup u} \qquad \textbf{oRep}\ \frac{x \overset{a}{\Rrightarrow} o}{\textbf{repeat}\ x \overset{a}{\Rrightarrow} o} \qquad \textbf{oTry}\ \frac{x \overset{a}{\Rrightarrow} o \qquad y \overset{a}{\Rrightarrow} u}{\textbf{try}\ x\ \textbf{unless}\ y \overset{a}{\Rrightarrow} o \cup u}$$

$$\textbf{oPush}_1\ \frac{x \overset{a}{\Rrightarrow} o \qquad x \overset{a}{\rightsquigarrow} \top}{x[A] \overset{a}{\Rrightarrow} o \cup A} \qquad \textbf{oPush}_1\ \frac{x \overset{a}{\Rrightarrow} o \qquad x \overset{a}{\not\rightsquigarrow} \top}{x[A] \overset{a}{\Rrightarrow} o}$$

Fig. 5. Rules for the output function o of the machine of PAR expressions

$$\textbf{Silent}\ \frac{x \overset{a}{\not\rightsquigarrow} \iota}{x \overset{a}{\longrightarrow} \textbf{silent}}.$$

This rule establishes that every expression becomes **silent** after declaring success or failure. This ensures the adequacy condition ($S1$). The special expression **silent** is defined by the following three rules:

$$\alpha\textbf{Silent}: \ \textbf{silent} \overset{a}{\rightsquigarrow} \iota \qquad \partial\textbf{Silent}: \ \textbf{silent} \overset{a}{\longrightarrow} \textbf{silent} \qquad \textbf{oSilent}: \ \textbf{silent} \overset{a}{\Rrightarrow} \varnothing$$

Output Function (Fig. 5)**.** The rules for output (**oEv**) and (**oSilent**) state that simple expressions generate no output. Rules (**oNeg**), (**oRep**) and (**oSeq**) state that the output is that of the evaluating subexpressions, while rules (**oSel**) and (**oTry**) combine the output from the subexpressions evaluated in parallel. The rules (**oPush$_1$**) and (**oPush$_2$**) govern how new output is added.

Example 6. The expression **repeat** (**a**;**try a**$[A]$ **unless b**), describes the behavior $[\![q_0]\!]$ shown in Fig. 2(a) (i.e., the behavior of state q_0 in machine M in Fig. 1.)

5 PAR Congruences and Output Equivalences

In this section we show that bisimilarity is the largest PAR congruence that refines output equivalence.

A PAR context (or simply a context) is a PAR expression with one special variable $\square$. The instantiation of context $C\langle\rangle$ with an expression x, denoted by $C\langle x\rangle$, corresponds to the resulting expression of substituting every occurrence of $\square$ by x in $C\langle\rangle$.

We say that a binary relation # between PAR expressions is a PAR congruence (or simply a congruence) if for every x, y, and every context $C\langle\rangle$, if $x \# y$ then $C\langle x\rangle \# C\langle y\rangle$. Examples of congruences include the empty relation, syntactic identity $\equiv$, and the universal relation PAR $\times$ PAR.[4] We say that a relation R *refines* a relation S if aRb implies aSb.

[4] In the Hidden Algebra line of research (see [5]) observations and experiments correspond to contexts of the language of the hidden specification (in our case this would be the language formed by o, α and ∂ in Definition 1). Note, on the other hand, that in this section we are reasoning about PAR congruences.

We first establish that if two states of arbitrary machines exhibit the same behavior then they are bisimilar.

Lemma 1. *If* $[\![m]\!] = [\![m']\!]$ *then* $m \approx m'$.

Proof. Consider the binary relations $R = \{\langle m, [\![m]\!]\rangle\}$ and $S = \{\langle m', [\![m']\!]\rangle\}$. A routine proof by coinduction shows that R and S are bisimulations, and therefore $R \circ S^{-1}$ is a bisimulation, which contains $\langle m, m'\rangle$ if $[\![m]\!] = [\![m']\!]$. □

Definition 7 (Output equivalence). *Two states* m *and* m' *are output equivalent, written* $m \sim m'$, *if* $\pi_1[\![m]\!] = \pi_1[\![m']\!]$.

The relation $\sim$ captures whether two states generate the same output when offered the same input. In practice, two states corresponding to output equivalent event-pattern reactive programs can be replaced by each other without modifying the observable behavior to the served component.

Unfortunately, replacing two output equivalent PAR expressions in arbitrary PAR contexts does not preserve output equivalence: consider the expressions **silent** and **a** (which are output equivalent since both generate the empty output) and the context $\square[A]$. Clearly, $\mathbf{silent}[A] \not\sim \mathbf{a}[A]$, as witnessed by the stream prefix a. Congruences that refine observational equivalence are important in practice too, since they allow to replace expressions as part of enclosing PAR programs while maintaining the behavior. Syntactic equivalence $\equiv$ is trivially a congruence that refines observational equivalence, but it is too fine for our purposes.

Theorem 4. *(1) Bisimilarity* $\approx$ *is a* PAR *congruence. (2) Bisimilarity is the largest* PAR *congruence that refines output equivalence.*

Proof. The proof of (1) is omitted. (2) Consider the relation $\#$ defined as: $x\#y$ precisely when, for every context $C\langle\rangle$, $C\langle x\rangle \sim C\langle y\rangle$; $\#$ clearly refines output equivalence and is itself a congruence, since the composition of contexts is a context. Moreover, every congruence S that refines output equivalence also refines $\#$ because if xSy then $C\langle x\rangle S C\langle y\rangle$, and then $C\langle x\rangle \sim C\langle y\rangle$ and $x\#y$. Hence, it is sufficient to show that $\#$ refines $\approx$. We show that $x\#y$ implies $[\![x]\!] = [\![y]\!]$. First, $x\#y$ implies $\pi_1[\![x]\!] = \pi_1[\![y]\!]$ by considering the empty context. Moreover, $x\#y$ also implies $\pi_2[\![x]\!] = \pi_2[\![y]\!]$. By contradiction, assume $\pi_2[\![x]\!] \neq \pi_2[\![y]\!]$, and consider the contexts $C_1 = \square; a[A]$ and $C_2 = \overline{\square}; a[A]$. Either $C_1\langle x\rangle \not\sim C_1\langle y\rangle$ or $C_2\langle x\rangle \not\sim C_2\langle y\rangle$, which contradicts $x\#y$. By Lemma 1 we can conclude that $x \approx y$. □

Indirectly, the previous proof provides an alternative definition of PAR bisimilarity, since $\#$ and $\approx$ are shown to be the same relation.

6 Proofs by Coinduction

The definitions and results from the previous two sections allow us to perform equational reasoning at the level of PAR expressions. Two PAR expressions x and

y that exhibit the same behavior cannot be distinguished by any experiment, and by Theorem 4, x and y are then also indistinguishable in any PAR context $C\langle\rangle$. This justifies the use of equations to represent that all their ground instances are bisimilar PAR expressions. Some examples of such equations are

$$\begin{array}{rr} x \mid y = y \mid x & x \,;(y\,; z) = (x\,; y)\,; z \\ x \mid (y \mid z) = (x \mid y) \mid z & \overline{x\,; y} = \overline{x} \mid x\,; \overline{y} \\ x \mid x = x & x\,; \textbf{try}\ y\ \textbf{unless}\ z = \textbf{try}\ x\,; y\ \textbf{unless}\ x\,; z \\ \overline{\overline{x}} = x & \textbf{repeat}\ x = x\,; \textbf{repeat}\ x \end{array}$$

Example 7. To illustrate the use of coinduction to show the validity of these equations we show the commutativity of the operator $|$, that is, for all PAR expressions x and y, $x|y \approx y|x$. It is sufficient to show that there is a bisimulation containing all pairs $\langle x \mid y, y \mid x\rangle$; we prove that $R = \{\langle x \mid y, y \mid x\rangle\} \cup \equiv$ is a bisimulation. Take arbitrary expressions x and y. First,

$$\begin{aligned} o_a(x \mid y) &= o_a x \cup o_a y = o_a(y \mid x), \text{ and} \\ \alpha_a(x \mid y) &= \alpha_a x \vee \alpha_a y = \alpha_a(y \mid x). \end{aligned}$$

Second, let $x' = \partial_a x$ and $y' = \partial_a y$. We split cases according to $\alpha_a x$ and $\alpha_a y$:

1. $\alpha_a x \neq \iota$ or $\alpha_a y \neq \iota$: in all these cases $\partial_a(x \mid y) \equiv \partial_a(y \mid x)$.
2. $\alpha_a x = \iota = \alpha_a y$. Here, $\partial_a(x \mid y) = x' \mid y'$ and $\partial_a(y \mid x) = y' \mid x'$. By definition of R, $(x' \mid y')R(y' \mid x')$.

Example 8. Let us also prove that for all expressions x and y,

$$\overline{x\,; y} = \overline{x} \mid (x\,; \overline{y}).$$

We show that $R = \{\langle \overline{x\,; y}, \overline{x} \mid (x\,; \overline{y})\rangle\} \cup \equiv$ is a bisimulation. First,

$$\begin{aligned} o_a(\overline{x\,; y}) &= o_a(x\,; y) = o_a x, \quad \text{and} \\ o_a(\overline{x} \mid (x\,; \overline{y})) &= o_a(\overline{x}) \cup o_a(x\,; \overline{y}) = o_a x \cup o_a x = o_a x. \end{aligned}$$

Now, let us consider all the cases for $\alpha_a x$:

1. $\alpha_a x = \bot$. Then, both $\alpha_a(\overline{x\,; y})$ and $\alpha_a(\overline{x}|(x;\overline{y}))$ become $\top$, and consequently both derivatives are **silent**.
2. $\alpha_a x = \iota$. Let $x' = \partial_a x$. First, both $\alpha_a(\overline{x\,; y})$ and $\alpha_a(\overline{x} \mid (x\,; \overline{y}))$ become ι. Also,
$$\begin{aligned} \partial_a(\overline{x\,; y}) &= \overline{x'\,; y}, \quad \text{and} \\ \partial_a(\overline{x} \mid (x\,; \overline{y})) &= \overline{x'} \mid (x'\,; \overline{y})), \end{aligned}$$
and then $\langle \partial_a(\overline{x\,; y}), \partial_a(\overline{x} \mid (x\,; \overline{y}))\rangle$ is in R.
3. $\alpha_a x = \top$. Then, again, $\alpha_a(\overline{x\,; y})$ and $\alpha_a(\overline{x} \mid (x\,; \overline{y}))$ become ι. Finally,
$$\begin{aligned} \partial_a(\overline{x\,; y}) &= \overline{y}, \quad \text{and} \\ \partial_a(\overline{x} \mid (x\,; \overline{y})) &= \overline{y}, \end{aligned}$$
which are related by $\equiv$, and hence by R.

7 Conclusions

Using coalgebraic techniques, we have built a framework for the study of languages to describe event-pattern reactive programs. This framework provides a convenient domain for the definition of the behavioral operational semantics of event-pattern reactive programs. Using this framework we have defined the formal semantics of PAR.

The semantics of event-pattern languages are most naturally defined compositionally. To enable such compositional semantics, we introduced a completion status, giving rise to the functor $\Gamma_1(X) = (\Sigma \to (\mathcal{O} \times \mathcal{C} \times X))$, rather than the simpler $\Gamma_2(X) = (\Sigma \to (\mathcal{O} \times X))$—which may have been expected to study synchronous maps from inputs to outputs.

Our results can be directly compared to other formalisms based on Γ_2 (like Moore and Mealy machines and interactive computation [6]), by simply hiding the completion component. (In fact, bisimulation in Γ_2 becomes output equality in Γ_1, $\sim$ as defined in Sect. 5.)

Some ongoing and future research includes:

Expressiveness. It is easy to show that every PAR expression has only a finite number of derivatives (for all possible input prefixes from Σ^*). Thus, all PAR behaviors can be expressed with finite memory. The converse is also true: every behavior that can be described with finite memory can also be described by a PAR expression. This result [17] parallels the correspondence between regular expressions and finite automata [11,12].

The syntax of PAR presented here is minimal in the sense that, by removing any one operator, expressive completeness is lost. In practice, though, it is useful to have more operators available. In [16] we introduced ECL, with a larger set of operators than PAR. We are studying the conciseness of specification and the complexity of analysis of these extensions.

Equational Reasoning. Section 6 presented some equalities that hold between the corresponding instances of both sides. We illustrated one such a proof using coinduction. Two important open problems are: (1) whether this proof technique can be automated—in other words— whether coinduction together with some other rules provides a complete proof system for PAR equivalences; (2) the existence of a finite list of PAR equations that form an axiomatization of bisimulation for PAR, which could lead to alternative decision procedures for checking (parametrized) equivalences.

References

1. Marcos Kawazoe Aguilera, Robert E. Strom, Daniel C. Sturman, Mark Astley, and Tushar Deepak Chandra. Matching events in a content-based subscription system. In *Symposium on Principles of Distributed Computing*, pages 53–61, 1999.
2. Jos C. M. Baeten and W. Peter Weijland. *Process Algebra*. Cambridge University Press, 1990.

3. Antonio Carzaniga, David S. Rosenblum, and Alexander L. Wolf. Design and evaluation of a wide-area event notification service. *ACM Transactions on Computer Systems*, 19(3):332–383, August 2001.
4. Corina Cîrstea. Semantic constructions from the specification of objects. *Theoretical Computer Science*, 260, 2001.
5. Joseph Goguen and Grant Malcolm. A hidden agenda. *Theoretical Computer Science*, 245(1), 2000.
6. Dina Q. Goldin. Persistent Turing Machines as a model of interactive computation. In *Foundations of Information and Knowledge Systems*, pages 116–135, Burg, Germany, February 2000.
7. Matthew Hennessy and Robin Milner. Algebraic laws for nondeterminism and concurrency. *Journal of the Association for Computer Machinery*, 32(1):137–161, January 1985.
8. C. Antony R. Hoare. *Communicating Sequential Processes.* Prentice-Hall, 1985.
9. Frank Hunleth, Ron Cytron, and Christopher Gill. Building customizable middleware using aspect oriented programming. In *Workshop on Advanced Separation of Concerns (OOPSLA'01)*, 2001.
10. Bart Jacobs and Jan J. M. M. Rutten. A tutorial on (co)algebras and (co)induction. *Bulletin of the European Association for Theoretical Computer Science*, 62:222–259, 1997.
11. Stephen C. Kleene. Representation of events in nerve nets and finite automata. In Claude E. Shannon and John McCarthy, editors, *Automata Studies*, number 34, pages 3–41. Princeton University Press, Princeton, New Jersey, 1956.
12. Robert F. McNaughton and H. Yamada. Regular expressions and state graphs for automata. *IEEE Transactions on Electronic Computers*, 9:39–47, 1960.
13. Robin Milner. *Communication and Concurrency.* Prentice-Hall, 1989.
14. Grigore Roşu. *Hidden Logic.* PhD thesis, University of California at San Diego, 2000.
15. Jan J. M. M. Rutten. Automata and coinduction (an exercise in coalgebra). In *CONCUR*, 1998.
16. César Sánchez, Sriram Sankaranarayanan, Henny B. Sipma, Ting Zhang, David Dill, and Zohar Manna. Event correlation: Language and semantics. In Rajeev Alur and Insup Lee, editors, *EMSOFT 2003*, pages 323–339. Spring-Verlag, 2003.
17. César Sánchez, Matteo Slanina, Henny B. Sipma, and Zohar Manna. Expressive completeness of an event-pattern reactive programming language. Submitted for publication.
18. Douglas C. Schmidt, David L. Levine, and Timothy H. Harrison. The design and performance of a real-time CORBA object event service. In *Proc. of OOPSLA'97*, 1997.
19. Bill Segall and David Arnold. Elvin has left the building: A publish/subscribe notification service with quenching. In *Queensland AUUG Summer Technical Conference, Brisbane, Australia*, 1997.
20. David Sharp. Reducing avionics software cost through component based product line development. In *Proc. of the Software Technology Conference*, 1998.

Complete Symbolic Reachability Analysis Using Back-and-Forth Narrowing

Prasanna Thati[1] and José Meseguer[2]

[1] Carnegie Mellon University, USA
thati@cs.cmu.edu
[2] University of Illinois at Urbana-Champaign, USA
meseguer@cs.uiuc.edu

Abstract. We propose a method called *back-and-forth narrowing* for solving reachability goals of the form $(\exists \overrightarrow{x}).t_1 \rightarrow^* t'_1 \wedge \ldots \wedge t_n \rightarrow^* t'_n$ in general term rewrite systems. The method is a complete semi-decision procedure in the sense that it is guaranteed to find a solution when one exists, but in general it may not terminate when there are no solutions. The completeness result is very general in that it makes no assumptions about the given term rewrite system. Specifically, the rewrite rules need *not* be linear, confluent, or terminating, and can even have extra-variables in the righthand side. Such generality is often essential while modeling concurrent systems or axiomatizing inference systems as rewrite rules, and in such applications back-and-forth narrowing can be used as a sound and complete technique for symbolic reachability analysis or as a deductive procedure for proving existential formulae.

1 Introduction

A concurrent or an inference system can be naturally expressed as a rewrite system $\mathcal{R} = (\Sigma, R)$, where Σ is a signature and R is a collection of rewrite rules. For a concurrent system terms represent states, and a rewrite rule $t \rightarrow t'$ is understood as a (parametric) local transition. For an inference system terms represent formulae, and rewrite rules specify basic inference steps. In this paper, we address the question of solving *reachability goals* in a rewrite system $\mathcal{R}$. By a reachability goal we mean an existentially quantified formula of the form

$$(\exists \overrightarrow{x})\ t_1 \rightarrow^* t'_1\ \wedge \ldots \wedge\ t_n \rightarrow^* t'_n$$

where each *source* t_i is a term with variables $Var(t_i) \subseteq \overrightarrow{x}$ specifying a possibly infinite set of initial configurations (namely all its instances by *ground substitutions*), and each *target* t'_i is a term with variables $Var(t'_i) \subseteq \overrightarrow{x}$ that represents likewise a possibly infinite set of configurations that we want to reach by a sequence of transitions starting from the corresponding source t_i. *Solutions* to this reachability problem can then be described by *substitutions* σ for which indeed we have, $\mathcal{R} \vdash \sigma(t_i) \rightarrow^* \sigma(t'_i)$ for $1 \leq i \leq n$. The meaning and interest of solving reachability goals such as the above is clear; it would serve as both a *symbolic*

J.L. Fiadeiro et al. (Eds.): CALCO 2005, LNCS 3629, pp. 379–394, 2005.

reachability analysis technique for concurrent systems, and, alternatively, as a *deductive procedure* for proving existential formulae in inference systems.

We propose a semi-decision procedure called *back-and-forth-narrowing* for solving reachability goals. This procedure is *complete* in the *solvability* sense in that it is guaranteed to find a solution when there is one. The procedure is very general in the sense that there are absolutely no assumptions on the given rewrite system $\mathcal{R}$. In particular, the rewrite rules in $\mathcal{R}$ need *not* be left or right linear, or confluent, or terminating, and can also have *extra variables* in the righthand side. This is to be contrasted with other approaches such as model-checking results for special classes of systems [4, 7, 20], or tree-automata based reachability analysis [8, 18, 3] where typically the rules and the goal are assumed to be linear. In some tree automata approaches [8, 18] non-linearity is dealt with using abstractions or conservative approximations of the reachability set; in contrast, back-and-forth narrowing is an *exact* procedure. A more detailed comparison with related work is presented in Section 7.

Back-and-forth-narrowing is a generalization of *narrowing*, a technique originally introduced as a complete method for generating all solutions of an equational unification problem. Specifically, narrowing was introduced for solving goals of the form $(\exists \overrightarrow{x})\ t_1 = t_1' \wedge \ldots \wedge t_n = t_n'$ in free algebras modulo a set of confluent and terminating equations used as rewrite rules [11, 12, 15]. Of course, in our new reachability setting, the meaning of a rewrite rule is changed from the previous meaning as an *equality* to a new meaning as a *transition* or *inference*. Further, the completeness of narrowing for equational unification critically depends on the confluence property of equations; an assumption which is done away with in our reachability setting. As a result of these generalizations, a naive extension of narrowing to the reachability setting turns about to be *incomplete* as shown in [14]. It is also shown in [14] that the naive narrowing procedure, however, is complete for certain restricted classes of rewrite theories such as those that are top-most or right-linear. We show, in this paper, that completeness can be regained for arbitrary rewrite systems by using back-and-forth narrowing.

Several applications of solving reachability goals using (naive) narrowing have been reported, especially in the area of verification of computer security protocols [2, 13, 14]. While the approach in [2, 13] is to use narrowing to symbolically search the reachable state space of a protocol, the approach in [14] is to use narrowing to solve appropriate existential formulae in the Dolev-Yao inference system [5] in order to discover attacks if any. These applications exploit the fact mentioned above that naive narrowing is complete for a restricted class of rewrite systems that is sufficient to model the protocols being considered. Our back-and-forth narrowing procedure would substantially expand the scope of these applications to cases where one needs completeness for general rewrite systems.

In the following section, we describe the essential idea behind back-and-forth narrowing at an intuitive level. We follow it up with a more formal treatment in Sections 3 to 6. We discuss related work in Section 7, and conclude in Section 8. We omit all proofs in this paper because of space constraints, but we will supply them in the forthcoming journal version.

2 The Basic Idea

The essential idea behind using narrowing for solving reachability goals is that a single narrowing sequence starting from a term t can be used to symbolically represent *many* rewrite sequences starting from instances of t. Specifically, for a term t and substitution σ, suppose $\sigma(t) \to t'$ by rewriting with the rule $l \to r$ at a non-variable position ω in t. Then clearly, l and $t|_\omega$ (the subterm of t at position ω) are unifiable. If ρ is the most general unifier, then we can show that $\rho(t) \to t''$ for some t'' by applying $l \to r$ at position ω, and there is a substitution η such that $t' = \eta(t'')$. This observation motivates the definition of the narrowing step $t \overset{\rho}{\rightsquigarrow} t''$; the intention is to use this narrowing step to symbolically represent several rewrite steps, one for each unifier σ of l and $t|_\omega$.

Building on the above idea, one can compose several narrowing steps to get a sequence that symbolically represents many underlying rewrite sequences. One can then hope to use narrowing sequences to search for solutions of reachability goals. Specifically, to solve a given goal $\exists \overrightarrow{x}.t_1 \to^* t_2$ we systematically explore the narrowing tree starting from t_1, and look for a narrowing sequence $t_1 \overset{\rho_1}{\rightsquigarrow} \dots \overset{\rho_n}{\rightsquigarrow} t_1'$ such that t_1' and t_2 are unifiable. If η is one such unifier then $\eta \circ \rho_n \circ \dots \rho_1$ is a solution. Unfortunately, although sound, this procedure is *not always complete* even in the solvability sense, i.e., it may fail to find a solution even when one exists. The crucial reason is that, by definition, narrowing can be performed only at non-variable positions, and therefore cannot account for rewrites that occur within the solution (i.e. under variable positions)[1]. Such "under-the-feet" rewrites can have non-trivial effects if the rewrite rules or the reachability goal are non-linear, and the rules are not confluent. Consider for example the rewrite rules: (i) $a \to b$, (ii) $a \to c$, (iii) $f(b,c) \to d$, and the reachability goal $\exists x.f(x,x) \to^* d$. The substitution $\{a/x\}$ is a solution, but the narrowing procedure returns no solutions since $f(x,x)$ can neither be narrowed further nor unified with d.

A natural question to ask is whether the simple narrowing procedure above is complete for specific classes of rewrite systems, or with respect to specific classes of solutions. Indeed, as shown in Section 5, the narrowing procedure above is *weakly* complete in that it can find all R-normalized solutions provided the rewrite rules have no extra variables in the righthand side (see Theorems 2 and 3). However, narrowing may not find solutions that are not normalized. More generally, in [14] we also identified several classes of rewrite systems for which the naive narrowing procedure can find *all* solutions, and applied these results to verify safety properties of cryptographic protocols.

In this paper, we establish a completeness result of a much broader scope by (i) generalizing the basic narrowing step through *linearization* of the term being narrowed, and (ii) using a *combination of forward and backward narrowing* with this generalized relation. Specifically, we account for under-the-feet rewrites by

[1] One could of course generalize the definition of narrowing to allow narrowing steps at variable positions. But that would make the narrowing procedure very inefficient since, in general, we will have to perform *arbitrary* instantiations of variables.

defining a narrowing step that is capable of "skipping" several such rewrites and capturing the first rewrite that occurs at a non-variable position. This is achieved by linearizing a term before narrowing it with a rule. The intermediate under-the-feet rewrites that have thus been skipped will be accounted for by extending the reachability goal with appropriate subgoals. For example, consider the goal $\exists x.\ f(x,x) \rightarrow^* d$ again. We (i) linearize the term $f(x,x)$ to, say, $f(x_1,x_2)$, (ii) narrow the linearized term with the rule $f(b,c) \rightarrow d$ and the unifier $\{b/x_1, c/x_2\}$, and (iii) extend the reachability goal with subgoals $x \rightarrow^* b$ and $x \rightarrow^* c$. This gives us the goal $\exists x.\ d \rightarrow^* d \wedge x \rightarrow^* b \wedge x \rightarrow^* c$.

Linearization alone does not help us regain completeness in general. For example, consider a goal $\exists \overrightarrow{x}.t \rightarrow^* t'$ where the solution σ is such that any rewrite sequence $\sigma(t) \rightarrow^* \sigma(t')$ is such that none of the rewrites occur at non-variable positions of t. But observe that if at least one of these rewrites occurs at a non-variable position in t', then we can narrow the right side t' in the *backward* direction, i.e. using R^{-1}, to obtain a simpler goal. For instance, in the goal $\exists x.\ d \rightarrow^* d \wedge x \rightarrow^* b \wedge x \rightarrow^* c$ above, backward narrowing twice gives us the goal $\exists x.\ d \rightarrow^* d \wedge x \rightarrow^* a \wedge x \rightarrow^* a$, which has the unifier (solution) $\{a/x\}$. In general, backward narrowing might in turn enable *forward* narrowing steps using R on the lefthand side, and so on, until we reach a point where all the rewrites occur under variable positions of both the lefthand and righthand sides. In this case, however, the lefthand and righthand sides are unifiable, and we are therefore done.

To keep the presentation simple at this point, we postpone a detailed example illustrating all of these until Section 6 (see Example 3). For the simple example considered above, however, note that just backward narrowing with R^{-1}, even without any linearization, gives us the solution as follows: $d \overset{id}{\leadsto} f(b,c) \overset{id}{\leadsto}\overset{id}{\leadsto} f(a,a)$. But as shown in Example 3, a combination of forward and backward narrowing is necessary, in that neither is complete by itself. In Theorems 5 and 6 we prove that with both the generalizations above we regain completeness in the solvability sense for arbitrary rewrite systems.

An important problem for back-and-forth narrowing to be effective in practice is to devise *strategies* that improve its efficiency. Otherwise, one would quickly face a combinatorial explosion in the number of possible narrowing sequences. When several back-and-forth narrowing derivations are possible for the same solution, the question is whether there is a preferred strategy and whether a standardization result is possible. Several *lazy* narrowing strategies that address these questions are known for special classes of rewrite systems [1], but extending these to back-and-forth narrowing is an open question and is beyond the scope of this paper. However, a few comments on a promising approach [6] are made in Section 8.

3 Background

A *signature* Σ is a ranked alphabet $\Sigma = \{\Sigma_n \mid n \in \mathbb{N}\}$, where Σ_n is a set of function symbols of arity n. A Σ-algebra is a set A together with a function

$f_A : A^n \to A$ for each $f \in \Sigma_n$. We assume an infinite set of variables X that are all different from constant symbols in Σ. We write T_Σ for the Σ-algebra of ground terms over Σ, and $T_\Sigma(X)$ for the Σ-algebra of terms with variables from the set X.

We use a finite sequence of positive integers, called a *position*, to denote an access path in a term. We let ω range over positions. For $t \in T_\Sigma(X)$ let $Var(t), Pos(t), FuPos(t)$ denote the set of variables, positions, and non-variable (or functional) positions in t, respectively. The root of a term is at position ϵ. We denote the subterm of t at position ω by $t|_\omega$.

A *substitution* is a mapping $\sigma : X \to T_\Sigma(X)$ which maps variables to terms, and which is different from the identity for only a finite subset $Dom(\sigma)$ of X. We denote the homomorphic extension of σ to $T_\Sigma(X)$ also by σ. The set of variables introduced by σ is $Ran(\sigma) = \cup_{x \in Dom(\sigma)} Var(\sigma(x))$. The restriction of a substitution σ to a set of variables V, is defined as $\sigma|_V(x) = \sigma(x)$ if $x \in V$, and $\sigma|_V(x) = x$ otherwise. We say that a substitution σ is *away* from a set of variables V if $Ran(\sigma) \cap V = \emptyset$. For substitutions σ, ρ such that $Dom(\sigma) \cap Dom(\rho) = \emptyset$ we define their union as

$$(\sigma \cup \rho)(x) = \begin{cases} \sigma(x) & \text{if } x \in Dom(\sigma) \\ \rho(x) & \text{if } x \in Dom(\rho) \\ x & \text{otherwise} \end{cases}$$

For a substitution σ that maps x_i to t_i for $1 \leq i \leq n$, we write $\{t_1/x_1, \ldots, t_n/x_n\}$ to denote σ. We denote the identity substitution by *id*.

The *subsumption* preorder $\ll$ on $T_\Sigma(X)$ is defined by $t \ll t'$ if there is a substitution σ such that $\sigma(t) = t'$; such a substitution σ is said to be a *match* from t to t'. For substitutions σ, ρ and a set of variables V we define $\sigma|_V = \rho|_V$ if $\sigma(x) = \rho(x)$ for all $x \in V$, and $\sigma|_V \ll \rho|_V$ if there is a substitution η such that $\rho|_V = (\eta \circ \sigma)|_V$.

A Σ-*equation* is an expression of the form $t = t'$. A *unifier* for the equation $t = t'$ is a substitution σ such that $\sigma(t) = \sigma(t')$. It is the case that, if t and t' are unifiable, then for any given finite set of variables V containing $W = Var(t) \cup Var(t')$, there is a most general unifier $\sigma = MGU(t = t', V)$ away from V such that (i) $Dom(\sigma) \subseteq W$, and (ii) $\sigma|_V \ll \rho|_V$ for any other unifier ρ of $t = t'$. This most general unifier σ is unique up to renaming of variables and can be computed by a unification algorithm [17].

A *rewrite rule* is an expression of the form $l \to r$, where $l, r \in T_\Sigma(X)$. An *(unconditional and unsorted) rewrite system* is a tuple $\mathcal{R} = (\Sigma, R)$ with Σ a signature, and R a set of rewrite rules. We write R^{-1} for the set that contains $l \to r$ if and only if $r \to l$ is in R. We define the *one-step rewrite relation* on $T_\Sigma(X)$ as follows: $t \to_R t'$ if there is an $\omega \in Pos(t)$, a rule $l \to r$ in R, and a substitution σ such that $t|_\omega = \sigma(l)$ and $t' = t[\omega \leftarrow \sigma(r)]$. We also write $t \xrightarrow{[\omega]}_R t'$ to make explicit the position at which the rewrite occurs. Note that $t \to_R t'$ if and only if $t' \to_{R^{-1}} t$. A term $t \in T_\Sigma(X)$ is called *R-irreducible* (or just *irreducible* if R is clear from the context) if there is no $t' \in T_\Sigma(X)$ such that $t \to_R t'$. For substitutions σ, ρ and a set of variables V we define $\sigma|_V \to_R \rho|_V$ if there is

$x \in V$ such that $\sigma(x) \to_R \rho(x)$ and for all other $y \in V$ we have $\sigma(y) = \rho(y)$. A substitution σ is called *R-normalized* if $\sigma(x)$ is irreducible for all x.

4 Reachability Goals

A *reachability goal* G is a conjunction of the form $\exists \overrightarrow{x}.t_1 \to^* t'_1 \wedge \ldots \wedge t_n \to^* t'_n$. To simplify notation, we will drop the existential quantification from now on, and we simply write $t_1 \to^* t'_1 \wedge \ldots \wedge t_n \to^* t'_n$. It is understood that the order of the subgoals $t_i \to^* t'_i$ in the expression is irrelevant, i.e., $\wedge$ is associative and commutative. We define $|G| = n$, and $Var(G) = \bigcup_i Var(t_i) \cup Var(t'_i)$. We write G^{-1} to denote the goal $t'_1 \to^* t_1 \wedge \ldots \wedge t'_n \to^* t_n$. A substitution σ is an *R-solution* of G (or just a solution of G when R is clear from the context) if $\sigma(t_i) \to^*_R \sigma(t'_i)$ for $1 \leq i \leq n$. Note that since $\sigma(t_i) \to^*_R \sigma(t'_i)$ if and only if $\sigma(t'_i) \to^*_{R^{-1}} \sigma(t_i)$, we have that ρ is an R-solution of G if and only if ρ is an R^{-1}-solution of G^{-1}. We denote the empty goal, i.e., for the case $n = 0$, by Λ, and define every substitution to be a solution of Λ. We call a goal G of the form $x_1 \to^* y_1 \wedge \ldots \wedge x_n \to^* y_n$, where all the lefthand sides and the righthand sides are variables, a *trivial* goal. Note that the substitution σ such that $\sigma(x_i) = \sigma(y_i) = z$ for some variable z, is a solution of this goal. We also define Λ to be a trivial goal.

Definition 1. *We define the rewrite relation on goals as follows.*

(Reduce)	$G \wedge t_1 \to^* t_2$	$\xrightarrow{[\omega]}_R$	$G \wedge t'_1 \to^* t_2$	*if* $t_1 \xrightarrow{[\omega]}_R t'_1$
(Eliminate)	$G \wedge t \to^* t$	$\xrightarrow{[\epsilon]}_R$	$G.$	

Note that in $G \xrightarrow{[\omega]}_R G'$, the position ω is not sufficient to determine the exact subgoal at which the rewrite happens. But we adopt this notation because it is sufficient for our purposes and it simplifies the presentation. Further, instead of $G \xrightarrow{[\omega]}_R G'$ we may simply write $G \to_R G'$.

Lemma 1. *σ is an R-solution of G if and only if $\sigma(G) \to^*_R \Lambda$.*

For a set of variables V containing $Var(G)$, we say that a set of substitutions $CSS(G, V)$ is a *complete* set of R-solutions of G away from V if: (i) every $\sigma \in CSS(G, V)$ is an R-solution of G, (ii) for each solution ρ of G there is a $\sigma \in CSS(G, V)$ such that $\sigma|_{Var(G)} \ll \rho|_{Var(G)}$, and (iii) for every $\sigma \in CSS(G, V)$, $Dom(\sigma) \subseteq Var(G)$ and $Ran(\sigma) \cap V = \emptyset$. We are interested in finding a complete set of R-solutions of a goal G in an (unconditional) rewrite system $\mathcal{R}$.

5 Narrowing: Soundness and Weak Completeness

In this section we show that narrowing provides a sound but only weakly complete procedure (in the sense made precise below) for computing the solutions of reachability goals. We introduced the main ideas in this Section in [14], but here we reformulate their technical presentation in a manner that allows a smooth extension to our more general back-and-forth narrowing procedure in the next section.

The essential idea behind narrowing is to *symbolically* represent the transition relation between terms as a narrowing relation between terms. Specifically, narrowing instantiates the variables in a term by the most general unifier that enables a rewrite with a given rule and a term position. This narrowing relation on terms is then extended to reachability goals by narrowing only the lefthand sides of the goals, while the righthand sides only accumulate substitutions. The idea is to repeatedly narrow the lefthand sides until each lefthand side unifies with the corresponding righthand side. The composition of the unifier with all the substitutions generated (in the reverse order) gives us a solution of the goal.

Definition 2 (narrowing of terms). *We define* $t \overset{\sigma}{\leadsto}_R t'$ *if there is* $\omega \in FuPos(t)$, *a rule* $l \to r$ *in* R *(assume* $Var(t) \cap Var(l,r) = \emptyset$*), such that for a set of variables* V *containing* $Var(t)$ *and* $Var(l,r)$ *and* $\sigma = MGU(t|_\omega = l, V)$, *we have* $t' = \sigma(t[\omega \leftarrow r])$.

Definition 3 (narrowing of goals). *The narrowing relation on goals is defined by the following two inference rules.*

$$\textbf{(Narrow)} \quad G \wedge t \to^* t' \overset{\sigma}{\leadsto}_R \sigma(G) \wedge t'' \to^* \sigma(t') \quad \text{if } t \overset{\sigma}{\leadsto}_R t'' \text{ and } \sigma \text{ is away from } Var(G,t,t')$$

$$\textbf{(Unify)} \quad G \wedge t \to^* t' \overset{\sigma}{\leadsto}_R \sigma(G) \quad \text{if } \sigma = MGU(t = t', Var(G,t,t'))$$

We write $G \overset{\sigma}{\leadsto}{}^*_R G'$ *if either* $G = G'$ *and* $\sigma = id$, *or there is a sequence of derivations* $G \overset{\sigma_1}{\leadsto}_R \ldots \overset{\sigma_n}{\leadsto}_R G'$ *such that* $\sigma = \sigma_n \circ \sigma_{n-1} \circ \ldots \circ \sigma_1$.

<u>**Soundness**</u>: We first consider the soundness problem. Following the idea in [11] we associate with each narrowing step between terms, a corresponding rewrite step. The proofs of the propositions below are easy.

Lemma 2. $t \overset{\sigma}{\leadsto}_R t'$ *implies* $\sigma(t) \to_R t'$.

Lemma 3. *If* $G \overset{\sigma}{\leadsto}_R G'$ *and* ρ *is a solution of* G' *then* $\rho \circ \sigma$ *is a solution of* G.

This gives us the following soundness theorem.

Theorem 1 (Soundness). *If* $G \overset{\sigma}{\leadsto}{}^*_R \Lambda$, *then* σ *is solution of* G.

<u>**Weak Completeness**</u>: The idea behind proving weak completeness is to associate with each rewrite step a corresponding narrowing step. It is possible to establish such a correspondence only under certain assumptions, and hence the weakness in the completeness. In the following, note that we assume that each rule $l \to r$ in R has no extra variables in its righthand side, i.e., $Var(r) \subseteq Var(l)$. However, we will *drop* this assumption in the following section where we consider the more general back-and-forth narrowing.

Lemma 4. *Let R be a set of rules with no extra variables in their righthand sides. Let ρ be an R-normalized substitution, and let V be a finite set of variables containing Var(t). Let $\rho(t) \rightarrow_R t'$ using the rule $l \rightarrow r$. Then there are σ, t'', η such that: (i) $t \stackrel{\sigma}{\rightsquigarrow}_R t''$ using the same rule, σ away from V, (ii) η is R-normalized, (iii) $\eta(t'') = t'$, and (iv) $\rho|_V = (\eta \circ \sigma)|_V$.*

The above lemma can be easily lifted to goals as follows.

Lemma 5. *Let R be a set of rules with no extra variables in their righthand sides. Let ρ be an R-normalized substitution, V be a finite set of variables containing Var(G), and let $\rho(G) \rightarrow_R G'$. Then, there are σ, G'', η such that: (i) $G \stackrel{\sigma}{\rightsquigarrow}_R G''$, σ away from V, (ii) η is R-normalized, (iii) $\eta(G'') = G'$, and (iv) $\rho|_V = (\eta \circ \sigma)|_V$.*

This gives us the following weak completeness result.

Theorem 2 (Weak Completeness). *Let R be a set of rewrite rules with no extra-variables in the righthand side, let ρ be an R-normalized solution of a reachability goal G, and let V be a finite set of variables containing Var(G). Then $G \stackrel{\sigma}{\rightsquigarrow}{}^*_R \Lambda$ for some σ away from V such that $\sigma|_V \ll \rho|_V$.*

We show below that Theorem 2 need not hold for substitutions ρ that are not R-normalized, and hence narrowing is only weakly complete.

A Weakly Complete Algorithm for Reachability Goals: A simple consequence of Theorems 1 and 2 is the following.

Theorem 3. *Let R be a set of rules with no extra-variables in the righthand side. Then for a finite set of variables V containing Var(G), the set of all substitutions $\sigma|_{Var(G)}$ such that $G \stackrel{\sigma}{\rightsquigarrow}{}^*_R \Lambda$ and σ is away from V, is a complete set of solutions of G away from V, with respect to R-normalized solutions.*

This theorem provides a general algorithm which builds a narrowing tree starting from G, to find all normalized solutions. Nodes in this tree correspond to goals, while edges correspond to one-step narrowing derivations. Since there can be infinitely long narrowing derivations, the algorithm has to expand the tree in a *fair* manner to cover each possible derivation.

Incompleteness of Narrowing: Narrowing is complete only with respect to normalized solutions. Specifically, it may not find solutions that are not normalized. We showed an example in the introduction where a reachability goal had a single non-normalized solution, but the narrowing procedure failed to find it. Here is another example.

Example 1. Let $\mathcal{R} = (\Sigma, R)$, where the signature Σ has unary function symbols s, f, g, and R has the following two rules: $s(x) \rightarrow s^2(x)$, and $f(s^2(x)) \rightarrow g(s(x))$. The reachability goal $G = f(x) \rightarrow^* g(x)$ has solutions $\sigma_k = \{s^k(y)/x\}$ for $k \geq 1$ (none of which is R-normalized). But narrowing returns only σ_2 as a solution, and it is not the case that $\sigma_2|_{\{x\}} \ll \sigma_1|_{\{x\}}$.

6 Back-and-Forth Narrowing

The main reason for the incompleteness of narrowing is that, since terms can be narrowed only at non-variable positions (see Definition 2), it is not possible to associate a narrowing step for the rewrite $\rho(t) \xrightarrow{[\omega]}_R t'$ where $\omega \notin FuPos(t)$. Such rewrites "under-the-feet" of t are possible if the substitution ρ is not normalized. This is precisely the reason for the assumption in Theorem 2 that the solution ρ of the goal G is normalized. Fortunately, it is possible to generalize the narrowing relation to one that, in some sense, also accounts for such under-the-feet rewrites.

Suppose ρ is a (not necessarily normalized) solution of the reachability goal $G = G_1 \wedge t_1 \to^* t_2$. Let

$$\rho(t_1) \xrightarrow{[\omega_1]}_R \dots u \xrightarrow{[\omega_k]}_R v \dots \xrightarrow{[\omega_n]}_R \rho(t_2) \quad (1)$$

and let k be such that $\omega_i \notin FuPos(t_1)$ for $1 \leq i < k$ and $\omega_k \in FuPos(t_1)$. Suppose we *linearize* the term t_1 by renaming each occurrence of a variable $x \in Var(t_1)$ to a distinct variable $x' \notin Var(G)$, and thereby obtain a term $\overline{t_1}$. Then, since all the rewrites in $\rho(t_1) \to^*_R u$ occur under-the-feet of t_1, i.e., at positions $\omega \notin FuPos(t_1)$, there is a substitution ρ' such that $\rho'(\overline{t_1}) = u$. Specifically, if a variable $x \in Var(t_1)$ is renamed to, say, $x_1, \dots, x_n$, in $\overline{t_1}$, then $\rho(x) \to^*_R \rho'(x_i)$ for $1 \leq i \leq n$. Now, as in Lemma 4, we can associate to the rewrite step $\rho'(\overline{t_1}) \xrightarrow{[\omega_k]}_R v$ a narrowing step $\overline{t_1} \overset{\sigma}{\rightsquigarrow}_R w$ for some σ and w.

The observation above motivates the definition of an extended narrowing relation on goals that effectively "skips" several under-the-feet rewrites and captures the first rewrite that occurs at a non-variable position in one of the lefthand sides of the goal. Specifically, in the generalized narrowing relation, to solve the goal $G = G_1 \wedge t_1 \to^* t_2$ above, we (i) linearize the lefthand side t_1 to $\overline{t_1}$, (ii) narrow the linearized term $\overline{t_1}$ as, say, $\overline{t_1} \overset{\sigma}{\rightsquigarrow}_R w$, and (iii) add to the resulting goal a subgoal H that accounts for the intermediate under-the-feet rewrites that have been skipped. Specifically, for each variable $x \in Var(t_1)$ whose occurrences are renamed to, say, $x_1 \dots x_n$, in $\overline{t_1}$, the subgoal H contains $x \to^* \sigma(x_1) \wedge \dots \wedge x \to^* \sigma(x_n)$. According to this extended narrowing relation, the goal G above narrows to the goal $G' = G_1 \wedge w \to^* t_2 \wedge H$[2]. Since G has a solution ρ as assumed above, it is the case that G' has a solution η such that $\eta|_{Var(G)} = \rho|_{Var(G)}$.

The above discussion applies in particular to the case where in the rewrite sequence (1) above, there is a k such that $\omega_k \in FuPos(t_1)$. Otherwise, there are two possibilities. First, if there is a k such that $\omega_k \in FuPos(t_2)$, then we can apply the above idea in the *backward* direction, i.e., we linearize the righthand side t_2 and narrow the resulting term using R^{-1}. This is justified by the observation that $t \to^*_R t'$ if and only if $t' \to^*_{R^{-1}} t$. Thus, we have a procedure that combines forward and backward reachability analysis. Of course, for this idea to work,

[2] Note that the subgoal G_1 is unchanged in the narrowing step. This is because the variables $x_1, \dots, x_n$ that are introduced during linearization of t_1 are fresh w.r.t G_1, and therefore the substitution σ has no effect on G_1 (see Definitions 4 and 5.)

unlike in Section 5, we have to *allow the rules in R to have extra variables in their righthand sides.* Finally, we are left with the case where $\omega_i \notin FuPos(t_1, t_2)$ for all $1 \leq i \leq n$. We note that in this case, $\rho(t_1)$ and $\rho(t_2)$ should be identical at all positions $\omega \in FuPos(t_1, t_2)$. This observation can be used to further instantiate variables in G, or to reduce G to a trivial goal.

Definition 4 (extended narrowing of terms). *For a term t, let $\overline{t}$ be a linearized form of t, where each occurrence of a variable $x \in Var(t)$ is renamed to a distinct fresh variable $y \notin Var(t)$. Further, suppose $\overline{t} \overset{\sigma}{\rightsquigarrow}_R t'$ for σ away from $Var(t)$. Then we define $t \twoheadrightarrow_R t'; H$, where H is the reachability goal such that if the occurrences of a variable $x \in Var(t)$ are renamed to, say, $x_1, \ldots, x_n$, then H includes the subgoal $x \rightarrow^* \sigma(x_1) \wedge \ldots \wedge x \rightarrow^* \sigma(x_n)$.*

For example, consider the rewrite system of the example in the Introduction. We have $f(x, x) \twoheadrightarrow_R d; (x \rightarrow^* b \wedge x \rightarrow^* c)$, using the rule $f(b, c) \rightarrow d$.

Definition 5 (back-and-forth narrowing of goals). *We define a back-and-forth narrowing relation on goals as a decorated relation of the form $G \overset{\sigma}{\twoheadrightarrow}_R G'$ defined as follows.*

(Narrow-left) $G \wedge t \rightarrow^* t'$
$\overset{id}{\twoheadrightarrow}_R \; G \wedge t'' \rightarrow^* t' \wedge H$ *if* $t \twoheadrightarrow_R t''; H$

(Narrow-right) $G \wedge t \rightarrow^* t'$
$\overset{id}{\twoheadrightarrow}_R \; G \wedge t \rightarrow^* t'' \wedge H^{-1}$ *if* $t' \twoheadrightarrow_{R^{-1}} t''; H$

(Decompose) $G \wedge f(t_1, \ldots, t_n) \rightarrow^* f(t'_1, \ldots, t'_n)$
$\overset{id}{\twoheadrightarrow}_R \; G \wedge t_1 \rightarrow^* t'_1 \wedge \ldots \wedge t_n \rightarrow^* t'_n$

(Match-left) $G \wedge x \rightarrow^* f(t_1, \ldots, t_n)$
$\overset{\sigma}{\twoheadrightarrow}_R \; \sigma(G) \wedge x_1 \rightarrow^* \sigma(t_1) \wedge \ldots \wedge x_n \rightarrow^* \sigma(t_n)$
if $x_i \notin Var(G, x, t_1, \ldots, t_n)$ *for* $1 \leq i \leq n$, *and* $\sigma = \{f(x_1, \ldots, x_n)/x\}$

(Match-right) $G \wedge f(t_1, \ldots, t_n) \rightarrow^* x$
$\overset{\sigma}{\twoheadrightarrow}_R \; \sigma(G) \wedge \sigma(t_1) \rightarrow^* x_1 \wedge \ldots \wedge \sigma(t_n) \rightarrow^* x_n$
if $x_i \notin Var(G, x, t_1, \ldots, t_n)$ *for* $1 \leq i \leq n$, *and* $\sigma = \{f(x_1, \ldots, x_n)/x\}$

(Unify) $G \wedge t \rightarrow^* t'$
$\overset{\sigma}{\twoheadrightarrow}_R \; \sigma(G)$ *if* $\sigma = MGU(t = t', Var(G, t, t'))$

For $t \twoheadrightarrow_R t''; H$ in the case **Narrow-left** *above, we impose the following additional condition. Suppose t is linearized to $\overline{t}$ and $\overline{t} \overset{\sigma}{\rightsquigarrow}_R t''$, then we require that the new variables introduced in linearizing t to $\overline{t}$ are fresh with respect to $Var(G, t, t')$, and the substitution σ is away from $Var(G, t, t')$. Similar conditions apply to* **Narrow-right**. *The relation $G \overset{\sigma}{\twoheadrightarrow}{}^*_R G'$ is defined by composing the substitutions of each step as expected.*

We cannot in general hope for a procedure that, given a goal G enumerates a complete set of solutions of G. For instance, consider the trivial goal $G = x_1 \to^* y_1 \wedge \ldots \wedge x_n \to^* y_n$. Enumerating a complete set of solutions of G is equivalent to enumerating a set S of tuples $(u_1, v_1, \ldots, u_n, v_n)$ such that (i) for each $(u_1, v_1, \ldots, u_n, v_n) \in S$ we have $u_i \to_R^* v_i$, and (ii) for each $s_1, t_1, \ldots, s_n, t_n$ such that $s_i \to_R^* t_i$ there is a $(u_1, v_1, \ldots, u_n, v_n) \in S$ and a substitution σ such that $s_i = \sigma(u_i)$ and $t_i = \sigma(v_i)$. We can systematically enumerate one such set S, namely, the set of all tuples $(u_1, v_1, \ldots, u_n, v_n)$ such that $u_i \to_R^* v_i$, but that would be extremely inefficient.

We will therefore give a procedure that is complete only as far as *solvability* of goals is concerned. Specifically, if a given goal G has a solution, then the procedure is guaranteed to find *some* solution of G. For example, for the trivial goal G above, the substitution σ such that $\sigma(x_i) = \sigma(y_i) = z$ will be returned as a solution. In addition, if we have a procedure that enumerates a complete set of solutions for trivial goals, we can combine it with the procedure for solvability to obtain a procedure that enumerates a complete set of solutions for any given goal G (see Theorem 6).

Examples: We now show a few examples where the narrowing procedure of Section 5 fails to find any solution, but back-and-forth narrowing succeeds.

Example 2. Consider the rewrite theory $\mathcal{R}$ and the reachability goal $G = f(x,x) \to^* d$ of Example 2 in Section 5. We have

$$
\begin{array}{lll}
f(x,x) \to^* d & \overset{id}{\twoheadrightarrow}_R \; f(x,x) \to^* f(b,c) & \textbf{(Narrow-right)} \\
& \overset{id}{\twoheadrightarrow}_R \; f(x,x) \to^* f(a,c) & \textbf{(Narrow-right)} \\
& \overset{id}{\twoheadrightarrow}_R \; f(x,x) \to^* f(a,a) & \textbf{(Narrow-right)} \\
& \overset{\{a/x\}}{\twoheadrightarrow}_R \; \Lambda & \textbf{(Unify)}
\end{array}
$$

Thus, back-and-forth narrowing finds the solution $\sigma = \{a/x\}$, whereas the narrowing procedure of Section 5 fails to find any solution. Here is another back-and-forth narrowing derivation that finds the same solution.

$$
\begin{array}{lll}
f(x,x) \to^* d & \overset{id}{\twoheadrightarrow}_R \; d \to^* d \wedge x \to^* b \wedge x \to^* c & \textbf{(Narrow-left)} \\
& \overset{id}{\twoheadrightarrow}_R \; x \to^* b \wedge x \to^* c & \textbf{(Unify)} \\
& \overset{id}{\twoheadrightarrow}_R \overset{id}{\twoheadrightarrow}_R \; x \to^* a \wedge x \to^* a & \textbf{(2} \times \textbf{Narrow-right)} \\
& \overset{\{a/x\}}{\twoheadrightarrow}_R \overset{id}{\twoheadrightarrow}_R \; \Lambda & \textbf{(2} \times \textbf{Unify)}
\end{array}
$$

Example 3. Here is an example that illustrates the use of **Decompose**, and **Match-left**. Consider the rewrite theory $\mathcal{R} = (\Sigma, R)$, where the signature Σ contains the constants a, b, c, a unary function symbol g, and two binary function symbols f, h, and the set R contains the following three rules

$$a \to h(b,c) \qquad b \to g(a) \qquad c \to h(b,c)$$

Consider the goal $G = f(x, y) \to^* f(g(y), h(x, y))$. Clearly, there is no narrowing derivation (in the sense of Section 5) starting from $f(x, y)$. But G has the solution $\sigma = \{g(a)/x, h(b, c)/y\}$ because

$$f(g(a), h(b, c)) \xrightarrow{[1.1]}_R f(g(h(b, c)), h(b, c)) \xrightarrow{[2.1]}_R f(g(h(b, c)), h(g(a), c))$$
$$\xrightarrow{[2.2]}_R f(g(h(b, c)), h(g(a), h(b, c))$$

Note that all the above rewrites occur under-the-feet of both the lefthand and righthand sides of G. The solution σ is found by back-and-forth narrowing as follows.

$$
\begin{array}{lr}
f(x, y) \to^* f(g(y), h(x, y)) & \\
\overset{id}{\twoheadrightarrow}_R\ x \to^* g(y) \ \wedge\ y \to^* h(x, y) & \textbf{(Decompose)} \\
\overset{\sigma_1}{\twoheadrightarrow}_R\ x_1 \to^* y \ \wedge\ y \to^* h(g(x_1), y) & \textbf{(Match-left)} \\
\overset{\sigma_2}{\twoheadrightarrow}_R\ x_1 \to^* h(y_1, y_2) \ \wedge\ y_1 \to^* g(x_1) \ \wedge\ y_2 \to^* h(y_1, y_2) & \textbf{(Match-left)} \\
\overset{id}{\twoheadrightarrow}_R\ x_1 \to^* h(y_1, y_2) \ \wedge\ y_1 \to^* g(x_1) \ \wedge\ y_2 \to^* c \ \wedge\ b \to^* y_1 \ \wedge\ c \to^* y_2 & \\
 & \textbf{(Narrow-right)} \\
\overset{\sigma_3}{\twoheadrightarrow}{}^*_R\ x_1 \to^* h(b, c) \ \wedge\ b \to^* g(x_1) & \textbf{(3} \times \textbf{Unify)} \\
\overset{id}{\twoheadrightarrow}_R\ x_1 \to^* h(b, c) \ \wedge\ g(a) \to^* g(x_1) & \textbf{(Narrow-left)} \\
\overset{\sigma_4}{\twoheadrightarrow}_R\ a \to^* h(b, c) & \textbf{(Unify)} \\
\overset{id}{\twoheadrightarrow}{}^*_R\ \Lambda & \textbf{(Narrow-left,Unify)}
\end{array}
$$

where $\sigma_1 = \{g(x_1)/x\}$, $\sigma_2 = \{h(y_1, y_2)/y\}$, $\sigma_3 = \{b/y_1, c/y_2\}$, and $\sigma_4 = \{a/x_1\}$. Thus, back-and-forth narrowing finds the solution $\sigma = (\sigma_4 \circ \sigma_3 \circ \sigma_2 \circ \sigma_1)|_{\{x,y\}}$, while the narrowing procedure of Section 5 doesn't.

Soundness: We now prove the soundness of back-and-forth narrowing of reachability goals. First, following is the analogue of Lemma 2 for the extended narrowing relation on terms.

Lemma 6. *If* $t \twoheadrightarrow_R t'; H$ *and* ρ *is a solution of* H*, then* $\rho(t) \to^*_R \rho(t')$*.*

The lemma above is lifted to goals as expected.

Theorem 4. *If* $G \overset{\sigma}{\twoheadrightarrow}_R G'$ *and* ρ *is a solution of* G' *then* $\rho \circ \sigma$ *is a solution of* G*.*

Completeness: Recall that in Section 5 the main idea behind establishing weak completeness of narrowing was to associate to each rewrite step on terms a corresponding narrowing step on terms (Lemma 4). To establish the completeness of back-and-forth narrowing, we generalize this idea to associate to a sequence of rewrites starting from $\rho(t)$, where all but the last rewrite occur at positions $\omega \notin FuPos(t)$, a single extended narrowing step starting from t. This is formalized in the following lemma. It is important to note that, unlike in Lemma 4, rewrite rules are now *allowed* to have extra variables in their righthand side[3].

[3] The no-extra-variable assumption is necessary in Lemma 4 to guarantee that η is R-normalized. This is in turn required for the assumption that ρ is R-normalized while inductively composing several applications of Lemma 4 to obtain Lemma 5. In contrast, Lemma 7 neither assumes ρ to be R-normalized, nor does it guarantee that η is R-normalized.

Lemma 7. *Let V be a finite set of variables containing $Var(t)$, and let $\rho(t) \xrightarrow{[\omega_1]}_R \ldots \xrightarrow{[\omega_n]}_R \xrightarrow{[\omega]}_R t'$ such that $\omega_i \notin FuPos(t)$ for $1 \leq i \leq n$ and $\omega \in FuPos(t)$. Then there are t'', H, η such that $t \twoheadrightarrow_R t''; H$, η is a solution of H, $\eta|_V = \rho|_V$, and $\eta(t'') = t'$.*

Lifting the above lemma to goals is a bit more complicated than its analogue, Lemma 5. Suppose ρ is a solution of G, and π is a rewrite sequence

$$\rho(G) \xrightarrow{[\omega_1]}_R G_1 \xrightarrow{[\omega_2]}_R \ldots \xrightarrow{[\omega_n]}_R \Lambda$$

We call π a *witness* for the solution ρ of G. Define the metrics $d(\pi) = \sum_{i=1}^{|\pi|} |\omega_i|$ and $\mu(\pi) = (|\pi|, d(\pi))$. Let $\preceq$ be the usual lexicographic ordering on pairs of natural numbers, i.e., $(m_1, n_1) \preceq (m_2, n_2)$ if $m_1 < m_2$, or $m_1 = m_2$ and $n_1 \leq n_2$. Define $(m_1, n_1) \prec (m_2, n_2)$ if $(m_1, n_1) \preceq (m_2, n_2)$ and $(m_1, n_1) \neq (m_2, n_2)$. Note that $\prec$ is a well-founded relation with $(0, 0)$ as the least element.

Lemma 8. *Let G be a non-trivial reachability goal, V a finite set of variables containing $Var(G)$, ρ a solution of G, and π a witness for the solution ρ. Then there are σ, η, G' such that $G \overset{\sigma}{\twoheadrightarrow}_R G'$, σ is away from V, $\rho|_V = (\eta \circ \sigma)|_V$, η is a solution of G', and there is a witness π' for η such that $\mu(\pi') \prec \mu(\pi)$.*

We are now ready to state the completeness of back-and-forth narrowing.

Theorem 5 (Completeness). *Let ρ be a solution of a reachability goal G, and let V be a finite set of variables containing $Var(G)$. Then there are σ and G' such that $G \overset{\sigma}{\twoheadrightarrow}_R^{*} G'$, σ is away from V, G' is a trivial goal, and there is a solution η of G' such that $\rho|_V = (\eta \circ \sigma)|_V$.*

A Complete Algorithm for Solvability of Reachability Goals:

Theorem 6. *Let V be a finite set of variables containing $Var(G)$, and let S be the set of all substitutions of the form $(\eta \circ \sigma)|_{Var(G)}$, where $G \overset{\sigma}{\twoheadrightarrow}_R^{*} G'$, σ is away from V, G' is a trivial goal, and $\eta \in CSS(G', V \cup Ran(\sigma) \cup Var(G'))$. Then S is a complete set of solutions of G away from V.*

Thus, if we are given a procedure for enumerating complete sets of solutions of trivial goals, then we also have a procedure for enumerating complete sets of solutions for any goal. In addition, since for the trivial goal $x_1 \rightarrow^* y_1 \wedge \ldots \wedge x_n \rightarrow^* y_n$, the substitution σ such that $\sigma(x_i) = \sigma(y_i) = z$ is a solution, it follows from Theorems 4 and 5 that we have a complete procedure for *solvability* of reachability goals. That is, if a given goal G has a solution, then the procedure finds some solution of G.

7 Related Work

Backward and forward narrowing may seem familiar in the context of equational unification [11, 12, 15, 16], where a unification goal $\exists \vec{x}.t_1 = t_2$ is transformed into the reachability goal $\exists \overrightarrow{x}.eq(t_1, t_2) \rightarrow^* tt$, and is then (naively) narrowed using $R \cup \{eq(t,t) \rightarrow tt\}$. Note that in the transformed goal one can narrow both the lefthand and righthand sides t_1 and t_2 using R, but both only in the *forward* direction[4]. Further, linearization is not necessary in the equational setting where under-the-feet rewrites are inconsequential due to the confluence assumption. But in a general setting where such assumptions are dropped, linearization becomes essential. In summary, equational unification procedure should *not* be confused with back-and-forth narrowing which is much more general. Specifically, the equational unification procedure just amounts to naive narrowing, and as shown by the examples in Sections 5 and 6, back-and-forth narrowing can solve goals which naive narrowing cannot.

Symbolic reachability analysis using narrowing is also reminiscent of tree-automata (TA) based techniques for reachability analysis [8, 18]. The nth unfolding of the narrowing tree roughly corresponds to the TA recognizing the states that are reachable within n steps. However, there are important differences between the two, which we highlight after briefly recalling the main TA based approaches. In the TA setting, given a rewrite system $\mathcal{R}$ and a regular tree language L, one considers the set $[\rightarrow^*_{\mathcal{R}}]L = \{t \in T_\Sigma \mid \exists u \in L \ s.t. \ u \rightarrow^*_{\mathcal{R}} t\}$. Then, given regular tree languages I and F, the reachability problem is posed as the question of whether the intersection $[\rightarrow^*_{\mathcal{R}}]I \cap F$ is nonempty. In general, $[\rightarrow^*_{\mathcal{R}}]I$ is not a regular tree language and this problem is undecidable. A first approach is to characterize classes of rewrite systems $\mathcal{R}$ for which, given any regular tree language L, the set $[\rightarrow^*_{\mathcal{R}}]L$ is also regular and we can effectively construct a tree automaton recognizing it if we are given a tree automaton recognizing L. Since the set of instances of a nonlinear term is not regular, some linearity assumptions are placed on $\mathcal{R}$ to characterize suitable classes (see [19, 18] for some of the most general classes known so far). A second, more generally applicable approach is to iteratively compute tree automata to recognize $[\rightarrow^n_{\mathcal{R}}]L$ (terms reachable from L in at most n steps). Since $[\rightarrow^*_{\mathcal{R}}]L = \cup_n[\rightarrow^n_{\mathcal{R}}]L$, this yields a semidecision procedure for reachability analysis provided each $[\rightarrow^n_{\mathcal{R}}]L$ is regular; for this again some linearity assumptions on $\mathcal{R}$ are needed, and in some approaches [10] non-linearity is dealt with by over approximations. A third related approach is to compute tree-automata-based *abstractions* that approximate the reachability set [8, 18].

In comparison with back-and-forth narrowing, the main differences have to do with the quite restricted assumptions on term rewriting systems required by TA approaches in order to ensure preservation of the *regularity* of the relevant sets of terms involved in the reachability analysis. By contrast, back-and-forth narrowing is a complete semidecision procedure for *arbitrary* rewrite systems; in particular, regularity-preserving restrictions on a term rewriting system are typically non-symmetric, whereas inverting the rules is part of the back-and-forth narrowing procedure. Under regularity-preserving conditions allowing the

[4] The idea behind the transformation is that for a confluent equational theory E, $\sigma(t_1) =_E \sigma(t_2)$ if and only if $\sigma(t_1) \rightarrow^*_E t$ and $\sigma(t_2) \rightarrow^*_E t$ for some term t. The proof of Lemma 8 should shed some light on the fact this idea is totally different from the one behind back-and-forth narrowing.

use of the first TA approach, the reachability problem is decidable, whereas back-and-forth narrowing is only a semidecision procedure. The third TA approach works by over-approximation, which ensures correctness of negative answers, but can result in false positives; instead, with back-and-forth narrowing a positive solution is always correct and is always found if there is one.

8 Conclusions and Future Work

We have presented back-and-forth narrowing as a semidecision procedure for solving reachability goals in unsorted and unconditional rewrite systems, and we have proved its completeness in the solvability sense. Although we have given an unsorted treatment using standard rewriting, our method can be extended to general order-sorted rewrite theories of the form (Σ, E, R) with equations E, under appropriate assumptions along the lines adopted in [14]. These assumptions include *pre-regularity* of Σ, that $E = \Delta \cup B$, where the equations Δ are confluent and terminating modulo B, and that Δ and R satisfy certain coherence properties relative to B. Such an extension, that we plan to document in a subsequent paper, will make our results available for many other systems.

Another important direction of research is to investigate efficient strategies for back-and-forth narrowing. Several lazy narrowing strategies are known in the functional-logic programming context [1, 9]. These strategies are all complete for special classes of rewrite systems, typically for left-linear and constructor-based systems. These assumptions are quite reasonable for functional-logic programming applications, but not so in non-equational contexts. In recent work with S. Escobar [6], we have proposed a lazy narrowing strategy called *natural narrowing* for general term rewrite systems that is complete in the weak sense, in that it is guaranteed to find all R-normalized solutions. We conjecture that natural narrowing can be extended to the back-and-forth setting so that completeness is regained even for non-normalized solutions. This problem will be dealt with in subsequent papers.

References

[1] S. Antoy, R. Echahed, and M. Hanus. A needed narrowing strategy. *Journal of the ACM*, 47(4):776–822, 2000.

[2] David Basin, Sebastian Modersheim, and Luca Vigano. Constraint differentiation: A new reduction technique for constraint-based analysis of security protocols. Technical Report TR-405, Swiss Federal Insititute of Technology, Zurich, May 2003.

[3] Ahmed Bouajjani and Tayssir Touili. Extrapolating tree transformations. In *Proc. 14th Int. Conf. on Computer Aided Verification (CAV'02)*, volume 2404 of *Lecture Notes in Computer Science*, 2002.

[4] O. Burkart, D. Caucal, F. Moller, and B. Steffen. Verification over Infinite States. In *Handbook of Process Algebra*, pages 545–623. Elsevier Publishing, 2001.

[5] D. Dolev and A. Yao. On the security of public key protocols. *IEEE Transaction on Information Theory*, 29(2):198–208, 1983.

[6] S. Escobar, J. Meseguer, and P. Thati. Natural narrowing for general term rewriting systems. In *International Conference on Rewriting Techniques and applications (RTA)*, 2005. also available at http://www.dsic.upc.es/users/elp/papers.html.
[7] Alain Finkel and Ph. Schnoebelen. Well-structured transition systems everywhere! *Theoretical Computer Science*, 256(1):63–92, 2001.
[8] T. Genet and F. Klay. Rewriting for cryptographic protocol verification. In *Automated Deduction—CADE-17*, volume 1831 of *Lecture Notes in Artificial Intelligence*, pages 271–290. Springer-Verlag, 2000.
[9] M. Hanus. The integration of functions into logic programming: From theory to practice. *Jounral of Logic Programming*, 19(20):583–628, 1994.
[10] Hiroyuki Seki Hitoshi Ohsaki and Toshinori Takai. ACTAS: A system design for associative and commutative tree automata theory. In *Proc. 5th Intl. Workshop on Rule-Based Programming (RULE 2004)*. Elsevier, ENTCS, 2004.
[11] J.M. Hullot. Canonical forms and unification. In W. Bibel and R. Kowalski, editors, 5th *Conference on Automated Deduction*, volume 87 of *Lecture Notes in Computer Science*, pages 318–334. Springer, 1980.
[12] Jean-Pierre Jouannaud, Claude Kirchner, and Helene Kirchner. Incremental construction of unification algorithms in equational theories. In 10th *International Colloquium on Automata, Languages and Programming*, volume 154 of *Lecture Notes in Computer Science*, pages 361–373. Springer, 1983.
[13] Catherine Meadows. The NRL protocol analyzer: An overview. *Journal of logic programming*, 26(2):113–131, 1996.
[14] José Meseguer and Prasanna Thati. Symbolic reachability analysis using narrowing and its application to analysis of cryptographic protocols. In *Workshop on Rewriting Logic and its Applications*, Electronic Notes in Theoretical Computer Science. Elsevier, 2004. To appear, also available at `http://osl.cs.uiuc.edu/docs/wrla04/main.ps`.
[15] A. Middeldorp and E. Hamoen. Counterexamples to completeness results for basic narrowing. In *Proceedings of the 3rd International Conference on Algebraic and Logic Programming*, Lecture Notes in Computer Science 632, pages 244–258, 1992.
[16] S. Okui, A. Middeldorp, and T. Ida. Lazy narrowing: Strong completeness and eager variable elimination. In *Proceedings of the 20th Colloquium on Trees in Algebra and Programming*, Lecture Notes in Computer Science 915, pages 394–408, 1995.
[17] G. E. Peterson and M. N. Wegman. Linear unification. *Journal of Computer and Systems Sciences*, 16:158–167, 1978.
[18] T. Takai. A verification technique using term rewriting systems and abstract interpretation. In *Proc. RTA 2004*, pages 119–133. Springer LNCS 3091, 2004.
[19] T. Takai, Y. Kaji, and H. Seki. Right-linear finite path overlapping term rewriting systems effectively preserve recognizability. In *Proc. RTA 2000*, pages 246–260. Springer LNCS 1833, 2000.
[20] P. Wolper and B. Boigelot. Verifying systems with infinite but regular state spaces. In *International Conference on Computer-Aided Verification*, volume 1427 of *Lecture Notes in Computer Science*, pages 88–97. Springer Verlag, 1998.

Final Sequences and Final Coalgebras for Measurable Spaces

Ignacio D. Viglizzo

Department of Mathematics, Indiana University,
Bloomington IN 47405, USA
igvigliz@indiana.edu

Abstract. A measure polynomial functor is a functor in the category Meas built up from constant measurable spaces, the identity functor and using products, coproducts and the probability measure functor Δ. In [1] it was proved that these functors have final coalgebras. We present here a different proof of that fact, one that uses the final sequence of the functor, instead of an ad hoc language. We also show how this method works for certain functors in Set and explore the connection with results in the literature that use the final sequence in other ways.

1 Introduction

Coalgebras for functors involving probability measures have been objects of interest, since they provide a framework for modelling probabilistic transition systems like those studied by Larsen and Skou [2], as noted by de Vink and Rutten in [3]. Previous work in this area has been limited to certain metric and topological spaces, while our results here apply to measure spaces in general.

In previous joint work with Lawrence S. Moss, [1], [4], we proved the existence of final coalgebras for measure polynomial functors (defined in section 2.3). Those final coalgebras were constructed as a set of sets of formulas in a language $\mathcal{L}(T)$ that depended on the functor T. The sets of formulas involved were *satisfied theories*, i.e., the sets of formulas satisfied by some point in some T-coalgebra. In this paper, we present a different proof of the same fact, but now we don't use any language. Instead, we look for our 'satisfied theories' inside the limit of the final sequence for the functor.

It is well known that for an ω^{op}-continuous functor T on the category of sets, the limit of the final sequence obtained by iterating the application of T to the terminal object 1 yields a final coalgebra for T. Our work presents some differences with this result. The main one is that we don't work (only) in the category Set, but also in the category Meas of measurable spaces and functions. Second, the functors we work with are not ω^{op} continuous. The functor Δ in Meas does not preserve these limits in general (see the next section).

The structure that the proof of the main result in section 3 follows is close to the one in [1], and the main idea was inspired by section 5 of [5]. We hope that the presentation here is more accessible than the one we gave previously, and can therefore make these tools available for a wider audience.

J.L. Fiadeiro et al. (Eds.): CALCO 2005, LNCS 3629, pp. 395–407, 2005.

2 Preliminaries

2.1 Measure Theory

Given a set M, a *σ-algebra* of sets is a non empty class Σ of subsets of M closed under finite intersections, complements and countable unions. A *measurable space* is a pair (M, Σ) where M is a set and Σ is a σ-algebra of subsets of M. The subsets of M which are in Σ are called *measurable sets.*

A *family of generators* $\mathcal{F}$ of a σ-algebra Σ is a family of subsets such that the smallest σ-algebra containing $\mathcal{F}$ is Σ. This is denoted by $\sigma(\mathcal{F}) = \Sigma$. A *π-system* is a family of subsets of a set M which is closed under finite intersections.

A function between measurable spaces $f : (M, \Sigma) \to (M', \Sigma')$ is said to be *measurable* if for every $E \in \Sigma', f^{-1}(E) \in \Sigma$. We will denote with Meas the category of measurable spaces and measurable functions.

A *measure* is a σ-additive function μ from the σ-algebra of a measurable space to $[0, \infty]$, such that $\mu(\emptyset) = 0$. If $\mu(M) = 1$, then it is called a *probability measure.* The following Lemma about probability measures is a consequence of Dynkin's π-λ-theorem and will be used later.

Lemma 1. *Suppose that μ_1, μ_2 are probability measures on $\sigma(\mathcal{F})$, where $\mathcal{F}$ is a π-system. If μ_1 and μ_2 agree on $\mathcal{F}$, then they agree on $\sigma(\mathcal{F})$.*

We will consider the operator Δ in Meas that assigns to each measurable space M the set ΔM of all the probability measures over M, endowed with the σ-algebra Σ_Δ generated by the sets of the form $\beta^p(E)$ where E is a measurable subset of M, p is any real number in the unit interval $[0, 1]$ and

$$\beta^p(E) = \{\mu \in \Delta M : \mu(E) \geq p\}. \tag{1}$$

As an easy consequence of the definition, we have that

Lemma 2. *If $f : M \to M'$ is a measurable function, then for all measurable $E \subseteq M'$ and $p \in [0, 1]$,*

$$\beta^p(f^{-1}(E)) = (\Delta f)^{-1}(\beta^p(E))$$

Using the above Lemma, it is easy to check that by defining $\Delta f : \Delta M \to \Delta M'$ as $\Delta f(\mu)(E) = \mu(f^{-1}(E))$ for every E measurable in M', Δf is measurable and Δ is an endofunctor in Meas.

Lemma 3. *(Heifetz and Samet, [5]) Let $\mathcal{F}$ be a boolean algebra of sets that generates the σ-algebra Σ on a measurable space M. Then the σ-algebra $\mathcal{F}_\Delta$ generated by the family of sets*

$$\{\beta^p(E) : E \in \mathcal{F} \text{ and } p \in [0, 1]\}$$

is the same as Σ_Δ.

Products and coproducts exist in Meas, and they can be constructed in a similar fashion to the corresponding constructions in Set (furthermore, Meas is a complete category, [6], page 71). The forgetful functor Meas $\to$ Set uniquely lifts products and coproducts. The σ-algebra of the product is the one generated by the "rectangles" formed by taking the cartesian product of measurable sets, while the σ-algebra of the coproduct is formed by taking (disjoint) unions of measurable sets in each of the summands.

Weak pullbacks are not preserved by Δ. A natural question that arises in this context is whether the functor Δ preserves weak pullbacks. Here we answer this negatively.

Pullbacks in Meas are constructed in a similar way as in Set. Given measurable functions $f : X \to Z$ and $g : Y \to Z$, one takes the set $P = \{\langle x, y\rangle : f(x) = g(y)\}$ endowed with the smallest σ-algebra that makes the projections measurable.

Now let $X = Y = Z$ be the interval $[0, 1]$ on the reals. Let

$$f(x) = \begin{cases} x \text{ if } x \neq 1/2 \\ 0 \text{ if } x = 1/2 \end{cases}$$

and $g(y) = 1 - y$. It is easy to check that f and g are measurable functions and $\{\langle x, y\rangle | f(x) = g(y)\} = \emptyset$.

Now consider $Q = \{q\}$, and functions $k : Q \to \Delta X, l : Q \to \Delta Y$ given by $k(q) = l(q)$ equal to the Lebesgue measure over $[0, 1]$. $(\Delta f)k(q) = (\Delta g)l(q)$ is again the Lebesgue measure, so the diagram commutes, but there is no measurable function $j : Q \to \emptyset$.

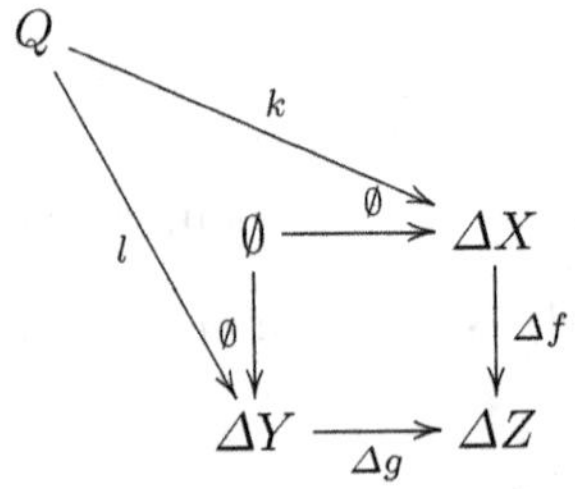

This counterexample can be "fixed" by changing f to be the identity on $[0, 1]$, so the pullback is not empty. Could this be a pathological case arising from the fact that two measurable functions can be essentially the same but differ on a set of measure zero? In other words, is there always a pair of functions $\hat{f}, \hat{g}$ so that $f = \hat{f}$ and $g = \hat{g}$ almost everywhere, but yet $\hat{f}$ and $\hat{g}$ have a non-empty pullback that can be weakly preserved by Δ? We show next that this is not the case.

We can assume that $\hat{f} = 1_{[0,1]}$ is the identity function on the interval $[0, 1]$ and $g = f$ on a set E of measure zero. So the pullback of f and g is $P = \{\langle x, x\rangle : x \in E\}$. Consider the functions $c_q : [0, 1] \to R$ defined as follows:

$$c_q(x) = \begin{cases} 1 + q \text{ if } x \in [0, 1/4) \cup (3/4, 1] \\ 1 - q \text{ if } x \in [1/4, 3/4] \end{cases}$$

Each c_q induces a probability measure on $[0,1]$ which we again denote c_q. For this, take $\int_F c_q(x)dx = (1+q)\lambda(F) - 2q\lambda(F \cap [1/4, 3/4])$ where λ is the Lebesgue measure over the reals and F is the measurable set for which we are calculating the measure.

Now let $Q = [0,1]$ and $k(q) = c_q = l(q)$. It is clear that $\Delta 1_{[0,1]} k(q) = c_q = (\Delta g)l(q)$. Furthermore, the functions k and l are measurable. Assume that there is a measurable function $j : Q \to \Delta P$ such that the diagram below commutes.

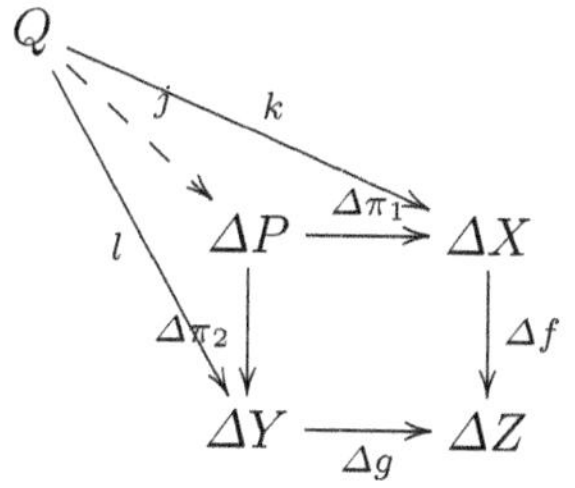

Then when we calculate $k(q)(E)$ we get: $\int_E c_q(x)dx = 0$, while on the other hand $\Delta\pi_1 j(q)(E) = j(q)(\pi_1^{-1}(E)) = j(q)(P) = 1$, contradiction.

2.2 The Final Sequence

Given an endofunctor T in a category with final object 1, we can consider the sequence:

$$1 \xleftarrow{\;!\;} T1 \xleftarrow{\;T!\;} T^2 1 \xleftarrow{\;T^2!\;} T^3 1 \longleftarrow \cdots$$

We will only consider countably many steps of this chain. It will be convenient here to describe briefly the limit of this ω^{op} sequence in Meas, called *inverse limit* or *projective limit.*

Notice that the function $! : T1 \to 1$ is surjective and measurable (assuming $T1 \neq \emptyset$) and it follows that $T^n!$ is also a surjective and measurable function. The inverse limit can then be constructed by taking the set

$$P = \{(x_n) \in \prod_n T^n 1 : \forall n \geq 0 (T^n!(x_{n+1}) = x_n)\}$$

and endowing it with the σ-algebra generated by the sets $\pi_n^{-1}(E)$ where E is a measurable subset of $T^n 1$ and π_n is the natural projection from $\prod_k T^k 1 \to T^n 1$ (see e.g. [7], Exercise 17.16).

Given a T-coalgebra (X, c), there is a natural way of obtaining a cone $(h_n^c : X \to T^n 1)$ mapping X to the final sequence. Let h_0^c be the unique map $!_X : X \to 1$, and given $h_n^c : X \to T^n 1$, let $h_{n+1}^c = Th_n^c \circ c$. Notice also that $h_n^c = T^n! \circ h_{n+1}^c$, as it's easy to prove by induction on n: for $n = 0$, $h_0^c = !_X = !_X \circ h_1^c$ and if $h_{n-1}^c = T^{n-1}! \circ h_n^c$, then $h_n^c = Th_{n-1}^c \circ c = T(T^{n-1}! \circ h_n^c) \circ c = T^n! \circ Th_n^c \circ c = T^n! \circ h_{n+1}^c$.

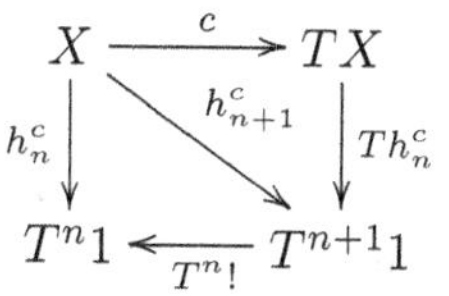

Definition 1. *Let $h^c : X \to P$ be the function defined by $\pi_n(h^c(x)) = h^c_n(x)$ for all $n \geq 0$. Now define $Z \subseteq P$ as the set of points that are in the image $h^c(X)$ for some T-coalgebra (X, c). Z is a measurable space with the structure it inherits from P.*

Lemma 4. *Given $f : (X, c) \to (Y, d)$ a T-coalgebra morphism, $h^d \circ f = h^c$.*

Proof. We prove by induction over n that $\pi_n h^d f = \pi_n h^c$, this is, $h^d_n \circ f = h^c_n$: $h^d_0 f = !_Y f = !_X = h^c_0$. Assuming that $h^d_n f = h^c_n$, we get that $h^d_{n+1} f = T h^d_n \circ d \circ f = T h^d_n \circ T f \circ c = T(h^d_n \circ f) \circ c = T h^c_n \circ c = h^c_{n+1}$. □

We will ignore the superscript of h whenever doing so does not lead to any confusion.

For any $n < m$, let $\tau_{mn} : T^m 1 \to T^n 1$ be defined by $\tau_{mn} = T^n! \circ T^{n+1}! \circ \ldots \circ T^{m-1}!$. Also let $\tau_{mm} = 1_{T^m 1}$.

$$T^n 1 \xleftarrow{T^n!} T^{n+1} 1 \xleftarrow{T^{n+1}!} \cdots \xleftarrow{T^{m-1}!} T^m 1, \qquad \tau_{mn} : T^m 1 \to T^n 1$$

ω^{op} sequences are not preserved by Δ. To show this we adapt a counterexample due to Andersen and Jessen [8] as it appears in [9]. Let $\{X_n\}_{n \in \omega}$ be a descending sequence of thick subsets of the real interval $[0, 1]$ such that $\cap X_n = \emptyset$. Thick sets have outer measure one, and each of these sets is endowed with a σ-algebra Σ_n consisting of all the intersections of Borel sets E with X_n. Let $\mu_n \in \Delta X_n$ be defined by $\mu_n(E \cap X_n) = \lambda(E)$, where λ is the Lebesgue measure. Since X_n is thick, μ_n is well defined.

Now let $Y_n = \prod_{k=0}^n X_k$, $f_n : Y_{n+1} \to Y_n$ the projection that forgets the last coordinate, and $S_n : X_n \to Y_n$ the application that sends x to $(x, x, \ldots, x)$. If $\nu_n \in \Delta Y_n$ is defined as $(\Delta S_n)\mu_n$, then one can check that $(\Delta f_n)\nu_{n+1} = \nu_n$ for all $n \geq 0$. The limit of the sequence of the Y_n is the infinite product $X = \prod_{n \in \omega} X_n$. Let $p_n : X \to Y_n$ be the evident projection.

If Δ preserved the limit, then there would exist a measure $\nu \in \Delta X$ such that for all $n \geq 0, (\Delta p_n)\nu = \nu_n$, since the sequence $\{\nu_n\}_{n \in \omega}$ belongs to the projective limit of the chain formed by the ΔY_n and functions Δf_n. But letting $E_n = p_n^{-1}(S_n(X_n))$ we get a sequence of measurable subsets of X such that $\nu(E_n) = 1$, and with empty intersection, a contradiction.

2.3 Measure Polynomial Functors

Definition 2. *The class of* measure polynomial functors *is the smallest class of functors on* Meas *containing the identity Id, the constant functor M for each measurable space M, and closed under binary products, coproducts and Δ.*

Definition 3. *Given a measure polynomial functor T we define the set of* ingredients of T, $\mathsf{Ing}(T)$ *as follows:* $\mathsf{Ing}(Id) = \{Id\}$; $\mathsf{Ing}(M) = \{M, Id\}$; $\mathsf{Ing}(U \times V) =$

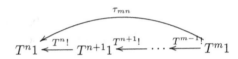

$\{U \times V\} \cup \mathsf{Ing}(U) \cup \mathsf{Ing}(V); \mathsf{Ing}(U + V) = \{U + V\} \cup \mathsf{Ing}(U) \cup \mathsf{Ing}(V)$ *and* $\mathsf{Ing}(\Delta S) = \{\Delta S\} \cup \mathsf{Ing}(S)$*. All the ingredients of* T *are measure polynomial functors, and* $\mathsf{Ing}(T)$ *is a finite set.*

The ingredients of a measure polynomial functor are defined in terms of the syntax with which the polynomial is presented.

Example 1. Let $T = Id + \Delta(Id \times M)$ for some fixed measurable space M. Then

$$\mathsf{Ing}(T) = \{Id, M, Id \times M, \Delta(Id \times M), T\}$$

For each ingredient S of T we will consider the ω^{op} chain

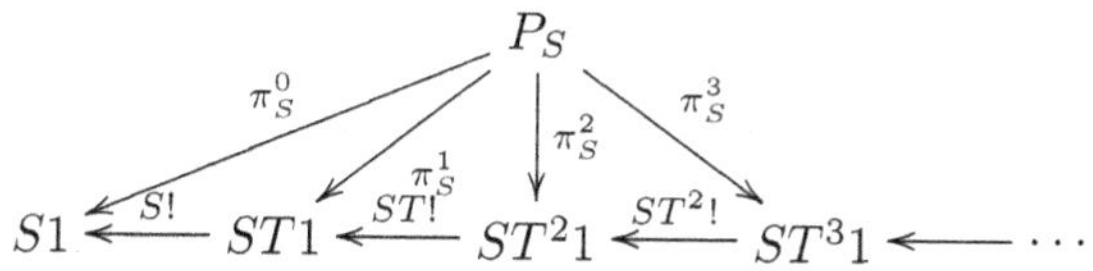

Let P_S the projective limit of this chain, constructed like P before, and let $\pi_S^n : P_S \to ST^n1$ be the corresponding projections. It is worth noting that unless the functor S is ω^{op}-continuous, π_S^n won't be in general equal to $S\pi_n$. For any given coalgebra (X, c), the cone $Sh_n : SX \to ST^n1$ induces a mapping $h_S : SX \to P_S$ such that $\pi_S^n h_S = Sh_n$.

Definition 4. *Just as when we defined* Z *in definition 1, let* Z_S *be the collection of all the points in* P_S *that are the image under* h_S^c *of some element in* SX *for some* T*-coalgebra* (X, c)*. Notice that we write* $Z_{Id} = Z$*.*

3 Final Coalgebras

The measurable space Z defined in the previous section will be the carrier of the final coalgebra for T. To define the structure map on Z, we will proceed in two stages. First, we define a map $\nu : Z \to Z_T$, and then we find a map r_T from Z_T to $T(Z)$ that will establish the result. The first part is taken care of in the following Lemma:

Lemma 5. *There exists a measurable map* $\nu : Z \to Z_T$*, such that for every coalgebra* (X, c)*,* $\nu hx = h_T cx$*.*

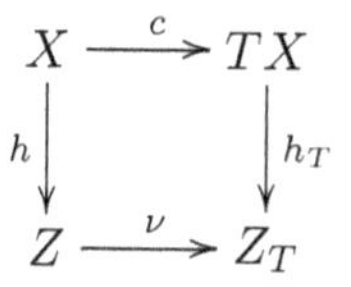

Proof. We define $\nu : Z \to P_T$ by $\pi^n_T \nu = \pi_{n+1}$ for all $n \geq 0$. We need to prove that $\pi^n_T \nu h = \pi^n_T h_T c$.

$$\begin{aligned}\pi^n_T h_T c &= T h_n c \\ &= h_{n+1} \\ &= \pi_{n+1} h \\ &= \pi^n_T \nu h\end{aligned}$$

This proves that $\nu h = h_T c$ and also that ν has codomain included in Z_T.

Next we prove that ν is measurable. Consider a measurable set $E_n \subseteq TT^n1$; since the sets of the form $E = (\pi^n_T)^{-1}(E_n)$ generate the σ-algebra on Z_T, it will be enough to prove that $\nu^{-1}(E)$ is measurable. But $\nu^{-1}(E) = \nu^{-1}(\pi^n_T)^{-1}(E_n) = (\pi^n_T \circ \nu)^{-1}(E_n) = (\pi^{n+1}_T)^{-1}(E_n)$, which we know to be measurable. □

To find the map $r_T : Z_T \to T(Z)$ satisfying the appropriate conditions, we need to work making reference to the structure of T. First, we introduce some auxiliary results.

Lemma 6. *If $(U \times V) \in \mathsf{Ing}(T)$, there is a measurable map $\langle \pi_1, \pi_2 \rangle : Z_{U\times V} \to Z_U \times Z_V$ so that for every T-coalgebra (X, c), $\langle \pi_1, \pi_2 \rangle h^c_{U\times V} = h^c_U \times h^c_V$ and for every $n \geq 0$, $(\pi^n_U \times \pi^n_V)\langle \pi_1, \pi_2 \rangle = \pi^n_{U\times V}$.*

Proof. Let $p^n_1 : (U \times V)T^n1 \to UT^n1$, $p^n_2 : (U \times V)T^n1 \to VT^n1$, $p_U : (U \times V)X \to UX$, $p_V : (U \times V)X \to VX$, $p_{Z_U} : Z_U \times Z_V \to Z_U$, $p_{Z_V} : Z_U \times Z_V \to Z_V$ be the natural projections.

We start by defining $\pi_1 : Z_{U\times V} \to P_U$ through $\pi^n_U(\pi_1(z)) = p^n_1(\pi^n_{U\times V}(z))$. From this definition it follows that $(\pi^n_U \times \pi^n_V)\langle \pi_1, \pi_2 \rangle = \pi^n_{U\times V}$.

To prove that $\langle \pi_1, \pi_2 \rangle h^c_{U\times V} = h_U \times h_V = \langle h_U p_U, h_V p_V \rangle$, we need to show that $\pi_1 h_{U\times V} = h_U p_U$ (and the corresponding equation for V). This will be proved once we prove that for all n, $\pi^n_U \pi_1 h_{U\times V} = \pi^n_U h_U p_U$.

Also, $\pi^n_U \pi_1 h_{U\times V} = p^n_1 \pi^n_{U\times V} h_{U\times V} = p^n_1 (U \times V) h_n = p^n_1 (U h_n \times V h_n) = U h_n p_U = \pi^n_U h_U p_U$. From this equation it follows that $\pi_1 : Z_{U\times V} \to Z_U$.

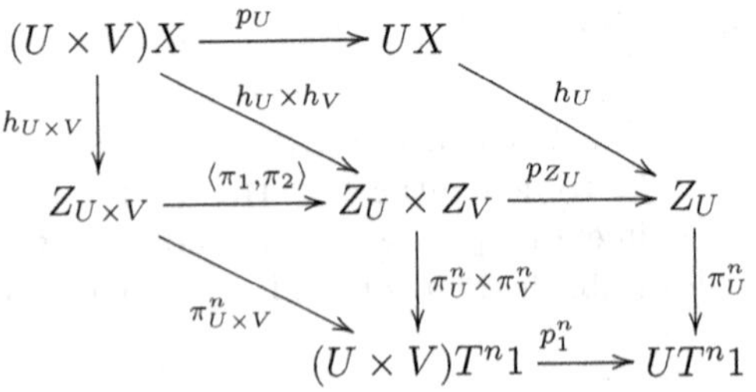

To prove that $\langle \pi_1, \pi_2 \rangle$ is measurable, it's enough to prove each of the components is measurable. We do it for π_1. Let E, E_n be as in the proof of Lemma 5. Then $\pi_1^{-1}(E) = \pi_1^{-1}(\pi^n_U)^{-1}(E_n) = (\pi^n_U \pi_1)^{-1}(E_n) = (p_1 \pi^n_{U\times V})^{-1}(E_n)$, which is measurable. □

Lemma 7. *If $(U+V) \in \mathsf{Ing}(T)$, there is a measurable map $\alpha : Z_{U+V} \to Z_U + Z_V$ so that for every T-coalgebra (X, c), $\alpha h^c_{U+V} = h^c_U + h^c_V$, and for all $n \geq 0$, $(\pi^n_U + \pi^n_V)\alpha = \pi^n_{U+V}$.*

Proof.

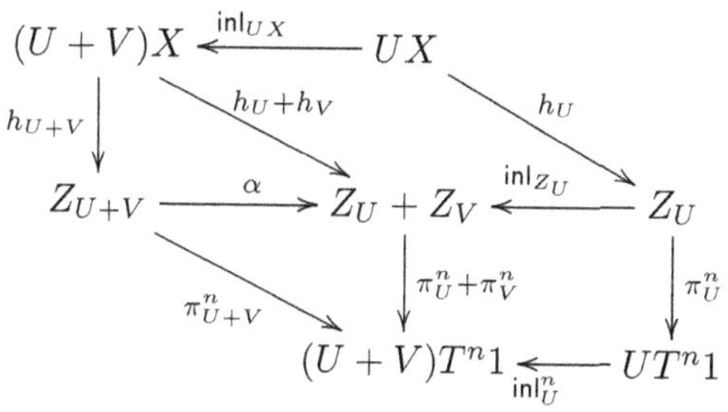

Since for each $z \in Z_{U+V}, z = h^c_{U+V}(x)$ for some $x \in (U+V)X$ for some coalgebra (X, c), we define α by $\alpha h_{U+V}(x) = (h^c_U + h^c_V)(x)$.

For every $n \geq 0$ we have:

$$\begin{aligned} \pi^n_{U+V} h_{U+V} &= (U+V)h_n \\ &= Uh_n + Vh_n \\ &= \pi^n_U h_U + \pi^n_V h_V \\ &= (\pi^n_U + \pi^n_V)(h_U + h_V) \end{aligned}$$

Now we can check α is well-defined. If (Y, d) is another coalgebra and $y \in (U+V)Y$ is such that $h^c_{U+V}(x) = h^d_{U+V}(y)$, then for all $n \geq 0$, $(\pi^n_U + \pi^n_V)(h^c_U + h^c_V)x = (\pi^n_U + \pi^n_V)(h^d_U + h^d_V)y$ so $(h^c_U + h^c_V)x = (h^d_U + h^d_V)y$, i.e., $\alpha h^c_{U+V}(x) = \alpha h^d_{U+V}(y)$.

It also follows from the computation above that $(\pi^n_U + \pi^n_V)\alpha h_{U+V} = (\pi^n_U + \pi^n_V)(h_U + h_V) = \pi^n_{U+V} h_{U+V}$.

To prove that α is measurable, consider $E = \mathsf{inl}_{Z_U}(\pi^n_U)^{-1}(E_n)$, with E_n a measurable subset of $UT^n 1$. Then $\alpha^{-1}(E) = \alpha^{-1}(\mathsf{inl}_{Z_U}(\pi^n_U)^{-1}(E_n)) = \alpha^{-1}(\pi^n_U + \pi^n_V)^{-1}\mathsf{inl}^n_U(E_n) = ((\pi^n_U + \pi^n_V)\alpha)^{-1}\mathsf{inl}^n_U(E_n) = (\pi^n_{U+V})^{-1}\mathsf{inl}^n_U(E_n)$ is a measurable set. Here we used the fact that $\mathsf{inl}_{Z_U}(\pi^n_U)^{-1}(E_n) = (\pi^n_U + \pi^n_V)^{-1}\mathsf{inl}^n_U(E_n)$, which is easy to verify. □

Lemma 8. *Let ΔS be an ingredient of T. Then there exists a measurable function $\epsilon : Z_{\Delta S} \to \Delta Z_S$ so that for every T-coalgebra (X, c) $\epsilon h^c_{\Delta S} = \Delta h^c_S$ and for every $n \geq 0$, $(\Delta \pi^n_S)\epsilon = \pi^n_{\Delta S}$.*

Proof. To define $\epsilon(z)$ for a given $z \in Z_{\Delta S}$, we start by doing it for the family $\mathcal{F}$ of sets of the form $E = (\pi^n_S)^{-1}(E_n)$ with E_n measurable in $ST^n 1$.

Given $z \in Z_{\Delta S}$, we define

$$\epsilon(z)(E) = \pi^n_{\Delta S}(z)(E_n)$$

It is worth remarking that this definition just depends on z. To check that it does not depend on the selection of n, consider (X, c) and $\mu \in \Delta SX$ so that $h_{\Delta S}\mu = z$.

$$\begin{aligned}\epsilon(z)(E) &= \pi^n_{\Delta S}(z)(E_n)\\ &= \pi^n_{\Delta S}(h_{\Delta S}(\mu))(E_n)\\ &= (\Delta S h_n)(\mu)(E_n)\\ &= \mu(Sh_n)^{-1}(E_n)\\ &= \mu(h_S)^{-1}(\pi^n_S)^{-1}(E_n)\\ &= \mu(h_S)^{-1}(E)\\ &= (\Delta h_S)(\mu)(E)\end{aligned}$$

The above equation not only proves the independence of the definition from the selection of n, but also that $\epsilon h_{\Delta S}\mu(E) = \Delta h_S \mu(E)$ for every $E \in \mathcal{F}$. Now we extend the definition of $\epsilon(z)$ to every measurable subset F of Z_S by letting $\epsilon(z)(F) = \Delta h_S \mu(F)$. We still need to check this is well defined. If $z = h^c_{\Delta S}(\mu) = h^d_{\Delta S}(\mu')$, then we know that $\epsilon h^c_{\Delta S}(\mu)$ and $\epsilon h^d_{\Delta S}(\mu')$ agree on all the elements of the family $\mathcal{F}$. Since for any $n \leq k, (\pi^n_S)^{-1}(E_n) \cap (\pi^k_S)^{-1}(E_k) = (\pi^k_S)^{-1}(\tau^{-1}_{kn} E_n \cap E_k)$, $\mathcal{F}$ is a π-system, by Lemma 1, the measures must agree on all measurable subsets.

To prove that ϵ is measurable, first notice that for any measurable subset E_n of $ST^n 1$, $(\Delta \pi^n_S \epsilon)(z)(E_n) = \epsilon(z)(\pi^n_S)^{-1}(E_n) = \pi^n_{\Delta S}(z)(E_n)$. By Lemma 3 it will be enough to prove that for a measurable subset $E_n \subseteq ST^n 1, \epsilon^{-1}\beta^p(\pi^n_S)^{-1}(E)$ is measurable.

$$\begin{aligned}\epsilon(\beta^p(\pi^n_S)^{-1}(E_n)) &= \epsilon^{-1}(\Delta\pi^n_S)^{-1}\beta^p(E_n)\\ &= (\Delta\pi^n_S\epsilon)^{-1}\beta^p(E_n)\\ &= (\pi^n_{\Delta S})^{-1}\beta^p(E_n)\end{aligned}$$

We know the set in the last line to be measurable. We used lemma 2 in the first line of the equation. □

Lemma 9. *There exists a measurable function $r_T : Z_T \to T(Z)$ so that for every (X, c), $r_T \circ h^c_T = Th^c$, and for every $n \geq 0, T\pi_n \circ r_T = \pi^n_T$.*

Proof. We will prove this by induction over the ingredients of T. This is, if $S \in \mathsf{Ing}(T)$, then there exists a measurable map $r_S : Z_S \to S(Z)$ such that $r_S h_S = Sh$ and for all $n \geq 0, S\pi_n \circ r_S = \pi^n_S$.

For $S = Id$, $r_{Id} = 1_Z$ is measurable and trivially satisfies the conditions. Notice that Z_{Id} is just Z.

For $S = M$, a constant functor, we let $r_M = \pi^0_M : Z_M \to M = M(1)$. Then $r_M h_M = \pi^0_M \langle 1_M \rangle_{n \geq 0} = 1_M = M(h)$, and $M\pi_n r_M = 1_M \pi^0_M = \pi^n_M$ for all $n \geq 0$.

Probability measures. We define $r_{\Delta S}$ as $\Delta r_S \circ \epsilon$.

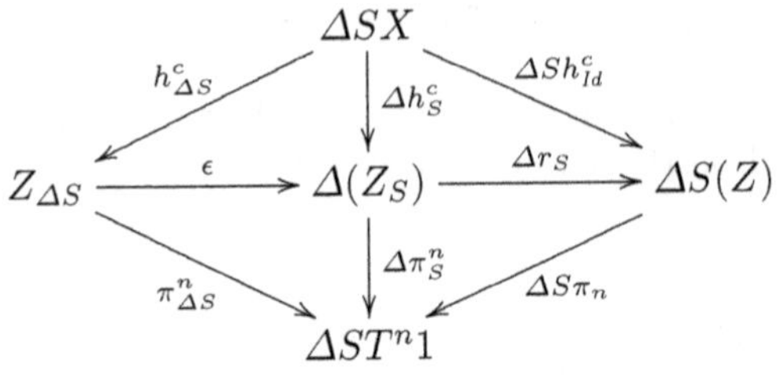

(2)

The triangles on the left commute by Lemma 8, and the ones on the right by the induction hypothesis. Hence the diagram commutes.

Products. The argument here is almost the same. Given r_U and r_V with the desired properties, we define $r_{U\times V}$ to be $(r_U \times r_V) \circ \langle \pi_1, \pi_2 \rangle$. One verifies that $r_{U\times V} h_{U\times V} = (U \times V)h$ with a figure similar as the one above (the changes in the objects are that ΔS is now $U \times V$, and $\Delta(Z_S)$ is $Z_U \times Z_V$; for the maps, the main change is that ϵ is now $\langle \pi_1, \pi_2 \rangle$), using Lemma 6.

Coproducts. We take r_{U+V} to be $(r_U + r_V) \circ \alpha$. We use the diagram from (2), replacing ΔS with $U + V$, and Lemma 8 with Lemma 7. □

Now we are ready to define $\gamma : Z \to T(Z)$ as

$$r_T \circ \nu : \ Z \to Z_T \to T(Z) \tag{3}$$

We shall show that (Z, γ) is a final T-coalgebra.

Lemma 10. *For each coalgebra (X, c), h^c is a morphism of coalgebras.*

Proof. Consider the diagram:

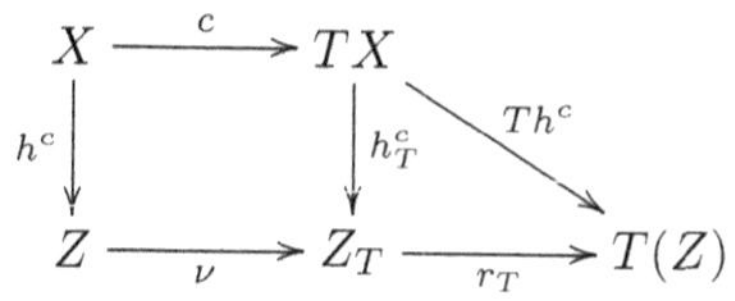

The square commutes by Lemma 5, and the triangle by Lemma 9. □

Lemma 11. $h^\gamma = 1_Z$.

Proof. It will be enough to prove that for each $n \geq 0$, $\pi_n h^\gamma = \pi_n$. For $n = 0$, we have that $\pi_0 h^\gamma = h_0 =! = \pi_0 1_Z$. Now assume that $h^\gamma_n = \pi_n h^\gamma = \pi_n$. Then,

$$\begin{aligned}
\pi_{n+1} h^\gamma &= h^\gamma_{n+1} \\
&= Th^\gamma_n \circ \gamma \\
&= T\pi_n \circ \gamma && \text{by inductive hypothesis} \\
&= T\pi_n \circ r_T \circ \nu && \text{by the definition of } \gamma \\
&= \pi^n_T \circ \nu && \text{by Lemma 9} \\
&= \pi_{n+1} && \text{by the definition of } \nu
\end{aligned}$$

□

Theorem 1. *(Z, γ) is a final coalgebra of T.*

Proof. Let (X, c) be a T-coalgebra. By Lemma 10, h^c is a coalgebra morphism. For the uniqueness, suppose that f is any morphism.

By Lemma 4, $h^\gamma \circ f = h^c$. But by Lemma 11, $h^\gamma = 1_Z$, so $f = h^\gamma \circ f = h^\gamma$. □

Example 2. Returning to the functor T of example 1, now we have that the structure map γ for Z is

$$\begin{array}{ccccccc} Z & \xrightarrow{\nu} & Z_T & \xrightarrow{\alpha} & Z + Z_{\Delta(Id \times M)} & \xrightarrow{1_Z + \epsilon} & Z + \Delta Z_{Id \times M} \\ \downarrow{\scriptstyle \gamma} & & & & & & \downarrow{\scriptstyle 1_Z + \Delta\langle \pi_1, \pi_2 \rangle} \\ Z + \Delta(Z \times M) & & & \xleftarrow{1_Z + \Delta(1_Z \times \pi^0_M)} & & & Z + \Delta(Z + Z_M) \end{array}$$

$$\gamma = (1_Z + \Delta((1_Z \times \pi^0_M)\langle \pi_1, \pi_2 \rangle)\epsilon)\alpha\nu.$$

4 Probabilistic Kripke Polynomial Functors in Set

The work from the previous section can also be carried out in Set, for probabilistic Kripke polynomial functors as introduced in [1]. These are functors built from the identity functor, constant functors for fixed sets, the finite, covariant power set functor denoted by $\mathcal{P}$, functions from a fixed set E, denoted by $\cdot^E$, and the discrete measure functor $\mathcal{D}$, that assigns to a set X the set of all functions $\mu : A \to [0,1]$ with finite support and such that $\sum_{a \in A} \mu(a) = 1$.

The set $\mathsf{Ing}(T)$ is defined similarly as done for measure polynomial functors, omitting the clause for Δ and adding $\mathsf{Ing}(\mathcal{P}S) = \{\mathcal{P}S\} \cup \mathsf{Ing}(S); \mathsf{Ing}(S^E) = \{S^E\} \cup \mathsf{Ing}(S); \mathsf{Ing}(\mathcal{D}S) = \{\mathcal{D}S\} \cup \mathsf{Ing}(S)$.

The required connecting maps are defined as follows: $\zeta : Z_{\mathcal{P}S} \to \mathcal{P}(Z_S)$

$$\zeta(z) \quad = \quad \{u \in Z_S | \forall n \geq 0 \ \pi^n_S(u) \in \pi^n_{\mathcal{P}S}(z)\}$$

$\eta : Z_{S^E} \to (Z_S)^E$ is such that for all $n \geq 0, e \in E$ and $z \in Z_S$,

$$\pi^n_S(\eta(z)(e)) = \pi^n_{S^E}(z)(e)$$

To define $\theta : Z_{\mathcal{D}S} \to \mathcal{D}(Z_S)$ one must first consider for each $z \in Z_{\mathcal{D}S}$ the set $S_z = \{u \in S_S | \forall n \geq 0 \ \pi^n_{\mathcal{D}S}(z)(\pi^n_S(u)) > 0\}$. Since this proves to be a finite set, there's a number n such that if $u, u' \in S_z$, and $u \neq u'$, then $\pi^n(u) \neq \pi^n(u')$. Let N be the first such number. Then

$$\theta(z)(u) = \begin{cases} \pi^N_{\mathcal{D}S}(z)(\pi^N_S(u)) & \text{if } u \in S_z \\ 0 & \text{otherwise.} \end{cases}$$

The functions defined above satisfy the conditions that make the following Lemma, analogous to Lemma 9, work.

Lemma 12. *There exists a function $r_T : Z_T \to T(Z)$ so that for every (X, c), $r_T \circ h^c_T = Th^c$, and for every $n \geq 0, T\pi_n \circ r_T = \pi^n_T$.*

Proof. (Sketch) The proof is much like the one of Lemma 9, defining $r_{\mathcal{P}S} = \zeta(\mathcal{P}r_S), r_{S^E} = \eta(r_S)^E$ and $r_{\mathcal{D}S} = \theta(\mathcal{D}r_S)$. □

The rest of the results of Section 3 are valid in Set, yielding the following Theorem:

Theorem 2. *Every probabilistic Kripke polynomial functors has a final coalgebra.*

5 Related Work and Open Questions

There exist in the literature other ways of obtaining final coalgebras from the final sequences for a functor. We briefly review them and point out similarities and differences with the approach presented here.

A simpler way of finding the coalgebra structure for Z in the category Set, is shown in Kurz and Pattinson, [10]. First, a coalgebra (C, δ) is constructed by taking a coproduct $(C, \delta) = \coprod_{z \in Z}(C^z, \delta^z)$ where (C^z, δ^z) is such that for some $c \in C^z, h^{\delta^z}(c) = z$. Then the map $h^\delta : C \to Z$ is surjective, and therefore has a right inverse $o : Z \to C$ which allows to define $\gamma' : Z \to T(Z)$ by letting $\gamma' = (Th^\delta)\delta o$.

$$\begin{array}{ccc} C & \xrightarrow{\delta} & TC \\ h^\delta \downarrow \uparrow o & & \downarrow Th^\delta \\ Z & \xrightarrow[\gamma']{} & TZ \end{array}$$

The function γ' is not necessarily the structure map of the final coalgebra; some more work would need to be done to show that Z is actually the carrier of the final coalgebra. This approach for finding γ' does not work in general in Meas, since surjective functions do not always have a measurable right inverse, and it is also less instructive about the structure of the final coalgebra than the construction presented here.

In [11], van Breugel et al. consider the functor $\mathbb{M}$ of *subprobabilities* over a measurable space. Given a countable set *Act*, coalgebras for $\mathbb{M}^{Act}$, the *Act*-fold product of $\mathbb{M}$ are models of labelled Markov processes. The construction of the final coalgebra of the functor $\mathbb{M}^{Act}$ is performed using the the connection of the final sequence for this functor with the one for the probabilistic powerdomain in topological spaces. Furthermore, they prove that the final coalgebra can be regarded as a Polish space.

In [12], Worrell shows a different way to get final coalgebras from final sequences in Set. One of the referees for this article has outlined how this method would also work in the category Meas. The idea is that the ω^{op} sequence can de extended beyond the limit. The key observation to develop this approach is that for the functor Δ, the connecting map $\Delta^\omega! : \Delta(P) \to P$ can be proved to be injective using the π-λ theorem.

In [1], we presented an application to Economics. To model beliefs of agents in a game with incomplete information, one wants to solve equations of the form $X = \Delta(M \times X)$ for some fixed measurable space M. In this way, an element $x \in X$ is identified with the probability distributions (beliefs) that x has on the state of nature, represented by the space M, and on the beliefs of the other players. A final coalgebra provides solutions for these equations up to isomorphisms.

The most interesting question at this point is what is the most general class of functors to which this method for constructing the final coalgebras can be applied. This question may have different answers in different categories. One is tempted to venture that it is related to the work by Worrell [12] and Adámek [13] on final sequences. This characterizations would prove the expressiveness of the languages introduced in [1].

References

1. Moss, L.S., Viglizzo, I.D.: Harsanyi type spaces and final coalgebras constructed from satisfied theories. Electronic Notes in Theoretical Computer Science **106** (2004) 279–295
2. Larsen, K.G., Skou, A.: Bisimulation through probabilistic testing. Inform. and Comput. **94** (1991) 1–28
3. deVink, E.P., Rutten, J.J.M.M.: Bisimulation for probabilistic transition systems: a coalgebraic approach. Theoretical Computer Science **221** (1999) 271–293 ICALP '97 (Bologna).
4. Moss, L.S., Viglizzo, I.: Final coalgebras for functors on measurable spaces. Information and Computation (2005) Accepted, under revision.
5. Heifetz, A., Samet, D.: Topology-free typology of beliefs. Journal of Economic Theory **82** (1998) 324–341
6. Giry, M.: A categorical approach to probability theory. In: Categorical aspects of topology and analysis. Volume 915 of Springer Lecture Notes in Mathematics. Springer, Berlin-New York (1982) 68–85
7. Kechris, A.S.: Classical Descriptive Set Theory. Springer-Verlag, New York (1995)
8. Andersen, E.S., Jessen, B.: On the introduction of measures in infinite product sets. Danske Vid. Selsk. Mat-Fys. Medd. **25** (1948) 8
9. Halmos, P.R.: Measure Theory. D, Van Nostrand Company, Inc., New York, N. Y. (1950) xi + 304 pp.
10. Kurz, A., Pattinson, D.: Definability, canonical models, compactness for finitary coalgebraic modal logic. In: CMCS 2002: coalgebraic methods in computer science (Berlin). (Volume 65 of Electron. Notes Theor. Comput. Sci.)
11. van Breugel, F., Mislove, M., Ouaknine, J., Worrell, J.: Domain theory, testing and simulation for labelled Markov processes. Theoret. Comput. Sci. **333** (2005) 171–197
12. Worrell, J.: Terminal sequences for accessible endofunctors. In: CMCS'99 Coalgebraic Methods in Computer Science (Amsterdam, 1999). Volume 19 of Electron. Notes Theor. Comput. Sci. Elsevier, Amsterdam (1999) 15 pp. (electronic)
13. Adámek, J.: On final coalgebras of continuous functors. Theoret. Comput. Sci. **294** (2003) 3–29 Category theory and computer science.

Bireachability and Final Multialgebras

Michał Walicki*

University of Bergen, Department of Informatics,
michal@ii.uib.no

Abstract. Multialgebras generalise algebraic semantics to handle non-determinism. They model relational structures, representing relations as multivalued functions by selecting one argument as the "result". This leads to strong algebraic properties missing in the case of relational structures. However, such strong properties can be obtained only by first choosing appropriate notion of homomorphism. We summarize earlier results on the possible notions of compositional homomorphisms of multialgebras and investigate in detail one of them, the outer-tight homomorphisms which yield rich structural properties not offered by other alternatives. The outer-tight homomorphisms are different from those obtained when relations are modeled as coalgebras and the associated congruence is the converse bisimulation equivalence. The category is co-complete but initial objects are of little interest (essentially empty). On the other hand, the category does not, in general, possess final objects for the usual cardinality reasons. The main objective of the paper is to show that Aczel's construction of final coalgebras for set-based functors can be modified and applied to multialgebras. We therefore extend the category admitting also structures over proper classes and show the existence of final objects in this category.

1 Introduction

In the tradition of algebraic specifications, nondeterminism has been modeled by means of multialgebras, that is, algebras where operations may return not only single elements but also sets thereof, e.g., [10,11,13,25,26]. Multialgebras, or variants of power structures, have been given some attention also in the mathematical community, e.g., [19,20,7,22,4,17], with the seminal work [14,15] which introduced them as "algebras of complexes" to represent relational structures and demonstrated representability of Boolean algebras with operators by such algebras. [3] gives a comprehensive overview. Some variants disallow empty result-sets, e.g., [7,24], but most do not. Then, applying the standard isomorphism

$$A_1 \times ... \times A_n \to \mathcal{P}(A) \;\simeq\; \mathcal{P}(A_1 \times ... \times A_n \times A), \tag{1}$$

one obtains another representation of relational structures, although with more algebraic properties, as will be observed below. This is the variant of multialgebras we will be using.

* Research partially supported by the Norwegian Research Council project MoSIS.

J.L. Fiadeiro et al. (Eds.): CALCO 2005, LNCS 3629, pp. 408–423, 2005.

The standard requirement put on a function $\phi : A \to B$ between two relational structures in order to obtain a homomorphism is preservation of all basic relations – for each relation symbol $R : R^A(a_1...a_n, a) \Rightarrow R^B(\phi(a_1)...\phi(a_n), \phi(a))$. This is extremely weak notion (e.g., such homomorphisms do not preserve even positive inclusions, the associated congruence is simply equivalence). Consequently, one finds in the literature numerous alternative, and stronger, requirements. In fact, the problem which we are addressing is that such proposals are *too numerous.* Of course, the choice of the notion of homomorphism can often depend on the specific context and need not be made uniformly once and for all. But it is not certain that the possibility of such a choice is itself a virtue rather than a nuisance (especially, if we compare to the tradition of universal total algebra with the unique and powerful notion of homomorphism).

In earlier work, [23], we have shown that, restricting the possible definitions of relational/multialgebraic homomorphisms to a reasonable and almost universally followed format, there are only nine choices which are compositional. Investigation of these categories showed that only few of them are finitely complete and cocomplete. From the point of view of the semantics of (algebraic) specifications, it is desirable that the model category possesses canonical (initial or final) objects of interest. Although we have investigated only the most generic situation of the whole category of all Σ-algebras for a given signature Σ, the canonical models (when existing) were of minimal relevance (basically, empty).

This paper addresses one of the earlier investigated categories which does not, in general, possess final objects. The reason for that is the same as the reason for which the categories of coalgebras for functors involving power-set do not possess final objects – the cardinality reasons which require one to step over to the proper classes (or limit the cardinality of power-set.) We show that, making that step, we obtain final multialgebras of quite interesting nature which, in some sense, are dual to final coalgebras. The homomorphisms of multialgebraic structures in the studied category carry a similar duality to the homomorphisms induced by the coalgebraic model of (binary) relations, while the associated congruence relations are converse bisimulations. The obtained category is cocomplete and we expect other positive results: it is, probably, complete; the homomorphisms have stronger preservation/reflection properties than the traditional (weak) ones; final objects can also be obtained for axiomatic theories. All these "probable" issues remain, however, for the future work. At the present, we only consider the existence of final objects and comment briefly their character.

Section 2 gives the basic definitions, summarises earlier results and signals some possible alternatives. Section 3 presents the category of interest, "outer-tight", focusing on the notion of its congruence – bireachability. It also describes the final objects (when these exist). Section 4 generalizes this category by allowing algebras over proper classes, shows its cocompleteness and the existence of final objects. The concluding section 5 lists some open problems, suggesting also improvements and further generalizations of the obtained results. The main aspects of central constructions are summarized as proof ideas – the complete proofs will be available in a forthcoming technical report.

2 Background

Multialgebras are many-sorted algebras where operations can return (possibly empty) sets of values rather than unique values. Following [8], (one-sorted) multialgebraic operation R on a set X can be seen as a dialgebra $R : F(X) \to \mathcal{P}(X)$ in the category $\mathsf{SET}^F_{\mathcal{P}}$, where functor F gives the source of the operation and $\mathcal{P}$ is the covariant existential-image power-set functor, i.e., sending a function $\phi : A \to B$ onto $\mathcal{P}(\phi)(X) = \{\phi(x) \mid x \in X\}$, for $X \subseteq A$. The variations in the definitions of homomorphisms to be encountered below could be then seen as variations of the morphisms of dialgebras (requiring, in addition, lax transformations). Less abstractly, we use the isomorphism (1.1), and view a multialgebra as a relational structure where, for each relation, one argument is designated as its "result" and used for composition with other relations.

Definition 2.1. *For a signature $\Sigma = \langle \mathcal{S}, \mathcal{F} \rangle$, a Σ-multialgebra M is given by:*

- *a (family of) carrier set(s) $|M| = \{s^M \mid s \in \mathcal{S}\}$,*
- *a function $R^M : s_1^M \times ... \times s_n^M \to \mathcal{P}(s^M)$ for each $R : s_1 \times ... \times s_n \to s \in \mathcal{F}$, with composition defined through additive extension to sets, i.e. $R^M(X_1, ..., X_n) = \bigcup_{x_i \in X_i} R^M(x_1, ..., x_n)$.*

The only structures addressed in the paper are multialgebras, so "multialgebra" and "algebra" will be used interchangeably. We assume a given signature with R ranging over all function/relation symbols.

Selection of the "result" argument corresponds, in a sense, to turning our considerations to binary relations with the additional operation of tupling the arguments. Composition of relations $R_1 : X_{11}...X_{1n} \to X_1, ..., R_k : X_{k1}...X_{kn} \to X_k$ and $R : X_1...X_k \to X$, corresponds to application of R to the tupling $\langle R_1...R_k \rangle$. We will freely switch between relational and functional notation, so the composition can be written as $R(R_1(x_1)...R_k(x_k))$ or $\langle R_1...R_k \rangle; R$. We write composition in diagrammatic order, $R; \phi$, resp. $\phi; R$, assuming implicitly ϕ to be binary (homomorphism or, strictly speaking, a tuple $\langle \phi_1, ..., \phi_{n+1} \rangle$ of unary functions, for each relevant argument/sort i.) The composition is, as just explained, an abbreviation for the multialgebraic one, i.e.:

$$\begin{array}{rl} \langle \langle a_1...a_n \rangle, b \rangle \in R; \phi & \iff \exists a : \langle \langle a_1...a_n \rangle, a \rangle \in R \wedge \langle a, b \rangle \in \phi_{n+1} \\ \textit{resp. } \langle \langle a_1...a_n \rangle, b \rangle \in \phi; R & \iff \exists b_1...b_n : \langle a_i, b_i \rangle \in \phi_i \wedge \langle \langle b_1...b_n \rangle, b \rangle \in R \end{array} \quad (2)$$

Having made these precautions, we will write things as if all relations were binary, algebras were one-sorted and homomorphisms simple functions (and not their families), but all considerations apply to the general case.

Selection of the "result" among the relational arguments leads to more algebraic structure reflected by homomorphisms. (In particular, derived operators of a multialgebra are analogous to those of classical algebra: for a signature Σ, the term structure T_Σ is itself a Σ-algebra, and preservation/reflection of Σ operations leads to the corresponding behaviour of the derived operators. For relational structures, derived operators are just boolean operators only very weakly related to the actual signature and not necessarily preserved by the homomorphisms preserving the basic relations. [5], V.3, p.203, considers this the reason for the subordinate role of homomorphisms in the study of relational structures.) However, the study of the obtained structure is not significantly simplified. As a matter of fact, the number of possible definitions of homomorphisms, congruences, etc. does not decrease. As the first step towards simplification of the rather complicated picture, we have earlier in [23] classified compositional homomorphisms of (relational structures modeled as) multialgebras and checked finite (co)completeness of the respective categories. We recall now these results in order to motivate our choice of the outer-tight homomorphisms.

Definition 2.3. *A definition $\Delta[\_]$ of a function $\phi : |A| \to |B|$ being a homomorphism of the multialgebraic structures $A \to B$ has the form:*

$$\Delta[\phi] \iff l_1[\phi]; R^A; r_1[\phi] \quad \bowtie \quad l_2[\phi]; R^B; r_2[\phi]$$

where $l[\_]$'s and $r[\_]$'s are relational expressions (using only relational composition and converse), and $\bowtie \in \{=, \subseteq, \supseteq\}$.

One can certainly consider other formats but most proposed definitions of homomorphisms conform to this one as, in particular, do all compositional definitions which we have ever encountered.

Definition 2.4. *A definition Δ is* compositional *iff for all $\phi : A \to B$, $\psi : B \to C$, we have $\Delta[\phi]$ & $\Delta[\psi] \Rightarrow \Delta[\phi;\psi]$, i.e.:*

$$l_1[\phi]; R^A; r_1[\phi] \bowtie l_2[\phi]; R^B; r_2[\phi] \quad \& \quad l_1[\psi]; R^B; r_1[\psi] \bowtie l_2[\psi]; R^C; r_2[\psi]$$
$$\Rightarrow \quad l_1[\phi;\psi]; R^A; r_1[\phi;\psi] \quad \bowtie \quad l_2[\phi;\psi]; R^C; r_2[\phi;\psi]$$

Theorem 2.5 ([23]). *A definition is compositional iff it is equivalent to one of:*

$$R^A;\phi \bowtie \phi;R^B \qquad \phi^-;R^A;\phi \rhd R^B \qquad \phi^-;R^A \rhd R^B;\phi^- \qquad R^A \rhd \phi;R^B;\phi^-$$

where $\bowtie \in \{=, \subseteq, \supseteq\}$ and $\rhd \in \{=, \supseteq\}$.

The following table summarises the naming conventions for the compositional cases. The name consists of two parts, the first (inner/left/...) indicating one of the four main cases in the theorem and the second (closed/tight/weak) the choice of the set relation. For the weak case there are no further distinctions, since all such cases are, in fact, equivalent. (They would not be equivalent if (homo)morphisms were relations – [6] analyses these four weak cases of "simulations", though without addressing the issue of (co)completeness.)

	$R^A;\phi \bowtie \phi;R^B$	$\phi^-;R^A;\phi \rhd R^B$	$\phi^-;R^A \rhd R^B;\phi^-$	$R^A \rhd \phi;R^B;\phi^-$
	inner	left	outer	right
closed: $\supseteq$	$\mathsf{MAlg}_{IC}(\Sigma)$	$\mathsf{MAlg}_{LC}(\Sigma)$	$\mathsf{MAlg}_{OC}(\Sigma)$	$\mathsf{MAlg}_{RC}(\Sigma)$
tight: $=$	$\mathsf{MAlg}_{IT}(\Sigma)$	$\mathsf{MAlg}_{LT}(\Sigma)$	$\mathsf{MAlg}_{OT}(\Sigma)$	$\mathsf{MAlg}_{RT}(\Sigma)$
weak: $\subseteq$	$\mathsf{MAlg}_{W}(\Sigma)$			

Table 1. Finite limits and co-limits in the categories of multialgebras

	initial	co-prod.	co-equal.	final	prod.	equal.
$\mathsf{MAlg}_W(\Sigma)$	+	+	+	+	+	+
$\mathsf{MAlg}_{IC}(\Sigma)$	–	–	–	+	–	–
$\mathsf{MAlg}_{IT}(\Sigma)$	–	–	+	–	–	–
$\mathsf{MAlg}_{LC}(\Sigma)$	–	–	+	+	–	–
$\mathsf{MAlg}_{LT}(\Sigma)$	–	–	+	–	–	–
$\mathsf{MAlg}_{OC}(\Sigma)$	+	+	–	+	–	+
$\mathsf{MAlg}_{OT}(\Sigma)$	+	+	+	+/–	?	+
$\mathsf{MAlg}_{RC}(\Sigma)$	+	+	+	+	+	+
$\mathsf{MAlg}_{RT}(\Sigma)$	+	–	–	–	–	+

The earlier results concerning finite (co)completeness of these categories are summarised in table 1.

The present paper addresses the category of outer-tight homomorphisms (the double row in the table) and, in particular, the position marked +/–. First, however, a few words about the possible alternatives.

Remark 2.6. *Viewing (binary) relations as coalgebras for the existential-image power-set functor, yields the homomorphism condition $R^A;\phi = \phi;R^B$, that is, the inner-tight homomorphisms. As we see from the table, the category $\mathsf{MAlg}_{IT}(\Sigma)$ has rather few (co)limits. This, of course, looks suspicious, since we know from [21] that any category of coalgebras over sets will be, at least, cocomplete. The difference is, however, due to the fact that although the homomorphism conditions are the same, the respective representations of relations are not:*

The absence of final objects is here due to the fact that the table addresses only categories based on sets. The non-existence of colimits is due to the algebraic character of operations, in particular, constants which correspond to predicates. For instance, for a signature with a single sort and constant $c :\to S$, the category $\mathsf{MAlg}_{IT}(\Sigma)$ has no initial multialgebra I – for any (in particular, empty) c^I there is no IT-homomorphism $\phi : I \to A$ making $\phi(c^I) = c^A$ when $|c^I| < |c^A|$. In a coalgebra, a (predicate) constant is an arrow $c : X \to \mathbf{2}$ and this enables one to achieve commutativity, $c^A;\phi = \phi;c^B$, also when $X = \emptyset$.

In fact, the meaning of the condition is different in the two cases: for coalgebras it requires equality of two functions while for multialgebras of two sets. As an example, take the carrier $X = \{1,2\}$ and one constant c. Let, in a multialgebra M, $c^M = \{1,2\}$, while in a coalgebra C, $c(1) = c(2) = \top$. Let $X' = \{1,2,3\}$ and $c^{M'} = \{1,2,3\}$ while in a coalgebra C', $c'(1,2,3) = \top$. Although both M and C, resp., M' and C' represent the same predicates, the inclusion $i : X \to X'$ is a coalgebraic homomorphism, since indeed $c;i = i;c'$, but it is not a multialgebraic IT-homomorphism since $i(c^M) = i(\{1,2\}) = \{1,2\} \neq \{1,2,3\} = c^{M'}$.

This might be taken as a suggestion that the multialgebraic representation of relations is not the most successful one. However, using coalgebras as models of relations is by no means straightforward. For the first, one has to decide on whether to use the functor $\mathcal{P}(X^n)$ or $\mathbf{2}^{(X^n)}$ – the difference in homomorphisms will be similar to that suggested in the above remark (between equality of sets and of functions). In either case one has to decide which power-set functor to use. Any choice involves sacrificing the pleasant and well understood behavior of polynomial functors. Additional complications arise if one wants to model many-sorted relations. (Although these are hardly theoretically demanding, they *are* complications, at least of the same order as in the case of many-sorted algebras.) Multialgebraic model, on the other hand, is in agreement with the traditional notion of relation/predicate as a subset. It deals with many-argument, as well as many-sorted, relations in the uniform and elementary way. In addition, one should remark that multialgebras were introduced not merely as representations of relational structures but of Boolean algebras with operators (central, if not always recognised, in modal logics, as Kripke-frames are such algebras) and, on the other hand, as a generalisation of algebraic semantics to handle nondeterminism (most common institutions can be naturally embedded into the institution of multialgebras, with weak homomorphisms as morphisms in the model categories, [16]). The investigation of homomorphisms arises from this background and is motivated primarily by the search for the interesting canonical objects (initial or final) for algebraic specifications with nondeterminism.

Now, weak homomorphisms are those which are most commonly used. Unfortunately, this is an extremely weak notion which is also reflected in its standard name. Although the initial objects exist, they are of little interest having all predicates and relations empty. Lifting existence of initial objects to the axiomatic classes depends, of course, on the language one wants to use, and this is by no means a clarified issue. Most approaches suggest, at least, the use of inclusions, but this again leads only to empty relations in the initial objects. Furthermore, even simplest formulae are not preserved. E.g., having two constants a, b interpreted in A as $\{1\}$, resp., $\{1,2\}$ makes $A \models a \subseteq b$. But the inclusion, which is a weak homomorphism, into B with $a^B = \{1,3\}$ and $b^B = \{1,2\}$ does not preserve this formula. Counterexamples can be easily found also when we restrict attention to preservation under homomorphic images. One way would be to design a specific syntax ensuring adequate restrictions of the model classes, as was done, for instance, with membership algebras, [18]. But this amounts to an application-oriented specialisation of the problem which we are not addressing here. (Similar remarks apply to the other (co)complete category $\mathsf{MAlg}_{RC}(\Sigma)$.)

The OT-homomorphisms seem to possess many desirable properties absent in other cases, especially that of weak homomorphisms. This paper characterizes final objects in the category $\mathsf{MAlg}_{OT}(\Sigma)$ and proves their existence. Now, the $+/-$ in the table 1 indicates that final objects can be constructed only in special cases. In general, they do not exist for the simple cardinality reasons. In the following section, we recall a series of basic facts about this category, and illustrate the character of final objects (when they exist). We also focus on the associated notion of congruence which can be seen as the converse bisimulation equivalence. Then, we will extend the category by allowing algebras with carriers being proper classes. In this category, final objects do exist, and we show it in the way analogous to that in which the corresponding fact is proven for the categories of coalgebras for "set-based" functors in [2].

3 The Category Outer-Tight

For $\Sigma = \langle \mathcal{S}, \mathcal{F} \rangle$, an OT-homomorphism, $\phi : A \to B$, is a (family of) function(s) $\phi_i : s_i^A \to s_i^B$, for each $s_i \in \mathcal{S}$, such that for every $R \in \mathcal{F}$:

$$\phi^-; R^A = R^B; \phi^-$$

in functional notation : $\forall b_1 ... b_n \in |B| : R^A(\phi_1^-(b_1)...\phi_n^-(b_n)) = \phi_{n+1}^-(R^B(b_1...b_n))$

which for constants specializes to : $c^A = \phi^-(c^B)$.

This requirement is strictly stronger than that of the weak homomorphism. Since we will be dealing exclusively with OT-homomorphisms, we will not qualify the name – saying "homomorphism", we will always mean an OT-homomorphism.

The following few facts are hardly surprising but they are used in later results.

Fact 3.1. *An OT-homomorphism ϕ is*

1) mono iff it is injective;
2) epi iff it is surjective;
3) iso iff it is bijective.

The following observation will not be referred to later on, but it is used in a couple of proofs of the results mentioned in the sequel.

Given $A, A' \in \mathsf{MAlg}_{OT}(\Sigma)$, A' is a subalgebra of A, $A' \sqsubseteq A$, iff the inclusion $|A'| \subset |A|$ is a homomorphism. (The categorical definition would not introduce any significant changes.) In general, an inclusion need not be a homomorphism. But the following fact holds.

Fact 3.2. *Inclusions between subalgebras of the same algebra are OT-homomorphisms. I.e., if $A_1 \sqsubseteq A$ and $A_2 \sqsubseteq A$ and $|A_2| \subset |A_1|$, then also $A_2 \sqsubseteq A_1$.*

The following fact ensures that the diagram of subalgebras is directed.

Fact 3.3. *For an algebra A and every set $X \subseteq |A|$, there is a smallest subalgebra $A_X \sqsubseteq A$ with $X \subseteq |A_X|$.*

Thus, if $A_1, A_2 \sqsubseteq A$, then there is also (a smallest) $A_3 \sqsubseteq A$, with $|A_1| \cup |A_2| \subseteq |A_3|$. In the proof, one extends appropriately the set X or, like in the classical case, verifies that intersection of subalgebras is a subalgebra.

3.1 Bireachability

In order for the quotient construction performed on a carrier of a (classical) Σ-algebra to yield a (quotient) Σ-algebra, the equivalence must be a Σ-congruence. However, for any (classical) algebra A and any equivalence $\sim$ on its carrier, the quotient $A/_{\sim}$, with operations collecting the possibly non-congruent results (i.e., defined by $R^{A/\sim}([a]) = \{[n] : n \in R^A(a'),\ a' \in [a]\}$), is a multialgebra, and the construction works in the same way if we start with a multialgebra, and not only a classical algebra. Defining the mapping $q : A \to A/_{\sim}$ by $q(a) = [a]$, the operations are obtained as $R^{A/\sim} = q^-; R^A; q$. In general, this mapping is only a weak homomorphism, just like the kernel of a weak homomorphism is, in general, only an equivalence. (This correspondence is perhaps the clearest expression of the weakness of this homomorphism notion.) OT-homomorphisms come along with a much stronger notion of a congruence.

Definition 3.4. *An equivalence* $\sim$ *on* A *is OT-congruence iff:* $\sim; R^A; \sim = \sim; R^A$

Whenever $\sim$ is an equivalence, the inclusion $\sim; R; \sim \subseteq \sim; R$ is equivalent to:

$$R; \sim \subseteq \sim; R. \tag{3}$$

Any equivalence satisfying this last condition is OT-congruence, since the opposite inclusion $\sim; R^A; \sim \supseteq \sim; R^A$ holds trivially for any reflexive $\sim$.

This characterisation of OT-congruence can be visualized as the converse of (bi)simulation. (Bi)simulation requires propagation of $\sim$ forward, while OT-congruence backward – we should be therefore allowed to call this relation "bireachability".[1] On the drawing, the dotted lines indicate the required existence implied by the regular lines:

(4)

Henceforth, we will use the words "bireachability" and "OT-congruence" as synonyms. The same meaning will be attached also to "congruence", unless the word is qualified in some other way.

Fact 3.7. *If* $\phi : A \to B$ *is OT then so is its kernel* $\sim_\phi$.

The opposite does not hold generally; even if the kernel of ϕ is OT, ϕ itself may be not. We have a slightly weaker claim.

[1] We are not addressing any details concerning bisimulations. For the sake of analogy, since OT-congruences are equivalences, it is most convenient to think of bisimulation defined as a symmetric simulation, rather than merely as a simulation with converse being also a simulation. Exact duality obtains between our bireachability and the equivalences satisfying the condition that for every $R : \sim; R^A; \sim = R^A; \sim$, i.e., IT-congruences or bisimulations in (3.6), referred to in remark 2.6. In [4] such equivalences were called "preserving the arguments" (as opposed to congruences "preserving the values"). In [9], the relation dual to mere simulation, without the requirement of equivalence, was called "opsimulation," but the name "biopsimulation" does not seem very appealing.

Fact 3.8. *If $\sim$ is a bireachability then the mapping $q : A \to A/_\sim$, $q(a) = [a]$, is an OT-epimorphism.*

This allows us to obtain epi-mono factorisation of morphisms in $\mathsf{MAlg}_{OT}(\Sigma)$.

Fact 3.9. *For every homomorphism $h : A \to B$ there is a (regular) epi $e : A \to Q$ and mono $m : Q \to B$ such that $h = e; m$.*

Bireachability on a Σ-multialgebra has itself a multialgebraic Σ-structure.

Definition 3.10. *Given a bireachability $\sim$ on an $A \in \mathsf{MAlg}_{OT}(\Sigma)$, we define $A^\sim \in \mathsf{MAlg}_{OT}(\Sigma)$:*

- $|A^\sim| = \{\langle a_1, a_2\rangle : a_1, a_2 \in |A| \wedge a_1 \sim a_2\}$, *and*
- $f^{A^\sim}(\langle a_1, b_1\rangle ... \langle a_n, b_n\rangle) = \{\langle x, y\rangle : x \in f^A(a_1...a_n) \wedge y \in f^A(b_1...b_n) \wedge x \sim y\}$, *which yields*
- *for constants* $c^{A^\sim} = \{\langle x, y\rangle : x, y \in c^A \wedge x \sim y\}$.

Fact 3.11. *Given a bireachability $\sim$ on A. 1) The two projections $\pi_1, \pi_2 : A^\sim \to A$, $\pi_i(\langle a_1, a_2\rangle) = a_i$ are OT-homomorphisms. 2) Moreover, $A/_\sim$ with the quotient homomorphism $q : A \to A/_\sim$ is their coequalizer.*

Maximal bireachability. Given a collection $C = \{\sim_i : i \in I\}$ of equivalences (on a set/algebra A), one defines their lub as the transitive closure of their union, i.e., $\sim\ = \bigvee_i \sim_i\ = (\bigcup_i \sim_i)^*$. Explicitly, $a \sim a'$ iff there exists a finite sequence $a = a_0 a_1 ... a_n = a'$ and a respective sequence of the equivalences from C, $\sim_1 \sim_2 ... \sim_n$, such that $a_i \sim_{i+1} a_{i+1}$ for all $0 \leq i < n$.

The same construction applies also to bireachabilities. The following lemma will be of crucial importance.

Lemma 3.12. *Given a collection $C = \{\sim_i : i \in I\}$ of bireachabilities on a multialgebra A, then $\sim\ = \bigvee_i \sim_i$ is a bireachability.*

Notice that the maximal bireachability need not be the standard unit relation. For instance, for the algebra $\begin{array}{cc} b_1 & b_2 \\ {\scriptstyle R}\uparrow & \\ a_1 & \end{array}$, the elements b_1 and b_2 cannot be related by any bireachability, according to the observation (3.5).

One verifies easily that the construction yields, in fact, the least upper bound – with respect to the subset relation – of the argument bireachabilities. Thus, the collection of all congruences on a multialgebra is a complete upper semilattice with the least element being identity, and so it is a complete lattice. (Greatest lower bounds are not, however, obtained as mere intersections.)

Fact 3.13. *Let $B \sqsubseteq A$, $\sim_A$ be a bireachability on A, and $\sim_B \subseteq \sim_A$ be restriction of $\sim_A$ to the carrier of B, i.e., $\sim_A \cap |B| \times |B|$. Then $\sim_B$ is bireachability on B.*

3.2 Final Objects in $\mathsf{MAlg}_{OT}(\Sigma)$

Final objects do not exist in $\mathsf{MAlg}_{OT}(\Sigma)$ due to the usual cardinality reasons. (A multialgebra for an operation $f : S \to S$ is essentially a coalgebra for the existential-image power-set functor.) As stated in the introduction $\mathsf{MAlg}_{OT}(\Sigma)$ is finitely cocomplete but the existence of final objects has been shown only for a very special case. We show here such a case mainly to illustrate the interesting features of the final objects.

Example 3.14. *Let $\Sigma = \langle\{s_1, s_2\}, \{c :\to s_1; f : s_1 \to s_2\}\rangle$. The final object Z in $\mathsf{MAlg}_{OT}(\Sigma)$ can be described as follows. (Expressions like "$\emptyset_1$" or "$fc\emptyset$" are simple names – mnemonic devices – not any sets or function applications.)*

- $s_1^Z = \{c, \emptyset_1\}$, $s_2^Z = \{fc, f\emptyset, fc\emptyset, \emptyset_2\}$
- $c^Z = \{c\}$ *and* $f^Z(c) = \{fc, fc\emptyset\}$, $f^Z(\emptyset_1) = \{f\emptyset, fc\emptyset\}$.

In words, each sort contains only elements needed to distinguish any combination of operations returning the elements of this sort. In s_1^Z it is enough with one element to interpret the constant. In addition, there is always a "junky" element not belonging to the result of any operation, $\emptyset_1$. s_2^Z contains one such element, $\emptyset_2$, as well as one element characteristic for (belonging only to) $f^Z(c) \ni fc$, one for $f^Z(\emptyset_1) \ni f\emptyset$ and one for $f^Z(c) \cap f^Z(\emptyset_1) \ni fc\emptyset$.

If we had two constants of sort s_1, we would obtain corresponding collection $\{c, d, cd, \emptyset_1\}$ in s_1^Z, while s_2^Z would now contain characteristic element for every possible $f^Z(x)$ when $x \in s_1^Z$, as well as for every intersection $\bigcap_{x \in X} f^Z(x)$ for every possible $X \subseteq s_1^Z$.

Viewing the set of results of any application, $f^Z(x)$, as the set of possible (or nondeterministic) observations of its argument x, the construction amounts to providing the minimal number of elements needed for every set of (every series of) observations to have its unique characteristic result.

The most general form of this construction can be obtained when signature does not contain any "loops". Call a signature "acyclic" if there is no derived operator t with target sort occurring also among the argument sorts.

Fact 3.15. *If Σ is acyclic then $\mathsf{MAlg}_{OT}(\Sigma)$ has final objects.*

We will now extend the category $\mathsf{MAlg}_{OT}(\Sigma)$ to allow for the existence of final objects without any restrictions on the signature. As in the case of coalgebras, we have to either impose some cardinality limits or else leave the set-based categories and allow algebras with proper classes as carriers. The former case leads to rather special conditions[2] and so we follow the later alternative.

[2] E.g., final objects can be obtained if algebras considered are such that every element of the carrier can be reached from at most finite number of other elements in at most finite number of ways – the "reachability" restriction which is, in a sense, dual to restricting the $\mathcal{P}$ functor to $\mathcal{P}^{fin}$ returning only finite sets.

4 The Category Outer-Tight with Classes

Given a Σ with sort symbols $\{s_1...s_n\}$, we allow algebras where carrier of each sort is a class. Likewise, operations and constants can return proper classes.[3] But we will need the assumption that

each such algebra is a colimit of its small subalgebras and, moreover, the category contains all algebras which are such colimits (over arbitrary and, possibly, large diagrams). (5)

Since colimit arrows are jointly epi (and, by fact 3.1, our epis are surjective) and the diagram of (small) subalgebras is directed (fact 3.3), the above assumption implies that:

for every algebra A and set $X \subseteq |A|$, there is a small subalgebra $sA \sqsubseteq A$ with $X \subseteq |sA|$. (6)

We denote this category $\mathsf{MAlg}^*_{OT}(\Sigma)$. (We will comment on more specific conditions which could replace (4.1) ensuring that all our constructions yield appropriate results in the concluding section.)

A bireachability R on an A which is a colimit of its small subalgebras A_i, is itself a colimit of its small subalgebras $R_i = R \cap |A_i| \times |A_i|$, i.e., $R \in \mathsf{MAlg}^*_{OT}(\Sigma)$. Lemma 3.12 applies unchanged when the collection is a proper class of small bireachabilities. Performing the same standard construction on the collection of *all small* bireachabilities on a given multialgebra yields the following lemma.

Lemma 4.3. $\forall A \in \mathsf{MAlg}^*_{OT}(\Sigma)$ *there exists a unique maximal bireachability* $\sim_A$.

The following easy technicality will be needed in the proof of the next lemma. (Notation follows the diagram below.)

Fact 4.4. *Let* $\{A_i : i \in I\}$ *be the class of small subalgebras of* A *(*A *being their colimit),* R *be a congruence on* A *and* R_i *the respective restrictions of* R *to* A_i*. Then the family of inclusions* $\{r_i : R_i \hookrightarrow R : i \in I\}$ *is jointly epi and, for every* $c : A \to C$*, if* $\forall i \in I : \pi_{i1}; a_i; c = \pi_{i2}; a_i; c$ *then* $\pi_1; c = \pi_2; c$.

The result which we will actually need is the following one.

Lemma 4.5. *Given an algebra* $A \in \mathsf{MAlg}^*_{OT}(\Sigma)$ *and a congruence* R *on* A*, the quotient* $A/_R$ *is a colimit of its small subalgebras.*

---

[3] This might cause some foundational worries since functions returning classes, and hence also indexed families of classes, are not legal objects in the most common class theory, NBG. This signals that we must use an alternative foundation, Grothendieck's hierarchy of universes being the natural candidate. We use the words "small"/"set" and "large"/"class" in the sense of membership in the lowest level $\mathcal{U}_1$ versus in some higher level $\mathcal{U}_i \setminus \mathcal{U}_1$ (for $i \geq 2$), respectively.

PROOF: We consider the following (schema of the) diagram:

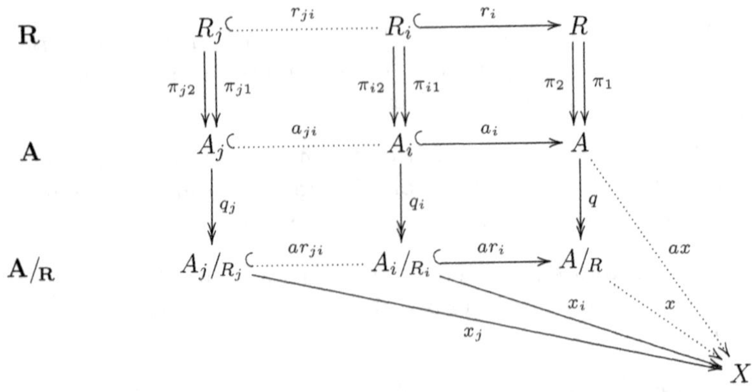

A, resp. **R**, stand for the whole diagrams consisting of the respective small subalgebras A_i of A and $R_i = R \cap |A_i| \times |A_i|$ (by Fact 3.13, $R_i \sqsubseteq R$) with the inclusion arrows a_{ji}, resp. r_{ji}. A with inclusions a_i is colimit of **A**. The collection of all r_i's, resp., all a_i's is jointly epi. All q_i's are epi.

The diagram $\mathbf{A}/_{\mathbf{R}}$ contains all quotient algebras $A_i/_{R_i}$ and inclusion arrows between them. Since for each $i : R_i = R \cap |A_i| \times |A_i|$, we have an inclusion $a_{ji} : A_j \hookrightarrow A_i$ iff $r_{ji} : R_j \hookrightarrow R_i$. But then, this implies the existence of a mono $ar_{ji} : A_j/_{R_j} \hookrightarrow A_i/_{R_i}$. For each $A_i/_{R_i}$, we can obtain an isomorphic algebra by replacing every element $[a]^{R_i}$ by $[a]^R$ (though $[a]^{R_i} \subseteq [a]^R$ and the inclusion can be proper, whenever $R(a_1, a_2)$ and $a_1, a_2 \in |A_i|$, then also $R_i(a_1, a_2)$). This – making all monos ar_i and ar_{ij} into inclusions – simplifies the argument below.

We want to show that $A/_R$ with all inclusions ar_i is colimit of $\mathbf{A}/_{\mathbf{R}}$. Obviously, for each (existing) ar_{ji} we do have that $ar_j = ar_{ji}; ar_i$, since all arrows are inclusions. So assume an X with arrows $x_i : A_i/_{R_i} \to X$ such that $x_j = ar_{ji}; x_i$ for all (relevant) i, j.

1. Since $q_j; ar_{ji} = a_{ji}; q_j$, we obtain that for all (relevant) $j, i : x_j = ar_{ji}; x_i \Rightarrow q_j; x_j = q_j; ar_{ji}; x_i = a_{ji}; q_i; x_i$. That is, X with $q_i; x_i$ is a commutative cocone over **A**. Since A is colimit of **A**, we obtain a unique arrow $ax : A \to X$ such that for all $i : q_i; x_i = a_i; ax$.

2. For every i, since $\pi_{i1}; q_i = \pi_{i2}; q_i$, so also $\pi_{i1}; q_i; x_i = \pi_{i2}; q_i; x_i$ and by 1, $\pi_{i1}; a_i; ax = \pi_{i2}; a_i; ax$. By Fact 4.4, we thus have $\pi_1; ax = \pi_2; ax$.

3. By Fact 3.11, $(A/_R, q)$ is coequalizer of π_1, π_2, and thus we obtain a unique arrow $x : A/_R \to X$ making $q; x = ax$. This is the arrow we are looking for:

4. Commutativity: $q_i; ar_i; x = a_i; q; x \overset{3.}{=} a_i; ax \overset{1.}{=} q_i; x_i$. But q_i is epi and so $ar_i; x = x_i$.

5. Uniqueness: assume another arrow $y : A/_R \to X$ with $ar_i; y = x_i$ for all i. Then also, $q_i; x_i = q_i; ar_i; y = a_i; q; y$ and thus, for every $i : a_i; q; y = a_i; q; x$. Since a_i are jointly epi, this means that $q; y = q; x$ and now, since q is epi, $x = y$.

□

4.1 Cocompleteness and Final Objects of $\mathsf{MAlg}^*_{OT}(\Sigma)$

The positive results for the category $\mathsf{MAlg}_{OT}(\Sigma)$ from table 1, generalise to the extended category $\mathsf{MAlg}^*_{OT}(\Sigma)$. We only mention the results needed in the construction of final objects, suggesting the constructions used in the proofs.[4]

Proposition 4.6. $\mathsf{MAlg}^*_{OT}(\Sigma)$ *has initial objects and coproducts.*

PROOF IDEA: Empty algebra is trivially an initial object.
Consider a class $\{A_i : i \in I\}$ of algebras. Its coproduct is the algebra C whose carrier is the disjoint union of the carriers of all A_i, with operations defined as:

$$f^C(\overline{x}) = \begin{cases} f^{A_i}(\overline{x}) & \text{if for all } x \in \overline{x} : x \in |A_i| \\ \emptyset & \text{otherwise} \end{cases}$$

and constants as disjoint unions: $c^C = \biguplus_i c^{A_i}$. The injections $\iota_i : A_i \to C$ are obviously OT, and C is colimit of small subalgebras (of all A_i's). □

Proposition 4.7. $\mathsf{MAlg}^*_{OT}(\Sigma)$ *has all coequalizers.*

PROOF IDEA: Given two arrows $\phi_1, \phi_2 : A \to B$, we start as usual by considering the equivalence closure $\sim$ on B of the relation $E = \{\langle \phi_1(a), \phi_2(a) \rangle : a \in |A|\}$. Equivalence classes induced by this relation are denoted $B_1, B_2 \ldots$. Assuming the global axiom of choice, we can choose the representatives $b_i \in B_i$, and the carrier of the coequalizer object C is the collection of such representatives. (We may occasionally write $[b_i]$ for B_i.)[5] Operations are defined by:

$$b_2 \in f^C(b_1) \iff B_2 \subseteq f^B(B_1)$$

which for constants specializes to: $b_i \in c^C \iff B_i \subseteq c^B$. The arrow $ce : B \to C$ is the usual $\forall x \in B_i : ce(x) = b_i$. By the definition of $\sim$, it makes $\phi_1; ce = \phi_2; ce$. Verification that it is OT and of the universality is rather lengthy and technical. □

We have thus shown that the assumption (4.1), according to which $\mathsf{MAlg}^*_{OT}(\Sigma)$ contains only those colimits of small algebras which happen to exist there, indeed is a category with *all* colimits. The main result is now obtained from the following lemma (with a straightforward proof).

Lemma 4.8. *For a given multialgebra A, let $\sim_A$ denote the maximal congruence on A (existing by Lemma 4.3). For any algebra B there is at most one homomorphism $B \to A/_{\sim_A}$.*

[4] Notice that due to the difference in the definition of homomorphism, cocompleteness of $\mathsf{MAlg}_{OT}(\Sigma)$ is not a special case of the general fact about dialgebras, according to which the category SET^F_G has all colimits preserved by the functor F.

[5] In case some of the equivalence classes B_i's are proper classes, we have to follow the trick of Dana Scott (quoted in [1], Appendix B) in order to obtain the quotient, i.e., to consider as B_i only its subset of the elements having the least possible rank in the cumulative hierarchy.

Theorem 4.9. $\mathsf{MAlg}^*_{OT}(\Sigma)$ *has final objects.*

PROOF: Let C be a coproduct of all small algebras in $\mathsf{MAlg}^*_{OT}(\Sigma)$ (which exists in $\mathsf{MAlg}^*_{OT}(\Sigma)$ by Fact 4.6). Let $\sim_C$ be the maximal bireachability on C (existing by Lemma 4.3), and let $Z = C/_{\sim_C}$. By Lemma 4.5, $Z \in \mathsf{MAlg}^*_{OT}(\Sigma)$.

For every small algebra $A \in \mathsf{MAlg}^*_{OT}(\Sigma)$, there is (at least one) morphism $A \to C$ and then, composing it with the quotient morphism $C \to Z$, exactly one (by lemma 4.8) morphism $a : A \to Z$.

Any other (large) $A \in \mathsf{MAlg}^*_{OT}(\Sigma)$ is colimit of its small subalgebras, with the components $\iota_i : A_i \to A$. Since there is also (exactly) one morphism $a_i : A_i \to Z$ for each small subalgebra $A_i \sqsubseteq A$, the colimit property yields a (unique) morphism $u : A \to Z$ (making $\iota_i; u = a_i$). But then, since there is such a morphism $u : A \to Z$ so, by lemma 4.8, it is unique. □

5 Conclusion

Multialgebras lie at the intersection of several research currents. They

- represent relations and, generally, Boolean algebras with operators;
- generalise traditional algebras, in particular,
- provide a fundamental instance of power structure construction;
- with one-argument operations, provide particular examples of coalgebras;
- but can also represent arbitrary coalgebras over polynomial functors (simply by considering the reversed coalgebra arrows);
- provide a specific and well-motivated example of dialgebras, [8].

The apparently poor algebraic structure and, on the other hand, a multiplicity of choices when generalising most of the standard notions might discourage investigation of multialgebras. We have argued that, as far as the notion of homomorphism is concerned, the number of choices is, after all, not so large and in fact limited to one, while further choices are mainly conditioned by this one. The category of multialgebras with outer-tight homomorphisms is cocomplete and the associated notion of congruence – bireachability – arises as the converse of the bisimulation equivalence.

We have shown that the category $\mathsf{MAlg}^*_{OT}(\Sigma)$ of multialgebras (admitting proper classes as carriers) possesses final objects with interesting structure which reflects the reachability relation in the way analogous to final coalgebras reflecting the similarity relation. We have considered only the class of all Σ-multialgebras and although we expect the existence of final objects can be lifted to (some) axiomatic classes, the possibility and scope of this lifting remain to be investigated. The question which still remains open is the existence of products which, intuitively, should be related (or even equal) to largest bireachability between the arguments. Attempts to construct counter-examples have failed and we are convinced that products do exist in $\mathsf{MAlg}^*_{OT}(\Sigma)$, but the claim and an explicit construction remain to be demonstrated. There remains also the open question concerning the more specific conditions, than those given in (4.1), on the actual algebras to be included in the category $\mathsf{MAlg}^*_{OT}(\Sigma)$. As can be seen from the proof of lemma 4.6, we must allow constants (unary predicates) to denote proper classes. We suspect that the following condition may be sufficient to ensure the existence of final objects and (co)completeness of the category: for every operation f in an algebra A : if $f^A(X)$ is a set then so is X. Sufficiency of this condition or, possibly, alternative fomulations remain to be investigated.

References

1. Peter Aczel. Non-well-founded sets. Technical Report 14, CSLI, 1988.
2. Peter Aczel and Nax Mendler. A final coalgebra theorem. In G. Goos and J. Hartmanis, editors, *Category Theory and Computer Science*, volume 389 of *LNCS*, pages 357–365. Springer, 1989.
3. Ivica Bošnjak and Rozália Madarász. On power structures. *Algebra and Discrete Mathematics*, 2:14–35, 2003.
4. Chris Brink. Power structures. *Algebra Universalis*, 30:177–216, 1993.
5. Paul M. Cohn. *Universal Algebra*, volume 6 of *Mathematics and Its Applications*. D.Reidel Publishing Company, 1965.
6. Willem-Paul de Roever and Kai Engelhardt. *Data Refinement: Model-Oriented Proof Methods and their Comparison*. Cambridge University Press, 1998.
7. George Grätzer. A representation theorem for multialgebras. *Arch. Math.*, 13, 1962.
8. Tatsuya Hagino. *A Categorical Programming Language*. PhD thesis, Department of Computer Science, University of Edinburgh, 1987.
9. Claudio Hermida. A categorical outlook on relational modalities and simulations. In Michael Mendler, Rajeev P. Goré, and Valeria de Paiva, editors, *Intuitionistic Modal Logic and Aplications*, volume 02-15 of *DIKU technical reports*, pages 17–34, July 26 2002.
10. Wim H. Hesselink. A mathematical approach to nondeterminism in data types. *ACM ToPLaS*, 10, 1988.
11. Heinrich Hußmann. Nondeterministic algebraic specifications and nonconfluent term rewriting. In *Algebraic and Logic Programming*. LNCS vol. 343, Springer, 1988.
12. Heinrich Hußmann. *Nondeterministic algebraic specifications*. PhD thesis, Fak. f. Mathematik und Informatik, Universitat Passau, 1990.
13. Heinrich Hußmann. *Nondeterminism in Algebraic Specifications and Algebraic Programs*. Birkhäuser, 1993. [revised version of [12]].
14. Bjarni Jónsson and Alfred Tarski. Boolean algebras with operators i. *American J. Mathematics*, 73:891–939, 1951.
15. Bjarni Jónsson and Alfred Tarski. Boolean algebras with operators ii. *American J. Mathematics*, 74:127–162, 1952.
16. Yngve Lamo. *The institution of multialgebras – a general framework for algebraic software development*. PhD thesis, Department of Informatics, University of Bergen, 2002.
17. Rozália Madarász. Remarks on power structures. *Algebra Universalis*, 34(2):179–184, 1995.

18. José Meseguer. Membership algebra as a logical framework for equational specification. In *In 12th International Workshop on Recent Trends in Algebraic Development Techniques (WADT'97)*, volume 1376 of *LNCS*, pages 18–61. Springer, 1998.
19. Günter Pickert. Bemerkungen zum homomorphie-begriff. *Mathematische Zeitschrift*, 53, 1950.
20. H.E. Pickett. Homomorphisms and subalgebras of multialgebras. *Pacific J. of Mathematics*, 21:327–342, 1967.
21. Jan J.M.M Rutten. Universal coalgebra: a theory of systems. *Theoretical Computer Science*, 249:3–80, 2000.
22. Dietmar Schweigert. Congruence relations on multialgebras. *Discrete Mathematics*, 53:249–253, 1985.
23. Michał Walicki, Adis Hodzic, and Sigurd Meldal. Compositional homomorphisms of relational structures. In R. Freivalds, editor, *Fundamentals of Computation Theory*, volume 2138 of *LNCS*. Springer, 2001.
24. Michał Walicki and Sigurd Meldal. A complete calculus for the multialgebraic and functional semantics of nondeterminism. *ACM ToPLaS*, 17(2), 1995.
25. Michał Walicki and Sigurd Meldal. Multialgebras, power algebras and complete calculi of identities and inclusions. In *Recent Trends in Data Type Specifications*, volume 906 of *LNCS*. Springer, 1995.
26. Michał Walicki and Sigurd Meldal. Algebraic approaches to nondeterminism – an overview. *ACM Computing Surveys*, 29(1), March 1997.

Parametrized Exceptions

Dennis Walter, Lutz Schröder, and Till Mossakowski

BISS, Department of Computer Science,
University of Bremen

Abstract. Following the paradigm of encapsulation of side effects via monads, the Java execution mechanism has been described by the so-called Java monad, encorporating essentially stateful computation and exceptions, which are heavily used in Java control flow. A technical problem that appears in this model is the fact that the return exception in Java is parametrized by the return value, so that method calls actually move between slightly different monads, depending on the type of the return value. We provide a treatment of this problem in the general framework of exception monads as introduced in earlier work by some of the authors; this framework includes generic partial and total Hoare calculi for abrupt termination. Moreover, we illustrate this framework by means of a verification of a pattern match algorithm.

1 Introduction

Many imperative languages allow for manipulating the control flow by means of exceptions. Particularly extensive use of this possibility is made in Java, where abnormally terminating statements are used e.g. in order to exit from loops or method calls. Exceptions therefore also play a prominent role in the design of program logics for Java.

Generally, imperative languages may be represented in standard higher order logic or in functional programming languages by encapsulating side effects as monads, a principle introduced by Moggi [9]. The Java exception mechanism has been modelled by the so-called Java monad [6], an instance of Moggi's exception monad transformer [8]. In previous work [16, 14, 15], we have developed monadic computational logics for generic side effects, and we have extended these logics with a generic treatment of exceptional termination that subsumes existing Hoare logics for abrupt termination [4, 5].

A technical problem that appears in the monadic modeling of the Java exception mechanism is the fact that the return exception in Java is parametrized by the value to be returned, so that method calls actually move between slightly different monads, depending on the type of the return value. In previous work [4], this problem has been treated by a workaround which involves storing the return value in a global variable. Here, we provide a more elegant solution in the shape of a monadic wrapper routine for method bodies that shifts the value of the return exception (which is now treated as a part of the exception, in accordance with the Java language specification [7]) into the value of the monadic computation.

J.L. Fiadeiro et al. (Eds.): CALCO 2005, LNCS 3629, pp. 424–438, 2005.

We also provide suitable Hoare rules for this wrapper. The generic framework is illustrated by means of a benchmark problem appearing also in [4, 6], the verification of a pattern match algorithm, thus also providing a first example application for the calculus presented in [15].

2 Monads for Imperative Programming and Specification

Following seminal work by Moggi [9], monads are being used in both semantics and programming to formalize and encapsulate side effects in an elegant, functional way; in particular, this idea is one of the central concepts of Haskell [11]. Intuitively, a monad associates to each type A a type TA of *computations* of type A; a function with side effects that takes inputs of type A and returns values of type B is, then, just a function of type $A \to TB$, also called a *(B-valued) program*. This approach abstracts away from particular notions of computation such as store, non-determinism, non-termination etc.; a surprisingly large amount of reasoning can in fact be carried out independently of the choice of such a notion.

Formally, there are three ingredients to a monad: the computation type constructor T, a lifting operation $\mathrm{ret} : a \to T\ a$ for each type a, and a binding operation $\_ \gg= \_ : T\ a \to (a \to T\ b) \to T\ b$ for all types a, b. The intuition behind these operations is that $\mathrm{ret}\ x$ is a computation without side-effects that just returns the value x (not to be confused with the return expcetions appearing further below), and that $p \gg= f$ executes the computation $p : T\ a$ and feeds the resulting values of type a into the program $f : a \to T\ b$; this is essentially sequential composition of statements. A *strong monad* additionally has an operation $t : a \times T\ b \to T\ (a \times b)$, called *(tensorial) strength*, for all types a, b. These data are governed by equational axioms requiring associativity of binding, neutrality of $\mathrm{ret}\, x$ w.r.t. binding, and compatibility of the strength with binding and lifting. A slight complication is caused by the fact that programs $a \to T\ b$ are in practice typically partial functions; a version of the monad axioms accommodating partial functions in such a way that typical monads such as the state monad of Example 1 below are actually subsumed (cf. discussions on this topic in [3]) is given in [16].

Example 1 ([9]). Computationally relevant monads on **Set** (all monads on **Set** are strong) include

- stateful computations with possible non-termination: $TA = (S \to ?\ (A \times S))$, where S is a fixed set of states and $\_ \to ?\ \_$ denotes the partial function type;
- (finite) non-determinism: $TA = \mathcal{P}_{\mathit{fin}}(A)$, where $\mathcal{P}_{\mathit{fin}}$ denotes the finite power set functor;
- exceptions: $TA = A + E$, where E is a fixed set of exceptions;
- interactive input: TA is the least fixed point of $\lambda\gamma.\, A + (U \to ?\ \gamma)$, where U is a set of input values;
- non-deterministic stateful computations: $TA = S \to \mathcal{P}_{\mathit{fin}}(A \times S)$, where, again, S is a fixed set of states.

As laid out in [18] and incorporated in the Haskell syntax, monads can be used to support an imperative style of notation: terms of the form

$$\text{do } x \leftarrow p;\ q$$

(where x may appear in q) are taken to abbreviate $p \gg= \lambda x.\, q$, and nested bindings do $x \leftarrow p_1$; do $y \leftarrow p_2$; $\ldots$ are denoted in the form do $x \leftarrow p_1; y \leftarrow p_2; \ldots$. Sequences of bindings $x_1 \leftarrow p_1; \ldots; x_n \leftarrow p_n$ will often be indicated by a bar notation $\bar{x} \leftarrow \bar{p}$ below. As indicated above, the notational coincidence with the usual sequential composition operator indeed conveys the right intuition, in particular in connection with variants of the above-mentioned state monad. Imperative control structures such as the while loop can be expressed as recursive programs in this notation; the definition of a slightly generalized loop construct *iter* where a return value is passed through the loop is shown in Fig. 1, along with a definition of the while loop by specialization to void results.

The background formalism used in Fig. 1 and in the further development is the wide-spectrum language HasCasl, which extends the standard algebraic specification language Casl by features aimed at providing support for modern functional programming languages. These features include higher order logic of partial functions, type class polymorphism in the style of Haskell, and general recursion over domains, specified in a bootstrap fashion as complete partial orders (cpo's) in much the same way as in HOLCF [12]. In the following, we briefly explain some of the less obvious syntactical features of HasCasl appearing in Fig. 1 and elsewhere; for further details, the reader is referred to [13, 17].

As indicated above, the treatment of recursion in HasCasl is via a type class *Cpo* of complete partial orders; type classes are just subsets of the syntactical universe of types. The types of total and partial continuous functions between cpo's a and b are denoted by $a \xrightarrow{cont} b$ and $a \xrightarrow{cont}? \, b$, respectively (in contrast to non-continuous function types $a \to b$ and $a \to ? \, b$). Operators containing type variables of a certain type class in their type are thereby declared to be polymorphic over that class; e.g. the *iter* operation of Fig. 1 is polymorphic over a so-called cpo monad T, i.e. a monad that allows lifting cpo structures on a to cpo structures on $T\ a$ in such a way that the monad operations become continuous. Here, the class *CpoMonad* is, like the more general class *Monad* of monads, a *constructor class*, i.e. a subset of the syntactical universe of type constructors. The specification of the class *CpoMonad* is imported by referencing the named specification CpoMonad (not shown here), which itself includes the specification of both cpo's and monads.

The keyword **program** invokes syntactical sugar for the fixed point operator on cpo's which allows writing programs in much the same style as e.g. in Haskell; in particular, the equation defining *iter* in Fig. 1 is an actual recursive definition rather than just a semantic equation. Program blocks in HasCasl can be automatically translated into Haskell by means of the Bremen heterogeneous tool set [10].

As indicated above, HasCasl incorporates a higher order logic of partial functions. Below, we will freely combine the monadic do-notation and higher order formulae, using e.g. monadic computations of truth values. A more detailed explanation of the required logical framework is given in [16]. We stress that the use of HASCASL serves mainly to drive home the point that our framework can be cast in a real specification language; the results as such remain valid essentially in any suitable higher order logic of partial functions, including standard set theory and topos logic.

spec ITERATION = CPOMONAD **and** BOOL **then**
    vars $T\colon CpoMonad;\ a\colon Cpo$
    op $iter\colon (a \overset{cont}{\longrightarrow} T\ Bool) \overset{cont}{\longrightarrow} (a \overset{cont}{\longrightarrow}?\, T\ a) \overset{cont}{\longrightarrow} a \overset{cont}{\longrightarrow}?\, T\ a$
    program *iter test f x* =

```
        do b ← test x
           if b then
              do y ← f x
                 iter test f y
              else return x
```

    op $while\ (b\colon T\ Bool)\ (p\colon T\ Unit)\colon T\ Unit = iter\ (\lambda x \bullet b)\ (\lambda x \bullet p)\ ()$

Fig. 1. The iteration control structure

In [16, 14], generic computational logics have been introduced that allow reasoning about the correctness of monadic programs. This includes a Hoare logic for partial correctness, a dynamic logic, and a Hoare logic for total correctness defined via dynamic logic. While partial Hoare logic works in principle over arbitrary monads, the semantics of dynamic logic (and hence also of total Hoare logic) is introduced axiomatically and works only for sufficiently well-behaved monads, including the monads of Example 1 and excluding e.g. the continuation monad; in the positive case we say that a monad *admits dynamic logic.*

A basic concept underlying all these logics is the notion of *deterministically side-effect free (dsef)* program engendered by the underlying monad. In the terminology of [2], p is dsef if p is central in the variety of discardable and copyable programs. Intuitively, discardability means that the state may be read, but not changed (this is in contrast to *stateless* programs, i.e. programs of the form ret a, which additionally do not access the state), and copyability amounts to determinism.

The monadic partial Hoare calculus is concerned with Hoare triples

$$\{\phi\}\ \bar{x} \leftarrow \bar{p}\ \{\psi\},$$

where $\bar{x} \leftarrow \bar{p}$ is a sequence of bindings for monadic programs p_i, ϕ and ψ are *formulae*, i.e. dsef computations of truth values, and ψ may mention the intermediate results x_i. The semantics of Hoare triples is defined in terms of equations between computations of truth values. In the monads of Example 1, this semantics instantiates as expected. For example, a Hoare triple $\{\phi\}\ x \leftarrow p\ \{\psi\}$ holds in the state monad iff, whenever ϕ holds in a state s, then ψ holds for x after successful execution of p from s with result x. In the non-deterministic state-monad, ψ must be satisfied for all possible pairs of results and post-states for p. A calculus for such Hoare triples may be derived directly from the definition of the semantics; the rules of the calculus include e.g. a sequencing rule

$$\text{(seq)}\quad \frac{\{\phi\}\ \bar{x} \leftarrow \bar{p}\ \{\psi\} \qquad \{\psi\}\ \bar{y} \leftarrow \bar{q}\ \{\chi\}}{\{\phi\}\ \bar{x} \leftarrow \bar{p}; \bar{y} \leftarrow \bar{q}\ \{\chi\}}$$

and, for cpo monads, a loop rule

$$\text{(iter)}\quad \frac{\{\phi\ x \wedge b\ x\}\ y \leftarrow p\ x\ \{\phi\ y\}}{\{\phi\ e\}\ y \leftarrow \mathit{iter}\ b\ p\ e\ \{\phi\ y \wedge \neg(b\ y)\}}\ (b \text{ dsef})$$

which generalizes the usual while rule.

Hoare triples $[\phi]\ \bar{x} \leftarrow \bar{p}\ [\psi]$ for *total* correctness are, as indicated above, defined using the diamond operator of monadic dynamic logic [16] and hence can be interpreted only over monads that admit dynamic logic, which however does not seem to be an overly serious restriction. As for partial Hoare triples, one can prove the usual set of rules from the definition of the semantics, including e.g. a rule for total correctness of loops which specializes to the usual total while rule.

3 Exception Monads and the Java Monad

In modern imperative languages, exceptions are often used as a means of manipulating the control flow, replacing explicit jumps. A maybe somewhat extreme example is Java, where common control statements such as `break`, `continue`, and `return` terminate abnormally, and where e.g. bodies of methods with non-void result type are in fact expressly forbidden to terminate normally. Hoare calculi for Java as developed e.g. in [4, 5] therefore need to accommodate correctness assertions for abnormal termination.

In the monadic setting, exceptions are modelled by Moggi's *exception monad transformer* [8] which, given a set E of exceptions, transforms a monad R into the associated exception monad $Ex\ E\ R = R\ (\_ + E)$ — i.e. $Ex\ E\ R$ models the side effects of R, extended by exceptions in E. A simple example is the *Java monad* [6] $Ex\ E\ ST$, where ST is the state monad of Example 1. In principle, the generic computational logics described in the previous section do apply to exception monads; in particular, $Ex\ E\ R$ admits dynamic logic if R does. However, since monadic computational logic as recalled in the previous section treats exceptional termination like non-termination in the sense that one has $\{\}\ p\ \{\bot\}$ if p throws an exception, the generic framework, like the concrete Hoare calculi of [4, 5], needs to be provided with an explicit treatment of abnormal termination.

Such a generic framework for exception monads has been introduced in [15]. It is obtained by first giving an equational characterization of exception monads in the above sense, and then combining this with the existing monadic Hoare logics as explained in Sect. 2. The equational description of an exception monad T is based on operations $catch : T\ a \to T\ (a + E)$ and $raise : E \to T\ a$ for each type a. Here, $raise\ e$ throws an exception, thus freezing the state (so that subsequent bindings are skipped until the exception is caught), and $catch\ p$ behaves like p if p terminates normally, and otherwise returns a thrown exception and unfreezes the state, i.e. resumes normal execution of statements. If $T = Ex\ E\ R$, then $catch\ p$, which for $p : Ex\ E\ R\ a$ is of type $Ex\ E\ R\ (a + E) = R\ ((a + E) + E)$, may be expressed in the do-notation for R as

$$\text{do } x \leftarrow p;\ case\ x\ of\ inl\ y \to \text{ret}\ (inl\ (inl\ y)) \mid inr\ e \to \text{ret}\ (inl\ (inr\ e)).$$

It is convenient to let $catch\ \bar{x} \leftarrow \bar{p}$ abbreviate $catch$ (do $\bar{x} \leftarrow \bar{p};\ \text{ret}\,\bar{x}$).

One can then define Hoare assertions that simultaneously cover normal and abnormal termination. A *partial extended Hoare assertion* $\{\phi\}\ \bar{x} \leftarrow \bar{p}\ \{\psi \parallel S\}$ abbreviates

$$\{\phi\}\ y \leftarrow (catch\ \bar{x} \leftarrow \bar{p})\ \{case\ y\ of\ inl\ \bar{x} \to \psi \mid inr\ e \to S\ e\},$$

thus stating that, if the program sequence $\bar{x} \leftarrow \bar{p}$ is executed in a state that satisfies ϕ, then if the execution terminates normally, the post-state and the result $\bar{x}$ satisfy ψ, and if the execution terminates abnormally with exception e, the post-state satisfies $S\ e$ (in particular note that $S\ e$ is a stateful formula, although as usual required to be dsef). The conditions ψ and S are referred to as the *normal* and the *abnormal postcondition*, respectively. Similarly, *total extended Hoare assertions* $[\phi]\ \bar{x} \leftarrow \bar{p}\ [\psi \parallel S]$ are defined using *catch* and standard total Hoare triples; the meaning of a total assertion is thus the conjunction of the partial assertion and termination of $catch\ \bar{x} \leftarrow \bar{p}$. A calculus for extended Hoare assertions is then easily derived from the standard Hoare calculus and the equational description of exception monads. The rules for partial extended Hoare assertions are shown in Figure 2, with the general fixed point rule (Y) of [15] replaced by its specialization to *iter* (which then gives rise to a *while* rule by further specialization). The rules for total extended Hoare assertions are largely the same, except for the loop rule, which becomes

$$\text{(iter-total)}\ \frac{\begin{array}{c}\_ < \_ : Pred\ (c \times c) \text{ is well-founded} \\ [\phi\ x \wedge b\ x \wedge (t\ x = z)]\ y \leftarrow p\ x\ [\phi\ y \wedge (t\ y < z) \parallel S]\end{array}}{[\phi\ e]\ y \leftarrow iter\ b\ p\ e\ [\phi\ y \wedge \neg(b\ y) \parallel S]}$$

(subject to the side condition that $t\ x : T\ c$ is dsef for all $x : a$). As explained in [15], this rule can be specialized to the *total exception while rule*

$$\frac{\begin{array}{c}[\phi \wedge b]\ p\ [\top \parallel \top] \\ \{\phi \wedge b \wedge t = z\}\ p\ \{\phi \wedge b \wedge t < z \parallel \top\} \\ \{\phi \wedge b\}\ p\ \{\top \parallel S\}\end{array}}{[\phi \wedge b]\ while\ b\ p\ [\bot \parallel S]}$$

Further below, we will use a slight variant of this rule, where the well-founded relation $<$ lives only on a subtype of c and the invariant ϕ guarantees that results of t are always in this subtype.

$$(\text{seq})\ \frac{\{\phi\}\ \bar{x} \leftarrow \bar{p}\ \{\psi \parallel S\} \quad \{\psi\}\ \bar{y} \leftarrow \bar{q}\ \{\chi \parallel S\}}{\{\phi\}\ \bar{x} \leftarrow \bar{p}; \bar{y} \leftarrow \bar{q}\ \{\chi \parallel S\}} \qquad (\text{ctr})\ \frac{\{\phi\}\ \ldots; x \leftarrow p; y \leftarrow q; \bar{z} \leftarrow \bar{r}\ \{\psi \parallel S\} \quad x \notin FV(\bar{r}) \cup FV(\psi)}{\{\phi\}\ \ldots; y \leftarrow (\text{do } x \leftarrow p;\ q); \ldots\ \{\psi \parallel S\}}$$

$$(\text{conj})\ \frac{\{\phi\}\ \bar{x} \leftarrow \bar{p}\ \{\psi_1 \parallel S_1\} \quad \{\phi\}\ \bar{x} \leftarrow \bar{p}\ \{\psi_2 \parallel S_2\}}{\{\phi\}\ \bar{x} \leftarrow \bar{p}\ \{\psi_1 \wedge \psi_2 \parallel S_1 \wedge S_2\}} \qquad (\text{disj})\ \frac{\{\phi_1\}\ \bar{x} \leftarrow \bar{p}\ \{\psi \parallel S\} \quad \{\phi_2\}\ \bar{x} \leftarrow \bar{p}\ \{\psi \parallel S\}}{\{\phi_1 \vee \phi_2\}\ \bar{x} \leftarrow \bar{p}\ \{\psi \parallel S\}}$$

$$(\text{wk})\ \frac{\{\phi\}\ \bar{x} \leftarrow \bar{p}\ \{\psi \parallel S\} \quad \phi' \Rightarrow \phi \quad \psi \Rightarrow \psi' \quad S \Rightarrow S'}{\{\phi'\}\ \bar{x} \leftarrow \bar{p}\ \{\psi' \parallel S'\}} \qquad (\text{stateless})\ \frac{}{\{\text{ret } \phi\}\ q\ \{\text{ret } \phi \parallel \lambda e.\ \text{ret } \phi\}}$$

$$(\text{dsef})\ \frac{p \text{ dsef}}{\{\phi\}\ x \leftarrow p\ \{\phi \wedge x = p \parallel \bot\}} \qquad (\text{catch})\ \frac{\{\phi\}\ \bar{x} \leftarrow \bar{p}\ \{\psi[\mathit{inl}\ \bar{x}/y] \parallel \lambda e.\ \psi[\mathit{inr}\ e/y]\}}{\{\phi\}\ y \leftarrow (\mathit{catch}\ \bar{x} \leftarrow \bar{p})\ \{\psi \parallel \bot\}}$$

$$(\text{raise})\ \frac{}{\{\phi\}\ \mathit{raise}\ e_0\ \{\bot \parallel \lambda e.\ (\phi \wedge e = e_0)\}} \qquad (\text{if})\ \frac{\{\phi \wedge b\}\ x \leftarrow p\ \{\psi \parallel S\} \quad \{\phi \wedge \neg b\}\ x \leftarrow q\ \{\psi \parallel S\}}{\{\phi\}\ x \leftarrow \mathit{if}\ b\ \mathit{then}\ p\ \mathit{else}\ q\ \{\psi \parallel S\}}$$

$$(\text{iter})\ \frac{\{\phi\ x \wedge b\ x\}\ y \leftarrow p\ x\ \{\phi\ y \parallel S\}}{\{\phi\ e\}\ y \leftarrow \mathit{iter}\ b\ p\ e\ \{\phi\ y \wedge \neg(b\ y) \parallel S\}}$$

Fig. 2. The generic Hoare calculus for partial exception correctness

It is shown in [15] that these generic Hoare calculi subsume the calculi of [4, 5] w.r.t. the treatment of exceptions (the work of [4, 5] covers also aspects of the Java class mechanism, which is not considered here).

Remark 2. The very simple exception mechanism laid out above can be used to capture also the more involved aspects of Java's treatment of exceptions. For instance, the monadic approach is also suitable for the modeling of side-effecting expressions. In the encoding, an expression needs to be decomposed into a sequence of bindings in a do-expression, where each binding corresponds to an application of a method. Hoare rules that actually work on unencoded expressions are easily designed using this observation, given a fixed order of evaluation for subexpressions.

Additionally, the following program *tryFinally* models Java's `try` statement with a `finally` block attached. It guarantees execution of the program q representing the finally block even in case of abnormal termination of the try block program p. It terminates normally if both p and q do. Otherwise, it raises the exception of p or q with q's exceptions overwriting those of p:

$$\begin{aligned} \mathit{tryFinally}\ p\ q = \mathrm{do}\ & x \leftarrow \mathit{catch}\ p \\ & \mathit{case}\ x\ \mathit{of}\ \mathit{inl}\ \_ \rightarrow q \mid \mathit{inr}\ e \rightarrow \mathrm{do}\, q;\ \mathit{raise}\ e \end{aligned}$$

(i.e. *tryFinally p q* corresponds to `try p finally q`; additional `catch` clauses can be coded in the obvious way). A corresponding Hoare rule is easily derived from existing rules of the calculus (cf. Fig. 2), making supplementary use of a rule for the case construct. The total Hoare rule shown below (the partial rule is analogous) in particular captures the fact that exceptions in the finally block dominate those of the try block. Moreover, it subsumes a rule stating that exceptions in p propagate beyond q supposing q itself does not raise exceptions.

$$(\text{try-finally})\ \frac{\begin{array}{c}[\phi]\ p\ [\chi \parallel R] \\ [\chi]\ q\ [\psi \parallel S] \\ [R\,e]\ q\ [S\,e \parallel S]\end{array}}{[\phi]\ \mathit{tryFinally}\ p\ q\ [\psi \parallel S]}\ (e \text{ fresh})$$

4 Parametrized Exceptions and Java Return Values

As indicated in the introduction, the translation of Java programs into the monadic framework faces the following technical problem. In Java, the only admissible way for a method to return a value is via a `return` statement, with the returned value as parameter. The `return` statement terminates abnormally, essentially raising an exception marked as a return exception and *containing* the return value. The rest of the method body is then skipped; the return exception is implicitly caught at the end of the method body, where the exceptional return value is turned into a normal result and normal execution is resumed. In the monadic framework, this means that the body p of a method m with type a of return values is of the type

$$p : Ex\ (E\ a)\ R\ b,$$

where $E\ a$ is a parametrized type of exceptions, with return exceptions carrying values of type a, and R is the underlying monad; the type b of 'normal' results does not really matter, since the method body is explicitly forbidden to terminate normally [7]. The standard *catch* function could be used to turn this into $\mathit{catch}\ p :$ $Ex\ (E\ a)\ R\ (b + E\ a)$; however, if the method call to m took place in the body of a further method with result type c, use of *catch* p would still lead to a type error since it is a computation in $Ex\ (E\ a)\ R$ rather than $Ex\ (E\ c)\ R$.

Possibly for this reason, the translation of method calls in the LOOP tool as described e.g. in [4] slightly deviates from the above procedure: return exceptions are treated as unparametrized, so that there is a monomorphic type E of exceptions. In order to pass the return value of a method to the caller, one then needs to side-step the exception mechanism, creating instead a new global variable in which the return value is stored at the time of execution of the `return` statement and from which it is later retrieved by the wrapper function of the method call. It is clear that this solution is not entirely satisfactory. We will now propose an alternative solution which conforms to the Java Language Specification.

We will work with a polymorphic datatype

$$E\ a = \mathit{MRet}\ a \mid \mathit{DropOff} \mid \dots$$

of exceptions, parametrized by the type a of return values. Here, *MRet* is the return exception carrying the return value, and *DropOff* ('dropped off end of method body') is a special exception to be raised when a non-void method terminates normally (i.e. never, since this case should according to [7] be caught at compile time — however, we will still need to insert some value into the corresponding case statements); the unmentioned further exceptions do not depend on a.

In our calculus, Java **return** statements are translated into the throwing of return exceptions, i.e. statements of the form $\mathit{raise}\ (\mathit{MRet}\ x)$, abbreviated by $\mathit{mret}\ x$. Similar to [4], every method body is protected by a wrapper function *mbody*, so that for each method m with body p one has $m = \mathit{mbody}\ p$. This wrapper function turns the abnormal state caused by a return exception back into a normal one; but in contrast to the existing *catch* operation it additionally allows shifting the type of return exceptions. This is to say that *mbody* needs to have the polymorphic type

$$\mathit{mbody} : \mathit{Ex}\ (E\ a)\ R\ b \rightarrow \mathit{Ex}\ (E\ c)\ R\ a$$

for all types a, b, c; the instantiation of c is then determined by the context of the monadic computation in which the method call appears.

In order to give a generic definition of *mbody*, we recall that, while we will usually want to write monadic programs directly in $\mathit{Ex}\ (E\ a)\ R$, we can also exploit our knowledge that $\mathit{Ex}\ (E\ a)\ R = R\ (\_ + E\ a)$, and program in the monadic notation for R. Thus, we can write

$$\begin{aligned}
&\mathit{mbody}\ p = \mathit{do}\\
&\quad x \leftarrow p;\\
&\quad \mathit{case}\ x\ \mathit{of}\\
&\qquad \mathit{inl}\ _ \ \rightarrow \mathrm{ret}\ (\mathit{inr}\ \mathit{DropOff})\\
&\qquad \mid \mathit{inr}\ e \ \rightarrow \mathit{case}\ e\ \mathit{of}\\
&\qquad\qquad\qquad \mathit{MRet}\ v \ \rightarrow \mathrm{ret}\ (\mathit{inl}\ v)\\
&\qquad\qquad\qquad \mid _ \ \rightarrow \mathrm{ret}\ (\mathit{inr}\ e)
\end{aligned}$$

(type checking e.g. in Isabelle or Haskell will detect that this is the do-notation in R rather than in $\mathit{Ex}\ (E\ a)\ R$).

In order to conduct proofs about programs involving *mbody*, we need to encapsulate its properties in suitable Hoare rules at the level of $\mathit{Ex}\ (E\ a)\ R$, i.e. in the Hoare calculi for abrupt termination. In both the total and the partial calculus, a single Hoare rule suffices to prove properties of programs that obey the above-mentioned restrictions, i.e. where method bodies never return values normally, but always via the *mret* statement. The partial Hoare rule is

$$\text{(mbody)}\ \frac{\{\phi\}\ x \leftarrow p\ \{\bot \parallel \lambda e.\ case\ e\ of\ MRet\ y \rightarrow \psi \mid e \rightarrow S\ e\}}{\{\phi\}\ y \leftarrow mbody\ p\ \{\psi \parallel S\}};$$

the total version is analogous (the only additional statement being that *mbody p* terminates, normally or abruptly, if p terminates). Given the definition of *mbody* above, the proof of these rules in the monadic Hoare calculus (without abrupt termination) for R is straightforward, recalling the proper definition of the binding operation for $Ex\ (E\ a)\ R$ in the do-notation for R.

Note that from these rules one can infer further intuitively expected properties. E.g. by choosing S such that $S\ (MRet\ x) = \bot$ for each x, one obtains that *mbody p* does not throw or let pass any return exceptions. Furthermore, one concludes that *mbody p* only raises a *DropOff* exception —indicating that p terminated normally— if either p indeed terminates normally or p itself raised a *DropOff* exception, by choosing S such that $S\ DropOff = \bot$. Further illustration of the use of the rules is given in the example proof in the next section.

5 Verification of a Pattern Match Algorithm

To give an idea of how the calculus described above is applied in the verification of programs exploiting abnormal termination mechanisms, we will now prove the correctness of a pattern match algorithm. This algorithm has already been used as an example for the application of the calculi of [4, 5]; the main point to be made here is that a concrete algorithm of this kind can indeed be verified in our generic framework.

The algorithm is implemented in an exception monad with dynamic references (implemented in our concrete algorithm as natural numbers). One therefore has to axiomatize additional operations on the monad (apart from ret and $\gg=$); the corresponding HasCASL specification is shown in Figure 3. The notation should be largely self-explanatory. The type *Logical* is HasCASL's built-in type of truth values. The imported specification ExceptionMonad defines the exception monad transformer Ex, which is used to generate a reference monad RE with exceptions over a loosely specified base monad R. This specification, in turn, imports the specification CpoMonad (cf. Section 2); the imported symbols include the higher order type constructor D which, given a monad T and a type a, extracts the type $D\ T\ a$ of deterministically side-effect free a-valued T-computations. The specification in Figure 3 extends the axiomatization of the dynamic reference monad in terms of 'normal' Hoare triples given in [14] by abnormal postconditions, which in most cases are $\bot$, asserting that the corresponding operations do not raise exceptions. An exception is the rule (new-distinct), which states that subsequent creation of references, with an arbitrary program p (which may raise exceptions) executed in between, produces distinct references.

A Haskell implementation of the pattern match algorithm (which does not really look all that different from the corresponding executable HasCASL specification) is shown in Figure 4. For syntactic reasons, the function `read` replaces the *-notation of Fig. 3. The infrastructure functions `mbody`, `mret`, and `raise` are implemented elsewhere as described above.

spec EXCREFERENCE = EXCEPTIONMONAD **then**
  var $a\colon Cpo$
  types $R\colon CpoMonad;\ Ref\ a, E\colon Flatcpo;$
  type $RE := Ex\ E\ R$
  ops $read\colon Ref\ a \xrightarrow{cont} D\ RE\ a;$
  $\_\_ := \_\_\colon Ref\ a \xrightarrow{cont} a \xrightarrow{cont} RE\ Unit$
  forall $x, y\colon a;\ r, s\colon Ref\ a$
  • $[]\ r := x\ [x = {*r} \parallel \bot]$ %(read-write)%
  • $[x = {*r} \wedge \neg r = s]\ s := y\ [x = {*r} \parallel \bot]$ %(read-write-other)%

spec DYNAMICEXCREFERENCE = REFERENCE **then**
  var $a, b\colon Type$
  op $new\colon a \xrightarrow{cont} RE\ (Ref\ a)$
  forall $x, y\colon a;\ t\colon Ref\ a;$
  $p\colon Ref\ a \to RE\ b;\ P\colon D\ RE\ Logical$
  • $[]\ r \leftarrow new\ x\ [x = {*r} \parallel \bot]$ %(read-new)%
  • $[y = {*t}]\ r \leftarrow new\ x\ [y = {*t} \vee t = r \parallel \bot]$ %(read-new-other)%
  • $[P]\ r \leftarrow new\ x;\ p\ r\ [\top \parallel \top] \Rightarrow$
  $[P]\ r \leftarrow new\ x;\ p\ r;\ s \leftarrow new\ y\ [\neg r = s \parallel \top]$ %(new-distinct)%

Fig. 3. Specification of the reference and the dynamic reference monad

We prove total correctness of the algorithm generically, i.e. without further assumptions on the underlying monad other than the axioms of Figure 3. For convenience, we make use of *existential equality* $\stackrel{e}{=}$ e.g. when comparing elements of lists. For instance, $a!!i \stackrel{e}{=} b!!i$ means that $a!!i$ and $b!!i$ are *defined* and equal, where $(!!) :: List\ a \to Int \to a$ is the indexing function for the list datatype, with $a!!i$ defined only for $0 \le i < len\ a$.

For the actual method body p, i.e. the argument of `mbody` in Figure 4, we claim that it terminates abnormally, raising either a return exception carrying as its value an index x that is the starting position of the first occurence of the pattern in the base string or a failure exception indicating that there is no occurrence of the pattern in the base string:

$$[]\ p\ [\bot \parallel \lambda e.\ case\ \ e\ of \quad MRet\ i \to MPOS\ i \wedge \forall j.\, MPOS\ j \Rightarrow i \le j \mid PatternNotFound \to \neg\exists i.\ MPOS\ i \mid \_ \to \bot] \tag{1}$$

The abnormal postcondition above will be denoted by *POST* below. Here, *MPOS* i states that the pattern is matched at position i in the base string:

$$MPOS\ i \equiv \forall j.\ 0 \le j < len\ pat \Rightarrow base!!(i+j) \stackrel{e}{=} pat!!j.$$

In order to apply the total exception while rule (cf. Sect. 3), we need to provide a loop invariant *INV* and a termination measure t. Putting

$$INV \equiv (\forall i.\ 0 \le i < {*r} \Rightarrow base!!({*s}+i) \stackrel{e}{=} pat!!i)\ \wedge \ \forall i.\ MPOS\ i \Rightarrow {*s} \le i$$

```
pmatch base pat = mbody (
  do r <- new 0
     s <- new 0
     while (ret ⊤)
           (do u <- read r
               v <- read s
               if u == len pat
                 then mret v
                 else if v + u == len base
                   then raise PatternNotFound
                   else if base!!(v+u) == pat!!u
                     then r := (u+1)
                     else do s := (v+1); r := 0)
  )
```

Fig. 4. Haskell implementation of the pattern match algorithm

(which implies $0 \leq {*r} \leq \mathit{len\ pat}$ and $0 \leq {*s} + {*r} \leq \mathit{len\ base}$) guarantees that the dsef term $t = (\mathit{len\ base} - {*s}, \mathit{len\ pat} - {*r})$ always yields results of type $\mathit{Nat} \times \mathit{Nat}$, on which we have the lexicographic ordering as a well-founded relation.

Establishing the invariant upon entrance into the loop is easy, since from the axioms given above,

$$[]\ r \leftarrow \mathit{new}\ 0; s \leftarrow \mathit{new}\ 0\ [{*s} = {*r} = 0 \wedge \neg(r = s) \parallel \bot] \tag{2}$$

can be derived by the rules (seq), (conj), (read-new-other) and (new-distinct). Inside the loop, there are essentially four branches, arising from three applications of the rule (if), so that the three premises of the total exception while rule are split into twelve proof goals. We discuss the proof of two of these goals in more detail.

Firstly, we show that the branch of successful termination ($\mathit{mret}\ v$) satisfies the third premise of the total exception while rule, i.e.

$$\{\mathit{INV} \wedge {*r} = u \wedge {*s} = v \wedge {*r} = \mathit{len\ pat}\}\ \mathit{mret}\ v\ \{\top \parallel \mathit{POST}\}.$$

From the precondition we infer $\mathit{MPOS}\ v \wedge \forall i.\, \mathit{MPOS}\ i \Rightarrow v \leq i$, a stateless formula. Noting that $\mathit{mret}\ v$ is just shorthand for $\mathit{raise}\ (\mathit{MRet}\ v)$, we may derive $\{\}\ \mathit{mret}\ v\ \{\bot \parallel \lambda e.\ e = (\mathit{MRet}\ v)\}$ by (raise). Thus, by (stateless), (conj), and (wk), we obtain

$$\{\mathit{INV} \wedge {*r} = u \wedge {*s} = v \wedge {*r} = \mathit{len\ pat}\}\ \mathit{mret}\ v$$
$$\{\bot \parallel \lambda e.\, e = \mathit{MRet}\ v \wedge \mathit{MPOS}\ v \wedge \forall i.\, \mathit{MPOS}\ i \Rightarrow v \leq i\}.$$

The formula in the abnormal postcondition implies POST, so that we are finished by another application of (wk).

Secondly, as part of the second premise of the total exception while rule in the third branch ($r := u+1$), we show that the termination measure t decreases, i.e. we derive the Hoare triple

$$\{INV \wedge {*r} = u \wedge {*s} = v \wedge \neg(u = len\ pat) \wedge \neg(u + v = len\ base) \wedge \\ z = (len\ base - v, len\ pat - u) \wedge \neg(r = s)\} \\ r := u + 1 \\ \{(len\ base - {*s}, len\ pat - {*r}) < z \parallel \top\} \quad (3)$$

By (read-write) and (read-write-other) one infers $\{\neg(r = s) \wedge v = {*s}\}\ r := u + 1\ \{v = {*s} \wedge u + 1 = {*r} \parallel \bot\}$, so that in (3) the value of $*s$ carries over from the pre- to the postcondition, while the value of $*r$ is increased exactly by one. Taken together, this forces the measure t to decrease strictly.

Having arrived at proving Formula (1) by composing (2) with the conclusion of the total exception while rule, we can then apply rule (mbody) to obtain the total correctness of the whole algorithm:

$$[]\ i \leftarrow mbody\ p\ [MPOS\ i \wedge \forall j.\, MPOS\ j \Rightarrow i \leq j \parallel \\ \lambda e.\, case\ e\ of\ \ PatternNotFound \rightarrow \neg\exists i.\, MPOS\ i \\ \mid \_ \rightarrow \bot].$$

6 Conclusion

The principle of encapsulation of side effects in monads can be applied to model the imperative aspects of realistic languages such as Java; in particular, the Java exception mechanism is accurately captured by the so-called Java monad. Generic program logics including partial and total Hoare logics can be formulated largely independently of the nature of specific side-effects, i.e. monads [14, 16]; specific Hoare logics for abnormal termination introduced as part of verification support frameworks for Java are also subsumed by generic logics [15].

Here, we have illustrated this principle by means of a 'benchmark' verification of a pattern match algorithm previously used also as a test case for existing specific Hoare logics, implemented in a loosely specified dynamic reference monad. The example has shown that the framework of [15], for which no example application had so far been provided, is able to deal with realistic examples making extensive use of abrupt termination.

A general technical problem with the monadic modelling of the Java termination mechanism that came up in the verification process is that Java return exceptions are of a polymorphic type, parametrized over the type of the return value, so that the Java monad must in fact be regarded as a 'polymorphic monad' — i.e. each method body is executed in the instance of the Java monad determined by its result type. In order to deal with this polymorphism, we have designed a generalized catch function to be implicitly wrapped around method bodies in the same style as in the translation implemented in the LOOP tool [4]. This wrapper function shifts return exceptions into regular monadic return values and converts the resulting computation to fit the ambient monad; this solution improves on the previous approach, which consisted in bypassing the exception mechanism by storing return values in global variables [4]. There is a natural Hoare rule for wrapped method bodies, so that method calls can be dealt with in the generic verification framework without further problems.

This work forms part of an ongoing effort to adapt the wide-spectrum language HASCASL to the specification of object-oriented programs, in particular in Java. Open problems include the modelling of the Java class mechanism in HASCASL and logical support for concurrency. There are indications that the latter may be integrated in a monadic framework by means of continuations [1]. This motivates the search for a program logic for the continuation monad, to which by the results of [16] the existing generic computational logics are not usefully applicable.

Acknowledgements

This work forms part of the DFG-funded project HasCASL2 (KR 1191/7-2).

References

[1] K. Claessen, *A poor man's concurrency monad*, J. Funct. Programming **9** (1999), 313–323.
[2] C. Führmann, *Varieties of effects*, Foundations of Software Science and Computation Structures, LNCS, vol. 2303, Springer, 2002, pp. 144–158.
[3] *The Haskell mailing list*, `http://www.haskell.org/mailinglist.html`, 2002.
[4] M. Huisman and B. Jacobs, *Java program verification via a Hoare logic with abrupt termination*, Fundamental Approaches to Software Engineering, LNCS, vol. 1783, Springer, 2000, pp. 284–303.
[5] B. Jacobs and E. Poll, *A logic for the Java Modeling Language JML*, Fundamental Approaches to Software Engineering, LNCS, vol. 2029, Springer, 2001, pp. 284–299.
[6] ———, *Coalgebras and Monads in the Semantics of Java*, Theoret. Comput. Sci. **291** (2003), 329–349.
[7] B. Joy, G. Steele, J. Gosling, and G. Bracha, *The Java language specification*, Addison-Wesley, 2000.
[8] E. Moggi, *An abstract view of programming languages*, Tech. Report ECS-LFCS-90-113, Univ. of Edinburgh, 1990.
[9] ———, *Notions of computation and monads*, Inform. and Comput. **93** (1991), 55–92.
[10] T. Mossakowski, *Heterogeneous specification and the heterogeneous tool set*, Habilitation thesis, University of Bremen, 2005.
[11] S. Peyton-Jones (ed.), *Haskell 98 language and libraries — the revised report*, Cambridge, 2003, also: J. Funct. Programming **13** (2003).
[12] F. Regensburger, *HOLCF: Higher order logic of computable functions*, Theorem Proving in Higher Order Logics, LNCS, vol. 971, 1995, pp. 293–307.
[13] L. Schröder and T. Mossakowski, HASCASL*: Towards integrated specification and development of functional programs*, Algebraic Methodology and Software Technology, LNCS, vol. 2422, Springer, 2002, pp. 99–116.
[14] ———, *Monad-independent Hoare logic in* HASCASL, Fundamental Aspects of Software Engineering, LNCS, vol. 2621, 2003, pp. 261–277.

[15] L. Schröder and T. Mossakowski, *Generic exception handling and the Java monad*, Algebraic Methodology and Software Technology, LNCS, vol. 3116, Springer, 2004, pp. 443–459.
[16] ———, *Monad-independent dynamic logic in* HasCasl, J. Logic Comput. **14** (2004), 571–619.
[17] L. Schröder, T. Mossakowski, and C. Maeder, HasCasl – *Integrated functional specification and programming. Language summary*, available at `http://www.informatik.uni-bremen.de/agbkb/forschung/formal_methods/CoFI/HasCASL`
[18] Philip Wadler, *How to declare an imperative*, ACM Computing Surveys **29** (1997), 240–263.

Property Preserving Redesign of Specifications*

Artur Zawłocki[1], Grzegorz Marczyński[1], and Piotr Kosiuczenko[2]

[1] Institute of Informatics, Warsaw University
[2] Department of Computer Science, University of Leicester

Abstract. In the traditional formal approach to system specification and implementation, the software development process consists of a number of refinement steps which transform the initial specification into its correct realisation. This idealised view can hardly capture common situations when a specification changes in a non-incremental way. An extra flexibility can be added to the development process by allowing for a *redesign of specifications*, in addition to refinement steps. In this paper, the notion of specification redesign is formalised for an arbitrary institution. Basic properties of redesign are investigated and the formalism is applied to provide a formal semantics for UML class diagram transformations. As examples, two refactoring patterns are described in terms of class diagrams and interpreted as redesigns of corresponding algebraic specifications.

1 Introduction

In the contemporary software engineering the phases of system specification and design occur in a series of interleaving steps. As the system specification changes due to a number of factors including changed or new client requirements, new technology enablers etc., an extensive re-engineering of the system specification and design is often needed. In the algebraic approach to the system specification ([AKKB99]) the progress of the software development process is often described in terms of refinement which by monotonicity assumption can not express the non-incremental changes to the system structure.

We perceive the signature of the system specification as a description of the system structure. Therefore changes of the system structure should be reflected by a changed specification over a new signature. We define in an institution-independent way (cf. [BG92]) the notion of *redesign of specifications* as an embedding of two specifications into an intermediate one over a "joint" signature. That intermediate specification determines the strength of the connection between the original and the new specifications.

Our definition of redesign of specifications has numerous applications. One of them is the ability to reason about the transformation of the class structure of the object-oriented systems.

* This research was supported by the EC 5th Framework project AGILE: Architectures for Mobility (IST-2001-32747).

J.L. Fiadeiro et al. (Eds.): CALCO 2005, LNCS 3629, pp. 439–455, 2005.

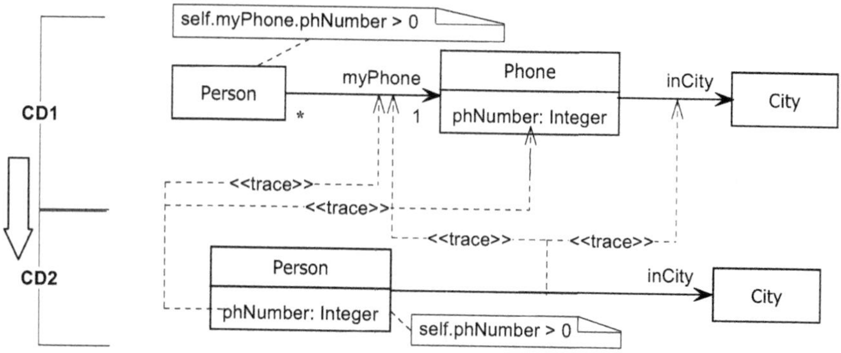

Fig. 1. An example of the Inline Class refactoring pattern

Object-oriented modelling languages provide textual and diagrammatic means for system specification (like UML, cf. [Obj03]). Class diagrams specify a common structure and relationships between objects. A well known approach to redesigning the object-oriented systems is the refactoring method ([Fow99]). It provides simple patterns to redesign the code and class structure in order to extend, improve and modify a system without altering its behaviour. Using our definition of *redesign class diagrams* (cf. Sect. 4) we make precise the otherwise ambiguous property of "preserving the system's behaviour".

The paper is organised as follows. In Sect. 2 we introduce our approach by means of an example – we argue that the *Inline Class* refactoring pattern (cf. [Fow99]) is indeed a redesign. Then (Sect. 3) we formally introduce a definition of a redesign diagram in an institution-independent way, describe a redesign of UML class diagrams (Sect. 4) and show (Sect. 5) an elaborate example — a formal proof that an application of the *Composite* design pattern ([GHJV95]) is a redesign. Finally we discuss related work (Sect. 6) and conclude (Sect. 7) the paper.

2 A Redesign Example – Inline Class Refactoring Pattern

In this section we present an example of *system redesign*, expressed on the level of UML class diagrams ([Obj03]). The redesign is done according to the *Inline Class* refactoring pattern ([Fow99]). This pattern allows us to join two classes, if one of them does not provide much functionality.

Let us consider class diagrams CD1 and CD2 shown on Fig. 1. The class Phone represents phone numbers, Person represents phone owners and City represents cities. Each person has a phone and every phone is located in a city. We attach an OCL constraint (cf. [Obj03]) to the class Person, saying that every person's number is larger than 0. The class Phone does not provide much functionality and is only used by the class Person. Therefore, we join those classes in CD2. The class City is not affected by the redesign.

We formalise CD1 and CD2 as algebraic specifications SP_1 and SP_2, respectively. In this case the encoding is self-explanatory. More details are provided in Sect. 4.

spec SP1 =
  sorts *Person*, *Phone*, *City*;
  ops *myPhone* : *Person* → *Phone*;
    inCity : *Phone* → *City*;
    phNumber : *Phone* → *Int*
  vars *p* : *Person*
  • *phNumber*(*myPhone*(*p*)) > *0*
end

spec SP2 =
  sorts *Person*, *City*;
  ops *inCity* : *Person* → *City*;
    phNumber : *Person* → *Int*
  vars *p* : *Person*
  • *phNumber*(*p*) > *0*
end

The axioms in SP_1 and SP_2 result from the formalisation of OCL constraints attached to CD1 and CD2, respectively.

We use a dependency relationship with the ⟨⟨trace⟩⟩ stereotype to relate elements of both diagrams. According to the convention introduced in [Kos05], classes with the same names in both diagrams are implicitly related by a ⟨⟨trace⟩⟩ relationship. We only need to draw the dashed arrows between attribute and link names. To represent the ⟨⟨trace⟩⟩ relationship at the formal level, we form a "joint" signature Σ consisting of the sum of sorts and a disjoint sum of operation symbols from Σ_1 and Σ_2 (symbols from Σ_i will be indexed with $_i$ in Σ, for $i \in \{1, 2\}$). The dependency relationship can be then translated to a set Φ of Σ-equations:

$$\forall p : Person \cdot phNumber_1(myPhone_1(p)) = phNumber_2(p)$$
$$\forall p : Person \cdot inCity_1(myPhone_1(p)) = inCity_2(p)$$

Intuitively, the first equation states that for any person p, the phone number obtained by evaluating *phNumber*(*myPhone*(*p*)) in the original system will be the same as the one obtained by evaluating *phNumber*(*p*) in the redesigned system. The second equation can be interpreted in a similar way.

SP_1 and SP_2 can be translated to the joint signature Σ simply by indexing all operation symbols in axioms with $_1$ and $_2$, respectively. After putting the translations together we add the equations from Φ to obtain the specification *SP* which can be treated as the encoding of the whole redesign diagram shown on Fig. 1:

spec SP =
  sorts *Person*, *Phone*, *City*;
  ops $myPhone_1$: *Person* → *Phone*;
    $inCity_1, inCity_2$: *Phone* → *City*;
    $phNumber_1$: *Phone* → *Int*;
    $phNumber_2$: *Person* → *Int*;
  vars *p* : *Person*;
  • $phNumber_1(myPhone_1(p)) > 0$

- $phNumber_2(p) > 0$
- $phNumber_1(myPhone_1(p)) = phNumber_2(p)$
- $inCity_1(myPhone_1(p)) = inCity_2(p)$

end

Notice that we can remove either the first or the second axiom from *SP* without changing the semantics of the specification, since Φ implies their equivalence.

The important property of *SP* is that it is a *conservative extension* of both SP_1 and SP_2. Roughly, this means that *SP* does not put any constraints on interpretations of $phNumber_1$, $myPhone_1$ and $inCity_1$ other than those resulting from the translation of SP_1, and similarly for the operations from Σ_2.

For the rest of this section let us adopt the usual notion of a model of a specification $\langle \Sigma, \Psi \rangle$ as a first-order many-sorted Σ-structure satisfying all sentences from Ψ. Conservativity of *SP* with respect to SP_1 means that any model M of SP_1 can be extended to a model of *SP* by interpreting $inCity_2$ as the composition of $myPhone^M$ and $inCity^M$ and interpreting $phNumber_2$ as the composition of $myPhone^M$ and $phNumber^M$. Such an extension then can be restricted to a model of SP_2 which is a "refactored" version of M. Conservativity of *SP* with respect to SP_2 is equivalent to the fact that every model of SP_2 can be obtained as a restriction of some model of *SP*.

3 A Formal Approach

As mentioned in the introduction, the redesign has to preserve the properties of the system being restructured. To make this statement precise, we must be able to compare the *semantics* of system descriptions. In the example above this involves encoding class diagrams in some specification language, translating and putting specifications together, as well as comparing specifications that use different signatures. Such operations can be carried out in an institutional framework. In this section we formally define a redesign of specifications in an institution-independent way.

3.1 Preliminaries

Let **Set** and **Cat** denote the category of all sets and the category of all categories, respectively. An *institution* (cf. [BG92]) is a tuple $\langle \mathbf{Sig}, \mathbf{Mod}, \mathbf{Sen}, \models \rangle$, where

- **Sig** is the category of *signatures*;
- $\mathbf{Mod} : \mathbf{Sig}^{op} \to \mathbf{Cat}$ is the *model functor*, assigning a category $\mathbf{Mod}(\Sigma)$ of Σ*-models* to every signature $\Sigma \in |\mathbf{Sig}|$ and a functor $\mathbf{Mod}(\sigma) : \mathbf{Mod}(\Sigma') \to \mathbf{Mod}(\Sigma)$ to every signature morphism $\sigma : \Sigma \to \Sigma'$;
- $\mathbf{Sen} : \mathbf{Sig} \to \mathbf{Set}$ is the *sentence functor* assigning a set $\mathbf{Sen}(\Sigma)$ of Σ*-sentences* to every $\Sigma \in |\mathbf{Sig}|$ and a σ*-translation* function $\mathbf{Sen}(\sigma) : \mathbf{Sen}(\Sigma) \to \mathbf{Sen}(\Sigma')$ to every $\sigma : \Sigma \to \Sigma'$;
- $\models$ is a family $\{\models_\Sigma\}_{\Sigma \in |\mathbf{Sig}|}$ of *satisfaction relations*, where $\models_\Sigma \subseteq |\mathbf{Mod}(\Sigma)| \times \mathbf{Sen}(\Sigma)$.

such that for any signature morphism $\sigma : \Sigma \to \Sigma'$ the functor $\mathbf{Mod}(\Sigma)$ and the translation function $\mathbf{Sen}(\sigma)$ preserve the satisfaction relation, that is, for any $\varphi \in \mathbf{Sen}(\Sigma)$ and $M' \in |\mathbf{Mod}(\Sigma')|$

$$\mathbf{Mod}(\sigma)(M') \models_\Sigma \varphi \quad \text{iff} \quad M' \models_{\Sigma'} \mathbf{Sen}(\sigma)(\varphi)$$

We write $M|_\sigma$ for $\mathbf{Mod}(\sigma)(M)$ and just $\sigma(\varphi)$ for $\mathbf{Sen}(\sigma)(\varphi)$.

Given an institution we can consider specifications as abstract objects, classified by signatures and defining classes of models.

That is, we require that operations *Sig* and $[\![\_]\!]$ be defined on the class of specifications so that, for every specification SP, $Sig(SP) \in |\mathbf{Sig}|$ and $[\![SP]\!] \subseteq |\mathbf{Mod}(Sig(SP))|$. Moreover, we require that the class of specifications is closed under the following *specification-building operations* ([ST88a]):

- For any $\Sigma \in |\mathbf{Sig}|$ and $\Phi \subseteq \mathbf{Sen}(\Sigma)$, a *presentation* $\langle \Sigma, \Phi \rangle$ is a specification with $Sig(\langle \Sigma, \Phi \rangle) = \Sigma$ and $[\![\langle \Sigma, \Phi \rangle]\!] = \{M \in |\mathbf{Mod}(\Sigma)| \mid M \models \Phi\}$.
- For any signature morphism $\sigma : \Sigma \to \Sigma'$ and a specification SP such that $Sig(SP) = \Sigma$, the *translation of SP along* σ is a specification $\sigma(SP)$ such that $Sig(\sigma(SP)) = \Sigma'$ and $[\![\sigma(SP)]\!] = \{M' \in \mathbf{Mod}(\Sigma') \mid M'|_\sigma \in [\![SP]\!]\}$.
- For any specifications SP_1, SP_2 such that $Sig(SP_1) = Sig(SP_2)$, the *union* $SP_1 \cup SP_2$ is a specification with $Sig(SP_1 \cup SP_2) = Sig(SP_1)$ and $[\![SP_1 \cup SP_2]\!] = [\![SP_1]\!] \cap [\![SP_2]\!]$.
- For any signature morphism $\sigma : \Sigma \to \Sigma'$ and a specification SP' such that $Sig(SP') = \Sigma'$, the *reduct of* SP' *along* σ is a specification $SP'|_\sigma$ such that $Sig(SP'|_\sigma) = \Sigma$ and $[\![SP'|_\sigma]\!] = \{M'|_\sigma \mid M' \in [\![SP']\!]\}$.

Specifications in an arbitrary institution form a category: a *specification morphism* (cf. Chap. 4 in [AKKB99]) $\sigma : SP \to SP'$ is a signature morphism $\sigma : Sig(SP) \to Sig(SP')$ such that for every model $M' \in [\![SP']\!]$, $M'|_\sigma \in [\![SP]\!]$. A specification morphism $\sigma : SP \to SP'$ is *conservative* if for every $M \in [\![SP]\!]$ there exists $M' \in [\![SP']\!]$ such that $M = M'|_\sigma$. A composition of conservative morphisms is also conservative.

By Prop. 4.22 in [AKKB99] (see also [BG92]), if the category **Sig** is finitely cocomplete and the class of specifications is closed with respect to the operations listed above, then the category of specifications is also finitely cocomplete and every pushout diagram is of the form

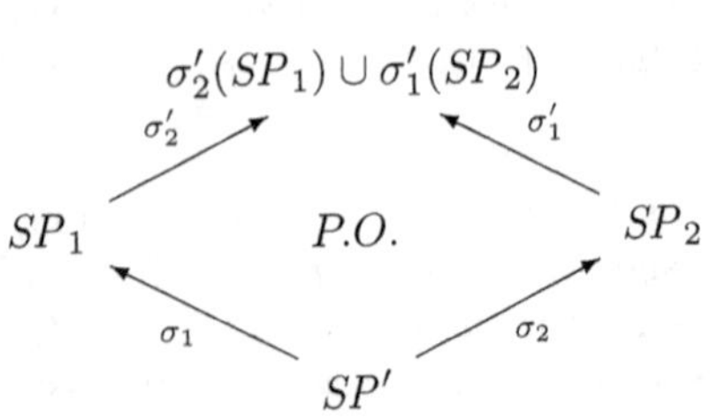

From now on we consider only institutions with a finitely cocomplete category of signatures and classes of specifications closed with respect to the specification-building operations.

Let D be a finite diagram of specifications. The functor **Mod** *preserves* a colimit of D if it maps a colimit in **Sig** of the corresponding diagram of signatures to a limit in **Cat**. We call such diagram D *an amalgamable diagram.*

3.2 An Abstract View of Redesign

Adopting an institutional semantics encourages the formulation of concepts on the level as general as possible, hence we look for a definition of redesign independent of the logical system used. We view the structure of the system as described simply by the signature of system specification. A redesign of the specification amounts then to expressing system properties using a different signature.

Let SP_1 and SP_2 be specifications. It would be too restrictive to require the existence of a specification morphism from SP_1 to SP_2 in order to consider the latter a redesign of the former. For instance, in the example of Sect. 2 there exists no signature morphism from Σ_1 to Σ_2. Instead, we require the existence of an intermediate specification SP and two conservative morphisms $\sigma_1 : SP_1 \to SP$ and $\sigma_2 : SP_2 \to SP$. Conservativity means that the specification SP does not put any restrictions on the interpretation of symbols from $Sig(SP_1)$ and $Sig(SP_2)$ besides those already present in SP_1 and SP_2. However, SP can also relate the symbols from the two signatures, for instance define the ones from $Sig(SP_2)$ in terms of those from $Sig(SP_1)$.

Definition 1 (Redesign). *Let SP be a specification. Let $\sigma_1 : Sig(SP_1) \to Sig(SP)$ and $\sigma_2 : Sig(SP_2) \to Sig(SP)$ be signature morphisms. SP_2 is a* redesign of SP_1 via σ_1, σ_2 *if $\sigma_1 : SP_1 \to SP$ and $\sigma_2 : SP_2 \to SP$ are conservative specification morphisms. In such a case we say, that*

$$SP_1 \xrightarrow{\sigma_1} SP \xleftarrow{\sigma_2} SP_2 \tag{1}$$

is a redesign diagram.

Let us observe that the specifications SP_1 and SP_2 play symmetric roles in the above definition: SP_2 is a redesign of SP_1 via σ_1, σ_2 iff SP_1 is a redesign of SP_2 via σ_2, σ_1.

Moreover, if the model functor preserves coproducts, for arbitrary SP_1 and SP_2 there is a redesign diagram with the coproduct SP_1+SP_2 as the intermediate specification. This redesign preserves a "minimal" behaviour, as components of the two signatures remain totally unrelated. We can require more properties to be preserved by strengthening the intermediate specification. For instance, if the invariant properties are expressed by a specification SP' with specification morphisms $\rho_1 : SP' \to SP_1$, $\rho_2 : SP' \to SP_2$, a pushout of the diagram

$$SP_1 \xleftarrow{\rho_1} SP' \xrightarrow{\rho_2} SP_2$$

is a redesign, provided the pushout morphisms are conservative.

Let us discuss briefly what are the properties preserved by a redesign of specifications. Given a specification SP we define the equivalence relation $\sim_{SP}$ on $\mathbf{Sen}(Sig(SP))$ as

$$\varphi \sim_{SP} \varphi' \quad \text{iff} \quad SP \models \varphi \iff SP \models \varphi'$$

We identify properties with abstraction classes of $\sim_{SP}$ and define *the set* $Prop(SP)$ *of SP-properties* as the quotient set $\mathbf{Sen}(Sig(SP))/_{\sim_{SP}}$. Observe that given the redesign diagram as in Def. 1, conservative specification morphisms σ_1 and σ_2 induce the injective functions (defined in an obvious way)

$$\sigma_1' : Prop(SP_1) \to Prop(SP), \qquad \sigma_2' : Prop(SP_2) \to Prop(SP)$$

Pairs of properties preserved by the redesign diagram are elements of the set

$$\{\langle p_1, p_2\rangle \mid \sigma_1'(p_1) = \sigma_2'(p_2),\ p_1 \in Prop(SP_1),\ p_2 \in Prop(SP_2)\}$$

We show that under certain assumptions on the model functor, redesigns can be composed "vertically".

Fact 2 (Category of redesigns). *If the functor* **Mod** *preserves all pushouts then redesigns form a category in which objects are specifications and a morphism from* SP_1 *to* SP_2 *is any redesign diagram of the form (1). Identity redesign diagram of a specification* SP *with a signature* Σ *is* $SP \xrightarrow{id_\Sigma} SP \xleftarrow{id_\Sigma} SP$. *The composition of* $SP_1 \xrightarrow{\sigma_1} SP' \xleftarrow{\sigma_2} SP_2$ *and* $SP_2 \xrightarrow{\sigma_2'} SP'' \xleftarrow{\sigma_3} SP_3$ *is defined using the pushout construction, as shown below.*

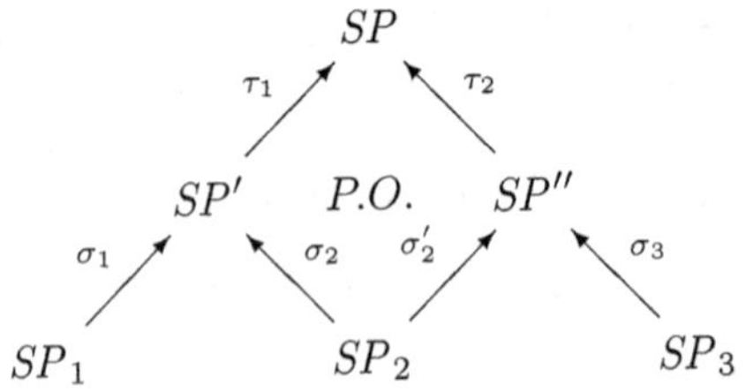

As pushouts are defined up to an isomorphism, for the above construction to work, the canonical choice of pushouts guaranteeing the associativity of morphism composition has to be given.

3.3 Redesigning Structured Specifications

Definition 1 allows us to decide whether a given diagram of specifications is a redesign. However, in a common scenario the specification SP_2 may not be known in advance: developers want to redesign a specification SP_1 to a new signature Σ_2. They relate the symbols of Σ_2 to those of $\Sigma_1 = Sig(SP_1)$ by means of a specification SP_{rel} over a "joint" signature Σ with signature morphisms $\sigma_1 : \Sigma_1 \to \Sigma$, $\sigma_2 : \Sigma_2 \to \Sigma$, such that $\sigma_1 : SP_1 \to \sigma_1(SP_1) \cup SP_{rel}$ is a conservative specification morphism. The problem now is to find a redesigned specification SP_2 over Σ_2, such that $\sigma_2 : SP_2 \to \sigma_1(SP_1) \cup SP_{rel}$ is a conservative specification morphism. Such a specification always exists, since we assumed that the class of specifications is closed with respect to reducts along signature morphisms.

Fact 3. *Let $\sigma_1 : SP_1 \to SP$ be a conservative specification morphism and let $\sigma_2 : \Sigma_2 \to Sig(SP)$ be a signature morphism. Then*

$$SP_1 \xrightarrow{\sigma_1} SP \xleftarrow{\sigma_2} SP|_{\sigma_2}$$

is a redesign diagram.

However, in many applications it is preferable to obtain SP_2 as a presentation. For instance, if SP_1 is a finite presentation resulting from encoding an UML class diagram with OCL constraints, we would also like SP_2 to consist of a list of axioms corresponding to OCL constraints for the redesigned specification. If SP_1 is obtained by application of specification-building operations, then we can use the next lemma to find SP_2 by following, to some extent, the structure of SP_1.

Lemma 4. *Let $SP_1 \xrightarrow{\sigma_1} SP \xleftarrow{\sigma_2} SP_2$ be a redesign diagram.*

i. (translation) Let $\tau : SP_1 \to SP_1'$ be a conservative specification morphism. If the following pushout diagram is amalgamable

$$\begin{array}{ccccc} & & SP' & & \\ & \overset{\sigma_1'}{\nearrow} & & \overset{\tau'}{\nwarrow} & \\ SP_1' & & P.O. & & SP \\ & \underset{\tau}{\nwarrow} & & \underset{\sigma_1}{\nearrow} & \\ & & SP_1 & & \end{array}$$

then

$$SP_1' \xrightarrow{\sigma_1'} SP' \xleftarrow{\sigma_2;\tau'} SP_2$$

is a redesign diagram.

ii. (union) Let SP_1' and SP_2' be specifications such that $Sig(SP_1') = Sig(SP_1)$ and $Sig(SP_2') = Sig(SP_2)$ and also

$$[\![SP \cup \sigma_1(SP_1')]\!] = [\![SP \cup \sigma_2(SP_2')]\!]$$

Then

$$SP_1 \cup SP_1' \xrightarrow{\sigma_1} SP \cup \sigma_1(SP_1') \xleftarrow{\sigma_2} SP_2 \cup SP_2'$$

is a redesign diagram.

iii. (coproduct) Let

$$SP_1' \xrightarrow{\sigma_1'} SP' \xleftarrow{\sigma_2'} SP_2'$$

be a redesign diagram. If the following three coproduct diagrams are amalgamable

$$SP_1 \to SP_1 + SP_1' \leftarrow SP_1'$$
$$SP \to SP + SP' \leftarrow SP'$$
$$SP_2 \to SP_2 + SP_2' \leftarrow SP_2'$$

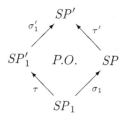

then

$$SP_1 + SP'_1 \xrightarrow{\rho_1} SP + SP' \xleftarrow{\rho_2} SP_2 + SP'_2$$

is also a redesign diagram, where ρ_1 *and* ρ_2 *are universal morphisms from coproducts* $SP_1 + SP'_1$ *and* $SP_2 + SP'_2$*, respectively.*

iv. (reduct) Let $\rho : \Sigma' \to Sig(SP_1)$ *be a signature morphism. Then*

$$SP_1|_{\rho} \xrightarrow{\rho;\sigma_1} SP \xleftarrow{\sigma_2} SP_2$$

is a redesign diagram.

The proof of the above lemma is straightforward.

The property (i) allows us to translate either the original or the redesigned specification via a conservative specification morphism and obtain a redesign diagram. This covers situations such as renaming symbols in a signature via an injective signature morphism or extending the specification with new symbols and axioms concerning only the new symbols. By (ii), both SP_1 and SP_2 can be enriched as long as the enriching parts "correspond to each other modulo SP", i.e. represent properties preserved by the redesign, using the terminology from Sect. 3.2. The property (iii) states that if the original specification is a disjoint sum of two components, each of the components can be redesigned separately in order to obtain a redesign of the sum. Finally, by (iv), either the original or the redesigned specification can be reduced along an arbitrary signature morphism.

The above lemma can be applied to the example of Sect. 2. By the property (ii) in order to conclude that the diagram in Fig. 1 describes a redesign it suffices to show that

$$\langle \Sigma_1, \emptyset \rangle \xrightarrow{\sigma_1} \langle \Sigma, \Phi \rangle \xleftarrow{\sigma_2} \langle \Sigma_2, \emptyset \rangle$$

is a redesign diagram, and then to check that

$$phNumber_1(myPhone_1(p)) > 0 \iff phNumber_2(p) > 0$$

follows from Φ.

4 Redesign of UML Class Diagrams

We apply the notions developed in previous sections for reasoning about transformations of UML class diagrams. The idea is to formalise such diagrams as specifications in the institution of CASL ([CoF04]) — a variant of order sorted, partial first-order logic — and then to generate an intermediate specification from dependency relationships between elements of the diagrams. Note that since the category of CASL signatures is finitely cocomplete such intermediate specification always exists. A transformation preserves essential system properties if it gives rise to a redesign diagram in the category of CASL specifications.

4.1 Formalising UML Class Diagrams

We represent UML class diagrams annotated with OCL constraints (cf. [Obj03]) as algebraic specifications in CASL institution following [BHTW99]. The only difference is that in order to make the presentation more readable we omit the concept of *environments* used there to represent methods with side-effects.

For a class diagram CD we create a specification SP. In the corresponding signature Σ each sort name corresponds to a class name from CD. These sorts represent collections of objects of that class.

Class inheritance is handled by the ordering on corresponding sorts. Unlike in [BHTW99] we require that the carrier of a sort corresponding to an abstract class is a disjoint union (up to an isomorphism) of carriers of sorts corresponding to its direct subclasses. Axioms guaranteeing this property for every abstract class are added to SP (as in the example in Sect. 5).

As we are aware that this kind of encoding does not permit to express overriding we refer the reader to e.g. [ACZ99] for a possible solution of this problem (see also [Mar04]).

All query methods (the ones that do not change system state) and attributes are encoded as functions with an additional first parameter representing the *self* object. To avoid all problems related to side effects, local object states, global system environment etc. that are not directly relevant to the problems described herein, we do not handle any non-query methods. For a similar reason we assume that the only OCL constraints contained in class diagrams are class invariants (i.e. there are no method pre- and postconditions). Translation of these OCL constraints to the CASL logic is straightforward using the method described in [BHTW99]. As in [Kos05] we call the function that takes an OCL annotated UML class diagram and produces a specification in the CASL institution the *translation function Trans*. Formally, since the whole class diagram can be described as an OCL sentence, *Trans* is a mapping of OCL terms to CASL formulas. Specifications SP_1 and SP_2 in Sec. 2 are example results of this translation.

In what follows we only consider diagrams consisting of two class diagrams, say CD1 and CD2, with trace relationships (dependencies marked with the ⟨⟨trace⟩⟩ stereotype) connecting corresponding components in both of them (Fig. 1 contains an example of such a diagram). Let us call such diagrams *class diagrams with traces*.

All UML trace dependencies have a mappingExpression attribute used to capture the relationship between elements linked by trace dependencies. In our examples only two values of mappingExpression are used — *composition* and *product*. For instance a trace dependency linking myPhone and inCity in CD1 with inCity in CD2 on Fig. 1 has mappingExpression set to *composition*, which means that inCity in CD2 is a composition of myPhone and inCity in CD1. In cases when the intended relationship is obvious, mappingExpression may be omitted from the class diagram.

4.2 Redesign Class Diagrams

Given a UML class diagram with traces CDT that consists of two class diagrams CD1 and CD2 such that CD2 results from some transformation of CD1 we would like to decide whether that transformation is a redesign. We assume that CD2 is already annotated with OCL constraints e.g. using some interpretation function (cf. Sect. 4.3). We use the translation *Trans* (described in Sect. 4.1) to represent CD1 as a specification SP_1 with a signature Σ_1, and CD2 as a specification SP_2 with a signature Σ_2.

Let X be a countable set of variables. Let us use traces connecting classes on CDT to generate a partial *sort mapping* $sm : Sorts(\Sigma_1) \rightarrow ?\ Sorts(\Sigma_2)$. We require that *sm* preserve the subsort relation and be injective (for the reasons described below). Similarly, using traces connecting methods and/or attributes and using the derivation strategy described by mappingExpression of each trace, we define a partial many sorted *term mapping* $tm : T_{\Sigma_1}(X) \rightarrow ?\ T_{\Sigma_2}(X)$ such that *sm* is a sort mapping associated with *tm*. The way of translation of trace endpoints to terms is straightforward (e.g. the composition of myPhone and phNumber is translated to a term $phNumber(myPhone(x))$ where $x \in Person$ as in Sect. 2)

The term mapping *tm* forms a connection between terms over two different signatures. We require the denotations of the corresponding terms to be equal. In general it is impossible to express such property as a sentence from either $\mathbf{Sen}(\Sigma_1)$ or $\mathbf{Sen}(\Sigma_2)$. Thus we construct a bigger signature Σ containing all that is needed to express the equality of terms being mapped by *tm*.

Let the signature Σ_{rts} describe the "rest of the system" i.e. translation of all UML model classes, attributes, methods etc. depicted neither in CD1 nor in CD2. We assume that Σ_{rts} is a part of both Σ_1 and Σ_2. As the sort mapping *sm* is functional and injective we can assume also that all sorts connected by *sm* are common (up to renaming) to both signatures. Thus let us define the signature Σ' as Σ_{rts} and additionally all sorts from $dom(sm)$ (Σ_{rts} and $dom(sm)$ are disjoint). The perfect candidate for a "joint" signature Σ is the pushout of Σ_1 and Σ_2 over the signature Σ'

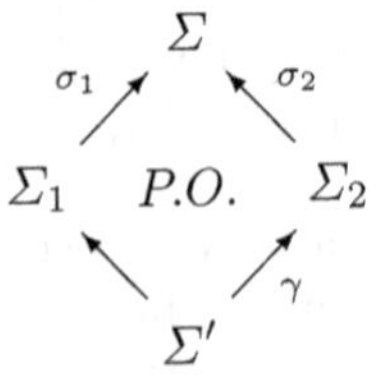

where γ is a signature morphism mapping all Σ_{rts} symbols to themselves and additionally mapping all sorts in $s \in dom(sm)$ to $sm(s)$. Since we require that *sm* preserves the subsort relation, γ is indeed a CASL subsorted signature morphism. Having Σ we can express the desired equality on terms. We define the following set of Σ-equations:

$$\Phi = \{\forall X \cdot \sigma_1(t) = \sigma_2(tm(t)) \mid t \in T_{\Sigma_1}(X), t \in dom(m)\}$$

Finally, we are able to formalise the requirements that need to be imposed on a class diagram to represent a redesign.

Definition 5 (Redesign class diagram). *Using the notation introduced above, the class diagram CDT is a* redesign class diagram *over the signature* Σ' *if translations of CD1 and CD2 to the* CASL *institution together with the specification* $\sigma_1(SP_1) \cup \langle \Sigma, \Phi \rangle$ *and the signature morphisms* σ_1, σ_2 *form the following redesign diagram (in the sense of Def. 1)*

$$SP_1 \xrightarrow{\sigma_1} \sigma_1(SP_1) \cup \langle \Sigma, \Phi \rangle \xleftarrow{\sigma_2} SP_2$$

The requirement that *sm* be injective is very sensible since every non-injective sort mapping leads to non-conservativity of σ_1.

4.3 Interpretation Functions

Interpretation functions (cf. [Kos05, Kos01]) are a partial solution to the problem of finding Σ_2-sentences equivalent to given Σ_1-sentences for the case of institutions with sentences containing term equalities. They can be very useful as a vehicle for an automatic transformation of OCL constraints when changes to class diagrams are performed.

An interpretation function is a partial function generated by a term mapping with an orthogonal domain (cf. [Kos05]). These functions have several useful properties. As stated in [Kos05] they preserve equational proofs, proofs using propositional tautologies, resolution rule and proofs by induction. From our perspective, the following property is the most important: Given a redesign class diagram like the one in Def. 5, the interpretation function $f : \mathbf{Sen}(\Sigma_1) \rightarrow ? \mathbf{Sen}(\Sigma_2)$ generated by the same term mapping *tm* as used in the definition of Φ, and a sentence $\phi_1 \in \mathrm{dom}(f)$, we have

$$[\![SP \cup \sigma_1(\phi_1)]\!] = [\![SP \cup \sigma_2(f(\phi_1))]\!]$$

where *SP* denotes $\sigma_1(SP_1) \cup \langle \Sigma, \Phi \rangle$.

Lemma 4, (ii), can then be applied in order to add ϕ_1 to SP_1 and $f(\phi_1)$ to SP_2.

We use an interpretation function to automatically transform the OCL constraint in the example of the next section.

5 An Elaborate Redesign Example – The Composite Pattern

The class diagram CD1 in Fig. 2 describes a *Directed Acyclic Graph (DAG)* data structure. Objects of the class A represent internal graph nodes, objects of the class B represent leaves. The OCL invariant of A guarantees that the structure is indeed a DAG (i.e. it doesn't contain a cycle).

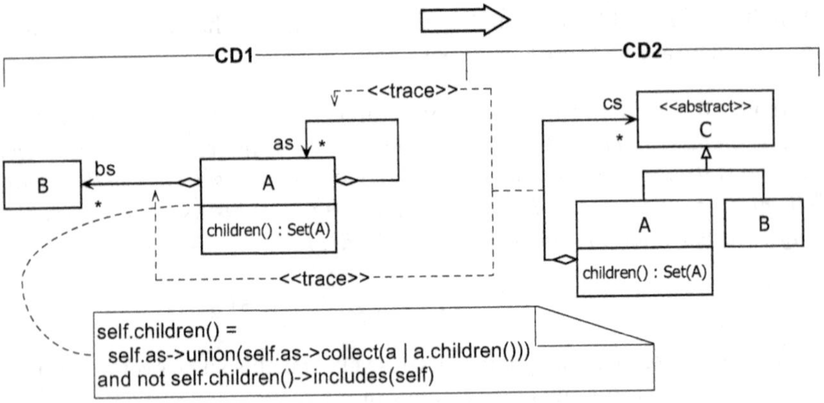

Fig. 2. An example of the Composite design pattern in UML

The class diagram CD2 is a result of the application of the *Composite design pattern* (cf. [GHJV95]) to the system described by CD1. Note the lack of OCL constraint describing an invariant of the class A in CD2. Thus CD2 does not necessarily describe a DAG. To fix the problem we use the interpretation function to generate the appropriate invariant (cf. Sect. 4.3 and [Kos05] for details). The following mapping

$$\texttt{x.as} \mapsto \texttt{x.cs}{-}{>}\ \texttt{select(a} \mid \texttt{a.isKindOf(A))}$$
$$\texttt{x.bs} \mapsto \texttt{x.cs}{-}{>}\ \texttt{select(b} \mid \texttt{b.isKindOf(B))}$$

is orthogonal and thus it could be extended (cf. [Kos05]) to the interpretation function that we use to transform the invariant of A in CD1

```
context A inv :
 self.children() = self.as-> union(self.as-> collect(a | a.children()))
 and not self.children()-> includes(self)
```

to the following invariant in CD2

```
context A inv :
  self.children() = (self.cs-> select(a | a.isKindOf(A)))-> union(
     (self.cs-> select(a | a.isKindOf(A)))-> collect(a | a.children()))
  and not self.children()-> includes(self)
```

To decide whether the above class diagram with an additional invariant for A in CD2 is a redesign class diagram we translate CD1 and CD2 augmented with an invariant to SP_1 and SP_2, respectively.

spec SP1 =
  sorts A, B
  ops $as : A \to Set[A]$;
    $bs : A \to Set[B]$;
    $children : A \to Set[A]$;
  vars $a, a' : A$
  • $a' \in children(a) \Leftrightarrow (a' \in as(a) \vee \exists a'' \bullet a'' \in as(a) \wedge a' \in children(a''))$
  • $a \notin children(a)$
end

spec SP2 =
  sorts $A, B, C; A \leq C; B \leq C$;
  ops $cs : A \to Set[C]$;
    $children : A \to Set[A]$;
  vars $a, a' : A$; $c : C$
  • $a' \in children(a) \Leftrightarrow (a' \in A \wedge (a' \in cs(a) \vee \exists a'' \bullet a'' \in cs(a) \wedge a' \in children(a'')))$
  • $a \notin children(a)$
  • $c \in A \Leftrightarrow c \notin B$
end

First two axioms of both specifications are translations of invariants of A. The last axiom in SP_2 says that C is a disjoint union of A and B (the encoding of the ⟨⟨abstract⟩⟩ stereotype on C).

The trace relationships (with the omitted mappingExpression attribute set to *product*, cf. Sect. 4.1) result in the following sort mapping *sm* and term mapping *tm*

$$sm = \{A \mapsto A; \; B \mapsto B\}$$

$$tm = \{as(x) \mapsto first(i(cs(x))); \; bs(x) \mapsto second(i(cs(x))); \; children(x) \mapsto children(x)\}$$

where i is an obvious isomorphism between $Set[C]$ and $Pair[Set[A], Set[B]]$ (justified by the requirement expressed by axioms in SP_2 saying that $C = A \uplus B$), *first* and *second* are product projections defined in the CASL standard library.

We use the procedure described in Sect. 4.2 to define a signature Σ' common to both SP_1 and SP_2, $\Sigma' = \{A, B\} \cup \Sigma_{rts}$, where Σ_{rts} is a signature with all standard sorts (e.g. integers, booleans, etc.) and operations on them (it was implicitly assumed to be a part of signatures of SP_1 and SP_2). Let Σ be a pushout of Σ_1 and Σ_2 over Σ'. The joint specification *SP* is the following:

spec SP =
  sorts $A, B, C; A \leq C; B \leq C$;
  ops $as_1 : A \to Set[A]$;
    $bs_1 : A \to Set[B]$;
    $cs_2 : A \to Set[C]$;
    $children_1, children_2 : A \to Set[A]$;
    $i : Set[C] \to Pair[Set[A], Set[B]]$
  vars $a, a' : A, b : B, c : C, s : Set[C]$
  • $as_1(a) = first(i(cs_2(a)))$
  • $bs_1(a) = second(i(cs_2(a)))$
  • $children_1(a) = children_2(a)$
  • $a' \in children_1(a) \Leftrightarrow (a' \in as(a) \vee \exists a'' \bullet a'' \in as(a) \wedge a' \in children_1(a''))$

- $a \notin children_1(a)$
- $c \in A \Leftrightarrow c \notin B$
- $(a\ is\_in\ first(i(s)) \Leftrightarrow a\ is\_in\ s) \wedge (b\ is\_in\ first(i(s)) \Leftrightarrow b\ is\_in\ s)$

end

The first three axioms are equalities resulting from the term mapping tm (the set Φ in Sect. 4.2). The last axiom defines the isomorphism i.

It is clear that $\sigma_1 : SP_1 \to SP$ and $\sigma_2 : SP_2 \to SP$ are specification morphisms. In the following we show that they are conservative.

First we prove this for σ_1. Let us assume $M_1 \in [\![SP_1]\!]$. We need to find a model $M \in [\![SP]\!]$ such that $M|_{\sigma_1} = M_1$. M can be constructed by putting $A^M = A^{M_1}$, $B^M = B^{M_1}$, $as_1^M = as^{M_1}$, and similarly for bs_1^M and $children_1^M$. Let $C^M = A^M \uplus B^M$. The function $cs_2^M : A^M \to P(C^M)$ is the only function that makes the following diagram in **Set** commute (we use the standard set-theoretical notation here)

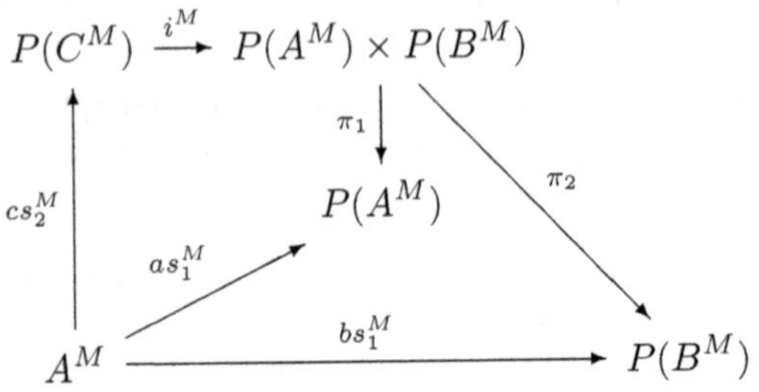

The existence of cs_2^M follows from the universal property of the product $P(A^M) \times P(B^M)$. Obviously $M = M|_{\sigma_1}$.

To show that σ_2 is conservative let us assume $M_2 \in [\![SP_2]\!]$. We need to find such $M \in [\![SP]\!]$ that $M|_{\sigma_2} = M_2$. As C is an abstract class i.e. $C^{M_2} = A^{M_2} \uplus B^{M_2}$, we construct M as M_2 and additionally interpret two functions $as_1 : A^M \to P(A^M)$ and $bs_1 : A^M \to P(B^M)$ as compositions of $cs_2; i$ with π_1 and π_2 respectively. Again is easy to see that $M_2 = M|_{\sigma_2}$.

5.1 Changing a Redesign Diagram

Imagine the situation that just after we had proved that the above class diagram is a redesign class diagram we discovered that actually the names of classes on CD2 were written incorrectly. They should have been (as it is in [GHJV95]) Component instead of C, Composite instead of A and Leaf instead of B and also there should have been additional method name in a class Component. We do not need to redo a proof that a transformation is a redesign. Since all above described changes are conservative, we can use[1] property (i) of Lemma 4 that any translation of SP_2 by a conservative specification morphism leads to a specification that is also a redesign of SP_1 (via the same intermediate specification).

[1] As it is easy to prove that the resulting pushout diagram is amalgamable in CASL.

6 Related Work

A number of approaches to redesigning UML class models exist already. One of them is *refactoring* (cf. [Fow99]), which provides simple patterns for code and class structure redesign to extend, improve and modify a system without altering its behaviour. Model transformations gain a lot of interest in recent time (cf. [MCG05]). Interpretation functions, used to formalise UML class digram transformations in [Kos01, Kos05] originate in abstract algebra. In [Tay73], an interpretation function transforms a single operation symbol into a complex term. Lano uses a form of interpretation function to define the notion of refinement for his Real Time Action Logic ([Lan95]). Graph transformations (cf. e.g. [GRPPS98]) may also be used to describe evolution of a specification.

Our notion of redesign is related to the concept of *implementation* (cf. e.g. [ONS96, ST88b]). The implementation of SP_1 by SP_2, as defined in [ONS96], is also parametrised by an intermediate specification. The main difference between the two notions is the constructive nature of the implementation: a *constructor* operation must be provided that transforms every model of SP_2 to a model of the intermediate specification that can be then reduced to the model of SP_1. A redesign diagram $SP_1 \xrightarrow{\sigma_1} SP \xleftarrow{\sigma_2} SP_2$ does not prescribe how to provide such operations, it merely guarantees that any *persistent constructors* ([ST88b]) from $[\![SP_2]\!]$ to $[\![SP]\!]$ would implement SP_1 by SP_2.

7 Conclusion and Future Work

In our paper we have defined a formal notion of the redesign of specifications. Our approach allows one to reason about such transformations of the system structure that are incomparable by means of a signature morphism. We have also shown (see Lemma 4) that, under certain assumptions, a structured specification can be redesigned in a step-by-step manner, to a specification structured similarly to the initial one.

As a practical application of our work we have presented a formalisation of the redesign of UML class diagrams. We have identified the conditions a class diagram must satisfy in order to describe a property preserving redesign.

While translating UML class diagrams to CASL we handled query methods only, did not care about objects local state, and did not support method overriding in subclasses. We are aware that in order to make our approach applicable in practice, we need not only to investigate all above-mentioned issues, but also to find a tool support basing on the number of currently available CASL tools. We plan to look closer at these problems in coming future.

We would like to thank Andrzej Tarlecki for his comments and suggestions.

References

[ACZ99] D. Ancona, M. Cerioli, and E. Zucca. A formal framework with late binding. In *Fundamental Approaches to Software Engineering, FASE'99*, volume 1577 of *LNCS*, pages 30–44. Springer, 1999.

[AKKB99] E. Astesiano, H.-J. Kreowski, and B. Krieg-Brückner, editors. *Algebraic Foundations of Systems Specification.* IFIP State-of-the-Art Report. Springer, 1999.

[BG92] R.M. Burstall and J.A. Goguen. Institutions: Abstract model theory for specification and programming. *Journ. of the ACM*, 39(1):95–146, 1992.

[BHTW99] M. Bidoit, R. Hennicker, F. Tort, and M. Wirsing. Correct realization of interface constraints with OCL. In R. France and B. Rumpe, editors, *UML'99: The Unified Modeling Language – Beyond the Standard.*, volume 1723 of *LNCS*, pages 399–415. Springer, 1999.

[CoF04] CoFI. CASL *Reference Manual*, volume 2960 (IFIP Series) of *LNCS*. Springer, 2004.

[Fow99] M. Fowler. *Refactoring: Improving the Design of Existing Code.* Addison-Wesley, 1999.

[GHJV95] E. Gamma, R. Helm, R. Johnson, and J. Vlissides. *Design Patterns: Elements of Reusable Object-Oriented Software.* Addison-Wesley, 1995.

[GRPPS98] M. Große-Rhode, F. Parisi-Presicce, and M. Simeoni. Refinements and modules for typed graph transformation systems. In J.L. Fiadeiro, editor, *Recent Trends in Algebraic Development Techniques, WADT'98*, LNCS, pages 138–151. Springer, 1998.

[Kos01] P. Kosiuczenko. Formal redesign of UML class diagrams. In A. Evans, R. France, A. Moreira, and B. Rumpe, editors, *Practical UML-Based Rigorous Development Methods - Countering or Integrating the eXtremists*, volume P-7 of *LNI*, pages 174–190. German Informatics Society, 2001.

[Kos05] P. Kosiuczenko. Redesign of UML class diagrams. Technical Report CS-05-01, University of Leicester, Department of Computer Science, 2005. `http://www.cs.le.ac.uk/people/pk82/RedesignTR.pdf`.

[Lan95] K. Lano. *Formal Object-Oriented Development.* Springer, 1995.

[Mar04] G. Marczyński. Specifications of internally dependent structures. Technical report, Warsaw University, Institute of Informatics. In preparation, 2004. `http://www.mimuw.edu.pl/~gmarc/papers/specidp04.pdf`.

[MCG05] T. Mens, K. Czarnecki, and P. Van Gorp. A taxonomy of model transformations. In J. Bezivin and R. Heckel, editors, *Language Engineering for Model-Driven Software Development*, number 04101 in Dagstuhl Seminar Proceedings. IBFI, Schloss Dagstuhl, Germany, 2005. `http://drops.dagstuhl.de/opus/volltexte/2005/11`.

[Obj03] Object Management Group. *Unified Modeling Language, version 1.5*, 2003. `http://www.omg.org/cgi-bin/doc?formal/03-03-01`.

[ONS96] F. Orejas, M. Navarro, and A. Sánchez. Algebraic implementation of abstract data types: a survey of concepts and new compositionality results. *Mathematical Structures in Computer Science*, 6(1):33–67, 1996.

[ST88a] D. Sannella and A. Tarlecki. Specifications in an arbitrary institution. *Information and Computation*, 76:165–210, 1988.

[ST88b] D. Sannella and A. Tarlecki. Toward formal development of programs from algebraic specifications: Implementations revisited. *Acta Informatica*, 25(3):233–281, 1988.

[Tay73] W. Taylor. Characterizing Malcev conditions. *Algebra Universalis*, 3:351–397, 1973. Springer, Berlin.

Author Index